PETROLEUM RELATED ROCK MECHANICS

DEVELOPMENTS IN PETROLEUM SCIENCE 72

PETROLEUM RELATED ROCK MECHANICS

Third Edition

ERLING FJÆR

RUNE MARTIN HOLT

PER HORSRUD

ARNE MARIUS RAAEN

RASMUS RISNES

ELSEVIER

Elsevier
Radarweg 29, PO Box 211, 1000 AE Amsterdam, Netherlands
The Boulevard, Langford Lane, Kidlington, Oxford OX5 1GB, United Kingdom
50 Hampshire Street, 5th Floor, Cambridge, MA 02139, United States

Library of Congress Cataloging-in-Publication Data
A catalog record for this book is available from the Library of Congress

British Library Cataloguing-in-Publication Data
A catalogue record for this book is available from the British Library

ISBN: 978-0-12-822195-2

Publisher: Candice Janco
Acquisitions Editor: Amy Shapiro
Editorial Project Manager: Michelle Fisher
Production Project Manager: Vijayaraj Purushothaman
Designer: Victoria Pearson

Typeset by VTeX

Contents

Biography *xv*
Foreword to the 1992 edition *xvii*
Preface to the third edition *xix*
Preface to the second edition *xxi*
Preface to the 1992 edition *xxiii*

1. Elasticity **1**
1.1. Stress 1
1.1.1. The stress tensor 4
1.1.2. Equations of equilibrium 6
1.1.3. Principal stresses in two dimensions 8
1.1.4. Mohr's stress circle 10
1.1.5. Principal stresses in three dimensions 11
1.1.6. Mohr's stress circles in three dimensions 12
1.1.7. Stress invariants 13
1.1.8. Deviatoric stresses 13
1.1.9. The octahedral stresses 16
1.2. Strain 16
1.2.1. The strain tensor and the strain invariants 20
1.2.2. Compatibility conditions 21
1.2.3. Principal strains 22
1.2.4. Plane strain and plane stress 23
1.3. Elastic moduli 24
1.4. Strain energy 28
1.5. Thermoelasticity 31
1.5.1. Thermal strain 31
1.5.2. Thermal stress 31
1.5.3. Stress–strain relation for linear thermoelasticity 31
1.5.4. Isothermal and adiabatic moduli 32
1.5.5. Example: thermal stresses in a constrained square plate 32
1.6. Poroelasticity 33
1.6.1. Suspension of solid particles in a fluid 34
1.6.2. Biot's poroelastic theory for static properties 35
1.6.3. Alternate form of the Biot constitutive equations 43
1.6.4. Effective stress 44
1.6.5. Specific storage 47
1.6.6. Pore volume compressibility and related topics 47
1.6.7. The Skempton coefficients 49
1.6.8. The undrained Poisson's ratio 50

1.6.9. The correspondence to thermoelasticity 52
1.6.10. Thermoporoelasticity 53
1.6.11. Other notation conventions 54
1.7. Anisotropy 54
1.7.1. Orthorhombic (orthotropic) symmetry 57
1.7.2. Transverse isotropy (TI) 60
1.7.3. Generalised bulk modulus 61
1.7.4. Anisotropic poroelasticity 62
1.8. Nonlinear elasticity 66
1.8.1. Stress–strain relations 66
1.8.2. The impact of cracks 69
1.8.3. Finite strain 71
1.9. Time-dependent effects 72
1.9.1. Consolidation 73
1.9.2. Thermal diffusivity 78
1.9.3. The Mandel–Cryer and Noordbergum effects 78
1.9.4. Creep 80
1.10. Further reading 85
References 86

2. Failure mechanics **89**
2.1. Basic concepts 89
2.1.1. Strength and related concepts 89
2.1.2. The failure surface 92
2.2. Tensile failure 94
2.3. Shear failure 95
2.3.1. The Mohr–Coulomb criterion 97
2.3.2. The Griffith criterion 101
2.4. Compaction failure 103
2.5. Failure criteria in three dimensions 106
2.5.1. Criteria independent of the intermediate principal stress 106
2.5.2. Criteria depending on the intermediate principal stress 108
2.6. Fluid effects 116
2.6.1. Pore pressure 116
2.6.2. Partial saturation 117
2.6.3. Chemical effects 120
2.7. Presentation and interpretation of data from failure tests 123
2.8. Beyond the yield point 125
2.8.1. Plasticity 126
2.8.2. Soil mechanics 134
2.8.3. Localisation 141
2.8.4. Liquefaction 142
2.9. Failure of anisotropic and fractured rocks 143

2.9.1. Intrinsic anisotropy and the failure surface 143
2.9.2. The plane of weakness model 144
2.9.3. Fractured rock 148
2.10. Stress history effects 149
2.10.1. Rate effects and delayed failure 150
2.10.2. Fatigue 151
References 151

3. Geological aspects of petroleum related rock mechanics 157
3.1. Underground stresses 157
3.2. Pore pressure 170
3.3. Sedimentological aspects 174
3.3.1. Grains and minerals 175
3.3.2. Pre-deposition and deposition 178
3.3.3. Post-deposition 179
3.4. Mechanical properties of sedimentary rocks 183
3.4.1. Sandstone 183
3.4.2. Carbonate 186
3.4.3. Shale 189
3.4.4. Rock salt 196
References 197

4. Stresses around boreholes. Borehole failure criteria 201
4.1. Stresses and strains in cylindrical coordinates 201
4.2. Stresses in a hollow cylinder 203
4.2.1. The equilibrium equations 203
4.2.2. Stress distributions with constant pore pressure 204
4.2.3. Stress distributions with varying pore pressure 207
4.2.4. Stress distributions with heat flow 211
4.2.5. Stress distributions in nonlinear formations 212
4.3. Elastic stresses around circular wells—the general solution 213
4.3.1. Transformation formulas 214
4.3.2. The general elastic solution 216
4.3.3. Borehole along a principal stress direction 217
4.4. Poroelastic time-dependent effects 219
4.4.1. Wellbore pressure invasion 220
4.4.2. Drillout-induced pore pressure changes 222
4.5. Borehole failure criteria 224
4.5.1. Vertical hole, isotropic horizontal stresses and impermeable borehole wall 224
4.5.2. Vertical hole, isotropic horizontal stresses and permeable borehole wall 228
4.5.3. Borehole along a principal stress direction 229

4.5.4. Borehole in a general direction 231
4.6. Elliptical borehole 232
4.6.1. Fracture initiation 235
4.6.2. Shear failure 236
4.7. Borehole in an anisotropic formation 237
4.8. Beyond failure initiation 238
4.8.1. A simple plasticity model 243
4.9. Cased borehole 250
4.10. Spherical coordinates 253
4.10.1. Basic equations 253
4.10.2. Stress distribution around a spherical cavity with no fluid flow 254
4.10.3. Stress distribution with fluid flow 256
References 256

5. Elastic wave propagation in rocks 259
5.1. The wave equation 259
5.2. P- and S-waves 262
5.3. Elastic waves in porous materials 265
5.3.1. Biot's theory of elastic wave propagation 266
5.3.2. Dispersion due to local flow 271
5.4. Attenuation 271
5.5. Anisotropy 277
5.5.1. The Christoffel equation 277
5.5.2. Weak anisotropy 280
5.6. Rock mechanics and rock acoustics 281
5.6.1. Static and dynamic moduli 281
5.6.2. Stress state and stress history 285
5.6.3. Additional effects 288
5.7. Reflections and refractions 294
5.7.1. Interface waves 298
5.8. Borehole acoustics 299
5.8.1. Borehole modes 302
5.8.2. Borehole alteration 308
5.9. Seismics 310
5.10. Acoustic emission 314
References 316

6. Rock models 319
6.1. Layered media 320
6.2. Models involving porosity only 323
6.3. Grain pack models 327
6.4. Models for cracks and other inclusions 335
6.4.1. Linear, isotropic models 336

6.4.2. Anisotropic models 339
6.4.3. Models accounting for interactions 343
6.4.4. Crack development in stressed rocks 348
6.5. Multicomponent models 353
6.6. Fractured rocks 357
6.6.1. Single fractures 357
6.6.2. Rocks with many fractures 361
6.7. Finite element analysis 364
References 366

7. Mechanical properties and stress data from laboratory analysis 373
7.1. Core samples for rock mechanical laboratory analysis 374
7.1.1. Core representativeness and size effects 374
7.1.2. Core alteration 375
7.1.3. Core handling 377
7.1.4. Preparation of test samples 378
7.2. Laboratory equipment 381
7.2.1. The load frame 382
7.2.2. The triaxial cell 384
7.2.3. Measurements of stresses and strains 385
7.2.4. Acoustic measurements 386
7.3. Laboratory tests for rock mechanical property determination 389
7.3.1. Stresses on cylindrical samples 389
7.3.2. Drained and undrained test conditions 390
7.3.3. Standard triaxial compression tests 391
7.3.4. Interpretation of elastic moduli from triaxial tests 392
7.3.5. Uniaxial (unconfined) compression tests 394
7.3.6. Hydrostatic tests and determination of poroelastic parameters 395
7.3.7. Triaxial testing of shales 399
7.3.8. Oedometer (K_0) test or uniaxial strain test (UST) 400
7.3.9. Stress path tests 402
7.3.10. Other triaxial failure tests 404
7.3.11. Hollow cylinder tests 405
7.3.12. Measurements on small samples 408
7.4. Laboratory tests for stress determination 409
7.4.1. Differential strain curve analysis 410
7.4.2. Anelastic strain recovery 412
7.4.3. Acoustic techniques 413
7.5. Index tests and other characterisation tests 414
7.5.1. Tensile strength indicators 415
7.5.2. Hardness measurements 416
7.5.3. Indentation and scratch tests 417
7.5.4. Punch tests 418

7.5.5. Fracture toughness 419
7.5.6. Specific shale characterisation tests 420
7.5.7. Shale-fluid interaction tests 422
References 423

8. Mechanical properties and in situ stresses from field data 429
8.1. Estimation of elastic parameters 429
8.1.1. Acoustic wireline logs 431
8.1.2. Acoustic logging while drilling 433
8.1.3. Acoustic measurements on drill cuttings 433
8.2. Estimation of strength parameters 434
8.2.1. Log data (wireline and MWD) 434
8.2.2. Drill cuttings measurements 435
8.2.3. Empirical correlations 435
8.2.4. Drilling data 436
8.3. Estimation of in situ stresses 437
8.3.1. Overburden stress (density log) 438
8.3.2. Horizontal stress directions (borehole logs) 439
8.3.3. Horizontal stress magnitude (acoustic logs) 442
8.3.4. Horizontal stress magnitude (fracture tests) 444
8.3.5. Other methods 454
8.3.6. Transferring stress data from offset wells 455
References 456

9. Stability during and after drilling 461
9.1. Unstable boreholes: symptoms, reasons and consequences 464
9.1.1. Tight hole/stuck pipe 464
9.1.2. Lost circulation 467
9.2. Rock mechanics analysis of borehole stability 469
9.2.1. Introduction 469
9.2.2. Lower mud weight limit 471
9.2.3. Upper mud weight limit 476
9.3. Time-delayed borehole failure 478
9.3.1. Establishment of pore pressure equilibrium 478
9.3.2. Temperature effects 482
9.3.3. Creep 483
9.4. Interaction between shale and drilling fluid 483
9.5. Borehole stability analysis for well design: incorporating effects of nonlinear elasticity, plasticity and rock anisotropy 487
9.6. Use of pressure gradients 492
9.6.1. Introduction 492
9.6.2. Depth reference and depth corrections 493
9.7. Beyond simple stability analysis 495

9.7.1. Field cases: the borehole stability problem in complex geology 495
9.7.2. Drilling below deep water 497
9.7.3. Surge and swab effects 499
9.7.4. Hole cleaning 499
9.7.5. Amount and quality of input data 500
9.8. Stability issues in different lithologies 502
9.8.1. Sandstone 502
9.8.2. Carbonate rock 503
9.8.3. Salt 504
9.8.4. Coal 506
9.9. Drilling in depleted reservoirs 506
9.10. Shale as a barrier 509
References 513

10. Solids production 519
10.1. Operational aspects of solids production 519
10.1.1. Consequences of solids production 519
10.1.2. Well completion and solids control 520
10.2. Sand 522
10.2.1. Necessary and sufficient conditions for sand production 523
10.2.2. Forces on a sand grain 523
10.2.3. Critical drawdown for cylindrical cavities 526
10.2.4. Stability and collapse of sand arches 536
10.2.5. Rate of produced sand 540
10.2.6. Sand transport 547
10.2.7. Sand prediction 549
10.3. Chalk 551
References 552

11. Mechanics of hydraulic fracturing 555
11.1. Conditions for tensile failure 557
11.2. Fracture initiation and formation breakdown 558
11.3. Fracture orientation, growth and confinement 565
11.3.1. Stress shadowing 570
11.3.2. The "wedge effect" 572
11.4. Fracture size and shape 574
11.5. Fracture closure 582
11.5.1. Estimation of σ_3 from shut-in/decline tests 583
11.5.2. Estimation of σ_3 from flowback tests 585
11.6. Thermal effects on hydraulic fracturing 587
11.7. Fracturing in unconventional reservoirs 590
11.8. Microseismic monitoring of fracturing 593
References 596

12. Reservoir geomechanics **601**
12.1. Compaction and subsidence 601
12.2. Modelling of reservoir compaction 603
12.2.1. Uniaxial reservoir compaction 603
12.2.2. The depleting sphere 605
12.2.3. Reservoir stress path 606
12.2.4. Beyond simple elastic theory 611
12.2.5. Time-delayed reservoir compaction 614
12.3. From compaction to subsidence 616
12.3.1. Geertsma's nucleus of strain model 616
12.3.2. Stress alteration in the overburden 626
12.4. Geomechanical effects on reservoir performance 632
12.4.1. Compaction drive 632
12.4.2. Stress effects on porosity 633
12.4.3. Stress effects on permeability 636
12.4.4. Geomechanics in reservoir simulation 643
12.4.5. Seismic reservoir monitoring 644
12.5. Well problems and reservoir geomechanics 648
12.5.1. Casing damage 649
12.5.2. Reservoir geomechanics as a tool to optimise drilling and production strategies 652
12.6. Some field cases: subsidence and induced seismicity 653
References 657

A. Rock properties **663**
References 674

B. SI metric conversion factors **675**

C. Mathematical background **677**
C.1. Introduction 677
C.2. Matrices 677
C.2.1. The transpose of a matrix 677
C.2.2. Symmetric matrix 678
C.2.3. Diagonal matrix 678
C.2.4. Matrix addition 678
C.2.5. Multiplication by a scalar 678
C.2.6. Matrix multiplication 678
C.2.7. The identity matrix 679
C.2.8. The inverse matrix 679
C.2.9. The trace of a matrix 680
C.2.10. Determinants 680
C.2.11. Systems of linear equations 681

C.2.12. Eigenvalues and eigenvectors 682
C.2.13. Similarity transforms and orthogonal transforms 683
C.3. Vectors and coordinate transforms 684
C.4. Tensors and coordinate transforms 685
C.5. Eigenvalues, eigenvectors and diagonalisation 686
C.6. Rotation of the coordinate system: the Euler angles 687
C.7. Examples 688
C.7.1. Rotation of the stress tensor 689
C.7.2. Inversion of an axis 689
C.7.3. Rotation and strain 690
C.7.4. Lode coordinates 690
C.8. Matrix invariants 691
C.9. Some trigonometric formulas 693
C.10. The Voigt notation spelled out 694
C.10.1. The Voigt mapping for the stress tensor 695
C.10.2. The Voigt mapping for the strain tensor 695
C.10.3. The Voigt mapping for the stiffness tensor 695
C.10.4. The Voigt mapping for the compliance tensor 695
C.10.5. The Voigt mapping for the Biot tensor 696
C.10.6. The Voigt mapping for the Skempton tensor 696
C.10.7. Direct rotation of the Voigt matrices 696
C.11. Elastic stability 697
C.12. The Einstein summing convention and other notation conventions 699
C.12.1. The Einstein summing convention 699
C.12.2. Kronecker's delta 700
C.12.3. Comma notation for partial derivatives 700
C.12.4. Operator notation for partial derivatives 701
References 701

D. Some relevant formulas 703
D.1. Elasticity 703
D.1.1. Stress invariants 703
D.1.2. Strain in spherical coordinates 704
D.1.3. Isotropic linear elastic stiffness tensor 705
D.1.4. Isotropic linear thermoporoelastic stress–strain law 705
D.1.5. The force balance equation 706
D.1.6. Transverse isotropy 706
D.2. Elastic wave propagation in rocks 708
D.2.1. Correction for non-laminar flow 708
D.2.2. Reflection, transmission and conversion coefficients at non-normal incidence 708
D.3. Rock models 709
D.3.1. Stresses at a crack tip 709

D.3.2. Self-consistent model for composite media 711
D.4. Solids production 712
D.4.1. Critical drawdown for turbulent flow 712
D.5. Subsidence 713
D.6. Permeability of tubes 715
D.7. Vector operators in cylindrical coordinates 716
References 717

E. Abbreviations **719**

F. List of symbols **721**

Index *733*

Biography

Erling Fjær

Chief Scientist at SINTEF and Adjunct Professor at NTNU (Norwegian University of Science and Technology).

Rune Martin Holt

Professor at NTNU (Norwegian University of Science and Technology) and Special Advisor to SINTEF.

Per Horsrud

Retired from the position as specialist in drilling related rock mechanics in Equinor.

Arne Marius Raaen

Scientist at SINTEF, previously with Statoil (Equinor).

Rasmus Risnes

Recognised as one of the pioneers in petroleum related rock mechanics and a professor at the University of Stavanger until he passed away in 2004.

Foreword to the 1992 edition

About 10 years ago, petroleum related rock mechanics was mostly confined to a few specific topics like hydraulic fracturing or drilling bit performance. Although a few precursors had already established the basics of what is now spreading out.

In fact, since that time, the whole petroleum industry has progressively realised that the state of underground stresses and its modification due to petroleum related operations, could have a significant impact on performances, in many different aspects of exploration and production. Therefore all those concepts need now to be presented in a simple but comprehensive way.

An engineering science, "Petroleum related rock mechanics" is also dependent on the variable and uncertain character of natural geological materials at depth. The limited availability of relevant data is also part of the problem. For that reason, it is essential that any potential user is aware of the high potential of the technique, together with the actual limitations.

This book should totally fulfil these needs. The reader will be provided with fundamentals and basics, but also with the techniques used in data acquisition, and eventually with a series of typical applications like wellbore stability, sand production or subsidence.

The book is mostly for students, geologists or engineers who want to know more about rock mechanics, and specifically rock mechanics applied to petroleum industry. It is well suited for instance, for drilling or mud engineers wanting to know more on the mechanical aspect of wellbore stability problems, for reservoir engineers who have to deal with stress related problems in their field, like compaction, stress dependent permeability or fracture injectivity. Operation geologists dealing with drilling in abnormal pressure zones will also benefit from this book.

Elf Aquitaine Norge, together with other Norwegian oil companies, has been supporting IKU for several years in their effort to build a strong group in rock mechanics. IKU has now a well-established team, whose competence is recognised on the international level. They have mostly been contributing in the field of acoustic wave propagation in rocks and sand production appraisal. They are also very active in research and consulting activities in the different fields of petroleum related rock mechanics.

Stavanger, April 1991
Alain Guenot

Preface to the third edition

More than a decade has passed since the last edition of this book. Petroleum related rock mechanics has developed significantly during this period, calling for an updated version of the book. This development includes new scientific results that enhances the capabilities of analyses and predictions within this area. The position of rock mechanics within the petroleum industry has also been strengthened. In particular, rock mechanical analyses and modelling have become vital for handling consequences of reservoir depletion, on near-well scale as well as reservoir scale where possible reactivation of fractures is a key topic. Rock mechanical analyses are also critical for prediction of possible fracturing during injection, not only for water injection as a means for enhanced productivity but also for CO_2 injection and waste disposal. Furthermore, while the main goals of petroleum related rock mechanics in the past were cost efficient drilling and improved production, plugging and abandonment is now becoming a subject of increasing importance. Thus applications of petroleum related rock mechanics extend beyond petroleum.

The basic structure of the book remains the same as for the previous editions. The text has been partly extended, partly improved or corrected, reflecting the extended and improved knowledge among the authors as well as the scientific community in general. The new or extended subjects include, amongst other things, anisotropic poroelasticity, creep, shale barriers, and fracturing in unconventional reservoirs. The collection of relevant formulas has been extended accordingly, both in the main text and in the appendices.

This revision of the book has been made by the original team of authors, except for Rasmus Risnes, who sadly passed away before the previous edition was published. Several friends and colleagues have given valuable support, advice and encouragement to our work, for which we are very grateful.

Trondheim, 15 June 2021

Erling Fjær
Rune Martin Holt
Per Horsrud
Arne Marius Raaen

Preface to the second edition

During the years that have passed since the first edition of this book, petroleum related rock mechanics has been well established as a significant supplier of premises and boundary conditions for the petroleum industry. More than ever, it is now recognised that engineers and geologists in the petroleum industry should possess a certain level of knowledge within rock mechanics. The need for a textbook like this is therefore even larger now than it was 15 years ago.

Although there are still a lot of uncovered areas within petroleum related rock mechanics, the topic has in many ways developed significantly over the later years. This is a natural consequence of the increased focus this area has gained. We are also proud to admit that the general knowledge of petroleum related rock mechanics among the authors of this book has increased since the first edition was prepared. Consequently, when the need for a new printing of the book appeared, we felt the need to revise the manuscript in order to account for this development.

The revision has been quite extensive for some parts of the manuscript. The basic structure of the book is however kept as it was, and parts of the text have only been subject to minor revisions. The major guideline for the work has been to update and add information where we felt it was relevant, while maintaining the concept of the book as an introduction to petroleum related rock mechanics as an engineering science. A consequence of this dual objective has been the introduction of a couple of new appendices, where more advanced mathematics and heavy formulas are presented in a compact form. This way, we hope that the main text shall remain easily accessible even for newcomers in the field, while at the same time the more complete formulas shall be available to the readers who need them.

The revision has been made by the same team of authors that wrote the first edition. In addition, several friends and colleagues have generously provided suggestions, advice and encouragement for the work. This support is greatly appreciated.

Trondheim, October 2004

Erling Fjær
Rune M. Holt
Per Horsrud
Arne M. Raaen
Rasmus Risnes

Preface to the 1992 edition

Systematic application of rock mechanics is quite new to the petroleum industry. Accordingly, the need for an introductory textbook for petroleum engineers and scientists has recently emerged. This need was felt by the authors when we started our research in this area, and it inspired us to develop the first version of this book as a manuscript for a two weeks continuing education course for petroleum engineers.

The first 6 Chapters deal with the fundamentals of rock mechanics. This includes theories of elasticity and failure mechanics, borehole stresses, and acoustic wave propagation. In addition, sedimentary rocks are viewed from the geological side as well as from the side of idealised mathematical modelling based on microstructure. For readers who wants to further extend their knowledge on rock mechanics, we suggest the book "Fundamentals of rock mechanics" by Jaeger and Cook as a continuation. Deeper insight into acoustic wave propagation in rocks can be achieved from e.g. the book "Acoustics of porous media" by Bourbie, Coussy and Zinszner or "Underground Sound" by White.

Chapters 7 and 8 are dedicated to the extremely important task of obtaining parameters that are relevant for rock mechanics field application, be it from laboratory tests or from analysis of field data like borehole logs. The last 4 Chapters discuss applications of rock mechanics in borehole stability, sand production, hydraulic fracturing and reservoir compaction/surface subsidence analyses.

It has also been our intention to make each chapter more or less self-contained, especially the chapters dealing with applications. Hopefully, this will make the book useful also to those who are interested only in one particular topic. The other chapters can then be used as support depending on the reader's previous knowledge. Notice, however, that the book is intended to be an introduction to petroleum related rock mechanics as an engineering science, rather than a "tool-box" for petroleum engineers.

We wish to thank Elf Aquitaine Norge and Fina Exploration Norway for the financial contributions which made it possible for us to write this book. In particular, we appreciate the positive feedback and encouragement provided by Alain Guenot of Elf. The skilful and patient support from Siri Lyng at IKU in preparing the manuscript for camera-ready quality is greatly appreciated. We also thank Eamonn F. Doyle for advice on our use of the English language.

Trondheim, May 1991

Erling Fjær
Rune M. Holt
Per Horsrud
Arne M. Raaen
Rasmus Risnes

CHAPTER 1

Elasticity

Most materials have an ability to resist and recover from deformations produced by forces. This ability is called elasticity. It is the foundation for all aspects of rock mechanics. The simplest type of response is one where there is a linear relation between the external forces and corresponding deformations. When changes in the forces are sufficiently small, the response is nearly always linear. Thus the theory of linear elasticity is fundamental for all discussions on elasticity.

The theory of elasticity rests on the concepts of stress and strain. These are defined in Sections 1.1 and 1.2. The linear equations relating stresses and strains are discussed in Section 1.3 for isotropic materials, and in Section 1.7 for anisotropic materials. Linear thermoelasticity is discussed in Section 1.5.

The region of validity for linear elasticity is often exceeded in practical situations. Some general features of nonlinear behaviour of rocks are described in Section 1.8.

Even when nonlinearity is important, the behaviour may still be linear over limited ranges of loading. The equations of linear elasticity then apply, provided that the reference point is taken properly into account. Hence, the linear elastic formulas have a broader applicability if expressed in terms of stress changes and associated strains.

In petroleum related rock mechanics, much of the interest is furthermore focused on rocks with significant porosity and permeability. The elastic theory for solid materials is not able to fully describe the behaviour of such materials, and the concept of poroelasticity has therefore to be taken into account. The elastic response of a rock material may also be time dependent, so that the deformation of the material changes with time, even when the external conditions are constant. The elastic properties of porous materials and time-dependent effects are described in Sections 1.6 and 1.9, respectively.

1.1 Stress

Consider the situation shown in Fig. 1.1. A weight is resting on the top of a pillar. Due to the weight, a force is acting on the pillar, whereas the pillar reacts with an equal but reversely directed force. The pillar itself is

Developments in Petroleum Science, Volume 72
ISSN 0376-7361, https://doi.org/10.1016/B978-0-12-822195-2.00010-3

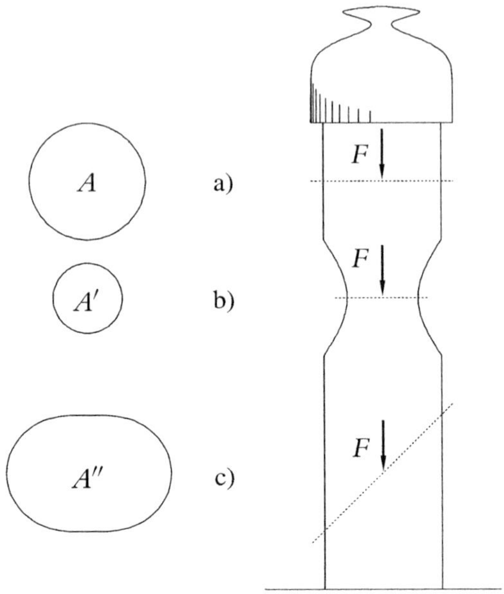

Figure 1.1 Illustration of forces and stress.

supported by the ground. Hence, the force acting at the top of the pillar must be acting through any cross-section of the pillar.

The area of the cross-section at a) is A. If the force acting through the cross-section is denoted F, then the *stress* σ at the cross-section is defined as

$$\sigma = \frac{F}{A} \tag{1.1}$$

The SI unit for stress is Pa (Pascal = N/m^2). In the petroleum industry, "oilfield" units like psi (pounds per square inch) are still extensively used, so that we need to be familiar with them. See Appendix B for an overview of some conversion factors.

The sign of the stress σ is not uniquely defined by the physics of the situation and has therefore to be defined by convention. In rock mechanics the common sign convention states that compressive stresses are positive (see e.g. Jaeger and Cook, 1979; Jaeger et al., 2007). The historical reason for this is that the stresses dealt with in rock mechanics are mostly compressive. The sign convention causes no problems when consistently used, but it is important to remember that the opposite sign convention is the

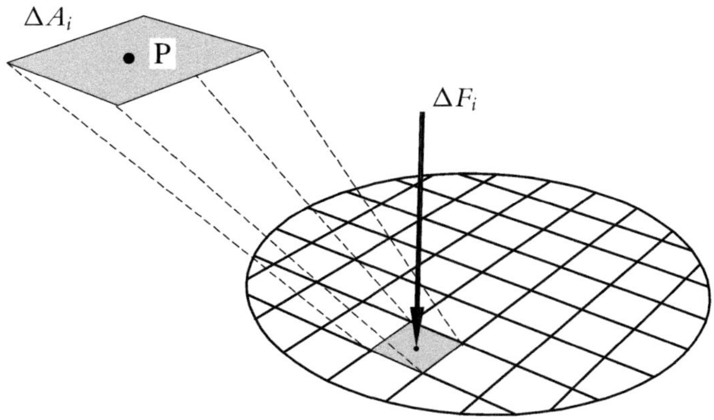

Figure 1.2 Local stress.

preferred choice in some other sciences involving elasticity and that it is also occasionally used in rock mechanics.

As Eq. (1.1) shows, the stress is defined by a force and a cross-section (or more generally, a surface) through which the force is acting. Consider the cross-section at b). The force acting through this cross-section is equal to the force acting through the cross-section at a) (neglecting the weight of the pillar). The area A' of the cross-section at b) is, however, smaller than A. Hence, the stress $\sigma = F/A'$ at b) is larger than the stress at a), that is, the stress depends on the position within the stressed sample. Going even further, we may divide the cross-section at a) into an infinite number of subsections ΔA, through which an infinitely small part ΔF of the total force F is acting (Fig. 1.2). The force ΔF may vary from one subsection to another. Consider a subsection i that contains a point P. The stress at the point P is defined as the limit value of $\Delta F_i/\Delta A_i$ as ΔA_i goes to zero, that is,

$$\sigma = \lim_{\Delta A_i \to 0} \frac{\Delta F_i}{\Delta A_i} \tag{1.2}$$

Eq. (1.2) defines the local stress at the point P within the cross-section at a), whereas Eq. (1.1) describes the average stress at the cross-section. When talking about the stress state at a point, we implicitly mean local stresses.

The orientation of the cross-section relative to the direction of the force is also important. Consider the cross-section at c) in Fig. 1.1, with area A''. Here the force is no longer normal to the cross-section. We may then decompose the force into one component F_n that is normal to the cross-

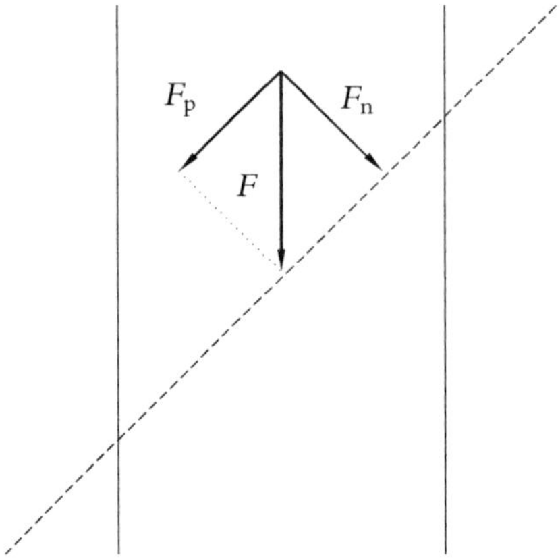

Figure 1.3 Decomposition of forces.

section and one component F_p that is parallel to the section (Fig. 1.3). The quantity

$$\sigma = \frac{F_n}{A''} \tag{1.3}$$

is called the *normal stress*, whereas the quantity

$$\tau = \frac{F_p}{A''} \tag{1.4}$$

is called the *shear stress*. Thus, there are two types of stresses, which may act through a surface, and the magnitude of each depends on the orientation of the surface.

1.1.1 The stress tensor

To give a complete description of the stress state at a point P within a sample, it is necessary to identify the stresses related to surfaces oriented in three orthogonal directions.

The stresses related to a surface normal to the x-axis may be denoted σ_x, τ_{xy} and τ_{xz}, representing the normal stress, the shear stress related to a force in y-direction, and the shear stress related to a force in the z-direction, respectively. Physically, there will be only one shear stress associated with this surface. However, the orientation of the shear stress has to be identified, and this is most conveniently done by identifying its y- and z-components, τ_{xy} and τ_{xz}. Similarly, the stresses related to a surface normal to the y-axis are denoted σ_y, τ_{yx} and τ_{yz}, whereas the stresses related to a surface normal to the z-axis are denoted σ_z, τ_{zx} and τ_{zy}. Thus there are all together nine stress components related to the point P:

$$\begin{pmatrix} \sigma_x & \tau_{xy} & \tau_{xz} \\ \tau_{yx} & \sigma_y & \tau_{yz} \\ \tau_{zx} & \tau_{zy} & \sigma_z \end{pmatrix} \tag{1.5}$$

Expression (1.5) is called the *stress tensor*. It gives a complete description of the stress state at the point P.

Associating the first index with the face normal and the second with the force direction is a matter of choice, similar to the sign convention. As a result, the opposite convention used in a number of works, for example, Landau and Lifshitz (1986). Further, due to the symmetry of the stress tensor (see below), the convention is of no practical importance.

It is sometimes convenient to denote the stress tensor by a single symbol, for instance, $\overleftrightarrow{\sigma}$. Thus $\overleftrightarrow{\sigma}$ implicitly means the collection of stress components given by (1.5). The stress tensor also has a definite physical meaning: if $\hat{r}$ is a unit vector, then the expression $|\overleftrightarrow{\sigma} \cdot \hat{r}|$ represents the total stress (normal and shear) in the direction of $\hat{r}$.

However, not all the nine components of the stress tensor are independent. Consider a small square of the (x, y) plane, as shown in Fig. 1.4. The stresses acting on the square are shown on the figure. The square is at rest, and hence no net translational or rotational force can act on it. While no translational force is already ensured, no rotational force requires that

$$\tau_{xy} = \tau_{yx} \tag{1.6}$$

Similarly, it may be shown that

$$\tau_{xz} = \tau_{zx} \tag{1.7}$$

and

$$\tau_{yz} = \tau_{zy} \tag{1.8}$$

Relations (1.6)–(1.8) are general, and they reduce the number of independent components of the stress tensor (1.5) to six.

Although being practical for many purposes, the notation used in (1.5) is not very convenient for theoretical calculations. For such purposes, the following notation is frequently used: both types of stresses (normal and

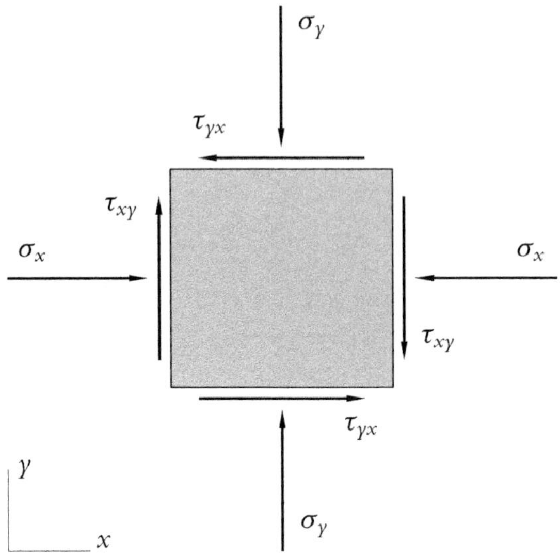

Figure 1.4 Stress components in two dimensions.

shear) are denoted σ_{ij}. The subscripts i and j may be any of the numbers 1, 2, 3, which represent the x-, y- and z-axis, respectively. The first subscript i identifies the axis normal to the actual surface, whereas the second subscript j identifies the direction of the force. Thus, from Fig. 1.4, we see that $\sigma_{11} = \sigma_x$, $\sigma_{13} = \tau_{xz}$, and so on. In this notation the stress tensor (1.5) becomes

$$\overleftrightarrow{\sigma} = \begin{pmatrix} \sigma_{11} & \sigma_{12} & \sigma_{13} \\ \sigma_{12} & \sigma_{22} & \sigma_{23} \\ \sigma_{13} & \sigma_{23} & \sigma_{33} \end{pmatrix} \tag{1.9}$$

where we have explicitly used the symmetry of the stress tensor. See Appendix C.4–C.7 for a discussion of how the tensor components change as a result of a change of coordinates.

1.1.2 Equations of equilibrium

Apart from forces acting on a surface of a body, there may also be forces acting on every part of the body itself. Such forces are called body forces. An example of a body force is gravity. We will denote by f_x, f_y, and f_z the components of the body forces per unit mass acting at the point x, y, z of a body. According to the sign convention, f_x is positive if it acts in the

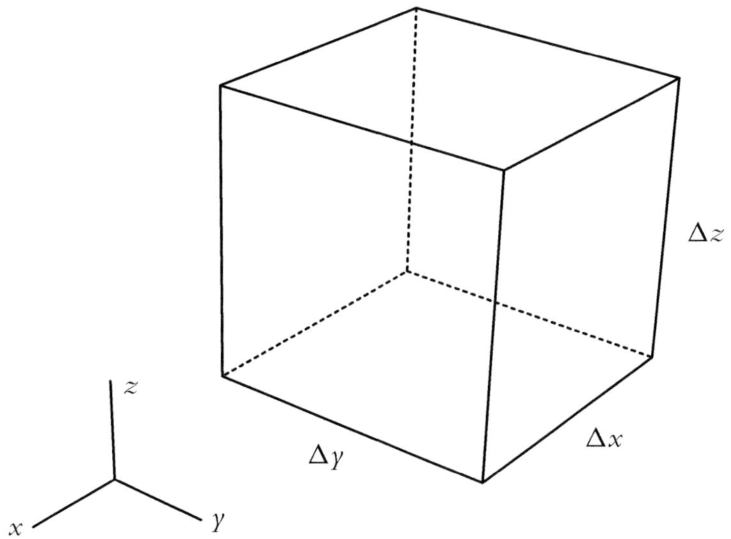

Figure 1.5

negative x-direction, and similarly for f_y and f_z. As an example, consider a small part of volume ΔV of a material with density ρ. If z is the vertical axis, then the body force due to gravity acting on this small volume is $\rho f_z \Delta V = \rho g \Delta V$, where g is the acceleration of gravity.

Body forces generally give rise to stress gradients. For instance, an element in a formation is not only subject to the gravity force, it also has to carry the weight of the formation above. Thus the total stress increases with increasing depth.

For a stressed body to remain at rest, it is required that all forces acting on the body cancel. This requirement produced a set of symmetry requirements for the stress tensor (Eqs. (1.6)–(1.8)). In addition, it produces a set of equations for the stress gradients. These equations are called the *equations of equilibrium*.

Consider the parallelepiped shown in Fig. 1.5. The forces acting on this body in the x-direction are as follows.

Normal forces:

$$-\sigma_x \Delta y \Delta z + \left(\sigma_x + \frac{\partial \sigma_x}{\partial x}\Delta x\right)\Delta y \Delta z \tag{1.10}$$

Shear forces:

$$-\tau_{yx}\Delta x\Delta z+\left(\tau_{yx}+\frac{\partial\tau_{yx}}{\partial y}\Delta y\right)\Delta x\Delta z \tag{1.11}$$

$$-\tau_{zx}\Delta y\Delta x+\left(\tau_{zx}+\frac{\partial\tau_{zx}}{\partial z}\Delta z\right)\Delta y\Delta x \tag{1.12}$$

Body forces:

$$\rho f_x\Delta x\Delta y\Delta z \tag{1.13}$$

Adding up Eqs. (1.10)–(1.13) and dividing by $\Delta x\Delta y\Delta z$, we find that the requirement for the forces in the x-direction to cancel is equivalent to

$$\frac{\partial\sigma_x}{\partial x}+\frac{\partial\tau_{yx}}{\partial y}+\frac{\partial\tau_{zx}}{\partial z}+\rho f_x=0 \tag{1.14}$$

Similarly, for the forces in the y- and z-directions, we find

$$\frac{\partial\sigma_y}{\partial y}+\frac{\partial\tau_{xy}}{\partial x}+\frac{\partial\tau_{zy}}{\partial z}+\rho f_y=0 \tag{1.15}$$

$$\frac{\partial\sigma_z}{\partial z}+\frac{\partial\tau_{xz}}{\partial x}+\frac{\partial\tau_{yz}}{\partial y}+\rho f_z=0 \tag{1.16}$$

Eqs. (1.14)–(1.16) are the equations of equilibrium in terms of stresses. Note that in the alternative notation (shown in Eq. (1.9) for the stresses with $x_1=x$, $x_2=y$, $x_3=z$), these equations take a particularly simple form

$$\sum_j\frac{\partial\sigma_{ji}}{\partial x_j}+\rho f_i=0 \tag{1.17}$$

1.1.3 Principal stresses in two dimensions

For special orientations of the coordinate system, the stress tensor has a particularly simple form. To reveal this form, we will initially study stresses in two dimensions. This is more than just an academic exercise; many problems of practical interest are effectively two-dimensional.

Consider the normal (σ) and shear (τ) stresses at a surface oriented normal to a general direction θ in the (x, y)-plane, as shown in Fig. 1.6. The triangle on the figure is at rest, so that no net forces act on it. Cancellation

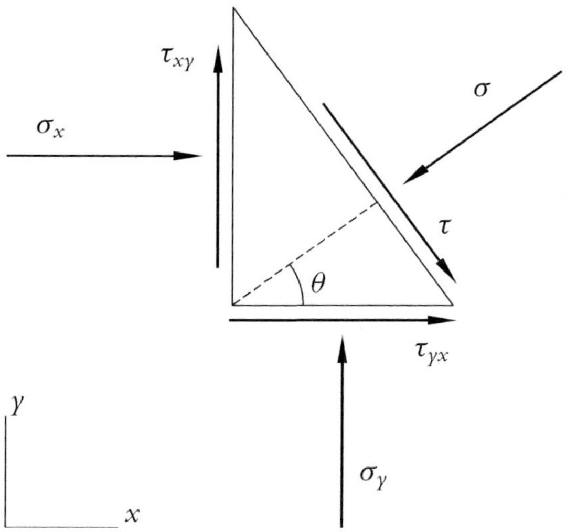

Figure 1.6 Force equilibrium on a triangle. The arrows shows the direction of the forces on the triangle, assuming that all the stress components are positive.

of forces implies that

$$\sigma = \sigma_x \cos^2\theta + \sigma_y \sin^2\theta + 2\tau_{xy} \sin\theta \cos\theta \tag{1.18}$$

$$= \frac{1}{2}\left(\sigma_x + \sigma_y\right) + \frac{1}{2}\left(\sigma_x - \sigma_y\right)\cos 2\theta + \tau_{xy}\sin 2\theta \tag{1.19}$$

$$\tau = \sigma_y \sin\theta\cos\theta - \sigma_x \cos\theta\sin\theta + \tau_{xy}\cos\theta\cos\theta - \tau_{yx}\sin\theta\sin\theta \tag{1.20}$$

$$= \frac{1}{2}\left(\sigma_y - \sigma_x\right)\sin 2\theta + \tau_{xy}\cos 2\theta \tag{1.21}$$

By proper choice of θ it is possible to obtain $\tau = 0$. From Eq. (1.21) we see that this happens when

$$\tan 2\theta = \frac{2\tau_{xy}}{\sigma_x - \sigma_y} \tag{1.22}$$

Eq. (1.22) has two solutions, θ_1 and θ_2. The two solutions correspond to two directions for which the shear stress τ vanishes. These two directions are called the *principal axes of stress.*

The corresponding normal stresses (σ_1 and σ_2) are called the *principal stresses* and are found by introducing Eq. (1.22) into Eq. (1.19):

$$\sigma_1 = \frac{1}{2}\left(\sigma_x + \sigma_y\right) + \sqrt{\tau_{xy}^2 + \frac{1}{4}\left(\sigma_x - \sigma_y\right)^2} \tag{1.23}$$

$$\sigma_2 = \frac{1}{2}\left(\sigma_x + \sigma_y\right) - \sqrt{\tau_{xy}^2 + \frac{1}{4}\left(\sigma_x - \sigma_y\right)^2} \tag{1.24}$$

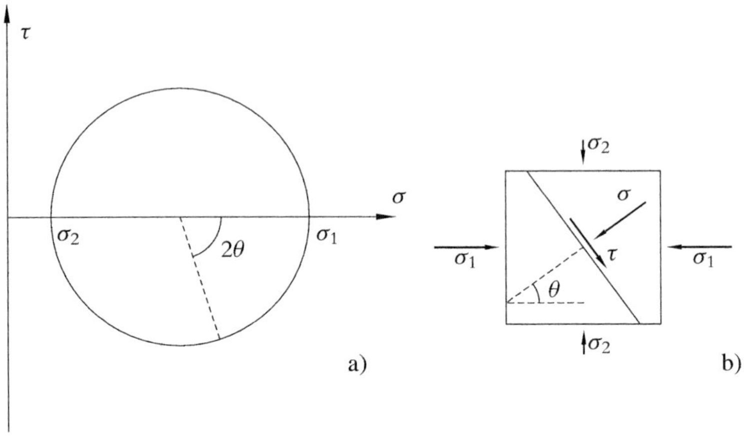

Figure 1.7 Mohr's circle.

It is convenient to choose the notation such that $\sigma_1 \geq \sigma_2$ Thus, in the direction θ_1, which identifies a principal axis, the normal stress is σ_1, and the shear stress is zero. In the direction θ_2, which identifies the other principal axis, the normal stress is σ_2, and the shear stress is zero. The principal axes are *orthogonal.*

1.1.4 Mohr's stress circle

It is often convenient to reorient the coordinate system so that the x-axis is parallel to the first principal axis and the y-axis parallel to the other. Then the stresses σ and τ in a general direction θ relative to the x-axis become, by Eqs. (1.19) and (1.21),

$$\sigma = \frac{1}{2}(\sigma_1 + \sigma_2) + \frac{1}{2}(\sigma_1 - \sigma_2)\cos 2\theta \tag{1.25}$$

$$\tau = -\frac{1}{2}(\sigma_1 - \sigma_2)\sin 2\theta \tag{1.26}$$

Plotting corresponding values of σ and τ in a diagram (Fig. 1.7a), we obtain a circle called Mohr's circle. The radius of the circle is $(\sigma_1 - \sigma_2)/2$, and the centre is at the point $(\sigma_1 + \sigma_2)/2$ on the σ-axis.

The stresses σ and τ in any direction θ (Fig. 1.7b) correspond to a point on Mohr's circle. It is seen from Fig. 1.7a that the largest absolute value for the shear stress is $(\sigma_1 - \sigma_2)/2$ and occurs for $\theta = \pi/4\,(45°)$ and $\theta = 3\pi/4\,(135°)$. Mohr's circle is a very useful tool in the analysis of conditions for rock failure, as will be seen in Chapter 2.

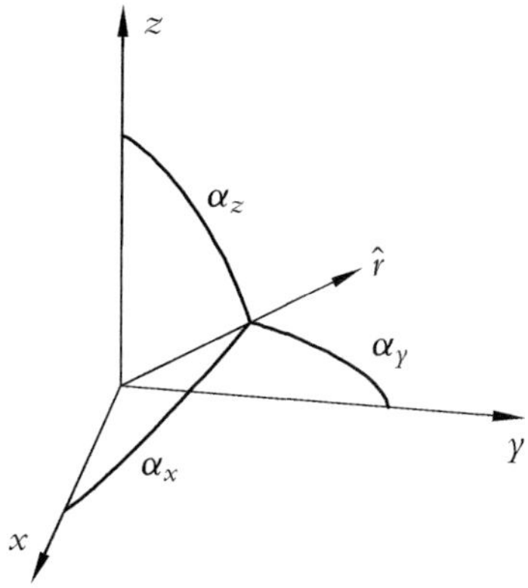

Figure 1.8 Direction cosines.

1.1.5 Principal stresses in three dimensions

Now moving to three dimensions, we first have to decide how to identify a direction in space. This can be done by the direction cosines (see also Appendix C.6):

$$l_x = \cos \alpha_x \tag{1.27}$$

$$l_y = \cos \alpha_y \tag{1.28}$$

$$l_z = \cos \alpha_z \tag{1.29}$$

The angles α_x, α_y and α_z are the angles between our chosen direction and the x-, y- and z-axes, respectively (Fig. 1.8). The vector $\hat{r} = (l_x, l_y, l_z)$ is a unit vector in the chosen direction. Note that we always have

$$l_x^2 + l_y^2 + l_z^2 = 1 \tag{1.30}$$

The principal stresses can be found by solving for σ the determinant equation (see Appendix C, in particular, Section C.2.12)

$$\begin{vmatrix} \sigma_x - \sigma & \tau_{xy} & \tau_{xz} \\ \tau_{xy} & \sigma_y - \sigma & \tau_{yz} \\ \tau_{xz} & \tau_{yz} & \sigma_z - \sigma \end{vmatrix} = 0 \tag{1.31}$$

The three solutions of this equation are the principal stresses σ_1, σ_2 and σ_3. The solutions are conventionally organised so that $\sigma_1 \geq \sigma_2 \geq \sigma_3$. The direction cosines l_{1x}, l_{1y} and l_{1z} identifying the principal axis corresponding to σ_1 are found by solving the equations

$$l_{1x}(\sigma_x - \sigma_1) + l_{1y}\tau_{xy} + l_{1z}\tau_{xz} = 0 \tag{1.32}$$

$$l_{1x}\tau_{xy} + l_{1y}(\sigma_y - \sigma_1) + l_{1z}\tau_{yz} = 0 \tag{1.33}$$

$$l_{1x}\tau_{xz} + l_{1y}\tau_{yz} + l_{1z}(\sigma_z - \sigma_1) = 0 \tag{1.34}$$

The principal axes corresponding to σ_2 and σ_3 are found similarly by substituting subscript 1 by 2 and 3, respectively, in Eqs. (1.32)–(1.34).

If the coordinate system is oriented so that the x-axis is parallel to the first, the y-axis parallel to the second, and the z-axis parallel to the third principal axis, then the stress tensor has the particularly simple form

$$\begin{pmatrix} \sigma_1 & 0 & 0 \\ 0 & \sigma_2 & 0 \\ 0 & 0 & \sigma_3 \end{pmatrix} \tag{1.35}$$

The stresses σ and τ in a general direction l_1, l_2, l_3 relative to this set of coordinate axes are determined by the equations

$$\sigma = l_1^2\sigma_1 + l_2^2\sigma_2 + l_3^2\sigma_3 \tag{1.36}$$

$$\sigma^2 + \tau^2 = l_1^2\sigma_1^2 + l_2^2\sigma_2^2 + l_3^2\sigma_3^2 \tag{1.37}$$

Starting with a full stress tensor (1.9), the corresponding expressions are

$$\begin{aligned} \sigma = {} & l_1(l_1\sigma_{11} + l_2\sigma_{12} + l_3\sigma_{13}) + l_2(l_1\sigma_{12} + l_2\sigma_{22} + l_3\sigma_{23}) \\ & + l_3(l_1\sigma_{13} + l_2\sigma_{23} + l_3\sigma_{33}) \end{aligned} \tag{1.38}$$

$$\begin{aligned} \sigma^2 + \tau^2 = {} & (l_1\sigma_{11} + l_2\sigma_{12} + l_3\sigma_{13})^2 + (l_1\sigma_{12} + l_2\sigma_{22} + l_3\sigma_{23})^2 \\ & + (l_1\sigma_{13} + l_2\sigma_{23} + l_3\sigma_{33})^2 \end{aligned} \tag{1.39}$$

1.1.6 Mohr's stress circles in three dimensions

Mohr's construction is, naturally, more complicated in three dimensions than in two dimensions and will not be treated in detail here. The basic features of the construction are shown in Fig. 1.9. If $l_x = 0$ (direction in the (y, z)-plane), then the stresses σ and τ are located on the small circle spanning from σ_3 to σ_2. If $l_z = 0$ (direction in the (x, y)-plane), then σ and τ are located on the circle spanning from σ_2 to σ_1, and, finally, if $l_y = 0$ (direction in the (x, z)-plane), then σ and τ are located on the large circle spanning from σ_3 to σ_1. For all other directions, σ and τ are located within the shaded areas.

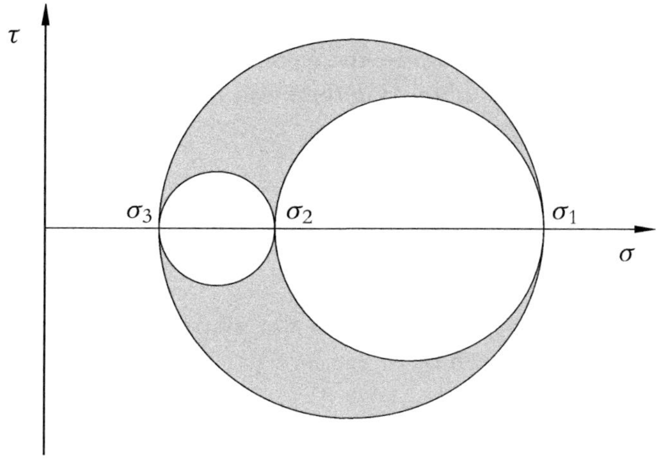

Figure 1.9 Mohr's construction in three dimensions.

1.1.7 Stress invariants

The stress tensor is a second-order tensor. When changing to a rotated set of coordinate axes, the components of the stress tensor change. However, as discussed in Appendix C, some properties of the stress tensor remain unchanged. The simplest of these is the mean normal stress

$$\overline{\sigma} = \left(\sigma_x + \sigma_y + \sigma_z\right)/3 \tag{1.40}$$

which equals 1/3 of the trace of the matrix (see Section C.2.9). The mean normal stress is thus an *invariant of stress.*

There also exist other stress combinations that are independent of the coordinate axes. Any combination of stress invariants will of course be a stress invariant as well. The commonly used stress invariants are

$$I_1 = \sigma_x + \sigma_y + \sigma_z \tag{1.41}$$

$$I_2 = -\left(\sigma_x\sigma_y + \sigma_y\sigma_z + \sigma_z\sigma_x\right) + \tau_{xy}^2 + \tau_{yz}^2 + \tau_{xz}^2 \tag{1.42}$$

$$I_3 = \sigma_x\sigma_y\sigma_z + 2\tau_{xy}\tau_{yz}\tau_{xz} - \sigma_x\tau_{yz}^2 - \sigma_y\tau_{xz}^2 - \sigma_z\tau_{xy}^2 \tag{1.43}$$

(See Appendix C.8 for some information about invariants.)

1.1.8 Deviatoric stresses

The mean normal stress $\overline{\sigma}$, defined in Eq. (1.40), essentially causes uniform compression or extension. Distortions, on the other hand, are essentially caused by the so-called deviatoric stresses. The deviatoric stress (also called stress deviator or stress deviation—the terminology is not consistent in the literature) is obtained by subtracting the mean normal stress from the normal stress components[1]:

$$\begin{pmatrix} s_x & s_{xy} & s_{xz} \\ s_{xy} & s_y & s_{yz} \\ s_{xz} & s_{yz} & s_z \end{pmatrix} = \begin{pmatrix} \sigma_x - \overline{\sigma} & \tau_{xy} & \tau_{xz} \\ \tau_{xy} & \sigma_y - \overline{\sigma} & \tau_{yz} \\ \tau_{xz} & \tau_{yz} & \sigma_z - \overline{\sigma} \end{pmatrix} \tag{1.44}$$

Invariants of stress deviation similar to the invariants of stress defined in Eqs. (1.41)–(1.43) are given by

$$J_1 = s_x + s_y + s_z = 0 \tag{1.45}$$

$$J_2 = -\left(s_x s_y + s_y s_z + s_z s_x\right) + s_{xy}^2 + s_{yz}^2 + s_{xz}^2 \tag{1.46}$$

$$J_3 = s_x s_y s_z + 2 s_{xy} s_{yz} s_{xz} - s_x s_{yz}^2 - s_y s_{xz}^2 - s_z s_{xy}^2 \tag{1.47}$$

The invariants J_1, J_2, J_3 (and combinations of them) are independent of the choice of coordinate axes. Invariants of stress deviation appear, for example, in failure criteria, since these must be independent of the choice of coordinate axes (for isotropic materials).

There are many different ways of writing the invariants of stress deviation. See Appendix D.1.1 for some useful expressions.

We will often encounter various variants of the stress invariants, in particular, the parameters q (known as the *von Mises stress* or the *generalised shear stress*[2]) and r, which are related to the basic invariants as

$$q = \sqrt{3J_2} = \sqrt{\frac{3}{2}\left[(\sigma_1 - \overline{\sigma})^2 + (\sigma_2 - \overline{\sigma})^2 + (\sigma_3 - \overline{\sigma})^2\right]} \tag{1.48}$$

$$r = \sqrt[3]{\frac{27}{2} J_3} = \sqrt[3]{\frac{27}{2}(\sigma_1 - \overline{\sigma})(\sigma_2 - \overline{\sigma})(\sigma_3 - \overline{\sigma})} \tag{1.49}$$

For a stress state in which two of the principal stresses are equal ($\sigma_2 = \sigma_3$), the expressions simplify to

$$q = |\sigma_1 - \sigma_3| \tag{1.50}$$

[1] The term deviatoric stress is sometimes used to express the difference between axial stress and confining pressure in laboratory tests. This use is not in accord with the definition.

[2] Beware that in some sciences the term *effective stress* is used for the *von Mises stress*. One may also encounter *effective von Mises stress* or *equivalent von Mises stress*. In this book, we restrict *effective stress* to the poroelastic version defined in Section 1.6.4.

and

$$r = \sigma_1 - \sigma_3 \tag{1.51}$$

The parameter q will be used extensively in Chapter 2. It gives a fair estimate of the maximum shear stress in the sample:

$$0.5 \leq \frac{\tau_{\max}}{q} \leq \frac{1}{\sqrt{3}} \approx 0.577 \tag{1.52}$$

The minimum occurs when two of the principal stresses are equal, whereas the maximum occurs when the intermediate principal stress is the average of the two others.

1.1.8.1 Geometric interpretation of the deviatoric stress invariants

The deviatoric stress invariants have a straightforward geometrical interpretation in principal stress space (see Section 2.1.2), as illustrated in Fig. 1.10. Eq. (1.48) is the equation of a circle centred on $\bar{\sigma}$, with the normal pointing along the hydrostatic axis $\sigma_1 = \sigma_2 = \sigma_3$. Thus the distance from a point $(\sigma_1, \sigma_2, \sigma_3)$ in principal stress space to the hydrostatic axis is

$$\sqrt{\frac{2}{3}}q = \sqrt{2J_2} \tag{1.53}$$

It can further be shown (see e.g. Chen and Han, 1988) that the angle ϑ_L, called the Lode angle (in honour of the Lode, 1926 paper), indicated in Fig. 1.10, is given by the invariants as

$$\cos(3\vartheta_L) = \left(\frac{r}{q}\right)^3 = \frac{3\sqrt{3}J_3}{2J_2^{3/2}} \tag{1.54}$$

(Note that since arccos is a multi-valued function, the Lode angle computed from Eq. (1.54) is not unique. If we choose the principal branch of arccos, then the result will be in the range 0° to 60° even if the actual stress state corresponds to another value.)

Using $z_L = I_1/\sqrt{3} = \sqrt{3}\bar{\sigma}$, $r_L = \sqrt{2J_2} = \sqrt{2/3}\,q$, we see that ($r_L$, ϑ_L, z_L) define a cylindrical coordinate system in which the z-axis points along the hydrostatic axis, and the new x (from which ϑ_L is measured) is determined by pointing along the projection of the original x-axis (i.e. the σ_1 axis) in the π-plane. These coordinates are known as the *Lode coordinates* or the *Haigh–Westergaard coordinates*.

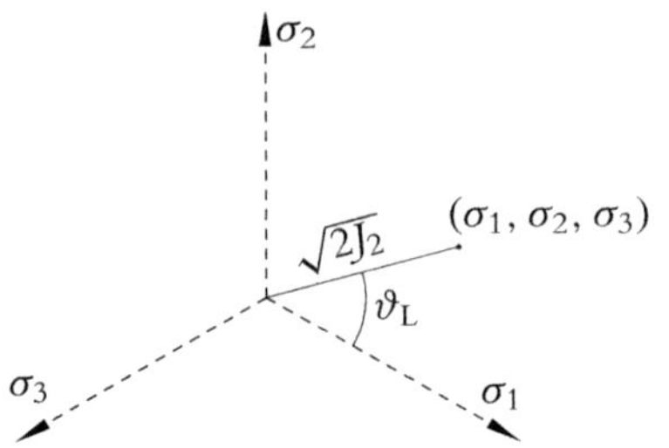

Figure 1.10 Geometrical interpretation of the deviatoric stress invariants in principal stress space. The dashed lines are the *projections* of the principal stress axes onto a deviatoric plane (i.e. a plane normal to the hydrostatic axis $\sigma_1 = \sigma_2 = \sigma_3$, also called the π-plane) passing through the point $(\sigma_1, \sigma_2, \sigma_3)$. The angle ϑ_L is called the Lode angle.

1.1.9 The octahedral stresses

A plane normal to the (1,1,1) direction in the principal stress space is called an octahedral plane, a π-plane or a deviatoric plane.

The normal stress on and shear stress in this plane are sometimes called the *octahedral normal stress* and the *octahedral shear stress*, and are given by

$$\sigma_{oct} = \frac{1}{3}(\sigma_1 + \sigma_2 + \sigma_3) = \overline{\sigma} = \frac{1}{3}I_1 \tag{1.55}$$

$$\tau_{oct} = \frac{1}{3}\sqrt{(\sigma_2 - \sigma_3)^2 + (\sigma_3 - \sigma_1)^2 + (\sigma_1 - \sigma_2)^2} = \sqrt{\frac{2}{3}J_2} = \frac{\sqrt{2}}{3}q \tag{1.56}$$

Note that the normal stress in the direction equally inclined to the principal stress axes is thus equal to the mean stress.

1.2 Strain

Consider a sample as shown in Fig. 1.11. The position of a specific particle within the sample is initially x, y, z. After the action of an external force, the position of this particle is shifted. We will denote the shift in the x-direction by u, the shift in the y-direction by v, and the shift in the z-direction by w. The quantities u, v and w are called the displacements of the particle. To make the signs of the *displacements* compatible with the signs of the stresses, as defined in Section 1.1, the displacements are taken to be positive when they are directed in the negative direction of the axes. Hence, the new position of the particle initially at x, y, z becomes

$$x' = x - u \tag{1.57}$$

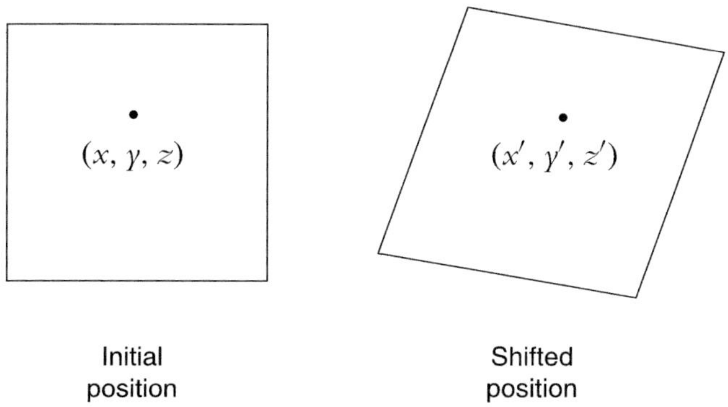

Figure 1.11 Sample deformation.

$$y' = y - v \tag{1.58}$$

$$z' = z - w \tag{1.59}$$

If the displacements u, v and w are constants, that is, they are the same for every particle within the sample, then the displacement is simply a *translation of a rigid body*. (Beware that with this definition, the displacement vector counter-intuitively points from the final position to the original position.)

Another simple form of displacements is the *rotation of a rigid body*. For a small rotation specified by $\vec{\omega}$, where the magnitude $|\vec{\omega}|$ gives the angle of rotation, whereas the direction of $\vec{\omega}$ gives the axis of rotation, the new position of the particle becomes

$$\vec{r}\,' = \vec{r} + \vec{\omega} \times (\vec{r} - \vec{r}_0) \tag{1.60}$$

where $\vec{r} = (x, y, z)$, $\vec{r}\,' = (x', y', z')$, and $\times$ denotes the vector product. The vector $\vec{r}_0$ is the centre of rotation, through which the axis of rotation goes.

If the relative positions of the particles within the sample are changed, so that the new positions cannot be obtained simply by a rigid translation and/or rotation of the sample, then the sample is said to be strained. Fig. 1.12 shows an example of a strained sample. The displacements related to the positions O and P are not equal. The contraction in the direction OP is given as

$$\varepsilon = \frac{L - L'}{L} = -\frac{\Delta L}{L} \tag{1.61}$$

Contraction belongs to a specific type of quantities known as *strains*, and ε as defined in Eq. (1.60) is called the *normal strain* corresponding to the point

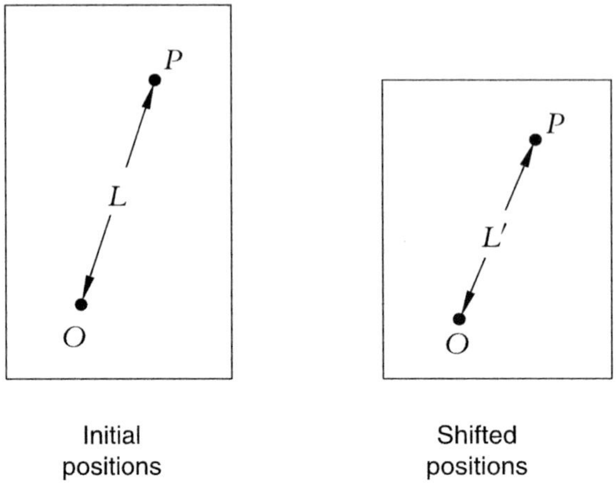

Figure 1.12 Deformation.

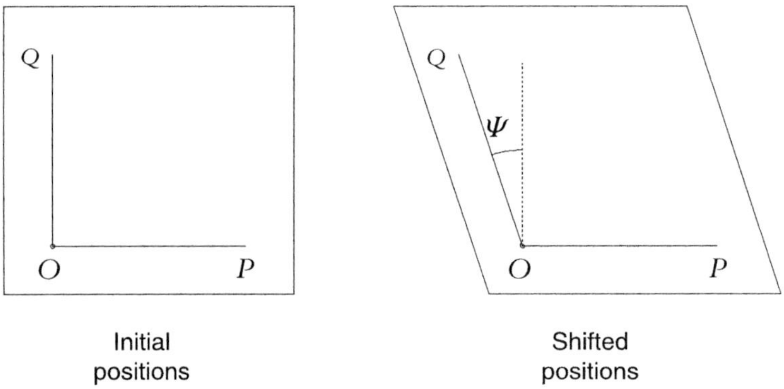

Figure 1.13 Shear deformation.

O in the direction OP. To comply with the sign convention for stresses, we require that the normal strain is positive for contraction. The other type of strains that may occur can be expressed by the change Ψ of the angle between two initially orthogonal directions (Fig. 1.13). The quantity

$$\Gamma = \frac{1}{2}\tan\Psi \tag{1.62}$$

is called the *shear strain* corresponding to the point O and the direction OP.

For many applications, we will only be dealing with *infinitesimal strains*, which implies that the strains ε and Γ are so small that their products and squares can be ignored, and we will make this approximation in the following. See Section 1.8 for a discussion of finite strains and nonlinear effects.

Now consider for a while the strains in two dimensions as shown in Fig. 1.14. The normal strain at x, in the x-direction, is given as

$$\begin{aligned}\varepsilon_x &= \frac{(x+\Delta x) - x - [(x+\Delta x - u(x+\Delta x)) - (x - u(x))]}{(x+\Delta x) - x} \\ &= \frac{u(x+\Delta x) - u(x)}{\Delta x}\end{aligned} \tag{1.63}$$

In the limit as $\Delta x \rightarrow 0$, we have

$$\varepsilon_x = \frac{\partial u}{\partial x} \tag{1.64}$$

Since the strains are small, we find for the shear strain corresponding to the x-direction:

$$\begin{aligned}\Gamma_{xy} &= \frac{1}{2}\tan\Psi \approx \frac{1}{2}\sin\Psi \\ &= -\frac{1}{2}\cos\left(\frac{\pi}{2} + \Psi\right) \\ &= -\frac{1}{2}\frac{\vec{P}_1' \cdot \vec{P}_2'}{\left|\vec{P}_1\right| \cdot \left|\vec{P}_2\right|}\end{aligned} \tag{1.65}$$

The vectors $\vec{P}_1$, $\vec{P}_1'$, $\vec{P}_2$, $\vec{P}_2'$ are found in Fig. 1.14. When $\Delta x \rightarrow 0$, $\Delta y \rightarrow 0$, and squares and products of the strains are neglected, we find that

$$\Gamma_{xy} = \frac{1}{2}\left(\frac{\partial u}{\partial y} + \frac{\partial v}{\partial x}\right) \tag{1.66}$$

It is clear from Eq. (1.66) that the shear strain Γ_{yx} corresponding to the y-direction is equal to Γ_{xy}.

To give a full description of the strain state at a point within a three-dimensional body, the elongations and shear strains corresponding to all three axes must be specified. In accordance with Eqs. (1.64) and (1.66), these strains are defined as

$$\varepsilon_x = \frac{\partial u}{\partial x} \tag{1.67}$$

$$\varepsilon_y = \frac{\partial v}{\partial y} \tag{1.68}$$

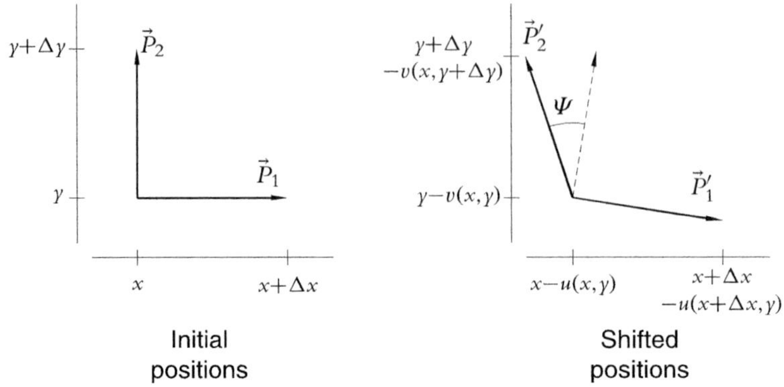

Figure 1.14 Parametrisation of shear deformation.

$$\varepsilon_z = \frac{\partial w}{\partial z} \tag{1.69}$$

$$\Gamma_{xy} = \Gamma_{yx} = \frac{1}{2}\left(\frac{\partial u}{\partial y} + \frac{\partial v}{\partial x}\right) \tag{1.70}$$

$$\Gamma_{xz} = \Gamma_{zx} = \frac{1}{2}\left(\frac{\partial u}{\partial z} + \frac{\partial w}{\partial x}\right) \tag{1.71}$$

$$\Gamma_{yz} = \Gamma_{zy} = \frac{1}{2}\left(\frac{\partial v}{\partial z} + \frac{\partial w}{\partial y}\right) \tag{1.72}$$

1.2.1 The strain tensor and the strain invariants

Similarly to Eq. (1.5), we may organise the strains (1.67)–(1.72) in a *strain tensor*

$$\overleftrightarrow{\varepsilon} = \begin{pmatrix} \varepsilon_x & \Gamma_{xy} & \Gamma_{xz} \\ \Gamma_{xy} & \varepsilon_y & \Gamma_{yz} \\ \Gamma_{xz} & \Gamma_{yz} & \varepsilon_z \end{pmatrix} \tag{1.73}$$

The numerical factor 1/2 in Eq. (1.67) and Eqs. (1.70)–(1.72) is necessary for the strain to have the properties of a mathematical tensor. It has been common, especially on older literature, to leave out the factor 1/2. The shear strain is then referred to as the *engineering shear strain*. Unfortunately, authors are not always explicit about their choice, but it is clearly vital to ascertain the convention before using formulas.

The trace of the strain tensor

$$\varepsilon_{\text{vol}} = \varepsilon_x + \varepsilon_y + \varepsilon_z \tag{1.74}$$

is identical to the *volumetric* strain, that is, the relative decrease in volume. The volumetric strain is independent on the choice of coordinate axes and is thus an *invariant of strain*.

Similarly to the stress invariants of Eqs. (1.42)–(1.43), it can be shown that the quantities

$$\mathcal{I}_2 = -(\varepsilon_x\varepsilon_y + \varepsilon_y\varepsilon_z + \varepsilon_z\varepsilon_x) + \Gamma_{xy}^2 + \Gamma_{yz}^2 + \Gamma_{xz}^2 \tag{1.75}$$

and

$$\mathcal{I}_3 = \varepsilon_x\varepsilon_y\varepsilon_z + 2\Gamma_{xy}\Gamma_{yz}\Gamma_{xz} - \varepsilon_x\Gamma_{yz}^2 - \varepsilon_y\Gamma_{xz}^2 - \varepsilon_z\Gamma_{xy}^2 \tag{1.76}$$

are also invariants of strain.

There also exists a mathematical notation for strains, similar to Eq. (1.9). In this notation, all strains are defined by

$$\varepsilon_{ij} = \frac{1}{2}\left(\frac{\partial u_i}{\partial x_j} + \frac{\partial u_j}{\partial x_i}\right) \tag{1.77}$$

The subscripts i and j may be any of the numbers 1, 2, 3, representing the x-, y- and z-axis, respectively. Thus $u_1 = u$, $u_2 = v$ and $u_3 = w$, whereas $x_1 = x$, $x_2 = y$ and $x_3 = z$. We then have $\varepsilon_{11} = \varepsilon_x$, $\varepsilon_{13} = \Gamma_{xz}$, and so on.

In this notation the strain tensor (1.73) becomes

$$\begin{pmatrix} \varepsilon_{11} & \varepsilon_{12} & \varepsilon_{13} \\ \varepsilon_{12} & \varepsilon_{22} & \varepsilon_{23} \\ \varepsilon_{13} & \varepsilon_{23} & \varepsilon_{33} \end{pmatrix} \tag{1.78}$$

Note that the strain is defined such that displacements that can be obtained by a rigid rotation of the sample shall be eliminated. Eq. (1.77) takes care of this, but only for small rotations. An extended definition that excludes any rotation is given in Appendix C.7.3.

1.2.2 Compatibility conditions

We note from the general definition of strain (Eq. (1.77)) that all strains are the derivatives (in various combinations) of the components of the displacement vector $\vec{u} = (u_1, u_2, u_3)$. Some useful expressions may be derived from this fact. For instance, we observe from Eqs. (1.74) and (1.77) that the volumetric strain ε_{vol} equals the divergence of $\vec{u}$, that is,

$$\varepsilon_{\text{vol}} = \nabla \cdot \vec{u} = -\frac{\mathrm{d}V}{V} \tag{1.79}$$

The minus sign is due to our sign convention for strains. Other relations can be obtained by comparing some of the second derivatives of the strains. We find, for example,

$$\frac{\partial^2 \varepsilon_x}{\partial y^2} + \frac{\partial^2 \varepsilon_y}{\partial x^2} = 2\frac{\partial^2 \Gamma_{xy}}{\partial x \partial y} \left(= \frac{\partial^3 u}{\partial x \partial y^2} + \frac{\partial^3 v}{\partial y \partial x^2} \right) \tag{1.80}$$

$$\frac{\partial^2 \varepsilon_x}{\partial z^2} + \frac{\partial^2 \varepsilon_z}{\partial x^2} = 2\frac{\partial^2 \Gamma_{xz}}{\partial x \partial z} \left(= \frac{\partial^3 u}{\partial x \partial z^2} + \frac{\partial^3 w}{\partial z \partial x^2} \right) \tag{1.81}$$

$$\frac{\partial^2 \varepsilon_z}{\partial y^2} + \frac{\partial^2 \varepsilon_y}{\partial z^2} = 2\frac{\partial^2 \Gamma_{zy}}{\partial z \partial y} \left(= \frac{\partial^3 w}{\partial z \partial y^2} + \frac{\partial^3 v}{\partial y \partial z^2} \right) \tag{1.82}$$

These three differential relations, together with three others that express $\partial^2 \varepsilon_x/(\partial y \partial z)$, $\partial^2 \varepsilon_y/(\partial x \partial z)$ and $\partial^2 \varepsilon_z/(\partial x \partial y)$ in terms of second derivatives of the shear strains, are known as the *compatibility conditions* for strain.

1.2.3 Principal strains

In Section 1.1, we saw that for some specific directions, the shear stress vanishes, so that for a specific orientation of the coordinate system (with axis parallel to the principal axes of stress), the stress tensor becomes particularly simple. The situation is similar for strains.

In two dimensions, it can be shown that the shear strain vanishes in the directions θ relative to the x-axis, which fulfil the equation

$$\tan 2\theta = \frac{2\Gamma_{xy}}{\varepsilon_x - \varepsilon_y} \tag{1.83}$$

Thus, in two dimensions, there are two orthogonal directions for which the shear strain vanishes. These directions are called the *principal axes of strain*. The elongations in the directions of the principal axes of strain are called the *principal strains*.

In three dimensions, there are three principal axes of strain. The principal strains are found by solution of the determinant equation

$$\begin{vmatrix} \varepsilon_x - \varepsilon & \Gamma_{xy} & \Gamma_{xz} \\ \Gamma_{xy} & \varepsilon_y - \varepsilon & \Gamma_{yz} \\ \Gamma_{xz} & \Gamma_{yz} & \varepsilon_z - \varepsilon \end{vmatrix} = 0 \tag{1.84}$$

The solutions are denoted ε_1, ε_2, ε_3. The direction cosines l_{1x}, l_{1y}, l_{1z} identifying the principal axis corresponding to ε_1 are found by solution of the equations

$$l_{1x}(\varepsilon_x - \varepsilon_1) + l_{1y}\Gamma_{xy} + l_{1z}\Gamma_{xz} = 0 \tag{1.85}$$

$$l_{1x}\Gamma_{xy} + l_{1y}(\varepsilon_y - \varepsilon_1) + l_{1z}\Gamma_{yz} = 0 \tag{1.86}$$

$$l_{1x}\Gamma_{xz} + l_{1y}\Gamma_{yz} + l_{1z}(\varepsilon_z - \varepsilon_1) = 0 \tag{1.87}$$

The principal axes corresponding to ε_2 and ε_3 are found similarly by substituting subscript 1 by 2 and 3, respectively. Eqs. (1.84) and (1.85)–(1.87) are seen to be equivalents of Eqs. (1.31) and (1.32)–(1.34) identifying the principal stresses and the principal axes of stress.

1.2.4 Plane strain and plane stress

In several practical applications, it is a good approximation to assume that all cross-sections along a given axis are in the same condition and that there is no displacement along the axis. This state of strain is called *plane strain.*

In the following, we will assume the unique axis to be the z-axis. The strain tensor for plane strain is then

$$\begin{pmatrix} \varepsilon_x(x, y) & \Gamma_{xy}(x, y) & 0 \\ \Gamma_{xy}(x, y) & \varepsilon_y(x, y) & 0 \\ 0 & 0 & 0 \end{pmatrix} \tag{1.88}$$

where all strain components are independent of z. The term *plane* of course refers to the fact that the strain is confined to a plane.

If we only have displacement along z, and this displacement is independent of z, then the strain state is referred to as *antiplane strain.* The strain tensor is then

$$\begin{pmatrix} 0 & 0 & \Gamma_{xz}(x, y) \\ 0 & 0 & \Gamma_{yz}(x, y) \\ \Gamma_{xz}(x, y) & \Gamma_{yz}(x, y) & 0 \end{pmatrix} \tag{1.89}$$

where again all components are independent of z.

A general situation in which the displacement along the z-axis is independent of z can be decomposed into a sum of plane strain and antiplane strain. This state of strain is referred to as a *generalised plane strain.* The strain tensor is

$$\begin{pmatrix} \varepsilon_x(x, y) & \Gamma_{xy}(x, y) & \Gamma_{xz}(x, y) \\ \Gamma_{xy}(x, y) & \varepsilon_y(x, y) & \Gamma_{yz}(x, y) \\ \Gamma_{xz}(x, y) & \Gamma_{yz}(x, y) & 0 \end{pmatrix} \tag{1.90}$$

Note, however, that the concept of generalised plane strain is not uniquely defined in the literature. Sometimes the term is used when ε_z rather than the displacement along z is independent of z. This leads to the strain tensor

$$\begin{pmatrix} \varepsilon_x(x,y) & \Gamma_{xy}(x,y) & \Gamma_{xz}(x,y) \\ \Gamma_{xy}(x,y) & \varepsilon_y(x,y) & \Gamma_{yz}(x,y) \\ \Gamma_{xz}(x,y) & \Gamma_{yz}(x,y) & \varepsilon_z(x,y) \end{pmatrix} \tag{1.91}$$

See Cheng (1998) for a thorough discussion.

Analogously, if all stress components are independent of z, and $\sigma_z = \tau_{xz} = \tau_{yz} = 0$ (still taking z to be the unique axis), then we call the situation *plane stress*. The stress tensor is then

$$\begin{pmatrix} \sigma_x(x,y) & \tau_{xy}(x,y) & 0 \\ \tau_{xy}(x,y) & \sigma_y(x,y) & 0 \\ 0 & 0 & 0 \end{pmatrix} \tag{1.92}$$

The term *generalised plane stress* is used when all stress components are independent of z, leading to the stress tensor

$$\begin{pmatrix} \sigma_x(x,y) & \tau_{xy}(x,y) & \tau_{xz}(x,y) \\ \tau_{xy}(x,y) & \sigma_y(x,y) & \tau_{yz}(x,y) \\ \tau_{xz}(x,y) & \tau_{yz}(x,y) & \sigma_z(x,y) \end{pmatrix} \tag{1.93}$$

1.3 Elastic moduli

The theory of linear elasticity deals with situations where there are linear relationships between applied stresses and resulting strains. Whereas most rocks do behave nonlinearly when subject to large stresses, their behaviour may normally be described by linear relations for sufficiently small changes in stress. Consider a sample of length L and cross-sectional area $A = D^2$ (Fig. 1.15). When the force F is applied on its end surfaces, the length of the sample is reduced to L'. The applied stress is then $\sigma_x = F/A$, and the corresponding elongation is $\varepsilon_x = (L - L')/L$ according to Eqs. (1.1) and (1.61). If the sample behaves linearly, then there is a linear relation between σ_x and ε_z, which we may write

$$\varepsilon_x = \frac{1}{E}\sigma_x \tag{1.94}$$

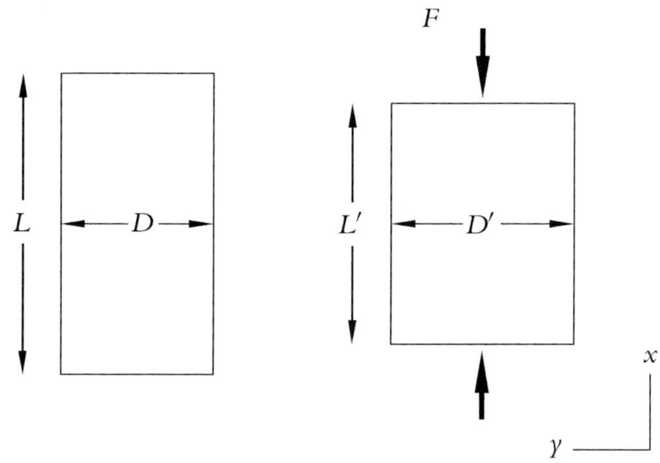

Figure 1.15 Deformation induced by uniaxial stress.

Eq. (1.94) is known as *Hooke's law*, whereas the coefficient E is called *Young's modulus* or simply the E-modulus. Young's modulus belongs to a group of coefficients called *elastic moduli*. It is a measure of the stiffness of the sample, that is, the resistance of the sample against being compressed by a uniaxial stress.

Another consequence of the applied stress σ_x (Fig. 1.15) is an increase in the width D of the sample. The lateral elongation is $\varepsilon_y = \varepsilon_z = (D - D')/D$. In general, $D' > D$, and thus ε_y and ε_z become negative. The ratio defined as

$$\nu = -\frac{\varepsilon_y}{\varepsilon_x} \tag{1.95}$$

is another elastic parameter, known as *Poisson's ratio*. It is a measure of lateral expansion relative to longitudinal contraction.

Eqs. (1.94) and (1.95), which relate one component of stress or strain to another, are defined by a specific state of stress, namely $\sigma_x \neq 0$ and $\sigma_y = \sigma_z = 0$. In general, each component of stress is a linear function of all components of strain, and vice versa.

Isotropic materials are materials whose response is independent of the orientation of the applied stress. For such materials, the principal axes of stress and the principal axes of strain always coincide. For isotropic materials, the general relations between stresses and strains may be written as

$$\sigma_x = (\lambda + 2G)\varepsilon_x + \lambda\varepsilon_y + \lambda\varepsilon_z \tag{1.96}$$

$$\sigma_y = \lambda\varepsilon_x + (\lambda + 2G)\varepsilon_y + \lambda\varepsilon_z \tag{1.97}$$

$$\sigma_z = \lambda\varepsilon_x + \lambda\varepsilon_y + (\lambda + 2G)\varepsilon_z \tag{1.98}$$

$$\tau_{yz} = 2G\Gamma_{yz} \tag{1.99}$$

$$\tau_{xz} = 2G\Gamma_{xz} \tag{1.100}$$

$$\tau_{xy} = 2G\Gamma_{xy} \tag{1.101}$$

The coefficients λ and G are elastic moduli, known as Lamé's parameters or moduli. The coefficient G is also known as the modulus of rigidity or the shear modulus. It is a measure of the resistance of a sample against shear deformation.

Another important elastic modulus is the bulk modulus K. It is defined as the ratio of hydrostatic stress σ_p relative to the volumetric strain $\varepsilon_{\rm vol}$ (Eq. (1.74)). For a hydrostatic stress state, we have $\sigma_p = \sigma_x = \sigma_y = \sigma_z$ and $\tau_{xy} = \tau_{yz} = \tau_{xz} = 0$. From Eqs. (1.96)–(1.98) we then find

$$K = \frac{\sigma_p}{\varepsilon_{\rm vol}} = \lambda + \frac{2}{3}G \tag{1.102}$$

where K is a measure of the resistance of a sample against hydrostatic compression; $C_{\rm b} = 1/K$ is known as the compressibility.

In the experiment (Fig. 1.15) defining Young's modulus and Poisson's ratio, the stress is uniaxial, that is, $\sigma_y = \sigma_z = \tau_{xy} = \tau_{xz} = \tau_{yz} = 0$. From Eqs. (1.96)–(1.98) we then find

$$E = \frac{\sigma_x}{\varepsilon_x} = G\frac{3\lambda + 2G}{\lambda + G} \tag{1.103}$$

$$\nu = -\frac{\varepsilon_y}{\varepsilon_x} = \frac{\lambda}{2(\lambda + G)} \tag{1.104}$$

From relations (1.102)–(1.104) it follows that when any two of the moduli E, ν, λ, G and K are defined, the remaining ones are fixed by these relations. Depending on which two of the moduli are known, special combinations of Eqs. (1.102)–(1.104) may be needed. Some of the most useful combinations are listed in Table 1.1.

Table 1.1 also includes some relations involving $H = \lambda + 2G$, the *uniaxial compaction modulus* or *oedometer modulus*, which is repeatedly used in later chapters. In the context of acoustics, H is referred to as the *plane wave modulus* or the *P-wave modulus*.

Table 1.1 Some relations between elastic moduli.

$E=3K(1-2\nu)$	$E=2G(1+\nu)$	$E=\dfrac{9KG}{3K+G}$	$E=G\dfrac{3\lambda+2G}{\lambda+G}$
$E=\dfrac{\lambda}{\nu}(1+\nu)(1-2\nu)$	$K=\lambda\dfrac{1+\nu}{3\nu}$	$K=\dfrac{2}{3}G\dfrac{1+\nu}{1-2\nu}$	$K=\lambda+\dfrac{2}{3}G$
$K=\dfrac{GE}{9G-3E}$	$K=\dfrac{H+2\lambda}{3}$	$H=\lambda+2G$	$H=K+\dfrac{4}{3}G$
$H=E\dfrac{1-\nu}{(1+\nu)(1-2\nu)}$	$H=2G\dfrac{1-\nu}{1-2\nu}$	$H=3K\dfrac{1-\nu}{1+\nu}$	$\nu=\dfrac{H-2G}{2(H-G)}$
$\nu=\dfrac{E-2G}{2G}$	$\nu=\dfrac{3K-2G}{2(3K+G)}$	$\nu=\dfrac{1}{2}\left(1-\dfrac{E}{3K}\right)$	$\nu=\dfrac{3K-H}{3K+H}$
$\nu=\dfrac{\lambda}{\lambda+H}$	$\dfrac{\lambda}{\lambda+G}=2\nu$	$\dfrac{G}{\lambda+G}=1-2\nu$	$\dfrac{\lambda+2G}{\lambda+G}=2(1-\nu)$
$\dfrac{3\lambda+2G}{\lambda+G}=2(1+\nu)$	$\dfrac{\lambda}{G}=\dfrac{2\nu}{1-2\nu}$	$\dfrac{3\lambda+4G}{\lambda+G}=2(2-\nu)$	$\dfrac{\lambda}{\lambda+2G}=\dfrac{\nu}{1-\nu}$

For sedimentary rocks, values for Poisson's ratio are usually found in the range 0.05–0.45, the median being around 0.2. For fluids, the rigidity G vanishes, which according to Eq. (1.104) implies that ν approaches 1/2. However, we may also encounter weak, porous rocks, where ν may approach zero. Some physical limits for the elastic moduli are discussed towards the end of Section 1.4.

The elastic moduli E, λ, G, H and K are measured in the same units as stress, for example, Pa, psi or bar. This follows from Eqs. (1.94), (1.96) and (1.102), and from Eq. (1.61), which shows that strain is dimensionless.

Typical values for the elastic moduli of some rocks are given in Appendix A. Note that the given values may change with the stress state. This will be further discussed in Section 1.8.

The stress–strain relations (1.96)–(1.101) are the fundamental equations for description of isotropic linear elastic materials. In many cases, however, it is convenient to have these equations on an alternative form, expressing the strains as functions of the stresses. Introducing expressions (1.103) and (1.104) for E and ν, this alternative form becomes

$$E\varepsilon_x=\sigma_x-\nu(\sigma_y+\sigma_z) \tag{1.105}$$

$$E\varepsilon_y=\sigma_y-\nu(\sigma_x+\sigma_z) \tag{1.106}$$

$$E\varepsilon_z=\sigma_z-\nu(\sigma_x+\sigma_y) \tag{1.107}$$

$$G\Gamma_{yz} = \frac{1}{2}\tau_{yz} \tag{1.108}$$

$$G\Gamma_{xz} = \frac{1}{2}\tau_{xz} \tag{1.109}$$

$$G\Gamma_{xy} = \frac{1}{2}\tau_{xy} \tag{1.110}$$

The stress–strain relations (1.96)–(1.101) may be written on a more compact form using the notation introduced in Eqs. (1.9) and (1.77) as

$$\sigma_{ij} = \lambda\varepsilon_{\text{vol}}\delta_{ij} + 2G\varepsilon_{ij} \tag{1.111}$$

where δ_{ij} is the Kronecker symbol (see Eq. (C.125)). See Appendix D.1.4 for some other useful ways of writing the formulas.

The concept of plane strain was introduced in Section 1.2.4. Consider again the situation in Fig. 1.15, but let us now assume that the body is constrained such that there is no strain in the z-direction, that is, it is in a state of plane strain. Introducing $\varepsilon_z = 0$ (and retaining $\sigma_y = 0$ from Fig. 1.15) in Eqs. (1.105)–(1.107), we find an equation corresponding to Eq. (1.94):

$$E' = \frac{\sigma_x}{\varepsilon_x} = \frac{E}{1-\nu^2} = \frac{2G}{1-\nu} \tag{1.112}$$

The quantity E' is called the *plane strain modulus* since it often shows up in plane strain problems. As an example, see Section 11.4. Note that we have $E \leq E' \leq H$, which means that the stiffness of the sample increases as it is increasingly confined in the lateral directions.

In fact, when applying the equations of elasticity to rock in the subsurface, we should always treat the stresses in the equations as stress changes. Strains in themselves (Section 1.2) represent changes in displacements or distortions of shape, and they are clearly related to changes in stress. In conventional solid mechanics, however, the reference stress state is often set to zero, whereas stresses in the Earth (Chapter 3) are far from zero. For compactness of notation, we will follow the solid mechanics tradition and leave out the Δ, especially in this chapter. It is then understood that the formulas apply to stress changes and associated strains.

1.4 Strain energy

A strained body possesses a potential energy which may be released during unloading. Consider a small cube of a material, with sides a, loaded uniaxially with the stress σ. The resulting elongation is $\varepsilon = \sigma/E$. The work done by increasing the stress from 0 to σ_1 is

$$\begin{aligned}
\text{work} &= \text{force} \cdot \text{distance} \\
&= \int_0^{\sigma_1} (a^2\sigma)(a\,\mathrm{d}\varepsilon) = a^3 \int_0^{\sigma_1} \sigma \frac{1}{E}\,\mathrm{d}\sigma \\
&= \frac{1}{2} a^3 \frac{\sigma_1^2}{E} = \frac{1}{2} a^3 E \varepsilon_1^2 = \frac{1}{2} a^3 \sigma_1 \varepsilon_1
\end{aligned} \tag{1.113}$$

where $\varepsilon_1 = \sigma_1/E$. As the stress state in this case is uniaxial, σ_1 is a principal stress, whereas ε_1 is a principal strain. When the other two principal stresses are non-zero, corresponding terms will add to the expression for the work. The work per unit volume (the potential energy per unit volume) then becomes

$$W = \frac{1}{2} (\sigma_1 \varepsilon_1 + \sigma_2 \varepsilon_2 + \sigma_3 \varepsilon_3) \tag{1.114}$$

W is called the *strain energy*.

A variety of expressions for the strain energy can be obtained by suitable substitutions for the principal stresses and/or the principal strains. Using Eqs. (1.96)–(1.98) to express the stresses in terms of the strains, we find that the strain energy (1.114) is equal to

$$W = \frac{1}{2} \left[(\lambda + 2G)(\varepsilon_1^2 + \varepsilon_2^2 + \varepsilon_3^2) + 2\lambda(\varepsilon_1\varepsilon_2 + \varepsilon_1\varepsilon_3 + \varepsilon_2\varepsilon_3) \right] \tag{1.115}$$

Note that Eq. (1.115) is quadratic in strain. As a result, the elastic moduli are sometimes referred to as *second-order elastic moduli*. In general, there will also be (normally smaller) energy contributions from higher-order combinations of strain; see Section 1.8.1. Coefficients similar to the elastic moduli above relating energy to cubic terms in strain are referred to as *third-order elastic moduli*.

Comparing to Eqs. (1.74) and (1.75), we may express the strain energy in terms of strain invariants:

$$W = \frac{1}{2} \left[(\lambda + 2G)\varepsilon_{\text{vol}}^2 + 4G\mathcal{I}_2 \right] \tag{1.116}$$

This allows writing the result for a general (non-principal) strain state:

$$\begin{aligned}
W = \frac{1}{2} \Big[&(\lambda + 2G)(\varepsilon_x^2 + \varepsilon_y^2 + \varepsilon_z^2) + 2\lambda(\varepsilon_x\varepsilon_y + \varepsilon_y\varepsilon_z + \varepsilon_z\varepsilon_x) \\
&+ 4G(\Gamma_{xy}^2 + \Gamma_{yz}^2 + \Gamma_{xz}^2) \Big]
\end{aligned} \tag{1.117}$$

Although convenient in transforming the equations, Eq. (1.116) is not ideal from a fundamental point of view, since a change in volumetric strain changes $\mathcal{I}_2$. Rewriting in terms of the deviatoric strain invariant $\mathcal{J}_2$ (in analogy with Eq. (D.5)), we find, using Table 1.1:

$$W = \frac{1}{2}\left[K\varepsilon_{\text{vol}}^2 + 4G\mathcal{J}_2\right] \tag{1.118}$$

which shows more clearly that the strain energy due to a volumetric change depends on the bulk modulus, whereas the strain energy from a shear deformation depends on the shear modulus.

Useful relations can be established by analysis of the strain energy. Taking the derivative of Eq. (1.117) with respect to ε_x and using Eq. (1.96), we find that

$$\frac{\partial W}{\partial \varepsilon_x} = \frac{1}{2}\left[2(\lambda + 2G)\varepsilon_x + 2\lambda(\varepsilon_y + \varepsilon_z)\right] = \sigma_x \tag{1.119}$$

Similar expressions connecting σ_y to ε_y and so on can also be established in the same way. We now observe, by taking the derivative of Eq. (1.119) with respect to ε_y, that it is possible to establish a set of expressions of the type

$$\frac{\partial \sigma_x}{\partial \varepsilon_y} = \frac{\partial^2 W}{\partial \varepsilon_y \partial \varepsilon_x} = \frac{\partial \sigma_y}{\partial \varepsilon_x} \tag{1.120}$$

This equation gives rise to a general symmetry restriction on the elastic moduli, which will be discussed later. (See Eq. (1.238).)

Going back to Eq. (1.113), we observe that Young's modulus E must be non-negative; otherwise, the system will be unstable ($E < 0$ implies that $\varepsilon_1 \to \infty$ will be energetically favourable). By considering other stress geometries we may similarly show that also the shear modulus G and the bulk modulus K must be non-negative. It follows from Table 1.1 that Poisson's ratio ν is then restricted to the region $-1 < \nu < 1/2$. (Note that these restrictions are derived assuming that the material is isotropic and linearly elastic.) Negative values are hardly ever seen in rocks. Theoretically, $\nu = -1$ would correspond to a material with finite shear but zero bulk modulus. Such materials are hypothetical, but negative Poisson's ratios are observed in several cases, like foams, cork and certain honeycomb structures.

Note that negative Poisson's ratios appear in a number of anisotropic materials, including crystals, but only in specific directions. Anisotropy gives rise to different bounds and also (as will be seen in Section 1.7) to several different Poisson's ratios. In the non-elastic regime, dilatancy may occur (Section 1.8, Chapter 2) leading to static Poisson's ratio larger than 0.5, but this is beyond the limits of linear elasticity.

1.5 Thermoelasticity

1.5.1 Thermal strain

Most materials expand or contract under a temperature change. Let us consider an elastic rod, which is free to expand. The initial temperature is T_i, and the temperature is changed to some other value T. The axial thermal strain resulting from the temperature change is then given by

$$\varepsilon_z = -\alpha_T (T - T_i) \tag{1.121}$$

where α_T is the coefficient of *linear thermal expansion*. The minus sign ensures that α_T is positive (for the normal cases where a temperature increase gives expansion). Some examples of the numerical values for thermal properties of rocks may be found in Appendix A.

When comparing the thermal expansion for rocks with that of fluids, it is important to be aware that for fluids, we often specify the coefficient of *volumetric thermal expansion*, $\alpha_{TV} = 3\alpha_T$.

1.5.2 Thermal stress

If the rod is constrained at the ends, such that it cannot change its length, a thermal stress will build up when the temperature increases. The magnitude of the thermal stress may be inferred by requiring that the thermal stress should give a strain of opposite sign and equal magnitude to the thermal strain computed from Eq. (1.121). From Eq. (1.94) we see that the thermal stress resulting from a temperature change $T - T_i$ for a rod that is fully constrained in one direction is

$$\sigma_z = -E\varepsilon_z = E\alpha_T (T - T_i) \tag{1.122}$$

1.5.3 Stress–strain relation for linear thermoelasticity

To take thermal effects into consideration, the stress–strain relations must be modified to take the thermal stress and strain into account.

Using the compact notation employed in Eq. (1.111) we may write

$$\sigma_{ij} = \lambda\varepsilon_{\text{vol}}\delta_{ij} + 2G\varepsilon_{ij} + 3\alpha_T K (T - T_i)\,\delta_{ij} \tag{1.123}$$

In terms of K and ν, this becomes

$$\sigma_{ij} = \frac{3K\nu}{1+\nu}\varepsilon_{\text{vol}}\delta_{ij} + 3K\frac{1-2\nu}{1+\nu}\varepsilon_{ij} + 3\alpha_{\text{T}}K(T-T_{\text{i}})\,\delta_{ij} \tag{1.124}$$

whereas in terms of E and ν, we have

$$\sigma_{ij} = \frac{E\nu}{(1+\nu)(1-2\nu)}\varepsilon_{\text{vol}}\delta_{ij} + \frac{E}{1+\nu}\varepsilon_{ij} + \frac{E}{1-2\nu}\alpha_{\text{T}}(T-T_{\text{i}})\,\delta_{ij} \tag{1.125}$$

Eqs. (1.123)–(1.125) may be derived from each other using the relations in Table 1.1. It is further straightforward to show that Eq. (1.125) is consistent with Eqs. (1.121) and (1.122) by specifying the appropriate strain and stress conditions. To get Eq. (1.121), we should assume all stresses to be zero, whereas for Eq. (1.122), we assume zero strain and non-zero stress in one direction, and zero stress and non-zero strains in the other two directions.

1.5.4 Isothermal and adiabatic moduli

To compute the sound velocity in a gas, it is important to use the adiabatic rather than the isothermal compressibility of the gas. The difference in velocity is typically several tens of percent.

In principle, there is a similar difference between the adiabatic and the isothermal elastic moduli in solids, but the magnitude of the difference is so small that it may always be safely neglected. We will therefore not discuss this further here, but refer the interested reader to, for example, Landau and Lifshitz (1986).

1.5.5 Example: thermal stresses in a constrained square plate

We consider a square plate, rigidly constrained in the x and y directions, but free to expand in the z direction. We write the stress–strain relation (1.123) in differential form using the constraints $\Delta\varepsilon_x = \Delta\varepsilon_y = \Delta\sigma_z = 0$. The result is

$$\Delta\sigma_x = \Delta\sigma_y = \lambda\Delta\varepsilon_z + 3\alpha_{\text{T}}K\Delta T \tag{1.126}$$

$$0 = (\lambda + 2G)\Delta\varepsilon_z + 3\alpha_{\text{T}}K\Delta T \tag{1.127}$$

which may be solved to give

$$\Delta\varepsilon_z = -\frac{3\alpha_{\text{T}}K\Delta T}{\lambda + 2G} \tag{1.128}$$

and

$$\Delta\sigma_x = 2G\frac{3\lambda + 2G}{\lambda + 2G}\alpha_T \Delta T \tag{1.129}$$

where we have used $K = \lambda + 2G/3$. In terms of E and ν, we have

$$\Delta\sigma_x = \Delta\sigma_y = \frac{\alpha_T E}{1-\nu}\Delta T \tag{1.130}$$

For $\alpha_T = 15 \cdot 10^{-6}\,\mathrm{K}^{-1}$, $\nu = 0.25$ and $E = 5\,\mathrm{GPa}$ (fairly typical values for weakly consolidated rock), this becomes

$$\frac{\Delta\sigma_x}{\Delta T} = 0.1\,\mathrm{MPa\ K}^{-1} \tag{1.131}$$

This shows that thermal stresses may be quite significant, and we will see in later chapters that there are several aspects of petroleum rock mechanics where temperature effects play an important role.

1.6 Poroelasticity

So far, we have treated rocks as if they were homogeneous solid materials. However, rocks are generally composite materials and hence inhomogeneous on a microscopic scale. The behaviour of rocks, such as their elastic responses, their failure stresses, and so on, depend, to a large extent, on the *non-solid* part of the materials. In this section, we will take into account the void space, which not only is essential for oil to be produced from a reservoir, but also plays an important role in rock mechanical behaviour. We will first consider a macroscopic description of porous and permeable media, which allows us to study both static and dynamic mechanical properties. This approach is based on the theory of Maurice A. Biot (Biot, 1941, 1955, 1956, 1962, and others).

In this section, we will give a brief and practical introduction to the theory. We will mainly limit ourselves to a description of an idealised porous material, which is microscopically homogeneous and isotropic. This approximation is sometimes referred to as the Gassmann limit (Berryman, 1992) of the Biot theory, in honour of Gassmann (1951). See however Section 6.5 for a brief introduction to the Brown–Korringa formulation.

For those desiring a more thorough exposition, we refer to the books by Wang (2000); Cheng (2016); Coussy (2004); Charlez (1991) and the papers by Detournay and Cheng (1993); Berryman (1992); Rice and Cleary

(1976). Zimmerman (1991) is also a very useful introduction to the mechanical properties of porous media.

1.6.1 Suspension of solid particles in a fluid

Let us first take a look at a very simple porous medium, in which the solid and fluid parts are deformed independently of each other. In practice, we may think of this medium as a suspension of solid particles in a fluid or, for instance, as a water-saturated, completely unconsolidated sand. If we place this mixture in a container, then the volumetric strain due to an external pressure σ_p is

$$\varepsilon_{\mathrm{vol}} = \frac{\sigma_p}{K_{\mathrm{eff}}} \tag{1.132}$$

where K_{eff} is the bulk modulus of the mixture. The total deformation must, however, equal the sum of the deformations of each component, weighted by the volume portion of each component:

$$\varepsilon_{\mathrm{vol}} = \frac{V_{\mathrm{s}}}{V_{\mathrm{tot}}}\varepsilon_{\mathrm{vol,s}} + \frac{V_{\mathrm{f}}}{V_{\mathrm{tot}}}\varepsilon_{\mathrm{vol,f}} \tag{1.133}$$

where subscripts s and f denote solid and fluid, respectively, and V_{tot} is the total volume. Now, we define porosity ϕ as the volume occupied by the fluid relative to the total volume:

$$\phi = \frac{V_{\mathrm{f}}}{V_{\mathrm{tot}}} \tag{1.134}$$

$$\frac{V_{\mathrm{s}}}{V_{\mathrm{tot}}} = 1 - \phi \tag{1.135}$$

The strains $\varepsilon_{\mathrm{vol,s}}$ and $\varepsilon_{\mathrm{vol,f}}$ are given by the bulk moduli of the solid (K_{s}) and the fluid (K_{f}), respectively, according to Eq. (1.102). Then Eq. (1.133) may be written as

$$\varepsilon_{\mathrm{vol}} = (1-\phi)\frac{\sigma_p}{K_{\mathrm{s}}} + \phi\frac{\sigma_p}{K_{\mathrm{f}}} \tag{1.136}$$

By combining Eqs. (1.132) and (1.136) we now find that the undrained[3] effective modulus of the suspension is given by

$$\frac{1}{K_{\mathrm{u}}} = \frac{1-\phi}{K_{\mathrm{s}}} + \frac{\phi}{K_{\mathrm{f}}} \tag{1.137}$$

[3] Since the medium is confined in a container, there is no way for the fluid to escape, which is called undrained conditions. This will be extensively discussed in the next paragraphs.

In terms of the corresponding compressibilities the equation is the weighted average

$$c_u = (1 - \phi)c_s + \phi c_f \tag{1.138}$$

The relations correspond to the Reuss average discussed in Section 6.2. In the context of acoustics, they are often referred to as Wood's equation (Wood, 1941). The Reuss average is also denoted as an isostress average, since the fluid and solid components experience the same stress. This is an example of a particularly simple porous material. We will now generalise it by taking into account the fact that rocks consist of a solid framework and a pore fluid which cannot be treated independently.

1.6.2 Biot's poroelastic theory for static properties

We will now consider an isotropic, porous and permeable medium consisting of two components, a solid and a fluid part. The properties of a porous medium are obviously different from the material it consists of. We will refer to the porous medium as the *frame*, the *framework* or sometimes the *skeleton*. We will refer to the material the porous medium consists of as the *solid constituent* material, possibly shortened to the *solid* material. Also, it may be referred to as the *grain* material.

Generally, in terms of indices on variables, "fr" will refer to the framework, whereas "s" will refer to the solid constituent; "p" will refer to the pore space, and "f" to the fluid. When referring to properties of the fluid-filled medium as a whole, such as volume V and strain ε_{vol}, we will generally drop indices. The *porosity* ϕ is as usual the fraction of void space relative to the total volume.

To describe a fluid-saturated porous medium, the stress–strain relations (1.96)–(1.101) must be extended by stress and strain parameters associated with the fluid. The appropriate stress parameter is the hydrostatic[4] *pore pressure* p_f, whereas Biot defined the additional strain parameter ζ as the "increment of fluid content".

Consider a porous material with pore volume V_p, fully saturated with a fluid. The fluid volume V_f is then equal to the pore volume initially. A

[4] The term "hydrostatic pressure" is sometimes used in a restricted sense, meaning the fluid pressure resulting from an overlying column of water. In this book, we use the term in a broader (and probably more widely used) sense, meaning the pressure in a static fluid irrespective of the source of the pressure. Further, we also use hydrostatic for the stress state in a solid when all the principal stresses are equal, that is, no shear stresses are present.

change in conditions may induce different changes in ΔV_p and ΔV_f, which implies that fluid is either absorbed or expelled by the porous material. The strain parameter ζ is given as

$$\zeta = \frac{\Delta V_\mathrm{f} - \Delta V_\mathrm{p}}{V} \tag{1.139}$$

where V is the total volume of the porous solid. According to our sign convention, ζ is positive if V_f increases or V_p decreases, that is, if the fluid is expelled.

The fluid volume is related to the pore pressure change Δp_f and the bulk modulus of the fluid as

$$\Delta p_\mathrm{f} = K_\mathrm{f}\varepsilon_{\mathrm{vol,f}} = -K_\mathrm{f}\frac{\Delta V_\mathrm{f}}{V_\mathrm{f}} \tag{1.140}$$

where the minus sign arises because an increase in pressure reduces the volume. Introducing Eq. (1.140) and using the relation $V_\mathrm{p} = V_\mathrm{f} = \phi V$, we have

$$\zeta = -\phi\left(\frac{\Delta V_\mathrm{p}}{V_\mathrm{p}} + \frac{\Delta p_\mathrm{f}}{K_\mathrm{f}}\right) \tag{1.141}$$

We now use $V = V_\mathrm{p} + V_\mathrm{s}$, where V_s is the volume of the solid constituent, which gives

$$\varepsilon_{\mathrm{vol}} = \phi\varepsilon_{\mathrm{vol,p}} + (1-\phi)\varepsilon_{\mathrm{vol,s}} \tag{1.142}$$

where $\varepsilon_{\mathrm{vol,p}} = -\Delta V_\mathrm{p}/V_\mathrm{p}$ and $\varepsilon_{\mathrm{vol,s}} = -\Delta V_\mathrm{s}/V_\mathrm{s}$ are the volumetric strain of the pore volume and the solid, respectively.

We may then rewrite Eq. (1.141) as

$$\zeta = \phi\left[\varepsilon_{\mathrm{vol}} + \frac{1-\phi}{\phi}(\varepsilon_{\mathrm{vol}} - \varepsilon_{\mathrm{vol,s}}) - \varepsilon_{\mathrm{vol,f}}\right] \tag{1.143}$$

$$= \phi\left[\varepsilon_{\mathrm{vol}} + (1-\phi)(\varepsilon_{\mathrm{vol,p}} - \varepsilon_{\mathrm{vol,s}}) - \varepsilon_{\mathrm{vol,f}}\right] \tag{1.144}$$

$$= \left[\varepsilon_{\mathrm{vol}} - (1-\phi)\varepsilon_{\mathrm{vol,s}}\right] - \phi\varepsilon_{\mathrm{vol,f}} \tag{1.145}$$

The structure of Eq. (1.141), with separate terms relating to the solid and fluid, remains. Considering a couple of limiting cases may give some insight. First, assuming the solid constituent to be incompressible ($\varepsilon_{\mathrm{vol,s}} = 0$), the solid contribution to ζ is equal to the volumetric strain, Eq. (1.143). This is because the change in total volumetric strain in this case is fully given by the change in pore volume, that is, $\Delta V = \Delta V_\mathrm{p}$.

Second, assume that $\varepsilon_{\mathrm{vol,p}} = \varepsilon_{\mathrm{vol,s}}$, which is a good approximation when the confining stress and the pore pressure are changed with the same amount; see the discussion of "unjacketed loading" later in this section. Then the solid contribution to ζ is reduced to $\phi\varepsilon_{\mathrm{vol}}$ since the relative change in pore volume is now equal to the relative change in total volume (see Eq. (1.144)).

Finally, Eq. (1.145) shows that in the general case the contribution from $\varepsilon_{\mathrm{vol}}$ is reduced by the porosity weighted strain in the solid constituent material.

The Biot theory is a *macroscopic theory*, in which we define both a fluid and a solid displacement for every point in space, irrespective of whether the underlying microscopic model has a pore or a grain at that point. The Biot solid displacement, which we will denote $\vec{u}_{\mathrm{s}}^{\mathrm{fr}}$, represents the displacement of the porous solid framework. It is in general different from the displacement of the solid material within a grain at the same point. For instance, the stiffness of the solid grain material may be so large that the displacement within each grain is uniform while the grains may move relative to each other, making $\vec{u}_{\mathrm{s}}^{\mathrm{fr}}$ non-uniform.

The saturated fluid network constitutes a framework similar (and perfectly complementary) to the solid framework. The Biot fluid displacement, here denoted $\vec{u}_{\mathrm{f}}^{\mathrm{fr}}$, represents the displacement of the fluid "framework". It is in general different from the displacement of the fluid within a pore. The relation between "fluid" and "fluid framework" is completely analogous to the relation between "solid" and "solid framework" described above.

The volumetric strain of the porous material is defined as the volumetric strain of the solid framework. This can be expressed as the divergence of the displacement vector (see Eq. (1.79)):

$$\varepsilon_{\mathrm{vol}} = \nabla \cdot \vec{u}_{\mathrm{s}}^{\mathrm{fr}} \tag{1.146}$$

Similarly, the divergence of the fluid displacement vector $\vec{u}_{\mathrm{f}}^{\mathrm{fr}}$ represents the volumetric strain of the fluid framework. The relative amount of fluid expelled from the porous material, which is represented by the strain parameter ζ, is given as the difference between the volumetric strains of the solid and fluid frameworks, scaled by the fraction of fluid in the porous material (i.e. ϕ):

$$\zeta = \phi \nabla \cdot \left(\vec{u}_{\mathrm{s}}^{\mathrm{fr}} - \vec{u}_{\mathrm{f}}^{\mathrm{fr}}\right) \tag{1.147}$$

Note that although $\vec{u}_s^{\rm fr}$ and $\vec{u}_f^{\rm fr}$ may appear as analogous quantities in the macroscopic model, they are fundamentally different in terms of physical interpretation on the microscopic level. Whereas $\vec{u}_s^{\rm fr}$ is a real, observable quantity, this is not the case for $\vec{u}_f^{\rm fr}$ unless the displacements are very small and shortlived. Instead, $\vec{u}_f^{\rm fr}$ can be derived indirectly from a combination of $\vec{u}_s^{\rm fr}$ and the fluid flux $\vec{q}_{\rm fl}$, which is also an observable quantity:

$$\vec{u}_f^{\rm fr} = \vec{u}_s^{\rm fr} - \frac{1}{\phi}\int \vec{q}_{\rm fl}\, {\rm d}t \tag{1.148}$$

where $\vec{q}_{\rm fl} = \vec{Q}/A$ and $\vec{Q}$ is the fluid volume per unit time flowing through the area A. The integral runs over the duration of the displacement. Note that a combination of Eqs. (1.147) and (1.148) may be used as an alternative definition of ζ (Section 1.9.1).

From the above discussion it is clear that $\nabla \cdot \vec{u}_f^{\rm fr}$ in general is different from the physical volumetric strain of the pore fluid $\varepsilon_{\rm vol,f}$. From Eq. (1.143) we infer

$$\nabla \cdot \vec{u}_f^{\rm fr} = \varepsilon_{\rm vol,f} - \frac{1-\phi}{\phi}(\varepsilon_{\rm vol} - \varepsilon_{\rm vol,s}) \tag{1.149}$$

Thus the criterion for equality is $\nabla \cdot \vec{u}_f^{\rm fr} = \varepsilon_{\rm vol,f}$ only if $\varepsilon_{\rm vol} = \varepsilon_{\rm vol,p} = \varepsilon_{\rm vol,s}$, that is, if the pore volume and the solid material deform identically (in other words, the porosity is constant).

The above can be restated in an alternative manner in view of the discussion following Eqs. (1.143)–(1.145). There we saw that the solid-related contribution to ζ depended on material parameters and the loading geometry. In Eq. (1.147) the solid deformation $\vec{u}_s^{\rm fr}$ appears in one specific way. This means that the variation due to loading path and so on is now taken care of by $\vec{u}_f^{\rm fr}$. Hence $\vec{u}_f^{\rm fr}$ is more than the fluid displacement; it also takes care of interaction with the solid.

The stress tensor $\overleftrightarrow{\sigma}$ represents the total external stress on a volume element attached to the solid framework. The volume element balances this stress partly by stresses in the solid and partly by the pore pressure $p_{\rm f}$. In accordance with the sign convention, all stresses, including the pore pressure, are positive in compression.

The presence of the pore fluid adds extra terms to the strain energy of the material. Hence the stress–strain relations (1.96)–(1.101) will also be modified. Biot (1962) showed how the linear stress–strain relations for this two-phase system can be expressed in terms of the strain parameters $\varepsilon_{\rm vol}$ and ζ, the stress tensor elements and the pore pressure $p_{\rm f}$:

$$\sigma_x = \lambda_u \varepsilon_{vol} + 2G\varepsilon_x - C\zeta \tag{1.150}$$

$$\sigma_y = \lambda_u \varepsilon_{vol} + 2G\varepsilon_y - C\zeta \tag{1.151}$$

$$\sigma_z = \lambda_u \varepsilon_{vol} + 2G\varepsilon_z - C\zeta \tag{1.152}$$

$$\tau_{yz} = 2G\Gamma_{yz} \tag{1.153}$$

$$\tau_{xz} = 2G\Gamma_{xz} \tag{1.154}$$

$$\tau_{xy} = 2G\Gamma_{xy} \tag{1.155}$$

$$p_f = C\varepsilon_{vol} - M\zeta \tag{1.156}$$

Written in the abbreviated form as Eq. (1.111), Eqs. (1.150)–(1.155) become

$$\sigma_{ij} = \lambda_u \varepsilon_{vol}\delta_{ij} + 2G\varepsilon_{ij} - C\zeta\delta_{ij} \tag{1.157}$$

where λ_u and G are the Lamé parameters of the porous material, and C and M are additional elastic moduli required to describe a two-phase medium. Note that C appears in both stress and pore pressure equations. Biot (1941) showed that this is a consequence of thermodynamic principles.

By letting $\zeta = 0$ in Eq. (1.157) we get Eq. (1.111); $\zeta = 0$ means that there is no fluid movement in the material, that is, the material is *undrained*. Thus the λ_u in Eq. (1.157) is *not* the λ of the *dry* porous medium, but the λ of the fluid-filled medium when the fluid is not allowed to move. This is discussed further later in this section.

Summation of Eqs. (1.150)–(1.152) gives

$$\overline{\sigma} = K_u \varepsilon_{vol} - C\zeta \tag{1.158}$$

where $\overline{\sigma}$ is defined by Eq. (1.40), and $K_u = \lambda_u + 2G/3$ is the bulk modulus of the porous rock in *undrained* condition.

We will now investigate how the elastic moduli K_u, C and M relate to the moduli of the constituents of the rock. First, imagine that we perform a "jacketed" test (see Fig. 1.16a): a porous medium is confined within an impermeable jacket and subjected to an external hydrostatic pressure σ_p.

The pore fluid is allowed to escape during loading (the test is therefore also referred to as a *drained* test), so that the pore pressure is kept constant, and hence the stress is entirely carried by the solid framework. From Eqs. (1.156) and (1.158) we then obtain

$$\frac{\sigma_p}{\varepsilon_{vol}} = K_u - \frac{C^2}{M} \stackrel{\text{def}}{=} K_{fr} \tag{1.159}$$

Since this test characterises the stiffness of the solid part of the rock, K_{fr} is called the bulk modulus of the framework or the frame modulus. Since

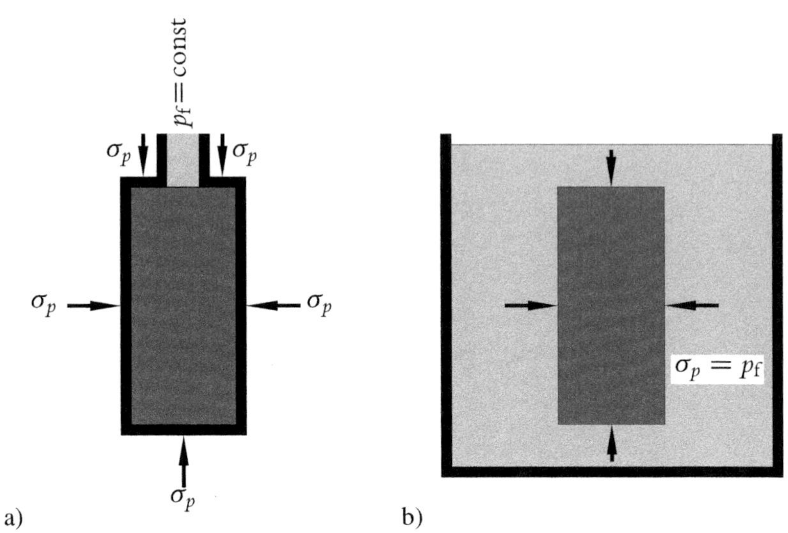

Figure 1.16 "Jacketed" (a) and "unjacketed" (b) test situations.

there are no shear forces associated with the fluid, we can directly identify the shear modulus of the porous system as the shear modulus of the framework, that is,

$$G = G_{fr} \tag{1.160}$$

Next, we proceed to an "unjacketed" test, as illustrated in Fig. 1.16b. The rock sample under investigation is here embedded in a fluid in such a way that the hydrostatic pressure on the sample is balanced by the pressure in the pores, that is, $p_f = \sigma_p$. Combining Eqs. (1.156) and (1.158), we now find that

$$\frac{\sigma_p}{\varepsilon_{vol}} = \frac{p_f}{\varepsilon_{vol}} = \frac{K_{fr}}{1 - \frac{C}{M}} \tag{1.161}$$

The loading $p_f = \sigma_p$ means that there is a uniform stress within the sample, which means that the rock framework deforms uniformly. Thus the volumetric strain of the total sample, the pore volume and the solid (grain) volume must be equal:

$$\frac{\Delta V_{tot}}{V_{tot}} = \frac{\Delta V_p}{V_p} = \frac{\Delta V_s}{V_s} \tag{1.162}$$

These equalities depend on the assumptions of microisotropy and microhomogeneity. Brown and Korringa (1975) showed that the assumptions may be relaxed by introducing one more parameter, such that the solid and pore volumes deform slightly differently under "unjacketed" loading. Detournay and Cheng (1993) suggest that neglecting the difference is a reasonable approximation: "*The error linked to the assumption is assumably modest, especially in view of the non-linear effects which are discussed next.*"

From $p_f = \sigma_p = -K_s(\Delta V_s / V_s)$, where K_s is the bulk modulus of the solid grains ($c_s = 1/K_s$ is often referred to as the *grain compressibility*), we hence find

$$\frac{\Delta V_s}{V_s} = \frac{\Delta V_p}{V_p} = \frac{\Delta V_{tot}}{V_{tot}} = -\varepsilon_{vol} = -\frac{p_f}{K_s} \tag{1.163}$$

The stress–strain response is therefore entirely given by the intrinsic elastic properties of the solid material. Comparing Eq. (1.163) to Eq. (1.161), we infer that

$$K_s = \frac{K_{fr}}{1 - \frac{C}{M}} \tag{1.164}$$

From Eqs. (1.141) and (1.163) it follows that

$$\zeta = \phi\left(\frac{1}{K_s} - \frac{1}{K_f}\right)p_f \tag{1.165}$$

On the other hand, the combination of Eqs. (1.156) and (1.158) with $p_f = \sigma_p$ gives

$$\zeta = \frac{C - K_u}{MK_u - C^2}p_f \tag{1.166}$$

By combining Eqs. (1.165) and (1.166) we obtain the relation

$$\phi\left(\frac{1}{K_s} - \frac{1}{K_f}\right) = \frac{C - K_u}{MK_u - C^2} \tag{1.167}$$

Although this equation was derived for a specific loading ("unjacketed test"), it is generally valid.

Eqs. (1.164) and (1.167), combined with the definition of K_{fr}, (Eq. (1.159)), allow us to express the elastic constants K_u, C and M in terms of the elastic moduli of the constituents of the rock (K_s and K_f) plus the porosity ϕ and the framework modulus K_{fr}. Defining

$$\alpha = \frac{C}{M} \tag{1.168}$$

(which we return extensively to in Section 1.6.4 and elsewhere), the results are

$$\frac{1}{M} = \frac{\alpha - \phi}{K_s} + \frac{\phi}{K_f} = \frac{(1-\alpha)(\alpha - \phi)}{K_{fr}} + \frac{\phi}{K_f} \tag{1.169}$$

$$C = \alpha M = \frac{K_f\left(1 - \frac{K_{fr}}{K_s}\right)}{\phi + \frac{K_f}{K_s}\left(1 - \phi - \frac{K_{fr}}{K_s}\right)} \tag{1.170}$$

$$K_u = K_{fr} + \alpha^2 M = K_{fr} + \alpha C = K_{fr} + \frac{K_f\left(1 - \frac{K_{fr}}{K_s}\right)^2}{\phi + \frac{K_f}{K_s}\left(1 - \phi - \frac{K_{fr}}{K_s}\right)} \tag{1.171}$$

Eqs. (1.169)–(1.171) may be summarised as

$$\frac{1}{M} = \frac{\alpha}{C} = \frac{\alpha^2}{K_u - K_{fr}} = \frac{\alpha - \phi}{K_s} + \frac{\phi}{K_f} \tag{1.172}$$

Eq. (1.171), which is called the Gassmann equation or the Biot–Gassmann equation, may alternatively be written as

$$\frac{K_u}{K_s - K_u} = \frac{K_{fr}}{K_s - K_{fr}} + \frac{1}{\phi}\frac{K_f}{K_s - K_f} \tag{1.173}$$

Note that in terms of the corresponding compressibilities the equation is slightly simpler:

$$\frac{1}{c_u - c_s} = \frac{1}{c_{fr} - c_s} + \frac{1}{\phi}\frac{1}{c_f - c_s} \tag{1.174}$$

Relations (1.169)–(1.171) do not give a clear physical meaning to each parameter. Better insight may be achieved by looking at a couple of limit cases, where the relations become simpler.

One case is a "hard" rock, where the frame is incompressible compared to the fluid. As $K_s \gg K_f$ in general, for this "stiff frame" case, we have

$$K_{fr}, G_{fr}, K_s \gg K_f \tag{1.175}$$

For porosities that are not too small (specifically, $\phi \gg (K_f/K_s^2)(K_s - K_{fr})$), Eqs. (1.169)–(1.171) then reduce to

$$\text{"Stiff frame"} \quad \begin{cases} K_u \approx K_{fr} \\ C \approx \frac{K_f}{\phi}\left(1 - \frac{K_{fr}}{K_s}\right) \\ M \approx \frac{K_f}{\phi} \end{cases} \tag{1.176}$$

We see that the bulk modulus K_u is here identified as the bulk modulus of the rock framework, whereas the constant M is entirely given by the properties of the pore fluid and the pore system.

The opposite limit is that of a "weak frame". For this case, we assume that

$$K_{fr}, G_{fr}, K_f \ll K_s \tag{1.177}$$

For porosities $\phi \gg K_f/K_s$, Eqs. (1.169)–(1.171) reduce to

$$\text{"Weak frame"} \quad \begin{cases} K_u & \approx & K_{fr} + \frac{K_f}{\phi} \\ C & \approx & M \approx \frac{K_f}{\phi} \end{cases} \tag{1.178}$$

In this case the bulk modulus K_u is influenced not only by the rock stiffness, but also by the fluid bulk modulus K_f. In the limiting case as $K_{fr} \to 0$ (suspension), $K_u = C = M \approx K_f/\phi$ are all given mainly by fluid properties. Disregarding the condition $K_s \gg K_f$ would in this case reproduce Eq. (1.137), which was obtained by simple physical arguments.

Note that the "stiff frame" and "weak frame" limits are rather extreme cases, which are mainly suited for illustrative purposes. For practical calculations, the complete expressions (1.169)–(1.171) should be used.

The theory outlined above contains two "unknown" parameters, which are not identified in terms of the properties of the rock constituents. These are the two elastic moduli of the framework, K_{fr}, and G_{fr}. We will see in Chapter 6 how microscopic theories can be used to estimate these moduli when further information about the rock structure is available. Empirically, the frame bulk modulus K_{fr} is found to be significantly smaller than K_s and to decrease strongly with the porosity.

1.6.3 Alternate form of the Biot constitutive equations

Using Eq. (1.156) to eliminate ζ from Eqs. (1.150)–(1.155) and introducing α from Eq. (1.168), we may write

$$\sigma_x - \alpha p_f = \lambda_{fr}\varepsilon_{vol} + 2G_{fr}\varepsilon_x \tag{1.179}$$

$$\sigma_y - \alpha p_f = \lambda_{fr}\varepsilon_{vol} + 2G_{fr}\varepsilon_y \tag{1.180}$$

$$\sigma_z - \alpha p_f = \lambda_{fr}\varepsilon_{vol} + 2G_{fr}\varepsilon_z \tag{1.181}$$

$$\tau_{yz} = 2G_{fr}\Gamma_{yz} \tag{1.182}$$

$$\tau_{xz} = 2G_{fr}\Gamma_{xz} \tag{1.183}$$

$$\tau_{xy} = 2G_{fr}\Gamma_{xy} \tag{1.184}$$

This may be written in compact form as

$$\sigma_{ij} = \lambda_{\text{fr}}\varepsilon_{\text{vol}}\delta_{ij} + 2G\varepsilon_{ij} + \alpha p_{\text{f}}\delta_{ij} \tag{1.185}$$

Summing Eqs. (1.179)–(1.181) gives

$$\overline{\sigma} - \alpha p_{\text{f}} = K_{\text{fr}}\varepsilon_{\text{vol}} \tag{1.186}$$

Notice that the elastic moduli are now the framework version. This follows directly from the algebra but is also obvious from the physics. Letting $\zeta = 0$ in Eqs. (1.150)–(1.152), we have an undrained situation, whereas $p_{\text{f}} = 0$ in Eqs. (1.179)–(1.181) is a drained situation. As before, $G_{\text{fr}} = G$.

1.6.4 Effective stress

If we define the *effective stress* $\overline{\sigma}'$ as

$$\overline{\sigma}' = \overline{\sigma} - \alpha p_{\text{f}} \tag{1.187}$$

then Eq. (1.186) becomes

$$\overline{\sigma}' = K_{\text{fr}}\varepsilon_{\text{vol}} \tag{1.188}$$

If we compare to Eq. (1.159), which may be rewritten as (using mean stress in place of hydrostatic external stress)

$$\overline{\sigma} = K_{\text{fr}}\varepsilon_{\text{vol}} \tag{1.189}$$

then we see that the right-hand sides of the equations are identical. Thus, when both external stress and pore pressure change, the deformation response to the combination $\overline{\sigma}' = \overline{\sigma} - \alpha p_{\text{f}}$ is the same as the response to the total external stress when the pore pressure is constant.

Physically, this means that the solid framework carries the part σ_p' of the total external stress σ_p, whereas the remaining part αp_{f} is carried by the fluid. The remaining pore pressure $(1-\alpha)p_{\text{f}}$ is counteracted by internal stresses in the solid. The parameter α is called the Biot coefficient, or sometimes the Biot–Willis coefficient, since it was first used by Biot and Willis (1957):

$$\alpha = \frac{C}{M} = 1 - \frac{K_{\text{fr}}}{K_{\text{s}}} \tag{1.190}$$

with the latter identity from Eq. (1.164). As mentioned above, K_{fr} is always smaller than K_{s}. Theoretically, the upper limit for K_{fr} is $(1-\phi)K_{\text{s}}$. (See

Section 6.2 for more information on bounds on elastic moduli of composite materials). The lower limit for K_{fr} is, of course, zero. Thus α is restricted to the region $\phi < \alpha \leq 1$. In unconsolidated or weak rocks, α is close to 1.

For a general stress state, the effective stress is given by

$$\sigma'_{ij} = \sigma_{ij} - \delta_{ij}\alpha p_f \tag{1.191}$$

where δ_{ij} is the Kronecker symbol; see Appendix C.12.2. Observe that only the normal effective stresses depend on the pore pressure.

The effective stress relations above were derived for rock deformation. The concept of effective stress often is applied to other physical quantities. It will be shown in Section 1.6.6 that porosity and solid volume changes according to Biot theory obey effective stresses with different coefficients. Permeability, wave velocities and electrical properties are found to have other effective stress coefficients. Hence we need to be careful when applying the term effective stress without specifying what property is addressed. Effective stress is not a single parameter characterising stress and pore pressure in the Earth.

From a formal point of view, the effective stress is related to whether a function of two variables can be written as a one-variable function of a linear combination of the two original arguments, that is, whether we can write a function

$$g(\sigma_p, p_f) \tag{1.192}$$

as

$$h(\sigma_p - cp_f) \tag{1.193}$$

If g is a linear function in both arguments, then this is obviously possible. It is equally clear that c depends on the details of g.

Note that for $p_f = 0$, Eqs. (1.156) and (1.190) give $\alpha = \zeta/\varepsilon_{vol}$. Then from Eq. (1.141) (with $p_f = 0$) and the definition of ε_{vol} we obtain

$$\alpha = \phi\frac{\Delta V_p/V_p}{\Delta V_{tot}/V_{tot}} = \frac{\Delta V_p}{\Delta V_{tot}} \tag{1.194}$$

which shows that α is a measure of the change in pore volume relative to the change in bulk volume at *constant pore pressure*. Thus, if the solid is incompressible and only the pore space is deformed, then $\alpha = 1$.

In addition to the bulk compressibility with respect to confining pressure at constant pore pressure $c_{fr} = 1/K_{fr}$, we may define the bulk compressibility with respect to pore pressure at constant confining pressure

$c_{bp} = 1/K_{bp}$. By letting $\overline{\sigma} = 0$ in Eq. (1.186) we may write

$$K_{bp} = -\frac{p_f}{\varepsilon_{vol}} = \frac{K_{fr}}{\alpha} \tag{1.195}$$

and thus we have

$$\alpha = \frac{K_{fr}}{K_{bp}} = \frac{c_{bp}}{c_{fr}} \tag{1.196}$$

which shows that the Biot coefficient α is the ratio of the bulk modulus at constant pore pressure to the bulk modulus at constant confining pressure.

By eliminating ζ between Eqs. (1.156) and (1.158) and setting $\varepsilon_{vol} = 0$ (see also Eq. (1.186)) we find

$$\alpha = \left(\frac{\overline{\sigma}}{p_f}\right)_{\varepsilon_{vol}=0} \tag{1.197}$$

which shows that α is the ratio of the mean stress change to the pore pressure change in an undrained test when the total volume is kept constant.

The effective stress concept was originally introduced in soil mechanics by Terzaghi in 1923 on an empirical basis. Terzaghi argued that

1. increasing the external hydrostatic pressure produces the same volume change of the material as reducing the pore pressure with the same amount,
2. the shear strength depends only on the difference between the normal stress σ and the pore pressure p_f.

These arguments lead to an effective stress law similar to Eq. (1.191), with $\alpha = 1$:

$$\sigma'_{ij} = \sigma_{ij} - \delta_{ij} p_f \tag{1.198}$$

For soils, this is a reasonable assumption. For rocks, however, the deviation of α from 1 should be taken into account.

Beware that the difference $\sigma_{ij} - \delta_{ij} p_f$ is sometimes referred to as the *net stress* (also in this book) or the *differential stress*; however, both these terms also have other meanings.

Terzaghi's statement 2 above implies that the effective stress, rather than the total stress, is determining whether the rock fails or not due to the external load. This is further discussed in Section 2.6.1.

1.6.5 Specific storage

The *specific storage* S is widely used in hydrogeology (see e.g. Wang, 2000) and is defined as the amount of fluid stored in the pore system per unit volume for a unit pressure increase.

It is thus given by

$$S = -\frac{\Delta\zeta}{\Delta p_{\rm f}} \tag{1.199}$$

From Eq. (1.156) it is clear that specific storage is not a unique quantity; it depends on the boundary conditions. Letting $\varepsilon_{\rm vol} = 0$ in Eq. (1.156), we see that specific storage at constant strain S_ε is given by

$$S_\varepsilon = \frac{1}{M} = \phi\left(\frac{1}{K_{\rm f}} - \frac{1}{K_{\rm s}}\right) + \frac{1}{K_{\rm s}}\left(1 - \frac{K_{\rm fr}}{K_{\rm s}}\right) = \frac{\phi}{K_{\rm f}} + \frac{\alpha - \phi}{K_{\rm s}} \tag{1.200}$$

where M is the Biot modulus, also referred to as the *storage modulus*.

To find the specific storage at constant stress, we set the stresses to zero in Eqs. (1.179)–(1.181), find $\varepsilon_{\rm vol}$ in terms of $p_{\rm f}$, and insert into Eq. (1.156) to find

$$S_\sigma = \frac{1}{M} + \frac{\alpha^2}{K_{\rm fr}} = \phi\left(\frac{1}{K_{\rm f}} - \frac{1}{K_{\rm s}}\right) + \frac{1}{K_{\rm fr}} - \frac{1}{K_{\rm s}} \tag{1.201}$$

Finally, the specific storage at uniaxial strain conditions is found by inserting $\sigma_x = \varepsilon_y = \varepsilon_z = 0$ in Eqs. (1.179)–(1.181) and proceeding as above:

$$S_0 = \frac{1}{M} + \frac{\alpha^2}{H_{\rm fr}} = \phi\left(\frac{1}{K_{\rm f}} - \frac{1}{K_{\rm s}}\right) + \frac{1}{K_{\rm s}}\left(1 - \frac{K_{\rm fr}}{K_{\rm s}}\right) + \frac{(1 - \frac{K_{\rm fr}}{K_{\rm s}})^2}{H_{\rm fr}} \tag{1.202}$$

Note that $S_\sigma > S_0 > S_\varepsilon$, that is, S becomes smaller as confinement increases.

1.6.6 Pore volume compressibility and related topics

The change of the pore volume as a result of the change in the pore pressure or the confining stress is of obvious interest in petroleum related rock mechanics.

By eliminating $\varepsilon_{\rm vol}$ from Eqs. (1.156) and (1.158), inserting the resulting expression for ζ into Eq. (1.141), and inserting for C, M and $K_{\rm u}$ from Eqs. (1.169)–(1.171), after some algebra, we find

$$\Delta V_{\rm p}/V_{\rm p} = -\frac{1}{\phi}\left(\frac{1}{K_{\rm fr}} - \frac{1}{K_{\rm s}}\right)\sigma_p + \frac{1}{\phi}\left(\frac{1}{K_{\rm fr}} - \frac{1+\phi}{K_{\rm s}}\right)p_{\rm f} \tag{1.203}$$

$$= -\frac{1}{\phi}\left(\frac{1}{K_{\text{fr}}} - \frac{1}{K_{\text{s}}}\right)\left[\sigma_p - \left(1 - \phi\frac{K_{\text{fr}}}{K_{\text{s}} - K_{\text{fr}}}\right)p_{\text{f}}\right] \tag{1.204}$$

This allows us to define the pore compressibility with respect to confining stress (with constant pore pressure) as

$$c_{\text{p}} = \frac{1}{K_{\text{p}}} = -\frac{1}{V_{\text{p}}}\left(\frac{\partial V_{\text{p}}}{\partial \sigma_p}\right)_{p_{\text{f}}} = \frac{1}{\phi}\left(\frac{1}{K_{\text{fr}}} - \frac{1}{K_{\text{s}}}\right) = \frac{1}{\phi}\frac{\alpha}{K_{\text{fr}}} \tag{1.205}$$

and the pore compressibility with respect to pore pressure (with constant confining stress) as

$$c_{\text{pp}} = \frac{1}{K_{\text{pp}}} = \frac{1}{V_{\text{p}}}\left(\frac{\partial V_{\text{p}}}{\partial p_{\text{f}}}\right)_{\sigma_p} = \frac{1}{\phi}\left(\frac{1}{K_{\text{fr}}} - \frac{1+\phi}{K_{\text{s}}}\right) = \frac{\alpha(1+\phi) - \phi}{\phi K_{\text{fr}}} = c_{\text{p}} - c_{\text{s}} \tag{1.206}$$

Note the different sign in the definitions, which ensure that the compressibilities are positive.

Defining the stress difference $\sigma_{\text{d}} = \sigma_p - p_{\text{f}}$, we see that the pore volume compressibility with respect to pore pressure at constant stress difference is

$$c_{\phi} = \frac{1}{K_{\phi}} = -\frac{1}{V_{\text{p}}}\left(\frac{\partial V_{\text{p}}}{\partial p_{\text{f}}}\right)_{\sigma_{\text{d}}} = \frac{1}{K_{\text{s}}} \tag{1.207}$$

where the last equality depends on the assumptions of microhomogeneity and microisotropy. Brown and Korringa (1975) allowed K_{ϕ} to differ from K_{s} to relax these conditions (see Section 6.5).

Zimmerman (1991) points out that an expression sometimes given for the pore compressibility, based on the weighted average $1/K_{\text{fr}} = \phi/K_{\text{p}} + (1-\phi)/K_{\text{s}}$, is incorrect.

We may find a related expression for the porosity by differentiating the definition $V_{\text{p}} = \phi V_{\text{tot}}$, which gives

$$\frac{\Delta\phi}{\phi} = \frac{\Delta V_{\text{p}}}{V_{\text{p}}} - \frac{\Delta V_{\text{tot}}}{V_{\text{tot}}} \tag{1.208}$$

Using the effective stress law $K_{\text{fr}}\varepsilon_{\text{vol}} = \sigma_p - \alpha p_{\text{f}}$ and Eq. (1.203), we find

$$\Delta\phi = -\frac{\alpha - \phi}{K_{\text{fr}}}(\sigma_p - p_{\text{f}}) = -\left(\frac{1-\phi}{K_{\text{fr}}} - \frac{1}{K_{\text{s}}}\right)(\sigma_p - p_{\text{f}}) \tag{1.209}$$

Finally, we may use the relation (see Eq. (1.142))

$$\frac{\Delta V_{\text{tot}}}{V_{\text{tot}}} = \phi\frac{\Delta V_{\text{p}}}{V_{\text{p}}} + (1-\phi)\frac{\Delta V_{\text{s}}}{V_{\text{s}}} \tag{1.210}$$

to derive the following expression for the deformation of the grain material (use e.g. Eq. (1.203) and the effective stress equation):

$$\frac{\Delta V_s}{V_s} = -\frac{1}{(1-\phi)K_s}(\sigma_p - \phi p_f) \tag{1.211}$$

Note that the grains expand if we increase the pore pressure while the confining stress is constant. Increasing the pore pressure means that the fluid carries more of the external loading, which means that the grains carry correspondingly less, and hence the grains expand.

Eq. (1.211) shows that the mean stress in the grains is

$$\overline{\sigma}_s = \frac{\sigma_p - \phi p_f}{1-\phi} \tag{1.212}$$

See Zimmerman (1991) for a simple sketch giving a geometrical justification of this result.

Eqs. (1.204), (1.209) and (1.211) all give an effective stress coefficient different from Biot's α defined in Eq. (1.190), in line with the discussion following Eq. (1.191).

1.6.7 The Skempton coefficients

An important characteristic of a porous medium is how the pore pressure responds to a change in the mean stress under *undrained* conditions.

For the elastic case, the response can be computed from the poroelastic equations. We add Eqs. (1.150)–(1.152), set $\zeta = 0$, and eliminate ε_{vol} using Eq. (1.156). The result is

$$\Delta p_f = \frac{C}{K_u}\Delta\overline{\sigma} \tag{1.213}$$

The Skempton B-coefficient is defined as

$$B = \frac{\Delta p_f}{\Delta\overline{\sigma}} = \frac{C}{K_u} = \frac{\frac{K_f}{\phi}\left(1-\frac{K_{fr}}{K_s}\right)}{\frac{K_f}{\phi}\left(1-\frac{K_{fr}}{K_s}\right) + K_{fr}\left(1-\frac{K_f}{K_s}\right)}$$
$$= \frac{\frac{1}{K_{fr}} - \frac{1}{K_s}}{\frac{1}{K_{fr}} - \frac{1}{K_s} + \phi\left(\frac{1}{K_f} - \frac{1}{K_s}\right)} = \frac{\frac{\alpha}{\phi}c_{fr}}{c_{pp} + c_f} \tag{1.214}$$

where the right-hand expressions are found by introducing K_u and C from Eqs. (1.171) and (1.170). It is clear from the formula that $B \le 1$.

Observe the similarity to Eq. (1.197). There the volume of the sample is kept constant, whereas above it is allowed to change freely.

Originally, Skempton (1954) defined the coefficients A and B according to

$$\Delta p_{\mathrm{f}} = B[\Delta\sigma_3 + A(\Delta\sigma_1 - \Delta\sigma_3)] \tag{1.215}$$

The form of this equation was chosen to be appropriate for triaxial tests (see Chapter 2). For a triaxial compression test, the change in mean stress may be written

$$\Delta\overline{\sigma} = \frac{1}{3}(\Delta\sigma_1 + 2\Delta\sigma_3) = \Delta\sigma_3 + \frac{1}{3}(\Delta\sigma_1 - \Delta\sigma_3) \tag{1.216}$$

which shows that B in Eq. (1.215) is the same as in Eq. (1.214).

Notice from Eq. (1.216) that for the isotropic elastic case, $A = 1/3$. The coefficient A was primarily introduced to characterise material inelasticity.

Neglecting K_{f} relative to K_{s} in Eq. (1.214) and introducing Biot's α, we have

$$B = \frac{K_{\mathrm{f}}}{K_{\mathrm{f}} + \frac{\phi}{\alpha} K_{\mathrm{fr}}} \tag{1.217}$$

In the weak frame limit, where K_{fr} can be neglected relative to K_{s}, we have

$$B = \frac{K_{\mathrm{f}}}{K_{\mathrm{f}} + \phi K_{\mathrm{fr}}} \tag{1.218}$$

Some examples of the variation of B (as computed from Eq. (1.214)) are shown in Fig. 1.17. Gas as the pore fluid will clearly give a low B except for very unconsolidated rocks.

Note the simple relation

$$\frac{K_{\mathrm{fr}}}{K_{\mathrm{u}}} = 1 - \alpha B \tag{1.219}$$

which follows from $B = C/K_{\mathrm{u}}$ and Eq. (1.171).

1.6.8 The undrained Poisson's ratio

It follows from Eq. (1.171) that the undrained bulk modulus is always larger than the drained one. From the expression (Table 1.1)

$$K = \frac{2}{3} G \frac{1 + \nu}{1 - 2\nu} \tag{1.220}$$

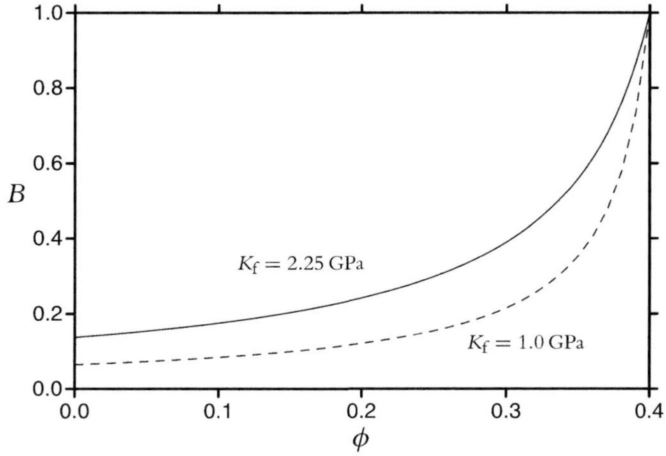

Figure 1.17 Examples of the variation of B with porosity for two different fluids. The full line corresponds to $K_f = 2.25$ GPa ("water"), and the dashed line to $K_f =$ 1 GPa ("oil"). The frame modulus K_{fr} is estimated from the critical porosity theory with $\phi_c = 40\,\%$ (see Section 6.2). In all cases, $K_s = 37.5$ GPa.

it then follows that the undrained Poisson's ratio is always larger than the drained counterpart.

Using the fact that G is independent of saturation, we may use Eq. (1.220) to relate Poisson's ratios to the bulk moduli. Using Eq. (1.219), we then find

$$\nu_u = \frac{3\nu_{fr} + \alpha B(1 - 2\nu_{fr})}{3 - \alpha B(1 - 2\nu_{fr})} \tag{1.221}$$

This equation is plotted for various values of αB in Fig. 1.18.

Inverted for αB, Eq. (1.221) becomes

$$\alpha B = 3\frac{\nu_u - \nu_{fr}}{(1 + \nu_u)(1 - 2\nu_{fr})} \tag{1.222}$$

which using Eq. (1.219) gives

$$\frac{K_u}{K_{fr}} = \frac{(1 + \nu_u)(1 - 2\nu_{fr})}{(1 + \nu_{fr})(1 - 2\nu_u)} \tag{1.223}$$

Since Eq. (1.223) is a direct consequence of $G_u = G_{fr}$, it is also easily inferred from Eq. (1.220).

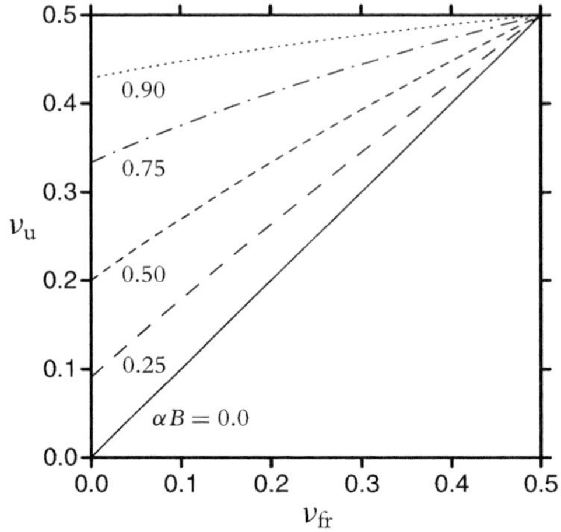

Figure 1.18 Undrained versus drained Poisson's ratio for various values of αB.

1.6.9 The correspondence to thermoelasticity

The equations governing poroelasticity are to some extent similar to the equations governing thermoelasticity. This implies that specific solutions to problems in one field may be used to solve the corresponding problems in the other (Geertsma, 1957).

In Eq. (1.185) the poroelastic stress–strain law was written as

$$\sigma_{ij} = \lambda_{fr}\varepsilon_{vol}\delta_{ij} + 2G\varepsilon_{ij} + \alpha p_f \delta_{ij} \tag{1.224}$$

Eq. (1.224) should be compared to the corresponding equation from thermoelasticity, Eq. (1.123):

$$\sigma_{ij} = \lambda\varepsilon_{vol}\delta_{ij} + 2G\varepsilon_{ij} + 3\alpha_T K (T - T_i)\,\delta_{ij} \tag{1.225}$$

This means that by making the identifications

$$p_f \leftrightarrow T - T_i \tag{1.226}$$

$$\alpha \leftrightarrow 3K\alpha_T = \frac{E}{1-2\nu}\alpha_T \tag{1.227}$$

solutions of problems in thermoelasticity (where a wide range of problems have been solved) may be directly applied to poroelasticity. The method is used, for example, by Haimson and Fairhurst (1967) in a classic paper on hydraulic fracturing.

There is one important point to note. Since stress or strain do not induce significant temperature changes, the temperature field is governed by the (decoupled) diffusion equation. Pore pressure, on the other hand, is of course directly coupled to stress (Eq. (1.156)), leading to coupled equations. There are, however, some important cases where decoupling occurs. These include all steady-state problems, the consolidation problem leading to Eq. (1.320), and the borehole problem for axisymmetric loading (see e.g. Detournay and Cheng, 1988, 1993; Wang, 2000).

1.6.10 Thermoporoelasticity

In the previous section, we discussed the similarities between poroelasticity and thermoelasticity. Clearly, a material may be subject to poroelastic and thermal loads at the same time, which calls for a generalisation of the basic equations.

Combining Eq. (1.123) and (1.185), we have

$$\sigma_{ij} = \lambda_{\mathrm{fr}}\varepsilon_{\mathrm{vol}}\delta_{ij} + 2G\varepsilon_{ij} + \alpha p_{\mathrm{f}}\delta_{ij} + 3\alpha_{\mathrm{T}}K_{\mathrm{fr}}\left(T - T_{\mathrm{i}}\right)\delta_{ij} \tag{1.228}$$

It follows that the average stress is given by

$$\overline{\sigma} = K_{\mathrm{fr}}\varepsilon_{\mathrm{vol}} + \alpha p_{\mathrm{f}} + 3\alpha_{\mathrm{T}}K_{\mathrm{fr}}\left(T - T_{\mathrm{i}}\right) \tag{1.229}$$

Assuming $\overline{\sigma} = p_{\mathrm{f}} = 0$, we find

$$\varepsilon_{\mathrm{vol}} = -3\alpha_{\mathrm{T}}\left(T - T_{\mathrm{i}}\right) \tag{1.230}$$

which shows that in this context, α_{T} is the linear thermal expansion coefficient for *drained* conditions.

Let us now consider undrained conditions. It is clear that if the linear thermal expansion coefficient of the fluid, α_{Tf}, is equal to that of the frame, α_{T}, then a temperature change will not change the pore pressure. The pressure effect must hence be proportional to the difference of the thermal expansion coefficients. Further, if the pore space was completely rigid, the pressure effect would simply be proportional to the bulk modulus of the fluid, K_{f}. However, this must be corrected for the change in pore volume as given by the pore volume compressibility c_{pp} of Eq. (1.206).

We thus have

$$p_{\mathrm{f}} = \frac{3(\alpha_{\mathrm{Tf}} - \alpha_{\mathrm{T}})(T - T_{\mathrm{i}})}{c_{\mathrm{pp}} + c_{\mathrm{f}}} \tag{1.231}$$

Using expression (1.214) for Skempton's B, we may write

$$p_{\mathrm{f}} = 3(\alpha_{\mathrm{Tf}} - \alpha_{\mathrm{T}})\phi B \frac{K_{\mathrm{fr}}}{\alpha}(T - T_{\mathrm{i}}) \tag{1.232}$$

Inserting into Eq. (1.229) with $\overline{\sigma} = 0$ and solving for $\varepsilon_{\mathrm{vol}}$, we obtain

$$\varepsilon_{\mathrm{vol}} = -3[\alpha_{\mathrm{T}} + \phi B(\alpha_{\mathrm{Tf}} - \alpha_{\mathrm{T}})](T - T_{\mathrm{i}}) \tag{1.233}$$

which shows that the *undrained* thermal expansion coefficient is given by

$$\alpha_{\mathrm{Tu}} = \alpha_{\mathrm{T}} + \phi B(\alpha_{\mathrm{Tf}} - \alpha_{\mathrm{T}}) \tag{1.234}$$

This formula was apparently first given, in a more general form without the assumptions of microhomogeneity and microisotropy, by McTigue (1986).

Eq. (1.232) shows how Eq. (1.156) is generalised for the thermal case:

$$p_{\mathrm{f}} = C\varepsilon_{\mathrm{vol}} - M\zeta + 3(\alpha_{\mathrm{Tf}} - \alpha_{\mathrm{T}})\phi B \frac{K_{\mathrm{fr}}}{\alpha}(T - T_{\mathrm{i}}) \tag{1.235}$$

1.6.11 Other notation conventions

We have seen that isotropic poroelasticity requires four independent moduli. In the initial Eqs. (1.150)–(1.156), we used λ_{u}, $G = G_{\mathrm{fr}}$, M and C. We also showed how these are related to the more physically understandable set consisting of the *undrained* bulk modulus K_{u}, the *drained* or *frame* bulk modulus K_{fr}, the solid grain bulk modulus K_{s} and the shear modulus $G = G_{\mathrm{fr}}$.

There are clearly many other possibilities; we refer to Detournay and Cheng (1993) for a thorough discussion.

We mention in particular the alternative used by Rice and Cleary (1976), Detournay and Cheng (1993) and others. They use the shear modulus, the *drained* and *undrained* Poisson's ratios. The fourth parameter is Skempton's B parameter.

1.7 Anisotropy

If the elastic response of a material depends on the orientation of the material for a given stress configuration, the material is said to be *anisotropic*. Thus the elastic moduli of an anisotropic material are different for different directions in the material.

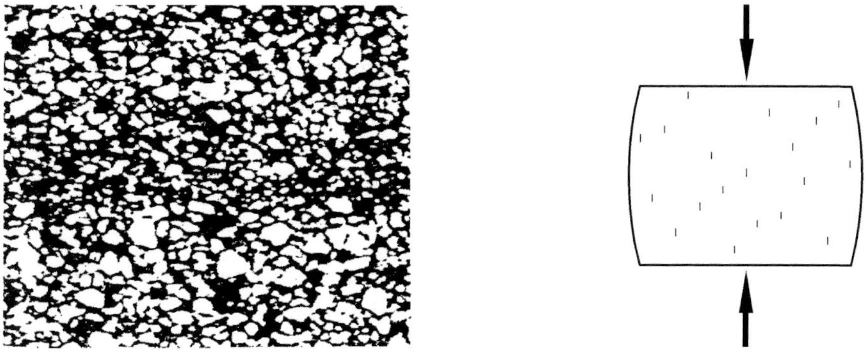

Figure 1.19 Illustration of intrinsic (lithological) and stress-induced anisotropy.

Most rocks are anisotropic to some extent. The origin of the anisotropy is always heterogeneities on a smaller scale than the volume under investigation, ranging from layered sequences of different rock types down to molecular configurations. Sedimentary rocks are created during a deposition process where the grains normally are not deposited randomly. For instance, in a river environment, there is a preferred direction (the direction of the streaming water), along which elongated or plane grains have a tendency to be oriented. Seasonal variations in the fluid flow rates may result in alternating microlayers of fine and coarser grain size distributions. The elastic moduli of rocks created under such conditions will be dependent on the orientation of the material, that is, they are anisotropic. Due to its origin, anisotropy of this type is said to be lithological or *intrinsic*.

Another important type is anisotropy *induced* by external stresses. The anisotropy is often caused by microcracks generated by a deviatoric stress and predominantly oriented normal to the lowest principal stress. The stiffness of uncemented grain contacts in unconsolidated or soft sandstone increases with increasing normal stress on the contacts, leading to macroscopic stiffness anisotropy. The microstructure causing lithological and stress-induced anisotropy is illustrated in Fig. 1.19.

In calculations on rock elasticity, anisotropy is often ignored. This simplification may be necessary rather than just comfortable, because, as we will see, an anisotropic description requires much more information about the material, information that may be not available. However, by ignoring anisotropy we may in some cases introduce large errors that invalidate the calculations.

For a general anisotropic material, each stress component is linearly related to every strain component by independent coefficients. In the mathematical notation used in Eqs. (1.9) and (1.78), this may be expressed as

$$\sigma_{ij} = \sum_{k,l} C_{ijkl}\varepsilon_{kl} \tag{1.236}$$

where C_{ijkl} are *elastic constants*, which are the components of the fourth-order *stiffness tensor* or *elastic tensor*. Since the indices i, j, k and l may each take the values 1, 2 or 3, there are all together 81 constants C_{ijkl}. However, some of these vanish, and others are equal by symmetry, so that the number of independent constants is considerably less. From Eqs. (1.6)–(1.8), (1.77) and (1.236) it may be deduced that

$$C_{ijkl} = C_{jikl} = C_{ijlk} = C_{jilk} \tag{1.237}$$

that is, interchanging the first index with the second one, or the third with the fourth, does not change the value of the constant. Furthermore, fulfilment of the equations of the type (1.120), which were derived from energy considerations, requires that

$$C_{ijkl} = C_{klij} \tag{1.238}$$

Relations (1.237) and (1.238) reduce the number of independent constants to 21. These 21 constants are not completely independent, since relations similar to those discussed towards the end of Section 1.4 for isotropy apply. See Appendix C.11.

The inverse relation of Eq. (1.236) may be written as

$$\varepsilon_{ij} = \sum_{k,l} S_{ijkl}\sigma_{kl} \tag{1.239}$$

where S_{ijkl} are the components of the fourth-order *compliance tensor*.

The same symmetry restrictions apply to S_{ijkl} as for C_{ijkl}, and the following inverse relationship is obeyed (δ is the Kronecker symbol, see Appendix C.12.2):

$$\sum_{k,l} C_{ijkl}S_{klmn} = \frac{1}{2}\left(\delta_{im}\delta_{jn} + \delta_{in}\delta_{jm}\right) \tag{1.240}$$

1.7.1 Orthorhombic (orthotropic) symmetry

Rocks can normally be described reasonably well by assuming that the material has three mutually perpendicular planes of symmetry. This is one of the *orthorhombic* symmetries.

Let us assume that the planes of symmetry are perpendicular to the coordinate axes x, y, z. From Eqs. (1.236) and (1.237), for the normal stress in x-direction, we have

$$\sigma_{11} = C_{1111}\varepsilon_{11} + C_{1122}\varepsilon_{22} + C_{1133}\varepsilon_{33} + 2C_{1112}\varepsilon_{12} + 2C_{1113}\varepsilon_{13} + 2C_{1123}\varepsilon_{23} \tag{1.241}$$

Due to the orthorhombic symmetry, Eq. (1.241) should look exactly the same when described in a coordinate system defined by $x' = x$, $y' = y$, $z' = -z$. However, in this new coordinate system, two of the strain components have changed sign (see Appendix C.7.2), namely $\varepsilon'_{13} = -\varepsilon_{13}$ and $\varepsilon'_{23} = -\varepsilon_{23}$. This implies that

$$C_{1113} = C_{1123} = 0 \tag{1.242}$$

By applying the same arguments on the remaining stress components, and on other orientations of the primed coordinate system, the number of independent constants C_{ijkl} is reduced to 9. These are C_{1111}, C_{2222}, C_{3333}, C_{1122}, C_{1133}, C_{2233}, C_{2323}, C_{1313}, C_{1212}.

In the compact notation, where σ_x is used instead of σ_{11} and so on, the constants C_{ijkl} have only two indices, I and J. In this so-called Voigt notation the indices ij are related to I as follows: 11→1, 22→2, 33→3, 23→4, 13→5 and 12→6. The Voigt notation is discussed in more detail in Appendix C.10.

The constants may then be written as

$$\begin{pmatrix} C_{11} & C_{12} & C_{13} & 0 & 0 & 0 \\ C_{12} & C_{22} & C_{23} & 0 & 0 & 0 \\ C_{13} & C_{23} & C_{33} & 0 & 0 & 0 \\ 0 & 0 & 0 & C_{44} & 0 & 0 \\ 0 & 0 & 0 & 0 & C_{55} & 0 \\ 0 & 0 & 0 & 0 & 0 & C_{66} \end{pmatrix} \tag{1.243}$$

The zeros represent the constants that vanish by the symmetry arguments of type (1.242). They are included to emphasise the matrix nature of the constants and also as a reminder that they do exist: for a material with orthorhombic symmetry, they only vanish when the planes of symmetry are perpendicular to the coordinate axes.

This means that for a general orientation, an imposed volumetric strain will generate shear stresses, and vice versa. As a consequence, the principal directions for stress and strain do not coincide in the general case.

Denoting the 6×6 matrix (1.243) by $\mathbf{C}$ and defining the 6×1 matrices $\boldsymbol{\sigma}$ and $\boldsymbol{\varepsilon}$ according to

$$\boldsymbol{\sigma} = \begin{pmatrix} \sigma_x \\ \sigma_y \\ \sigma_z \\ \tau_{yz} \\ \tau_{xz} \\ \tau_{xy} \end{pmatrix} \qquad \boldsymbol{\varepsilon} = \begin{pmatrix} \varepsilon_x \\ \varepsilon_y \\ \varepsilon_z \\ 2\Gamma_{yz} \\ 2\Gamma_{xz} \\ 2\Gamma_{xy} \end{pmatrix} \tag{1.244}$$

we can write the equations of type (1.236) as the matrix product

$$\boldsymbol{\sigma} = \mathbf{C} \cdot \boldsymbol{\varepsilon} \tag{1.245}$$

Eq. (1.245) is in reality six equations. Written in an explicit form in the Voigt notation, these equations look as

$$\sigma_x = C_{11}\varepsilon_x + C_{12}\varepsilon_y + C_{13}\varepsilon_z \tag{1.246}$$

$$\sigma_y = C_{12}\varepsilon_x + C_{22}\varepsilon_y + C_{23}\varepsilon_z \tag{1.247}$$

$$\sigma_z = C_{13}\varepsilon_x + C_{23}\varepsilon_y + C_{33}\varepsilon_z \tag{1.248}$$

$$\tau_{yz} = 2C_{44}\Gamma_{yz} \tag{1.249}$$

$$\tau_{xz} = 2C_{55}\Gamma_{xz} \tag{1.250}$$

$$\tau_{xy} = 2C_{66}\Gamma_{xy} \tag{1.251}$$

These stress–strain relations generally describe most types of rocks.

The matrix $\mathbf{C}$ is called the *stiffness matrix*, and its components C_{IJ} are called *elastic constants*. The inverse of the stiffness matrix, $\mathbf{S} = \mathbf{C}^{-1}$, is called the *compliance matrix*.

It follows from Eq. (1.245) that the compliance matrix relates strains to stresses in the following way:

$$\boldsymbol{\varepsilon} = \mathbf{S} \cdot \boldsymbol{\sigma} \tag{1.252}$$

The constants (1.243) describe the elastic properties of any linear elastic material with orthorhombic or higher symmetry. Thus they may also describe an isotropic rock. Comparing Eqs. (1.96)–(1.101) to (1.246)–(1.251), we find that for an isotropic material,

$$C_{11} = C_{22} = C_{33} = \lambda + 2G \tag{1.253}$$

$$C_{12} = C_{13} = C_{23} = \lambda \tag{1.254}$$

$$C_{44} = C_{55} = C_{66} = G \tag{1.255}$$

For an isotropic material, the linear elastic properties are completely described when any *two* of the elastic moduli λ, G, ν, E or K are identified. To give a complete description of an anisotropic rock, all the nine constants of (1.243) must be identified. This is not easily achieved in practical situations.

The elastic moduli do no longer have unique values for anisotropic materials. Since the elastic properties are different in different directions, the values for E (Eq. (1.103)) and ν (Eq. (1.104)) may obviously vary according to the direction of the applied stress.

As an example, consider the uniaxial stress state defining Young's modulus and Poisson's ratio (Fig. 1.15). In this example, $\sigma_y = \sigma_z = 0$ and $\tau_{xy} = \tau_{xz} = \tau_{yz} = 0$. The stress–strain relations (1.246)–(1.251) then become

$$\sigma_x = C_{11}\varepsilon_x + C_{12}\varepsilon_y + C_{13}\varepsilon_z \tag{1.256}$$

$$0 = C_{12}\varepsilon_x + C_{22}\varepsilon_y + C_{23}\varepsilon_z \tag{1.257}$$

$$0 = C_{13}\varepsilon_x + C_{23}\varepsilon_y + C_{33}\varepsilon_z \tag{1.258}$$

$$0 = 2C_{44}\Gamma_{yz} \tag{1.259}$$

$$0 = 2C_{55}\Gamma_{xz} \tag{1.260}$$

$$0 = 2C_{66}\Gamma_{xy} \tag{1.261}$$

Solving Eqs. (1.257) and (1.258) for $\nu = -\varepsilon_y/\varepsilon_x$, we find

$$\nu = -\frac{\varepsilon_y}{\varepsilon_x} = \frac{C_{12}C_{33} - C_{13}C_{23}}{C_{22}C_{33} - C_{23}^2} \tag{1.262}$$

whereas for $\nu = -\varepsilon_z/\varepsilon_x$, we find (by interchanging indices 2 and 3)

$$\nu = -\frac{\varepsilon_z}{\varepsilon_x} = \frac{C_{13}C_{22} - C_{12}C_{23}}{C_{22}C_{33} - C_{23}^2} \tag{1.263}$$

Thus the value of Poisson's ratio depends not only on the direction of the applied stress, but also on the direction in which lateral expansion is measured.

1.7.2 Transverse isotropy (TI)

A special type of symmetry, which is relevant for many types of rocks, is full rotational symmetry around one axis. Rocks possessing such symmetry are said to be *transversely isotropic*, often abbreviated TI. It implies that the elastic properties are equal for all directions within a plane, but different in the other directions. This extra element of symmetry reduces the number of independent elastic constants to 5.

Assuming that the x- and y-directions are equivalent and the z-direction is the different one, we may rotate the coordinate system any angle around the z-axis without altering the elastic constants. For this to be possible, it is required that $C_{11} = C_{22}$, $C_{13} = C_{23}$, $C_{12} = C_{11} - 2C_{66}$ and $C_{44} = C_{55}$. Then the stiffness matrix for a transversely isotropic material having the z-axis as the unique axis is

$$\mathbf{C} = \begin{pmatrix} C_{11} & C_{11} - 2C_{66} & C_{13} & 0 & 0 & 0 \\ C_{11} - 2C_{66} & C_{11} & C_{13} & 0 & 0 & 0 \\ C_{13} & C_{13} & C_{33} & 0 & 0 & 0 \\ 0 & 0 & 0 & C_{44} & 0 & 0 \\ 0 & 0 & 0 & 0 & C_{44} & 0 \\ 0 & 0 & 0 & 0 & 0 & C_{66} \end{pmatrix} \tag{1.264}$$

Transverse isotropy is normally considered to be a representative symmetry for horizontally layered sedimentary rocks. Stress-induced anisotropy may often be described by transverse isotropy as well. Thus, for geophysicists, transverse isotropy is probably the most important type of symmetry next to isotropy.

The TI compliance matrix may be written as

$$\mathbf{S} = \begin{pmatrix} \frac{1}{E_h} & -\frac{\nu_{hh}}{E_h} & -\frac{\nu_{vh}}{E_v} & 0 & 0 & 0 \\ -\frac{\nu_{hh}}{E_h} & \frac{1}{E_h} & -\frac{\nu_{vh}}{E_v} & 0 & 0 & 0 \\ -\frac{\nu_{vh}}{E_v} & -\frac{\nu_{vh}}{E_v} & \frac{1}{E_v} & 0 & 0 & 0 \\ 0 & 0 & 0 & \frac{1}{G_v} & 0 & 0 \\ 0 & 0 & 0 & 0 & \frac{1}{G_v} & 0 \\ 0 & 0 & 0 & 0 & 0 & \frac{2(1+\nu_{hh})}{E_h} \end{pmatrix} \tag{1.265}$$

By assuming loading in the x-, y- and z-directions, respectively, we find that E_h is Young's modulus for loading in the plane and E_v is Young's modulus for loading along the unique axis. Similarly, ν_{vh} is Poisson's ratio for loading along the unique axis, whereas ν_{hh} is Poisson's ratio for loading in the plane and observing the deformation in the plane. Note that for loading in the plane and observing the out-of-plane deformation, Poisson's ratio is

$$\nu_{\text{hv}} = \nu_{\text{vh}} \frac{E_{\text{h}}}{E_{\text{v}}} \tag{1.266}$$

Thus there are three different Poisson's ratios, but two of them are related via the Young's moduli. Finally, G_{v} is the out of plane shear modulus.

Since the matrices in Eqs. (1.264) and (1.265) are inverses, it is in principle straightforward to derive the relations between the parameters. For convenience, we give the formulas in Appendix D.1.6.

1.7.3 Generalised bulk modulus

Applying isotropic stress to an isotropic material gives isotropic strain, and vice versa. This is not so in an anisotropic material: Isotropic (hydrostatic) stress results in anisotropic strain, and applying isotropic strain gives anisotropic stress.

We may define the bulk modulus as a ratio of the mean stress to the volumetric strain, but the relation is no longer unique. To see this, assume the strain tensor in Eq. (1.236) to be given by

$$\varepsilon_{ij} = \frac{1}{3}\delta_{ij}\varepsilon_{\text{vol}} \tag{1.267}$$

The mean stress is then given by

$$\overline{\sigma} = \frac{1}{9}\sum_{i,j} C_{iijj}\varepsilon_{\text{vol}} \tag{1.268}$$

and we may define

$$K^{\varepsilon} = \frac{1}{9}\sum_{i,j} C_{iijj} \tag{1.269}$$

where K^{ε} may be called the isostrain bulk modulus. Note that with the stiffness matrix in Voigt form, K^{ε} is the average of the nine elements in the upper left quarter.

Similarly, we may use a hydrostatic stress given by

$$\sigma_{ij} = \overline{\sigma}\delta_{ij} \tag{1.270}$$

in the corresponding compliance expression (1.239). We then find

$$\varepsilon_{\mathrm{vol}} = \sum_{i,j} S_{iijj}\overline{\sigma} \tag{1.271}$$

and may define the isostress bulk modulus

$$K^{\sigma} = \frac{1}{\sum_{i,j} S_{iijj}} \tag{1.272}$$

The moduli K^{σ} and K^{ε} are in general unequal. As an example, the expressions for transverse isotropy are (see also Section D.1.6)

$$K^{\varepsilon} = \frac{1}{9}\left[C_{33} + 4(C_{11} - C_{66} + C_{13})\right] = \frac{1}{9}\frac{(1-\nu_{\mathrm{hh}})E_{\mathrm{v}} + 2E_{\mathrm{h}}(1+2\nu_{\mathrm{vh}})}{1-\nu_{\mathrm{hh}} - 2\nu_{\mathrm{vh}}^2\frac{E_{\mathrm{h}}}{E_{\mathrm{v}}}} \tag{1.273}$$

and

$$K^{\sigma} = \frac{(C_{11} - C_{66})C_{33} - C_{13}^2}{C_{11} + C_{33} - C_{66} - 2C_{13}} = \frac{E_{\mathrm{v}}}{1 + 2(1-\nu_{\mathrm{hh}})\frac{E_{\mathrm{v}}}{E_{\mathrm{h}}} - 4\nu_{\mathrm{vh}}} \tag{1.274}$$

These expressions are equal only if

$$\frac{E_{\mathrm{v}}}{1-\nu_{\mathrm{vh}}} = \frac{E_{\mathrm{h}}}{1-\nu_{\mathrm{hh}}} \tag{1.275}$$

Thus, even for transverse isotropy, $K^{\varepsilon} = K^{\sigma}$ is a special case, although it is always fulfilled for isotropy.

For a numerical example, Heyliger et al. (2003) measured the moduli of single crystal quartz. Although quartz is quite anisotropic (trigonal symmetry; see Section 6.5), the difference between the bulk moduli is small ($K^{\varepsilon} = 37.9\,\mathrm{GPa}$, $K^{\sigma} = 37.5\,\mathrm{GPa}$). (The TI expressions are also valid for quartz.)

The moduli K^{ε} and K^{σ} represent symmetric loading in strain or stress, respectively. By going to less symmetric loadings, both lower and higher generalised bulk moduli will occur; see Section C.11. Eqs. (C.120) and (C.121) are valid also for quartz and give $K^{\mathrm{min}} = 27.3\,\mathrm{GPa}$ and $K^{\mathrm{max}} = 39.2\,\mathrm{GPa}$.

1.7.4 Anisotropic poroelasticity

Anisotropy also influences the poroelastic properties. Gassmann (1951); Biot (1955) gave early treatments of the subject. Cheng's excellent tutorial

(Cheng, 1997) gives a more modern notation. We refer to Cheng's paper for details and derivations, and reproduce a few main results here. As for the isotropic case, we restrict ourselves to materials that are microscopically homogeneous and isotropic. See the discussion in Cheng (1997).

The poroelastic stress–strain equations may be inferred by considering Eq. (1.236) and Eqs. (1.179)–(1.184). We find

$$\sigma_{ij} - \alpha_{ij} p_{\mathrm{f}} = \sum_{k,l} C^{\mathrm{fr}}_{ijkl} \varepsilon_{kl} \tag{1.276}$$

Eq. (1.156) is generalised to

$$p_{\mathrm{f}} = M\left(\sum_{i,j} \alpha_{ij}\varepsilon_{ij} - \zeta\right) \tag{1.277}$$

The Biot coefficient is now a second-order tensor; in principle coupling also shear strain to the pore pressure. The anisotropic Biot coefficient may be written as

$$\alpha_{ij} = \delta_{ij} - \frac{\sum_k C^{\mathrm{fr}}_{ijkk}}{3K_{\mathrm{s}}} \tag{1.278}$$

and the average is given by

$$\overline{\alpha} = \frac{1}{3}\sum_i \alpha_{ii} = 1 - \frac{K^{\varepsilon}_{\mathrm{fr}}}{K_{\mathrm{s}}} \tag{1.279}$$

The parameter M is still given by Eq. (1.169), but with α replaced by the mean from Eq. (1.279):

$$\frac{1}{M} = \frac{\overline{\alpha} - \phi}{K_{\mathrm{s}}} + \frac{\phi}{K_{\mathrm{f}}} \tag{1.280}$$

The Biot parameter C is now a second-order tensor given by $\alpha_{ij}M$.

The Skempton's B also becomes a tensor defined by

$$\Delta p_{\mathrm{f}} = \frac{1}{3}\sum_{i,j} B_{ij}\Delta\sigma_{ij} \tag{1.281}$$

where B_{ij} may be written as

$$B_{ij} = 3M\sum_{kl} S^{\mathrm{u}}_{ijkl}\alpha_{kl} = \frac{3\sum_k S^{\mathrm{fr}}_{ijkk} - \delta_{ij}\frac{1}{K_{\mathrm{s}}}}{\left(\frac{1}{K^{\sigma}_{\mathrm{fr}}} - \frac{1}{K_{\mathrm{s}}}\right) + \phi\left(\frac{1}{K_{\mathrm{f}}} - \frac{1}{K_{\mathrm{s}}}\right)} \tag{1.282}$$

with the average given by

$$\overline{B}=\frac{1}{3}\sum_{i}B_{ii}=\frac{\left(\frac{1}{K_{\text{fr}}^{\sigma}}-\frac{1}{K_{\text{s}}}\right)}{\left(\frac{1}{K_{\text{fr}}^{\sigma}}-\frac{1}{K_{\text{s}}}\right)+\phi\left(\frac{1}{K_{\text{f}}}-\frac{1}{K_{\text{s}}}\right)} \tag{1.283}$$

As for isotropy, $\overline{B}\leq 1$.

Since B is a symmetric second-order tensor, it may be diagonalised by a rotation of the coordinate system (in full analogy with the determination of principal stresses). For isotropy, all three diagonal elements are equal, for transverse isotropy, there are two unique values, whereas in the general case, there are three.

To relate to Skempton's original formulation(1.215), we assume a triaxial stress state and find

$$\Delta p_{\text{f}}=\overline{B}\left(\Delta\sigma_3+\frac{B_{11}}{3\overline{B}}(\Delta\sigma_1-\Delta\sigma_3)\right) \tag{1.284}$$

Comparing to Eq. (1.215), this shows that

$$A=\frac{B_{11}}{3\overline{B}} \tag{1.285}$$

Thus, for an anisotropic material, the *elastic* A is expected to differ from 1/3 and to vary with sample orientation.

To find the undrained moduli, we may let $\zeta=0$ in Eq. (1.277) and insert into Eq. (1.276). The result is

$$C_{ijkl}^{\text{u}}=C_{ijkl}^{\text{fr}}+\alpha_{ij}\alpha_{kl}M \tag{1.286}$$

It follows that the undrained isostrain bulk modulus is given by

$$K_{\text{u}}^{\varepsilon}=K_{\text{fr}}^{\varepsilon}+\overline{\alpha}^2 M \tag{1.287}$$

which shows that with a redefinition of some parameters, Eq. (1.172) is essentially unchanged:

$$\frac{1}{M}=\frac{\overline{\alpha}}{\overline{C}}=\frac{\overline{\alpha}^2}{K_{\text{u}}^{\varepsilon}-K_{\text{fr}}^{\varepsilon}}=\frac{\phi}{K_{\text{f}}}+\frac{\overline{\alpha}-\phi}{K_{\text{s}}} \tag{1.288}$$

Similarly, to find the undrained compliances, we may insert the effective stress law and Eq. (1.281) into Eq. (1.239), which gives

$$S_{ijkl}^{\mathrm{u}} = S_{ijkl}^{\mathrm{fr}} - \frac{1}{3}\left(\sum_{m,n} S_{ijmn}^{\mathrm{fr}} \alpha_{mn}\right) B_{kl} \tag{1.289}$$

from which we find the undrained isostress bulk modulus as

$$\frac{1}{K_{\mathrm{u}}^{\sigma}} = \frac{1}{K_{\mathrm{fr}}^{\sigma}} - \left(\frac{1}{K_{\mathrm{fr}}^{\sigma}} - \frac{1}{K_{\mathrm{s}}}\right)\overline{B} \tag{1.290}$$

1.7.4.1 Voigt form

Rewriting the expressions for the Biot coefficient and undrained stiffness matrix to Voigt form, we find

$$\alpha_i = \delta_i - \frac{1}{3K_{\mathrm{s}}} \sum_{j=1}^{3} C_{ij}^{\mathrm{fr}} \tag{1.291}$$

and

$$C_{ij}^{\mathrm{u}} = C_{ij}^{\mathrm{fr}} + \alpha_i \alpha_j M \tag{1.292}$$

where $\delta_i = 1$ if $i = 1, 2, 3$ and $\delta_i = 0$ if $i = 4, 5, 6$.

We observe that requirement for non-zero α_4, α_5 and α_6 is that the upper right (or by symmetry, the lower left) quarter of the stiffness matrix is non-zero. Under the same conditions, the undrained shear stiffnesses are different from the drained. This is related to the fact that, in contrast to isotropy, a pure shear stress may generate a volumetric strain.

By similar arguments it is clear that the B tensor does not depend on the lower right quarter of the compliance matrix.

1.7.4.2 Transverse isotropy

In a transversely isotropic material with the unique axis along the z-axis, the Biot tensor is diagonal and given by

$$\alpha_{\mathrm{h}} = \alpha_{11} = \alpha_{22} = 1 - \frac{E_{\mathrm{h,fr}}(1 + \nu_{\mathrm{h,fr}})}{3K_{\mathrm{s}}\left(1 - \nu_{\mathrm{h,fr}} - 2\nu_{\mathrm{v,fr}}^2 \frac{E_{\mathrm{h,fr}}}{E_{\mathrm{v,fr}}}\right)} \tag{1.293}$$

$$\alpha_{\mathrm{v}} = \alpha_{33} = 1 - \frac{E_{\mathrm{v,fr}}(1 - \nu_{\mathrm{h,fr}}) + 2E_{\mathrm{h,fr}}\nu_{\mathrm{v,fr}}}{3K_{\mathrm{s}}\left(1 - \nu_{\mathrm{h,fr}} - 2\nu_{\mathrm{v,fr}}^2 \frac{E_{\mathrm{h,fr}}}{E_{\mathrm{v,fr}}}\right)} \tag{1.294}$$

and the Skempton B is

$$B_{\mathrm{h}} = B_{11} = B_{22} = \frac{\frac{K_{\mathrm{f}}}{\phi}\left[3(1-\nu_{\mathrm{h,fr}}) - \frac{E_{\mathrm{v,fr}}}{K_{\mathrm{s}}}\right]}{2\frac{K_{\mathrm{f}}}{\phi}(1-\nu_{\mathrm{h,fr}}) - \frac{E_{\mathrm{h,fr}}}{E_{\mathrm{v,fr}}}\frac{K_{\mathrm{f}}}{\phi}\left(\frac{E_{\mathrm{v,fr}}}{K_{\mathrm{s}}} + 4\nu_{\mathrm{v,fr}} - 1\right) + E_{\mathrm{h,fr}}\left(1 - \frac{K_{\mathrm{f}}}{K_{\mathrm{s}}}\right)} \tag{1.295}$$

$$B_{\mathrm{v}} = B_{33} = \frac{\frac{K_{\mathrm{f}}}{\phi}\frac{E_{\mathrm{h,fr}}}{E_{\mathrm{v,fr}}}\left[3\left(1-2\nu_{\mathrm{h,fr}}\right) - \frac{E_{\mathrm{v,fr}}}{K_{\mathrm{s}}}\right]}{2\frac{K_{\mathrm{f}}}{\phi}(1-\nu_{\mathrm{h,fr}}) - \frac{E_{\mathrm{h,fr}}}{E_{\mathrm{v,fr}}}\frac{K_{\mathrm{f}}}{\phi}\left(\frac{E_{\mathrm{v,fr}}}{K_{\mathrm{s}}} + 4\nu_{\mathrm{v,fr}} - 1\right) + E_{\mathrm{h,fr}}\left(1 - \frac{K_{\mathrm{f}}}{K_{\mathrm{s}}}\right)} \tag{1.296}$$

For transverse isotropy, there are thus two independent Skempton B coefficients. Taking the z-axis as the unique, this may be written

$$\begin{pmatrix} B_{\mathrm{h}} & 0 & 0 \\ 0 & B_{\mathrm{h}} & 0 \\ 0 & 0 & B_{\mathrm{v}} \end{pmatrix} \tag{1.297}$$

To relate this to Skempton A for loading at an angle θ to the unique axis, we transform the B tensor according to the formalism in Appendix C.6. The result for the 33-component is

$$B_{33} = \frac{B_{\mathrm{v}} + B_{\mathrm{h}}}{2} + \frac{B_{\mathrm{v}} - B_{\mathrm{h}}}{2}\cos 2\theta \tag{1.298}$$

To find A, we then use Eq. (1.285) (remembering the change of the unique axis) and find

$$A = \frac{(B_{\mathrm{v}} + B_{\mathrm{h}}) + (B_{\mathrm{v}} - B_{\mathrm{h}})\cos 2\theta}{6\overline{B}} = \frac{B_{\mathrm{h}}\sin^2\theta + B_{\mathrm{v}}\cos^2\theta}{3\overline{B}} \tag{1.299}$$

See Holt et al. (2018).

1.8 Nonlinear elasticity

1.8.1 Stress–strain relations

For a linear elastic material, there is always a constant relationship between the applied stress and the resulting strain, regardless the magnitude of the stress and the strain. The stress–strain relation (1.94) for such a material is therefore a straight line, as shown in Fig. 1.20a. The elastic modulus corresponding to this stress–strain pair is the slope of the curve.

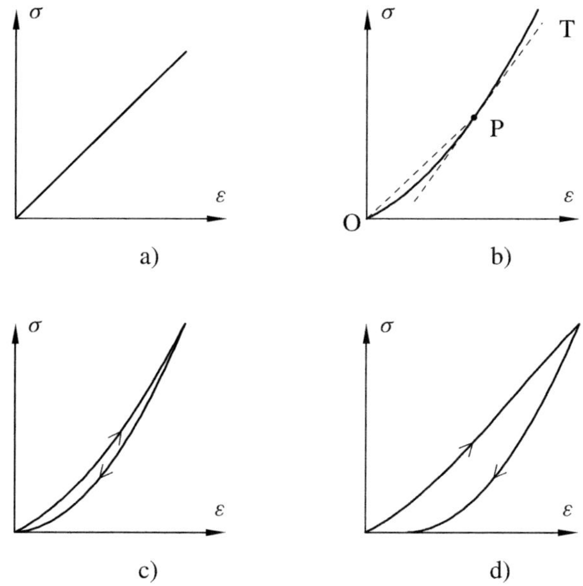

Figure 1.20 Stress–strain relations for a) linear elastic material. b) perfectly elastic material, c) elastic material with hysteresis and d) material suffering permanent deformation.

Any material not obeying a linear stress–strain relation is said to behave nonlinearly. For a nonlinear elastic material, the stress–strain relation may be written as

$$\sigma = E_1\varepsilon + E_2\varepsilon^2 + E_3\varepsilon^3 + \cdots \tag{1.300}$$

Recalling that σ and ε generally are tensors, it is clear that nonlinear elasticity may be very complicated mathematically.

The coefficient E_1 corresponds to the elastic moduli introduced in Section 1.3 and is hence a fourth-order tensor in the general case. E_2 represents a sixth-order tensor, whose components are commonly referred to as third-order elastic moduli; see Section 1.4. For isotropy, E_2 corresponds to three independent moduli, whereas the number for transverse isotropy is 9 (see e.g. Clayton, 2011).

Nonlinear behaviour may have various causes and appears in many different ways. Fig. 1.20b shows one example. This material has a nonlinear stress–strain relation, since the ratio of stress to strain is not the same for all stresses. The relation is, however, identical for the loading and unloading process. Such materials are said to be perfectly elastic.

For nonlinear stress–strain relations, the elastic modulus is no longer uniquely defined, not even for a specific stress level. As shown in Fig. 1.20b, the modulus related to the point P on the curve may either be identified as the slope of the line OP (*secant modulus*), or it may be identified as the slope of the tangent PT (*tangent modulus*).

If the correct definition for the moduli is used, then the linear form of the stress–strain relations may be used far beyond the initial linear region. For example, Hooke's law (Eq. (1.94)) may be written in its original form

$$\sigma = E_{\text{sec}}(\varepsilon)\varepsilon \tag{1.301}$$

where $E_{\text{sec}}(\varepsilon)$ is the secant value of Young's modulus. Alternatively, the relation may be written in a differential form

$$\Delta\sigma = E_{\text{tan}}(\varepsilon)\Delta\varepsilon \tag{1.302}$$

where $\Delta\sigma$ and $\Delta\varepsilon$ represent differential increments in stress and strain, respectively, and $E_{\text{tan}}(\varepsilon)$ is the tangent value of Young's modulus. Note, however, that both $E_{\text{tan}}(\varepsilon)$ and $E_{\text{sec}}(\varepsilon)$ depend on the strain ε. This complicates the use of the equations. By comparing Eqs. (1.301) and (1.302) we observe that the relation between the secant modulus and the tangent modulus is

$$E_{\text{sec}}(\varepsilon) = \frac{1}{\varepsilon}\int_0^{\varepsilon} E_{\text{tan}}(\varepsilon')\,\mathrm{d}\varepsilon' \tag{1.303}$$

The stress–strain relation shown in Fig. 1.20c is commonly observed in rocks. The unloading path is different from the loading path. This effect is called *hysteresis*. For materials behaving like this, the work done during loading is not entirely released during unloading, that is, a part of the strain energy dissipates in the material. Elastic moduli related to the unloading path are called *unloading moduli*.

If, as in Fig. 1.20c, the strain vanishes when the stress returns to zero, the material is said to be *elastic*. If not, as in Fig. 1.20d, the material has suffered a permanent deformation during the loading/unloading cycle. For sufficiently large stresses, many rocks enter a phase where permanent deformation occurs, yet the material is still able to resist loading (i.e. the slope of the stress–strain curve is still positive). The material is then said to be *ductile*. The point where the transition from elastic to ductile behaviour occurs is called the *yield point*.

In Eqs. (1.301)–(1.303) the elastic modulus has been expressed as a function of strain. Since there is a relation between the applied stress and the resulting strain, we might have expressed the modulus as a function of stress rather than strain. In general, the elastic moduli depend on all components of stress (or all components of strain). For example, Young's modulus may depend on the confining pressure, as shown in Fig. 2.3.

1.8.2 The impact of cracks

Cracks occur in all types of rocks, and they generally have a large impact on the elastic properties of the material. For some materials, like poorly consolidated sedimentary rocks, it may be difficult to imagine the presence of cracks. However, weak or failing grain contacts have much of the same impact on elastic parameters as cracks have and may for modelling purposes often be considered as such.

The occurrence of cracks in rocks produce various types of nonlinear behaviour. We will here look at a couple of examples. Consider first the situation shown in Fig. 1.21a. The stressed sample contains a crack oriented with its face normal to the stress σ_x. Since no stress can be transferred across the crack itself, the effective Young's modulus E_{eff} of the sample will be reduced:

$$\frac{\sigma_x}{\varepsilon_x} = E_{\mathrm{eff}} = E(1 - \xi\, Q) \tag{1.304}$$

Here E is Young's modulus of the material without cracks, ξ is called the *crack density* and is a function of the size and number of cracks, and Q is a coefficient depending on the shape and orientation of the crack. (Micromechanical models for calculation of Q and equivalent parameters are described in Section 6.4.) As the stress is increased, the strain ε_x also increases. A part of the strain increase is due to closure of the crack. At a certain stress level σ_x^c the crack is closed. At stresses above this point, ξ will vanish, and $E_{\mathrm{eff}} \to E$ according to Eq. (1.304). The stress–strain relation for this sample is then as shown in Fig. 1.21b.

For a material containing many cracks of different sizes and closure stresses, the stress–strain relation may look as in Fig. 1.20b. Such closure of cracks as stress increases may explain the typical feature that elastic constants of rocks normally increase with increasing hydrostatic pressure.

Now consider a material containing a *closed* crack with its face oriented at an angle relative to the stress σ_x (Fig. 1.22a). Due to friction, the closed crack will be able to transfer shear stress τ up to a certain level τ_c given by

$$\tau_c = S_0 + \mu\sigma \tag{1.305}$$

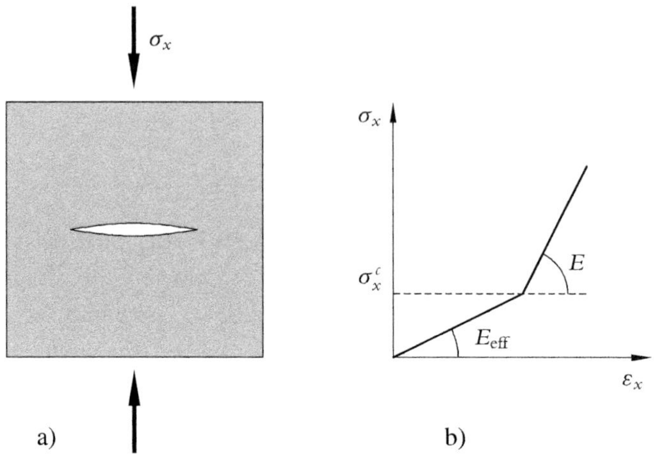

Figure 1.21 Nonlinear stress–strain relation due to crack closure.

Here σ is the stress normal to the crack face, S_0 is the inherent shear strength of the closed crack, and μ is the coefficient of friction. When the shear stress τ across the crack exceeds τ_c, the crack surfaces slip and slide relative to each other. Then τ is reduced, and it may either vanish (local damage), or the crack surfaces may again stick to each other, and τ increases from a lower level (point A in Fig. 1.23a). The sliding of the crack surfaces will result in an additional strain $\Delta\varepsilon$ (point A in Fig. 1.23b).

Upon unloading, τ is reduced and may eventually become equal to $-\tau_c$. A reverse sliding will then take place (point B in Figs. 1.23a and 1.23b). The stress–strain relation for the entire sample (Fig. 1.23b) is seen to have a hysteresis. Also, the sample is seen to have suffered a permanent deformation, since the strain does not go to zero when the stress vanishes.

Real rocks normally contain cracks or crack-like defects of many sizes and different orientations. The resulting stress–strain relation is typically as shown in Fig. 1.20d.

In Fig. 1.22b the sliding of the crack surfaces is accompanied with the opening of two other cracks oriented with their faces normal to the direction of the compression stress σ. Opening of cracks with this orientation relative to applied uniaxial stress is a typical feature occurring at high shear stresses. A consequence of such crack opening is an inelastic increase in the volume of the stressed material. This phenomenon is known as *dilatancy*. Another consequence is a stress-induced mechanical anisotropy, as

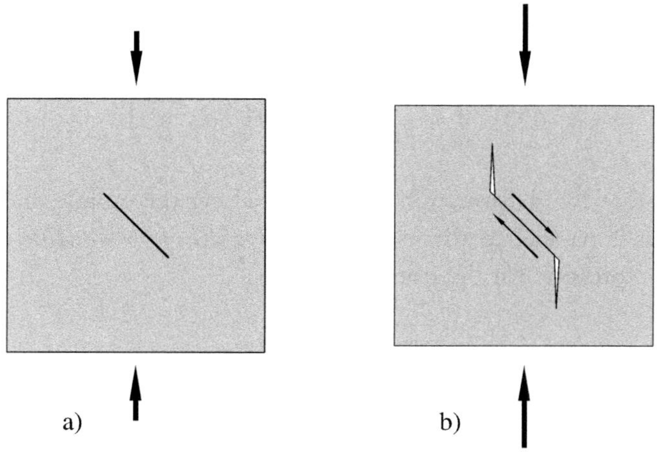

Figure 1.22 Material with sliding crack.

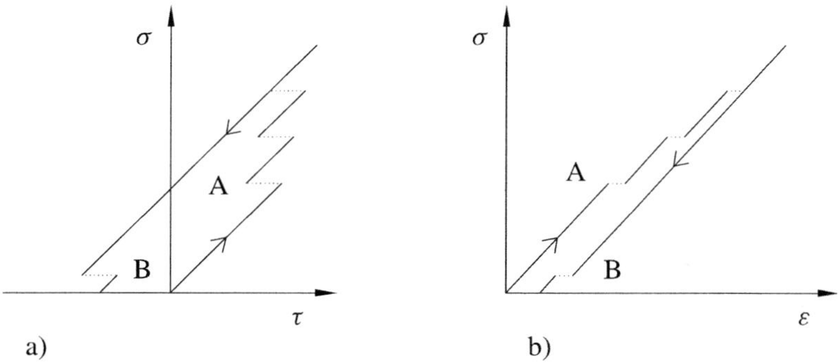

Figure 1.23 Material with a sliding crack. a) Normal stress versus shear stress across the crack surface. b) Stress–strain relation for the material.

described in the previous section. The anisotropy may, for instance, be observed by acoustical techniques (see Section 5.5).

1.8.3 Finite strain

A rock sample of length L that is subjected to a length alteration ΔL suffers a normal strain of $\varepsilon = -\Delta L/L$ according to the infinitesimal strain assumption (Eq. (1.63)). However, for finite strain, this definition is not unique. Consider that we obtain the length alteration ΔL through a series of n steps and calculate the strain for each step, taking into account that the length of the sample changes for every step in the process. In the limit as $n \to \infty$, we then find that the normal strain is given as

$$\varepsilon = -\int_{L}^{L+\Delta L} \frac{\mathrm{d}L'}{L'} = -\ln\left(1 + \frac{\Delta L}{L}\right) \tag{1.306}$$

Assuming that the deformation is uniform along the sample and that the sample axis is parallel to the x-axis, we have $\Delta L/L = -\partial u/\partial x$. Thus the complete expression for the normal strain is

$$\varepsilon_x = -\ln\left(1 - \frac{\partial u}{\partial x}\right) \tag{1.307}$$

This is called the *true* strain or *natural* strain.

Eq. (1.307) converges to Eq. (1.64) in the limit $\partial u/\partial x \to 0$, which is the infinitesimal strain limit. For finite strain, the difference may be significant. For the extreme case that $\Delta L \to -L$, Eq. (1.307) approaches the intuitively correct limit $\varepsilon_x \to \infty$, whereas the infinitesimal strain Eq. (1.64) approaches unity.

Although rocks may suffer large strains, the *elastic part* of the strain is usually small and may be described in the infinitesimal strain limit. This part of the strain is associated with stress changes in accordance with Hooke's law. The remaining, nonelastic part is governed by other rules, as described in Section 2.8.1. In general, finite strain is a complex subject and will not be further discussed here. See, for instance, Rudnicki (2014) for a broader discussion.

1.9 Time-dependent effects

So far, we have assumed that any change in applied stress is followed instantaneously by the corresponding deformation. Quite often, however, it is observed that the deformation of rocks continues for a long time after a change in the applied stress. The time-dependent effects can be divided into two groups, *consolidation* and *creep*. Consolidation is due to pore pressure gradients induced by a change in the stress state and the fact that it takes time to re-establish pore pressure equilibrium. Creep is related to visco-elastic behaviour of the solid framework. In the following, consolidation and creep are discussed separately. In practice, however, it is sometimes difficult to distinguish between genuine creep and consolidation effects.

1.9.1 Consolidation

Consolidation theory describes the transient process, where pore pressure equilibrium is re-established after a change in the stress state. This process involves a flow of the pore fluid through the porous rock. Viscous flow in porous rocks is described by Darcy's law, which states that the fluid flow rate $\vec{Q}$ (fluid volume per unit time flowing through a surface) is proportional to the pore pressure gradient ∇p_{f}:

$$\vec{q}_{\mathrm{fl}} = \frac{\vec{Q}}{A} = -\frac{k}{\eta_{\mathrm{f}}} \nabla p_{\mathrm{f}} \tag{1.308}$$

Here A is the surface through which the fluid flows, η_{f} is the dynamic viscosity of the fluid, and k is the permeability of the rock. Normally, k is measured in the unit Darcy, defined as the permeability that gives a flowrate of one centimetre per second of a fluid with viscosity one centipoise for a pressure gradient of one atmosphere per cm. This means that we have $1\,\mathrm{D} = 0.9869 \cdot 10^{-12}\,\mathrm{m}^2$ since 1 atmosphere is equal to 101325 Pa.[5] Permeabilities of reservoir rocks may vary from the low milliDarcy region up to several Darcy.

Eq. (1.308) basically describes stationary flow of fluid in a homogeneous pore pressure gradient field. However, the equation may be further evaluated to make it applicable for description of transient processes. The fluid flux $\vec{q}_{\mathrm{fl}}$ expresses the difference between the solid and fluid displacement rates, that is,

$$\vec{q}_{\mathrm{fl}} = \phi \left(\frac{\partial \vec{u}_{\mathrm{s}}^{\mathrm{fr}}}{\partial t} - \frac{\partial \vec{u}_{\mathrm{f}}^{\mathrm{fr}}}{\partial t} \right) \tag{1.309}$$

By comparing Eq. (1.309) with the definition (1.147) of the strain parameter ζ we find that the divergence of the flux $\vec{q}_{\mathrm{fl}}$ is proportional to the time derivative of ζ, that is,

$$\nabla \cdot \vec{q}_{\mathrm{fl}} = \frac{\partial \zeta}{\partial t} \tag{1.310}$$

[5] In some sciences, including soil mechanics and ground water research, an alternative definition for permeability with unit m/s is used. The relation to our definition is

$$k(\text{hydrosciences}) = k(\text{petroleum sciences})\gamma_{\mathrm{w}}/\eta_{\mathrm{w}}$$

where γ_{w} and η_{w} are the specific weight and viscosity of water, respectively.

Note that this equation may also serve as a definition of ζ and be used as an alternative starting point for the poroelastic theory (Rice and Cleary, 1976; Detournay and Cheng, 1993).

By introducing Eq. (1.308) we have

$$\frac{\partial \zeta}{\partial t} = -\frac{k}{\eta_f}\nabla^2 p_f \tag{1.311}$$

Eq. (1.311) describes transient fluid flow in a porous rock and also flow in non-homogeneous pore pressure gradient fields.

We may eliminate ζ from Eq. (1.311) by using Eq. (1.156) to find an equation involving p_f and ε_{vol}:

$$\frac{k}{\eta_f}\nabla^2 p_f = \frac{1}{M}\frac{\partial p_f}{\partial t} - \alpha\frac{\partial \varepsilon_{vol}}{\partial t} \tag{1.312}$$

Alternatively, we may find an equation involving p_f and $\overline{\sigma}$ by combining Eqs. (1.311), (1.156) and (1.158):

$$\frac{k}{\eta_f}\nabla^2 p_f = \frac{\alpha}{K_{fr}B}\frac{\partial p_f}{\partial t} - \frac{\alpha}{K_{fr}}\frac{\partial \overline{\sigma}}{\partial t} \tag{1.313}$$

where Skempton's B was defined in Eq. (1.214).

It is clear that the equations in general involve a coupling between the pore pressure and the strain or stress of the solid material. The pore pressure will obey an uncoupled equation only in special cases. For a thorough discussion, see, for example, Wang (2000); Detournay and Cheng (1993).

We will now consider an example of decoupled consolidation by considering a simple experiment: a porous material is confined with no lateral movement ($u_x = u_y = 0$) in a vertical column with an impermeable bottom (at $z = 0$) and a highly permeable piston on the top (at $z = h$). The bottom of the sample is fixed, that is, $u_z(z{=}0) = 0$. Initially, at $t = 0$, a vertical stress $\sigma_z = \sigma_0$ is applied by the piston. The pore pressure is now a function of both time t and height z, that is, $p_f = p_f(t, z)$. The boundary conditions are

$$p_f = 0 \quad \text{at } z = h \tag{1.314}$$

$$\frac{\partial p_f}{\partial z} = 0 \quad \text{at } z = 0 \tag{1.315}$$

The first condition is a result of the piston permeability causing the same pore pressure on the inside as on the outside. The second condition is due to the requirement of no flow through the bottom ($\partial \vec{u}_{\mathrm{s}}^{\mathrm{fr}}/\partial t = \partial \vec{u}_{\mathrm{f}}^{\mathrm{fr}}/\partial t$; see Eq. (1.309)).

Recalling that $u_x = u_y = 0$ and writing $\lambda_{\mathrm{u}} + 2G = H_{\mathrm{u}}$, we find that Eq. (1.152) now becomes

$$\sigma_0 = H_{\mathrm{u}}\varepsilon_z - C\zeta \tag{1.316}$$

whereas Eq. (1.156) becomes

$$p_{\mathrm{f}} = C\varepsilon_z - M\zeta \tag{1.317}$$

Using Eqs. (1.316) and (1.317) to eliminate ζ, we obtain an expression for the vertical strain ε_z:

$$\varepsilon_z = \frac{M\sigma_0 - Cp_{\mathrm{f}}}{H_{\mathrm{u}}M - C^2} = \frac{\sigma_0 - \frac{C}{M}p_{\mathrm{f}}}{H_{\mathrm{u}} - \frac{C^2}{M}} = \frac{\sigma_0 - \alpha p_{\mathrm{f}}}{H_{\mathrm{fr}}} \tag{1.318}$$

where $H_{\mathrm{fr}} = H_{\mathrm{u}} - C^2/M$ is defined in analogy with Eq. (1.159).

Eq. (1.318) shows that the vertical strain depends on the pore pressure, and hence it will change during the time it takes for the pore pressure to reach equilibrium. It is straightforward to find the initial and final deformation of the column. At the instant the load is applied, before the pore fluid has time to move, we observe the *undrained* stiffness of the material. Letting $\zeta = 0$ in Eq. (1.316), we see that the immediate strain is $\varepsilon_z = \sigma_0/H_{\mathrm{u}}$.

At late times, when the pore pressure has dissipated to zero, we observe the *drained* stiffness, which from Eq. (1.318) gives $\varepsilon_z = \sigma_0/H_{\mathrm{fr}}$. The difference and hence the magnitude of the consolidation or time-dependent strain is

$$\varepsilon_z = \sigma_0\left(\frac{1}{H_{\mathrm{fr}}} - \frac{1}{H_{\mathrm{u}}}\right) \tag{1.319}$$

The time dependence of the pore pressure is governed by Eqs. (1.311)–(1.313). Differentiating Eq. (1.318) and inserting into Eq. (1.312), we find the following differential equation for p_{f}:

$$\frac{\partial p_{\mathrm{f}}}{\partial t} = \frac{k}{\eta_{\mathrm{f}}}\frac{H_{\mathrm{u}}M - C^2}{H_{\mathrm{u}}}\frac{\partial^2 p_{\mathrm{f}}}{\partial z^2} = D_{\mathrm{h}}\frac{\partial^2 p_{\mathrm{f}}}{\partial z^2} \tag{1.320}$$

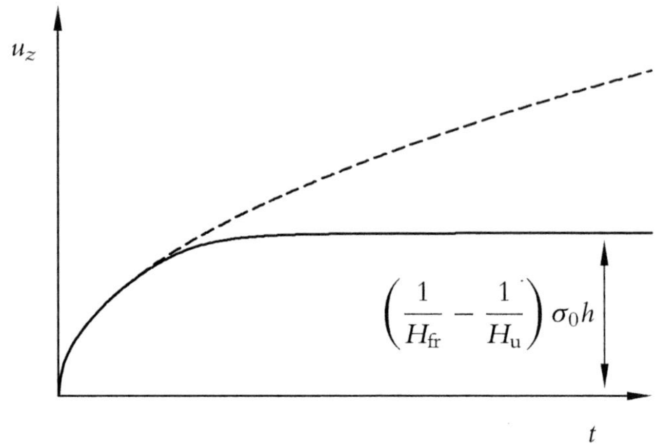

Figure 1.24 Settlement caused by consolidation as a function of time. The solid curve represents the settlement of a column of finite height, while the dashed curve represents the settlement for an infinitely high column. (After Biot, 1941.)

This equation is a diffusion equation. The diffusion constant, the *hydraulic diffusivity* is

$$D_h = \frac{k}{\eta_f}\frac{H_u M - C^2}{H_u} = \frac{k}{\eta_f}M\frac{H_{fr}}{H_u} = \frac{k}{\eta_f}\left(\frac{\phi}{K_f} + \frac{\alpha - \phi}{K_s} + \frac{\alpha^2}{K_{fr} + \frac{4}{3}G_{fr}}\right)^{-1} \tag{1.321}$$

The constant D_h is also called the *consolidation coefficient*.

Eq. (1.320) and the boundary conditions (1.314) and (1.315) fully determine the pore pressure $p_f(t, z)$. Biot (1941) discussed the solution of the problem and found that the settlement u_z can be written in terms of an infinite series. He also provided an approximate solution, which applies to the settlement of an infinitely high column ($h \to \infty$) or to the initial settlement ($t \to 0$) of a column of finite height:

$$u_z(t) \approx \frac{2}{\sqrt{\pi}}\sigma_0\left(\frac{1}{H_{fr}} - \frac{1}{H_u}\right)\sqrt{D_h t} \tag{1.322}$$

This result is illustrated in Fig. 1.24.

By assuming $K_{fr}, G_{fr} \ll K_s$ we find that the diffusion constant D_h is approximately given by

$$D_h \approx \frac{kK_f}{\eta_f \phi}\left[1 + \frac{K_f}{\phi\left(K_{fr} + \frac{4}{3}G_{fr}\right)}\right]^{-1} \tag{1.323}$$

In the "stiff frame" limit $K_{\mathrm{fr}}, G_{\mathrm{fr}} \gg K_{\mathrm{f}}/\phi$, and we find that the hydraulic diffusivity is governed by the permeability and the elastic properties of the fluid:

$$D_{\mathrm{h}} \to \frac{k}{\eta_{\mathrm{f}}} \frac{K_{\mathrm{f}}}{\phi} \tag{1.324}$$

On the other hand, if $K_{\mathrm{fr}}, G_{\mathrm{fr}} \ll K_{\mathrm{f}}/\phi$, then the hydraulic diffusivity is governed more by the elastic properties of the framework:

$$D_{\mathrm{h}} \to \frac{k}{\eta_{\mathrm{f}}} \left(K_{\mathrm{fr}} + \frac{4}{3} G_{\mathrm{fr}} \right) \tag{1.325}$$

Physically, D_{h} is a measure of how far (l_{D}) a pore pressure disturbance can propagate during a given time τ_{D}. The length l_{D} is called the *diffusion length* and relates to τ_{D} and D_{h} as

$$l_{\mathrm{D}}^2 = D_{\mathrm{h}} \tau_{\mathrm{D}} \tag{1.326}$$

It is an important result that the characteristics of the fluid flow, here represented by the diffusion constant D_{h} and the diffusion length l_{D}, generally depend not only on the fluid parameters, but also on the elastic properties of the rock itself. This is particularly significant for weak rocks, as Eq. (1.325) shows.

The consolidation discussed above is based on an elastic theory, which means that the behaviour is reversible. This is often not the case in a practical situation, in particular, not in sedimentary rocks at high pressure. In such cases, inelasticity of the rock framework should be taken into account.

Note that if we insert ζ from Eq. (1.199) into Eq. (1.311), then we get

$$\frac{\partial p_{\mathrm{f}}}{\partial t} = \frac{k}{\eta_{\mathrm{f}} S} \nabla^2 p_{\mathrm{f}} \tag{1.327}$$

The full decoupling is only apparent, since S in general depends on strain. The relation between the hydraulic diffusivity and specific storage is seen to be

$$D_{\mathrm{h}} = \frac{k}{\eta_{\mathrm{f}} S} \tag{1.328}$$

which underlines that D_{h} (like S) depends on the boundary conditions.

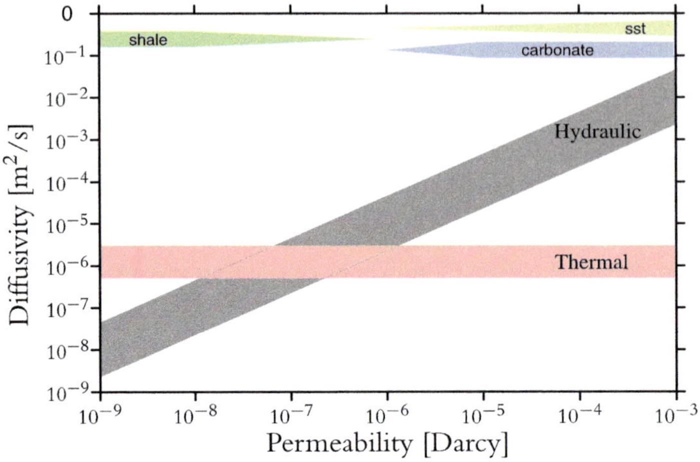

Figure 1.25 Illustration of typical ranges of hydraulic (grey) and thermal (light red) diffusivities.

1.9.2 Thermal diffusivity

Eq. (1.320) is similar to the well-known 1-D thermal diffusion equation

$$\frac{\partial T}{\partial t} = D_{\mathrm{T}} \frac{\partial^2 T}{\partial z^2} \tag{1.329}$$

where the thermal diffusivity is given by

$$D_{\mathrm{T}} = \frac{\kappa_{\mathrm{T}}}{\rho C_{\mathrm{r}}} \tag{1.330}$$

Here C_{r} is the heat capacity, κ_{T} is the thermal conductivity, and ρ is the density.

The diffusivities D_{h} and D_{T} are both expressed in length squared over time and can be directly compared. If D_{h} is greater than D_{T}, then thermally induced pore pressure will dissipate quickly, whereas if it is smaller, temperature changes will lead to sustained pore pressure changes.

Whereas D_{h} spans many orders of magnitude, D_{T} for rock varies relatively little. This is illustrated in Fig. 1.25.

1.9.3 The Mandel–Cryer and Noordbergum effects

Published independently (Mandel, 1953; Cryer, 1963; Verruijt, 1969), the *Mandel–Cryer effect* and the *Noordbergum effect* are manifestations in various

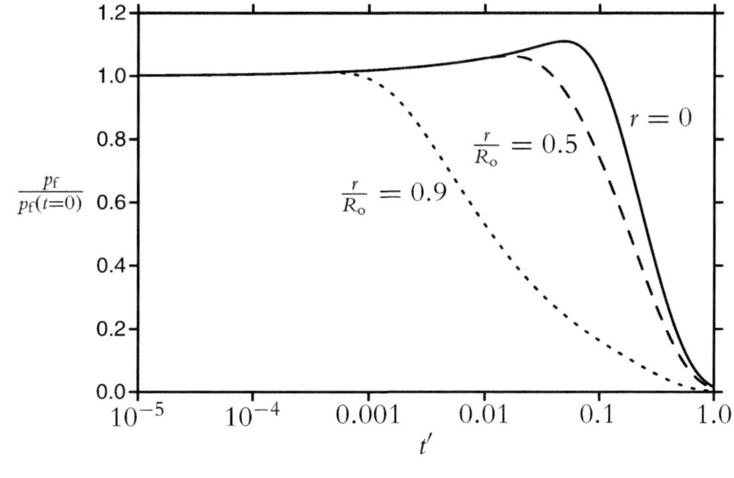

Figure 1.26 Pressure development in an infinite cylinder that is subjected to a radial stress at $t = 0$. The pore pressure is kept at zero at the outside. $t' = tD_h/R_o^2$. The figure was computed using $\nu_u = 0.4$ and $\nu_{fr} = 0.2$.

geometries of a fundamental poroelastic effect: poroelastic coupling in certain cases gives a non-monotonous development of pore pressure, which does not happen when the coupling is absent.

We will illustrate the effect by a variant described by Detournay and Cheng (1993). We refer to that paper for the mathematics and the details, and here show a figure demonstrating the effect. See also Abousleiman et al. (1996); Wang (2000); Cheng (2016).

We consider a cylinder in plane strain, which at time zero is subjected to a radial stress. This induces a pore pressure, which is allowed to drain radially as the pore pressure at the surface is kept at zero.

As the pore pressure falls in the outer part of the sample, this part becomes less stiff (going from undrained towards drained response). As a result, the inner part must carry more of the load, and hence the pore pressure increases before diffusion makes it fall.

Fig. 1.26 shows the pore pressure development at the centre, at $r/R = 0.5$, and at $r/R = 0.9$. An initial increase is observed, most strongly at the centre.

While Mandel and Cryer studied model systems, the Noordbergum[6] effect is a real-world phenomenon: As the pumps were started and pressure

[6] Noordbergum (Frisian: Noardburgum) is a small village in the vicinity of the wells where the effect was first reported.

decreased in one water well, a transient increase in pressure was seen in observation wells.

1.9.4 Creep

Creep is a time-dependent deformation that may occur in materials under constant stress. Creep originates from visco-elastic effects in the solid framework, and thus creep may, unlike consolidation, occur in both dry and saturated rocks.

Deformation related to creep follows in addition to the instantaneous deformation, that is,

$$\varepsilon_{\text{total}}(t) = \varepsilon_{\text{inst}} + \varepsilon_{\text{cr}}(t) \tag{1.331}$$

where $\varepsilon_{\text{total}}$ is the total strain, $\varepsilon_{\text{inst}}$ is the instantaneous elastic strain, and ε_{cr} is the creep strain. Thus $\varepsilon_{\text{cr}}(t) \to 0$ as $t \to 0$. The development of creep strain implies that there is an ongoing deformation of the rock even when the stress state is kept constant. Alternatively, if the deformation is fixed (i.e. $\varepsilon_{\text{total}}(t) = 0$), then the elastic strain will have to decrease simultaneously with the increase in creep strain. A reduction in elastic strain implies a reduction in stress according to Hooke's law. Thus, if deformation is fixed, then creep induces a time-dependent stress reduction known as *stress relaxation* proportional to the creep strain.

There are three stages of creep following a change in the stress state. First, there is a region where the rate of the time-dependent deformation decreases with time (Fig. 1.27). This is called *transient* (or primary) creep.

In the next stage the deformation rate is constant. This is called *steady-state* (or secondary) creep. Finally, the deformation rate may increase with time. This is called *accelerating* (or tertiary) creep. This stage leads rapidly to failure.

The actual creep behaviour of a rock depends on the magnitude of the applied stress. For low or moderate stresses, the material may stabilise after a period of transient creep. For high stresses, the material may rapidly run through all three stages of creep and finally fail. The intermediate stress regime, where the material fully develops each stage of creep, may be small and hard to find in practice (Fig. 1.28).

It is assumed that cracks propagating in a stable, quasi-static manner, known as *subcritical crack growth*, play an important role for creep in rocks. A dominant mechanism for subcritical crack growth seems to be *stress corrosion*, which is a chemical reaction between rock and pore fluid at the

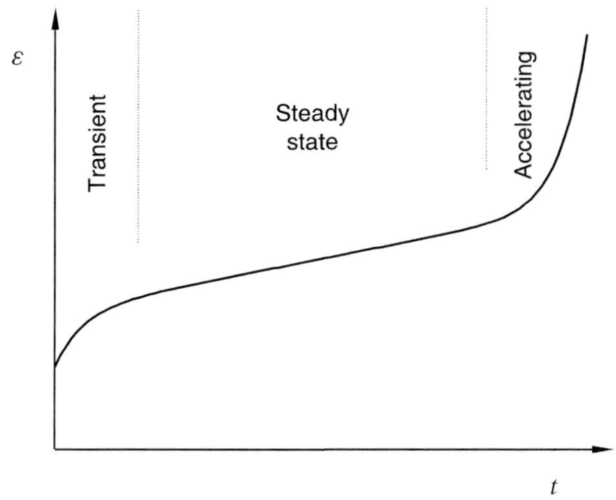

Figure 1.27 Strain versus time for a creeping material.

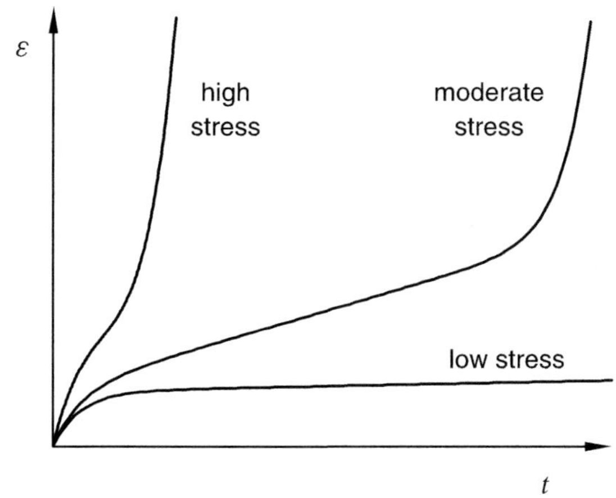

Figure 1.28 The development of creep for different values of the applied stress.

crack tips. Transient creep may be seen as subcritical crack growth at a decaying rate, with little or no interaction between the cracks. This process may disappear gradually as the crack growth rate vanishes, which typically happens at low stress levels. At higher stress levels the crack density may eventually reach a threshold where crack interactions become significant, leading to increasing crack growth rate and accelerating creep. This suggests that steady-state creep may be just an inflection between transient and accelerating creep (Brantut et al., 2013).

The fact that creep may lead to failure means that a rock that is loaded to a level somewhat below its ultimate strength may fail after some time if the load is maintained. This effectively reduces the long-term strength of the rock. The effect will be further discussed in Section 2.10.

Various mathematical models, with varying degree of sophistication, are being used to describe creep strain. Transient creep is often described by a power law model (see Louchet and Duval (2009) for an overview):

$$\varepsilon_{\mathrm{cr}} \propto t^n \qquad \text{with} \qquad 0 < n < 1 \tag{1.332}$$

or by a logarithmic expression (Lomnitz, 1956)

$$\varepsilon_{\mathrm{cr}} \propto \ln(1 + at) \tag{1.333}$$

Note that neither Eq. (1.332) nor Eq. (1.333) is valid as $t \to \infty$, and hence predictions into the far future based on these models are of limited value.

For steady-state creep, the creep strain increases linearly with time:

$$\varepsilon_{\mathrm{cr}} \propto t \tag{1.334}$$

that is, the creep rate $\dot{\varepsilon}_{\mathrm{cr}} \equiv \partial \varepsilon_{\mathrm{cr}} / \partial t$ is simply a constant.

Creep is a molecular process, which implies that it is strongly dependent on mineralogy and temperature. The temperature dependence of the creep rate is given by the Arrhenius relation

$$\dot{\varepsilon}_{\mathrm{cr}} \propto \mathrm{e}^{-\frac{Q}{RT}} = \mathrm{e}^{-\frac{T_a}{T}} \tag{1.335}$$

where T is the absolute temperature, R (8.31 J mol^{-1} K^{-1}) is the molar gas constant, Q is an activation energy, and $T_a = Q/R$ is the corresponding temperature. Typical values for Q are in the range 10–100 kJ/mol, corresponding to T_a in the range 1000–10000 K.

The creep rate increases as T_a becomes lower. However, the temperature sensitivity increases as T_a increases; T_a in the range 5000–10000 K coarsely corresponds to a doubling in creep rate per 10 K.

The creep rate also strongly depends on the stress level. This is often expressed as

$$\dot{\varepsilon}_{\mathrm{cr}} \propto \left(\frac{\sigma}{\sigma_0}\right)^m \tag{1.336}$$

or

$$\dot{\varepsilon}_{\rm cr} \propto {\rm e}^{\frac{\sigma}{\sigma_0}} \tag{1.337}$$

where σ_0 is a characteristic stress. The exponent m is typically in the range from 1 to 8. These expressions may be applicable for stress variations within a limited range. The dependency on stress is however more complex in general as discussed below and further in Section 2.10.

Rheological models are often used to describe creep in rocks. These are mathematical models designed for other physical systems; however, they provide some insight, and the described behaviour resembles observations on rocks to some extent. The models are based on combinations of linear elements obeying Hooke's law (Eq. (1.94))

$$\sigma = \kappa\varepsilon \qquad \text{spring element, Fig. 1.29a} \tag{1.338}$$

viscous elements obeying the stress–strain relation

$$\sigma = \chi\frac{\partial \varepsilon}{\partial t} \qquad \text{dashpot element, Fig. 1.29b} \tag{1.339}$$

and slider elements

$$\begin{array}{lll} \varepsilon = 0 & \text{for} & \sigma \le \sigma_{\rm y} \\ \sigma = \sigma_{\rm y} & & \text{otherwise} \end{array} \qquad \text{slider element, Fig. 1.29c} \tag{1.340}$$

Here κ is a spring constant, χ is a coefficient of viscosity, and $\sigma_{\rm y}$ is a yield stress.

One of these models (the "Burgers substance") is shown in Fig. 1.30. The model describes both instantaneous strain, transient creep, and steady-state creep. Consider, for instance, a stress path as shown in Fig. 1.31a. For $t < 0$, the system is unstrained. At $t = 0$ the stress is increased to σ^*, whereas at $t = t_{\rm c}$ the stress is released. The resulting strain (shown in Fig. 1.31b) is found to be

$$\varepsilon = \begin{cases} 0 & \text{for} \quad t < 0 \\ \dfrac{\sigma^*}{\kappa_2} + \dfrac{\sigma^*}{\kappa_1}\left(1 - {\rm e}^{-t/t_1}\right) + \dfrac{\sigma^*}{\chi_2}t & \text{for} \quad 0 < t < t_{\rm c} \\ \dfrac{\sigma^*}{\kappa_1}\left({\rm e}^{t_{\rm c}/t_1} - 1\right){\rm e}^{-t/t_1} + \dfrac{\sigma^*}{\chi_2}t_{\rm c} & \text{for} \quad t > t_{\rm c} \end{cases} \tag{1.341}$$

Here $t_1 = \chi_1/\kappa_1$ is the time constant of the transient creep.

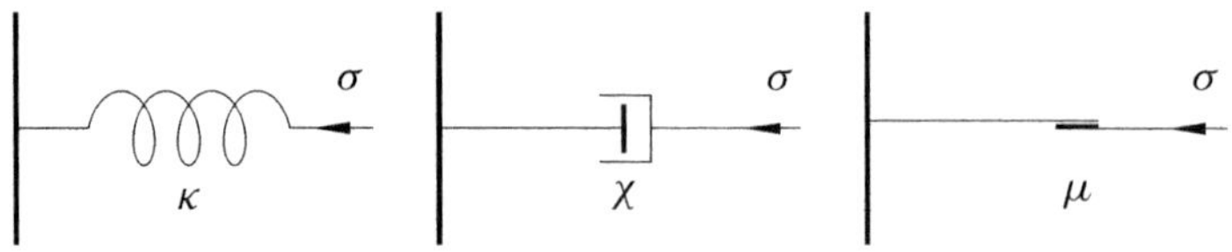

Figure 1.29 Building elements in visco-elastic models. a) Spring element. b) Dashpot element. c) Slider element.

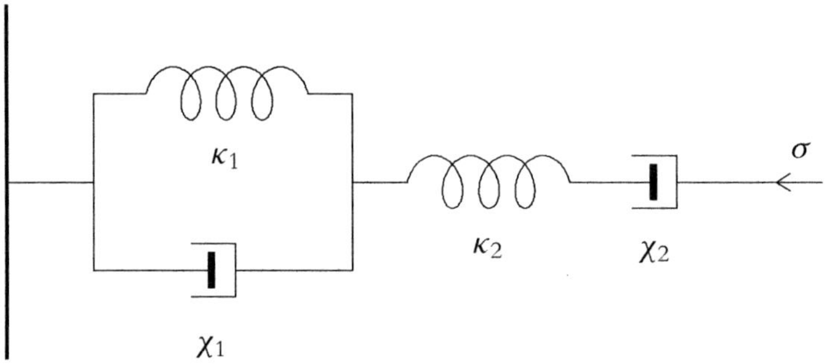

Figure 1.30 The Burgers substance. The Burgers substance can be seen as consisting of two simpler substances, the Maxwell substance (a spring and a dashpot in series) and the Kelvin substance (a spring and a dashpot in parallel).

The instantaneous strain is seen to be σ^*/κ_2, the steady-state creep velocity is σ^*/χ_2, and the permanent strain is $\sigma^* t_c/\chi_2$, resulting from the stress path. Thus the permanent strain is seen to arise from the steady state creep. Due to this term, the system has a "memory" of its stress history.

As mentioned above, it takes a certain stress level to activate the various stages of creep. This feature can be taken care of by including slider elements (Fig. 1.29c) in the rheological model. For instance, if a slider element is coupled in parallel with the second (rightmost) dashpot element in the Burgers model (Fig. 1.30), then steady-state creep will only occur if $\sigma^* > \sigma_y$, and the steady-state creep rate will be $(\sigma^* - \sigma_y)/\kappa_2$. The unloading path is unaffected by the slider element; however, the permanent strain is as before given by the accumulated steady-state creep during $0 < t < t_c$ (i.e. $= (\sigma^* - \sigma_y)t_c/\kappa_2$ if $\sigma^* > \sigma_y$, $= 0$ if $\sigma^* < \sigma_y$). See Section 2.10 for a further discussion of the impact of stress level.

It is often assumed that creep is proportional to the deviatoric stresses in a material, whereas hydrostatic stress alone will not produce creep effects. Thus creep effectively reduces the shear modulus and Young's modulus

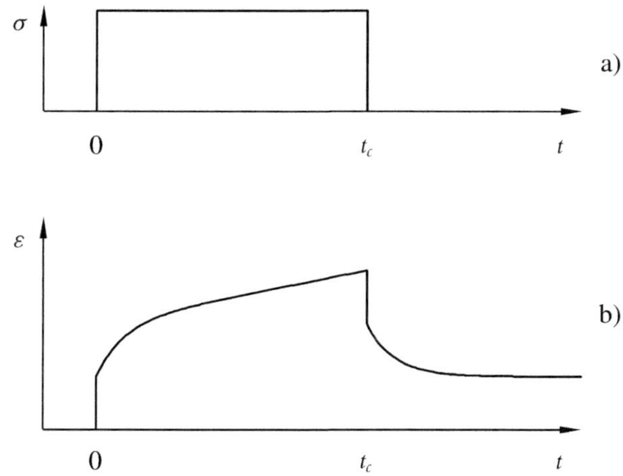

Figure 1.31 The response of the Burgers substance. a) Applied stress versus time. b) Strain versus time.

of a material (Fig. 1.32, left), whereas the bulk modulus is not affected by it. It follows from Table 1.1 that Poisson's ratio should increase due to creep. Note, however, that there may be local shear stresses in a sample subject to external hydrostatic loading due to inhomogeneities in the material microstructure. Therefore a material may creep even under hydrostatic external loads, and this favours a decrease in Poisson's ratio due to creep. Thus the impact of creep on Poisson's ratio is not unique and may even change with time under constant conditions (Fig. 1.32, right).

Note that our knowledge about creep is largely based on laboratory observations and that most creep models are designed and calibrated to describe observations in laboratory tests. Such tests are usually conducted under idealised boundary conditions, like constant stress or fixed strain. In a field situation, a piece of rock is interacting with its surroundings in such a way that the boundary conditions are rarely idealised. Furthermore, there is often a coupling between creep and consolidation as well. This has to be taken into account when analysing rock behaviour in field conditions.

1.10 Further reading

Elasticity has application in many sciences outside rock mechanics, and the number of books is vast. Rice (1999) lists some highlights, among them

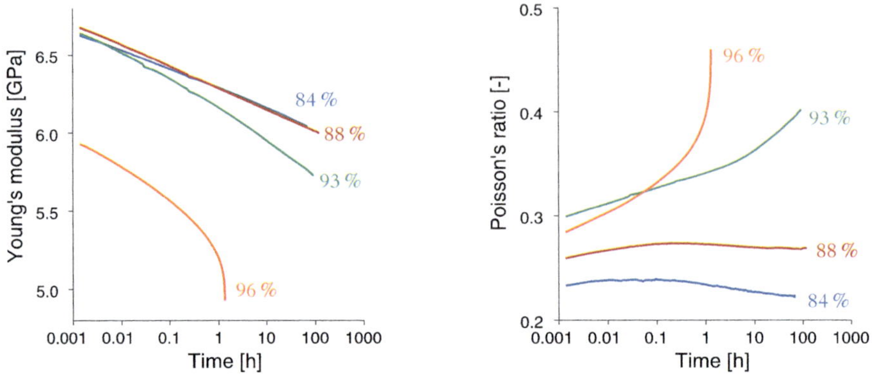

Figure 1.32 Young's modulus and Poisson's ratio (secant moduli) as functions of time during an inserted period with constant stress in a triaxial test. The figure shows results from four different tests at 5 MPa confining stress on dry Castlegate sandstone. The labels show the axial stress at the hold period, relative to the expected peak stress value in a regular test. Courtesy of SINTEF.

Love (1944); Timoshenko and Goodier (1970); Landau and Lifshitz (1986), which are all considered classics.

Modern presentations of the mechanics of continuous media may be found, for example, in Malvern (1969); Rice (1999); Rudnicki (2014).

Accounts of the history of poroelasticity are found in de Boer and Ehlers (1988); de Boer (1992, 2000). Skempton (1960) gave an overview of the development of the effective stress concept (see also de Boer and Ehlers, 1990).

References

Abousleiman, Y., Cheng, A.H.-D., Cui, L., Detournay, E., Roegiers, J.-C., 1996. Mandel's problem revisited. Géotechnique 46, 187–195. https://doi.org/10.1680/geot.1996.46.2.187.

Berryman, J.G., 1992. Effective stress for transport properties of inhomogeneous porous rock. J. Geophys. Res. 97 (B12), 17409–17424. https://doi.org/10.1029/92JB01593.

Biot, M.A., 1941. General theory of three-dimensional consolidation. J. Appl. Phys. 12, 155–164. https://doi.org/10.1063/1.1712886.

Biot, M.A., 1955. Theory of elasticity and consolidation for a porous anisotropic solid. J. Appl. Phys. 26, 182–185. https://doi.org/10.1063/1.1721956.

Biot, M.A., 1956. General solutions of the equations of elasticity and consolidation for a porous material. J. Appl. Mech. 23, 91–96.

Biot, M.A., 1962. Mechanics of deformation and acoustic propagation in porous media. J. Appl. Phys. 33, 1482–1498. https://doi.org/10.1063/1.1728759.

Biot, M.A., Willis, D.G., 1957. The elastic coefficients of the theory of consolidation. J. Appl. Mech. 24, 594–601.

Brantut, N., Heap, M.J., Meredith, P.G., Baud, P., 2013. Time-dependent cracking and brittle creep in crustal rocks: a review. J. Struct. Geol. 52, 17–43. https://doi.org/10.1016/j.jsg.2013.03.007.

Brown, R.J.S., Korringa, J., 1975. On the dependence of the elastic properties of a porous rock on the compressibility of the pore fluid. Geophysics 40 (4), 608–616. https://doi.org/10.1190/1.1440551.

Charlez, P.A., 1991. Rock Mechanics. Volume I. Theoretical Fundamentals. Éditions Technip, Paris.

Chen, W.F., Han, D.J., 1988. Plasticity for Structural Engineers. Springer-Verlag, New York.

Cheng, A.H.-D., 1997. Material coefficients of anisotropic poroelasticity. Int. J. Rock Mech. Min. Sci. 34 (2), 199–205. https://doi.org/10.1016/S0148-9062(96)00055-1.

Cheng, A.H.-D., 1998. On generalized plane strain poroelasticity. Int. J. Rock Mech. Min. Sci. 35, 183–193. https://doi.org/10.1016/S0148-9062(97)00327-6.

Cheng, A.H.-D., 2016. Poroelasticity. Springer, Berlin.

Clayton, J.D., 2011. Nonlinear Mechanics of Crystals. Solid Mechanics and Its Applications, vol. 177. Springer.

Coussy, O., 2004. Poromechanics. John Wiley & Sons Ltd., Chichester, England. ISBN 0-470-84920-7.

Cryer, C.W., 1963. A comparison of the three-dimensional consolidation theories of Biot and Terzaghi. Q. J. Mech. Appl. Math. 16, 401–412. https://doi.org/10.1093/qjmam/16.4.401.

de Boer, R., 1992. Development of porous media theories – a brief historical review. Transp. Porous Media 9, 155–164. https://doi.org/10.1007/BF01039634.

de Boer, R., 2000. Theory of Porous Media. Highlights in Historical Development and Current State. Springer, Berlin. ISBN 3-540-65982-X.

de Boer, R., Ehlers, W., 1988. A historical review of the formulation of porous media theories. Acta Mech. 74, 1–8. https://doi.org/10.1007/BF01194337.

de Boer, R., Ehlers, W., 1990. The development of the concept of effective stresses. Acta Mech. 83, 77–92. https://doi.org/10.1007/BF01174734.

Detournay, E., Cheng, A.H.-D., 1988. Poroelastic response of a borehole in a non-hydrostatic stress field. Int. J. Rock Mech. Min. Sci. Geomech. Abstr. 25, 171–182. https://doi.org/10.1016/0148-9062(88)92299-1.

Detournay, E., Cheng, A.H.-D., 1993. Fundamentals of poroelasticity. In: Hudson, J.A. (Ed.), Comprehensive Rock Engineering, vol. 2. Pergamon Press, Oxford, pp. 113–171.

Gassmann, F., 1951. Über die Elastizität poröser Medien. Vierteljahrsschr. Nat.forsch. Ges. Zür. 96, 1–23.

Geertsma, J., 1957. A remark on the analogy between thermoelasticity and the elasticity of saturated porous media. J. Mech. Phys. Solids 6 (1), 13–16. https://doi.org/10.1016/0022-5096(57)90042-X.

Haimson, B., Fairhurst, C., 1967. Initiation and extension of hydraulic fractures in rocks. Soc. Pet. Eng. J. 7, 310–318. https://doi.org/10.2118/1710-PA.

Heyliger, P., Ledbetter, H., Kim, S., 2003. Elastic constants of natural quartz. J. Acoust. Soc. Am. 114 (2), 644–650. https://doi.org/10.1121/1.1593063.

Holt, R.M., Bakk, A., Stenebråten, J.F., Bauer, A., Fjær, E., 2018. Skempton's A – a key to man-induced subsurface pore pressure changes. ARMA18-0949 In: 52nd US Rock Mechanics/Geomechanics Symposium. Seattle, WA, USA.

Jaeger, J.C., Cook, N.G.W., 1979. Fundamentals of Rock Mechanics, third edition. Chapman and Hall, London.

Jaeger, J.C., Cook, N.G.W., Zimmerman, R.W., 2007. Fundamentals of Rock Mechanics, fourth edition. Blackwell Publishing.

Landau, L.D., Lifshitz, E.M., 1986. Theory of Elasticity, third edition. Pergamon Press.

Lode, W., 1926. Versuche über den Einfluß der mittleren Hauptspannung auf das Fließen der Metalle Eisen, Kupfer und Nickel. Z. Phys. 36, 913–939. https://doi.org/10.1007/BF01400222.
Lomnitz, C., 1956. Creep measurements in igneous rocks. J. Geol. 64 (5), 473–479. https://doi.org/10.1086/626379.
Louchet, F., Duval, P., 2009. Andrade creep revisited. Int. J. Mater. Res. 100, 1433–1439. https://doi.org/10.3139/146.110189.
Love, A.E.H., 1944. A Treatise on the Mathematical Theory of Elasticity, fourth edition. Cambridge University Press, Cambridge.
Malvern, L.E., 1969. Introduction to the Mechanics of a Continuous Medium. Prentice-Hall, Englewood Cliffs, NJ.
Mandel, J., 1953. Consolidation des sols (étude mathématique). Géotechnique 3 (7), 287–299. https://doi.org/10.1680/geot.1953.3.7.287.
McTigue, D.F., 1986. Thermoelastic response of fluid-saturated porous rock. J. Geophys. Res. 91 (B9), 9533–9542. https://doi.org/10.1029/JB091iB09p09533.
Rice, J.R., 1999. Foundations of solid mechanics. In: Meyers, M.A., Armstrong, R.W., Kirchner, H. (Eds.), Mechanics and Materials: Fundamentals and Linkages. Wiley, New York.
Rice, J.R., Cleary, M.P., 1976. Some basic stress diffusion solutions for fluid-saturated elastic porous media with compressible constituents. Rev. Geophys. Space Phys. 14 (2), 222–241. https://doi.org/10.1029/RG014i002p00227.
Rudnicki, J.W., 2014. Fundamentals of Continuum Mechanics. Wiley, New York.
Skempton, A.W., 1954. The pore pressure coefficients *A* and *B*. Géotechnique 4, 143–147. https://doi.org/10.1680/geot.1954.4.4.143.
Skempton, A.W., 1960. Significance of Terzaghi's concept of effective stress. Terzaghi's discovery of effective stress. In: Bjerrum, L., Casagrande, A., Peck, R.B., Skempton, A.W. (Eds.), From Theory to Practice in Soil Mechanics. John Wiley & Sons, New York, pp. 42–53.
Timoshenko, S.P., Goodier, J.N., 1970. Theory of Elasticity. McGraw-Hill Kogakusha Ltd., Tokyo.
Verruijt, A., 1969. Elastic storage of aquifers. In: Dewiest, R.J.M. (Ed.), Flow Through Porous Media. Academic Press Inc., New York.
Wang, H.F., 2000. Theory of Linear Poroelasticity. Princeton University Press, Princeton.
Wood, A.B., 1941. A Textbook of Sound, second edition. G. Bell and Sons Ltd., London.
Zimmerman, R.W., 1991. Compressibility of Sandstones. Elsevier Science Publishers B.V., Amsterdam.

CHAPTER 2

Failure mechanics

2.1 Basic concepts

When a piece of rock is subject to sufficiently large stresses, a failure of some kind will occur. This implies that the rock changes its shape permanently and possibly also falls apart. The condition is accompanied with a reduced ability to carry loads. Rock failure is an important phenomenon also for petroleum related rock mechanics, as it is the origin of severe problems such as borehole instability and solids production. It is therefore useful to be able to predict under which conditions a rock is likely to fail.

In this chapter, we discuss the most elementary and well-known models for rock failure. We should keep in mind, however, that these are only simplified descriptions of real rock behaviour. Rock failure is a complex process, which is still not fully understood. Much of the framework used to handle rock failure is therefore based on convenient mathematical descriptions of observed behaviour, rather than on derivations from basic laws of physics. Some of the concepts used here, including the concept of failure, may thus be poorly defined and sometimes not very relevant, as we will see below.

For most of the chapter, we will assume that rocks are homogeneous and isotropic. Anisotropy is discussed in Section 2.9, whereas some consequences of inhomogeneities are briefly discussed in Sections 2.5 and 2.8.

2.1.1 Strength and related concepts

The stress level at which a rock typically fails is commonly called the *strength* of the rock. Obviously, as "stress level" is not a uniquely defined parameter, neither is strength. Rock strength is therefore a meaningful parameter only when the stress geometry—that is: the type of test, in a laboratory setting—is also specified. We will soon describe the impact of stress geometry on rock strength. First, we will however take a look at a couple of the most important tests used to measure rock strength, the *uniaxial* and *triaxial* tests, to illustrate the complexity of rock failure and to introduce some basic concepts. Rock mechanical testing procedures for strength measurements will be described in more detail in Chapter 7.

Developments in Petroleum Science, Volume 72
ISSN 0376-7361, https://doi.org/10.1016/B978-0-12-822195-2.00011-5

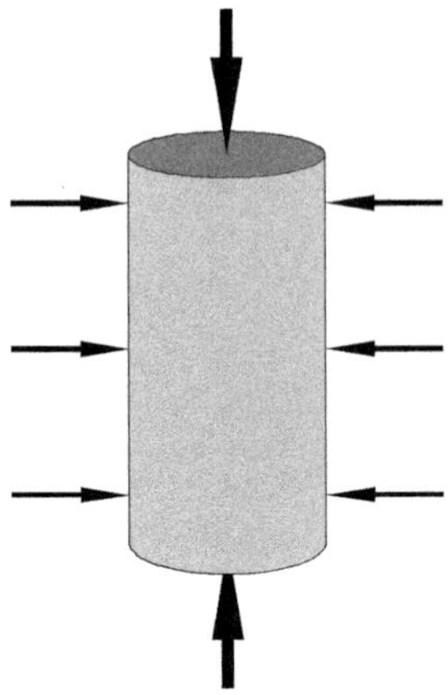

Figure 2.1 Typical test specimen for a uniaxial or triaxial test. A typical sample diameter for petroleum applications is 38 mm (1 1/2″).

Fig. 2.1 illustrates a typical test specimen, a cylinder with length-to-diameter ratio 2:1. A pair of pistons applies (axial) stress to the end faces of the cylinder, whereas a confining oil bath provides a stress of possibly different magnitude to the circumference. It is normally assumed that the stress state within the specimen is homogeneous. If the confining stress is zero, then we have a uniaxial stress test (also called unconfined compression test). When the test is performed with a non-zero confining pressure, a so-called triaxial test is performed.

Fig. 2.2 shows a typical result from a uniaxial test. The applied axial stress (denoted σ_z) is plotted as a function of the axial strain (ε_z) of the sample. Several important concepts are defined in the figure:

Elastic region: The rock deforms elastically. If the stress is released, then the specimen will return to its original state.

Yield point: The point beyond which permanent changes will occur. The sample will no longer return to is original state upon stress relief.

Uniaxial (unconfined) compressive strength (UCS): The peak stress.

Ductile region: A region in which the sample undergoes permanent deformation without loosing the ability to support load.

Brittle region: A region in which the specimen's ability to withstand stress decreases rapidly as deformation is increased.

A triaxial test is usually performed by increasing the axial and confining loads simultaneously until a prescribed hydrostatic stress level is reached. Then the confining pressure is kept constant while the axial load is increased

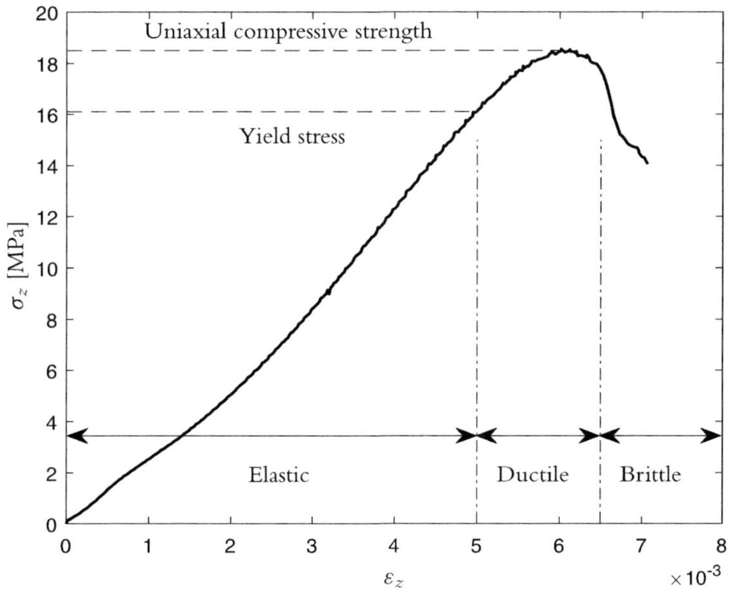

Figure 2.2 Stress versus deformation in a uniaxial compression test. In some tests the ductile region can be very small.

until failure occurs. The axial loading is normally applied such that it gives a constant axial deformation rate.

For a triaxial test, it is customary to plot the difference between the axial stress and the confining pressure σ_r versus the axial deformation. We then obtain a curve that looks similar to Fig. 2.2. However, the behaviour may be quite different at the higher stress levels. Fig. 2.3 illustrates results from triaxial tests with various confining pressures. It is seen that for the higher confining pressures, the specimen's ability to support load is not lost, although its stiffness is clearly reduced. This is further discussed in Section 2.8.

So far, we have not given a precise definition of failure, mainly because it is rather difficult to give a general definition. For the uniaxial test shown in Fig. 2.2, a seemingly unambiguous definition of failure may be given, corresponding to the peak stress point on the curve. For the higher confining pressures in Fig. 2.3, we may on the other hand define failure at some point where the slope of the stress–strain curve changes. This may not seem totally appropriate, however, since the specimen still supports increasing load after it has failed. Disregarding these problems, we will in the following sections discuss failure criteria assuming that a consistent definition of fail-

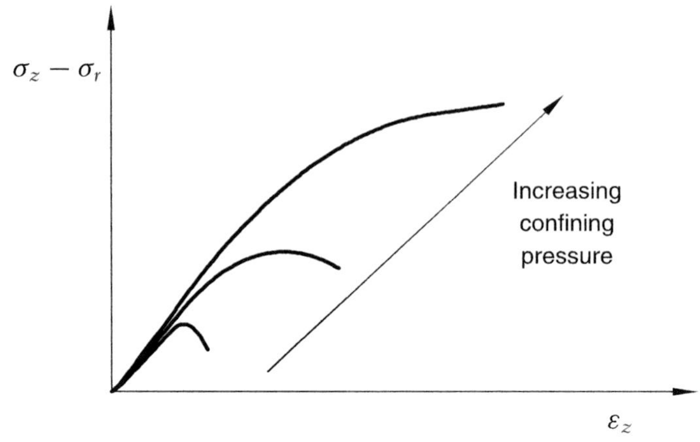

Figure 2.3 Triaxial testing: typical influence of the confining pressure on the shape of the differential stress (axial stress minus confining pressure) versus axial strain curves.

ure exists. Later, in Section 2.8, on post-failure behaviour described by the theory of plasticity, we will be a little more precise, at least for problems that can be described within the framework of that theory.

The most common mode of failure observed in uniaxial and triaxial tests is shear failure. This failure mode is caused by excessive shear stress. Another failure mode is tensile failure, which is caused by excessive tensile stress. Finally, pore collapse is a failure mode that is normally observed in highly porous materials, usually caused by excessive hydrostatic stress. The various failure modes are discussed in Sections 2.2–2.4.

The concept of rock failure is associated with the state of the solid framework. Thus the stresses that cause failure are the *effective stresses* felt by the framework. The effective stress concept was introduced in Section 1.6.4, where we also introduced the notation σ' to distinguish the effective stress from the total stress σ. Note however that the relation specifying the effective stress in a failure criterion is in general different from relation (1.187), which specifies the effective stress for deformation of a linearly elastic material. This will be further discussed in Section 2.6.1.

2.1.2 The failure surface

From the considerations above we see that a rock fails when the stress exceeds a certain limit, whereas it remains intact (more or less, as we will see later) as long as the stress is lower than this limit. We also see that this limit depends on the total stress state, not only the stress in one direction.

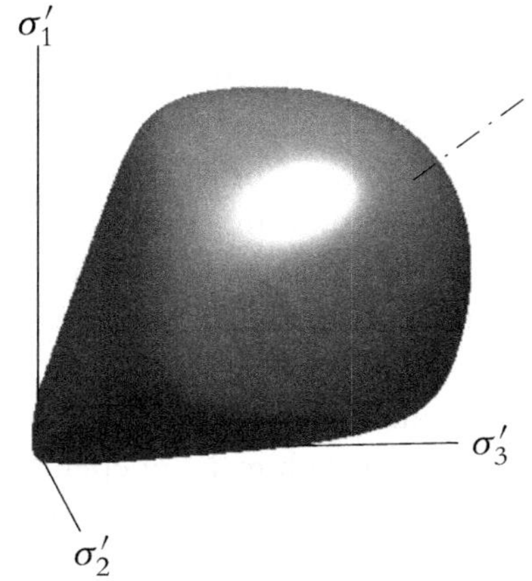

Figure 2.4 Schematic picture of a failure surface in principal stress space. The dash-dot line represents the hydrostatic axis. Note that the conventional relation $\sigma_1' \geq \sigma_2' \geq \sigma_3'$ has been abandoned in this figure, to illustrate that the failure surface is closed: The rock is supposed to fail at some stress level for any ratios between the principal stresses.

A graphical representation of this in an abstract "stress space" turns out to be quite useful.

Consider a test specimen of an isotropic material subject to a certain stress state described by the three principal stresses $(\sigma_1', \sigma_2', \sigma_3')$. We may represent the stress state graphically as a point in the *principal stress space*, that is, the space spanned by the $(\sigma_1', \sigma_2', \sigma_3')$ axes. (The principal stress space is also referred to as the *Haigh–Westergaard stress space*; see e.g. Chen and Han, 1988.) Imagine that the specimen is taken to failure by increasing the principal stresses in some manner and that the point of failure is plotted in the principal stress space. If this procedure could be repeated an infinite number of times, following different stress paths, we would get an infinite number of failure points in the stress space. We assume that these points form a continuous surface, which we call the *failure surface*. A schematic illustration of such a surface is shown in Fig. 2.4.

The failure surface may be described by the equation

$$f(\sigma_1', \sigma_2', \sigma_3') = 0 \tag{2.1}$$

In this simplified picture the assumption is that the rock is intact at stress states inside the failure surface, whereas it fails for any stress state outside. This does not imply that any stress state outside the failure surface is inaccessible, since the rock may under suitable conditions be able to support increasing load after failure, as mentioned above.

It is difficult to draw a surface in three dimensions. Therefore the failure surface is often presented by cross-sections instead. The most common of such representations are cross-sections in a *π-plane*. These planes are normal to the hydrostatic axis (also called hydrostat), that is, the axis where $\sigma_1' = \sigma_2' = \sigma_3'$. Figs. 2.16 and 2.18 are examples of π-plane cross-sections of failure surfaces.

The existence of the failure surface is by no means obvious. For instance, its existence implies that failure is independent of stress gradients and of stress history for stress states on the inner side of the surface. This approximation is not always fulfilled, as we will see later in this chapter. Also, we have already seen that failure does not necessarily occur abruptly at the failure point, but may be a gradual process that changes rate at the failure surface.

In the following, we will first consider the failure surface in two-dimensional stress space, focusing on the impact of the largest and smallest principal stresses. These configurations are relatively simple to handle both intuitively and mathematically, and are also most commonly used for practical applications. The corresponding failure criteria in the three-dimensional stress space are discussed in Section 2.5.

2.2 Tensile failure

Tensile failure occurs when the effective tensile stress across some plane in the sample exceeds a critical limit. This limit is called the *tensile strength*; it is given the symbol T_0 and has the same unit as stress. The tensile strength is a characteristic property of the rock. Most sedimentary rocks have a rather low tensile strength, typically only a few MPa or less. In fact, it is a standard approximation for several applications that the tensile strength is zero.

A sample that suffers tensile failure typically splits along one—or very few—fracture planes, as illustrated in Fig. 2.5. Thus tensile failure is a highly localised and inhomogeneous process. The fracture planes often originate from pre-existing cracks, oriented more or less normal to the direction of the tensile stress. The highest probability for further damage of the rock is at the perimeter of the largest of these cracks, and hence the largest crack(s) will grow increasingly faster than the other and rapidly split the sample. We will see later (in Section 6.4.4) that the tensile strength is very sensitive to the presence of cracks in the material.

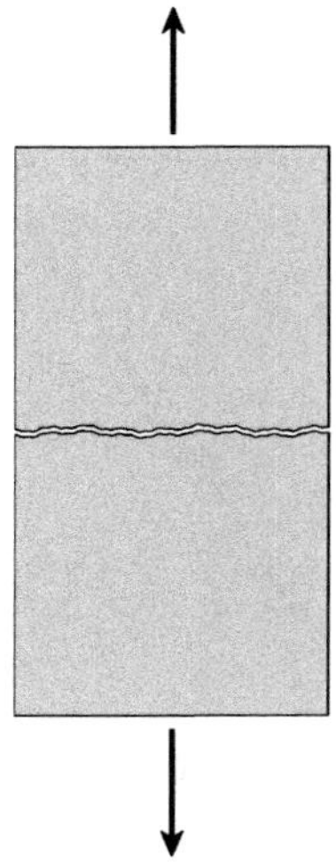

Figure 2.5 Tensile failure.

The failure criterion, which specifies the stress condition for which tensile failure will occur and identifies the location of the failure surface in the principal stress space, is given as

$$\sigma' = -T_0 \tag{2.2}$$

For isotropic rocks, the conditions for tensile failure will always be fulfilled first for the lowest principal stress, so that the tensile failure criterion becomes

$$\sigma_3' = -T_0 \tag{2.3}$$

2.3 Shear failure

Shear failure occurs when the shear stress along some plane in the sample is sufficiently high. Eventually, a fault zone will develop along the failure plane, and the two sides of the plane will move relative to each other in a frictional process, as shown in Fig. 2.6.

The frictional force that acts against the relative movement of two bodies in contact depends on the force that presses the bodies together. It is therefore reasonable to assume that the critical shear stress τ_{max}, for which shear failure occurs, depends on the normal stress σ' acting over the failure

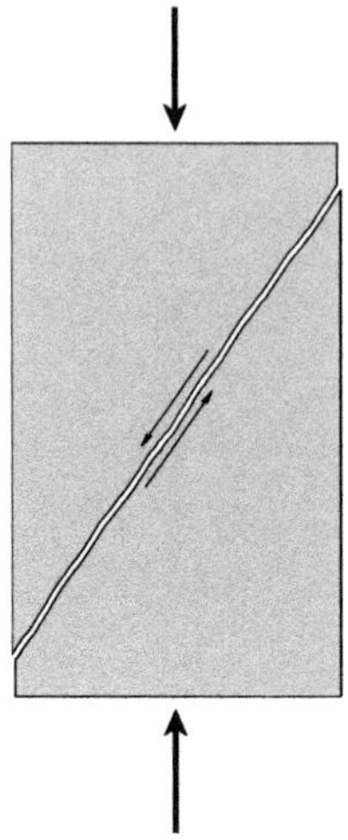

Figure 2.6 Shear failure.

plane, that is,

$$|\tau_{\max}| = f(\sigma') \tag{2.4}$$

This assumption is called Mohr's hypothesis.

In the (σ', τ)-plane, Eq. (2.4) describes a line that separates a "safe region" from a "failure" region, and we may consider Eq. (2.4) as a representation of the failure surface in the (σ', τ)-plane. The line is sometimes referred to as the failure line or the failure envelope. An example is shown in Fig. 2.7, where we have also indicated the three principal stresses and Mohr's circles connecting them. It was explained in Section 1.1.6 that for a given set of principal stresses, all possible combinations of τ and σ' lie within the area in between the three circles (i.e. the shaded area of Fig. 2.7).

The stress state of Fig. 2.7 represents a safe situation, as no plane within the rock has a combination of τ and σ' that lies above the failure line. Assume now that σ_1' is increased. The circle connecting σ_1' and σ_3' will expand and eventually touch the failure line. The failure criterion is then fulfilled for some plane(s) in the sample, and the sample fails. Note that the value of the intermediate principal stress σ_2' has no influence on this situation. Since σ_2' by definition lies within the range (σ_3', σ_1'), it does not affect the outermost of Mohr's circles, and hence it does not affect the failure. Thus, pure shear failure, as defined by Mohr's hypothesis, depends

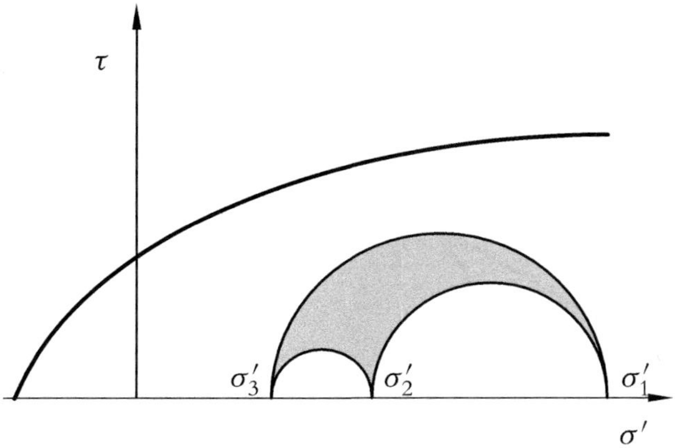

Figure 2.7 Failure line, as specified by Eq. (2.4), in the shear stress–normal stress diagram. Also shown are the Mohr's circles connecting the principal stresses $(\sigma_1', \sigma_2', \sigma_3')$.

only on the minimum and maximum principal stresses and not on the intermediate stress.

By choosing specific forms of the function $f(\sigma')$ of Eq. (2.4), various criteria for shear failure are obtained. The simplest possible choice is a constant. The resulting criterion is called the Tresca criterion. The criterion simply states that the material will yield when a critical level of shear stress is reached:

$$\tau_{\max} = \frac{1}{2}(\sigma_1' - \sigma_3') = S_0 \tag{2.5}$$

S_0 is the *inherent shear strength* (also called *cohesion*) of the material. In a Mohr $\tau - \sigma'$ plot the Tresca criterion appears simply as a straight horizontal line.

2.3.1 The Mohr–Coulomb criterion

A more general and frequently used criterion is the Mohr–Coulomb criterion, which is based on the assumption that $f(\sigma')$ is a linear function of σ':

$$|\tau| = S_0 + \mu\sigma' \tag{2.6}$$

Here μ is the *coefficient of internal friction*. The latter term is clearly chosen by analogy with sliding of a body on a surface, which to the first approximation is described by Amontons' law

$$\tau = \mu\sigma' \tag{2.7}$$

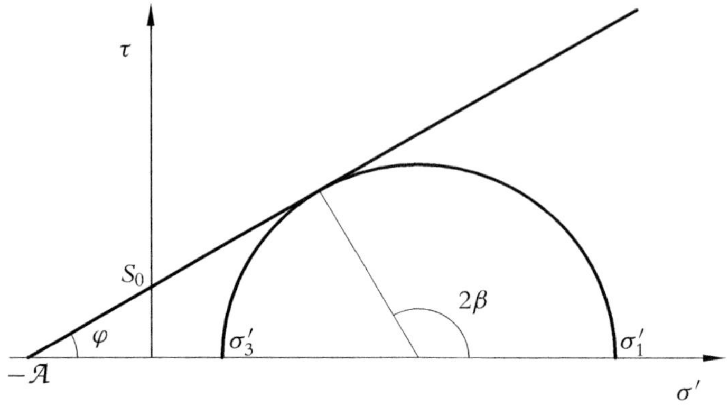

Figure 2.8 Mohr–Coulomb criterion in (σ', τ)-space. Also shown is Mohr's circle corresponding to a critical stress state.

In Fig. 2.8, we have drawn the Mohr–Coulomb criterion and Mohr's circle that touches the failure line. The angle φ defined in the figure is called the *angle of internal friction* (or *friction angle*) and is related to the coefficient of internal friction by

$$\tan \varphi = \mu \tag{2.8}$$

Note that the Tresca criterion can be considered as a particular case of the Mohr–Coulomb criterion with $\varphi = 0$.

The intersection point between the Mohr–Coulomb failure line and the normal stress axis is of no practical interest in itself, as the point is inaccessible due to tensile failure. However, for some purposes, it is convenient to make use of the parameter $\mathcal{A}$ defined as the distance from the intersection point to the origin (see Fig. (2.8)). The parameter is called the *attraction*. The attraction is related to the other Mohr–Coulomb parameters by

$$\mathcal{A} = S_0 \cot \varphi \tag{2.9}$$

Fig. 2.8 also shows the angle 2β, which gives the position of the point where Mohr's circle touches the failure line. It can be seen from the figure that the shear stress at this point of contact is

$$|\tau| = \frac{1}{2}(\sigma_1' - \sigma_3') \sin 2\beta \tag{2.10}$$

whereas the normal stress is

$$\sigma' = \frac{1}{2}(\sigma_1' + \sigma_3') + \frac{1}{2}(\sigma_1' - \sigma_3')\cos 2\beta \tag{2.11}$$

Also, we see that β and φ are related by

$$\varphi + \frac{\pi}{2} = 2\beta \tag{2.12}$$

Since β is the angle for which the failure criterion is fulfilled, it gives the orientation of the failure plane (see Fig. 1.7), and is hence called the *failure angle*. From Eq. (2.12) we have that

$$\beta = \frac{\pi}{4} + \frac{\varphi}{2} \tag{2.13}$$

The allowable range for φ is from 0° to 90° (in practice the range will be smaller and centred around approximately 30°), and hence it is clear that β may vary between 45° and 90°. We may conclude that the failure plane is always inclined at an angle smaller that 45° to the direction of σ_1'. Fig. 2.9 shows schematically how the failure planes may be oriented in a rock described by the Mohr–Coulomb criterion.

One important point to note is that β is given solely by φ, which is a constant in the Mohr–Coulomb criterion. Thus the orientation of the failure plane is independent of the confining stress. This is a special feature of the Mohr–Coulomb criterion. Experiments often show that the failure angle decreases with increasing confining pressure, in particular, at low confining pressures.

Introducing expressions (2.10) and (2.11) for σ' and τ into the failure criterion (2.6), we find

$$\frac{1}{2}(\sigma_1' - \sigma_3')\sin 2\beta = S_0 + \mu\left[\frac{1}{2}(\sigma_1' + \sigma_3') + \frac{1}{2}(\sigma_1' - \sigma_3')\cos 2\beta\right] \tag{2.14}$$

Replacing β and μ by φ, according to Eqs. (2.8) and (2.13), we obtain

$$\frac{1}{2}(\sigma_1' - \sigma_3')\cos\varphi = S_0 + \frac{1}{2}(\sigma_1' + \sigma_3')\tan\varphi - \frac{1}{2}(\sigma_1' - \sigma_3')\tan\varphi\sin\varphi \tag{2.15}$$

Multiplying with $2\cos\varphi$ and rearranging, we find

$$(\sigma_1' - \sigma_3')(\cos^2\varphi + \sin^2\varphi) = 2S_0\cos\varphi + (\sigma_1' + \sigma_3')\sin\varphi \tag{2.16}$$

$$\sigma_1'(1 - \sin\varphi) = 2S_0\cos\varphi + \sigma_3'(1 + \sin\varphi) \tag{2.17}$$

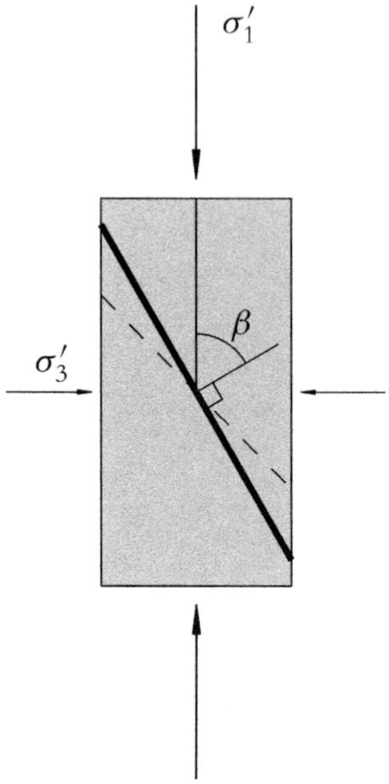

Figure 2.9 Orientation of the failure plane relative to the largest principal stress. The thick solid line shows the failure plane for a friction angle of 30°. The dashed line shows the maximum inclination of the failure plane relative to σ_1' according to the Mohr–Coulomb criterion.

$$\sigma_1' = 2S_0 \frac{\cos\varphi}{1-\sin\varphi} + \sigma_3' \frac{1+\sin\varphi}{1-\sin\varphi} \tag{2.18}$$

Fig. 2.10 is a plot of this relation in the (σ_1', σ_3')-plane. Again, we have a linear relationship with a positive intercept on the σ_1'-axis, quite similar to the plot in Fig. 2.8. The angle γ in the (σ_1', σ_3')-plane is related to φ by

$$\tan\gamma = \frac{1+\sin\varphi}{1-\sin\varphi} \tag{2.19}$$

or

$$\sin\varphi = \frac{\tan\gamma - 1}{\tan\gamma + 1} \tag{2.20}$$

An expression for the uniaxial compressive strength C_0 is obtained by putting $\sigma_3' = 0$ in Eq. (2.18), giving

$$C_0 = 2S_0 \frac{\cos\varphi}{1-\sin\varphi} = 2S_0 \tan\beta \tag{2.21}$$

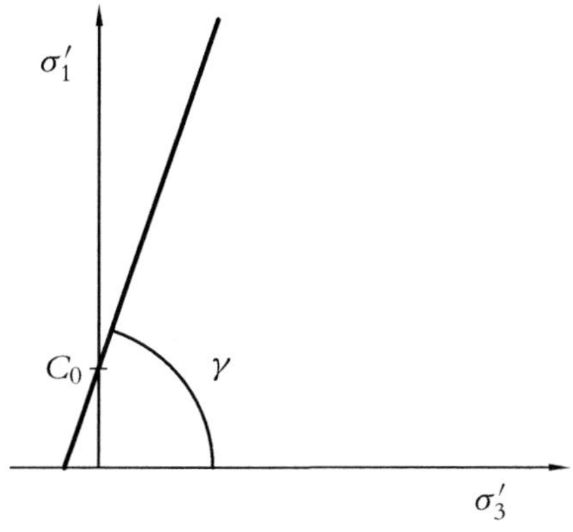

Figure 2.10 Mohr–Coulomb criterion in the (σ_1', σ_3')-plane.

The last equality here is derived from Eq. (2.12). It must be emphasised that the above expression is only valid if the failure mechanism under uniaxial stress is shear failure. This may not be the case even when shear failure occurs at rather low confining stresses.

Making use of Eqs. (2.21) and (2.12), we note that Eq. (2.18) may be written in the following simple way:

$$\sigma_1' = C_0 + \sigma_3' \tan^2 \beta \tag{2.22}$$

This formulation of the Mohr–Coulomb criterion will be extensively used later in this book.

We might perhaps consider computing the tensile strength T_0 by putting $\sigma_1' = 0$ in Eq. (2.22). However, this is not a valid approach, since the Mohr–Coulomb criterion describes shear failure whereas tensile failure is a different failure mode.

2.3.2 The Griffith criterion

Griffith (1921) developed a failure criterion based on a study of elliptical microcracks in a two-dimensional model. When the tensile stress at the tip of the crack exceeds a certain value characteristic of the material, the crack will grow, and the failure process is initiated. The theory is scaled in terms of the uniaxial tensile strength T_0, and the resulting failure criterion can be written

$$(\sigma_1' - \sigma_3')^2 = 8T_0(\sigma_1' + \sigma_3') \qquad \text{if } \sigma_1' + 3\sigma_3' > 0 \tag{2.23}$$

$$\sigma'_3 = -T_0 \qquad \text{if } \sigma'_1 + 3\sigma'_3 < 0 \tag{2.24}$$

In a principal stress plot the criterion is represented by a parabola ending in a straight line. This is illustrated in Fig. 2.11a.

The uniaxial compressive strength C_0 is given by Eq. (2.23) as

$$C_0 = 8T_0 \tag{2.25}$$

It is seen that the ratio between the uniaxial compressive strength and the tensile strength is here given as a fixed number. This ratio of 8 appears to be reasonable compared to experimental values, which often are in the range of 10–15. However, it is clear that fitting the criterion to experimental data may sometimes be difficult, since the criterion only has one free parameter.

In (σ', τ) coordinates the Griffith criterion is given by only one equation

$$\tau^2 = 4T_0(\sigma' + T_0) \tag{2.26}$$

(See, for instance, Jaeger and Cook (1979) for the transition from Eqs. (2.23) and (2.24) to Eq. (2.26)).

Eq. (2.26) shows that the criterion is a parabola also in a (σ', τ) plot, as illustrated in Fig. 2.11b. Unlike the Mohr–Coulomb criterion, the Griffith criterion is seen to have a steeper slope at low confining pressures. This feature is typical for experimental observations.

It is often observed that the Griffith criterion may give a reasonably good description of failure at low confining stresses, whereas a straight line Mohr–Coulomb criterion gives a better description at higher confining stresses. This has led to the development of a modified Griffith criterion (Brace, 1960; McClintock and Walsh, 1962). This criterion is based on the idea that the microcracks will close at sufficiently high pressure levels, resulting in a transition towards frictional behaviour corresponding to the Mohr–Coulomb criterion. If we assume that the stress required to close the cracks is so small that it can be neglected, then the modified Griffith criterion is simply the Griffith criterion for tensile conditions ($\sigma' < 0$) coupled to the Mohr–Coulomb criterion for compressive conditions ($\sigma' > 0$). The critical shear stress $\tau = 2T_0$ predicted by the Griffith criterion at $\sigma' = 0$ (see Eq. (2.26)) is used for the cohesion S_0 in the Mohr–Coulomb part of the criterion. A result from this modified theory is that the ratio between uniaxial compressive strength and the tensile strength is given by

$$\frac{C_0}{T_0} = \frac{4}{\sqrt{\mu^2 + 1} - \mu} \tag{2.27}$$

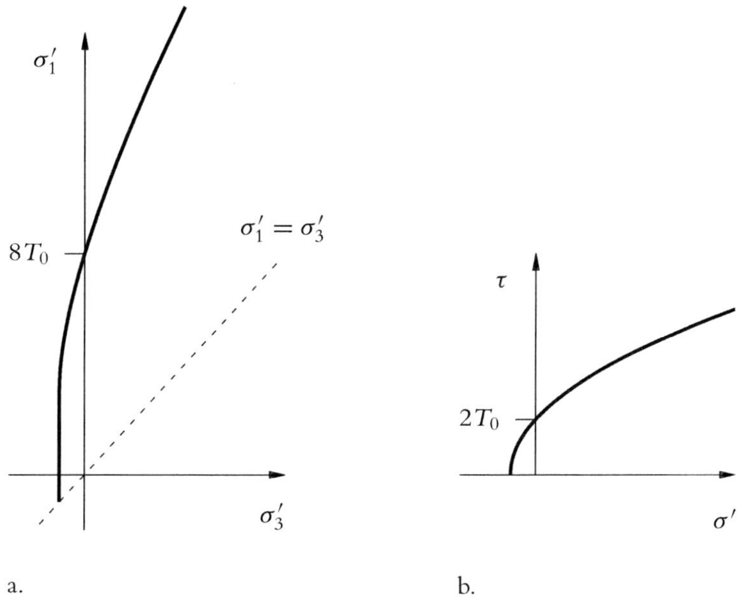

Figure 2.11 The Griffith criterion. a. Principal stress plot. b. (σ', τ) plot.

where μ is the coefficient of internal friction related to the Mohr–Coulomb part of the criterion.

2.4 Compaction failure

Pore collapse is a failure mode that is normally observed only in high-porosity materials, where the grain skeleton forms a relatively open structure. When the material is compressed, grains may loosen or break and then be pushed or twisted into the open pore space, resulting in a closer packing of the material. The process is called compaction. This deformation mode is schematically illustrated in Fig. 2.12.

In sandstones where the size of the pores is of the same order of magnitude as the size of the grains, pore collapse typically consists in reorientation of the grains to better fill the void spaces, as indicated in Fig. 2.12. This is particularly relevant for high-porosity chalks, where the size of the individual grains may be an order of magnitude smaller than the dimensions of the pore space.

Pore collapse may occur under pure hydrostatic loading. Microscopically, however, failure will be due to local excessive shear forces acting

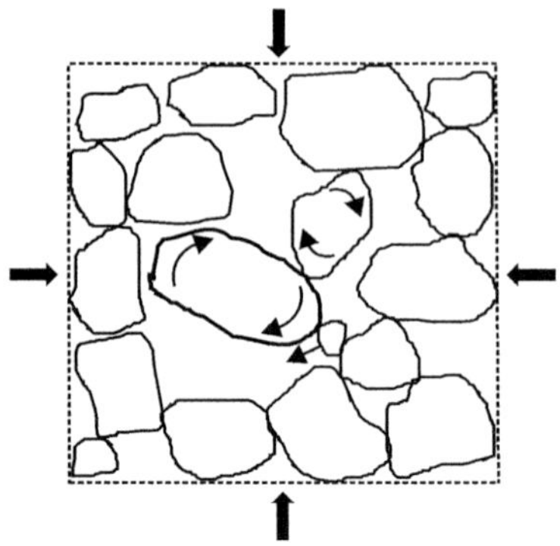

Figure 2.12 Grain reorientation resulting in a closer packing.

through grains and grain contacts. From this point of view, pore collapse may be regarded as distributed shear failure within the material.

Another failure mechanism that may occur under hydrostatic loading is grain crushing. If the stresses are sufficiently high, then the grains may be partly crushed at the grain contacts, and splitting of the grains may result. Either way, these local failure mechanisms represent permanent damage of the rock framework and cause yielding, with associated reduction in the stiffness of the rock. This type of failure also occurs to some extent under non-hydrostatic stress conditions and may be observed in triaxial tests at high confining pressure (see Fig. 2.3; high confining pressure). The process is then referred to as shear-enhanced compaction.

In the principal stress space, this type of failure is represented by an "end cap" that closes the failure surface at high stresses, as seen on Fig. 2.13. An elliptical form is often used for such an end cap (DiMaggio and Sandler, 1971; Wong et al., 1997):

$$\frac{1}{(1-\gamma)^2}\left(\frac{\overline{\sigma}'}{p^*}-\gamma\right)^2+\frac{1}{\delta^2}\left(\frac{q}{p^*}\right)^2=1 \tag{2.28}$$

where $\overline{\sigma}'$ is the mean effective stress (see Eq. (1.40)), q is a deviatoric stress invariant (see Eq. (1.48)), and p^*, γ (≈ 0.5) and δ (≈ 0.5–0.7) are constants; p^* is called the critical effective pressure (or crushing pressure) of the rock. Eq. (2.28) represents a failure criterion for compaction. Note that the equation predicts failure at $\overline{\sigma}'=p^*$ if the stress is hydrostatic ($q=0$).

Compaction leads to a denser structure, and the rock will still be able to carry load. As the structure becomes increasingly dense, the load car-

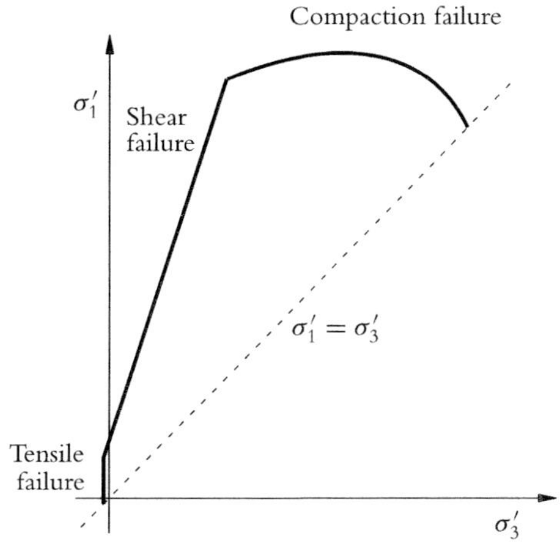

Figure 2.13 Location of the various failure modes in principal stress space, as defined by Eq. (2.3) for tensile failure, by Eq. (2.22) for shear failure, and by Eq. (2.29) (assuming that $\sigma_2' = \sigma_3'$) for compactive yield. For real rocks, the transition between the various failure modes is smoother than shown here.

rying capacity may even increase. Hence Eq. (2.28) represents essentially a compactive yield surface. This differs from tensile failure and unconfined shear failure, where the load carrying capacity is completely lost after failure. Compaction is also a more homogeneous failure mode, however localisation may occur even for compaction. This is further discussed in Section 2.8.

Boutéca et al. (2000) argued that the following simple form of Eq. (2.28) (obtained by choosing $\gamma = 0$ and $\delta = 1$) may be an acceptable approximation for many rocks:

$$\bar{\sigma}'^2 + q^2 = p^{*2} \tag{2.29}$$

It is to be expected that the critical effective pressure p^* decreases with increasing porosity. From theoretical considerations Zhang et al. (1990) derived the relation

$$p^* \propto \phi^{-\frac{3}{2}} \tag{2.30}$$

Given that the uniaxial compressive strength C_0 of a rock also depends on the porosity to some extent, we may furthermore expect a certain degree of correlation between p^* and C_0. Boutéca et al. (2000) estimated p^* to be 6–7 times larger than C_0 for a set of sandstones with porosity in the range 15–25 %.

Note that a transition from the principal stress space to the (σ', τ)-plane is not trivial for the end cap. Any point on the failure line in Fig. 2.13 that lies above the end point $\sigma_1' = \sigma_3' = p^*$ on the $\sigma_1' = \sigma_3'$ line represents a circle in the (σ', τ)-plane that extends beyond the (infinitesimally small) circle corresponding to the collapse point $(\sigma' = p^*, \tau = 0)$. Thus there is no line in the (σ', τ)-plane representing a boundary for Mohr's circles for collapse failure in the same way as for shear failure.

2.5 Failure criteria in three dimensions

The stresses in the underground formations and around wells will generally be three-dimensional in the sense that the principal stresses will have different values. It is therefore necessary to extend the two-dimensional failure criteria to deal with the more general case.

This also brings up the question of the role of the intermediate principal stress. With the development of equipment for true triaxial testing, it has been possible to approach this question experimentally. The general result from such tests is that under shear conditions the intermediate stress plays a role, although minor compared to the effect of the other stresses.

2.5.1 Criteria independent of the intermediate principal stress

Most of the failure criteria used in rock mechanics were developed before the role of the intermediate principal stress was clear. As the real influence of the intermediate principal stress is relatively small (compared to the significance of the other two), such criteria are still valuable approximations and will still be used.

From Fig. 2.10 we are able to infer how the Mohr–Coulomb failure surface looks like in the three-dimensional principal stress space $(\sigma_1', \sigma_2', \sigma_3')$. We will here temporarily abandon the conventional relation that $\sigma_1' \geq \sigma_2' \geq \sigma_3'$ to illustrate the entire surface. First, we consider the situation where σ_2' is the intermediate principal stress. The two cases $\sigma_1' > \sigma_3'$ and $\sigma_1' < \sigma_3'$ can be represented by two lines that are symmetric around the line $\sigma_1' = \sigma_3'$, as shown in Fig. 2.14. The figure shows the projection onto the (σ_1', σ_3')-plane of the part of the surface for which σ_2' is intermediate. Similar projections can be made onto the (σ_1', σ_2')-plane for the case where σ_3' is intermediate and onto the (σ_2', σ_3')-plane for the case where σ_1' is intermediate. The complete failure surface is indicated in Fig. 2.15. The figure shows that

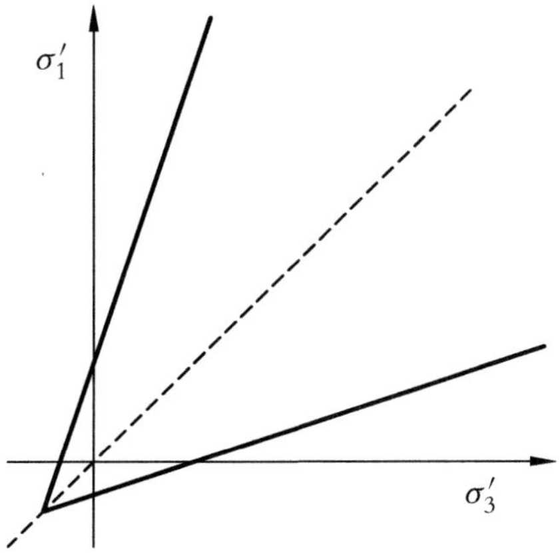

Figure 2.14 Projections onto the (σ_1', σ_3')-plane of the parts of the failure surface for which σ_2' is the intermediate principal stress. Note the symmetry about the projection of the hydrostatic axis (dashed line).

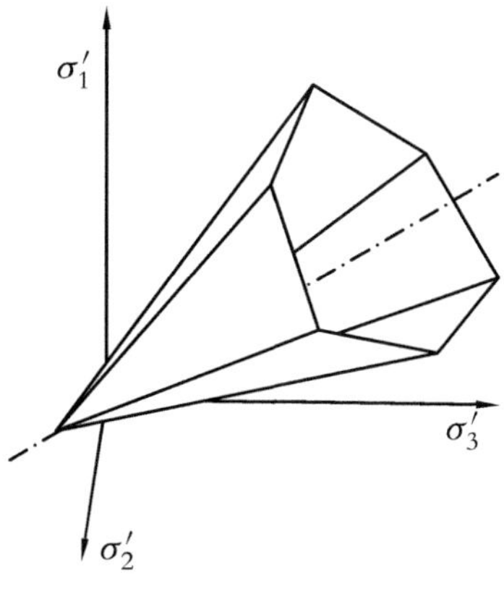

Figure 2.15 The Mohr–Coulomb failure surface in principal stress space.

opposite pairs of surface sections project onto the (σ_1', σ_3')-, (σ_2', σ_3')- and (σ_1', σ_2')-planes, respectively, along lines as indicated on Fig. 2.14.

The irregular hexagonal pyramid shape of the surface reflects that according to the Mohr–Coulomb criterion, failure is independent of the intermediate principal stress. The surface is not differentiable at the corners, a fact that may cause problems in numerical calculations involving the criterion. The cross-section of the Mohr–Coulomb failure surface in a π-plane (see Section 2.1.2) is shown in Fig. 2.16. It is seen to be an irregular hexagon with sharp corners and threefold symmetry.

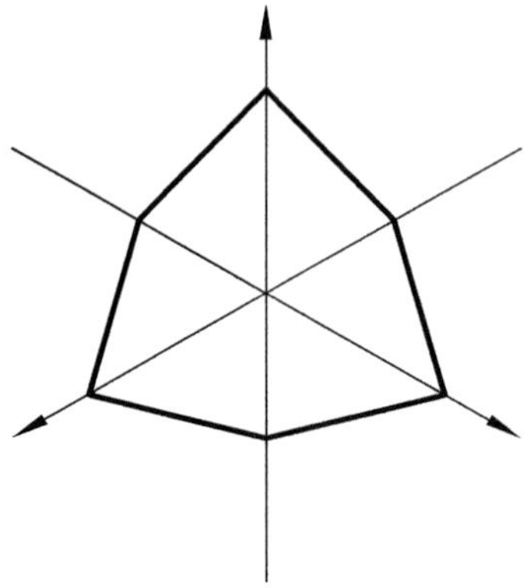

Figure 2.16 Cross section of the Mohr–Coulomb criterion in a π-plane. The arrows represent projections of the principal stress axes onto the plane. The friction angle $\varphi = 30°$.

The Tresca criterion also has a six-sided cross-section in the π-plane, but for this criterion, the section will be a regular hexagon. As the friction angle is zero, the failure surface will be parallel to the hydrostatic axis.

A three-dimensional failure surface can be obtained for the Griffith criterion in a similar way as for the Mohr–Coulomb criterion. The three pairs of planes ending in a sharp corner for the Mohr–Coulomb criterion (Fig. 2.15) is for the Griffith criterion replaced by three pairs of singly curved parabolic surfaces ending in a plane section.

The failure surface constructions presented here demonstrate one basic feature of the failure surface in three dimensions, namely the three-fold symmetry of the π-plane cross-section. However, they suffer from the fact that they ignore the influence of the intermediate principal stress. Other extensions to three dimensions have therefore been proposed, as discussed below.

2.5.2 Criteria depending on the intermediate principal stress

Mounting experimental evidence has shown that the intermediate principal stress σ_2' has a significant—although moderate—impact on the strength of several rock types (see, for instance, Colmenares and Zoback, 2002, for an overview). Typically, it is found that rocks are stronger when $\sigma_1' > \sigma_2' > \sigma_3'$ than for the situations where $\sigma_2' = \sigma_1'$ or $\sigma_2' = \sigma_3'$.

As mentioned in the previous paragraph, the failure surfaces constructed simply on the basis of two-dimensional criteria also possess sharp corners,

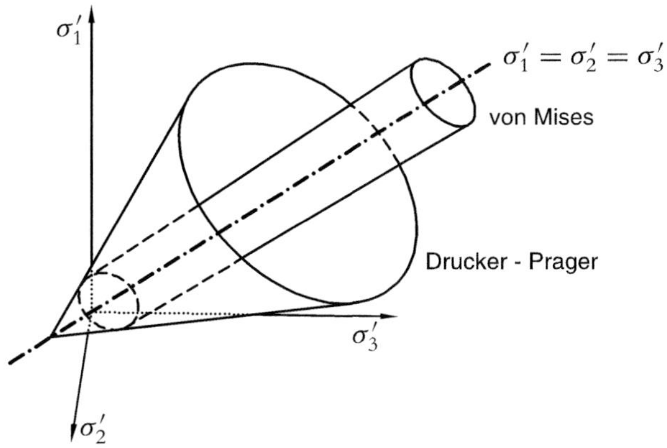

Figure 2.17 The von Mises criterion and the Drucker–Prager criterion in principal stress space.

which cause problems when the criteria are used for numerical modelling. Various other failure criteria that include the intermediate principal stress have therefore been proposed.

One simple solution is to implement rotational symmetry for the π-plane cross-section. This approach has no physical foundation; however, it is mathematically attractive and is the basis for some of the most commonly used criteria shown below. (Alternative formulations of these criteria, in terms of invariants, are given in Section 2.7.)

One of these criteria is the *von Mises criterion*, which can be written

$$(\sigma_1' - \sigma_2')^2 + (\sigma_1' - \sigma_3')^2 + (\sigma_2' - \sigma_3')^2 = C^2 \tag{2.31}$$

where C is a material parameter related to cohesion. In the principal stress space, this criterion is represented by a cylinder centred around the hydrostatic axis, as illustrated in Fig. 2.17. The criterion is seen to be identical to the Tresca criterion for $\sigma_2' = \sigma_1'$ or $\sigma_2' = \sigma_3'$. Like the Tresca criterion, the von Mises criterion describes a shear failure mechanism where the failure condition is independent of the stress level in the material. The von Mises criterion is commonly used to describe yield in metals. It has however very limited applications for rocks.

The corresponding generalisation of the Mohr–Coulomb criterion is the *Drucker–Prager criterion*. It can be formulated as

$$(\sigma_1' - \sigma_2')^2 + (\sigma_1' - \sigma_3')^2 + (\sigma_2' - \sigma_3')^2 = C_1(\sigma_1' + \sigma_2' + \sigma_3' + C_2)^2 \tag{2.32}$$

where C_1 and C_2 are material parameters related to internal friction and cohesion. In the principal stress space, this corresponds to a regular cone, as illustrated in Fig. 2.17.

Murrel (1963) introduced the *extended Griffith criterion*, which degenerates to the original Griffith criterion (Eqs. (2.23)–(2.24)) in two dimensions. In the principal stress space, it is represented by a paraboloid terminated by a pyramid, expressed as

$$(\sigma_1' - \sigma_2')^2 + (\sigma_1' - \sigma_3')^2 + (\sigma_2' - \sigma_3')^2 = 24T_0(\sigma_1' + \sigma_2' + \sigma_3') \tag{2.33}$$

and ending on the planes

$$\sigma_1' = -T_0, \quad \sigma_2' = -T_0, \quad \sigma_3' = -T_0 \tag{2.34}$$

The extended Griffith criterion predicts that the relation between uniaxial compressive strength and tensile strength is given by

$$C_0 = 12T_0 \tag{2.35}$$

This value of C_0/T_0 is typical for many rocks. Also for this criterion, the cross-section in the π-plane is given by a circle, except for the conical part at negative stresses.

Al-Ajmi and Zimmerman (2005) introduced the *Mogi–Coulomb criterion* for creation of a fracture plane:

$$\frac{1}{3}\sqrt{(\sigma_1' - \sigma_2')^2 + (\sigma_2' - \sigma_3')^2 + (\sigma_1' - \sigma_3')^2} = a + \frac{1}{2}(\sigma_1' + \sigma_3')b \tag{2.36}$$

This is a linear version of the more general criterion introduced by Mogi (1971) for creation of a fracture plane. Note that for $\sigma_2' = \sigma_3'$, which is the situation for a conventional triaxial test, criterion (2.36) coincides with the Mohr–Coulomb criterion (2.22).

An empirical failure criterion based on actual observations of the behaviour of soils was formulated by Lade (1977). A modified version of this criterion was presented by Ewy (1999) :

$$\frac{I_1'^3}{I_3'} - 27 = \eta_L \tag{2.37}$$

where I_1' and I_3' are modified representations of the first and third stress invariants defined as

$$I_1' = (\sigma_1' + S_L) + (\sigma_2' + S_L) + (\sigma_3' + S_L) \tag{2.38}$$

$$I_3' = (\sigma_1' + S_L)(\sigma_2' + S_L)(\sigma_3' + S_L) \tag{2.39}$$

where S_L is a material parameter related to the cohesion and the friction angle of the rock,

$$S_L = \frac{S_0}{\tan \varphi} \tag{2.40}$$

whereas η_L is related only to the internal friction:

$$\eta_L = 4 \tan^2 \varphi \frac{9 - 7 \sin \varphi}{1 - \sin \varphi} \tag{2.41}$$

In the principal stress space the criterion has the form of a convex triangularly shaped cone.

The modified Lade criterion is a simple criterion, which appears to account for the influence of the intermediate principal stress on shear strength in a realistic way. As such, this is a useful criterion for practical calculations. The formulation based on I_1' and I_3' (Eq. (2.37)) may however be somewhat impractical, as it requires solution of a cubic equation. An alternative formulation is given in Eqs. (2.42)–(2.45).

2.5.2.1 π-plane representation

The Drucker–Prager, von Mises and Murrel criteria are circular in the π-plane, whereas the modified Lade, Mogi–Coulomb and Mohr–Coulomb criteria have a three-fold symmetry. It is clear that for an isotropic material, the failure surface must at least have a three-fold symmetry in the π-plane. This means that in general, the failure surface in principal stress space may be given by specifying the mean stress and the radius in the π-plane as a periodic function of the angle $3\vartheta_L$ (where ϑ_L is the Lode angle defined in Eq. (1.54) and Fig. 1.10). Fig. 2.18 shows the shape of the π-plane cross-sections of the various failure criteria.

The radial distance to a point on the failure surface is given by Eq. (1.53). Defining two general functions f_1 and f_2, we may thus write a general failure criterion as

$$\sqrt{J_2} = f_1(\cos 3\vartheta_L) f_2(\overline{\sigma}') \tag{2.42}$$

where the cosine function ensures the required symmetry in the π-plane, and the factor 2 of Eq. (1.53) has been absorbed in f_1 and f_2. The separation of ϑ_L and $\overline{\sigma}'$ means that the shape of the π-plane cross-section is independent of $\overline{\sigma}'$.

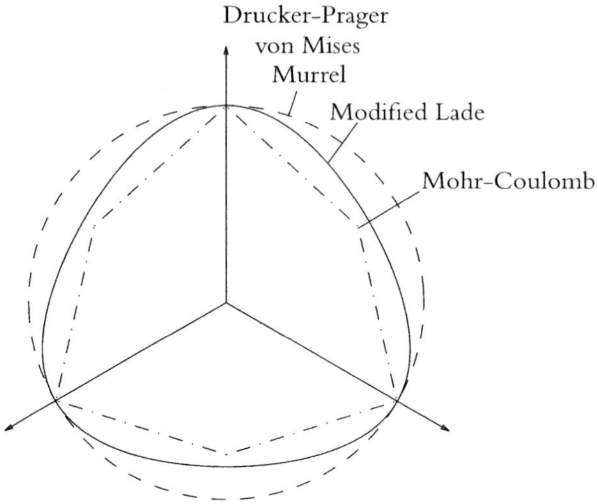

Figure 2.18 Characteristic shape of a π-plane cross-section for some failure criteria. The friction angle $\varphi = 22.5°$. The arrows represent projections of the principal stress axes onto the plane. Note that the criteria have been scaled so that they coincide at the intercepts with the projections of the principal stress axes.

The choice of the cosine in Eq. (2.42) is convenient, since $\cos 3\vartheta_L$ can be expressed in terms of invariants (Eq. (1.54)), and does not represent a limitation since any periodic function may be expressed by a series in $\cos 3\vartheta_L$.

The function f_1 cannot be chosen completely arbitrarily, since both empirical evidence (see e.g. Lade, 1993) and theoretical considerations (see e.g. Chen and Han, 1988) indicate that the failure surface should be convex. If we choose f_1 to be a constant, Eq. (2.42) represents the von Mises criterion if f_2 is also a constant, the Drucker–Prager criterion if f_2 is linear in $\overline{\sigma}'$, and the extended Griffith criterion if f_2 is proportional to the square root of $\overline{\sigma}'$.

A representation of the modified Lade criterion can be obtained by expressing I_1' and I_3' in terms of $\overline{\sigma}'$, q and ϑ_L (see Section 1.1.8), and defining a new variable $x = q/(S_L + \overline{\sigma}')$ (note that $\sqrt{J_2} = x(S_L + \overline{\sigma}')/\sqrt{3}$). Introducing these expressions into Eq. (2.37) gives a cubic equation in x, which can be solved explicitly (Viète's solution) to give

$$f_1 = \frac{3}{2\cos 3\vartheta_L}\left(1 - 2\cos\left[\frac{1}{3}\arccos\left(1 - \frac{2\eta_L \cos^2 3\vartheta_L}{27 + \eta_L}\right) \pm \frac{\pi}{3}\right]\right) \tag{2.43}$$

$$\simeq \sqrt{\frac{3\eta_{\rm L}}{27+\eta_{\rm L}}}\frac{1-\frac{2}{9}\sqrt{\frac{3\eta_{\rm L}}{27+\eta_{\rm L}}}\cos 3\vartheta_{\rm L}}{1-\frac{1}{3}\sqrt{\frac{3\eta_{\rm L}}{27+\eta_{\rm L}}}\cos 3\vartheta_{\rm L}} \tag{2.44}$$

$$f_2 = \frac{1}{\sqrt{3}}(S_{\rm L} + \overline{\sigma}') \tag{2.45}$$

Expression (2.44), obtained by one step Newton iteration with the $\vartheta_{\rm L}$-independent solution as starting point, is a good approximation for all realistic friction angles and may be the most practical form, since the two solutions (+ and −) given by Eq. (2.43) have to be patched together to make the solution continuous in $\vartheta_{\rm L}$. (+ applies for $-\pi/6 < \vartheta_{\rm L} < \pi/6 + n\pi/3$, whereas − applies for $\pi/6 < \vartheta_{\rm L} < \pi/2 + n\pi/3$. The expressions are undefined for $\cos 3\vartheta_{\rm L} = 0$, but the approximate solution is exact at these points. The third root (not shown) never gives a physical solution.)

The parameter $\eta_{\rm L}$ determines the shape of the cross-section in the π-plane. As the value of $\eta_{\rm L}$ increases, the cross-sectional shape changes from circular to triangular with smoothly rounded edges.

A general simple form for f_1 may be convenient for practical purposes. We may, for instance, choose

$$f_1 = k + (1-k)\cos 3\vartheta_{\rm L} \tag{2.46}$$

where k is a number. Choosing $k = 1$ gives a circular cross-section. Choosing k to make the uniaxial and biaxial ($\sigma_2' = \sigma_1'$) strengths equal would require

$$k = \frac{1}{2}\left(1 + \frac{3-\sin\varphi}{3+\sin\varphi}\right) \tag{2.47}$$

This relation gives $k = 6/7 \approx 0.857$ for $\varphi = 30°$ (note that the convexity criterion requires $k \geq 10/11 \approx 0.909$). Eqs. (2.46) and (2.47), in combination with

$$f_2 = C_0\frac{1-\sin\varphi}{3-\sin\varphi} + 2\frac{\sin\varphi}{3-\sin\varphi}\overline{\sigma}' \tag{2.48}$$

may be used as a rough approximation for the Mohr–Coulomb criterion.

The Mohr–Coulomb criterion may also be formulated in terms of the invariants $\vartheta_{\rm L}$ and $\overline{\sigma}'$ (the relations between these invariants and the principal stresses are given in Appendix D.1.1). The expression

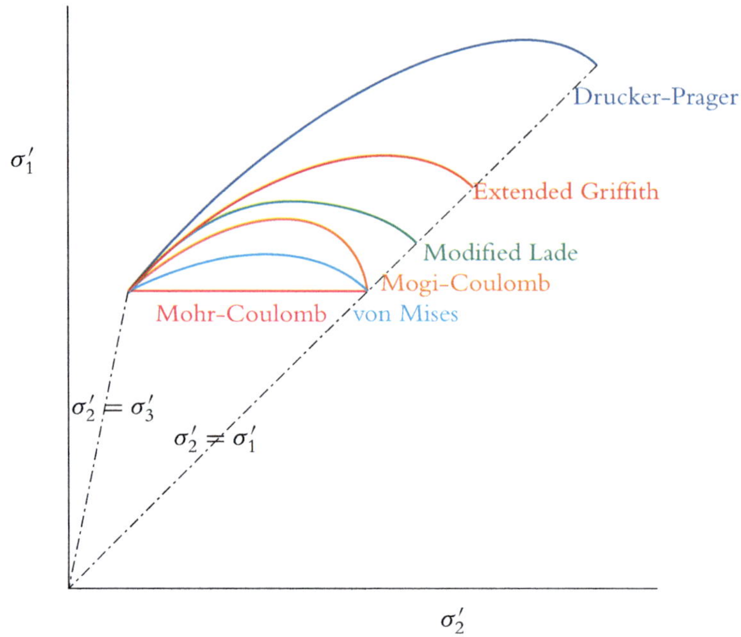

Figure 2.19 Characteristic shape of various failure criteria in a (σ'_2, σ'_1) cross-section. The friction angle $\varphi = 25°$, whereas $\sigma'_3/C_0 = 0.4$. The criteria have been scaled so that they coincide for $\sigma'_2 = \sigma'_3$.

$$f_1 = \sqrt{3}\left(\cos\vartheta_L + \sqrt{3}\sin\vartheta_L \frac{1+\sin\varphi}{3-\sin\varphi}\right)^{-1} \tag{2.49}$$

in combination with Eq. (2.48) is valid for $0 \le \vartheta_L \le \pi/3$. By this formulation, however, we loose the explicit three-fold symmetry of Eq. (2.42). This is not a major concern, since using arccos in Eq. (1.54) maps all stress states to ϑ_L in the correct range.

For papers using the formulation presented here, see, for example, van Eekelen (1980) or Nordal et al. (1989). This representation is convenient for generation of π-plots like Fig. 2.18.

The cross-sections shown in Fig. 2.18 represent paths in the principal stress space where all three principal stresses changes while the mean stress remains constant. To get a clearer picture of the impact of the intermediate principal stress, it is useful to also consider the failure envelope in a plot where σ'_1 varies as a function of σ'_2 while σ'_3 remains constant, as shown in Fig. 2.19.

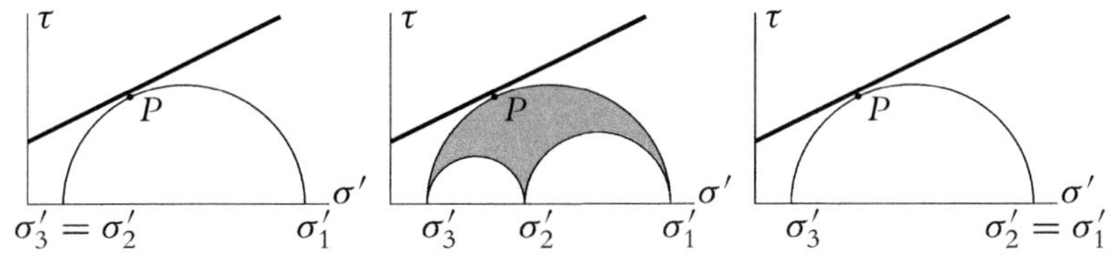

Figure 2.20 (σ', τ) plots with Mohr's circles and failure line for three different stress states.

2.5.2.2 Physical explanations

Wiebols and Cook (1968) proposed a model that offers a physical explanation to the impact of the intermediate principal stress on rock strength. They considered the shear strain energy associated with microcracks in the material. Activation of such a crack, in terms of sliding between the crack surfaces, will occur when a frictional criterion like the Mohr–Coulomb criterion (Eq. (2.6)) is fulfilled locally for this crack. The shear and normal stresses controlling this process are given by the relative orientation of the crack and the principal stresses (see Eq. (1.36)). The situation is illustrated graphically in Fig. 2.20, showing three different stress states: $\sigma_1' > \sigma_2' = \sigma_3'$, $\sigma_1' > \sigma_2' > \sigma_3'$ and $\sigma_1' = \sigma_2' > \sigma_3'$. We assume that there exists a large number of such cracks with random orientation in the material. For simplicity, we also assume that they all have the same activation threshold, so that the failure line shown in the figure is the same for all cracks. Thus, for every possible stress state represented by a point on (or within) Mohr's circle(s) in the (σ', τ) plot, there exist some corresponding cracks that are activated if the point is above the failure line.

The fundamental postulate of Wiebols and Cook is that the material fails when the total strain energy associated with the activated cracks reaches a limit value, which in our simplified picture implies that the material fails when a sufficient number of cracks have been activated. When σ_1' is increased, the cracks with an orientation corresponding to point P in each diagram will obviously be activated first. For the two cases $\sigma_1' > \sigma_2' = \sigma_3'$ and $\sigma_1' = \sigma_2' > \sigma_3'$, there is a large number of cracks corresponding to point P and its vicinity due to the rotational symmetry of the stress state in both cases. Hence the total strain energy associated with these cracks will soon reach the limit value when the (σ_1', σ_3')-circle crosses the failure line. For the case $\sigma_1' > \sigma_2' > \sigma_3'$, the stress states are distributed in the shaded area, and most of these cracks are therefore further away from the failure line. Hence the $\sigma_1' - \sigma_3'$-circle has to expand further before a sufficient number of cracks is activated so that the rock fails. As a consequence, the rock is stronger for the case $\sigma_1' > \sigma_2' > \sigma_3'$ than for the cases $\sigma_1' > \sigma_2' = \sigma_3'$ and $\sigma_1' = \sigma_2' > \sigma_3'$.

An alternative explanation was proposed by Fjær and Ruistuen (2002). They based their analysis on the classical assumption (see Section 2.3) that the rock will fail when the shear stress across a potential failure plane fulfils a failure criterion, like, for instance, the Mohr–Coulomb criterion. Any imaginary plane through a rock sample is a potential failure plane, and hence any point representing a possible stress state in the (σ', τ) plot (Fig. 2.20) also represents at least one possible failure plane.

Fjær and Ruistuen assumed that there is a variation in the strength of each potential failure plane due to the natural heterogeneity of rocks. This implies that there is a finite probability for a plane to fail even if its corresponding point in the Mohr's circle diagram lies below the failure line, and there is correspondingly a finite probability for the plane not to fail even if its corresponding point lies above the failure line. For the high-symmetry cases $\sigma_1' > \sigma_2' = \sigma_3'$ and $\sigma_1' = \sigma_2' > \sigma_3'$ of Fig. 2.20, there is a large number of planes corresponding to point P, which reaches the failure line first, and hence there is a high probability that at least one of these planes will fail early. For the low-symmetry case $\sigma_1' > \sigma_2' > \sigma_3'$, nearly all of the points representing these planes have moved away from the failure line, and the probability for early failure is correspondingly reduced. The net effect of this is that the rock appears as stronger—statistically—for the low-stress symmetry case $(\sigma_1' > \sigma_2' > \sigma_3')$ than for the cases of higher stress symmetry $(\sigma_1' > \sigma_2' = \sigma_3'$ and $\sigma_1' = \sigma_2' > \sigma_3')$.

2.6 Fluid effects

2.6.1 Pore pressure

In our discussion of failure criteria the pore pressure has so far appeared only indirectly through the effective stresses. The effective stresses are thought to represent the forces transmitted through the rock skeleton, which in turn causes the deformation of the material while the remaining parts of the total stresses are carried by the pore fluid. As the pore pressure is equal in all directions, it will affect only the normal stresses. The shear stresses that are due to differences in the principal stresses will be unaffected. In a (σ', τ) plot the effect of increasing the pore pressure while the total stresses are kept constant is moving Mohr's circles to the left and closer to the shear and tensile failure lines, as illustrated in Fig. 2.21. Thus increasing pore pressure may destabilise a rock with respect to shear and tensile failure.

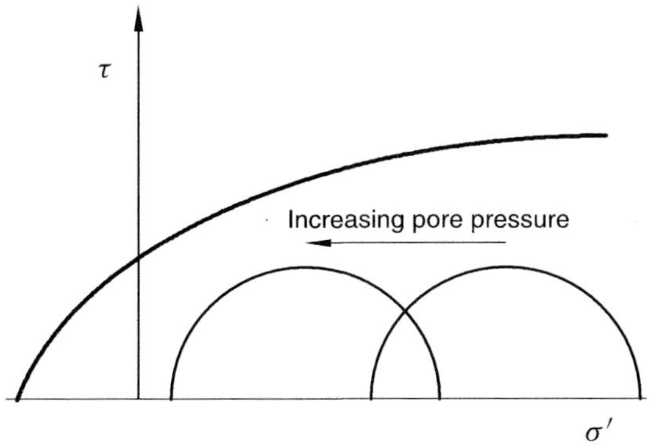

Figure 2.21 Mohr's circle and failure lines: the effect of increasing pore pressure.

The effective stress concept, as given by Eq. (1.187) in terms of the Biot constant α, was derived for linear poroelasticity and is not directly applicable for a rock at failure. It is, however, generally accepted that Terzaghi's definition of effective stress (see Section 1.6.4)

$$\sigma' = \sigma - p_f \tag{2.50}$$

appears to be the most relevant definition for effective stress to be used in failure criteria (Detournay and Cheng, 1988; Boutéca and Guéguen, 1999). Short of any straightforward theoretical derivation for this expression, it should be considered primarily as an empirical relation. However, the absence of an effective stress coefficient different from unity may be attributed to failure initiation being controlled by local stress concentrations, which are proportional to the difference between far-field stresses and pore pressure (Garg and Nur, 1973; Jaeger et al., 2007).

We may find an argument for this observation by observing that in the vicinity of failure the rock is softening (that is, the slope of the stress–strain curve is strongly reduced; see Fig. 2.2). Note however that the stiffness in triaxial tests is given by Young's modulus, and hence it is governed to a large extent by shear stiffness. It is not obvious, however, that Eq. (2.50) is the best definition for the effective yield stress, as the rock is supposed to behave elastically until it yields.

2.6.2 Partial saturation

Even unconsolidated sand may have some degree of cohesion. Any child that has visited a sandy beach on a sunny day has experienced the magic

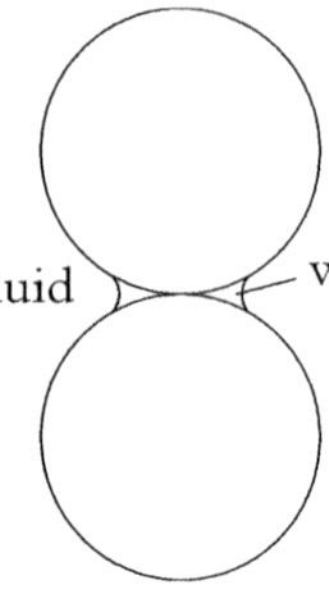

Figure 2.22 Schematic illustration of the distribution of fluids at a grain contact.

of moist sand, which can be shaped and reshaped to make all kinds of fantastic figures. When the tide comes and floods the beach, the magic is broken, and the figures disintegrate into loose sand. The "magic" is caused by the fact that moist sand is *partially* water saturated. This means that there is a meniscus of water at every grain contact, as shown schematically in Fig. 2.22. It is energetically favourable for the grain–water–air system to maintain this constellation, and hence a certain force is needed to rip the grains apart. This acts as a cohesion, giving the moist sand a finite shear and tensile strength. The strength is fully recovered after a reorganisation of the grains, and hence the moist sand can be reshaped indefinitely. When the sand becomes completely dry or fully water saturated, the meniscus constellation is destroyed, and the cohesion and strength are gone.

The effects of partial saturation occur whenever the pore space is filled with at least two immiscible fluids, like, for instance, oil and water. Normally, it is energetically favourable that one of the fluids (the wetting fluid) stays in contact with the solid material while the other (non-wetting) fluid is shielded from the solid to some degree, giving a constellation as illustrated in Fig. 2.22. This so-called capillary effect produces a difference in the pressure for the two fluids given as

$$p_{\mathrm{cp}} = p_{\mathrm{nw}} - p_{\mathrm{we}} \tag{2.51}$$

where p_{we} is the pressure in the wetting fluid, and p_{nw} is the pressure in the non-wetting fluid. The capillary pressure p_{cp} depends on the interfacial tension γ between two fluids and the radius of the capillary, typically the pore size r:

$$p_{\mathrm{cp}} = \frac{2\gamma}{r} \tag{2.52}$$

The surface tension of water is about 0.072 N/m at room temperature, and the interfacial tension for oil against water $\gamma_{\text{oil-water}}$ is typically 0.05 N/m. The value of p_{cp} will be 0.1–1 MPa in chalk with sub μm pore sizes, whereas it will be several MPa for shale with nm pores.

The capillary pressure p_{cp} is also called capillary suction: As the figure shows, the shape of the meniscus indicates that the pressure in the non-wetting phase exceeds that in the wetting phase. This means, in the case of air at atmospheric conditions, that the water pressure has to be considered as negative in order for p_{cp} to be positive; hence the term "suction".

The magnitude of the capillary suction depends on the type of fluids, the condition of the solid surface (which determines the degree of wettability for the various fluids) and the size of the pore at the point where the two fluid phases meet. The wetting fluid will always tend to fill the smallest pores, so that the contact between the fluid phases will move to larger pores when the degree of saturation for the wetting fluid is increased. Thus p_{cp} varies from one pore to the next and falls off rapidly with the degree of saturation for the wetting fluid.

The capillary suction has some effect on the effective stresses in the rock, and we may define a generalised effective stress (Bishop, 1959)

$$\sigma' = \sigma - \alpha(p_{\text{nw}} - S_{\text{we}} p_{\text{cp}}) \tag{2.53}$$

where S_{we} is the degree of saturation of the wetting fluid. The term $S_{\text{we}} p_{\text{cp}}$ typically has a peak at a low value for S_{we} (≈ 0.1 or less) and vanishes for $S_{\text{we}} = 0$ and $S_{\text{we}} = 1$. It is normally quite small, however, typically less than 1 MPa, even at its peak value, and can in most cases be ignored with respect to the effective stresses.

The capillary suction also affects material properties that can be related to the intergranular cohesion of the rock (Forsans and Schmitt, 1994; Papamichos et al., 1997). Forsans and Schmitt matched the effect of water saturation (between 20 % and 65 %) on the unconfined strength with experimental data for Tournemire Shale by calculating the capillary suction from the pore size distribution. In this range of saturation the strength dropped by a factor of 3 from the lowest to the highest saturation. No data was available for full saturation, but the model predicted very low strength in that case. For lower saturations, water may be grain coating more than contributing to the cohesion, and the capillary model becomes irrelevant. Significant capillary effect on both the strength and stiffness of chalk has been demonstrated experimentally (Brignoli et al., 1995), as illustrated in Fig. 2.23.

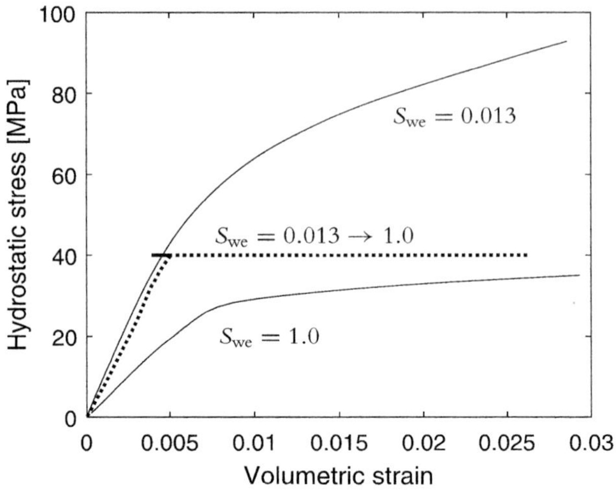

Figure 2.23 Hydrostatic compression tests at different levels of water saturation on Pietra Leccese chalk. In the test labelled $S_{we} = 0.013 \rightarrow 1.0$ the sample was saturated at $S_{we} = 0.013$ until the stress had reached 40 MPa, then the saturation was changed to $S_{we} = 1.0$ while the stress was kept constant. After Brignoli et al. (1995).

Rozhko (2020) developed the ideas of capillarity further, primarily for low-frequency dynamic elastic properties of partially saturated porous media (see Section 5.6.3). He accounted explicitly for interfacial tension, wettability (through the contact angle) and dual porosity and permeability by incorporating microcracks. In particular, the water sensitivity of the shear modulus was demonstrated and related to water sensitivity of strength through the impact of shear modulus on fracture toughness. In this model, increased surface tension implies softening and weakening (reduced strength), which may explain why water softens rock more than oil. Note that oil in a water-wet rock is predicted to cause higher strength than water in the same rock in oil-wet conditions. Closing microcracks by applied stress was also shown to reduce fluid sensitivity.

2.6.3 Chemical effects

The previous paragraph addressed partially saturated rock. Fluid sensitivity is however commonly observed also in fully monophasic saturated rocks, where there is no capillary effect. It may be argued (e.g. Rozhko, 2020) that, particularly for low permeability rock, 100 % saturation may be difficult to reach, even in controllable laboratory experiments. Nevertheless, there are also reasons for fluid sensitivity in absence of capillarity.

Most experimental and theoretical work along these lines has been related to wave propagation (Chapter 5). This implies undrained conditions, which obviously leads to fluid sensitivity as expressed by the Biot–Gassmann equation (1.171). Furthermore, high-frequency ultrasonic and sonic frequencies may give higher stiffnesses than low-frequency/quasi-static data because of dispersion, which gives the impression of stiffening by the fluid, even if the material is softened in the static limit.

There is however a number of published static experiments with sandstone, limestone, chalk and shale that show water softening. Vutukuri et al. (1974) reported measured tensile strength of 11 % porosity Indiana Limestone saturated with different liquids such as water, glycerine, nitro-benzene and alcohols. They found a monotonous decrease in strength with increasing surface tension of the saturating fluid. It is tempting to compare these results with the predictions of Rozhko's model above, but remember that the capillary-based approach is only valid in case full saturation was not achieved.

Experimental studies by Risnes et al. (2003) have shown that the chemical activity of the pore water seems to have an impact on the strength of chalk. In a set of $\kappa' = 0.9$ tests (see Section 7.3.9) on Liège chalk, they found the relation

$$\frac{\Delta\sigma_y}{\Delta a_w} = -6\,\text{MPa} \tag{2.54}$$

between the yield stress σ_y and the water activity a_w. Thus a low salt concentration in the pore water (which corresponds to a high water activity) makes the chalk weaker. During water injection, with partial saturation, capillary effects are also likely to contribute.

There are several observations of water weakening in sandstones (e.g. Baud et al., 2000; Duda and Renner, 2013; David et al., 2015), covering a wide range of porosities and strengths. Effects are similar to what was seen in chalk: Strength is reduced for water-saturated compared to dry specimens. Oil-saturated samples of Sherwood sandstone (David et al., 2015) are shown to be weaker than dry but stronger than water-saturated ones.

The most common explanation for water sensitivity of fully saturated rocks is based on the reduction of solid surface energy (see Section 6.4.4) in presence of water. Griffith's failure criterion (see Sections 2.5 and 6.4) relates failure initiation to strain energy exceeding the surface free energy, defined as the work needed to produce new surface. Parks (1984) summarised the impact of water on quartz and pointed to the huge difference between the surface energy of truly dry versus moist and immersed quartz surfaces. The surface energy drops when water molecules are adsorbed to the surface and reaches a further reduced value when the surface is fully covered by water. Note, as seen in experiments by Li and de Bruyn (1966), that the reduction in surface energy increases with increasing pH above the point of zero charge (which is when the electrical potential across the water–solid interface is zero) and that adding salt (reducing chemical activity) further reduces the surface energy. This aligns with the observations of Risnes et al. (2003), provided that calcite surfaces react to changes in chemical activity in a similar way as quartz.

Røyne et al. (2011) used a double torsion technique and verified the reduction of surface energy in calcite by exposing specimens to glycol–water mixtures with water concentrations from 0 to 1. Baud et al. (2000) related their observation to the ratio of surface energy for the solid–liquid interface to that for a solid–solid interface ($c = w_{sw}/w_s$) and found the brittle strength of the tested sandstone to change with $c^{1/2}$, whereas the yield stress for compactive failure varied with $c^{3/2}$.

The alteration of surface energy is thought to be responsible for water-softening observed in rock physics experiments. This effect is violating the assumption that the stiffness of the solid framework is not affected by the fluid, which is one of the basic assumptions behind the Biot–Gassmann equation (Tittmann et al., 1980; Murphy et al., 1984; Tutuncu and Sharma, 1992). It is however important to agree on the meaning of "dry" reference samples: If the sample is oven-dried and evacuated, then the initial high surface energy prevails, whereas in the field, samples are never vacuum dry. Hence, for practical purposes, the dry laboratory reference sample should be a sample kept at irreducible water saturation.

Duda and Renner (2013) found no significant impact of clay on the water sensitivity in their triaxial tests. Their samples contained as much as 15 % clay, primarily illite. On the other hand, Papamichos et al. (2000) observed that dry or oil-saturated Red Wildmoor sandstone has a uniaxial compressive strength of about 14 MPa, whereas it essentially turns into loose sand if it is saturated with fresh water. The reason for this is that this rock consists of quartz grains cemented together only by clay minerals (largely smectite), which dissolve in water.

The surface energy mechanism is still valid in clay and shale. Indeed, because of the narrow (nanometre size) pores, adsorbed or bound water will play a much more important role than in sand and chalk/limestone. Additional effects of clay swelling/shrinkage associated with osmotic and ionic exchange mechanisms complicate their water sensitivity, which nevertheless is of large practical importance, in particular, for drilling. The water sensitivity of clay and shale will be discussed further in Chapters 3 and 9.

Some abundant components in rocks, like salts and some clay and carbonate minerals are dissolvable in water. Over geological time the solid minerals in a rock establish a chemical equilibrium with the pore water, which implies that the minerals dissolve and precipitate at the same rate. If the pore fluid is changed, for instance, during drilling or production, then the chemical equilibrium may be disturbed, and a net dissolution or deposition of minerals may occur. This may have a strong effect on the rock properties; typically, a reduction in strength of 30–100 % is seen in many rocks due to deterioration of the cement (Broch, 1974).

The solubility of minerals in the pore water may also be affected by the acidity (pH-value) of the water, as well as its temperature and pressure. Thus it is to be expected that the strength of some rocks may be sensitive to changes in these parameters too if the changes are given sufficient time to act.

2.7 Presentation and interpretation of data from failure tests

To determine the mechanical properties of a rock, normally, a series of triaxial compression tests are performed at different confining pressures. From the stress–strain diagrams both elastic properties and failure or yield data are obtained. The strength data resulting from such a test series will consist of pairs of corresponding yield (or failure) and confining stresses. This set of failure data can be displayed in different types of diagrams.

It may be confusing that the word failure is often used in a somewhat general sense, not distinguishing between failure and yield. An argument for this practice is that yield can be considered as the onset of failure. However, when actual failure data are presented, it should be checked if the data refer to yield or failure in the strict sense.

Three types of plots are commonly used to present failure data: the principal stress plots (Fig. 2.10), the (σ', τ) plots (Fig. 2.8) and the (p', q) plots, which will be described below.

In a principal stress plot the stress path of a given test can be traced as a line, whereas failure is represented by the point where the stress path crosses the failure surface. The principal stress plot is suitable for mapping the entire failure surface, including all types of failure, as demonstrated in Fig. 2.13.

A (σ', τ) plot displays effectively how the shear stress and the normal stress across a plane in a stressed rock varies with the orientation of the plane relative to the orientation of the principal stress axes. It is therefore well suited for identifying critical shear stresses and for illustrating orientational effects, like Fig. 2.20. As explained in Section 2.4, the (σ', τ) plot is not well suited for displaying collapse.

The (p', q) plot essentially displays the maximum shear stress versus the mean effective stress. It is based on the generalised shear stress q (Section 1.1.8),

$$q = \frac{1}{\sqrt{2}}\sqrt{(\sigma_1' - \sigma_2')^2 + (\sigma_2' - \sigma_3')^2 + (\sigma_1' - \sigma_3')^2} \tag{2.55}$$

and the parameter p', which is identical to the mean effective stress $\overline{\sigma}'$,

$$p' = \frac{1}{3}(\sigma_1' + \sigma_2' + \sigma_3') = \overline{\sigma}' \tag{2.56}$$

Both q and p' are stress invariants (see Sections 1.1.7 and 1.1.8). Under triaxial conditions, $\sigma_2' = \sigma_3'$, which implies that $q = \sigma_1' - \sigma_3'$.

As for the principal stress plot, the stress path of a given test can be traced as a line in the (p', q) plot, and failure occurs where the stress path crosses the failure line. The (p', q) plot corresponding to Fig. 2.13 is shown in Fig. 2.24. Note that the transition from Fig. 2.13 to Fig. 2.24 is not unique since both q and p' depend on σ_2'.

Some of the three-dimensional failure criteria are simpler to express in terms of q and p' than in terms of the principal stresses, for instance,

– the von Mises criterion (Eq. (2.31)):

$$q^2 = \frac{1}{2}C^2 \tag{2.57}$$

– the Drucker–Prager criterion (Eq. (2.32)):

$$q^2 = \frac{1}{2}C_1^2(3p' + C_2)^2 \tag{2.58}$$

– the modified Lade criterion (Eq. (2.37) with $f_1(\vartheta_L, \eta_L)$ given by Eq. (2.43) (note the similarity with Eq. (2.58)):

$$q = f_1(\vartheta_L, \eta_L)(S_L + p') \tag{2.59}$$

– the extended Griffith criterion (Eq. (2.33)):

$$q^2 = 36T_0p' \tag{2.60}$$

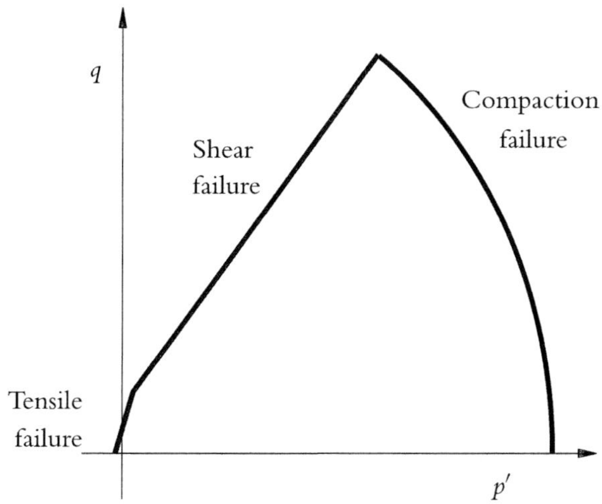

Figure 2.24 Location of the various failure modes in the (p', q) plane, as defined by Eq. (2.3) for tensile failure, by Eq. (2.22) for shear failure and by Eq. (2.29) for compactive yield, assuming that $\sigma_2' = \sigma_3'$.

– the compactive yield criterion (Eq. (2.29)):

$$p'^2 + q^2 = p^{*2} \tag{2.61}$$

The (p', q) plots are standard for plotting of failure surfaces in soil mechanics (see Section 2.8.2) and have a growing popularity also in rock mechanics.

2.8 Beyond the yield point

In the previous sections, we studied various failure criteria and their failure surfaces in the principal stress space without clearly expressing what we mean by failure. This is partly due to the fact that failure is not very well defined. Although the failure point can be defined relatively precisely for the uniaxial test in Fig. 2.2, the situation is much less clear for the higher confining pressures in Fig. 2.3. Here we have a gradual transition to a material with decreased stiffness, but still with an increasing capacity to carry load as the strain increases. Thus the material cannot be considered as completely failed, although it has been significantly altered at this stage. Whether the alteration is critical or not depends on the actual situation. For example, the practical criteria for a road tunnel and a wellbore may be quite different.

Whereas no change in profile is acceptable for a road tunnel, we may well live with a borehole with breakouts in many cases (see e.g. Guenot, 1989).

The remarks above illustrate the need for a post-yield description linked to the definition of failure and the failure criterion. To properly describe post-yield behaviour, we need a description treating the rock as an inhomogeneous medium penetrated by interacting faults and cracks.

Compared to a continuum model, such a description would, of course, be very complicated. It is therefore reasonable to look for a continuum theory, which can at least model some post-yield behaviour with some precision, although the physical interpretation may be less precise. One possible candidate is the *theory of plasticity*, which is widely used in the description of metals. An introduction to plasticity is given in Section 2.8.1.

Beyond the yield point the solid framework of the rock is gradually destroyed, and it may become increasingly relevant to compare its behaviour to unconsolidated materials. A model derived for the description of soils is presented in Section 2.8.2.

At some point in the failure process, the description of the rock as homogeneous is clearly not valid, even on a macroscopic scale, as exemplified by the formation of shear planes in a triaxial test. This calls for other types of modelling. Models addressing localisation effects at failure are briefly discussed in Section 2.8.3.

2.8.1 Plasticity

Plasticity is a concept describing non-elastic deformation of a material. Unlike elastic deformation, plastic deformation is not recovered when the load causing the deformation is released. Rocks stressed beyond the yield point typically suffer such deformations, and it is often relevant to describe post-yield behaviour of rocks by the concepts of plasticity.

The theory of plasticity is designed to model *ductile* behaviour, that is, behaviour in which the material can sustain a load comparable to the failure load well beyond failure, and no attempt should be made to model brittle failure with this theory.

The theory of plasticity is based on four major concepts:

1. *Plastic strain.* The total strain increment associated with a stress increment is assumed to consist of an elastic part and a plastic part:

$$d\varepsilon_{ij} = d\varepsilon_{ij}^{e} + d\varepsilon_{ij}^{p} \tag{2.62}$$

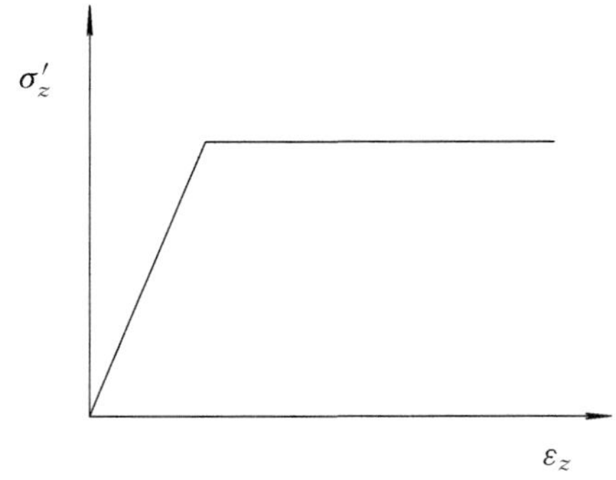

Figure 2.25 Stress–strain curve for a linearly elastic – ideally plastic material.

where $d\varepsilon_{ij}^{e}$ is related to the stress increment by conventional elasticity theory and will vanish when the stress is released. The plastic strain $d\varepsilon_{ij}^{p}$ is a permanent deformation and will remain when the stress is relieved.

2. *A yield criterion.* In Section 2.1, we defined yield as the point at which irreversible changes occur in the rock. Thus the yield point represents the onset of plastic deformation. Clearly, yield can be defined more precisely than failure. A yield criterion is similar to the failure criteria defined in Sections 2.2–2.4 and defines the surface in the stress space where plasticity is initiated.
3. *A flow rule.* The flow rule describes how the plastic strains develop for a given loading situation.
4. *A hardening rule.* We mentioned in Section 2.1.1 that a rock under certain conditions may sustain increasing load after the initial failure. This is described by the hardening rule. The hardening (or, alternatively, the softening) may be interpreted as a change of the yield surface in the principal stress space. This can be described by changing Eq. (2.1) to

$$f(\sigma_1', \sigma_2', \sigma_3', \kappa) = 0 \tag{2.63}$$

where κ is a parameter describing the hardening effects.

An *ideally plastic* material is a material that can endure infinite plastic strain without change in the stress level. Fig. 2.25 shows schematically the stress–strain diagram for a linearly elastic, an ideally plastic material. After the initial elastic phase, the material deforms at constant stress.

2.8.1.1 Plastic flow

The function of the flow rule is describing the development of the plastic strain increments. The basic assumption concerning plastic flow dates back to Saint-Venant in the 19th century. It states that

$$d\varepsilon_{ij}^{p} = d\lambda\, h_{ij}(\sigma'_{kl}) \tag{2.64}$$

where λ is a scalar not specified by the flow rule. It is seen that the h_{ij} are functions of the stress components. There are two main implications of Eq. (2.64). First, it states that the direction of plastic flow is determined by the *stress state* and is not influenced by the stress increments or by stress gradients. Second, the magnitude of the plastic strain is not unique. Referring to Fig. 2.25, this is quite trivial for an ideally plastic material, since the yield stress is maintained for any magnitude of the plastic strain.

The assumption that the plastic strains are independent of stress increments is intuitively understandable in the following simple example. Consider an object resting on some surface, and assume that a force nearly overcoming the rest friction between the object and the surface is applied. Now a small force increment is applied at some angle to the primary force. If the increment has a component in the direction of the primary force, then the rest friction will be overcome, and the object will start to slide in the direction of the primary force, irrespective of the direction of the force increment.

Eq. (2.64) puts some restrictions on the plastic behaviour; however, it is far from a complete description, which requires a specification of the functions h_{ij}. Some help is found from the fact that plastic deformation is a dissipative process, which implies that

$$\sum_{ij} \sigma'_{ij} d\varepsilon_{ij}^{p} \geq 0 \tag{2.65}$$

A significant simplification results from the assumption of von Mises (1928) that the h_{ij} can be derived as the gradient of a function g in the stress space:

$$d\varepsilon_{ij}^{p} = d\lambda \frac{\partial g}{\partial \sigma'_{ij}} \tag{2.66}$$

The function g is called the *plastic potential* and must of course be chosen such that Eq. (2.65) is obeyed. Although the assumption of a plastic potential reduces the need for a specification of six functions h_{ij} of six variables to one function g of six variables, it is by no means sufficient to completely specify plastic flow.

One solution to the problem is Drucker's definition (Drucker, 1950) of a stable, work hardening material. Such a material is defined by a more strict version of Eq. (2.65):

$$\sum_{ij} \mathrm{d}\sigma'_{ij}\mathrm{d}\varepsilon^{\mathrm{p}}_{ij} \geq 0 \tag{2.67}$$

Note that although Eq. (2.65) is a thermodynamic law, Eq. (2.67) is not, and thus the consequences derived from Eq. (2.67) need not be obeyed by all materials. From Eq. (2.67) Drucker found that the plastic potential g is identical to the function f describing the yield surface, that is,

$$\mathrm{d}\varepsilon^{\mathrm{p}}_{ij} = \mathrm{d}\lambda \frac{\partial f}{\partial \sigma'_{ij}} \tag{2.68}$$

Thus, once the yield surface is specified, so is the plastic flow, a gratifying result indeed. However, we will see that some of the consequences of Eq. (2.68) are not always fulfilled by rock materials, so that we may need to return to the more general assumption (2.66).

Since in Eq. (2.68) the plastic flow is derived from the yield criterion, it is called an associated flow rule. The flow rule derived from the more general plastic potential g is conversely called a non-associated flow.

2.8.1.2 Associated flow

In this section, we will discuss the associated plastic flow, mainly in conjunction with the Mohr–Coulomb criterion. Fig. 2.26 shows a principal stress plot with a Mohr–Coulomb failure line (see Fig. 2.10). We assume that this line represents the yield surface. According to Drucker (1950), it thus also represents the plastic potential as described above. Consider that a material is brought to the stress state represented by point A on the figure. Any small increment of the stresses that brings the stress state above the line will result in a plastic deformation.

The axes of plastic strain increments coincide with the principal stress axes. According to Eq. (2.68), the plastic strain increment is parallel to the gradient of the plastic potential. This implies that the direction of the plastic flow is parallel to the normal of the yield surface at point A, as indicated by the arrow on the figure. (The gradient of a function is always normal to

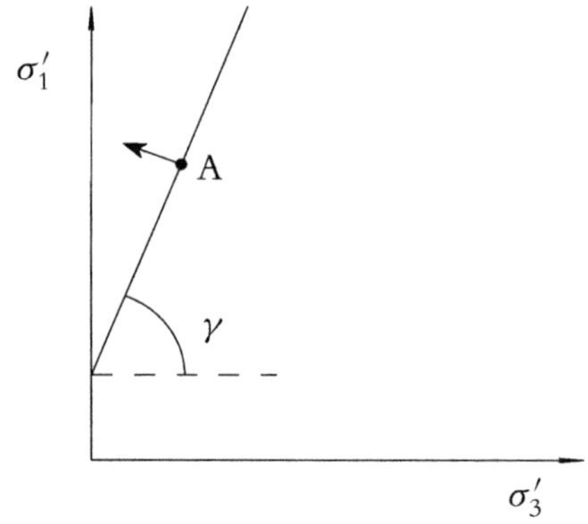

Figure 2.26 Associated plastic flow for the Mohr–Coulomb criterion in a principal stress plot.

the equisurface of that function.) From the figure we see that

$$d\varepsilon_1^p = d\lambda \cos\gamma \tag{2.69}$$

$$d\varepsilon_3^p = -d\lambda \sin\gamma \tag{2.70}$$

and therefore

$$d\varepsilon_3^p + d\varepsilon_1^p \tan\gamma = 0 \tag{2.71}$$

The Mohr–Coulomb criterion does not depend on the intermediate principal stress σ_2', and hence the failure surface is normal to the (σ_1', σ_3')-plane, and therefore $d\varepsilon_2^p = 0$. Then the volumetric strain increment $d\varepsilon_{vol}^p = d\varepsilon_1^p + d\varepsilon_2^p + d\varepsilon_3^p$ is

$$d\varepsilon_{vol}^p = d\varepsilon_1^p(1 - \tan\gamma) \tag{2.72}$$

For the Mohr–Coulomb criterion, γ is greater than 45°, and therefore

$$d\varepsilon_{vol}^p \le 0 \tag{2.73}$$

This means that the volume *increases* (recall our sign convention from Chapter 1). This effect is called *dilatancy*. Referring to Fig. 2.25, we note that for an ideally plastic material, this effect occurs while the mean stress is constant. This is different from a linearly elastic material, for which a change in volume only occurs as a result of a change in the mean stress.

In the literature, we will encounter plots in which the direction of plastic flow is indicated as a normal to the yield criterion in a (σ', τ) plot, as

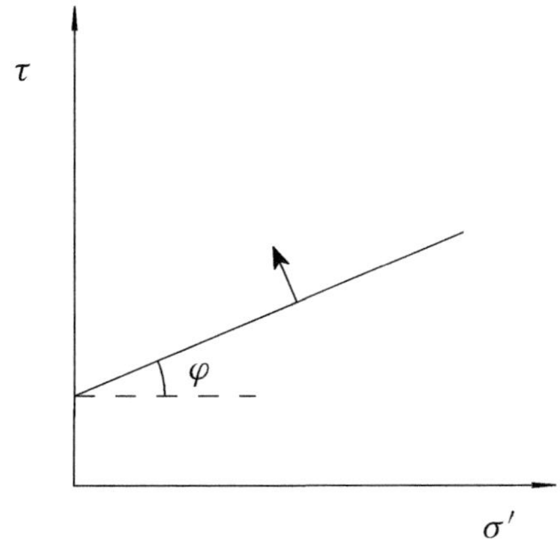

Figure 2.27 Associated flow for the Mohr–Coulomb criterion in a shear stress–normal stress plot.

shown in Fig. 2.27. Since the interpretation of this at first glance may be unclear, we will briefly discuss it here. Eq. (2.16) may be written as

$$\frac{\sigma_1' - \sigma_3'}{2\cos\varphi} = S_0 + \frac{1}{2}(\sigma_1' + \sigma_3')\tan\varphi \tag{2.74}$$

Therefore a plot giving the stress deviation $(\sigma_1' - \sigma_3')$ scaled by $1/(2\cos\varphi)$ versus $((\sigma_1' + \sigma_3')/2)$ will look exactly like Fig. 2.27. Hence, to find the plastic strains, we may interpret the y axis of Fig. 2.27 as $(\varepsilon_1^p - \varepsilon_3^p)/(2\cos\varphi)$, and the x axis as $(\varepsilon_1^p + \varepsilon_3^p)/2$. It can thus be concluded that the inclination of the arrow in Fig. 2.27 determines the dilatancy of the material as follows:

1. If the arrow is tilted to the left ($\varphi \geq 0$), then the material is positively dilatant.
2. If the arrow is vertical ($\varphi = 0$), then the material does not change volume (incompressible plastic flow).
3. If the arrow is tilted to the right ($\varphi \leq 0$), then the material is negatively dilatant or contractant.

Thus a normal Mohr–Coulomb criterion describes a dilatant plastic flow, whereas the Tresca criterion describes an incompressible plastic flow.

Experimentally, we often observe some dilatation but rarely to the degrees predicted by the Mohr–Coulomb criterion and associated plastic flow. This indicates that the assumptions leading to the associated flow do not always hold. Also, the restriction given by Eq. (2.67) means that the stress–strain curve is a monotonically increasing curve, again a consequence of associated plastic flow that is by no means always fulfilled.

2.8.1.3 Non-associated flow

A non-associated plastic flow may occur if the plastic potential is not identical to the yield surface. It is a convenient model to use for control on dilatancy without changing the yield criterion.

For plastic flow associated with the Mohr–Coulomb criterion, the plastic potential can be written as

$$f(\sigma_1', \sigma_3') = \sigma_1' - C_0 - \sigma_3' \tan\gamma = \sigma_1' - C_0 - \sigma_3' \frac{1+\sin\varphi}{1-\sin\varphi} \tag{2.75}$$

A non-associated yield criterion is found by choosing a different angle in Eq. (2.75), that is,

$$g(\sigma_1', \sigma_3') = \sigma_1' - C_0 - \sigma_3' \frac{1+\sin\Psi}{1-\sin\Psi} \tag{2.76}$$

where Ψ is called the *dilatancy angle*. From the discussion in the previous section we know that we will have dilatant, incompressible, or contractant flow depending on whether Ψ is larger than, equal to, or smaller than zero. Note that the allowable range for Ψ is restricted, since Eq. (2.65) has to be fulfilled.

The implicit definition of Ψ as expressed by Eq. (2.76) is linked to the Mohr–Coulomb-type flow criterion. An alternative, explicit definition of the dilatancy angle (Vermeer and de Borst, 1984) states that

$$\sin\Psi = -\frac{d\varepsilon_{vol}^{p}}{d\gamma^{p}} \tag{2.77}$$

where ε_{vol}^{p} is the plastic volumetric strain, and γ^{p} is the generalised plastic shear strain. (In the example above, $d\varepsilon_{vol}^{p} = d\varepsilon_1^{p} + d\varepsilon_3^{p}$, whereas $d\gamma^{p} = |d\varepsilon_1^{p} - d\varepsilon_3^{p}|$.) Note that dilation is a function of plastic strain, and hence the dilatancy angle will change as the rock deforms (Detournay, 1986).

2.8.1.4 Hardening

According to Eq. (2.63), hardening can be described by a change in the yield surface as a function of a parameter κ, which is, in some way, related to the plastic strains. This is shown schematically in Fig. 2.28. In the figure, we define the initial yield surface and a current yield surface, that is, the yield surface after some plastic straining. For weak rocks, the initial yield surface can be quite small, just reaching out to the highest stress level that the rock

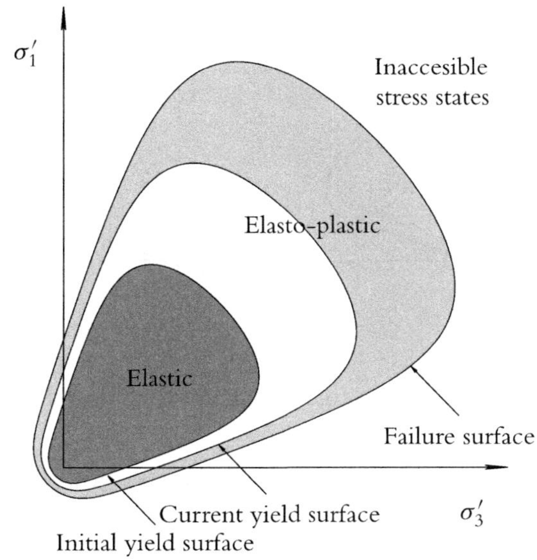

Figure 2.28 Sketch of hardening in stress space.

has experienced before. The yield point is therefore reached very early when loading the material for the first time, such that most of the loading curve involves plastic straining. During the loading process, the current yield surface expands so that the highest stress that the rock has experienced is just included. Upon unloading and reloading the yield surface will remain unaffected until the stress state again crosses the surface. Thus, if the material is subjected to several identical loading cycles, then only the first loading sequence will involve plastic straining, and the stiffness of the rock will be correspondingly smaller in that sequence.

Also shown in Fig. 2.28 is the failure surface, which is defined as the surface that separates accessible states from inaccessible states. Note that for an ideally plastic material, the initial yield surface and the failure surface coincide. The use of only one parameter κ to describe hardening is, of course, a simplification, since there are six plastic strain components.

There are two common ways to relate κ to the plastic strain. One way is to assume that κ is a function of the total plastic strain. This is called *strain hardening* and can be expressed as

$$\kappa = \kappa\left(\int_S \mathrm{d}\varepsilon_{ij}^{\mathrm{p}}\right) \tag{2.78}$$

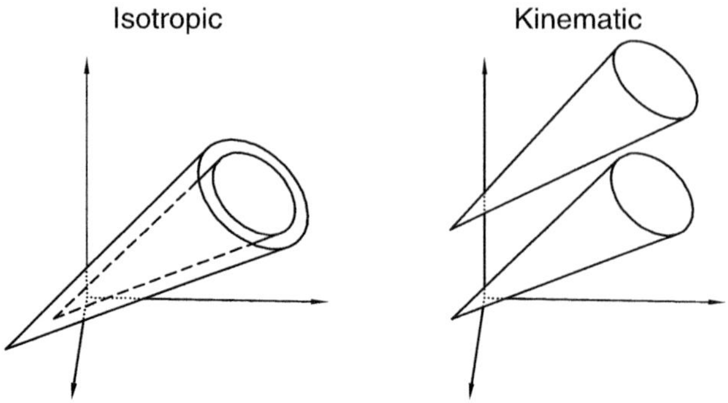

Figure 2.29 Isotropic and kinematic hardening in principal stress space.

where $\int_S$ symbolises integration over the stress path. Another way is relating κ to the total plastic work:

$$\kappa = \kappa\left(\int_S \sigma'_{ij} d\varepsilon^p_{ij}\right) \tag{2.79}$$

This is called *work hardening*.

Hardening is commonly decomposed into two main modes, isotropic hardening and kinematic hardening. Fig. 2.29 illustrates the difference between these concepts. *Isotropic* hardening means that the yield surface expands (or shrinks) in a uniform way about the hydrostatic axis. *Kinematic* hardening consists of a translation of the failure surface in stress space. In practice, hardening must be described by a combination of the modes or an even more complicated behaviour, in which different parts of the yield surface deform in different ways.

Before concluding, we will mention one important effect concerning hardening, which is normally observed experimentally. The *Baushinger effect* states that if a material is subjected to a given plastic strain in one direction, given a yield stress σ'_1, then the corresponding yield stress found when the specimen is later loaded in the opposite direction will be smaller than σ'_1. We see that the Baushinger effect may be obeyed by kinematic hardening, but not by isotropic hardening.

2.8.2 Soil mechanics

The failure mechanics described so far in this chapter was primarily developed for hard rocks. Weak sedimentary rocks are however intermediate between hard rocks and soils, and it has been found that a mechanical model developed for soils may be applied with success in some weakly cemented sedimentary rocks, such as chalk (Jones and Leddra, 1989) and shale (Marsden et al., 1989; Nakken et al., 1989). Here we will give an introduction to the main concepts of a mechanical model called "critical state soil mechanics". For more details and a more comprehensive description, see, for instance, Atkinson and Bransby (1978); Head (1984); Wood (1990).

Soil mechanics has been developed for systems with no or little cement between the individual grains. Clay is a material that fits into this description. In this section, we will take a look at the behaviour of a clay under isotropic compression and define some concepts commonly used in soil mechanics.

The *void ratio* (or *voids ratio*) e is the volume of voids V_{void} relative to the volume of the solid grains V_{solid} in the material:

$$e = \frac{V_{\text{void}}}{V_{\text{solid}}} \tag{2.80}$$

The *specific volume* υ is the total volume (grains + voids) divided by the volume of the solid grains, that is,

$$\upsilon = \frac{V_{\text{solid}} + V_{\text{void}}}{V_{\text{solid}}} = 1 + e \tag{2.81}$$

The specific volume and the voids ratio are related to the porosity ϕ by

$$\phi = \frac{e}{1+e} = \frac{\upsilon - 1}{\upsilon} \tag{2.82}$$

In a *drained test* the fluid pressure in the sample is controlled at a given value by allowing the fluid to enter or leave the sample (see Section 1.6). An *undrained test* is a test where no fluid is allowed to enter or escape from the sample during the test. The fluid pressure in the sample will change during an undrained test.

Fig. 2.30 schematically shows the result of isotropic loading of a clay under drained conditions. The plot shows the specific volume υ versus the logarithm of effective mean stress $p' = p - p_{\text{f}}$ (= the Terzaghi effective stress, see Section 1.6.4). Note that we use the soil mechanics convention of denoting the mean stress by p ($= \bar{\sigma}$ in Section 1.1.7). The reason for using logarithmic scale on the pressure axis is that the loading paths are close to straight lines in this case. As a sample is loaded, the specific volume will decrease along the line 1–2–3–4 in the figure. Assume now that we start with an identical sample and load it up to the point 2 and then unload it. The

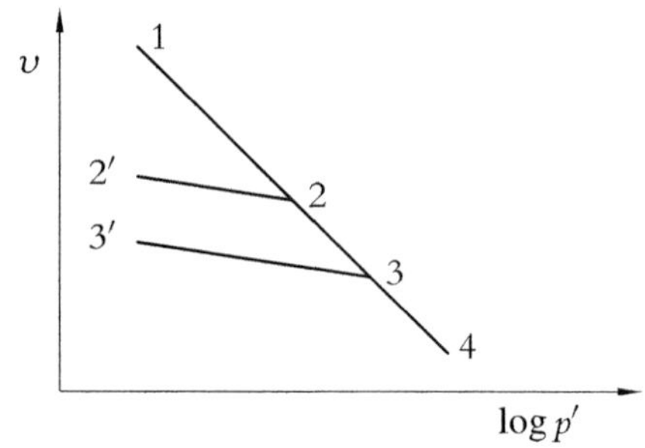

Figure 2.30 Isotropic compression of clay, schematically.

unloading path follows the line 2–2′, which has a lower slope. Reloading again from 2′, the path follows the line 2′–2 up to the point 2, and thereafter it follows the line of the first sample towards point 3. Unloading again after reaching point 3, the sample follows the line 3–3′, which has a slope equal to that of 2–2′.

The description above is idealised. In practice the lines will not be exactly straight, there will be some hysteresis along the lines 2–2′ and 3–3′, and the slopes of 2–2′ and 3–3′ will differ somewhat. Still, the description gives a good picture of the overall behaviour of clays. A main point to be learned is the following: the stiffness of a sample that is loaded above the highest pressure it has been exposed to before is lower than the stiffness of a sample that has seen higher pressures than the present state.

A sample that is at the highest stress it has ever experienced is said to be *normally consolidated*. The sample described above was normally consolidated whenever it was on the line 1–4 (Fig. 2.30). A sample that has been subjected to higher stresses before is said to be *overconsolidated*. The sample above was overconsolidated when it was on one of the lines 2–2′ or 3–3′ (Fig. 2.30). The highest stress a sample has been exposed to is commonly called the *preconsolidation stress*. The *overconsolidation ratio* of a sample is defined as the highest experienced stress divided by the current stress.

Triaxial tests on sandstones are usually run on drained samples, and plots showing stress versus strain (like Figs. 2.2 and 2.3) are used to present the results. Due to the low permeability, triaxial tests on a material like clay are usually run on undrained samples. It is also customary to analyse the test results in terms of the effective mean stress p' and the generalised shear stress q defined in Section 2.7.

Fig. 2.31 schematically shows a typical test on a normally consolidated clay. The loading path curves to the left, which means that the mean ef-

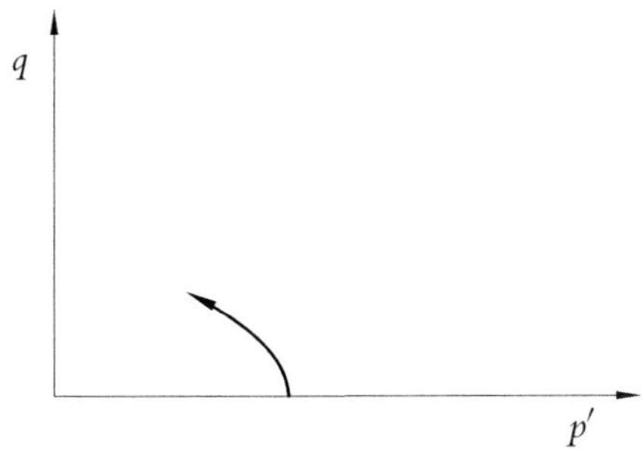

Figure 2.31 Loading path for an undrained test on a normally consolidated clay in the (p', q)-plane.

fective stress is decreasing. Thus the pore pressure increases throughout the test. For strongly overconsolidated samples, we expects the path to curve to the right as the sample approaches failure.

Typical behaviour during undrained triaxial testing for normally consolidated clays and strongly overconsolidated clays can be summarised as follows.

Normally consolidated clay:

1. No peak stress in the stress–strain diagram, that is, the shear stress does not fall below the previous value as the strain increases.
2. The sample contracts throughout the loading.
3. The loading path in a (p', q) plot curves to the left, that is, the pore pressure increases.

Strongly overconsolidated clay:

1. A peak is normally observable in the stress–strain diagram, that is, at some point the shear stress falls below the previous values as the strain increases.
2. The sample contracts initially, then dilates as failure is approaching.
3. The loading path in a (p', q) plot curves to the right near failure, that is, the pore pressure decreases.

The normally consolidated clays and strongly overconsolidated clays behave quite differently. We will now describe each of them in more detail.

2.8.2.1 Normally consolidated clays

Fig. 2.32 illustrates the stress paths for three *undrained* tests on samples of a normally consolidated clay, starting from three different isotropic compression levels. The arrows show the paths of the tests from the start of the

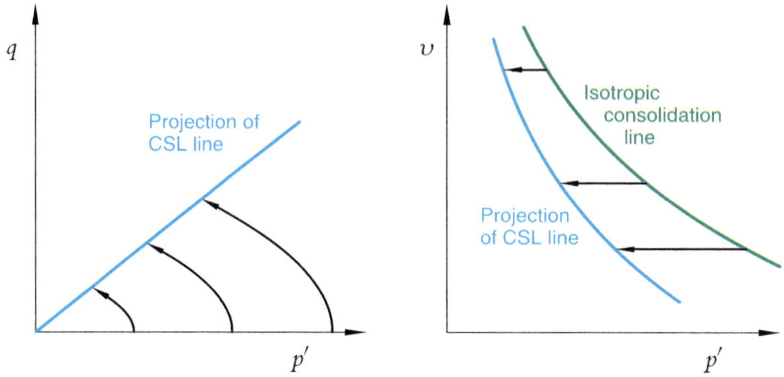

Figure 2.32 Undrained triaxial compression tests from three different isotropic compression levels on samples from a normally consolidated clay. CSL is the critical state line.

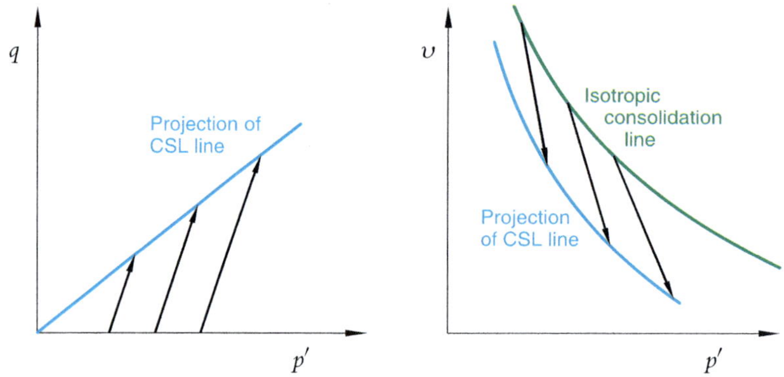

Figure 2.33 Drained triaxial compression tests from three different isotropic compression levels on samples from a normally consolidated clay.

triaxial phase to the failure state, that is, to the point where the stress–strain curve flattens out and large shear strains may occur at no increase in the shear stress. The loading paths in the (p', υ) plot are horizontal, since there is no change in specific volume during an undrained test. Note that the ends of the paths seem to fall on continuous "failure" lines in both plots.

Fig. 2.33 shows the corresponding stress paths for *drained* tests. Again, the end points are seen to fall on continuous lines. It has been found that the continuous failure lines defined by the ends of the tests in Figs. 2.32 and 2.33 are indeed the same lines to a fair approximation. If we plot the test results in a three-dimensional plot with (υ, p', q) as the axes, then we find that the failure lines in the (p', q)- and (p', υ)-planes are projections of a

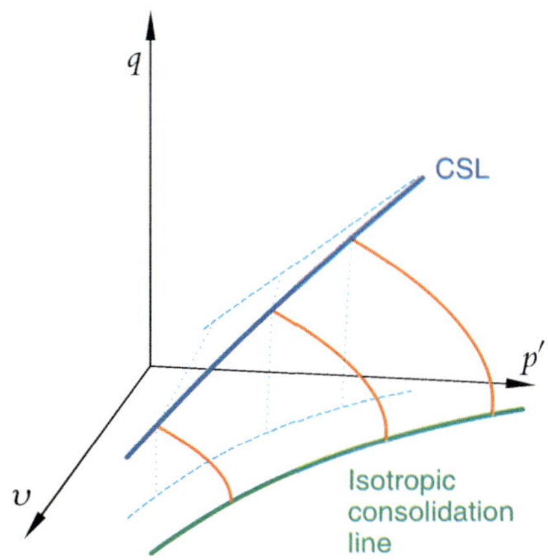

Figure 2.34 The Roscoe surface in the (υ, p', q)-space.

line in this three-dimensional space, the *critical state line* (CSL). The critical state line with its projections onto the (p', q)- and (p', υ)-planes are shown in Fig. 2.34. The critical state represents an ultimate state where large shear strains may occur at no change in shear stress.

A large number of drained tests on identical samples, starting from different isotropic compression levels, will define a surface in the (υ, p', q)-space. Similarly, undrained tests also define a surface in this space. Since the two surfaces have the critical state line and the isotropic consolidation line in common, it seems reasonable to assume that the two surfaces may in fact be the same. This assumption has been confirmed with reasonable precision from experiments. The surface is called the Roscoe surface. Its location in the (υ, p', q)-space is a characteristic property of the actual material. All tests on a normally consolidated sample of this material—drained, undrained or intermediate—will follow a path on the Roscoe surface and end at the critical state line at failure. The Roscoe surface is shown in Fig. 2.34.

2.8.2.2 Overconsolidated clays

For a normally consolidated clay, there is a unique relationship between the specific volume and the effective stress, defined by the normal consolidation line in the (p', υ)-plane. For an overconsolidated clay, there may be a range of voids ratios for a given effective stress, depending on the preconsolidation stress from which the material was unloaded. This means that whereas an undrained test on a normally consolidated material traverses a unique path

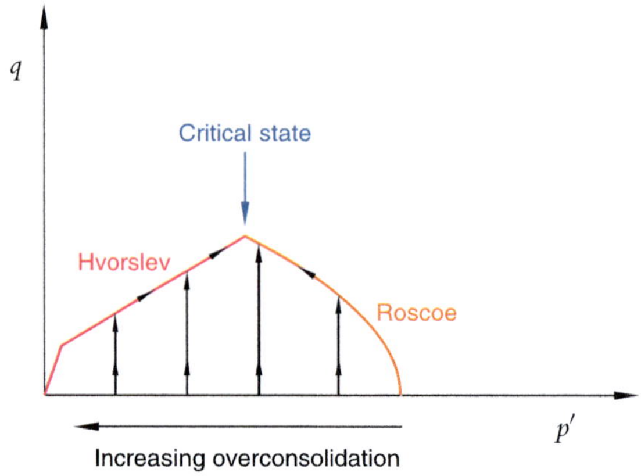

Figure 2.35 Undrained tests on samples with various overconsolidation ratios for a given specific volume, projected onto the (p', q)-plane. The change in slope for low p' corresponds to tensile stress states.

in the (υ, p', q)-space if the starting specific volume is given, there will be a range of possible paths for an overconsolidated material.

We now imagine that we do a series of undrained tests on a clay for a given specific volume but start at different effective hydrostatic stress levels corresponding to different overconsolidation ratios. Fig. 2.35 shows the (idealised) results of these tests, projected onto the (p', q)-plane. A slightly overconsolidated material moves more or less vertically until it reaches the Roscoe surface and then follows this surface until it reaches the critical state. (The curve is initially vertical only to the extent that the Skempton B coefficient is close to 1; otherwise, the curve tilts slightly to the right; see Section 1.6.7.) For strongly overconsolidated specimens, the movement is again vertical in the initial phases of the experiment. The path does not reach the Roscoe surface, but changes direction upon reaching a limiting curve, which like the Roscoe surface passes through the critical state. This limiting curve is nearly a straight line. In the figure, we have assumed that the sample cannot sustain tensile stresses, and the change on slope at low p' reflects this.

If similar tests are done for the same material at various specific volumes, we will find that they map out a surface in the (υ, p', q)-space, the Hvorslev surface. This surface plays a similar role for overconsolidated samples as the Roscoe surface does for normally (and slightly over-)consolidated samples. It is a limiting surface for all stress paths, not only undrained tests. The

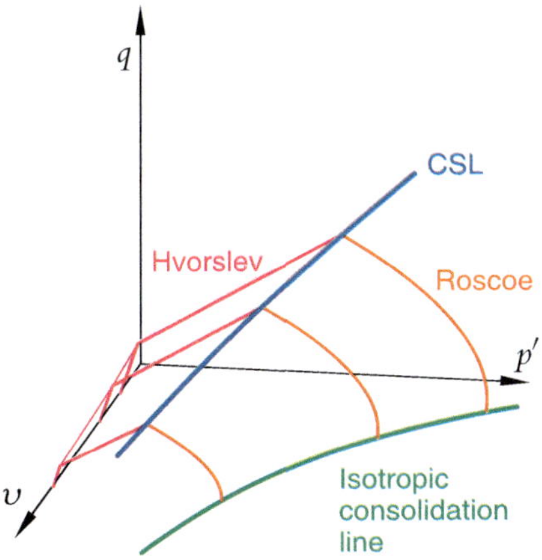

Figure 2.36 The Hvorslev and Roscoe surfaces in the (υ, p', q)-space.

Hvorslev and Roscoe surfaces are shown in Fig. 2.36. Note that they are joined together at the critical state line.

The behaviour of the material at the Hvorslev surface is less predictable than at the Roscoe surface. If the material behaved uniformly, then it would follow the Hvorslev surface up to the critical state line. However, overconsolidated samples often behave in a non-uniform way due to localisation effects such as the formation of shear bands, as discussed earlier in this chapter. This means that different parts of the sample take different loads, and thus the recorded behaviour is not indicative of the intrinsic material properties. As a result, the sample does not end up at the ultimate critical state as it would have if it had behaved uniformly.

2.8.3 Localisation

Complete failure of a rock sample usually implies that the sample breaks up into several pieces and falls apart. This implies that the rock at some point in the failure process ceases to behave as a homogeneous material even on the macroscopic scale. It also implies that failure is complete only within a few restricted areas of the rock sample, like fracture planes developed under tensile failure, and narrow shear bands developed under shear failure.

The formation and growth of localised failure zones occur when it is energetically favourable for cracks or other (pre-existing or randomly induced) defects to grow and interact with nearby defects. For tensile failure and unconfined shear failure, this is usually an unstable situation that rapidly leads to macroscopic failure of the sample. However, for failure under compressive stresses, the situation may lead to non-uniform deformation in a planar band within the sample, whereas the deformation outside the band remains homogeneous, and the material remains stable.

Rudnicki and Rice (1975) described the conditions for localisation of deformation into planar bands. The formation of such bands can be considered as a constitutive instability, closely linked to the plastic hardening of the material. It is found that for a material obeying associated plastic flow (see Section 2.8.1), the formation of shear bands can only occur after the peak stress point. For materials obeying non-associated plastic flow, such bands may also be formed prior to the peak stress point, at least for stress configurations close to pure shear.

At lower stress levels the rock sample is supposed to follow one unique path of uniform deformation. When the conditions for the formation of a localised deformation band are fulfilled, there are at least two possible orientations of the band. Thus further deformation of the rock sample will follow one out of several equivalent paths. This spontaneous splitting of the deformation path is called *bifurcation*, and the point where the conditions for localised deformation are first fulfilled is called the bifurcation point (see Vardoulakis and Sulem, 1995).

Other forms of non-uniform deformation are also possible, for instance, surface buckling (Biot, 1965; Vardoulakis, 1984), which implies that the material spontaneously changes its shape when the conditions for such deformation are fulfilled. It has been demonstrated (Mollema and Antonellini, 1996; Olsson, 1999; Olsson and Holcomb, 2000) that also compaction can be localised, in the so-called *compaction bands*. Such bands may act as barriers for fluid flow and may therefore be of importance under given conditions by changing flow paths or altering pore pressure gradients and thereby also the effective stresses. (See also Borja and Aydin, 2004; Fossen et al., 2007.)

2.8.4 Liquefaction

Failure of a highly porous rock results in a volume reduction and a corresponding reduction in the pore space. If the rock fails under undrained conditions, the pore space reduction will lead to increasing pore pressure. This implies reduced effective stresses and a corresponding reduction in confinement. In extreme situations—for a poorly consolidated (or previously damaged) rock and a sufficient increase in pore pressure—the grains are pulled apart, and the pore fluid is taking over the entire load, turning the rock into a suspension of uncemented particles. This is a highly unstable condition, which may result in very large deformations driven by the static shear stress.

This extreme condition is called liquefaction. It is sometimes observed in connection to earthquakes and is highly noticeable due to its sudden appearance and quick development, and the large distance the liquefied material may move (see e.g. Kramer, 1996; Jefferies and Been, 2016).

Liquefaction may occur if the mean effective stress $\overline{\sigma}' = \overline{\sigma} - p_\mathrm{f}$ becomes tensile, that is, if

$$\overline{\sigma} < p_\mathrm{f} \tag{2.83}$$

Note that tensile failure may occur if the smallest effective principal stress becomes tensile, which is a condition more easily fulfilled.

Liquefaction is usually a short-lived condition. The liquefied rock is highly permeable so that pore fluid may bleed off to the surroundings, allowing the pore pressure to drop accordingly. When Eq. (2.83) is no longer fulfilled, the effective confinement of the grain framework reappears, and the rock regains shear strength.

2.9 Failure of anisotropic and fractured rocks

In the discussion so far, it has been assumed that the material properties are isotropic, so that the strength is independent of the orientation of the applied stresses. In reality, this may not be the case. Isotropy is often assumed just for simplicity.

We will here distinguish between intrinsic anisotropy and structural anisotropy. Intrinsic anisotropy implies that an otherwise homogeneous material has different mechanical properties in different directions. Structural anisotropy is associated with localised discontinuities like planes of weakness or fractures. Several failure criteria have been proposed to describe anisotropic strength, including mathematical generalisations of isotropic criteria, and empirical criteria where cohesion and friction angle are allowed to vary with the orientation. We will focus on a third group of criteria, where strength anisotropy is assumed to originate from physical mechanisms associated with fractures or bedding planes.

2.9.1 Intrinsic anisotropy and the failure surface

For isotropic materials, the principal axes of strain and the principal axes of stress coincide. For anisotropic materials this is not the case in general. Still it is possible, at least in principle, to determine a failure surface in the stress space by performing experiments along different stress paths. However, such a failure surface will depend on the orientation of the anisotropic material relative to the principal stress axes. There is no longer a unique failure surface that characterises the material behaviour. The concept of a failure surface is thus less convenient for visualising failure properties.

2.9.2 The plane of weakness model

The plane of weakness model (Jaeger, 1960) is a simple approach to strength anisotropy. The model assumes that the inherent strength is the same in all directions, except for one set of parallel planes where the strength is lower. Since the bedding planes in sedimentary rocks may be planes of weakness, the model has a physical basis, and it is therefore quite important in spite of its simplicity. Obviously, the model also applies to a set of parallel fracture planes.

Assume that we run a series of triaxial tests on a material with a set of parallel planes of weakness. According to Mohr–Coulomb-type failure criteria, it is clear that the weak planes have no effect on strength if we choose the axis of the plug normal or parallel to the planes, since we have no shear stress on the weak planes in these cases. (Recall though that the assumption of shear failure may not be correct—see the discussion of uniaxial compressive and tensile strength in Section 2.3.2.) It is also clear that for some intermediate orientations, we expect the weak planes to fail at a lower stress than expected for the intact material.

Consider a (σ', τ) plot for this material, as illustrated in Fig. 2.37. The material has two failure criteria—one ordinary, isotropic criterion and one for the weak planes—and correspondingly two failure lines. The weak plane criterion is given by the cohesion $S_{0\mathrm{w}}$ and the friction angle φ_w. The corresponding failure angle is given as

$$\beta_\mathrm{w} = \frac{\pi}{4} + \frac{\varphi_\mathrm{w}}{2} \tag{2.84}$$

If the stress state in the rock sample is such that Mohr's circle touches the failure line for the weak planes, as shown in Fig. 2.37, then the material will fail only if the sample is oriented so that the angle θ between the major principal stress and the normal to the weak planes is equal to β_w. If the sample has a different orientation, it can take a higher value for σ_1' and a correspondingly a higher shear stress. Then the situation is as shown in Fig. 2.38, where Mohr's circle intersects the weak plane failure line in two points. At this stress level the sample will fail only if $\theta = \beta_1$ or $\theta = \beta_2$. For

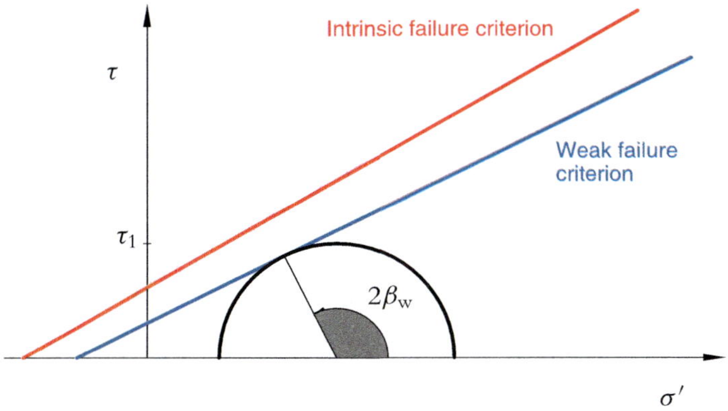

Figure 2.37 (σ', τ) plot for a material containing a plane of weakness. The illustrated stress configuration represents the lowest strength possible for any orientation of this material.

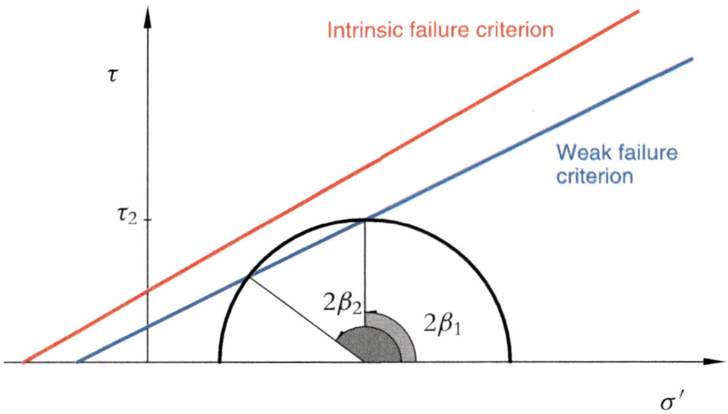

Figure 2.38 (σ', τ) plot for a material containing a plane of weakness. The illustrated stress configuration is accessible only for some orientations of the material.

any orientation within the range $\beta_1 < \theta < \beta_2$, the sample will fail along the weak planes at a lower stress level.

Finally, if the stress state is such that Mohr's circle touches the failure line for the isotropic criterion, as shown in Fig. 2.39, then the sample will fail for any orientation of the weak planes, except for orientations where $\beta_{\min} < \theta < \beta_{\max}$, in which case the sample will already have failed along the weak planes at a lower stress level.

For the stress configuration considered here ($\sigma'_2 = \sigma'_3$), we can express the two failure criteria as follows (see Eqs. (2.14)–(2.17)):

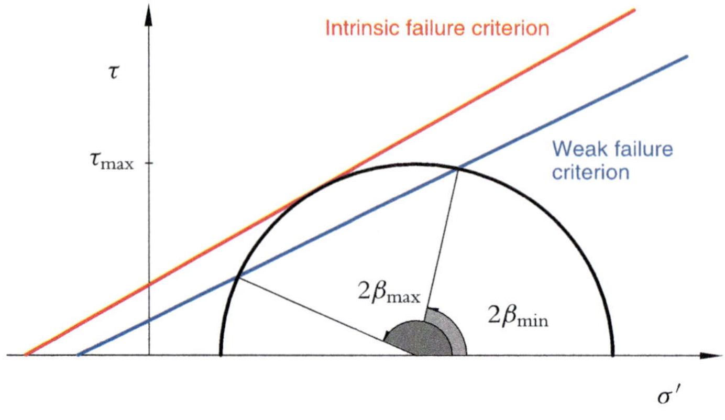

Figure 2.39 (σ', τ) plot for a material containing a plane of weakness. The illustrated stress configuration represents the highest strength possible for any orientation of this material.

Isotropic failure criterion (cohesion S_0, friction angle φ)

$$\sigma_1' - \sigma_3' = 2\frac{S_0 \cos\varphi + \sigma_3' \sin\varphi}{1 - \sin\varphi} \tag{2.85}$$

Weak plane failure criterion

$$\sigma_1' - \sigma_3' = 2\frac{S_{0w} \cos\varphi_w + \sigma_3' \sin\varphi_w}{\sin 2\theta \cos\varphi_w - (\cos 2\theta + 1)\sin\varphi_w} = \frac{S_{0w} \cos\varphi_w + \sigma_3' \sin\varphi_w}{\sin(\theta - \varphi_w)\cos\theta} \tag{2.86}$$

The criterion that predicts the lowest strength for a given orientation θ is always the relevant criterion. Fig. 2.40 illustrates how the strength varies with the orientation of the sample for such a material.

Observe that Eq. (2.86) diverges for $\theta = \varphi_w$. This is because the shear stress–normal stress trajectory moves parallel to the failure criterion in this case. In other words, as loading increases the shear stress, the normal stress increases sufficiently to counteract it. The same applies for smaller θ, where the point moves away from the failure criterion. (The negative algebraic solutions for $\theta < \varphi_w$ are clearly unphysical.) In the limit $\varphi_w = 0$, the above is the same as the fact that loading normal to the weak planes will not give weak plane failure.

Note that for a general, three-dimensional stress state, the shear and normal stresses over the weak planes are given by Eqs. (1.36)–(1.37), and the weak plane failure criterion becomes much more complicated. This means that in general, weak plane failure depends on all three principal

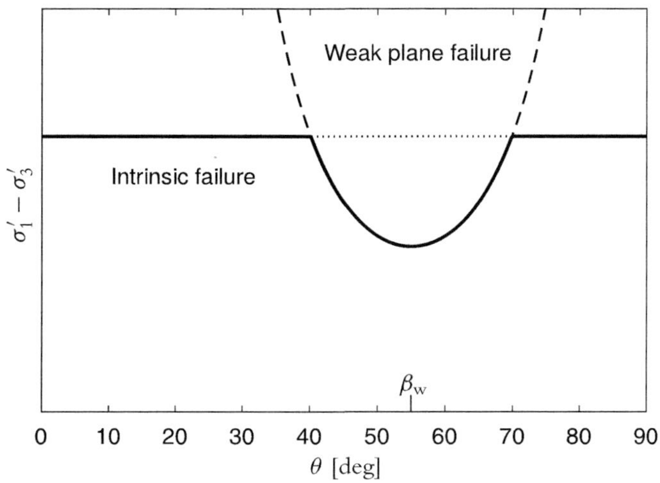

Figure 2.40 Sketch of failure stress as function of the angle θ between the major principal stress and the normal to the weak planes.

stresses even in a Mohr–Coulomb context. This is because the failure plane is fixed, in contrast to intrinsic failure, where the "freedom to choose" the orientation of the failure plane eliminates the influence of the intermediate principal stress.

Experimental data often show a smoother dependence on orientation than predicted by the plane of weakness model as shown in Fig. 2.40. Fjær and Nes (2014) suggested that this discrepancy could be explained by weak patches within the weak planes. The weak patches are unable to transfer the same shear stresses as the surrounding rock, and hence the surrounding rock has to take a larger portion of the total load. This reduces the effective strength both for weak plane failure and for intrinsic failure, depending on orientation. In this "patchy weakness model" the intrinsic failure criterion (corresponding to Eq. (2.85)) is given by

$$\sigma_1' - \sigma_3' = 2\frac{S_0 \cos\varphi + \sigma_3' \sin\varphi}{1 - \sin\varphi}(1 - \eta_p \sin^2 2\theta) \tag{2.87}$$

whereas the weak plane failure criterion (corresponding to Eq. (2.86)) is given by

$$\sigma_1' - \sigma_3' = 2\frac{S_{0w} \cos\varphi_w + \sigma_3' \sin\varphi_w}{\sin 2\theta \cos\varphi_w - (\cos 2\theta + 1)\sin\varphi_w}(1 - \eta_p \sin^2 2\theta) \tag{2.88}$$

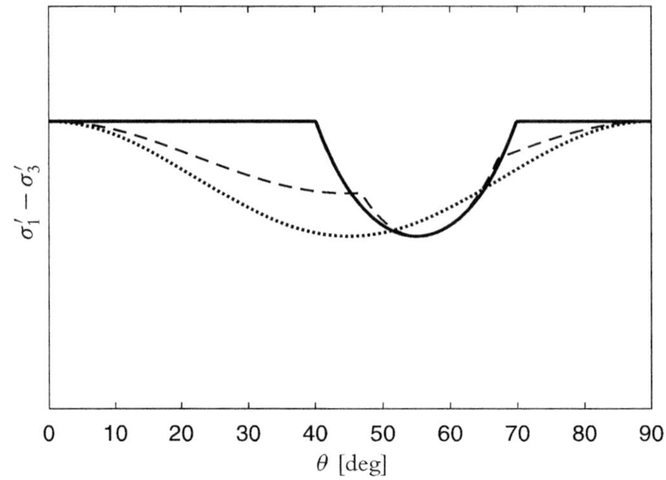

Figure 2.41 Failure stress as function of the angle θ for various forms of the patchy weakness criterion (cf. Fig. 2.40). Solid line: $\eta_p = 0$; dashed line: $\eta_p > 0$; dotted line: $\eta_p > 0$ while S_{0w} and/or φ_w are sufficiently large so that failure is not triggered along the weak plane.

The *patchiness parameter* η_p is a measure of the amount of weak patches. Note that η_p tends to decline with increasing σ_3'. The plane of weakness model is a limit case of the patchy weakness model (for $\eta_p = 0$). Fig. 2.41 illustrates how anisotropic strength may take various forms according to the patchy weakness model.

2.9.3 Fractured rock

The behaviour of a fractured rock depends both on the properties of the fractures and on the properties of the intact rock. Generally, fractured rocks will be much weaker than the corresponding intact rock, as the resistance against shear failure is considerably less for an already existing fracture.

To predict the behaviour of a fractured rock, it may be necessary to apply numerical simulation methods to the actual fracture system. Alternatively, effective rock properties may be assigned to a representative volume that is much larger than the fracture spacing.

Hoek and Brown (1980) derived an empirical failure criterion for fractured rocks:

$$\sigma_1' = \sigma_3' + \sqrt{m_b C_0 \sigma_3' + s C_0^2} \tag{2.89}$$

where C_0 is the uniaxial compressive strength of intact (unfractured) rock, and m_b and s are constants depending on the rock properties and the fracture system. The uniaxial compressive strength of the fractured rock is given by

$$C_{0f} = \sqrt{sC_0^2} \tag{2.90}$$

Obviously, $s = 1$ for intact rock, so that $1 - s$ represents the degree of fracturing. A comparison of the Hoek–Brown criterion to the Mohr–Coulomb criterion (Eq. (2.22)) indicates that m_b has some relation to the internal friction (see Eqs. (2.13) and (2.8)); however, there is no direct correspondence between Eqs. (2.89) and (2.22).

A typical application of this criterion implies that the value m_i is determined for m_b by a set of triaxial tests on intact parts of the rock, whereas s is estimated by visual inspection, following a specific characterisation scheme (Hoek and Brown, 1997). The actual value of m_b is obtained as

$$m_b \approx m_i s^{0.321} \tag{2.91}$$

For highly fractured rocks, a generalised version of the Hoek–Brown criterion applies (Hoek and Brown, 1997):

$$\sigma_1' = \sigma_3' + C_0 (m_b \frac{\sigma_3'}{C_0} + s)^a \tag{2.92}$$

where $s \to 0$ and $a \to 0.65$. Note that Eq. (2.92) is identical to Eq. (2.89) when $a = 0.5$.

2.10 Stress history effects

In our discussion of failure, we have in this chapter mostly assumed that the rock will fail when it has been brought to a stress state that fulfils a failure criterion, regardless *how* it was brought to this state. We have seen, however, that the yield surface may move as a result of plastic deformation (Section 2.8.1) and that the presence of fractures—which may be the result of previous loads—reduces the strength of the rock (Section 2.9.3). It was also described in Section 1.9.4 that creep may eventually lead to failure. These examples tell us that not only the current stress state, but also the stress history may have an effect on rock failure.

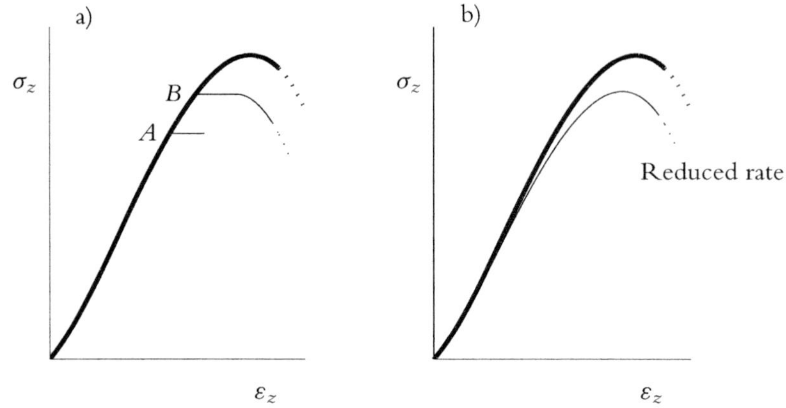

Figure 2.42 Schematic illustration of a triaxial test, showing the effects of: a) creep, b) strain rate.

2.10.1 Rate effects and delayed failure

It was shown in Section 1.9.4 that a rock may deform continuously under a constant shear load. The effect is more significant the closer the actual load is to the shear strength of the rock.

Consider a rock sample that has been tested in a standard triaxial test (Fig. 2.42a). In a second test on an identical sample, we stop loading at some level A before the peak stress of the standard test is reached and maintain the stress at this level. When the stress reached level A, the rock may already have suffered some damage, so that on a microscopic scale, some parts of the rock are intact while other parts are not. While the stress is kept constant, the rock creeps, and the increasing shear deformation implies that the shear load on the intact parts of the rock increases, and the areas of failure are growing. Thus parts of the deformation and corresponding damage that occurred at stress levels above A in the standard test occurs while the stress is maintained at A in the second test, only delayed.

When the loading is stopped at a relatively low level, like A, the creeping process declines and eventually stops after a limited amount of delayed deformation. For a sample that is brought to a sufficiently high stress level (like B, Fig. 2.42a) before the loading is stopped, the creeping process is unstable, and it will after a while accelerate and result in failure (accelerating creep, see Section 1.9). The time t_{fail} from the constant stress level σ_0 has been established until failure occurs depends on σ_0, as expressed by Main (2000):

$$t_{\text{fail}} \propto \sigma_0^{-n} \tag{2.93}$$

The fact that all of the damage that may occur at a given stress level does not occur immediately points directly to an effect of loading rate on rock strength. Clearly, if the rock is loaded at a lower rate, then more damage may occur per unit of stress increase, and the rock will essentially be weaker (Fig. 2.42b). Based on a model for subcritical crack growth supported by experimental data, Sano et al. (1981) derived a relation between peak stress $\sigma_{\max}$ and loading rate $\dot{\sigma}$:

$$\sigma_{\max} \propto \dot{\sigma}^{1/(n+1)} \tag{2.94}$$

The "stress corrosion index" n, which appears in both Eqs. (2.93) and (2.94), is typically in the range 30–60 for sandstone, whereas significantly higher values have been reported for limestone. Thus the peak stress under constant loading rate only weakly depends on the loading rate, whereas the time to failure under constant stress depends strongly on the stress level.

Note that effects of loading rate may also be seen in hydrostatic loading, even if creep is primarily thought to be related to shear deformation. This can be ascribed to local fluctuations in the stresses that will occur in a heterogeneous material like a porous rock.

2.10.2 Fatigue

Excessive deformation at a limited stress level, as described in the previous paragraph, may be provoked and accelerated if the stress is released and reapplied over and over again. For each unloading–reloading cycle, a little more damage is added, and the rock sample deforms correspondingly. Eventually, the rock may be destroyed by *fatigue* failure. The number of cycles required to bring a rock to failure increases dramatically if the peak stress level for the loading cycles is reduced.

If the peak stress of the loading cycles is kept sufficiently low, then the rock may not fail even after a very large number of cycles. Still, the cyclic loading may have affected the strength of the rock, so that in a subsequent failure test the rock will fail at a stress level different from its normal strength. Typically, cyclic loading will reduce the strength of a rock sample; however, increased strength due to cyclic loading is also reported (Ray et al., 1999).

References

Al-Ajmi, A.M., Zimmerman, R.W., 2005. Relation between the Mogi and the Coulomb failure criteria. Int. J. Rock Mech. Min. Sci. 42 (3), 431–439. https://doi.org/10.1016/j.ijrmms.2004.11.004.

Atkinson, J.H., Bransby, P.L., 1978. The Mechanics of Soils. An Introduction to Critical State Soil Mechanics. McGraw-Hill, London.

Baud, P., Zhu, W., Wong, T-f., 2000. Failure mode and weakening effect of water on sandstone. J. Geophys. Res. 105 (B7), 16371–16389. https://doi.org/10.1029/2000JB900087.

Biot, M.A., 1965. Mechanics of Incremental Deformations. Wiley, New York.

Bishop, A.W., 1959. The principle of effective stress. Tekn. Ukebl. 39, 859–863.

Borja, R.I., Aydin, A., 2004. Computational modeling of deformation bands in granular media. I. Geological and mathematical framework. Comput. Methods Appl. Mech. Eng. 193, 2667–2698. https://doi.org/10.1016/j.cma.2003.09.019.

Boutéca, M., Guéguen, Y., 1999. Mechanical properties of rocks: pore pressure and scale effects. Oil Gas Sci. Technol. – Rev. IFP 54 (6), 703–714. https://doi.org/10.2516/ogst:1999060.

Boutéca, M.J., Sarda, J.-P., Vincké, O., 2000. Constitutive law for permeability evolution of sandstones during depletion. SPE58717. In: SPE International Symposium on Formation Damage. Lafayette, LA, USA.

Brace, W.F., 1960. An extension of Griffith theory of fracture to rocks. J. Geophys. Res. 65 (10), 3477–3480. https://doi.org/10.1029/JZ065i010p03477.

Brignoli, M., Papamichos, E., Santarelli, F.J., 1995. Capillary effects in sedimentary rocks: application to reservoir water-flooding. In: Proc. 35th U.S. Symposium on Rock Mechanics. Lake Tahoe, CA, USA, pp. 619–625.

Broch, E., 1974. The influence of water on some rock properties. In: Proc. 3rd Congress, ISRM. Part A, pp. 33–38.

Chen, W.F., Han, D.J., 1988. Plasticity for Structural Engineers. Springer-Verlag, New York.

Colmenares, L.B., Zoback, M.D., 2002. A statistical evaluation of intact rock failure criteria constrained by polyaxial test data for five different rocks. Int. J. Rock Mech. Min. Sci. 39, 695–729. https://doi.org/10.1016/S1365-1609(02)00048-5.

David, C., Dautriat, J., Sarout, J., Delle Piane, C., Menéndez, B., Macault, R., Bertauld, D., 2015. Mechanical instability induced by water weakening in laboratory fluid injection tests. J. Geophys. Res. 120 (B6), 4171–4188. https://doi.org/10.1002/2015JB011894.

Detournay, E., 1986. Elastoplastic model of a deep tunnel for a rock with variable dilatancy. Rock Mech. Rock Eng. 19, 99–108. https://doi.org/10.1007/BF01042527.

Detournay, E., Cheng, A.H.-D., 1988. Poroelastic response of a borehole in a non-hydrostatic stress field. Int. J. Rock Mech. Min. Sci. Geomech. Abstr. 25, 171–182. https://doi.org/10.1016/0148-9062(88)92299-1.

DiMaggio, F.L., Sandler, I.S., 1971. Material model for granular soils. J. Eng. Mech. Div. Am. Soc. Civ. Eng. 97, 935–950.

Drucker, D.C., 1950. Some implications of work hardening and ideal plasticity. Q. Appl. Math. 7, 411–418. https://doi.org/10.1090/qam/34210.

Duda, M., Renner, J., 2013. The weakening effect of water on the brittle failure strength of sandstone. Geophys. J. Int. 192 (3), 1091–1108. https://doi.org/10.1093/gji/ggs090.

Ewy, R.T., 1999. Wellbore-stability predictions by use of a modified Lade criterion. SPE Drill. Completion 14, 85–91. https://doi.org/10.2118/56862-PA.

Fjær, E., Nes, O.-M., 2014. The impact of heterogeneity on the anisotropic strength of an outcrop shale. Rock Mech. Rock Eng. 47, 1603–1611. https://doi.org/10.1007/s00603-014-0598-5.

Fjær, E., Ruistuen, H., 2002. Impact of the intermediate principal stress on the strength of heterogeneous rock. J. Geophys. Res. 107 (B2), 2032. https://doi.org/10.1029/2001JB000277.

Forsans, T.M., Schmitt, L., 1994. Capillary forces: the neglected factor in shale instability studies? In: Rock Mechanics in Petroleum Engineering. Delft, Netherlands.

Fossen, H., Schultz, R.A., Shipton, Z.K., Mair, K., 2007. Deformation bands in sandstone: a review. J. Geol. Soc. Lond. 164, 755–769. https://doi.org/10.1144/0016-76492006-036.
Garg, S.K., Nur, A., 1973. Effective stress laws for fluid-saturated porous rocks. J. Geophys. Res. 78 (26), 5911–5921. https://doi.org/10.1029/JB078i026p05911.
Griffith, A.A., 1921. The phenomena of rupture and flow in solids. Phil. Trans. R. Soc. A 221, 163–198. https://doi.org/10.1098/rsta.1921.0006.
Guenot, A., 1989. Borehole breakouts and stress fields. Int. J. Rock Mech. Min. Sci. Geomech. Abstr. 26, 185–195. https://doi.org/10.1016/0148-9062(89)91968-2.
Head, K.H., 1984. Manual of Soil Laboratory Testing, vols. 1–3. ELE International Ltd, London.
Hoek, E., Brown, E.T., 1980. Underground Excavations in Rocks. The Institution of Mining and Metallurgy, London.
Hoek, E., Brown, E.T., 1997. Practical estimates of rock strength. Int. J. Rock Mech. Min. Sci. 34 (8), 1165–1186. https://doi.org/10.1016/S1365-1609(97)80069-X.
Jaeger, J.C., 1960. Shear failure of anisotropic rocks. Geol. Mag. 97, 65–72. https://doi.org/10.1017/S0016756800061100.
Jaeger, J.C., Cook, N.G.W., 1979. Fundamentals of Rock Mechanics, third edition. Chapman and Hall, London.
Jaeger, J.C., Cook, N.G.W., Zimmerman, R.W., 2007. Fundamentals of Rock Mechanics, fourth edition. Blackwell Publishing.
Jefferies, M., Been, K., 2016. Soil Liquefaction. A Critical State Approach, second edition. CRC Press.
Jones, M.E., Leddra, M.J., 1989. Compaction and flow of porous rocks at depth. In: Maury, V., Fourmaintraux, D. (Eds.), Rock at Great Depth. Balkema, Rotterdam, pp. 891–898.
Kramer, S.L., 1996. Geotechnical Earthquake Engineering. Prentice-Hall, New Jersey.
Lade, P.V., 1977. Elasto-plastic stress-strain theory for cohesionless soil with curved yield surfaces. Int. J. Solids Struct. 13, 1019–1035. https://doi.org/10.1016/0020-7683(77)90073-7.
Lade, P.V., 1993. Rock strength criteria: the theories and the evidence. In: Hudson, J.A. (Ed.), Comprehensive Rock Engineering. Principles, Practice & Projects, vol. 1. Pergamon Press, London, pp. 255–284.
Li, H.C., de Bruyn, P.L., 1966. Electrokinetic and adsorption studies on quartz. Surf. Sci. 5 (2), 203–220. https://doi.org/10.1016/0039-6028(66)90082-3.
Main, I.G., 2000. A damage mechanics model for power-law creep and earthquake aftershock and foreshock sequences. Geophys. J. Int. 142 (1), 151–161. https://doi.org/10.1046/j.1365-246x.2000.00136.x.
Marsden, J.R., Wu, B., Hudson, J.A., Archer, J.S., 1989. Investigation of peak rock strength behaviour for wellbore stability prediction. In: Maury, V., Fourmaintraux, D. (Eds.), Rock at Great Depth. Balkema, Rotterdam, pp. 753–760.
McClintock, F.A., Walsh, J.B., 1962. Friction on Griffith cracks under pressure. In: Proc. Fourth U.S. Nat. Congress of Appl. Mech., pp. 1015–1021.
Mogi, K., 1971. Fracture and flow of rocks under high triaxial compression. J. Geophys. Res. 76 (5), 1255–1269. https://doi.org/10.1029/JB076i005p01255.
Mollema, P.N., Antonellini, M.A., 1996. Compaction bands: a structural analog for anti-mode I cracks in aeolian sandstone. Tectonophysics 267, 209–228. https://doi.org/10.1016/S0040-1951(96)00098-4.
Murphy III, W.F., Winkler, K.W., Kleinberg, R.L., 1984. Contact microphysics and viscous relaxation in sandstones. In: Johnson, D.L., Sen, P.N. (Eds.), Physics and Chemistry of Porous Media. Am. Inst. Phys., pp. 176–190.
Murrel, S.A.F., 1963. A criterion for brittle fracture of rocks and concrete under triaxial stress and the effect of pore pressure on the criterion. In: Fairhurst, C. (Ed.), Rock Mechanics. Pergamon Press, Oxford, pp. 563–577.

Nakken, S.J., Christensen, T.L., Marsden, R., Holt, R.M., 1989. Mechanical behaviour of clays at high stress levels for wellbore stability applications. In: Maury, V., Fourmaintraux, D. (Eds.), Rock at Great Depth. Balkema, Rotterdam, pp. 141–148.

Nordal, S., Jostad, H.P., Kavli, A., Grande, L., 1989. A coulombian soil model applied to an offshore platform. In: Proc. 12th Int. Conf. Soil Mechanics and Foundation Engineering, vol. 4. Rio de Janeiro, pp. 471–474.

Olsson, W.A., 1999. Theoretical and experimental investigation of compaction bands in porous rock. J. Geophys. Res. 104 (B4), 7219–7228. https://doi.org/10.1029/1998JB900120.

Olsson, W.A., Holcomb, D.J., 2000. Compaction localization in porous rock. Geophys. Res. Lett. 27 (21), 3537–3540. https://doi.org/10.1029/2000GL011723.

Papamichos, E., Brignoli, M., Santarelli, F.J., 1997. An experimental and theoretical study of a partially saturated collapsible rock. Mech. Cohes.-Frict. Mater. 2 (3), 251–278. https://doi.org/10.1002/(SICI)1099-1484(200001)5:1<1::AID-CFM74>3.0.CO;2-H.

Papamichos, E., Tronvoll, J., Vardoulakis, I., Labuz, J.F., Skjærstein, A., Unander, T.E., Sulem, J., 2000. Constitutive testing of Red Wildmoor sandstone. Mech. Cohes.-Frict. Mater. 5 (1), 1–40. https://doi.org/10.1002/(SICI)1099-1484(200001)5:1<1::AID-CFM74>3.0.CO;2-H.

Parks, G.A., 1984. Surface and interfacial free energies of quartz. J. Geophys. Res. 89 (B6), 3997–4008. https://doi.org/10.1029/JB089iB06p03997.

Ray, S.K., Sarkar, M., Singh, T.N., 1999. Effect of cyclic loading and strain rate on the mechanical behaviour of sandstone. Int. J. Rock Mech. Min. Sci. 36 (4), 543–549. https://doi.org/10.1016/S0148-9062(99)00016-9.

Risnes, R., Haghighi, H., Korsnes, R.I., Natvik, O., 2003. Chalk–fluid interactions with glycol and brines. Tectonophysics 370, 213–226. https://doi.org/10.1016/S0040-1951(03)00187-2.

Rozhko, A.Y., 2020. Effects of pore fluids on quasi-static shear modulus caused by pore-scale interfacial phenomena. Geophys. Prospect. 68 (2), 631–656. https://doi.org/10.1111/1365-2478.12864.

Rudnicki, J.W., Rice, J.R., 1975. Conditions for the localization of deformation in pressure-sensitive dilatant materials. J. Mech. Phys. Solids 23 (6), 371–394. https://doi.org/10.1016/0022-5096(75)90001-0.

Røyne, A., Bisschop, J., Dysthe, D.K., 2011. Experimental investigation of surface energy and subcritical crack growth in calcite. J. Geophys. Res. 116 (B4), 1–10. https://doi.org/10.1029/2010JB008033.

Sano, O., Ito, I., Terada, M., 1981. Influence of strain rate on dilatancy and strength of Oshima granite under uniaxial compression. J. Geophys. Res. 86 (B10), 9299–9311. https://doi.org/10.1029/JB086iB10p09299.

Tittmann, B.R., Clark, V.A., Richardson, J.M., Spencer, T.W., 1980. Possible mechanism for seismic attenuation in rocks containing small amounts of volatiles. J. Geophys. Res. 85 (B10), 5199–5208. https://doi.org/10.1029/JB085iB10p05199.

Tutuncu, A.N., Sharma, M.M., 1992. The influence of fluids on grain contact stiffness and frame moduli in sedimentary rocks. Geophysics 57 (12), 1571–1582. https://doi.org/10.1190/1.1443225.

van Eekelen, H.A.M., 1980. Isotropic yield surfaces in three dimensions for use in soil mechanics. Int. J. Numer. Anal. Methods Geomech. 4 (1), 89–101. https://doi.org/10.1002/nag.1610040107.

Vardoulakis, I., 1984. Rock bursting as a surface instability problem. Int. J. Rock Mech. Min. Sci. Geomech. Abstr. 21 (3), 137–144. https://doi.org/10.1016/0148-9062(84)91531-6.

Vardoulakis, I., Sulem, J., 1995. Bifurcation in Geomechanics. Blackie Academic & Professional, Glasgow.

Vermeer, P.A., de Borst, R., 1984. Non-associated plasticity for soils, concrete and rocks. Heron 29, 1–64.
von Mises, R., 1928. Mechanik der plastischen Formänderung von Kristallen. Z. Angew. Math. Mech. 8 (3), 161–185. https://doi.org/10.1002/zamm.19280080302.
Vutukuri, V.S., Lama, R.D., Saluja, S.S., 1974. Mechanical Properties of Rocks, vols. I–IV. Trans Tech Publications, Clausthal.
Wiebols, G.A., Cook, N.G.W., 1968. An energy criterion for the strength of rock in polyaxial compression. Int. J. Rock Mech. Min. Sci. 5 (6), 529–549. https://doi.org/10.1016/0148-9062(68)90040-5.
Wong, T.-f., David, C., Zhu, W., 1997. The transition from brittle faulting to cataclastic flow in porous sandstones: mechanical deformation. J. Geophys. Res. 102 (B2), 3009–3025. https://doi.org/10.1029/96JB03281.
Wood, D.M., 1990. Soil Behaviour and Critical State Soil Mechanics. Cambridge University Press, New York.
Zhang, J., Wong, T.-f., Davis, D.M., 1990. Micromechanics of pressure-induced grain crushing in porous rocks. J. Geophys. Res. 95 (B1), 341–352. https://doi.org/10.1029/JB095iB01p00341.

CHAPTER 3

Geological aspects of petroleum related rock mechanics

From a practical rock mechanics view point, the present state of a rock and its present mechanical properties are of interest. The long and in many cases complicated process from an initial state as a loose sediment to the present state as a rock will however clearly affect rock mechanical behaviour.

A sedimentary basin may be exposed to sedimentary subsidence, sea-level changes and tectonic forces creating repeated cycles of elevation and depression, in addition to erosion, changes in sedimentary environment, changes in sedimentation rate, solution and precipitation of cementing material etc. All these effects complicate the geological description of the sedimentary basin. These geological activities and events affect not only current rock mechanical properties, but also current boundary conditions in terms of in situ stresses and pore pressure.

Therefore knowledge of geological processes is valuable in rock engineering. Although on a totally different time scale and length scale, geological processes are often comparable with events in rock mechanics laboratory testing, which means that such tests may be used to enhance our understanding of geological phenomena.

The purpose of this chapter is to give an introduction to geological aspects that are of particular importance in petroleum related rock mechanics.

3.1 Underground stresses

Normally, an underground formation has to carry the weight of the overlying formations. The vertical stress at the bottom of a homogeneous column of height z is $\sigma_v = \rho g z$, where ρ is the density of the material, and g is the acceleration of gravity. If the density varies with depth, then the vertical stress at depth D becomes

$$\sigma_v = \int_0^D \rho(z) g \, dz \tag{3.1}$$

Note that the z-axis is here pointing vertically downwards, with $z = 0$ corresponding to the Earth surface. The average density of sediments in the

Developments in Petroleum Science, Volume 72
ISSN 0376-7361, https://doi.org/10.1016/B978-0-12-822195-2.00012-7

overburden is between 1.8 and 2.2 g/cm^3, so as a rough number, the vertical stress increases downwards with about 20 MPa/km (typically, 1 psi/ft).

The sedimentary rocks encountered during oil well drilling and production are porous and hence contain fluids. We refer to the pore pressure at depth D as normal if it is given by the weight of a fluid column above, that is, in analogy to Eq. (3.1) the normal pore pressure is

$$p_{\mathrm{fn}} = \int_0^D \rho_{\mathrm{f}}(z) g \, \mathrm{d}z \tag{3.2}$$

The pore fluid density in case of brine with sea water salinity is in the range 1.03–1.07 g/cm^3, so the pore pressure increase with depth is roughly 10 MPa/km (0.45 psi/ft). The effective vertical stress σ_{v}' is then also increasing with approximately 10 MPa/km. In many important cases, however, the pore pressure deviates from the normal value p_{fn}. We will return to these so-called abnormal pore pressures and their possible origins in Section 3.2.

The underground stress state consists of the three mutually orthogonal principal stresses plus the pore pressure. It is very common (and convenient) in the oil industry to assume that the vertical stress is a principal stress. This is reasonable at large depth within a homogeneous Earth, in areas that have not been exposed to tectonic activity or are relaxed in the sense that there are no remnant stresses from previous tectonic activity. The vertical stress is governed by gravity, which has a unique direction, pointing towards the centre of the Earth. So it is not unreasonable to assume that vertical is a principal stress direction. Let us however be aware that there will be cases where this is not fulfilled, such as near the surface: Because the surface is stress free, the principal stress directions at and near it will be governed by the surface topography. In the case of a strongly sloping surface, even at depth, the principal stress directions may be far from the vertical–horizontal directions. Also, near heterogeneities, such as inclusions or faults, near underground openings such as boreholes (Chapter 4), or near depleting reservoirs (Chapter 12), principal stress directions will differ from the vertical–horizontal orientation.

Below we will consider the vertical stress as a principal stress. To begin with, assume a relaxed area where the horizontal stress is induced simply as a result of the vertical stress. For simplicity, think of a fluid: The overburden causes a vertical stress (pressure) in the fluid, but at the same time an equal horizontal stress (pressure). In a rock the ability to resist shear stresses causes the horizontal stress σ_h in general to be different from the vertical stress. We write (in terms of effective stresses)

$$\sigma'_h = K'\sigma'_v \tag{3.3}$$

The ratio K' of the effective horizontal and vertical stresses may vary significantly. At shallow depths (0–150 m), it may vary from 1 to 10 or even higher, whereas values from 0.2 to 1.5 may be found at larger depths. Bjørlykke and Høeg (1997) pointed out that chemical compaction increases in importance at depths below 2–3 km and will contribute to horizontal stresses by altering the trend seen from pure mechanical compaction above this level.

It has been suggested that, with time, $K' \to 1$, so that the stress state becomes hydrostatic with the magnitude given by the weight of the overburden according to Eq. (3.1). This is called *Heim's rule*, after work by the Swiss geologist Albert Heim in the 19th century. Heim suggested that the mechanism causing this development is creep. This state of stress is referred to as *lithostatic*.[1]

The transition towards lithostacy is normally a very slow process, and the lithostatic state of stress is rare in the lithosphere (Engelder, 1993). Some examples of underground stress illustrating non-lithostatic stress states are shown in Figs. 3.12 and 9.13.

The idea that creep brings the stress state towards lithostacy is however considered sound, and it is therefore not likely to find a general rule for predicting horizontal stress at depth, unless very sophisticated models taking long-time stress history into account are developed.

We now consider two very simple models for estimating the in situ stress state. First, assume that the formation under consideration is laterally constrained, that is, there is no horizontal strain under the process of rock formation. Furthermore, we assume that the rock during this process behaves according to the theory of linear elasticity. Then from Eqs. (1.105)–(1.106) with $\varepsilon_x = \varepsilon_y = 0$ we find

$$\sigma'_h = \frac{\nu_{fr}}{1 - \nu_{fr}}\sigma'_v \tag{3.4}$$

[1] In the literature, the term lithostatic stress is sometimes defined as the vertical stress being given by Eq. (3.1), without reference to the horizontal stresses. In this book, we follow Jaeger and Cook (1979); Engelder (1993) and others, and take lithostatic to imply that the stress state is also hydrostatic.

In this particular case the coefficient K' is given the name K_o. In a fluid, where Poisson's ratio is 1/2, $K_o = 1$. For a rock with $\nu_{fr} = 1/3$, $K_o = 1/2$.

There are many reasons to apply the relationship above with great care. We assumed linear elasticity, and we assumed zero lateral movement. In reality, we also assumed that the elastic properties of the rock have been constant throughout the whole process of rock formation. This is clearly not true. An approach based on a complete stress history analysis was presented by Warpinski (1989), incorporating variations in mechanical properties over time. Consolidation, diagenesis, changes in pore pressure due to gas generation, temperature gradients, and various tectonic and thermal episodes may be incorporated in this model. Visco-elasticity appeared, for the cases studied, to be more relevant for stresses in shale than in sandstone. Sone and Zoback (2014) fitted laboratory data on shale gas reservoir rock to a visco-elastic model, which was then used to estimate relaxation of in situ horizontal stresses over geological time.

Another simple approach is to say that the rock has been, and still is, in a critical state within the Earth, that is, it should obey some kind of a failure criterion. Leaning on ideas of self-organised criticality (see e.g. the book by Bak, 1996), this may be a reasonable assumption, in particular, in areas of active tectonics. Choosing the Mohr–Coulomb criterion (2.18) or (2.22) as a criterion for active faulting, the ratio between the minimum and the maximum effective stress should be (for the case where the unconfined strength C_0 is negligible)

$$\sigma_3' \leq \frac{1 - \sin\varphi}{1 + \sin\varphi}\sigma_1' = \frac{1}{\tan^2\beta}\sigma_1' \tag{3.5}$$

If the vertical stress is the maximum principal stress, then (3.5) predicts the following limiting value for K':

$$K' = \frac{1 - \sin\varphi}{1 + \sin\varphi} \tag{3.6}$$

If the friction angle is 30°, then $K' = 1/3$. A lower friction angle will result in a higher value for K'.

These are very simple examples of models for horizontal stress estimation. In reality, as mentioned above, horizontal stresses are difficult to assess from mathematical models. The most direct method of obtaining horizontal stress is to try to measure it, for instance, by a fracturing test of the formation (see Sections 8.3 and 11.5). Empirical relations based on hydraulic fracture testing exist, and an example is given in Section 8.3.4.4.

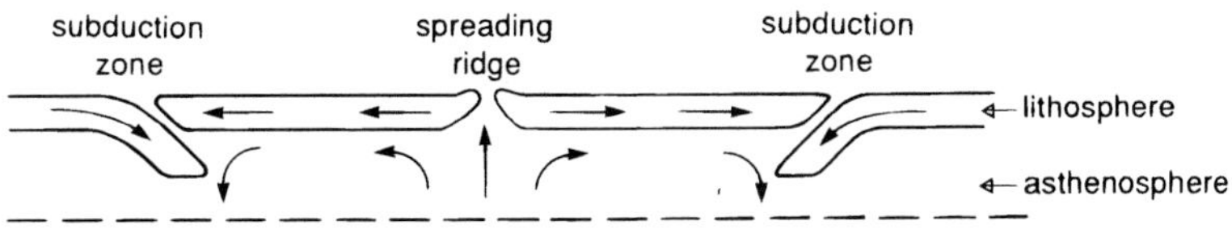

Figure 3.1 Schematic illustration of tectonic plate movement.

The principal horizontal stresses are in general not equal, contrary to what has been anticipated in Eqs. (3.3)–(3.6). We will in the continuation refer to the maximum horizontal stress as σ_H and to the minimum horizontal stress as σ_h. The main reason for horizontal stress anisotropy is *tectonic* stresses. The terms tectonic activity and tectonic stresses relate back to the theory of tectonophysics. The Earth's crust consists of a number of discrete tectonic plates. These are extensive (a few hundred to thousands of kilometres across) but thin (15–200 km thick) plates that move about the Earth's surface as rigid bodies. Fig. 3.1 shows two types of plate boundaries, *spreading ridge* (two plates moving away from each other) and *subduction zone* (two plates moving towards each other, and one plate subducts under the other).

A third type of plate boundary is a transform fault, where two plates slide past each other. Extensive deformation (faulting, earthquakes) occurs along plate boundaries where the plates interact. Tectonic activity refers to all forms of breaking and bending in this outer layer of the Earth. A tectonic stress component in one direction may also be accompanied by stress components in other directions. Consider for example a situation where there is a tectonic component of the stress $\Delta\sigma_x'$ in the x-direction. If there is no horizontal displacement in the y-direction ($\varepsilon_y = 0$), and no vertical restriction, the accompanying stresses $\Delta\sigma_y'$ and $\Delta\sigma_z'$ would become (according to linear elastic theory, Eqs. (1.105)–(1.107))

$$\Delta\sigma_y' = \nu_{\text{fr}}\Delta\sigma_x' \tag{3.7}$$

$$\Delta\sigma_z' = 0 \tag{3.8}$$

The total stresses in the region are then given by superimposing these stresses on the stresses given in Eqs. (3.1) and (3.4). Previous tectonic activity and associated effects are thus of major concern when discussing underground stresses, as both magnitudes and principal directions are affected.

Fig. 3.2 shows a World Stress Map, where principal stress orientations obtained from various techniques (primarily earthquake focal mechanisms

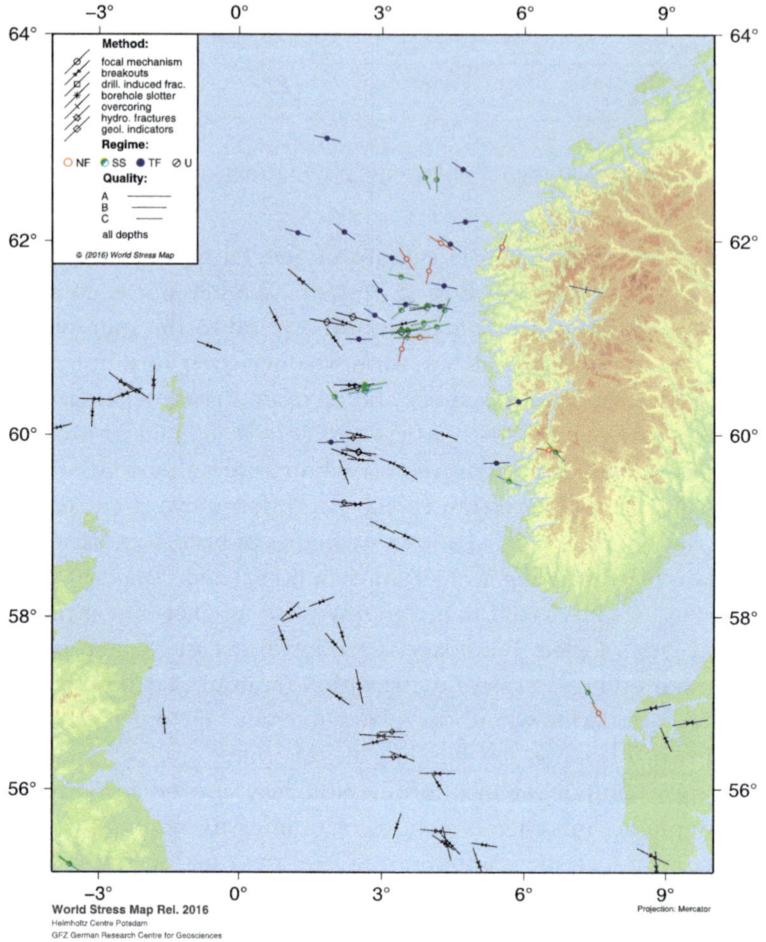

Figure 3.2 Stress map from the World Stress Map Project. Lines indicate the direction of σ_H, and line colours indicate tectonic regime (normal fault, strike-slip, thrust fault). (Heidbach et al., 2016, available online at www.world-stress-map.org).

and borehole breakouts; see Section 8.3.2) are indicated. Maps of selected geographical areas can be found at http://www.world-stress-map.org.

As mentioned above, the horizontal stresses may become very large at shallow depths. This may be because of residual stresses originating in the previous history of the rock or structural stresses (caused by large scale inhomogeneities).

An example of how residual stresses may originate is illustrated in Fig. 3.3. This figure shows how relative block movement exposes one

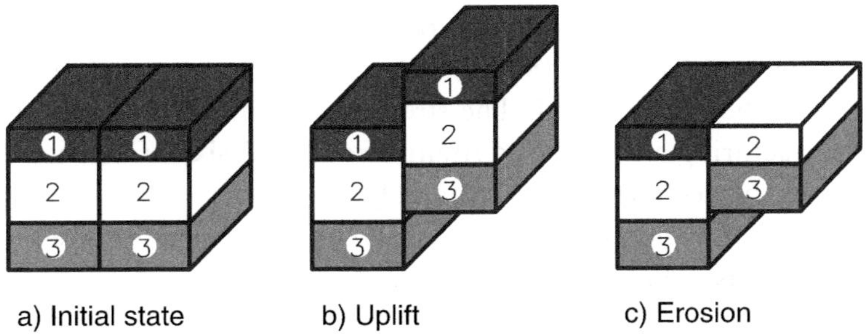

Figure 3.3 Residual stresses due to uplift and erosion.

block to erosion. The uplifted block will now, at the same depth as the neighbouring block, have a different stress history. If erosion takes place relatively rapidly, then the higher stresses that existed previously will not decay as rapidly as the erosion takes place. If the rock behaves as a visco-elastic material, then the stresses will at some stage die away. If the rock yields however, then the stresses may not relax back to normal stresses at that depth.

The process of uplift will in addition result in cooling and shrinkage of the formation. Depending on the stiffness of the rock, this can reduce the stress in the formation significantly (see Section 1.5). Hence it is also possible to experience stresses that are lower than expected for a normally consolidated rock.

The maximum stress a rock has ever been exposed to is often referred to as the paleostress. In soil mechanics (Section 2.8.2) the term *preconsolidation stress* is used. The uplifted and unloaded block in Fig. 3.3 has previously experienced higher effective stresses. Such a sediment is in soil mechanics said to be *overconsolidated* (Section 2.8.2). Examples of structural stresses are stress fields below slopes or mountains. The structural inhomogeneities affect both stress magnitudes and directions.

The Ekofisk field in the North Sea provides a good example of how geological events can explain the present state of stress. Ekofisk is the largest of several chalk reservoirs in the southern part of the Norwegian sector of the North Sea. The chalks are of Maastrichtian (Upper Cretaceous) and Danian (Lower Tertiary) age, and the field has a dome-like structure of elliptical shape. The chalk has a relatively high porosity (around 30% average) but a low matrix permeability (around 1 milliDarcy). The reservoir is however highly fractured, providing a much (around two orders of magnitude) higher reservoir permeability. This fracturing is believed to be a result of stresses induced by the growth of an underlying salt dome (Byrd, 1975). Teufel and Farrell (1990) presented results from in situ stress measurements in the Ekofisk field. The stress measurements were made using hydraulic fracturing and anelastic strain recovery (ASR) of oriented cores (more about this technique in Section 7.4).

Fig. 3.4 shows the azimuth of the maximum horizontal stress determined from ASR. On the crest the maximum horizontal stress tends to be oriented sub-parallel to the long axis of the ellipse, whereas on the flanks the maximum horizontal stress tends to be oriented perpendicular to the structural contours of the dome. Teufel and Farrell also looked at the distribution of natural fractures and found that a radial fracture pattern existed, aligned closely with the direction of the maximum horizontal stress on the flanks of the structure. Teufel and Farrell assumed the principal stress directions to lie in the horizontal and vertical planes. This is a reasonable assumption, since the areal extent of the dome is large compared to its curvature.

Since the structure of the field obviously affects the horizontal stress directions, it is however reasonable to infer that principal stress directions will not be completely horizontal and vertical at all locations. Rather, they will tend to have one principal direction normal to the dome surface. This implies that on the crest, principal stress directions are likely to be vertical and horizontal, whereas they will become tilted on the flanks of the dome.

There is a close relationship between structural geology (see e.g. Fossen (2016)) and rock mechanics. Rock mechanics laboratory experiments allow us to study processes in small scale, which are similar to those of the large scale of structural geology. For instance, a brittle shear failure in a laboratory specimen is a miniature analogue of a *fault*. Faults are by definition shear fractures with relative displacement along the plane of the fracture. The surface of the fracture is the fault plane, which can be described by its dip and strike; see Fig. 3.5. *Strike* is defined by a horizontal line formed by intersection of a horizontal plane and the tilted layer and is given in degrees relative to a compass direction. The *dip-angle* is the angle between the tilted layer and a horizontal plane. Brittle behaviour leading to fault formation is characteristic of rocks subjected to low confining pressure, that is, in some respect close to the surface of the Earth.

A tensile, brittle failure in a laboratory test produces a miniature *joint*. A joint is characterised by no displacement along (parallel to) the fracture plane. The rock on each side of the tensile fracture is moved apart, perpendicular to the failure plane.

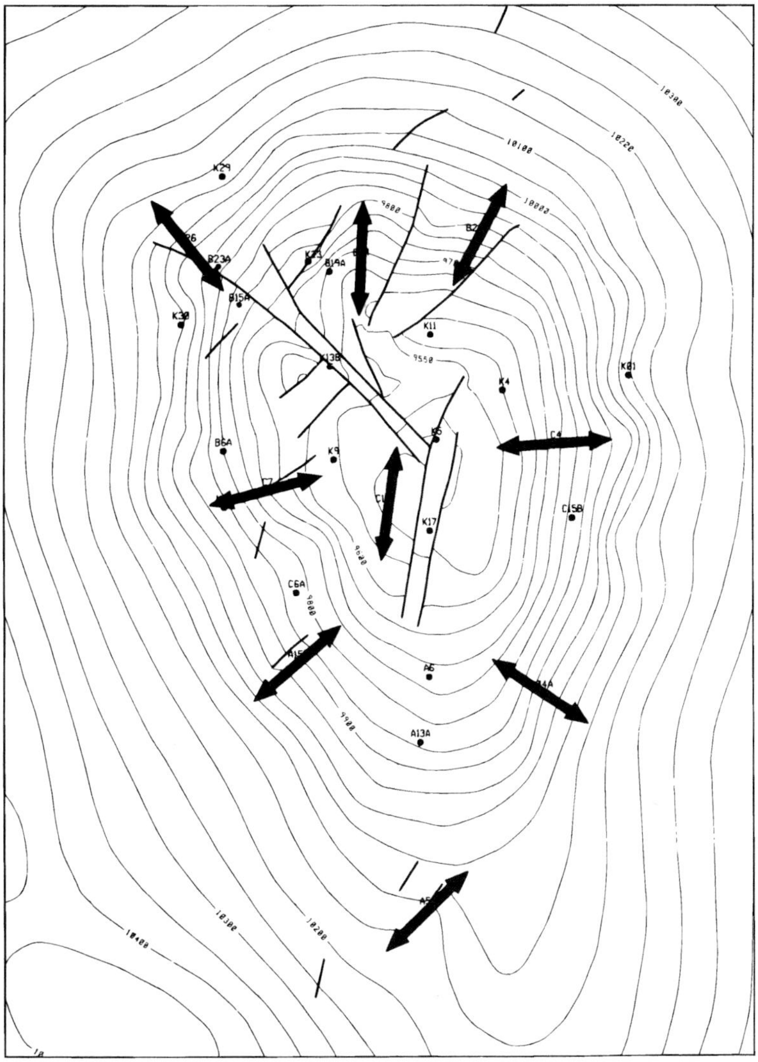

Figure 3.4 Structure contour map for the top of the Ekofisk formation showing the azimuth of the maximum horizontal stress determined from anelastic strain recovery measurements. The crest of the structure is at a depth of approximately 2.9 km (9500 ft), and contour intervals are 15.2 m (50 ft). (From Teufel and Farrell (1990); with permission from the authors.)

The various types of faults can be classified (Anderson, 1951; Twiss and Moores, 1992) according to well-known theories of shear failure (e.g. Mohr-Coulomb). According to these theories, fracturing will take place in

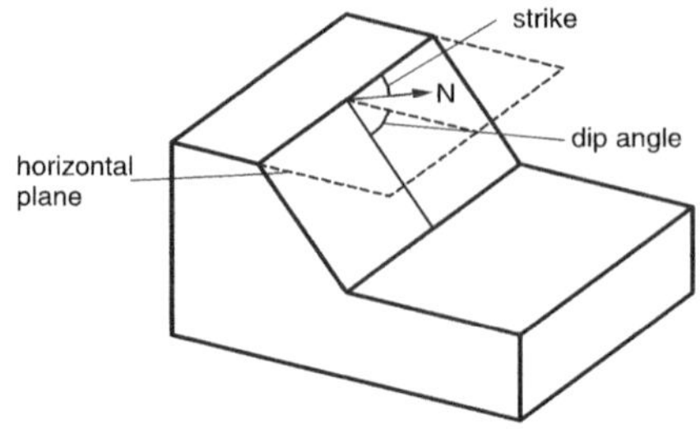

Figure 3.5 Illustration of dip and strike.

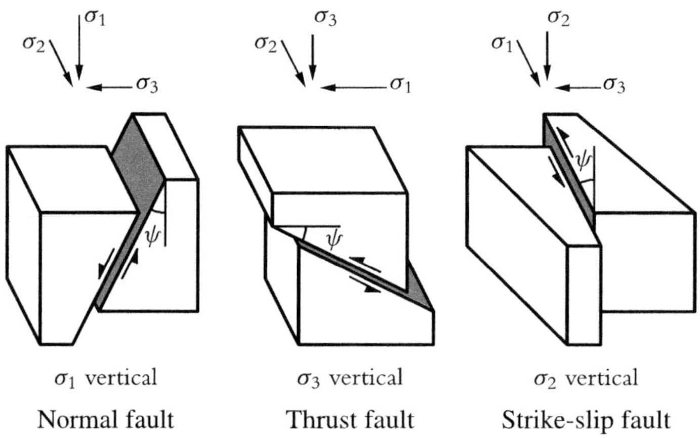

Figure 3.6 Illustration of fault types.

one or both pairs of conjugate planes, which are parallel to the direction of the intermediate principal stress and are both at equal angles of less than 45° to the direction of the maximum principal stress. Some of the most common types of faulting are illustrated in Fig. 3.6, assuming that one principal direction is vertical.

Normal faulting occurs when the maximum principal stress σ_1 is vertical, and the dip is hence larger than 45° (usually around 60°). If the minimum principal stress σ_3 is vertical, then the hanging wall is moving upwards, the dip is less than 45° (usually around 30°), and a thrust fault is formed. Thrust faults with very shallow dips (less than 10°) are also called overthrusts. If the intermediate principal stress σ_2 is vertical, then vertical failure planes are created. This is referred to as strike-slip faulting.

With reference to Fig. 3.6, ψ is the angle between the maximum principal stress and the failure plane. This angle is then related to the failure

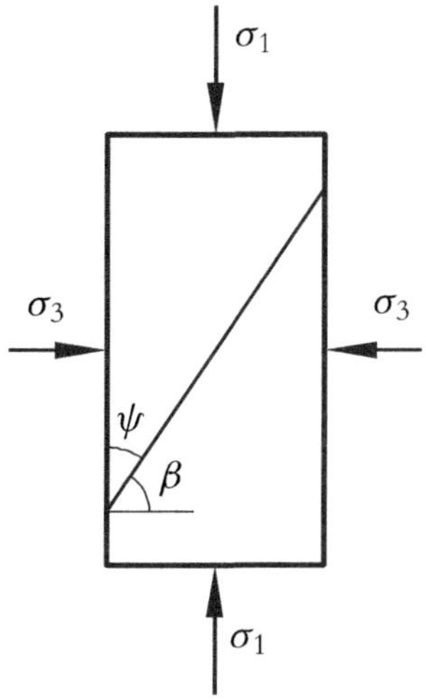

Figure 3.7 Relation between ψ and β.

angle β such that (see Fig. 3.7)

$$\psi = \pi/2 - \beta \tag{3.9}$$

Laboratory testing of sand and sandstone commonly yields values of the failure angle in the range 55°–70°, corresponding to ψ values in the range 35°–20°. According to the Mohr–Coulomb failure criterion (see Section 2.3.1), shear fracturing will take place when

$$\sigma_1' = C_0 + \sigma_3' \tan^2 \beta \tag{3.10}$$

where C_0 is the unconfined (uniaxial) compressive strength of the rock.

The type of fault is dependent on the relative values of the principal stresses. Whether faulting will take place, and at what angle, is however also dependent on the failure parameters of the rock.

To illustrate this further, let us consider some simple examples, using the Mohr–Coulomb criterion for shear fracturing.

Consider an unconsolidated sand with zero unconfined strength (typical of shallow sediments). A typical failure angle of the sand is 60° corresponding to ψ=30°. The Mohr–Coulomb criterion in this case (note the use of effective stresses) reduces to

$$\sigma_1' = 3\sigma_3' \tag{3.11}$$

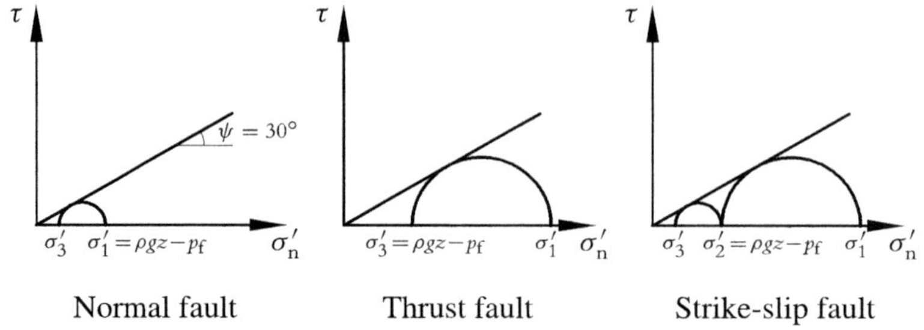

Figure 3.8 Stresses causing faulting illustrated by Mohr's circles.

We assume principal stresses to be vertical and horizontal, with the total vertical stress equal to $\rho g z$. If the vertical stress is the largest principal stress (see Fig. 3.8a), then the effective horizontal stress must be as low as one third of the effective vertical stress for normal faulting to occur. For thrust faulting, the effective horizontal stress must be three times larger than the effective vertical stress (see Fig. 3.8b). For strike-slip faulting, the vertical is the intermediate principal stress. The two horizontal stresses are limited as follows: If the maximum horizontal stress is equal to the vertical, then the minimum effective horizontal stress must be $\sigma'_h = (\rho g z - p_f)/3$. If the minimum horizontal stress equals the vertical, then the maximum effective horizontal stress must be $\sigma'_H = 3(\rho g z - p_f)$.

The above analysis may be visualised in a single plot. Let us assume that the vertical stress is a principal stress and has been determined to 50 MPa. Let us further assume, as above, that the stress state is limited by activation of cohesionless faults with a given angle of friction. By considering in turn all pairs of principal stresses we may map out the allowed states of stress in the (σ_H,σ_h)-plane.

The resulting *stress polygon* (Zoback et al., 1986, 2003; Zoback, 2007) is shown for two different pore pressures and two different friction angles in Fig. 3.9.

When using such methods for estimating or constraining in situ stresses by back-calculation, it is important to be aware of the uncertainties involved. Both the model(s), assumptions and the input data used will contribute to the uncertainty. In some cases the conclusion can even be fundamentally wrong; see Section 8.3.5.

There are however also examples of faults that cannot directly be identified from the three types shown in Fig. 3.6. One such example is a so-called oblique slip, where there are both vertical and horizontal movements; see

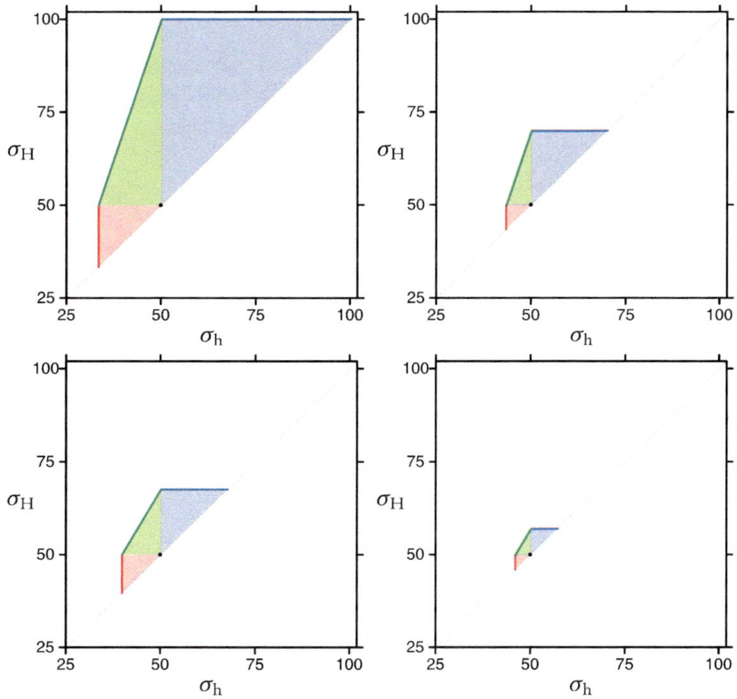

Figure 3.9 Stress polygon, assuming $\sigma_V = 50\,\text{MPa}$. Upper left: Normal pressure, $\varphi = 30°$. Upper right: 60 % overpressure, $\varphi = 30°$. Lower left: Normal pressure, $\varphi = 15°$. Lower right: 60 % overpressure, $\varphi = 15°$. Red: Normal fault, Green: Strike-slip, Blue: Thrust-fault.

Fig. 3.10. The most obvious reason for this fault direction is that the principal stresses are not aligned vertically and horizontally. In a formation that has not been fractured by previous tectonic activity, the fault direction is mainly controlled by principal stress directions. The fault direction may however be disturbed if the rock has been fractured previously. Fault movement can then be partly controlled by the previous fault direction and partly by present principal stress directions. Complex fracture systems are generated in this way.

Rocks buried deep in the Earth, at high confining stress and high temperature, tend to exhibit a more ductile behaviour when exceeding the elastic limit. This plastic flow can involve change both in shape and in volume. This may result in folding of the rock; see Fig. 3.11.

Available techniques for measurement and estimation of in situ stress directions and magnitudes are discussed further in Section 8.3. The most

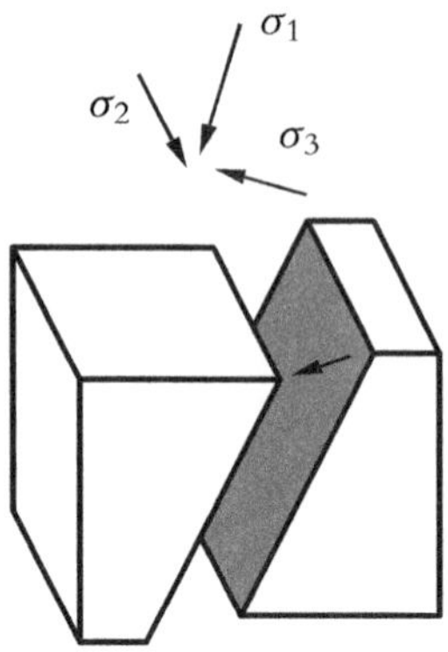

Figure 3.10 Oblique slip, normal fault.

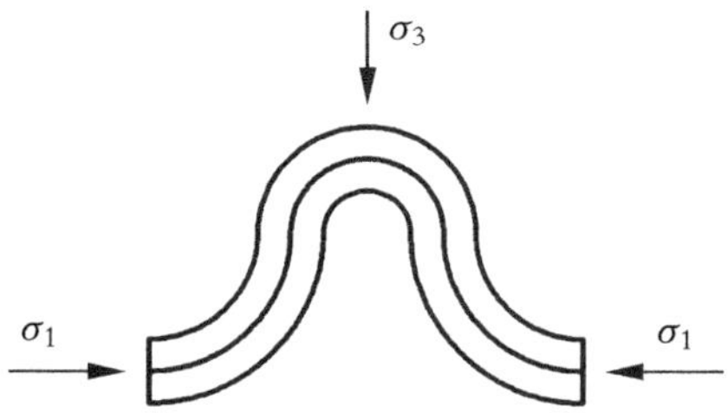

Figure 3.11 Illustration of compressional folding, cross-section.

direct measurements are those involving hydraulic fracturing, where detection of fracture closure pressure yields information about stress magnitudes. Borehole imaging and/or borehole shape measurements (by e.g. a 4-arm caliper) may reveal information about fracture or drilling-induced breakout directions and hence stress directions (Section 8.3). There is also a number of core-based techniques (anelastic strain recovery, wave velocity anisotropy, differential strain or wave velocity analysis, acoustic emission), which can be related to stress magnitudes and directions (Section 7.4). For large-scale stress fields, earthquake focal mechanisms represent a valuable source of information.

3.2 Pore pressure

Pore pressure is an important parameter in any rock mechanics study of porous, fluid-filled rock systems. The pore fluid will carry part of the total stresses applied to the system, thus relieving the rock matrix from part of the load. The effective stress as defined by Terzaghi is equal to the total stress minus the pore pressure. This effective stress concept was introduced in soil mechanics in 1923 on an empirical basis. It has later been refined by Biot; see Section 1.6 for further details.

There is overwhelming evidence that porous, saturated and permeable rocks obey an effective stress law. Both strain, given by the stress–strain relationship (constitutive equation), and yield or failure of the rock is controlled by effective rather than total stresses.

Therefore, when studying borehole stability during drilling, rock stability during production, and compaction/subsidence, knowledge of the pore pressure in the various formations is extremely important.

Pore pressure will develop in a saturated formation as sediments are buried. If the pore fluid can escape and migrate to the surface at about the same rate as the rate of compaction, then a normal pore pressure gradient is maintained, given by the weight of the fluid column above (Eq. (3.2)).

There are however several cases where the pore pressure within a zone has a value different from the expected normal pore pressure. Usually, the pore pressure will in such cases be higher than the normal, and the zone is referred to as *abnormally pressured* or *overpressured*. High pore pressures in a reservoir will of course make the field more prolific. On the other hand, overpressured formations are a potential hazard during drilling. Furthermore, borehole stability problems are often encountered in overpressured shales.

Briefly stated, abnormal pore pressure (overpressure) has three main causes (Osborne and Swarbrick, 1997; Yassir and Addis, 2002):

1. The rate of sedimentation and compaction being higher than the rate of fluid expulsion and migration (disequilibrium compaction, undercompaction). A prime example of this is the thick Tertiary sequence in the Gulf of Mexico.
2. Tectonic loading that leads to undrained shear stress with associated pore pressure development.
3. Pore fluid generation or expansion by thermal or chemical processes.

Due to the low permeability of compacted clays, shaly zones can easily become overpressured. Shale permeabilities are typically in the nanoDarcy range, but may be even lower (Swan et al. (1989); Chenevert and Sharma (1991); Katsube et al. (1991); van Oort et al. (1996), and others). Actually, laboratory measurements representative of in situ shale permeability are very scarce, because of the experimental difficulties involved in measuring such low permeabilities and because shale cores when retrieved from depth are altered in such a way that the permeability is likely to be overestimated from experiment. Anyway, shale permeability is low enough that a thick shale formation may not be able to expel fluid at the same rate as it is compacted. Sands that are embedded within such shale bodies are also likely to become overpressured.

Rapid sedimentation is obviously another possible reason for the development of overpressures. Abnormal pressures generated by disequilibrium compaction tend to decline in the course of geologic time. As pointed out by Osborne and Swarbrick (1997), this may explain why overpressure is more common in the younger Tertiary than Paleozoic sequences.

As an example, consider a shale with a permeability of 0.01 nanoDarcy, porosity of 25 %, and uniaxial compaction modulus of 3 GPa, saturated with a brine of bulk modulus 2.5 GPa and viscosity 1 cP. The characteristic diffusion time will be l_D^2/D_h, where D_h is the diffusion constant given by Eq. (1.323), and l_D is the characteristic diffusion length, say the thickness of the shale layer. The characteristic time for establishment of pore pressure equilibrium in a 100 m thick shale layer using these values turns out to be about 30 million years, which is on the geological time scale. Osborne and Swarbrick (1997) suggested that disequilibrium compaction is the primary mechanism responsible for overpressure in shale sequences.

As pointed out by Yassir and Addis (2002), there is a strong correlation between occurrence of overpressure and compressional tectonics. According to Eq. (1.215), pore pressure will increase with increasing shear stress in a manner proportional to Skempton's A-parameter. Consider as an example a situation where the horizontal stress σ_H is the major and the vertical stress (given by the weight of the overburden) is the minor principal stress. Keeping σ_v constant during tectonic activity, the pore pressure change is

$$\Delta p_f = AB\Delta\sigma_H \tag{3.12}$$

For normal consolidation processes, A may exceed 1, whereas B is close to 1, leading to very high pore pressure, possibly exceeding the vertical stress. In overconsolidated or well-cemented low-porosity rock on the other hand, shear loading is associated with dilatancy, leading to pore pressure reduction. This mechanism is therefore expected to cause abnormally high pore pressure preferably in young sediments. Note that tectonic activity (e.g. salt tectonics) will result in sustained abnormal pore pressure only if the system remains closed and does not fracture.

Another possible source of overpressure is the uplift-erosion process shown in Fig. 3.3. If the rock maintains its pore pressure after the uplift, it will be abnormally pressurised compared with its neighbouring formations at the same depth. Often, however, faulting is associated with both tectonic and uplift processes, and in such cases the pore pressure build-up would be only transient.

As mentioned above, generation or expansion of pore fluid may lead to overpressure, primarily in shale sections. This may be a result of increase in temperature (aquathermal pressuring). It may be caused by hydrocarbon (kerogen or gas) generation. It may also be caused by free water released during transformation of smectite to illite. This transition is temperature dependent, requiring temperatures of 70–90 °C, corresponding to depths of 2–3 km in areas of average geothermal gradient.

When the pore pressure increases due to fluid expansion/generation, there will be an associated horizontal stress increase. The relation between horizontal stress and pore pressure is hence not unique but depends on the overpressure mechanism (Yassir and Addis, 2002).

Yet another class of mechanisms is associated with fluid movement caused by, for example, density differences between liquids and gases or possibly by osmotic potentials. There is however no evidence that these mechanisms play any significant role in practice.

Fig. 3.12 presents typical pore pressure gradient curves from two North Sea fields, the Gullfaks field and the Hod field. Hod is a chalk reservoir (Cretaceous) in the Valhall area of the Norwegian sector of the North Sea. Gullfaks is a sandstone reservoir (Jurassic) located further north. Both parts of Fig. 3.12 include the estimated fracture pressure, taken from leak-off and mini-frac tests (see Section 8.3). The Hod field is characterised by a high initial pore pressure. Close to the top of the reservoir at 2500 m, the pore pressure is about 44.7 MPa. The difference between the overburden stress and the pore pressure is quite small at reservoir depth. Thus the effective vertical stress is very low.

In such a situation the net grain-to-grain stress is small, and unless the strength of the cement between the grains is significant, particles may easily be mobilised. Production problems have been experienced in the North Sea chalk fields, materialising both as inflow of solids and casing collapse (see Chapter 10 for analysis of particle production).

Both fields are characterised by a rapid increase of the pore pressure gradient just above the reservoir (top of Gullfaks is at approximately 1850 m). This is relatively typical in many of the North Sea reservoirs. The effective vertical stress in the Gullfaks field is approximately 6 MPa.

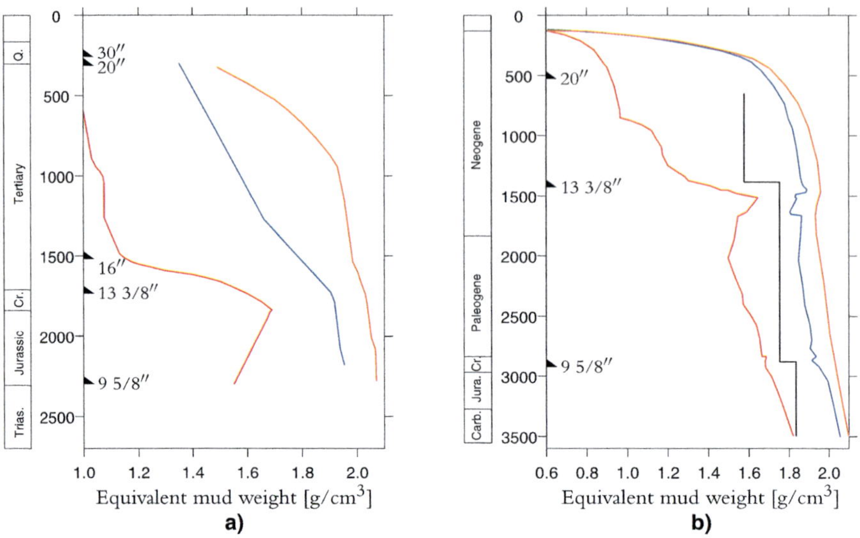

Figure 3.12 Pressure gradient (red), fracture gradient (blue) and overburden gradient (orange), given as equivalent mud weight: a) Gullfaks (Courtesy of Equinor), b) Valhall-Hod area (Courtesy of Aker BP). The depth is in metres relative to the drill floor. In the right hand plot the mud weight is shown in black.

Pore pressures above the reservoir, in the low-permeability zones, are normally estimated from trendlines and deviations from expected trendlines in wireline log readings. In many cases the accuracy of these methods is not good enough, and stability problems during drilling may be a consequence (see also Chapter 9). In the reservoir, where the rock has a higher permeability, more direct and accurate pore pressure measurement methods can be applied, thus improving the quality of the estimates. This is also a possibility in permeable intervals in the overburden.

3.3 Sedimentological aspects

The mechanical properties of a rock normally refer to constants in the constitutive equation, which the rock is assumed to obey. A linearly elastic and perfectly brittle isotropic rock would hence be described by five mechanical parameters: two elastic parameters (e.g. Young's modulus and Poisson's ratio; Section 1.3) and three strength parameters (e.g. friction angle and uniaxial compressive strength when applying the Mohr–Coulomb failure criterion, plus the tensile strength; see Chapter 2).

Table 3.1 Grain-size scale for sediments. In geology, the so-called phi-scale is commonly used. This scale is defined by phi $= -\log_2(d_g)$, where d_g is the grain diameter in mm (see e.g. Pettijohn et al., 1987).

Grain diameter range [mm]	phi-scale	Term
> 256	< -8	Boulder
64–256	−6 to −8	Cobble
4–64	−2 to −6	Pebble
2–4	−1 to −2	Granule
1/16–2	4 to −1	Sand
1/256–1/16	8 to 4	Silt
<1/256	> 8	Clay

The present properties of a sedimentary rock are determined by the entire process from erosion of rock fragments to transportation, deposition, compaction and lithification. Some knowledge of these processes is therefore valuable when trying to assess the mechanical properties of a rock.

3.3.1 Grains and minerals

Some definitions related to grains and granular materials are appropriate at this stage:

Grain size is a measure of the diameter of the grain; see Table 3.1. According to this classification scheme, the grain size determines the classes of sedimentary rocks.

Grain shape involves both the roundness (angularity of corners) and sphericity (proximity to a spherical shape).

Grain sorting is a measure of the range of grain sizes (grain-size distribution). A rock containing a wide range of grain sizes is said to be poorly sorted, whereas well sorted implies a narrow distribution.

The packing of grains is important since it affects porosity and permeability. The packing depends both on grain size, shape and sorting. Round grains will have a smaller friction angle than more angular grains at the same porosity. A poorly sorted sand will have a higher friction angle than a well-sorted sand.

There are several rock-forming mineral groups: silicates (e.g. feldspar, clay, mica), carbonates (calcite, dolomite), oxides, sulphides, sulphates and phosphates.

The silicate group is the most important rock-forming mineral group, making up more than 90 % of the Earth's crust. The basic unit of the silicates is the SiO_4^{-4} silicon tetrahedron. Various structures are formed by linking tetrahedral units together by sharing oxygen. Feldspar is the most important single silicate mineral. The three main groups of feldspars are potassium feldspars ($KAlSi_3O_8$), sodium feldspars ($NaAlSi_3O_8$) and calcium feldspars ($CaAl_2Si_2O_8$). These are formed by Al substituting for Si, thus allowing additional or different cations to enter the structure.

Although chemically an oxide, silica (SiO_2), is closely related to silicates, quartz, which is one of the minerals in the silica group, is the most common mineral in sandstone, ranging from 65 % to practically 100 %. The feldspar content in sandstone is typically 10–15 %, and mica and clay minerals may also be found.

The most important minerals in shaly rocks are clay minerals (also referred to as argillaceous minerals, after the Greek word for clay), which are composed of layers of sheet-shaped crystals. Two basic structural units exist. One type of sheet is built from silica tetrahedra linked together in a hexagonal structure with chemical formula $Si_4O_{10}^{4-}$. The second type is an octahedral sheet in which silicon is replaced by cations like Al^{3+} (this sheet is named gibbsite, with basic formula $Al_2(OH)_6$), Mg^{2+} (named brucite, $Mg_3(OH)_6$), Fe^{3+} or Fe^{2+}, surrounded by six hydroxide groups). Within a given sheet, ions different from the dominant ones may be present (isomorphic substitution), giving rise to possible charge deficiency. Usually, mainly because of oxygen and hydroxide at the surfaces, these have negative excess charge and will therefore attract cations.

The basic sheets are combined to form layered clay minerals. The most common groups of clay minerals are kaolinite, illite and smectite. Mixed layer variations also exist for these groups. Kaolinite is a two-layer mineral (often referred to as a 1:1 layer silicate) consisting of alternate layers of gibbsite and silicon tetrahedron sheets. A basic unit is 0.7 nm thick, whereas typical composite crystals may be 70–100 layers, have six-fold symmetry and are flake shaped, up towards the micrometre range in lateral extent. The successive layers are strongly bonded to each other with a hydrogen bonding, and therefore water is not permitted to enter in between layers. Mineral density is 2.6–2.7 g/cm^3.

Smectite is a 2:1 layer mineral, composed of a central gibbsite sheet embedded between two silicon tetrahedron sheets with a combined thickness of about 1 nm. Isomorphous substitutions are common both within the central octahedral sheet and in the tetrahedral sheet. The bonding between two silica sheets connecting the unit layers is very weak (through van der Waals' forces). In smectite, this permits water and exchangeable ions to enter between the platelets, leading to a swelling capacity: At surface conditions, smectite minerals may absorb up to 10 times their own weight in water ("swelling clays"). Calcium and sodium are common ions in the interlayer space. Smectite crystals are normally not more than 2–5 nm thick in hydrated state, with an average diameter of about 1 μm. Normally, the basal spacing is found to be near 1.5 nm, indicating that two monolayers of water are accommodated within the mineral. The specific surface area associated with the interlayer space is 500–700 m^2/g. Mineral density varies between 2.0 and 2.6 g/cm^3.

Montmorillonite is a name often used synonymously with smectite or as a common term for expandable clay minerals. In principle, this is a member of the smectite group with geographical origin in Montmorillon, France. Another well-known smectitic clay is Bentonite, from Wyoming in the USA. Vermiculite, which is hydrated muscovite (a mica mineral), is also classified as smectite but has much less swelling capacity than the clay minerals mentioned above.

Illite is another 2:1 layer clay mineral, formed by weathering of feldspars, degradation of muscovite, and transformation of smectite to illite at depth. Some of the silicon atoms in smectite are replaced by aluminium, causing a negative charge, which is balanced by potassium ions that provide bonding between the silica tetrahedron sheet. This bond is much stronger than in smectite, preventing hydration, but it is considerably weaker than in kaolinite. The basal layer thickness of illite is 1 nm. Mineral density is 2.6–2.9 g/cm^3.

The water near the surfaces of clay crystals is different from free water. This was suggested by Derjaguin and Churayev (1971). Molecular dynamics simulations (e.g. Skipper et al., 1991; Karaborni et al., 1996; Park and Sposito, 2002) show that bound water in the intralayer space and adsorbed water on surfaces have ordered, crystal-like structures; see also Section 3.4.3.

Chlorite is often also referred to as a clay mineral. It has a 2:1 sandwich structure but with an additional brucite-like (where the dominant ion may be magnesium, iron or nickel) layer embedded between the tetrahedral layers. It has the characteristic flake shape but does not have any swelling capacity. Density may be from 2.6 and up to 3.3 g/cm^3.

The basic unit of carbonates is $(CO_3)^{2-}$. Common carbonate minerals are calcite ($CaCO_3$), aragonite (which is a polymorph of calcite) and dolomite ($CaMg(CO_3)_2$). Calcite is the main component of carbonate sediments (chalk, limestone).

3.3.2 Pre-deposition and deposition

Clastic sedimentary rocks originate from some sort of pre-existing rock. The pre-existing rock is broken into fragments by weathering (mechanical and/or chemical) and erosion. Then it is transported by water (rivers, tides) or wind to another site, where it is deposited. This accumulation of fragments from pre-existing rocks form what is called a clastic sediment. The most common types of clastic sediments are conglomerate, sandstone, siltstone and shale. Other types of sediments are chemical sediments (e.g. precipitation of salt crystals due to evaporation) and organic sediments (e.g. shells and skeletons from sea organisms forming a chalk or limestone). Note that, for instance, a limestone can be both clastic and organic in origin.

Transport, deposition and sediment accumulation may take place in a variety of depositional environments. These various environments will give rise to different distributions of grain size and grain shape, which may affect the mechanical properties of the rock millions of years later. Particles may be transported by air (aeolian), water (fluvial) or ice (glacial transport). Deposition occurs if the particle flow velocity is lower than a settling velocity, being proportional to the square of particle size and the density contrast between particle and suspending fluid, and inversely proportional to fluid viscosity.

When transported by air (aeolian transport), the particles have a very small buoyancy, and the viscosity of the air is low. Particles transported by winds are therefore typically more fine-grained than particles transported by water.

In water, such as in a river or a marine environment, particles can be transported in suspension (suspended load) or along the bottom (bed load). Sediments deposited in running water are termed alluvial (poorly sorted, emerging from a high energy environment) or fluvial (well-sorted, emerging from a low-energy environment). As long as the concentration of particles in the suspension is low, the water can be regarded as a Newtonian fluid obeying ordinary hydrodynamic laws. As the concentration increases, the properties of the suspension (density, viscosity) will start to deviate significantly from those of pure water. Such a suspension can start to flow as a heavy liquid due to the difference in density between the suspension and the surrounding water. This is known as a turbidity current. This kind of sediment transport is believed to be an important mechanism when considering transport of sediments from deltas and shallow areas, down the continental shelf and into the deep oceans.

Grain flow is another mechanism of sediment transport, resulting from grain collisions that contribute to keep the grains from concentrating close to the bottom. Grain flow will have a low content of clay matrix material. As the contents of silt and clay increases, debris flow and eventually mudflow develops. Debris flow and mudflow sediments are often very poorly sorted.

In the shallow beach zone, deposition will often result in well-sorted coarse-grained rocks with high porosity. The grains will tend to be well rounded due to the constant action of waves, currents and tides. Further down, the grain size decreases, and the sorting is poorer. Clastic deposits of terrestrial origin, resulting from stream transport into the sea, can also be well sorted, but normally contain more clay than the beach deposits. The delta is the zone where fluvial and marine processes meet. If the marine activity is low, then the fluvial depositions in the delta will be maintained. If the marine activity is high, then the deposits will eventually accumulate as a marine sediment. In a delta front deposit, fresh water, sea water and pore water from previous sediments meet. Water from these different sources will be of completely different composition and form an environment of active precipitation and dissolution. This can lead to cementation (silica, carbonate) of the sediment.

Although carbonate rocks can originate from clastic sediments, the primary source of carbonates is from plants and animals, which make use of carbonates in some way. Carbonate sediments will be fine-grained (grain size in the micrometre-range), and the final properties of the carbonate rock will depend very much on the post-depositional processes (diagenesis), since the carbonates are very strongly influenced by pore fluid chemistry and temperature/pressure. Carbonates may also contain larger grains (e.g. fragments of fossils) of sand particle size or larger. In the same way as for sand deposits, the carbonate is more likely to be coarse grained and well sorted in high-energy shallow waters.

3.3.3 Post-deposition

Sediment deposition is followed by a lengthy process of transformation into a rock (sand to sandstone, clay to shale, silt to siltstone). As the sedimentation process continues, an overburden of younger sediments is deposited. This extra weight on top of a sediment causes the sediment to compact. Compaction will result in a closer packing of the grains, thus reducing both porosity and permeability. The friction angle will increase as a result of compaction; see Fig. 3.13a. Compaction itself is however not sufficient

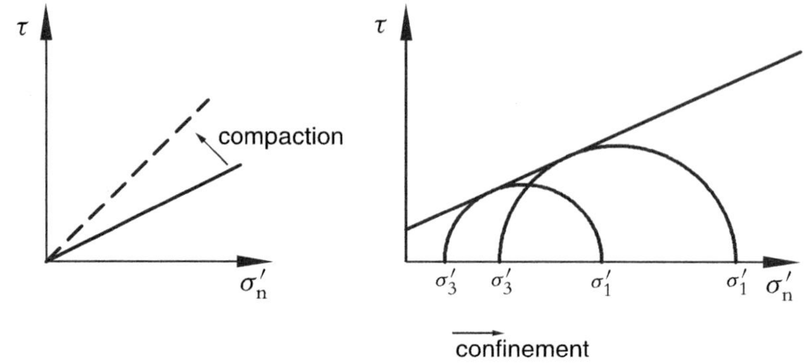

Figure 3.13 Effects of compaction on strength properties. Left: Increase of friction angle as a result of compaction. Right: Increase of failure stress as a result of confinement.

to create a massive, cemented rock, as closer packing alone does not affect the cohesion. The sediment will however act as if it is stronger when the weight of the overburden increases, since the friction angle is always larger than zero. The increased overburden will result in increased horizontal stresses when the sediment is confined horizontally. This will have the same effect as a confining pressure in the laboratory, and the sediment can tolerate larger shear stresses before it fails. This is illustrated in Fig. 3.13b.

For the transformation to be complete, processes by which grains are cemented together are required. These physical and chemical post-depositional processes are referred to as diagenesis. The parts of the diagenetic processes that contribute to harden the rock are also referred to as lithification.

The term consolidation is used with different meanings in the literature. In soil mechanics, it refers to time-dependent deformation due to dissipation of pore pressure (see Biot's theory of consolidation in Section 1.9.1). In this context, poorly consolidated basically means poorly compacted. In other applications (as well as in this book), terms like "poorly consolidated" and "unconsolidated" tend to refer more to the mechanical strength of the rock. The reason for this is that there is usually some correlation between the amount of diagenetic cementation and the effective stresses developing in a formation due to compaction, especially in sand deposits.

Physical changes, in addition to compaction, include changes brought about by tectonic activity. With respect to diagenetic effects, the chemical changes are however the most important. Cementation is mainly the result of precipitation of a binding agent. When considering chemical processes, the interaction between the solid grain minerals (especially the surface minerals) and the pore fluid is the basic process. As only the pore water is free to move in the sediment (at least as long as porosity and permeability are maintained), these processes are closely linked to changes in pore water chemistry. Precipitation of calcite onto the surface of sand grains is one such example of chemically induced cementation.

Carbonate (especially calcite) and silica are the minerals that account for most of the cementation of sandstones. If the contact stress between quartz grains gets sufficiently high, then pressure solution may occur. The idea is that dissolution at the grain contact points liberates SiO_2 to the pore water. The SiO_2 is then reprecipitated as quartz overgrowth because the pressure in the pore water is lower, resulting in silica cementation. Both calcite and quartz cementation lead to strong sandstones. If the grains are held together by clay minerals, then the sandstone will be weak.

In carbonate sediments, cementation is mainly precipitation of carbonate minerals. Primary cementation will often consist of aragonite or calcite with a high (up to 30%) magnesium content. These calcium carbonates are metastable and will after some thousands of years transform to the more stable form, low-magnesium-calcite, through dissolution and precipitation. Dolomitisation involves conversion of calcium carbonate and magnesium to dolomite:

$$2CaCO_3 + MgCl_2 \rightarrow CaMg(CO_3)_2 + CaCl_2$$

This process requires a certain ratio of Mg/Ca to take place. This required ratio is reduced as sediments are buried and the temperature increases because the magnesium ions then become less hydrated.

Dolomitisation, as most other processes of precipitation and cementation, often leads to reduced porosity as the void space is fully or partially filled up. This increases the rock strength in most cases but also reduces the potential of the rock as a reservoir formation.

Cementation processes however, as we have seen, require increased pressures or some throughflow of pore water. If these conditions are not met, cementation will come to a halt. Some of the North Sea reservoirs can exemplify this. Both the sandstone and chalk reservoirs are characterised by abnormally high pore pressures (see Fig. 3.12). This has prevented pressure solution, and an efficient cap rock has prevented further circulation of pore fluid through the reservoir rock. This has helped maintain a very high porosity, 30–50% in the most prolific chalk and around 30% in the sandstone reservoirs. This has however also resulted in very low cohesive strengths, and some formations are more or less uncemented, almost like soil. The cohesive strength in situ is then mainly a result of capillary forces and hence very low. This has resulted in significant production problems, such as production of sand particles and flow of chalk material into wells (see Chapter 10).

As a general statement, the degree of cementation and hence the strength of a given rock increases with depth of burial due to increased compaction, pressure solution and other diagenetic processes. This general trend is however often broken, either due to lithological changes or due to conditions that may retard the diagenetic processes, such as the abnormal pore pressure mentioned above.

By volume, sandstones and carbonates make up around 15 % and 10 %, respectively, of sedimentary rocks. The most abundant sedimentary rock is however shale (~75 %). Most of the overburden, which has to be drilled through to get down to the reservoir, consists of shale. These shale sections are often associated with stability problems during drilling (see Chapter 9) and are therefore of interest in petroleum rock mechanics.

Diagenetic processes in clay are mainly a result of reactions between the clay minerals and the pore water. Unstable minerals react with the pore water, depositing new and more stable minerals. For this process to continue, there must be a continuous flow of pore water through the sediment; if not, then the process will stop. In fine-grained sediments like clay, the permeability is very low (nanoDarcy and below), and equilibrium between pore water and minerals will be established. Clays will therefore not be much altered, and the diagenetic processes that take place will be very slow.

One example of diagenesis in clay is the transition of smectite to mixed layer illite/smectite and illite (See also Section 3.2). This transition requires temperatures corresponding to depths of 2–3 km to take place and is accompanied by release of water to free pore water.

Due to its plate-like structure, an uncompacted and soft clay (as a soil) will normally have a low friction angle. As compaction and consolidation takes place, and porosity and fluid content decrease, the shale will develop a higher friction angle. Shale, like any other sedimentary rock, is however anisotropic by nature. Often transversely isotropic symmetry is assumed, with the symmetry axis normal to bedding. This will affect permeability, elastic properties and strength parameters of the shale (Section 3.4).

Figure 3.14 Thin section image of Castlegate sandstone (courtesy of SINTEF).

3.4 Mechanical properties of sedimentary rocks

Here we give a brief overview of the mechanical properties of sedimentary rocks of interest to the petroleum industry, namely sandstones, carbonates, and shales. The main difference between these rocks is grain (or pore) size, ranging from 0.1–1 mm in sands down to the nanometre range in shales. This affects petrophysical characteristics, such as the permeability, which has a profound influence on mechanical behaviour, in particular, in time-dependent phenomena. Also, the relevant laws of physics controlling the force transmission at microscale are different in the nanometre and millimetre environments. However, in all cases, we deal with assemblies of bonded particles, and the gross behaviour of the different rocks is largely seen more similar than different.

We have also included a brief description of rock salt, which impacts amongst other seismic interpretation and drilling operations in many sedimentary basins.

3.4.1 Sandstone

A thin section image of a sandstone is shown in Fig. 3.14. The dominant grain sizes of sandstone are typically 0.01–1 mm. Pore sizes are the same order of magnitude but slightly smaller. This gives permeabilities ranging from microDarcy to several Darcy.

The predominant matrix mineral is quartz. Although the term sand may be used also when the grains are carbonate, the term sandstone is commonly restricted to siliceous grains (see e.g. Pettijohn et al., 1987). The cement may be quartz as well (in the case of pressure solution), it may be carbonate (especially calcite) or clay minerals.

The texture may vary from isotropic to anisotropic. Intrinsic (lithological) anisotropy is caused by microlamination (bedding), induced, for instance, by seasonal variations during deposition. This typically gives al-

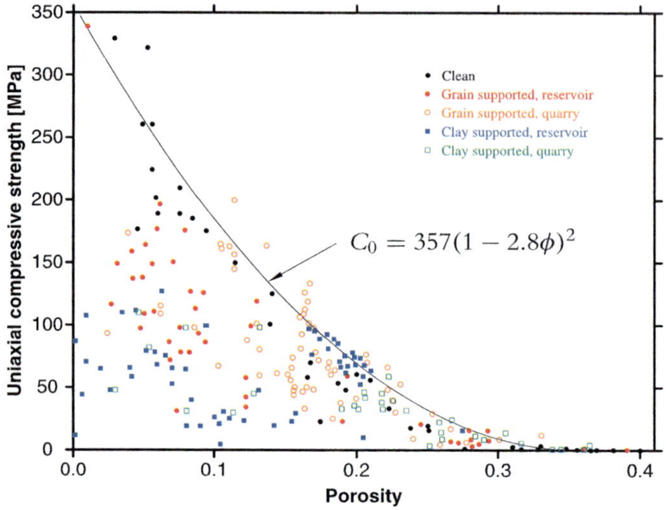

Figure 3.15 Strength of sandstones. After Plumb (1994).

ternating layers of coarse and fine grains. Anisotropy may also be caused by deposition of flat grains with their long axis parallel to each other.

An unconfined test with a sandstone, or a triaxial test at a low confining pressure, typically shows nonlinear stress–strain behaviour during initial loading, with stiffening as the stress is increased. This nonlinearity can be attributed to pre-existing microcracks within the core, generated by stress release during coring (for cores from depth; see Section 7.1.2) or caused by weathering (for outcrop or near surface cores). For weak sandstones, where many grain contacts are not cemented, the grain contact itself is a nonlinear element (cf. Hertzian contact theory; Chapter 6).

It is not possible to state any number that characterises the strength or stiffness of a typical sandstone—a "typical" rock simply does not exist. We find sandstones with porosities ranging from less than 5 % up to 40 %. The strength varies with porosity, as illustrated by Fig. 3.15 (from Plumb, 1994).

Plumb found an empirical correlation for the unconfined strength of very clean sandstones:

$$C_0 = 357(1 - 2.8\phi)^2 \qquad (\phi < 0.357) \qquad (3.13)$$

with C_0 given in MPa and ϕ as a fraction.

This represents an upper bound to the unconfined strength. For sandstones containing clay, the strength falls well below the trend suggested

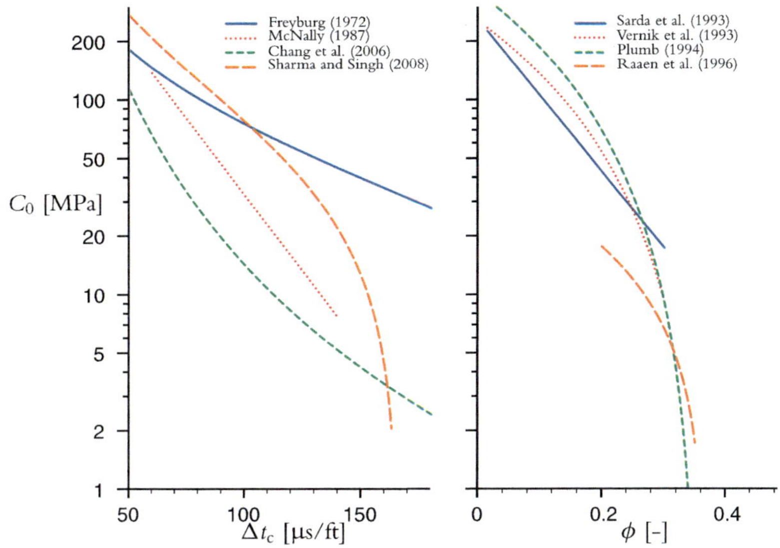

Figure 3.16 A set of empirical relations for prediction of sandstone strength. The formulas are given in Table A.9. Note that the strength-axis is on log scale. References: (Freyburg, 1972; McNally, 1987; Chang et al., 2006; Sharma and Singh, 2008; Sarda et al., 1993; Vernik et al., 1993; Plumb, 1994; Raaen et al., 1996).

by this equation. Note the similarity between Eq. (3.13) and the critical porosity model discussed in Section 6.2.

A collection of empirical relations linking unconfined strength of sandstone to sonic travel time Δt_c and porosity ϕ, respectively, are listed in Table A.9. The relations are plotted in Fig. 3.16.

Young's modulus shows a strong correlation with strength. Deere and Miller (1966) investigated 82 sandstones from 18 different locations in USA. They found the static Young's modulus to be proportional to the unconfined strength C_0 but with significant scatter. Most data presented by Deere and Miller and elsewhere give a proportionality ratio E/C_0 between 100 and 400.

The tensile strength T_0 is not often measured directly on sandstone cores. Values are normally derived from indirect measurements, such as Brazilian tests (see Section 7.5). In most cases, C_0/T_0 is found to be between 5 and 20 for sandstones. Note however that unconfined data are often not very reliable because of pre-existing cracks or flaws in the rocks tested.

When a sandstone starts to yield at low confining pressure, the behaviour is often found dilatant. There is a brittle–ductile transition: Above a certain confining pressure, the rock fails in a ductile manner, and plasticity theory is required to describe the constitutive material behaviour. The failure envelope is often matched to a linear criterion, although it is usually found to be nonlinear, with a decreasing friction angle at high confining stresses. The typical friction angle is near 30° but may vary typically between 20° and 40°. The higher friction angles are found for low porosities and low clay contents.

Sandstones are found to exhibit an end cap in their failure envelope (i.e. material failure under high isotropic stresses). This is often associated with grain crushing but may also result from collapse of the pore structure by grain reorganisation. Localised "compaction bands" of crushed material have been observed in triaxial tests with high-porosity sandstone at high confining pressure (Olsson and Holcomb, 2000). The stress threshold for grain crushing depends on particle size distribution and particle shape, and on the degree of cementation. Data on sandstones between 15 and 25% porosity by Wong et al. (1997) show (Boutéca et al., 2000) that the hydrostatic yield stress is 6–7 times the unconfined strength.

3.4.2 Carbonate

Carbonate rocks are a class of sedimentary rocks that consist primarily of carbonate minerals. Limestone is composed mainly of the minerals calcite or aragonite, both different crystal forms of $CaCO_3$. The other major carbonate rock is dolomite (also known as dolostone), where the main mineral is dolomite $CaMg(CO_3)_2$. Carbonates are typically very heterogeneous, especially with respect to porosity. A high percentage of the porosity can be made up of small pores on a micron scale. However, a secondary porosity system is often present, consisting of fractures and vugs. This complicates the description of the rock, since many properties, including mechanical properties, then depend on the scale of observation. Upscaling based on log data and/or core data is not trivial.

Some empirical correlations for carbonate strength are given in Table A.8. The relations are plotted in Fig. 3.17. In the remainder of this section, we focus on weaker chalk (a variety of limestone), primarily found in deposits in Western Europe.

A microscope image of chalk is shown in Fig. 3.18. Chalk particles originate as skeletons of algae that are called coccospheres, with a typical initial size of 30 µm. During burial the coccospheres are crushed, and most particles (and pores) of present chalk are in the range of a few (1–10) µm, with associated matrix permeabilities between micro- and milliDarcy. The

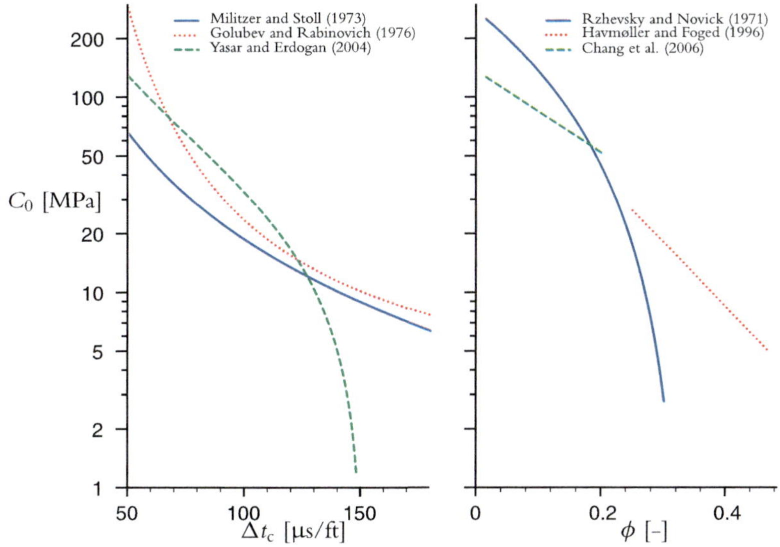

Figure 3.17 A set of empirical relations for prediction of carbonate strength. The formulas are given in Table A.8. Note that the strength-axis is on log scale. References: (Militzer and Stoll, 1973; Golubev and Rabinovich, 1976; Yasar and Erdogan, 2004; Rzhevsky and Novick, 1971; Havmøller and Foged, 1996; Chang et al., 2006).

Figure 3.18 SEM image of Liège outcrop chalk (from Risnes (2001); with permission from the publisher).

predominant mineral is calcite, which has a bulk modulus of 74 GPa and a shear modulus of 27.5 GPa (see Appendix A). In addition to calcite, chalks contain silica and clay minerals.

Chalk porosity may be as high as 70 %. In normally pressured areas, chalk porosity is typically less than 10 % at depths larger than 2000 m. In North Sea reservoirs however, chalk porosities of 15–50 % are found at depths of 2500–3500 m because of overpressure. These reservoirs are also naturally fractured, leading to high reservoir scale permeabilities in the 100 milliDarcy range.

North Sea chalk has been widely studied because of the prominent chalk reservoirs (Ekofisk, Eldfisk, Valhall, Tommeliten and others). Andersen (1995) gives a comprehensive summary of these studies. Havmøller and Foged (1996) compiled a large amount of North Sea reservoir and outcrop chalk data to establish correlations between mechanical properties and porosity (see also Engstrøm, 1992). The overall trends they found (for North Sea chalk) can be summarised in the following equations:

$$C_0 = 174\,\mathrm{e}^{-7.57\phi} \tag{3.14}$$

$$\frac{C_0}{T_0} \sim 8 \tag{3.15}$$

$$E = 225\,\mathrm{e}^{-11.2\phi} \tag{3.16}$$

$$H = 136\,\mathrm{e}^{-9.29\phi} \tag{3.17}$$

where C_0 and T_0 are given in MPa, E and H (uniaxial compaction modulus) are given in GPa, and ϕ is given as a fraction.

As for sandstone, chalk has a failure surface with a brittle shear failure at low confining pressures and also with an end cap at high stresses. Friction angles are typically between 10° and 30°, with a decreasing trend as a function of increasing porosity. Risnes (2001) pointed out that compressive, tensile and end cap failure modes appear to be governed by one failure surface. The end cap is associated with pore collapse, a phenomenon which may significantly reduce the pore space in high porosity chalks. An example of a typical stress–strain curve for chalk under hydrostatic compression (as e.g. given by Dahou et al. (1995)) is shown in Fig. 3.19. The stress reaches a plateau after an initial elastic phase. This plateau represents pore collapse. After pore collapse, the material regains some stiffness if it is continued to be strained.

From Havmøller and Foged (1996) the onset of yielding (pore collapse) in hydrostatic tests is given as

$$\sigma_{\mathrm{y}} = 435\,\mathrm{e}^{-8.31\phi} \tag{3.18}$$

where σ_{y} is given in MPa, and ϕ is given as a fraction. In uniaxial compaction the yield stress is

$$\sigma_{\mathrm{y}}^{K_0} = 363\,\mathrm{e}^{-7.36\phi} \tag{3.19}$$

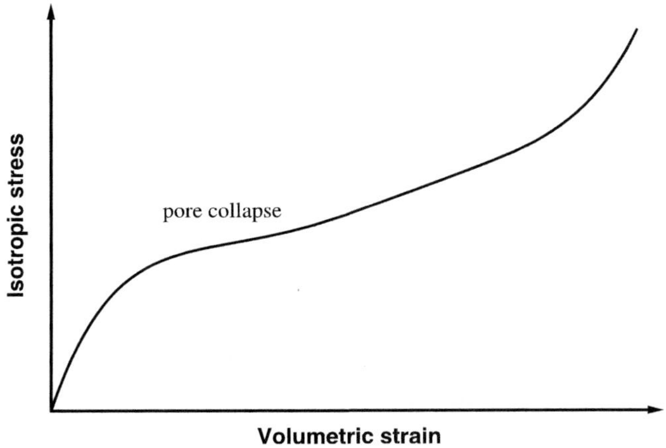

Figure 3.19 Typical stress–strain curve for chalk under hydrostatic compression.

The uniaxial compaction modulus in the plastic regime is

$$H_{pl} = 47.6\,e^{-12.1\phi} \tag{3.20}$$

Because of its significance as a reservoir rock in the North Sea, chalk has been widely studied and characterised. Less comprehensive laboratory studies have been performed with more competent carbonate rocks (e.g. Yasar and Erdogan (2004) and references therein). Yasar and Erdogan studied limestone, dolomite and (metamorphous) marble, and established correlations between P-wave velocity (see Section 5.2) versus strength and Young's modulus:

$$C_0 = 31.5v_p - 63.7 \tag{3.21}$$

$$E = 10.7v_p - 18.7 \tag{3.22}$$

where C_0 is given in MPa, E is given in GPa, and v_p is given in km/s.

3.4.3 Shale

A common geological definition of shale is a fine-grained sedimentary rock that forms from compaction of silt and clay-size mineral particles ("mud") and that exhibits lamination and fissility. Shale is hence defined as a mudstone, but unlike other mudstones, it is anisotropic and fissile, expressing the presence of weak planes along which it may split. The fine-grained nature of shale results in low permeability. With the development of unconventional resources, the word *shale* has become used in engineering language to describe rock with so low permeability that fracturing is required for profitable production of hydrocarbons. Many of those rocks do not qualify as shales in geological terms.

In conventional hydrocarbon reservoirs, shales are common source rocks. Unconventional reservoirs are often produced from the source rock itself. The main role of shales in conventional petroleum engineering is as part of the overburden, where they form the top seal and represent the dominant part of rock that is drilled through to reach the target reservoir. Most of these overburden shales are usually geologically young, which means that they are clay rich and have not undergone significant diagenesis.

Shales may be classified according to their mineralogical composition into silicious, carbonaceous, argillaceous and "mixed" shales (Passey et al., 2010). The so-called unconventionals may belong to any of these groups, with lower clay content shales having better reservoir quality (higher permeability) and usually being easier to stimulate by fracturing. Conventional overburden shales tend to belong to the argillaceous group.

Scanning Electron Images of a North Sea overburden shale are shown in Fig. 3.20. Shales consist to a large extent of clay minerals. From the rock mechanical viewpoint, it is natural to define such a shale as a rock in which clay minerals constitute a load-bearing framework. In practice, this means that the clay content needs to be higher than about 40 %. As demonstrated in Fig. 3.20, shale texture is strongly anisotropic. This is often manifested as a plane of weakness along which the shale easily splits. It means that all non-scalar physical quantities will be anisotropic, a fact which is often neglected in practical rock mechanical analysis due to lack of data.

Because of the abundance of clay minerals, pore sizes in argillaceous shale are very small, typically between 5 and 25 nm. A small pore size is also typical for other categories of shales, partly because of finite contents of clay and silt and partly because of diagenetic processes that have reduced the porosity and pore sizes. In addition, the clay minerals contain adsorbed water on their surfaces as well as bound water inside the stacks of platelets. This water is structurally ordered, has finite shear stiffness and high viscosity and reduces the flow active part of the pore space (Holt and Kolstø, 2017, and references therein).

Shales have, because of the narrow pore space, very large specific surface areas. The mineral surfaces are negatively charged and attract cations from the pore water. Hydrated ions are often attached to the mineral sur-

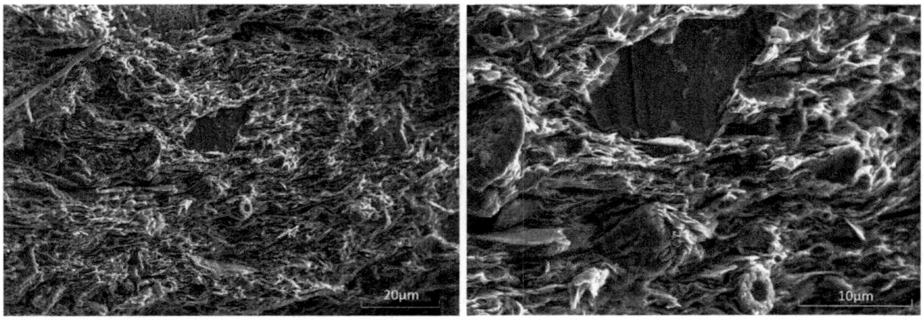

Figure 3.20 SEM images of an overburden North Sea shale, with 40 % porosity and 80 % clay content. Smectite is the most abundant clay mineral. Two different magnifications are shown. Notice that preferred mineral orientation leading to anisotropy is seen in both images. The larger (blackish) object seen in both pictures is mica or illite. Courtesy of SINTEF.

faces. Shale porosity may vary from very low (a few %) to quite high (up to 70 %). Even with the highest porosities, permeability remains very low. The nanometre size pores lead to laboratory measured permeabilities in the nanoDarcy range. Even lower values may be expected for shale under in situ conditions.

The adsorbed and bound water associated with clay minerals means that it is difficult to measure and define the elastic properties of the solid material contained in shales. The content of adsorbed or bound water depends strongly on the conditions under which the shale is measured with respect to for instance humidity and temperature. Thus a single value for the grain compressibility or the grain bulk modulus K_s does not exist.

Various approaches have been used to estimate a representative value of K_s for shales, including theoretical models (Alexandrov and Ryzhova, 1961; Katahara, 1996), extrapolation of wave velocity measurements on shales (Tosaya, 1982; Han, 1986; Castagna et al., 1995, and others summarised in Mavko et al., 1998) or clay powders (Vanario et al., 2003) to zero porosity, and theoretical inversion of velocity measurements on epoxy impregnated shale samples (Wang et al., 2001). Values scatter widely—the extrapolation techniques give bulk modulus between 5 and 25 GPa and shear modulus ranging from 4 to 10 GPa. These numbers clearly depend on which type of clay mineral (kaolinite, smectite, illite) is dominant and in particular on the adsorbed or bound water present within minerals and on mineral surfaces. When sufficiently large crystals are available, such as in the case of muscovite, we may measure directly the stiffness of the solid material in a dry state (no bound water). Such data give for this particular mineral (Carmichael, 1984; Alexandrov and Ryzhova, 1961; Katahara, 1996) a bulk modulus of 51 GPa and a shear modulus of 32 GPa. Note however that this mineral is strongly anisotropic, with moduli in the crystal symmetry plane being 4–6 times larger than moduli representing normal to plane stiffnesses.

The low permeability makes shale a very special rock material to study. It means that rock mechanical tests become extremely time-consuming, since pore pressure equilibration is slow. It also means that cores, when taken out of the Earth, always will have a tendency to be damaged by tensile failure occurring during core retrieval: The core would have to be retrieved at an extremely low rate to prevent overpressure inside. This means that shale cores may be incompletely saturated when at surface conditions. Such a damaged core is clearly not representative of the shale in situ. Re-saturation of a shale core can not be done as in standard petrophysics analysis. Capillary forces may lead to further damage of the shale. Experimental experience (Horsrud et al., 1998; Ewy, 2018; see also Section 7.3) is that the best results are obtained when a pore pressure is enforced (or measured as a response to applied stress).

The difficulties in testing and specimen preparation are discussed further in Chapter 7. It means however that good laboratory test results are scarce and have to be judged carefully in light of the laboratory procedures applied. Also, because shale is not a reservoir rock in conventional reservoirs, it has not been studied to the same extent as other sedimentary rocks.

Lashkaripour and Dusseault (1993) collected a large set of shale data from published literature and in-house studies. The majority of shales tested had porosity below 20%. They found unconfined strength to increase with decreasing porosity and proposed the relation

$$C_0 = 1.00\,\phi^{-1.14} \tag{3.23}$$

where C_0 is given in MPa, and ϕ is given as a fraction.

Their data set also showed strength and stiffness to be related. The ratio between Young's modulus and unconfined strength E/C_0 is typically around 200. The compressive strength is typically 10–15 times higher than the tensile strength. Note that the saturation conditions of these shales during testing are not known for all of them, so the data are likely to be biased by cores that may have dehydrated from their preserved state and hence may show higher strength than if they were fully saturated. This may indeed be the case with several data in the literature from before 2000, as pointed out by Dewhurst et al. (2015). Obviously, some of these data may still have relevance for gas shale studies, although less valuable than if their level of saturation is documented.

Horsrud (2001) studied preserved and fully saturated North Sea shales, most of them with higher porosities (30–55 %). He confirmed a proportionality between the undrained Young's modulus (measured at 50 % of peak shear stress) and unconfined strength with a proportionality constant of ~150. These data show good correlation of strength with porosity, but the correlations are different from those found for lower porosity shales by Lashkaripour and Dusseault. Both works demonstrated that P-wave velocity measurements show a good correlation to shale strength, which is a valuable result when relating shale strength to sonic measurements on drill cuttings or from log/seismic data. Horsrud (2001) found the relation

$$C_0 = 0.77\, v_p^{2.93} \tag{3.24}$$

where C_0 is given in MPa, and v_p is given in km/s.

Due to its plate-like structure, an uncompacted and soft clay (like soil) will normally have a low friction angle. As compaction and consolidation takes place, and porosity and fluid content decrease, the shale will develop a higher friction angle. Horsrud also showed that the friction angle in most tests with high porosity shale was very low, typically 10°–20°.

Dewhurst et al. (2015) presented results from multi-stage triaxial tests with a large number of preserved and fully saturated shales from different parts of the globe. Most porosities were below 15 %, but in establishing their correlations, they used additional data from other sources (including those of Horsrud, 2001). The best correlations for unconfined strength (in MPa) were found versus porosity (as a fraction)

$$C_0 = 1.98\, \phi^{-1.03} \tag{3.25}$$

and versus the product of cationic exchange capacity (CEC) (in centimol per kg, which is the same as milliequivalents/100 g for monovalent ions) and clay content V_{cl} (as fraction):

$$C_0 = 43.33(\mathrm{CEC} \cdot V_{cl})^{-1.13} \tag{3.26}$$

Dewhurst et al. also found a less accurate correlation between unconfined strength and P-wave velocity (normal to bedding) (written with v_p in km/s):

$$C_0 = 0.14 v_p^{3.93} + 3.53 \tag{3.27}$$

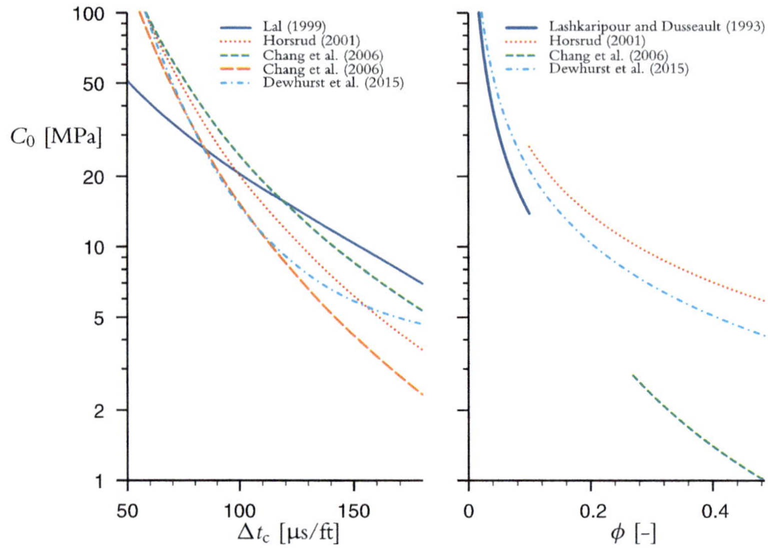

Figure 3.21 A set of empirical relations for prediction of shale strength. The formulas are given in Table A.7. Note that the strength-axis is on log scale. References: (Lal, 1999; Horsrud, 2001; Chang et al., 2006; Dewhurst et al., 2015; Lashkaripour and Dusseault, 1993).

They did however, for all their correlations, indicate lower and upper bounds to their data, demonstrating the scatter, which is normal for naturally heterogeneous rocks. They also note that it is difficult to find good correlations for the friction angle but point to the specific surface area and dielectric coefficient measurements as possible attributes (also noted by Leung and Steiger, 1992).

A collection of empirical relations linking unconfined strength of shales to sonic travel time Δt_c and porosity ϕ, respectively, are listed in Table A.7. The relations are plotted in Fig. 3.21.

As mentioned above, the plate-like structure of shale also leads to anisotropy. Fig. 3.22 shows two samples loaded to failure, one with the largest principal stress direction normal to the bedding (a) and one parallel to the bedding (b). Samples with the maximum load parallel to the bedding will often have a higher failure angle than samples with maximum load normal to the bedding. With equal confining pressures, the sample with maximum load normal to the bedding will require a higher maximum stress to fail. This general trend is reasonable to expect, since the bedding planes will represent possible planes of weakness. The strength anisotropy is also likely to influence borehole failure, in particular, for deviated holes.

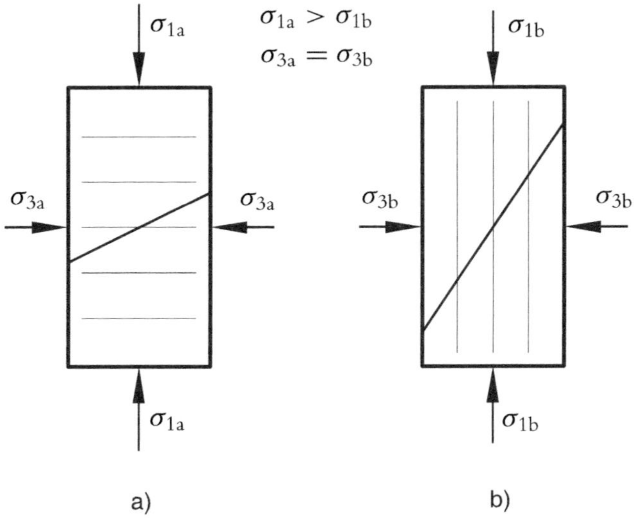

Figure 3.22 Samples loaded to failure: a) Maximum load normal to bedding, b) Maximum load parallel with bedding.

The degree of anisotropy may however vary significantly, depending on both depositional environment and post-depositional processes. Tectonic activities resulting in fracturing of the rock may enhance or disturb this general trend, depending on the direction of the induced fractures.

The anisotropy of shale has been addressed in far more detail by the rock physics than the rock mechanics community. In rock physics, elastic moduli are the target parameters, and as we will see in Chapter 5, the dynamic moduli (derived from density and wave velocities) in general exceed the static moduli. There are several reasons for this, including difference in strain amplitudes and in measurement frequencies. Dynamic moduli are however measures of the undrained response, which are most commonly also derived from static shale testing (Chapter 7), and dynamically measured anisotropies may—depending on the source of anisotropy—be indicative of anisotropy in static properties.

Horne (2013) collected velocity data from cores, logs and also VSPs (Vertical Seismic Profiling) for more than 1000 different shale formations and made a statistical evaluation of the key parameters. Note that there is no filtering based on possible differences in core quality, core preparation, or differences in frequency or length scale. He interpreted the data in terms of the expected transversely isotropic (TI) symmetry for shales with five independent elastic stiffnesses. He correlated P-wave and S-wave anisotropies and pointed to inherent correlations between different anisotropic moduli. These are probably rooted in textural similarities caused by, for example, a spatially ordered distribution of elongated clay mineral platelets. We will return to this in Section 6.6. For practical purposes, we may utilise correlations to reduce the number of parameters required in a practical situation where not all five elastic coefficients are available. Equally important is that a default assumption of isotropy may be bypassed by selecting a base case with a given non-zero anisotropy.

3.4.4 Rock salt

Rock salt is precipitated from sea water and may occur in the Earth as extensive salt beds or interstratified with, for example, sedimentary rocks. The mineralogical composition of natural rock salts varies from very homogeneous (99 % halite; NaCl) to heterogeneous mineral associations. In many areas, salt domes are found, such as beneath the Ekofisk field in the North Sea, where the underlying salt has a strong impact on the reservoir stresses (see Section 3.1). Salt may also be found above reservoirs, such as in the Gulf of Mexico area and offshore Brazil. Sometimes, salt is found to impose drilling problems. Salt has very low permeability and is therefore of interest for long-term storage of hazardous waste.

Salt grains (or crystals) can be between 1 and 50 mm in size. Virgin rock salt is usually characterised by very low porosity (< 0.5–1.0 %), which in some cases may be less than 0.1 %. A significant portion of the pore volume may occur as closed voids containing gas, brine or both. Pore sizes are in the nanometre to micrometre range. Permeability of virgin rock salt in the Earth is probably in the nanoDarcy range or lower (Cosenza and Ghoreychi, 1993). Ultra low permeability of natural intact rock salt enables us to hold this rock impermeable in many practical situations. The negligible permeability of rock salt is also attributed to healing processes and creep taking place under in situ conditions (Horseman, 1988).

A practical problem of measuring porosity and permeability is the solubility of rock salt in the liquids usually used in laboratory routine work. Therefore organic fluids or inert gas is often used for permeability tests. Laboratory measured permeabilities and porosities may be much larger than those representative for field conditions.

The value of Young's modulus in rock salt as obtained in a conventional static test is rate-sensitive. To reduce the effect of rate sensitivity, Young's modulus is usually measured during unloading-reloading paths, yielding E-

values of 10–30 GPa for various types of rock salt. Poisson's ratio ranges between 0.15 and 0.4 being 0.2–0.3 on the average (Hansen et al., 1984).

Some rock salt types have tight cementation and are quite competent, whereas others are loosely cemented and can be crushed by hand pressure. Uniaxial compressive strength C_0 typically ranges from about 15 MPa to 35 MPa. Tensile strength T_0 varies from less than 1 MPa to 2–3 MPa. Low resistance against tensile stresses is one of the characteristic features of rock salt. The ratio C_0/T_0 can be above 20 (Silberschmidt and Silberschmidt, 2000). The angle of internal friction ranges from 40° to 65°. Confining pressure remarkably increases the ductility of rock salt. Axial strain measured at failure in the confined regime can reach 10–25 % (Lux and Rokahr, 1984).

The plastic behaviour of rock salt is linked to very significant creep behaviour. This phenomenon can be explained microscopically by a dislocation glide mechanism (Munson and Wawersik, 1991; Fokker and Kenter, 1994) and can be modelled macroscopically in analogy with time-dependent metal plasticity. The amount of creep strain increases with increasing deviatoric stress and increases strongly with increasing temperature.

References

Alexandrov, K.S., Ryzhova, T.V., 1961. Elastic properties of rock-forming minerals. II. Layered silicates. Bull. USSR Acad. Sci. Geophys. Ser. 9, 1165–1168.

Andersen, M.A., 1995. Petroleum Research in North Sea Chalk. RF — Rogaland Research, Stavanger, Norway.

Anderson, E.M., 1951. The Dynamics of Faulting and Dyke Formation with Applications to Britain. Oliver and Boyd, Edinburgh.

Bak, P., 1996. How Nature Works. The Science of Self-Organized Criticality. Springer-Verlag, New York.

Bjørlykke, K., Høeg, K., 1997. Effects of burial diagenesis on stresses, compaction and fluid flow in sedimentary basins. Mar. Pet. Geol. 14 (3), 267–276. https://doi.org/10.1016/S0264-8172(96)00051-7.

Boutéca, M.J., Sarda, J.-P., Vincké, O., 2000. Constitutive law for permeability evolution of sandstones during depletion. SPE58717. In: SPE International Symposium on Formation Damage. Lafayette, LA, USA.

Byrd, W.D., 1975. Geology of the Ekofisk field, offshore Norway. In: Petroleum and the Continental Shelf of Northwest Europe. John Wiley & Sons, New York, pp. 439–445.

Carmichael, R.S., 1984. Handbook of Physical Properties of Rocks, vol. III. CRC Press.

Castagna, J.P., Han, D.-H., Batzle, M.L., 1995. Issues in rock physics and implications for DHI interpretation. Lead. Edge 14 (8), 883–885. https://doi.org/10.1190/1.1437178.

Chang, C., Zoback, M.D., Khaksar, A., 2006. Empirical relations between rock strength and physical properties in sedimentary rocks. J. Pet. Sci. Eng. 51, 223–237. https://doi.org/10.1016/j.petrol.2006.01.003.

Chenevert, M.E., Sharma, A.K., 1991. Permeability and effective pore pressure of shales. SPE Drill. Completion 8 (1), 28–34. https://doi.org/10.2118/21918-PA.

Cosenza, Ph., Ghoreychi, M., 1993. Coupling between mechanical behavior and transfer phenomena in salt. In: Ghoreychi, M., Bérest, P., Hardy Jr., H.R., Langer, M. (Eds.), Proc. 3rd Conference on the Mechanical Behavior of Salt. Trans Tech Publications, Clausthal-Zellerfeld, pp. 285–307.

Dahou, A., Shao, J.F., Bederiat, M., 1995. Experimental and numerical investigations on transient creep of porous chalk. Mech. Mater. 21 (2), 147–158. https://doi.org/10.1016/0167-6636(95)00004-6.

Deere, D.U., Miller, R.P., 1966. Engineering classification and index properties for intact rock. Techn. Report AFWL-TR-65-116.

Derjaguin, B.V., Churayev, N.V., 1971. Investigation of the properties of water II. J. Colloid Interface Sci. 36 (4), 415–426. https://doi.org/10.1016/0021-9797(71)90376-6.

Dewhurst, D.N., Sarout, J., Delle Piane, C., Siggins, A.F., Raven, M.D., 2015. Empirical strength prediction for preserved shales. Mar. Pet. Geol. 67, 512–525. https://doi.org/10.1016/j.marpetgeo.2015.06.004.

Engelder, T., 1993. Stress Regimes in the Lithosphere. Princeton University Press, Princeton.

Engstrøm, F., 1992. Rock mechanical properties of Danish North Sea chalk. In: Proc. 4th North Sea Chalk Symposium. Deauville, France, 21–23 September.

Ewy, R.T., 2018. Practical approaches for addressing shale testing challenges associated with permeability, capillarity and brine interactions. Geomech. Energy Environ. 14, 3–15. https://doi.org/10.1016/j.gete.2018.01.001.

Fokker, P.A., Kenter, C.J., 1994. The micro mechanical description of rocksalt plasticity. SPE/ISRM28117. In: Proc. Eurock '94. Delft, Netherlands, pp. 705–713.

Fossen, H., 2016. Structural Geology, 2 edition. Cambridge University Press, Cambridge, United Kingdom.

Freyburg, E., 1972. Der Untere und Mittlere Buntsandstein SW-Thüringen in seinen gesteinstechnischen Eigenshaften. Ber. Dtsch. Ges. Geol. Wiss. A; Berlin 17, 911–919.

Golubev, A.A., Rabinovich, G.Y., 1976. Resultaty primeneniya apparatury akusticheskogo karotazha dlya opredeleniya prochnostnych svoistv gornykh porod na mestorozhdeniyach tverdych iskopaemykh. Prikl. Geofiz. Moskva 73, 109–116 (in Russian).

Han, D.-H., 1986. Effects of porosity and clay content on acoustic properties of sandstones and unconsolidated sediments. PhD thesis. Stanford University.

Hansen, F.D., Mellegard, K.D., Senseny, P.E., 1984. Elasticity and strength of ten natural rock salts. In: Hardy Jr., H.R., Langer, M. (Eds.), Proc. 1st Conference on the Mechanical Behavior of Salt. Trans Tech Publications, Clausthal-Zellerfeld, pp. 71–83.

Havmøller, O., Foged, N., 1996. Review of rock mechanics data for chalk. In: Proc. 5th North Sea Chalk Symposium. Reims, France.

Heidbach, O., Rajabi, M., Reiter, K., Ziegler, M., WSM Team, 2016. World Stress Map Database Release 2016. GFZ Data Services.

Holt, R.M., Kolstø, M.I., 2017. How does water near clay mineral surfaces influence the rock physics of shales? Geophys. Prospect. 65 (6), 1615–1629. https://doi.org/10.1111/1365-2478.12503.

Horne, S.A., 2013. A statistical review of mudrock elastic anisotropy. Geophys. Prospect. 61 (4), 817–826. https://doi.org/10.1111/1365-2478.12036.

Horseman, S.T., 1988. Moisture content — a major uncertainty in storage cavity closure prediction. In: Hardy Jr., H.R., Langer, M. (Eds.), Proc. 2nd Conference on the Mechanical Behavior of Salt. Trans Tech Publications, Clausthal-Zellerfeld, pp. 53–68.

Horsrud, P., 2001. Estimating mechanical properties of shale from empirical correlations. SPE Drill. Completion 16 (2), 68–73. https://doi.org/10.2118/56017-PA.

Horsrud, P., Sønstebø, E.F., Bøe, R., 1998. Mechanical and petrophysical properties of North Sea shales. Int. J. Rock Mech. Min. Sci. 35, 1009–1020. https://doi.org/10.1016/S0148-9062(98)00162-4.

Jaeger, J.C., Cook, N.G.W., 1979. Fundamentals of Rock Mechanics, third edition. Chapman and Hall, London.

Karaborni, S., Smit, B., Heidug, W., Urai, J., van Oort, E., 1996. The swelling of clays: molecular simulations of the hydration of montmorillonite. Science 271, 1102–1104. https://doi.org/10.1126/science.271.5252.1102.

Katahara, K.W., 1996. Clay mineral elastic properties. In: SEG Ann. Tech. Conf. Abstracts, pp. 1691–1694.

Katsube, T.J., Best, M.E., Mudford, B.S., 1991. Petrophysical characteristics of shales from the Scotian shelf. Geophysics 56, 1681–1689. https://doi.org/10.1190/1.1442980.

Lal, M., 1999. Shale stability: drilling fluid interaction and shale strength. SPE54356. In: SPE Asia Pacific Oil and Gas Conference and Exhibition. Jakarta, Indonesia.

Lashkaripour, G.R., Dusseault, M.D., 1993. A statistical study of shale properties: relationships among principal shale properties. In: Li, K.S., Lo, S-C.R. (Eds.), Probabilistic Methods in Geotechnical Engineering. Balkema, pp. 195–200.

Leung, P.K., Steiger, R.P., 1992. Dielectric constant measurements: a new, rapid method to characterize shale at the wellsite. IADC/SPE23887. In: Proc. 1992 IADC/SPE Drilling Conf. New Orleans, LA, USA, pp. 401–408.

Lux, K.H., Rokahr, R., 1984. Laboratory investigations and theoretical statements as a basis for the design of caverns in rock salt formations. In: Hardy Jr., H.R., Langer, M. (Eds.), Proc. 1st Conference on the Mechanical Behavior of Salt. Trans Tech Publications, Clausthal-Zellerfeld, pp. 275–310.

Mavko, G., Mukerji, T., Dvorkin, J., 1998. The Rock Physics Handbook. Cambridge University Press, New York (Note: Third edition 2020).

McNally, G.H., 1987. Estimation of coal measures rock strength using sonic and neutron logs. Geoexploration 24, 381–395. https://doi.org/10.1016/0016-7142(87)90008-1.

Militzer, H., Stoll, R., 1973. Einige Beiträge der Geophysik zur Primärdatenerfassung im Bergbau. Neue Bergbautechnik 3, 21–25.

Munson, D.E., Wawersik, W.R., 1991. Constitutive modelling of salt behavior — state of the technology. In: Proc. 7th ISRM Congress. Aachen, Germany.

Olsson, W.A., Holcomb, D.J., 2000. Compaction localization in porous rock. Geophys. Res. Lett. 27 (21), 3537–3540. https://doi.org/10.1029/2000GL011723.

Osborne, M.J., Swarbrick, R.E., 1997. Mechanisms for generating overpressure in sedimentary basins: a reevaluation. AAPG Bull. 81 (6), 1023–1041. https://doi.org/10.1306/522B49C9-1727-11D7-8645000102C1865D.

Park, S.-H., Sposito, G., 2002. Structure of water adsorbed on a mica surface. Phys. Rev. Lett. 89, 085501. https://doi.org/10.1103/PhysRevLett.89.085501.

Passey, Q.R., Bohacs, K.M., Esch, W.L., Klimentidis, R., Sinha, S., 2010. From oil-prone source rock to gas-producing shale reservoir – geologic and petrophysical characterization of unconventional shale-gas reservoirs. SPE131350. In: CPS/SPE International Oil and Gas Conference and Exhibition. Beijing, China.

Pettijohn, F.J., Potter, P.E., Siever, R., 1987. Sand and Sandstone, second edition. Springer-Verlag.

Plumb, R.A., 1994. Influence of composition and texture on the failure properties of clastic rocks. SPE/ISRM28022. In: Proc. Eurock '94. Delft, Netherlands. Balkema, pp. 13–20.

Raaen, A.M., Hovem, K.A., Jøranson, H., Fjær, E., 1996. FORMEL: a step forward in strength logging. SPE36533. In: Proc. 1996 SPE Ann. Techn. Conf. & Exh. Denver, CO, USA, pp. 439–445.

Risnes, R., 2001. Deformation and yield in high porosity outcrop chalk. Phys. Chem. Earth, Part A, Solid Earth Geod. 26, 53–57. https://doi.org/10.1016/S1464-1895(01)00022-9.

Rzhevsky, V., Novick, G., 1971. The Physics of Rocks. MIR Publishers, Moscow.

Sarda, J.-P., Kessler, N., Wicquart, E., Hannaford, K., Deflandre, J.-P., 1993. Use of porosity as a strength indicator for sand production evaluation. SPE26454. In: SPE 68th Ann. Tech. Conf. & Exh. Houston, TX, USA.

Sharma, P.K., Singh, T.N., 2008. A correlation between P-wave velocity, impact strength index, slake durability index and uniaxial compressive strength. Bull. Eng. Geol. Environ. 67, 17–22. https://doi.org/10.1007/s10064-007-0109-y.

Silberschmidt, V.G., Silberschmidt, V.V., 2000. Analysis of cracking in rock salt. Rock Mech. Rock Eng. 33, 53–70. https://doi.org/10.1007/s006030050004.

Skipper, N.T., Refson, K., McConnell, J.D.C., 1991. Computer simulation of interlayer water in 2:1 clays. J. Chem. Phys. 94, 7434–7445. https://doi.org/10.1063/1.460175.

Sone, H., Zoback, M.D., 2014. Viscous relaxation model for predicting least principal stress magnitudes in sedimentary rocks. J. Pet. Sci. Eng. 124, 416–431. https://doi.org/10.1016/j.petrol.2014.09.022.

Swan, G., Cook, J., Bruce, S., Meehan, R., 1989. Strain rate effects in Kimmeridge Bay shale. Int. J. Rock Mech. Min. Sci. Geomech. Abstr. 26, 135–149. https://doi.org/10.1016/0148-9062(89)90002-8.

Teufel, L.W., Farrell, H.E., 1990. In situ stress and natural fracture distribution in the Ekofisk field, North Sea. In: Third North Sea Chalk Symp. Copenhagen, Denmark.

Tosaya, C.A., 1982. Acoustical properties of clay bearing rocks. PhD thesis. Stanford University.

Twiss, R.J., Moores, E.M., 1992. Structural Geology. W.H. Freeman & Co.

van Oort, E., Hale, A.H., Mody, F.K., Sanjit, R., 1996. Transport in shales and the design of improved water-based shale drilling fluids. SPE Drill. Completion 11 (3), 137–146. https://doi.org/10.2118/28309-PA.

Vanario, T., Prasad, M., Nur, A., 2003. Elastic properties of dry clay mineral aggregates, suspensions and sandstones. Geophys. J. Int. 155, 319–326. https://doi.org/10.1046/j.1365-246X.2003.02046.x.

Vernik, L., Bruno, M., Bovberg, C., 1993. Empirical relations between compressive strength and porosity of siliciclastic rocks. Int. J. Rock Mech. Min. Sci. Geomech. Abstr. 30 (7), 677–680. https://doi.org/10.1016/0148-9062(93)90004-W.

Wang, Z., Wang, H., Cates, M.E., 2001. Effective elastic properties of solid clays. Geophysics 66 (2), 428–440. https://doi.org/10.1190/1.1444934.

Warpinski, N.R., 1989. Elastic and viscoelastic calculations of stresses in sedimentary basins. SPE Form. Eval. 4, 522–530. https://doi.org/10.2118/15243-PA.

Wong, T.-f., David, C., Zhu, W., 1997. The transition from brittle faulting to cataclastic flow in porous sandstones: mechanical deformation. J. Geophys. Res. 102 (B2), 3009–3025. https://doi.org/10.1029/96JB03281.

Yasar, E., Erdogan, Y., 2004. Correlating sound velocity with the density, compressive strength and Young's modulus of carbonate rocks. Int. J. Rock Mech. Min. Sci. 41 (5), 871–875. https://doi.org/10.1016/j.ijrmms.2004.01.012.

Yassir, N., Addis, M.A., 2002. Relationships between pore pressure and stress in different tectonic settings. In: Pressure Regimes in Sedimentary Basins and Their Prediction, vol. 76. American Association of Petroleum Geologists, pp. 79–88.

Zoback, M.D., 2007. Reservoir Geomechanics. Cambridge University Press, Cambridge. ISBN 978-0-521-77069-09.

Zoback, M.D., Mastin, L., Barton, C., 1986. In-situ stress measurements in deep boreholes using hydraulic fracturing, wellbore breakouts, and Stonely wave polarization. In: Stephansson, Ove (Ed.), International Symposium on Rock Stress and Rock Stress Measurements. Stockholm, Sweden. International Society for Rock Mechanics and Rock Engineering, pp. 289–299.

Zoback, M.D., Barton, C.A., Brudy, M., Castillo, D.A., Finkbeiner, T., Grollimund, B.R., Moos, D.B., Peška, P., Ward, C.D., Wiprut, D.J., 2003. Determination of stress orientation and magnitude in deep wells. Int. J. Rock Mech. Min. Sci. 40, 1049–1076. https://doi.org/10.1016/j.ijrmms.2003.07.001.

CHAPTER 4

Stresses around boreholes. Borehole failure criteria

Underground formations are always in a stressed state, mostly due to overburden and tectonic stresses (see Chapter 3). When a well is drilled into a formation, stressed solid material is removed. The borehole wall is then supported only by the fluid pressure in the hole. As this fluid pressure generally does not match the in situ formation stresses, there will be a stress redistribution around the well. This may lead to deviatoric stresses greater than the formation can support, and failure may result. Knowledge of the stresses around a well is therefore essential for discussions of well problems.

4.1 Stresses and strains in cylindrical coordinates

To examine the stresses in the rock surrounding a borehole, we need to express the stresses and strains in cylindrical coordinates. The stresses at a point P identified by the coordinates r, θ, z are denoted σ_r, σ_θ, σ_z, $\tau_{r\theta}$, τ_{rz} and $\tau_{\theta z}$. The stresses in a plane perpendicular to the z-axis are indicated in Fig. 4.1(a).

The relations between the stresses in cylindrical and Cartesian coordinates are as follows (compare to Eqs. (1.18)–(1.21); see also Section C.7.1):

$$\sigma_r = \tfrac{1}{2}(\sigma_x + \sigma_y) + \tfrac{1}{2}(\sigma_x - \sigma_y)\cos 2\theta + \tau_{xy}\sin 2\theta \tag{4.1}$$

$$\sigma_\theta = \tfrac{1}{2}(\sigma_x + \sigma_y) - \tfrac{1}{2}(\sigma_x - \sigma_y)\cos 2\theta - \tau_{xy}\sin 2\theta \tag{4.2}$$

$$\sigma_z = \sigma_z \tag{4.3}$$

$$\tau_{r\theta} = \tfrac{1}{2}(\sigma_y - \sigma_x)\sin 2\theta + \tau_{xy}\cos 2\theta \tag{4.4}$$

$$\tau_{rz} = \tau_{xz}\cos\theta + \tau_{yz}\sin\theta \tag{4.5}$$

$$\tau_{\theta z} = \tau_{yz}\cos\theta - \tau_{xz}\sin\theta \tag{4.6}$$

The corresponding relations between the strains are obtained by replacing the stress parameters by the corresponding strains:

$$\varepsilon_r = \tfrac{1}{2}(\varepsilon_x + \varepsilon_y) + \tfrac{1}{2}(\varepsilon_x - \varepsilon_y)\cos 2\theta + \Gamma_{xy}\sin 2\theta \tag{4.7}$$

$$\varepsilon_\theta = \tfrac{1}{2}(\varepsilon_x + \varepsilon_y) - \tfrac{1}{2}(\varepsilon_x - \varepsilon_y)\cos 2\theta - \Gamma_{xy}\sin 2\theta \tag{4.8}$$

Developments in Petroleum Science, Volume 72
ISSN 0376-7361, https://doi.org/10.1016/B978-0-12-822195-2.00013-9

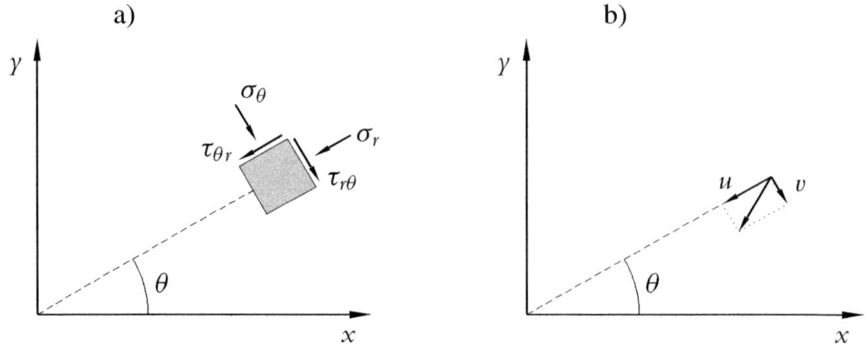

Figure 4.1 Stresses and displacements in cylindrical coordinates.

$$\varepsilon_z = \varepsilon_z \tag{4.9}$$

$$\Gamma_{r\theta} = \tfrac{1}{2}(\varepsilon_y - \varepsilon_x)\sin 2\theta + \Gamma_{xy}\cos 2\theta \tag{4.10}$$

$$\Gamma_{rz} = \Gamma_{xz}\cos\theta + \Gamma_{yz}\sin\theta \tag{4.11}$$

$$\Gamma_{\theta z} = \Gamma_{yz}\cos\theta - \Gamma_{xz}\sin\theta \tag{4.12}$$

The symbols u, v, w are often used for displacements also in cylindrical coordinates, but now redefined as u being the displacement in the radial direction, v the displacement in the tangential direction and w in the axial direction. This redefinition is illustrated in Fig. 4.1(b). The relations between strains and displacements are

$$\varepsilon_r = \frac{\partial u}{\partial r} \tag{4.13}$$

$$\varepsilon_\theta = \frac{u}{r} + \frac{1}{r}\frac{\partial v}{\partial \theta} \tag{4.14}$$

$$\varepsilon_z = \frac{\partial w}{\partial z} \tag{4.15}$$

$$\Gamma_{r\theta} = \frac{1}{2r}\left(\frac{\partial u}{\partial \theta} - v\right) + \frac{1}{2}\frac{\partial v}{\partial r} \tag{4.16}$$

$$\Gamma_{rz} = \frac{1}{2}\left(\frac{\partial w}{\partial r} + \frac{\partial u}{\partial z}\right) \tag{4.17}$$

$$\Gamma_{\theta z} = \frac{1}{2}\left(\frac{1}{r}\frac{\partial w}{\partial \theta} + \frac{\partial v}{\partial z}\right) \tag{4.18}$$

Hooke's law (see Eqs. (1.96)–(1.101)) has the same form in cylindrical as in Cartesian coordinates (for a porous and permeable formation, effective stresses should be used):

$$\sigma_r' = (\lambda_{\text{fr}} + 2G_{\text{fr}})\varepsilon_r + \lambda_{\text{fr}}\varepsilon_\theta + \lambda_{\text{fr}}\varepsilon_z \tag{4.19}$$

$$\sigma_\theta' = \lambda_{\text{fr}}\varepsilon_r + (\lambda_{\text{fr}} + 2G_{\text{fr}})\varepsilon_\theta + \lambda_{\text{fr}}\varepsilon_z \tag{4.20}$$

$$\sigma_z' = \lambda_{\text{fr}}\varepsilon_r + \lambda_{\text{fr}}\varepsilon_\theta + (\lambda_{\text{fr}} + 2G_{\text{fr}})\varepsilon_z \tag{4.21}$$

$$\tau_{r\theta} = 2G_{\text{fr}}\Gamma_{r\theta} \tag{4.22}$$

$$\tau_{rz} = 2G_{\text{fr}}\Gamma_{rz} \tag{4.23}$$

$$\tau_{\theta z} = 2G_{\text{fr}}\Gamma_{\theta z} \tag{4.24}$$

The equations of equilibrium (1.14)–(1.16) may also be expressed in cylindrical coordinates:

$$\frac{\partial \sigma_r}{\partial r} + \frac{1}{r}\frac{\partial \tau_{\theta r}}{\partial \theta} + \frac{\partial \tau_{zr}}{\partial z} + \frac{\sigma_r - \sigma_\theta}{r} + \rho f_r = 0 \tag{4.25}$$

$$\frac{1}{r}\frac{\partial \sigma_\theta}{\partial \theta} + \frac{\partial \tau_{r\theta}}{\partial r} + \frac{\partial \tau_{z\theta}}{\partial z} + \frac{2\tau_{r\theta}}{r} + \rho f_\theta = 0 \tag{4.26}$$

$$\frac{\partial \sigma_z}{\partial z} + \frac{\partial \tau_{rz}}{\partial r} + \frac{1}{r}\frac{\partial \tau_{\theta z}}{\partial \theta} + \frac{\tau_{rz}}{r} + \rho f_z = 0 \tag{4.27}$$

Note that these equations apply to the total stresses.

4.2 Stresses in a hollow cylinder

The hollow cylinder model, as sketched in Fig. 4.2, is a simple example of a borehole in a stressed formation. The model is important in itself, as laboratory tests concerning well stability often are carried out on such samples. The hollow cylinder model also provides a model for vertical wells through formations where the horizontal stresses are equal.

4.2.1 The equilibrium equations

The infinite hollow cylinder has full rotational symmetry about the axis of the cylinder and full translational symmetry along the axis. As a starting point, we assume that the cylinder is loaded with an axial stress σ_{v}, and we will derive expressions for the stresses in the cylinder as it is loaded with an internal pressure p_{w} and an external stress $\sigma_{r\text{o}}$. The external stresses on the cylinder are at all times normal and independent of θ and z. Therefore the cylindrical coordinate axes (with the z-axis along the cylinder axis) also represent the principal stress directions.

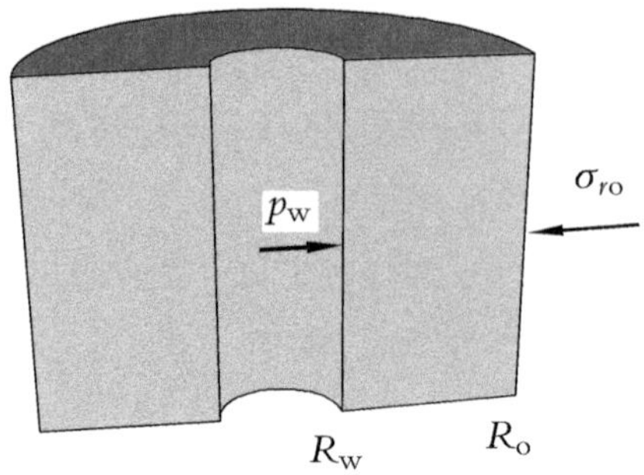

Figure 4.2 Section of the hollow cylinder model.

It is then clear that the only deformation will be in the radial direction and that there will be no variation along the axis. Thus the model will be in plane strain (see Section 1.2.4), and $\varepsilon_z = 0$. (If we consider a finite height cylinder, e.g., in a laboratory test, the plane strain will apply only if the top and bottom platens are fixed and if end effects can be ignored.)

Ignoring body forces, the equations of equilibrium (4.25)–(4.27) are simplified to the single equation

$$\frac{d\sigma_r}{dr} + \frac{\sigma_r - \sigma_\theta}{r} = 0 \tag{4.28}$$

We now replace the stress with the radial displacement u using Eqs. (4.19)–(4.20) (Hooke's law), the definition of effective stress (Eq. (1.191)), and the strains in terms of u from Eqs. (4.13) and (4.14). The result is

$$\frac{d^2u}{dr^2} + \frac{1}{r}\frac{du}{dr} - \frac{u}{r^2} + \frac{\alpha}{\lambda_{fr} + 2G_{fr}}\frac{dp_f}{dr} = 0 \tag{4.29}$$

When this equation is solved, the radial and tangential strain can be determined, and hence the stresses can be found using Hooke's law.

4.2.2 Stress distributions with constant pore pressure

For constant pore pressure, the displacement equation (4.29) reduces to

$$\frac{d^2u}{dr^2} + \frac{1}{r}\frac{du}{dr} - \frac{u}{r^2} = \frac{d}{dr}\left(\frac{du}{dr} + \frac{u}{r}\right) = \frac{d}{dr}\left(\frac{1}{r}\frac{d(ru)}{dr}\right) = 0 \tag{4.30}$$

The sum in the parenthesis in the middle can be recognised as the sum of radial and tangential strain, which thus is seen to be constant. Together with the plane strain condition $\varepsilon_z = 0$, this means that for the present case, the elastic rearrangement of the stresses around a wellbore does not result in any volumetric changes. Further, from Hooke's law (Eq. (1.102)) it follows that the mean stress is constant.

The expression

$$u = C_1 r + \frac{C_2}{r} \tag{4.31}$$

is the general solution of the displacement equation (4.30) with integration constants C_1 and C_2.

The radial and tangential strains are given by

$$\varepsilon_r = \frac{\mathrm{d}u}{\mathrm{d}r} = C_1 - \frac{C_2}{r^2} \tag{4.32}$$

$$\varepsilon_\theta = \frac{u}{r} = C_1 + \frac{C_2}{r^2} \tag{4.33}$$

Substitution of the strain expressions (4.32)–(4.33) into Eq. (4.19) gives

$$\sigma_r - \alpha p_{\mathrm{f}} = (2\lambda_{\mathrm{fr}} + 2G_{\mathrm{fr}})C_1 - 2G_{\mathrm{fr}}\frac{C_2}{r^2} \tag{4.34}$$

where p_{f} is the constant pore fluid pressure.

Before matching the boundary conditions, it is convenient to assemble all the r-independent terms in one constant and the r^{-2} terms in another:

$$\sigma_r = C_1' + \frac{C_2'}{r^2} \tag{4.35}$$

This small mathematical trick simplifies subsequent calculations and also makes it immediately clear that p_{f} and the elastic constants will not appear in the final answer.

Similarly, we find for the tangential stress

$$\sigma_\theta = C_1' - \frac{C_2'}{r^2} \tag{4.36}$$

The two integration constants may now be found from the boundary conditions for the radial stress

$$\sigma_r = p_{\mathrm{w}} \quad \text{for} \quad r = R_{\mathrm{w}} \tag{4.37}$$

$$\sigma_r = \sigma_{r\mathrm{o}} \quad \text{for} \quad r = R_{\mathrm{o}} \tag{4.38}$$

where we write p_{w} and σ_{ro} for the radial stresses at the inner and outer boundaries, respectively (see Fig. 4.2).

The result is

$$C_1' = \frac{R_{\mathrm{o}}^2 \sigma_{ro} - R_{\mathrm{w}}^2 p_{\mathrm{w}}}{R_{\mathrm{o}}^2 - R_{\mathrm{w}}^2} \tag{4.39}$$

$$C_2' = -\frac{R_{\mathrm{o}}^2 R_{\mathrm{w}}^2}{R_{\mathrm{o}}^2 - R_{\mathrm{w}}^2} \left(\sigma_{ro} - p_{\mathrm{w}} \right) \tag{4.40}$$

Note that as $R_{\mathrm{o}} \to \infty$, we have $C_1 = \sigma_{\mathrm{o}}/(2\lambda_{\mathrm{fr}} + 2G_{\mathrm{fr}})$, thats is, $u \to \infty$ as $r \to \infty$. This is because our reference is zero stress, and hence imposing the in situ stresses gives infinite deformation at infinity. If we alternatively use the in situ stress state as reference, then we can immediately simplify the equations since $C_1 = 0$ because radial deformation must be bounded.

The radial and tangential stresses in a hollow porous cylinder can then be written

$$\sigma_r = \frac{R_{\mathrm{o}}^2 \sigma_{ro} - R_{\mathrm{w}}^2 p_{\mathrm{w}}}{R_{\mathrm{o}}^2 - R_{\mathrm{w}}^2} - \frac{R_{\mathrm{o}}^2}{R_{\mathrm{o}}^2 - R_{\mathrm{w}}^2} \frac{R_{\mathrm{w}}^2}{r^2} \left(\sigma_{ro} - p_{\mathrm{w}} \right) \tag{4.41}$$

$$\sigma_\theta = \frac{R_{\mathrm{o}}^2 \sigma_{ro} - R_{\mathrm{w}}^2 p_{\mathrm{w}}}{R_{\mathrm{o}}^2 - R_{\mathrm{w}}^2} + \frac{R_{\mathrm{o}}^2}{R_{\mathrm{o}}^2 - R_{\mathrm{w}}^2} \frac{R_{\mathrm{w}}^2}{r^2} \left(\sigma_{ro} - p_{\mathrm{w}} \right) \tag{4.42}$$

Note that the sum of the radial and tangential stresses is a constant independent of r. Since the mean stress is constant (see the discussion following Eq. (4.30)), this means that the axial stress is constant.

When these hollow cylinder expressions are applied to well conditions, we may assume that $R_{\mathrm{o}} \gg R_{\mathrm{w}}$. Assuming a vertical borehole, we write σ_{h} (the horizontal stress) instead of σ_{ro} and find

$$\sigma_r = \sigma_{\mathrm{h}} - (\sigma_{\mathrm{h}} - p_{\mathrm{w}}) \frac{R_{\mathrm{w}}^2}{r^2} = \left(1 - \frac{R_{\mathrm{w}}^2}{r^2} \right) \sigma_{\mathrm{h}} + \frac{R_{\mathrm{w}}^2}{r^2} p_{\mathrm{w}} \tag{4.43}$$

$$\sigma_\theta = \sigma_{\mathrm{h}} + (\sigma_{\mathrm{h}} - p_{\mathrm{w}}) \frac{R_{\mathrm{w}}^2}{r^2} = \left(1 + \frac{R_{\mathrm{w}}^2}{r^2} \right) \sigma_{\mathrm{h}} - \frac{R_{\mathrm{w}}^2}{r^2} p_{\mathrm{w}} \tag{4.44}$$

$$\sigma_z = \text{const} \tag{4.45}$$

Fig. 4.3 shows an illustration of the stress distribution around a borehole as predicted by the hollow cylinder model, assuming that the borehole pressure is well below the far-field stress. Note that the tangential stress (commonly referred to as the *hoop stress*) is significantly increased in the near borehole region. The effect that the stress in the vicinity of an inho-

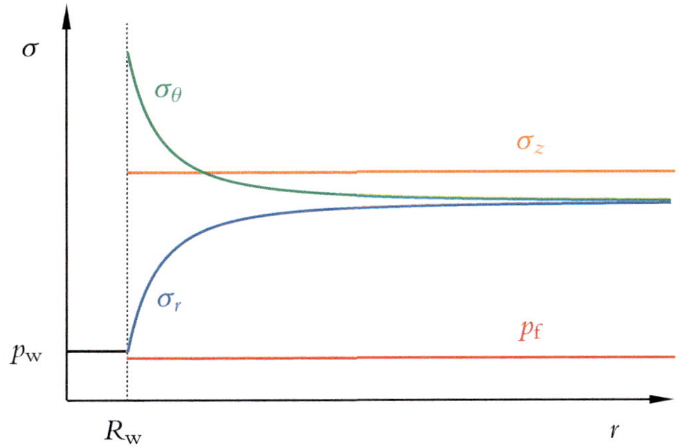

Figure 4.3 Stresses around a borehole in a linear elastic formation.

mogeneity (in our case the hole) rises above the far-field stress is commonly referred to as *stress concentration.*

The radial displacement u can be determined by calculating C_1 and C_2 from C_1' and C_2' and inserting into Eq. (4.31). Assuming that $R_o \gg R_w$, the result is

$$u = \frac{\sigma_h - \alpha p_f}{2\lambda_{fr} + 2G_{fr}} r + \frac{\sigma_h - p_w}{2G_{fr}} \frac{R_w^2}{r} \tag{4.46}$$

Note that u given in Eq. (4.46) is the displacement relative to the state where $\sigma_h = p_w = 0$. This expression is relevant for a hollow cylinder test in the laboratory.

In a field situation, it is convenient to use the in situ stress state as the reference. Then the displacement due to the drillout is seen to be given by

$$u_d = \frac{\sigma_h - p_w}{2G_{fr}} \frac{R_w^2}{r} \tag{4.47}$$

Eq. (4.47) shows that a reduction in well pressure results in a positive radial displacement, which corresponds to a reduction of the borehole radius, as expected.

4.2.3 Stress distributions with varying pore pressure

The displacement equilibrium equation (4.29) is readily integrated to give

$$\frac{1}{r}\frac{d}{dr}(ru) + \frac{\alpha}{\lambda_{fr} + 2G_{fr}} p_f = 2C_1 \tag{4.48}$$

where $2C_1$ is an integration constant. (The factor 2 will disappear in the next steps.)

The solution of this equation is

$$u = C_1 r + \frac{C_2}{r} - \frac{\alpha}{\lambda_{\mathrm{fr}} + 2G_{\mathrm{fr}}} \frac{1}{r} \int_{R_\mathrm{w}}^{r} r' p_\mathrm{f} \, \mathrm{d}r' \tag{4.49}$$

where C_2 is a second integration constant.

The stresses are found by proceeding in the same way as for the constant pressure case, that is, by computing the strains using Hooke's law and matching boundary conditions.

Defining

$$\Delta p_\mathrm{f}(r) = p_\mathrm{f}(r) - p_\mathrm{fo} \tag{4.50}$$

we can write the result

$$\begin{aligned} \sigma_r = & \frac{R_\mathrm{o}^2 \left(r^2 - R_\mathrm{w}^2\right) \sigma_{r\mathrm{o}} + R_\mathrm{w}^2 \left(R_\mathrm{o}^2 - r^2\right) p_\mathrm{w}}{r^2 \left(R_\mathrm{o}^2 - R_\mathrm{w}^2\right)} \\ & + \frac{2\eta}{r^2} \left(\int_{R_\mathrm{w}}^{r} r' \Delta p_\mathrm{f}(r') \, \mathrm{d}r' - \frac{r^2 - R_\mathrm{w}^2}{R_\mathrm{o}^2 - R_\mathrm{w}^2} \int_{R_\mathrm{w}}^{R_\mathrm{o}} r' \Delta p_\mathrm{f}(r') \, \mathrm{d}r' \right) \end{aligned} \tag{4.51}$$

$$\begin{aligned} \sigma_\theta = & \frac{R_\mathrm{o}^2 \left(r^2 + R_\mathrm{w}^2\right) \sigma_{r\mathrm{o}} - R_\mathrm{w}^2 \left(R_\mathrm{o}^2 + r^2\right) p_\mathrm{w}}{r^2 \left(R_\mathrm{o}^2 - R_\mathrm{w}^2\right)} \\ & - \frac{2\eta}{r^2} \left(\int_{R_\mathrm{w}}^{r} r' \Delta p_\mathrm{f}(r') \, \mathrm{d}r' - r^2 \Delta p_\mathrm{f}(r) \right. \\ & \left. + \frac{r^2 + R_\mathrm{w}^2}{R_\mathrm{o}^2 - R_\mathrm{w}^2} \int_{R_\mathrm{w}}^{R_\mathrm{o}} r' \Delta p_\mathrm{f}(r') \, \mathrm{d}r' \right) \end{aligned} \tag{4.52}$$

$$\begin{aligned} \sigma_z = & \sigma_\mathrm{v} + 2\nu_{\mathrm{fr}} \left(\sigma_{r\mathrm{o}} - p_\mathrm{w}\right) \frac{R_\mathrm{w}^2}{R_\mathrm{o}^2 - R_\mathrm{w}^2} + 2\eta \Delta p_\mathrm{f}(r) \\ & - 4\eta \frac{\nu_{\mathrm{fr}}}{R_\mathrm{o}^2 - R_\mathrm{w}^2} \int_{R_\mathrm{w}}^{R_\mathrm{o}} r' \Delta p_\mathrm{f}(r') \, \mathrm{d}r' \end{aligned} \tag{4.53}$$

where we have introduced the *poroelastic stress coefficient*

$$\eta = \frac{G_{\mathrm{fr}}}{\lambda_{\mathrm{fr}} + 2G_{\mathrm{fr}}} \alpha = \frac{1 - 2\nu_{\mathrm{fr}}}{2(1 - \nu_{\mathrm{fr}})} \alpha \tag{4.54}$$

It is clear from the definition that $0 < \eta < 0.5$ (assuming that $\nu > 0$).

Assuming that $R_\mathrm{o} \gg R_\mathrm{w}$ and $\sigma_{r\mathrm{o}} = \sigma_\mathrm{h}$, the equations simplify to

$$\sigma_r = \left(1 - \frac{R_\mathrm{w}^2}{r^2}\right) \sigma_\mathrm{h} + \frac{R_\mathrm{w}^2}{r^2} p_\mathrm{w} + \frac{2\eta}{r^2} \int_{R_\mathrm{w}}^{r} r' \Delta p_\mathrm{f}(r') \, \mathrm{d}r' \tag{4.55}$$

$$\sigma_\theta = \left(1 + \frac{R_\mathrm{w}^2}{r^2}\right)\sigma_\mathrm{h} - \frac{R_\mathrm{w}^2}{r^2}p_\mathrm{w} - \frac{2\eta}{r^2}\int_{R_\mathrm{w}}^{r} r'\Delta p_\mathrm{f}(r')\,\mathrm{d}r' + 2\eta\Delta p_\mathrm{f}(r) \tag{4.56}$$

$$\sigma_z = \sigma_\mathrm{v} + 2\eta\Delta p_\mathrm{f}(r) \tag{4.57}$$

The stresses at the borehole wall are seen to be

$$\sigma_r = p_\mathrm{w} \tag{4.58}$$

$$\sigma_\theta = 2\sigma_\mathrm{h} - p_\mathrm{w} - 2\eta\left(p_\mathrm{fo} - p_\mathrm{f}(R_\mathrm{w})\right) \tag{4.59}$$

$$\sigma_z = \sigma_\mathrm{v} - 2\eta\left(p_\mathrm{fo} - p_\mathrm{f}(R_\mathrm{w})\right) \tag{4.60}$$

where $p_\mathrm{f}(R_\mathrm{w})$ is the reservoir fluid pressure at the wellbore wall. If there is open communication between the well fluid and the formation, then $p_\mathrm{f}(R_\mathrm{w})$ is equal to the well pressure p_w. If there is a mud cake, then $p_\mathrm{f}(R_\mathrm{w})$ is the fluid pressure behind the mud cake.

From a physical point of view, this result can be explained as follows: If the well pressure is higher than the reservoir fluid pressure, then fluids are injected, and the pore fluid pressure is increased around the well. This will give a tendency for the material to expand and the stresses increase. If on the other hand the well is on production, then the well pressure is lower than the reservoir pressure. The lower fluid pressure around the well will make the formation shrink and hence the corresponding reduction in the tangential and axial stress.

Note that the wall stresses are independent of the details of the far-field radial variation in $p_\mathrm{f}(r)$, and only the difference between the value and the value at the borehole wall appears in the equations.

4.2.3.1 The superposition principle

Since the governing equations are linear, the final expressions can be seen as a sum of different basic loads, which are uncoupled. For the hollow cylinder, these basic loads are the external radial stress, the borehole pressure and the pore pressure.

As an illustration, consider Eqs. (4.43) and (4.44) for $p_\mathrm{w} = 0$:

$$\sigma_r = \left(1 - \frac{R_\mathrm{w}^2}{r^2}\right)\sigma_\mathrm{h} \tag{4.61}$$

$$\sigma_\theta = \left(1 + \frac{R_\mathrm{w}^2}{r^2}\right)\sigma_\mathrm{h} \tag{4.62}$$

To go back to the case with well pressure, we may let $\sigma_h \to \sigma_h - p_w$ and add a uniform stress p_w. This gives the far-field stress σ_h and wellbore stress p_w, and Eqs. (4.43) and (4.55) result.

This explains why the well pressure term $p_w R_w^2/r^2$ is the same in Eqs. (4.43) and (4.55). Further, we expect the well pressure term to be the same if we introduce a more complicated far-field stress, for example, corresponding to a deviated hole in an anisotropic stress field. See Eq. (4.85).

4.2.3.2 Radial flow

A special case is obtained if the pressure variation is given by stationary radial flow into (or out of) the well. We will first study the case of a hollow cylinder with outer radius R_o and look for a stationary solution, that is, we assume that all transients have died away.

The fluid pressure can then be calculated from Darcy's law (see Section 1.9.1)

$$\frac{Q_o}{2\pi rh} = \frac{k}{\eta_f}\frac{dp_f}{dr} \tag{4.63}$$

where Q_o is the constant flow rate, h is the height of the cylinder, k is the permeability, and η_f is the fluid viscosity. Note that Q_o is here defined to be positive for flow into the well.

Solving the equation above and assuming a constant pressure p_{fo} at the outer radius R_o, the pressure is given by the well-known equation

$$p_f = p_w + \frac{p_{fo} - p_w}{\ln\frac{R_o}{R_w}}\ln\frac{r}{R_w} = p_{fo} + \frac{p_{fo} - p_w}{\ln\frac{R_o}{R_w}}\ln\frac{r}{R_o} \tag{4.64}$$

Inserting this result into Eqs. (4.51)–(4.53), we find the equations published by Risnes et al. (1982):

$$\begin{aligned}\sigma_r = \sigma_h &+ (\sigma_h - p_w)\frac{R_w^2}{R_o^2 - R_w^2}\left[1 - \left(\frac{R_o}{r}\right)^2\right]\\ &- (p_{fo} - p_w)\eta\left\{\frac{R_w^2}{R_o^2 - R_w^2}\left[1 - \left(\frac{R_o}{r}\right)^2\right] + \frac{\ln(R_o/r)}{\ln(R_o/R_w)}\right\}\end{aligned} \tag{4.65}$$

$$\begin{aligned}\sigma_\theta = \sigma_h &+ (\sigma_h - p_w)\frac{R_w^2}{R_o^2 - R_w^2}\left[1 + \left(\frac{R_o}{r}\right)^2\right]\\ &- (p_{fo} - p_w)\eta\left\{\frac{R_w^2}{R_o^2 - R_w^2}\left[1 + \left(\frac{R_o}{r}\right)^2\right] + \frac{\ln(R_o/r) - 1}{\ln(R_o/R_w)}\right\}\end{aligned} \tag{4.66}$$

$$\sigma_z = \sigma_v + 2\nu_{fr}\left(\sigma_h - p_w\right)\frac{R_w^2}{R_o^2 - R_w^2} - (p_{fo} - p_w)\eta\left\{\frac{2\nu_{fr}R_w^2}{R_o^2 - R_w^2} + \frac{2\ln(R_o/r) - \nu_{fr}}{\ln(R_o/R_w)}\right\} \tag{4.67}$$

A common assumption in reservoir engineering is that the reservoir pressure p_{fo} is constant at all times at a drainage radius R_e. (We underline, however, that in reality, true stationary flow will never be reached in an infinite formation.)

If the drainage radius R_e can be taken to be much larger than the well radius, then the equations are simplified to

$$\sigma_r = \sigma_h - (\sigma_h - p_w)\left(\frac{R_w}{r}\right)^2 + (p_{fo} - p_w)\eta\left[\left(\frac{R_w}{r}\right)^2 - \frac{\ln(R_e/r)}{\ln(R_e/R_w)}\right] \tag{4.68}$$

$$\sigma_\theta = \sigma_h + (\sigma_h - p_w)\left(\frac{R_w}{r}\right)^2 - (p_{fo} - p_w)\eta\left[\left(\frac{R_w}{r}\right)^2 + \frac{\ln(R_e/r) - 1}{\ln(R_e/R_w)}\right] \tag{4.69}$$

$$\sigma_z = \sigma_v - (p_{fo} - p_w)\eta\frac{2\ln(R_e/r) - \nu_{fr}}{\ln(R_e/R_w)} \tag{4.70}$$

These equations can be used as an approximation to a producing well. Eq. (4.70) depends on the plane strain condition, which is valid for an infinitely long borehole. In a real case the flow will be in a reservoir of limited height, and hence a solution based on generalised plane stress (see Section 1.2.4) would be valid when the depleted zone extends far into the reservoir, and the full vertical stress is transmitted to the reservoir. In practice, there will be some arching of the overburden preventing the maintenance of the full vertical stress, and hence the true solution will be somewhere in between.

4.2.4 Stress distributions with heat flow

When a well is drilled, the drilling mud will alter the temperature of the surrounding formation. When the temperature changes, the formation expands or shrinks, and the stresses will change. (See Section 1.5 for the basic formulas of thermoelasticity.)

From the correspondence between poroelasticity and thermoelasticity (see Section 1.6.9 and Eqs. (4.58)–(4.60)), we may immediately write the expressions for the stresses at the wellbore wall:

$$\sigma_r = p_{\rm w} \tag{4.71}$$

$$\sigma_\theta = 2\sigma_{\rm h} - p_{\rm w} + \frac{E_{\rm fr}}{1-\nu_{\rm fr}}\alpha_{\rm T}(T_{\rm w} - T_{\rm o}) \tag{4.72}$$

$$\sigma_z = \sigma_{\rm v} + \frac{E_{\rm fr}}{1-\nu_{\rm fr}}\alpha_{\rm T}(T_{\rm w} - T_{\rm o}) \tag{4.73}$$

where $T_{\rm w}$ is the temperature in the well, and $T_{\rm o}$ is the reservoir temperature. The thermally induced changes in stress affect both tangential and axial stresses. The radial stress is of course equal to the well pressure regardless of any temperature changes.

4.2.5 Stress distributions in nonlinear formations

The results in Section 4.2.2 as displayed in Fig. 4.3 showed that there is a rapid increase in the stress deviation towards the borehole wall. These results were based on the assumption of linear elasticity, that is, that the elastic moduli of the rock are independent of the stress state. However, elastic moduli of rocks are normally stress dependent to some extent (see Section 1.8). Thus it is reasonable to assume that the elastic moduli close to the wellbore wall will differ from those of the virgin formation due to the stress alteration in the vicinity of the borehole.

Santarelli et al. (1986); Santarelli (1987) modelled this effect by assuming that Young's modulus depends on the minor principal stress σ_r as $E(\sigma_r) = E_0\sigma_r^a$. Here E_0 and a are empirical constants; normally, $0 < a < 1$, and E_0 may be interpreted as the value of Young's modulus as measured in uniaxial compression tests. Based on this assumption, they found the following expressions for σ_r and σ_θ corresponding to Eqs. (4.43)–(4.44):

$$\sigma_r = \sigma_{\rm h}\left\{\left[\left(\frac{p_{\rm w}}{\sigma_{\rm h}}\right)^{1-a} - 1\right]\left(\frac{R_{\rm w}}{r}\right)^N + 1\right\}^{\frac{1}{1-a}} \tag{4.74}$$

$$\sigma_\theta = \frac{N}{1-a}\sigma_{\rm h}\left(\frac{\sigma_r}{\sigma_{\rm h}}\right)^a + M\sigma_r \tag{4.75}$$

where

$$N = \frac{1}{1-\nu_{\rm fr}}[(1-2\nu_{\rm fr})(1-a)+1] \tag{4.76}$$

$$M = \frac{\nu_{\rm fr}(1-a)-1}{(1-\nu_{\rm fr})(1-a)} = \frac{1-a-N}{1-a} \tag{4.77}$$

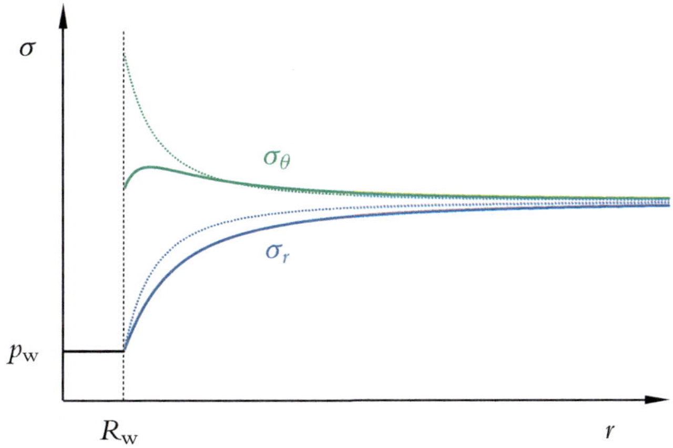

Figure 4.4 Stresses around a borehole in a formation with pressure-dependent elastic properties. The dotted lines show the linear elastic solution.

The implications for the stress distribution around the borehole are significant. Fig. 4.4 shows the variation of stresses as functions of radial distance as calculated by Eqs. (4.74)–(4.77) using $a = 0.5$ and $\nu_{\mathrm{fr}} = 0.2$. Also, the vertical stress is reduced close to the wellbore, as σ_z is linked to σ_r and σ_θ through Hooke's law (although it is no longer linear).

An important consequence of the non-constant E-modulus is the relative reduction of σ_θ near the wellbore wall. This effect is most pronounced for low borehole pressures, for which it may even happen that the difference $\sigma_\theta - \sigma_r$ is larger at some distance into the formation than at the borehole wall. Thus the commonly observed effect that Young's modulus increases with increasing confining pressure has the consequence that the stress deviation near the borehole wall at low borehole pressures is reduced compared to the linear elastic solution.

4.3 Elastic stresses around circular wells—the general solution

In the previous sections, we studied the simple example of a vertical borehole in a formation with isotropic horizontal stress. We will now proceed to describe the general elastic situation: the borehole is deviated, and the horizontal stress is anisotropic.

We will assume that the principal stresses are the vertical stress σ_v, and the major and minor horizontal stresses are σ_H and σ_h. (The generalisation to the case where the vertical stress is not a principal stress is trivial.)

As an introduction, a brief historical overview may be appropriate: The stress distribution around a circular hole in an infinite plate in one-dimensional tension was published by Kirsch (1898).[1] The Kirsch formulas easily generalise to a vertical borehole with unequal far-field stress, so that this solution is also often referred to as the Kirsch equations.

The "standard", easily available reference in petroleum rock mechanics literature for the full solution is Bradley (1979), who quoted a report by Fairhurst (1968). However, according to Peška and Zoback (1995), the equations were first published by Hiramatsu and Oka (1962).

There appears to be a sign error in part of the $\tau_{r\theta}$ expression in Bradley's paper. This error has propagated in the literature, for instance, in the first edition of this book and in several other works. An easily available reference that appears to be correct is Hiramatsu and Oka (1968).

4.3.1 Transformation formulas

The in situ principal stresses define a coordinate system, which we denote (x', y', z'), as indicated in Fig. 4.5. We take σ_v to be parallel to z', σ_H to be parallel to x', and σ_h to be parallel to y'.

We introduce a second coordinate system (x, y, z) such that the z-axis points along the axis of the hole, the x-axis points towards the lowermost radial direction of the hole, and the y-axis is horizontal (see Fig. 4.5).

A transform from (x', y', z') to (x, y, z) can be obtained in two operations (see Fig. 4.6):

1. a rotation a around the z'-axis,
2. a rotation i around the y-axis.

The transformation can be described mathematically by the direction cosines, where $l_{ij'}$ is the cosine of the angle between the i-axis and the

[1] The famous paper of Ernst Gustav Kirsch (1841–1901) is an example of a classic paper that is probably more often cited than read. As a result, many variants of the title and citation details exist in the literature, some of which suggest that the writer does not understand the German language. It is therefore a pleasure to bring the correct citation here. By the way, the original paper does not reveal Kirsch' first name or initials. The Kirsch paper is not a good starting point for those wishing to the study the derivation of the Kirsch equations, as the paper relies on detailed references to a contemporary German textbook. We suggest that the interested reader should instead study Chapter 8 of Jaeger et al. (2007), Chapter 10 of Jaeger and Cook (1979) or Chapter 5 of Charlez (1991).

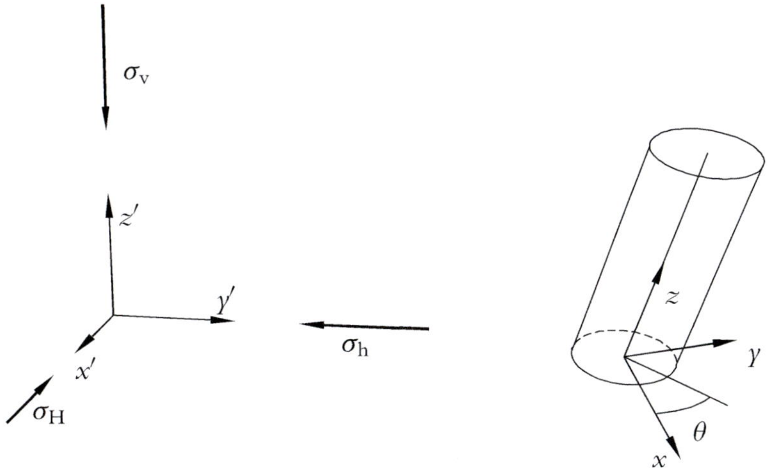

Figure 4.5 Coordinate system for a deviated borehole.

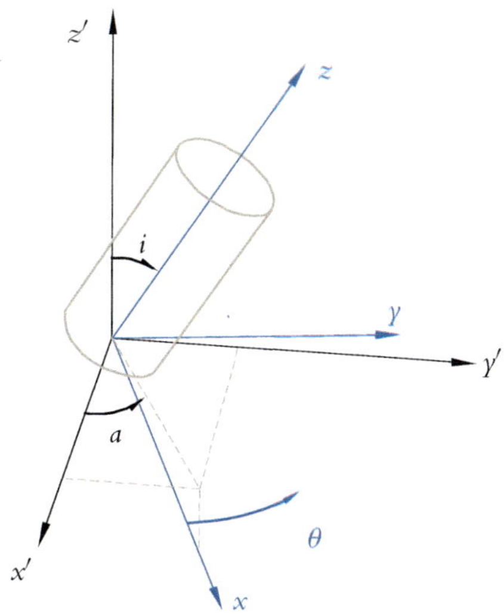

Figure 4.6 The transformation geometry.

j'-axis. With reference to Fig. 4.6, it is straightforward to derive the following expressions:

$$\begin{array}{lll} l_{xx'} = \cos a \cos i & l_{xy'} = \sin a \cos i & l_{xz'} = -\sin i \\ l_{yx'} = -\sin a & l_{yy'} = \cos a & l_{yz'} = 0 \\ l_{zx'} = \cos a \sin i & l_{zy'} = \sin a \sin i & l_{zz'} = \cos i \end{array} \tag{4.78}$$

(An alternate formulation, using the Euler angles, is discussed in Appendix C.6.)

Expressed in the (x, y, z) coordinate system, the formation stresses σ_{H}, σ_{h} and σ_{v} become

$$\sigma_x^{\mathrm{o}} = l_{xx'}^2 \sigma_{\mathrm{H}} + l_{xy'}^2 \sigma_{\mathrm{h}} + l_{xz'}^2 \sigma_{\mathrm{v}} \tag{4.79}$$

$$\sigma_y^{\mathrm{o}} = l_{yx'}^2 \sigma_{\mathrm{H}} + l_{yy'}^2 \sigma_{\mathrm{h}} + l_{yz'}^2 \sigma_{\mathrm{v}} \tag{4.80}$$

$$\sigma_z^{\mathrm{o}} = l_{zx'}^2 \sigma_{\mathrm{H}} + l_{zy'}^2 \sigma_{\mathrm{h}} + l_{zz'}^2 \sigma_{\mathrm{v}} \tag{4.81}$$

$$\tau_{xy}^{\mathrm{o}} = l_{xx'} l_{yx'} \sigma_{\mathrm{H}} + l_{xy'} l_{yy'} \sigma_{\mathrm{h}} + l_{xz'} l_{yz'} \sigma_{\mathrm{v}} \tag{4.82}$$

$$\tau_{yz}^{\mathrm{o}} = l_{yx'} l_{zx'} \sigma_{\mathrm{H}} + l_{yy'} l_{zy'} \sigma_{\mathrm{h}} + l_{yz'} l_{zz'} \sigma_{\mathrm{v}} \tag{4.83}$$

$$\tau_{zx}^{\mathrm{o}} = l_{zx'} l_{xx'} \sigma_{\mathrm{H}} + l_{zy'} l_{xy'} \sigma_{\mathrm{h}} + l_{zz'} l_{xz'} \sigma_{\mathrm{v}} \tag{4.84}$$

The superscript o on the stresses denote that these are the virgin formation stresses.

4.3.2 The general elastic solution

The solutions corresponding to Eqs. (4.43)–(4.45) are found assuming plane strain normal to the borehole axis.

Again, it is convenient to express the stresses in terms of cylindrical polar coordinates r, θ and z, where r represents the distance from the borehole axis, θ is the azimuth angle relative to the x-axis, and z is the position along the borehole axis (see Fig. 4.6). The stress solutions can be written

$$\begin{aligned}\sigma_r = {} & \frac{\sigma_x^{\mathrm{o}} + \sigma_y^{\mathrm{o}}}{2}\left(1 - \frac{R_{\mathrm{w}}^2}{r^2}\right) + \frac{\sigma_x^{\mathrm{o}} - \sigma_y^{\mathrm{o}}}{2}\left(1 + 3\frac{R_{\mathrm{w}}^4}{r^4} - 4\frac{R_{\mathrm{w}}^2}{r^2}\right)\cos 2\theta \\ & + \tau_{xy}^{\mathrm{o}}\left(1 + 3\frac{R_{\mathrm{w}}^4}{r^4} - 4\frac{R_{\mathrm{w}}^2}{r^2}\right)\sin 2\theta + p_{\mathrm{w}}\frac{R_{\mathrm{w}}^2}{r^2}\end{aligned} \tag{4.85}$$

$$\begin{aligned}\sigma_\theta = {} & \frac{\sigma_x^{\mathrm{o}} + \sigma_y^{\mathrm{o}}}{2}\left(1 + \frac{R_{\mathrm{w}}^2}{r^2}\right) - \frac{\sigma_x^{\mathrm{o}} - \sigma_y^{\mathrm{o}}}{2}\left(1 + 3\frac{R_{\mathrm{w}}^4}{r^4}\right)\cos 2\theta \\ & - \tau_{xy}^{\mathrm{o}}\left(1 + 3\frac{R_{\mathrm{w}}^4}{r^4}\right)\sin 2\theta - p_{\mathrm{w}}\frac{R_{\mathrm{w}}^2}{r^2}\end{aligned} \tag{4.86}$$

$$\sigma_z = \sigma_z^{\mathrm{o}} - \nu_{\mathrm{fr}}\left[2(\sigma_x^{\mathrm{o}} - \sigma_y^{\mathrm{o}})\frac{R_{\mathrm{w}}^2}{r^2}\cos 2\theta + 4\tau_{xy}^{\mathrm{o}}\frac{R_{\mathrm{w}}^2}{r^2}\sin 2\theta\right] \tag{4.87}$$

$$\begin{aligned}\tau_{r\theta} = {} & \frac{\sigma_y^{\mathrm{o}} - \sigma_x^{\mathrm{o}}}{2}\left(1 - 3\frac{R_{\mathrm{w}}^4}{r^4} + 2\frac{R_{\mathrm{w}}^2}{r^2}\right)\sin 2\theta \\ & + \tau_{xy}^{\mathrm{o}}\left(1 - 3\frac{R_{\mathrm{w}}^4}{r^4} + 2\frac{R_{\mathrm{w}}^2}{r^2}\right)\cos 2\theta\end{aligned} \tag{4.88}$$

$$\tau_{\theta z} = (-\tau_{xz}^{o} \sin\theta + \tau_{yz}^{o} \cos\theta)\left(1 + \frac{R_w^2}{r^2}\right) \tag{4.89}$$

$$\tau_{rz} = (\tau_{xz}^{o} \cos\theta + \tau_{yz}^{o} \sin\theta)\left(1 - \frac{R_w^2}{r^2}\right) \tag{4.90}$$

The borehole influence is given by terms in r^{-2} and r^{-4}, which vanish rapidly with increasing r. Thus, for large r, the borehole influence vanishes, and the equations degenerate to Eqs. (4.1)–(4.6).

The solutions depend on the angle θ (see Fig. 4.5) indicating that the stresses vary with position around the wellbore. Generally, the shear stresses are non-zero. Thus σ_r, σ_θ and σ_z are not principal stresses for arbitrary orientations of the well.

At the wellbore wall the equations are simplified to

$$\sigma_r = p_w \tag{4.91}$$

$$\sigma_\theta = \sigma_x^{o} + \sigma_y^{o} - 2(\sigma_x^{o} - \sigma_y^{o}) \cos 2\theta - 4\tau_{xy}^{o} \sin 2\theta - p_w \tag{4.92}$$

$$\sigma_z = \sigma_z^{o} - \nu_{fr}\left[2(\sigma_x^{o} - \sigma_y^{o}) \cos 2\theta + 4\tau_{xy}^{o} \sin 2\theta\right] \tag{4.93}$$

$$\tau_{r\theta} = 0 \tag{4.94}$$

$$\tau_{\theta z} = 2(-\tau_{xz}^{o} \sin\theta + \tau_{yz}^{o} \cos\theta) \tag{4.95}$$

$$\tau_{rz} = 0 \tag{4.96}$$

These are the equations used in linear elastic analysis of borehole stability.

The stress solutions as given by Eqs. (4.85)–(4.90) and Eqs. (4.91)–(4.96) are the stress solutions for non-porous materials or for porous material with constant pore pressure. Due to the superposition principle, pore pressure effects may simply be added. Thus, if we assume a radially symmetric pressure distribution around the borehole, then the pore pressure terms from Eqs. (4.55)–(4.57) or Eqs. (4.68)–(4.70) may be used directly.

4.3.3 Borehole along a principal stress direction

For a hole along a principal stress direction, the general equations (4.85)–(4.90) simplify considerably. We here give the resulting equations for a *vertical borehole*, but the expressions for a horizontal borehole along a principal stress are easily obtained by interchanging the far-field stresses:

$$\sigma_r = \frac{\sigma_H + \sigma_h}{2}\left(1 - \frac{R_w^2}{r^2}\right) + \frac{\sigma_H - \sigma_h}{2}\left(1 + 3\frac{R_w^4}{r^4} - 4\frac{R_w^2}{r^2}\right)\cos 2\theta + p_w\frac{R_w^2}{r^2} \tag{4.97}$$

$$\sigma_\theta = \frac{\sigma_H + \sigma_h}{2}\left(1 + \frac{R_w^2}{r^2}\right) - \frac{\sigma_H - \sigma_h}{2}\left(1 + 3\frac{R_w^4}{r^4}\right)\cos 2\theta - p_w\frac{R_w^2}{r^2} \tag{4.98}$$

$$\sigma_z = \sigma_v - 2\nu_{fr}(\sigma_H - \sigma_h)\frac{R_w^2}{r^2}\cos 2\theta \tag{4.99}$$

$$\tau_{r\theta} = -\frac{\sigma_H - \sigma_h}{2}\left(1 - 3\frac{R_w^4}{r^4} + 2\frac{R_w^2}{r^2}\right)\sin 2\theta \tag{4.100}$$

$$\tau_{rz} = \tau_{\theta z} = 0 \tag{4.101}$$

Note that it is clear from the general definitions that θ is in this case measured relative to the direction of the *major* horizontal stress.

At the borehole wall the equations simplify to

$$\sigma_r = p_w \tag{4.102}$$

$$\sigma_\theta = \sigma_H + \sigma_h - 2(\sigma_H - \sigma_h)\cos 2\theta - p_w \tag{4.103}$$

$$\sigma_z = \sigma_v - 2\nu_{fr}(\sigma_H - \sigma_h)\cos 2\theta \tag{4.104}$$

$$\tau_{r\theta} = \tau_{\theta z} = \tau_{rz} = 0 \tag{4.105}$$

The sum of the normal stresses is

$$\sigma_r + \sigma_\theta + \sigma_z = \sigma_v + \sigma_H + \sigma_h - 2(1 + \nu_{fr})(\sigma_H - \sigma_h)\frac{R_w^2}{r^2}\cos 2\theta \tag{4.106}$$

which shows that unlike the situation for equal horizontal stresses (see the discussion following Eq. (4.42)), the drillout leads to a change in the mean stress. The average change (along a circular path concentric with the hole) is zero but consists of an increase in mean stress in the direction of the minor horizontal stress and of a decrease in the major horizontal stress direction. See Fig. 4.7 for an illustration. Thus, immediately after drillout, we expect a pore pressure increase along the σ_h-direction and a pore pressure decrease along the σ_H-direction (since a change in mean stress leads to a volumetric strain).

Eq. (4.103) shows that the tangential stress at the borehole wall varies between the maximum value

$$\sigma_{\theta,\max} = 3\sigma_H - \sigma_h - p_w \tag{4.107}$$

and the minimum value

$$\sigma_{\theta,\min} = 3\sigma_h - \sigma_H - p_w \tag{4.108}$$

where the maximum value occurs in the direction of σ_h and the minimum value in the direction of σ_H. The variation in tangential stress around the borehole is illustrated in Fig. 4.8.

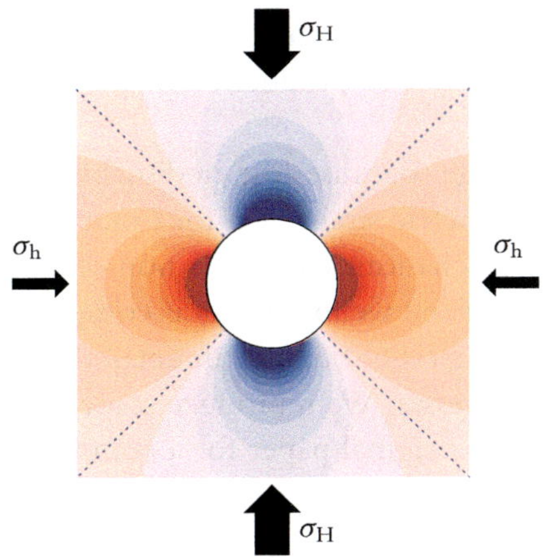

Figure 4.7 Change in mean stress due to the drillout of a borehole in anisotropic far-field stresses. Red is increase in mean stress, blue is decrease in mean stress. The dotted lines show the contours of no change.

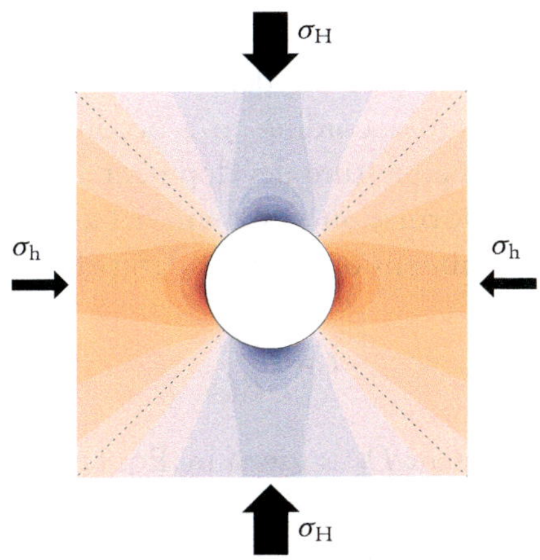

Figure 4.8 Change in tangential stress (relative to the isotropic case) due to the drill-out of a borehole in anisotropic far-field stresses. Red is increase in tangential stress, blue is decrease in tangential stress. The dotted lines show the contours of no change.

4.4 Poroelastic time-dependent effects

In a non-porous, linear elastic material the stress redistribution upon drilling or well pressure change is immediate. In a porous and permeable medium, however, there are time effects due to the fact that the pore pressure propagates with a limited rate (see Section 1.9.1).

There are two main sources of the poroelastic time effects:

1. Pore pressure change due to production or the invasion of the wellbore fluid.
2. Pore pressure change because the redistribution of the formation stresses gives a volumetric strain; see, for example, Eq. (4.106).

Whereas the first mechanism requires a non-sealing wellbore wall, the second will occur even if the wellbore wall is completely sealed.

Detournay and Cheng (1988) discuss in detail the case of a vertical hole with unequal far-field stresses. We reproduce some of their results here but strongly recommend the original paper for a complete discussion. See also Chapter 2 of Charlez (1997).

4.4.1 Wellbore pressure invasion

We first consider case 1, in which the pore pressure is changed by communication with the wellbore pressure. Let us assume that initially $p_{\mathrm{w}} = p_{\mathrm{fo}}$, that we have a permeable borehole wall and that at time zero the well pressure is changed by Δp_{w}. Assuming an axisymmetric pressure profile, the pore pressure is decoupled from the deformation, and the pressure inside the formation will develop according to a decoupled diffusion equation similar to Eq. (1.320). The pressure development as a function of time and radial distance into the formation is shown in Fig. 4.9. The parameter t' labelling the curves is a dimensionless time defined by

$$t' = t\frac{D_{\mathrm{h}}}{R_{\mathrm{w}}^2} \tag{4.109}$$

where the diffusion constant D_{h} is given by Eq. (1.321). For an 8 1/2″ diameter hole, $t' = 1$ corresponds to milliseconds or less for a high-permeable sand and up to several days for a tight shale.

Once the pore pressure profile is known, the tangential stress may be computed by Eq. (4.56). If we assume that as time tends to infinity, the well pressure change penetrates deeply into the formation, then we see from Eq. (4.56) that the contribution to the tangential stress from the pore pressure change is

$$\Delta\sigma_\theta(t \to \infty) = \eta\Delta p_{\mathrm{w}}\left(1 + \frac{R_{\mathrm{w}}^2}{r^2}\right) \tag{4.110}$$

This is of course in addition to the elastic contribution as given by the first two terms of Eq. (4.56) (or by Eq. (4.44)).

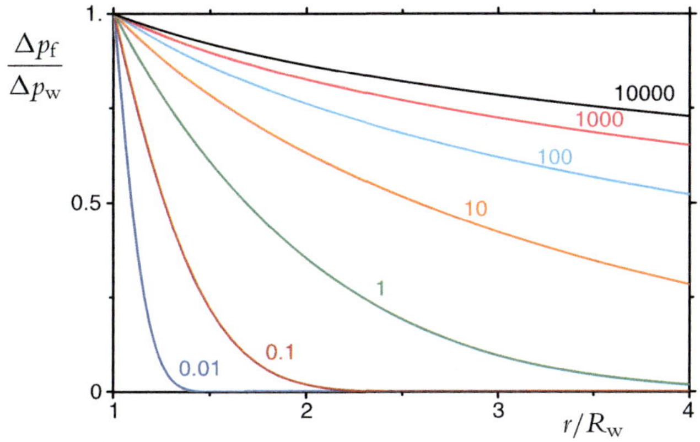

Figure 4.9 Well pressure penetration. The curves are labelled by the dimensionless time t' (see Eq. (4.109)). After Detournay and Cheng (1988).

Note that the assumptions leading to Eq. (4.110) are different from those leading to Eq. (4.69). The latter assumes a steady flow for large times, whereas the former assumes that the pore pressure far into the formation has become equal to the well pressure for large time. Both cases are approximations that are not fulfilled in practice.

We also see from Eq. (4.59) that at the borehole wall the change in tangential stress is

$$\Delta\sigma_\theta = 2\eta\Delta p_w \tag{4.111}$$

at all times.

There are no explicit analytical solutions of the diffusion equation for the current problem, and hence it must be solved by numerical methods.

In Fig. 4.10, we show the time development of the pore pressure induced contribution to the tangential stress. Let us assume that we consider production, that is, Δp_w is negative. We see that at the borehole wall the tangential stress is reduced relative to the elastic case at all times. However, for short times, we have an increase in the tangential stress in the near wellbore region. The region of increased stress propagates outwards with time.

The origin of this increase in tangential stress may be understood by considering that the reduced tangential stress near the borehole must be compensated by an increase of tangential stress further away. The position

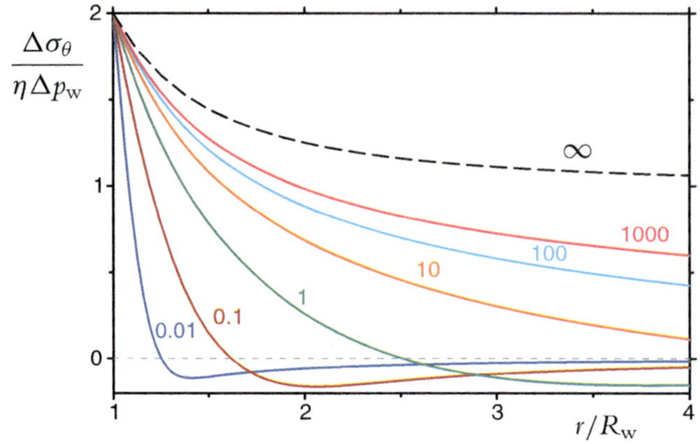

Figure 4.10 Tangential stress due to well pressure penetration. The curves are labelled by the dimensionless time t'. After Detournay and Cheng (1988).

of the region of increased tangential stress follows the position the pressure front. Hence it propagates outwards and broadens with time.

4.4.2 Drillout-induced pore pressure changes

We saw (e.g. in Eq. (4.106)) that when the far-field horizontal stresses are unequal, the drillout leads to a volumetric strain and hence to a change in pressure, which will dissipate with time due to fluid flow. This will lead to time-dependence of the near wellbore stresses, also when the wellbore wall is completely sealing.

The pore pressure change depends both on the radial position and on the azimuth. Hence there will be both radial and tangential fluid flows as a result of the drillout. Since the diffusion distance is shortest close to the borehole, pore pressure will reach equilibrium fastest here. A partly or fully open wellbore wall will contribute further to this.

As a result, the near wellbore behaviour will be drained, whereas the response deeper into the formation will be undrained for short times. Since the drained moduli are smaller than the undrained, the stress concentration in the near wellbore region will initially be shielded relative to its full elastic value. This is similar to the shielding seen for nonlinear formation, or plastic formations; see Sections 4.2.5 and 4.8. Obviously, the effect will be strongest when there is a significant difference between drained and undrained parameters, that is, for soft materials.

As the pore pressure imbalance dissipates, the shielding disappears, and the elastic stress concentration is re-established. Detournay and Cheng

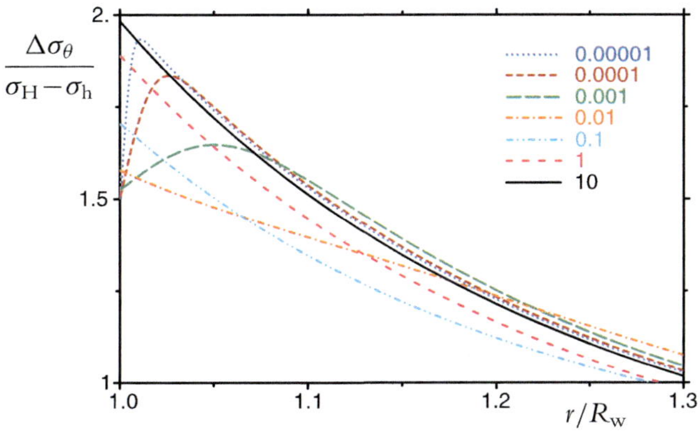

Figure 4.11 Time development of tangential stress (induced by anisotropic horizontal stresses) in the direction of minimum horizontal stress. The different curves correspond to different t' as shown by the legend. After Detournay and Cheng (1988).

(1988) suggested that this may be a mechanism of delayed failure for boreholes in tight formations.

Fig. 4.11 shows the variation in tangential stress as a function of radius in the minimum stress direction for various dimensionless times (Eq. (4.109)). The plot has been generated for $\nu_u = 0.4$, $\nu_{fr} = 0.2$ and Skempton's $B = 0.8$. (An example of a set of parameters in our notation that gives these values is $\phi = 0.15$, $K_{fr} = 4\,\text{GPa}$, $G_{fr} = 4\,\text{GPa}$ and $K_s = 37.5\,\text{GPa}$).

We underline that the figure shows only the tangential stress due to the deviatoric loading, that is, the stress corresponding to the $\cos 2\theta$-term in Eq. (4.98).

Fig. 4.12 shows the variation in tangential stress at the wellbore wall as a function of time, again in the direction of the minimum horizontal stress. The initial reduction in stress concentration is evident.

For a numerical example, assume that $p_w = 20\,\text{MPa}$, $\sigma_H = 40\,\text{MPa}$ and $\sigma_h = 30\,\text{MPa}$. Then Fig. 4.12 shows that $\sigma_\theta \approx 15\,\text{MPa}$ initially and $\sigma_\theta \approx 20\,\text{MPa}$ at large times. The total tangential stress is found by adding the contribution from the mean far-field stress: $\sigma_\theta = \sigma_H + \sigma_h - p_w = 50\,\text{MPa}$. Thus the tangential stress at the borehole wall (in the direction of the minor stress) increases from 65 MPa at early times to 70 MPa at large times.

In addition to the poroelastic time effects, there are also time effects due to temperature change and chemical processes. See Chapter 9 for further discussion.

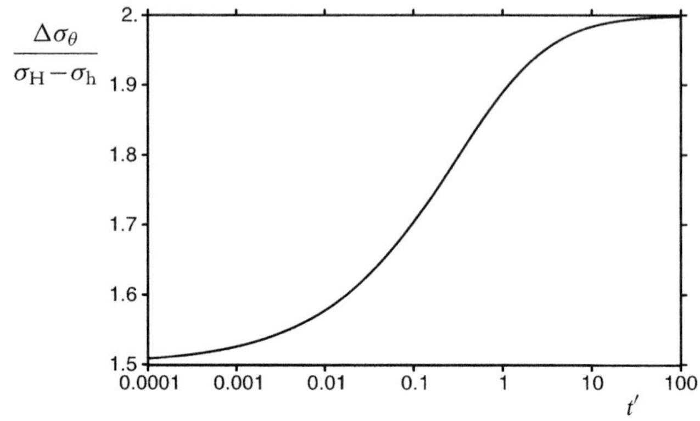

Figure 4.12 Time development of tangential stress (induced by anisotropic horizontal stresses) at the borehole wall in the direction of the minimum horizontal stress. After Detournay and Cheng (1988).

4.5 Borehole failure criteria

As shown in the previous sections, there may be large stress deviations in the formation close to the borehole when the borehole pressure differs from the formation stress. If the stress deviation somewhere exceeds the failure criterion for the rock, then the rock fails. We will associate this situation with the term "borehole failure" in this section. Correspondingly, "borehole failure criterion" means the boundary conditions for which borehole failure occurs. In this chapter, we focus on shear and tensile failure.

The consequences of borehole failure according to this definition are normally borehole deformations of some kind. Note, however, that such deformations are not necessarily dramatic from an operational point of view (see Chapter 9). Thus "borehole failure" as defined here should by no means be taken as a synonym for a lost well. Development of borehole failure beyond failure initiation is discussed in Section 4.8.

4.5.1 Vertical hole, isotropic horizontal stresses and impermeable borehole wall

For a material that behaves linearly elastic, the largest stress differences occur at the borehole wall, and hence rock failure is expected to initiate there.

We start with the simplest case, a vertical borehole with constant pore pressure and isotropic far-field horizontal stresses. We assume that the wellbore is lined with a perfect mud cake, which means that the pore pressure is not influenced by the well pressure.

Within these assumptions, according to Eqs. (4.43)–(4.45), the principal stresses at the borehole wall are

$$\sigma_r = p_{\mathrm{w}} \tag{4.112}$$

$$\sigma_\theta = 2\sigma_{\mathrm{h}} - p_{\mathrm{w}} \tag{4.113}$$

$$\sigma_z = \sigma_{\mathrm{v}} \tag{4.114}$$

There are several conditions for which the borehole may fail, depending on the relative magnitudes of the principal stresses. Let us first assume that we are lowering p_{w} so that $\sigma_r = p_{\mathrm{w}}$ becomes the smallest principal stress. We see from Eqs. (4.113) and (4.114) that depending on the relative magnitude of σ_{h} and σ_{v}, either σ_θ or σ_z will become the largest principal stress at the borehole wall.

Consider first the situation where $\sigma_\theta > \sigma_z > \sigma_r$ at the borehole wall. According to the Mohr–Coulomb criterion (Eq. (2.22)), failure will occur when

$$\sigma'_\theta = C_0 + \sigma'_r \tan^2\beta \tag{4.115}$$

Using an effective stress coefficient equal to 1 for failure (see Section 2.6.1), we find that the failure criterion for the borehole becomes

$$p_{\mathrm{w,min}} = \frac{2\sigma_{\mathrm{h}} + p_{\mathrm{f}}(\tan^2\beta - 1) - C_0}{1 + \tan^2\beta} = p_{\mathrm{f}} + \frac{2(\sigma_{\mathrm{h}} - p_{\mathrm{f}}) - C_0}{1 + \tan^2\beta} \tag{4.116}$$

Thus, if the well pressure falls below the value given by Eq. (4.116), then shear failure will occur at the borehole wall.

Next, assume that σ_{h} is smaller relative to σ_{v}, such that σ_z is the largest principal stress at failure. Proceeding as above, the failure criterion becomes

$$p_{\mathrm{w}} = \frac{\sigma_{\mathrm{v}} + p_{\mathrm{f}}(\tan^2\beta - 1) - C_0}{\tan^2\beta} = p_{\mathrm{f}} + \frac{\sigma_{\mathrm{v}} - p_{\mathrm{f}} - C_0}{\tan^2\beta} \tag{4.117}$$

In practice, we need to calculate both cases and pick the criterion that gives the largest borehole pressure at failure.

To map the region of mechanical stability for a borehole, we need to consider all six permutations of the three principal stresses σ_r, σ_θ and σ_z. The resulting equations are summarised in Table 4.1. The various conditions for borehole failure are shown graphically in Fig. 4.13. However, cases d, e and f are mainly of academic interest, since they imply a wellbore pressure higher than the overburden stress, a condition that is usually unacceptable in drilling. This may however be different for unequal far-field stresses. See discussion below.

The conditions constitute a polygon enframing a region where the borehole is stable with respect to shear failure; if subject to stress states outside the polygon, then the borehole will fail.

Table 4.1 Conditions for shear failure in vertical boreholes with isotropic far-field horizontal stresses and impermeable borehole wall.

Case	$\sigma_1 \geq \sigma_2 \geq \sigma_3$	Borehole failure occurs if
a	$\sigma_\theta \geq \sigma_z \geq \sigma_r$	$p_w \leq p_f + \dfrac{2(\sigma_h - p_f) - C_0}{1 + \tan^2\beta}$
b	$\sigma_z \geq \sigma_\theta \geq \sigma_r$	$p_w \leq p_f + \dfrac{\sigma_v - p_f - C_0}{\tan^2\beta}$
c	$\sigma_z \geq \sigma_r \geq \sigma_\theta$	$p_w \geq p_f + 2(\sigma_h - p_f) - \dfrac{\sigma_v - p_f - C_0}{\tan^2\beta}$
d	$\sigma_r \geq \sigma_z \geq \sigma_\theta$	$p_w \geq p_f + \dfrac{2(\sigma_h - p_f)\tan^2\beta + C_0}{1 + \tan^2\beta}$
e	$\sigma_r \geq \sigma_\theta \geq \sigma_z$	$p_w \geq p_f + (\sigma_v - p_f)\tan^2\beta + C_0$
f	$\sigma_\theta \geq \sigma_r \geq \sigma_z$	$p_w \leq p_f + 2(\sigma_h - p_f) - (\sigma_v - p_f)\tan^2\beta - C_0$

In addition to the shear failure criteria, we must consider that σ'_θ at the borehole wall becomes negative if the well pressure is sufficiently large, according to Eq. (4.113). If $\sigma'_\theta < -T_0$, where T_0 is the tensile strength of the material, then tensile failure will occur at the borehole wall. This adds an additional criterion for borehole failure to the list:

$$p_{w,max}^{frac} = 2\sigma_h - p_f + T_0 \tag{4.118}$$

According to this criterion, tensile failure will occur at the borehole wall if the well pressure is increased above the value given by Eq. (4.118). Borehole failure of this kind is called hydraulic fracturing, which is a topic we will return to in Chapters 8, 9 and 11.

Note that the criterion in Eq. (4.118) applies if the full stress concentration around the borehole as predicted by Eq. (4.113) exists. This requires a

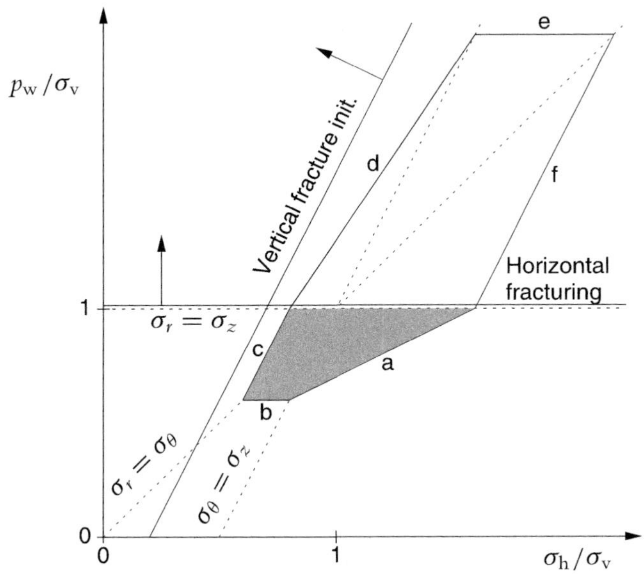

Figure 4.13 Graphical representation of conditions for borehole failure. The Mohr–Coulomb failure criterion with $C_0 = 0$, $p_f = 0.4\sigma_v$ and $\tan^2\beta = 3$ is assumed. The polygon will grow in all directions if C_0 is non-zero. In particular, if $T_0 = 0$, then the line c will coincide with the vertical fracturing line if $C_0 = \sigma_v - p_f$. (After Guenot, 1987.)

perfectly circular hole and a linear elastic material. In practice, both these criteria will be only partly fulfilled, and hence the real limit for hydraulic fracturing will occur at a lower value for p_w.

When the fracture has propagated away from the borehole, it may continue to propagate if p_w is higher than approximately $\sigma_h + T_0$. (See further discussion in Section 11.3.) This corresponds to the diagonal line labelled $\sigma_r = \sigma_\theta$ in Fig. 4.13 (for the case $T_0 = 0$).

Fig. 4.13 is a graphical representation of the failure criteria discussed, assuming that $C_0 = 0$, $p_f = 0.4\sigma_v$ and $\tan^2\beta = 3$. The shaded region is bounded by the shear failure criteria, whereas the oblique line labelled "Vertical fracture init." represents hydraulic fracturing.

The horizontal line labelled "Horizontal fracturing" represents another failure criterion, which needs to be taken into account: If there is any pre-existing fracture or flaw in the borehole wall, then the wall pressure will act with a vertical component on the formation there. A horizontal fracture will then grow if $\sigma_v - p_w$ is less than $-T_0$, that is, if $p_w > \sigma_v + T_0$.

4.5.2 Vertical hole, isotropic horizontal stresses and permeable borehole wall

If the borehole wall is permeable, the pore pressure at the borehole wall is equal to the well pressure. This means that we must use p_{w} rather than p_{f} when computing the effective stresses. In addition, we must take into account the change in the total stresses induced by the varying pore pressure, as given by Eqs. (4.58)–(4.60).

Thus the stresses at the wellbore wall are

$$\sigma_r = p_{\mathrm{w}} \tag{4.119}$$

$$\sigma_\theta = 2\sigma_{\mathrm{h}} - p_{\mathrm{w}} + 2\eta(p_{\mathrm{w}} - p_{\mathrm{fo}}) \tag{4.120}$$

$$\sigma_z = \sigma_{\mathrm{v}} + 2\eta(p_{\mathrm{w}} - p_{\mathrm{fo}}) \tag{4.121}$$

where η is the poroelastic stress coefficient (see Eq. (4.54)), and p_{fo} is the far-field pore pressure. Recall that the last equation follows from the plane strain assumption and is hence strictly valid only for an infinitely long borehole.

Inserting these expressions into the Mohr–Coulomb criterion and using p_{w} when computing the effective stress, we find that for $\sigma_\theta > \sigma_z > \sigma_r$,

$$p_{\mathrm{w,min}} = \frac{2\sigma_{\mathrm{h}} - 2\eta p_{\mathrm{fo}} - C_0}{2 - 2\eta} = \frac{2\sigma_{\mathrm{h}} - \alpha\frac{1-2\nu_{\mathrm{fr}}}{1-\nu_{\mathrm{fr}}} p_{\mathrm{fo}} - C_0}{2 - \alpha\frac{1-2\nu_{\mathrm{fr}}}{1-\nu_{\mathrm{fr}}}} \tag{4.122}$$

Note that the equation contains Biot's α because of the elastic redistribution of the stresses due to the pore pressure profile. However, we still use an effective stress coefficient equal to unity in the failure criterion. Also, observe that since the minor effective stress is zero, the failure angle does not appear in Eq. (4.122).

Similarly, for $\sigma_z > \sigma_\theta > \sigma_r$, we find

$$p_{\mathrm{w,min}} = \frac{\sigma_{\mathrm{v}} - 2\eta p_{\mathrm{fo}} - C_0}{1 - 2\eta} = \frac{\sigma_{\mathrm{v}} - \alpha\frac{1-2\nu_{\mathrm{fr}}}{1-\nu_{\mathrm{fr}}} p_{\mathrm{fo}} - C_0}{1 - \alpha\frac{1-2\nu_{\mathrm{fr}}}{1-\nu_{\mathrm{fr}}}} \tag{4.123}$$

Since σ_r' is always equal to zero, it is of little interest to study the cases corresponding to c,d,e,f in Table 4.1. The remaining criteria to be considered are thus the hydraulic fracturing criteria, $\sigma_\theta' = -T_0$ and $\sigma_z' = -T_0$. We find for vertical fracturing, in agreement with Haimson and Fairhurst (1967),

$$p_{\mathrm{w,max}}^{\mathrm{frac}} = \frac{2\sigma_{\mathrm{h}} - 2\eta p_{\mathrm{fo}} + T_0}{2 - 2\eta} = \frac{2\sigma_{\mathrm{h}} - \alpha\frac{1-2\nu_{\mathrm{fr}}}{1-\nu_{\mathrm{fr}}} p_{\mathrm{fo}} + T_0}{2 - \alpha\frac{1-2\nu_{\mathrm{fr}}}{1-\nu_{\mathrm{fr}}}} \tag{4.124}$$

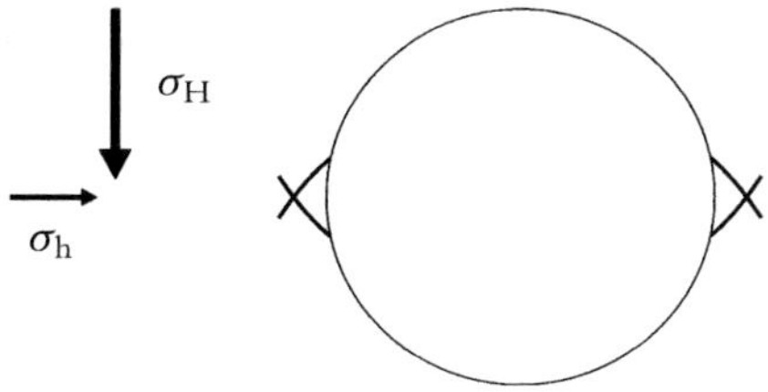

Figure 4.14 Location of shear failure initiation at the wall of a vertical hole with anisotropic horizontal stresses.

and for horizontal fracturing,

$$p_{w,max}^{frac} = \frac{\sigma_v - 2\eta p_{fo} + T_0}{1 - 2\eta} = \frac{\sigma_v - \alpha \frac{1-2\nu_{fr}}{1-\nu_{fr}} p_{fo} + T_0}{1 - \alpha \frac{1-2\nu_{fr}}{1-\nu_{fr}}} \tag{4.125}$$

Again, this last equation assumes plane strain conditions and cannot be directly applied to pressurisation of a limited section of a borehole. Hence the equation has limited practical applicability and is mainly of academic interest.

4.5.3 Borehole along a principal stress direction

The stresses at the wall of a vertical hole with unequal far-field stresses and constant pore pressure, were given in Eqs. (4.102)–(4.105). Since all shear stresses vanish in this case, σ_r, σ_θ and σ_z are the principal stresses and may be used directly in the failure criterion.

From Eq. (4.103) and Fig. 4.8 it is clear that the maximum tangential stress occurs in the direction of the minimum horizontal stress, and hence shear failure at the borehole wall will initiate in the direction of the minimum horizontal stress; see Fig. 4.14.

Comparing to Eq. (4.113), we see that the critical well pressure when $\sigma_\theta > \sigma_z > \sigma_r$ (case a of Table 4.1) may be found by substituting $2\sigma_h$ with $3\sigma_H - \sigma_h$ into Eq. (4.116):

$$p_{w,min} = p_f + \frac{3\sigma_H - \sigma_h - 2p_f - C_0}{1 + \tan^2\beta} \tag{4.126}$$

Similarly, by comparing Eq. (4.104) to Eq. (4.114) we see that for $\sigma_z > \sigma_\theta > \sigma_r$ (case b of Table 4.1), we have

$$p_{w,min} = p_f + \frac{\sigma_v + 2|\nu_{fr}|(\sigma_H - \sigma_h) - p_f - C_0}{\tan^2\beta} \tag{4.127}$$

Note that we need to take the absolute value of Poisson's ratio. This is because the largest value of σ_z in Eq. (4.104) occurs for $\theta = 0°$ for negative ν_{fr} and for $\theta = 90°$ for positive ν_{fr}.

The tensile failure criterion (4.118) is also modified if the horizontal stress is anisotropic. In this case, failure will occur at the positions where σ_θ is smallest, and hence fracturing will occur in the direction of maximum horizontal stress; see Fig. 4.8. The criterion becomes

$$p_{w,max}^{frac} = 3\sigma_h - \sigma_H - p_f + T_0 \tag{4.128}$$

Here we see that an increase in the largest horizontal stress reduces the upper stability limit for the well pressure. Thus anisotropy in the formation around a borehole reduces the region where the borehole is stable.

Observe that if σ_H is sufficiently large, the pressure given by Eq. (4.128) may be smaller than the criterion for fracture growth outside the influence of the borehole, which is $p_w \approx \sigma_h + T_0$. Eq. (4.128) may thus be the criterion for initiation of a fracture that does not propagate beyond the stress concentration around the borehole. See Section 11.2 for a more detailed discussion.

It is important to observe that with unequal far-field stresses the tangential stress may vary considerably along the perimeter of the borehole. In some cases, it may go from being the largest stress in the direction of σ_h to the smallest in the direction of σ_H. Hence it is important to not only check Eq. (4.126), but also all cases discussed in Table 4.1.

An example is shown in Fig. 4.15 illustrating the large variation in the effective tangential stress. Fig. 4.16 shows the mobilised friction coefficient, that is, the magnitude of the required friction coefficient for a stable hole relative to the actual value. The mobilised friction coefficient was calculated assuming a fixed attraction $\mathcal{A}$.

The curve is above 1.0 for several θ-values, which means an unstable hole. Looking only at Eq. (4.126), we would conclude that increasing the mud weight (to 223 bar) will help. However, this makes the situation worse at $\theta = 0°$ (in the direction of σ_H). The correct conclusion is therefore that this borehole is unstable with respect to shear failure irrespective of mud weight.

The above equations were derived for a vertical borehole. It is however trivial to adapt them to a horizontal borehole along the principal stress direction by appropriately interchanging the far-field stresses. As an example, for a horizontal hole along σ_H, we must interchange σ_v and σ_H (assuming that $\sigma_v > \sigma_h$).

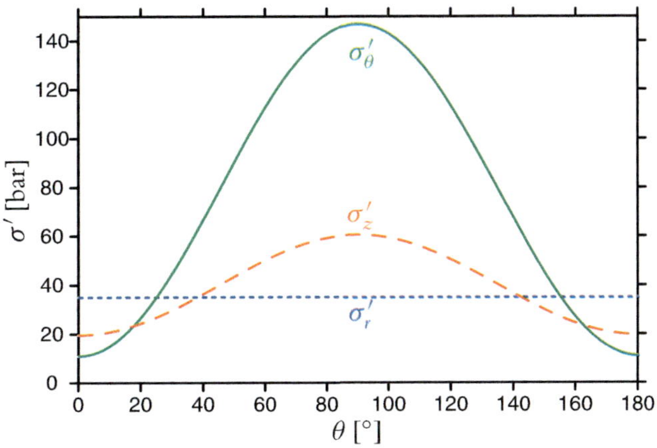

Figure 4.15 Effective stresses σ'_r (short dash), σ'_θ (full line) and σ'_z (long dash) along the perimeter of a vertical borehole. Parameters are $\nu_{\mathrm{fr}} = 0.3$, $\sigma_{\mathrm{H}} = 254\,\mathrm{bar}$, $\sigma_{\mathrm{h}} = 220\,\mathrm{bar}$, $\sigma_{\mathrm{v}} = 220\,\mathrm{bar}$, $p_{\mathrm{f}} = 180\,\mathrm{bar}$ and $p_{\mathrm{w}} = 215\,\mathrm{bar}$.

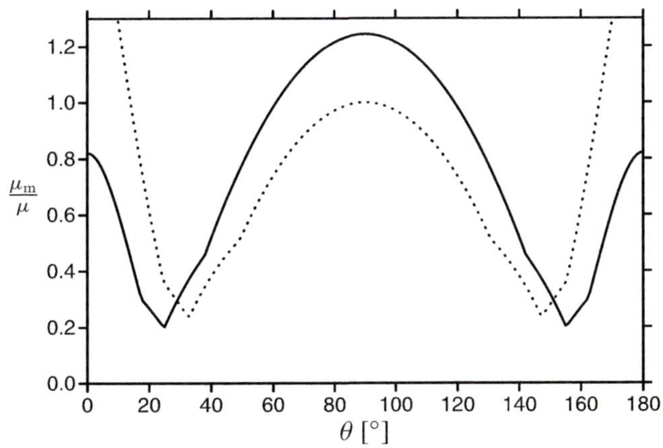

Figure 4.16 Mobilised friction along the perimeter of the vertical borehole of Fig. 4.15. Additional parameters are $C_0 = 10\,\mathrm{MPa}$ and $\varphi = 30°$. The dotted line shows the curve for $p_{\mathrm{w}} = 223\,\mathrm{bar}$.

4.5.4 Borehole in a general direction

In the general case, it is not possible to obtain closed-form expressions for the critical well pressure, and we must resort to numerical solution methods.

In brief, the procedure for establishing the lowest possible well pressure is as follows:

1. From the known far-field stresses compute the stresses in the borehole coordinate system according to Eqs. (4.79)–(4.84).
2. Compute the stresses at the borehole wall according to Eqs. (4.91)–(4.96).
3. Subdivide the interval $0 < \theta < \pi$ into a number of sub-intervals (depending on the accuracy required). For each sub-interval, perform the following:
   - Diagonalise the stress tensor at the borehole wall to find the principal stresses.
   - Find the critical well pressure by inserting the principal stresses into the failure criterion.
4. The lowest admissible p_w is the largest critical well pressure found from the iteration in the step above.

4.6 Elliptical borehole

The analysis so far has assumed perfectly circular holes. In practice, there will be deviations from this. Failure of the near wellbore material is a mechanism for this. Mechanical action from the drill string is another, especially in deviated holes.

Deviation from a circular cross section changes the stress concentration around the borehole and hence influences both fracturing and shear failure. The simplest deviation from a circular hole is an elliptical hole. We will here consider a few results for an elliptical hole aligned along the principal stress direction.

Jaeger et al. (2007) give an expression for the tangential stress at the wall of an empty elliptical hole oriented at an angle β_h to the minor transversal principal stress (See also Pollard, 1973; Jaeger and Cook, 1979).

In the following, we assume the hole to be vertical, but the change to a hole along an arbitrary principal stress is trivial. The geometry is illustrated in Fig. 4.17.

In terms of the *ellipse flattening*

$$f = \frac{a - b}{a} \tag{4.129}$$

where a and b are the major and minor semi-axes, the expression for the tangential stress at the wall of an empty hole is

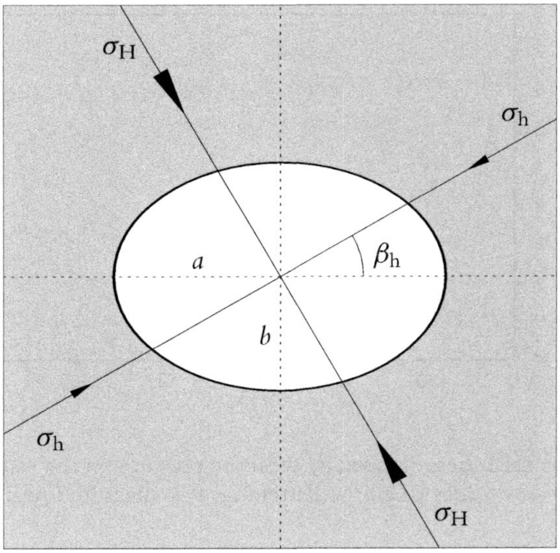

Figure 4.17 Sketch of a vertical elliptical borehole. The long diameter $2a$ of the ellipsis is inclined at an angle β_h to the minor horizontal stress.

$$\sigma_\theta = \frac{2(1-f)(\sigma_h+\sigma_H) + (\sigma_H-\sigma_h)\left[(2-f)^2\cos 2(\beta_h-\theta) - (2-f)f\cos 2\beta_h\right]}{2-2f+f^2-(2-f)f\cos 2\theta} \tag{4.130}$$

where the elliptical coordinate θ is measured relative to the major semi-axis.

We now use the simple method discussed following Eq. (4.62) to proceed to the case with pressure in the borehole. Here we give an expression for $\beta_h = 0$, that is, the borehole elongation is towards the minor stress, as is the case with breakouts:

$$\sigma_\theta(r = R_w) = \frac{A + B\cos 2\theta}{C} \tag{4.131}$$

where

$$A = 2\frac{1-f}{2-f}(\sigma_H + \sigma_h) - f(\sigma_H - \sigma_h) - \frac{2-2f-f^2}{2-f}p_w \tag{4.132}$$

$$B = (2-f)(\sigma_H - \sigma_h) - fp_w \tag{4.133}$$

$$C = \frac{2}{2-f} - f - f\cos 2\theta \tag{4.134}$$

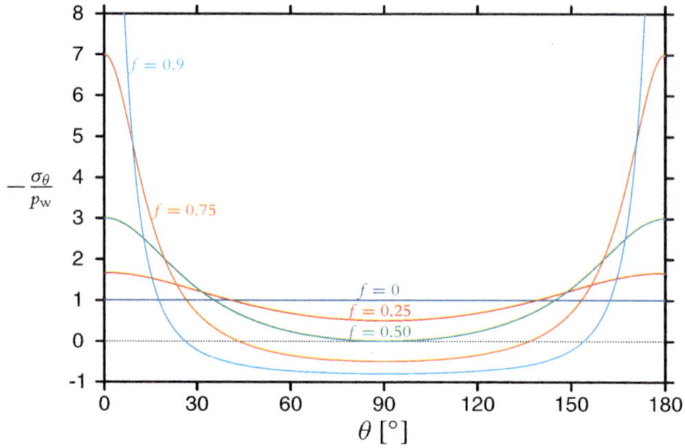

Figure 4.18 Tangential stress induced by wellbore pressure on the wall of an elliptical borehole, for various values of ellipse flattening. θ is measured relative to the long axis of the ellipse.

The contribution to σ_θ from the well pressure is found by subtracting Eq. (4.130) from Eq. (4.131). The result must clearly be independent of β_h and the stresses and is

$$\Delta\sigma_\theta(p_w, r = R_w) = -\frac{2 - 2f - f^2 + (2 - f)f\cos 2\theta}{2 - 2f + f^2 - (2 - f)f\cos 2\theta} p_w \tag{4.135}$$

Aadnøy and Angell-Olsen (1995) took a different route, which resulted in a rather complex expression. It can however be shown that their Eq. (4) is equivalent to Eq. (4.135).

For a circular hole, a change in well pressure leads to a uniform change in the tangential stress at the wall by the same magnitude but opposite signal, see, for example, Eq. (4.44). For an elliptical hole, this still holds for the average change, but the induced tangential stress varies around the periphery.

As expected, the change is largest where the curvature radius is smallest. For $f = 0.5$, the change is exactly zero for $\theta = \pi/2$, that is, for this direction (which we assumed to be the maximum stress direction in Eq. (4.131)), the borehole pressure does not influence fracturing.

As f increases above 0.5, we get an increasing part of the hole perimeter where the sign is swapped: Increasing p_w reduces σ_θ. This is compensated by a stronger effect where the borehole curvature is large. This is illustrated for a few f in Fig. 4.18.

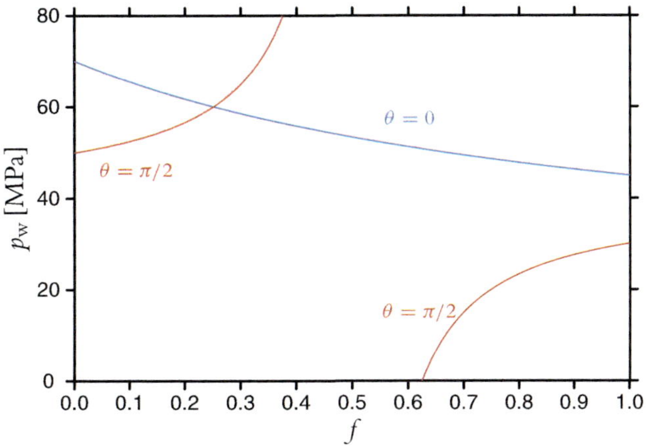

Figure 4.19 Fracture initiation at $\theta = 0°$ and $\theta = 90°$ as a function of ellipse flattening, assuming that $\sigma_H = 45$ MPa, $\sigma_h = 40$ MPa, $p_f = 25$ MPa and $T_0 = 0$.

4.6.1 Fracture initiation

With our stress assumptions, the natural direction for the hydraulic fracture is $\theta = \pi/2$. It is evident from Fig. 4.18 that as f increases, fracture initiation will switch to where the stress concentration is largest, that is, at $\theta = 0$ or $\theta = \pi$. We should therefore consider the fracturing criteria for θ equal to 0 and $\pi/2$.

The results follow from Eq. (4.131)[2] using the fact that the effective tangential stress is equal to the tensile strength $-T_0$ at fracture initiation:

$$p_w^{\text{frac}}(\theta = 0) = \frac{(3 - f)\sigma_H - (1 - f)(\sigma_h + p_f - T_0)}{1 + f} \tag{4.136}$$

$$p_w^{\text{frac}}(\theta = \pi/2) = \frac{(3 - 2f)\sigma_h - \sigma_H - p_f + T_0}{1 - 2f} \tag{4.137}$$

Fig. 4.19 shows Eqs. (4.136)–(4.137) for $\sigma_H = 45$ MPa, $\sigma_h = 40$ MPa, $p_f = 25$ MPa and $T_0 = 0$.

For a circular hole, fracture initiation is expected at $\theta = \pi/2$ for well pressure $p_w = 50$ MPa; see Eq. (4.128). Then as f increases, fracture initiation at $\theta = \pi/2$ becomes more difficult, since the shape of the hole reduces the stress concentration there. On the other hand, at $\theta = 0$ the hole shape increases the stress concentration, and from about $f = 0.25$ fracture initiation takes place at $\theta = 0, \pi$ for the parameters chosen.

[2] There is a misprint in Eq. (13a) of Aadnøy and Angell-Olsen (1995).

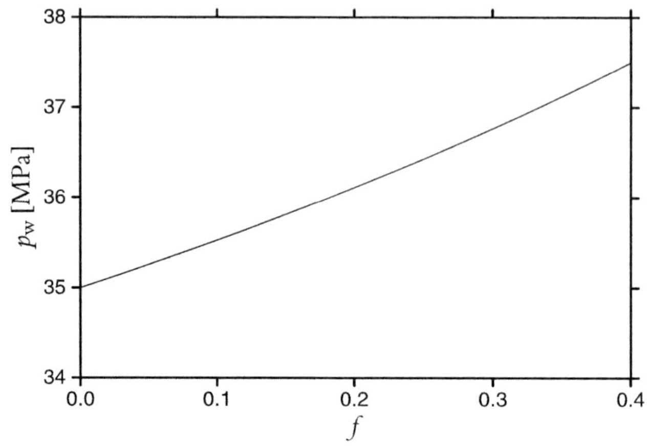

Figure 4.20 Shear initiation at $\theta = 0$ as a function of ellipse flattening, assuming that $\sigma_H = 45$ MPa, $\sigma_h = 40$ MPa, $p_f = 25$ MPa and $T_0 = 0$.

Note that the $\theta = \pi/2$ curve diverges for $f = 0.5$. This is because σ_θ is then independent of p_w. As f increases above 0.5, the sign of the well pressure contribution to σ_θ at $\theta = \pi/2$ is reversed, and there is in principle fracturing at low well pressure. However, this is typically below normal well pressure in petroleum wells.

4.6.2 Shear failure

For shear failure, we must in principle consider all the cases in Table 4.1. We will here limit ourselves to case a), that is, where the tangential stress is the largest principal stress, and the well pressure is the smallest. From Fig. 4.18 it is clear that reducing the well pressure increases the tangential pressure most strongly at $\theta = 0$, so we expect shear failure at $\theta = 0$.

The minimum allowable well pressure is found to be (assuming the Mohr–Coulomb criterion in terms of the UCS C_0 and failure angle β)

$$p_{w,\min} = \frac{\left[(3-f)\sigma_H - (1-f)(C_0 + p_f + \sigma_h)\right]\cos^2\beta + (1-f)p_f \sin^2\beta}{1 + f\cos 2\beta} \tag{4.138}$$

which simplifies to Eq. (4.126) for $f = 0$.

Fig. 4.20 shows a plot of minimum well pressure to avoid shear failure as a function of f. (We limit the range to make it less likely that other stress combinations are of importance.) There is a moderate increase as f increases, indicating that an elliptical hole is less stable than a circular hole.

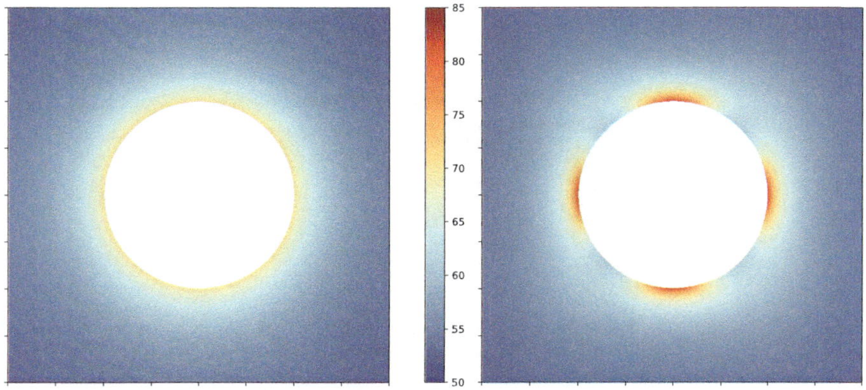

Figure 4.21 Total tangential stress around a horizontal borehole aligned along the minimum horizontal stress. $\sigma_V = \sigma_H = 50\,\text{MPa}$, $p_w = 30\,\text{MPa}$. Left: isotropic formation. Right: transversely isotropic formation where the vertical shear modulus G_V was reduced by 50 %.

4.7 Borehole in an anisotropic formation

In the preceding discussion, we have considered anisotropic stresses but assumed full isotropy in the elastic parameters. Many relevant rocks, particularly shale, are sufficiently anisotropic that making this assumption is not satisfactory.

Amadei (1983) presented a full anisotropic elastic solution based on Lekhnitskii (1963). Amadei's formulas have been implemented and verified by several groups, among them Ong and Rogiers (1993); Gaede (2011); Karpfinger et al. (2011); Jia and Schmitt (2014); Setiawan and Zimmerman (2018); Fang (2018); Asaka and Holt (2020). Raaen et al. (2019) highlighted some technical issues that are important for a successful implementation.

Increasing the number of elastic parameters obviously complicates matters. Further, a full anisotropic description also requires anisotropic strength. Raaen et al. (2019) emphasised the need for using a consistent anisotropic Skempton B.

Here we restrict the discussion to a basic example, which underlines that the effects are significant. We consider a horizontal hole aligned along the minimum horizontal stress with equal vertical and maximum horizontal stresses; see Fig. 4.21.

The left-hand part of the figure shows the basic isotropic solution with symmetric increase of the tangential stress to 70 MPa at the borehole wall;

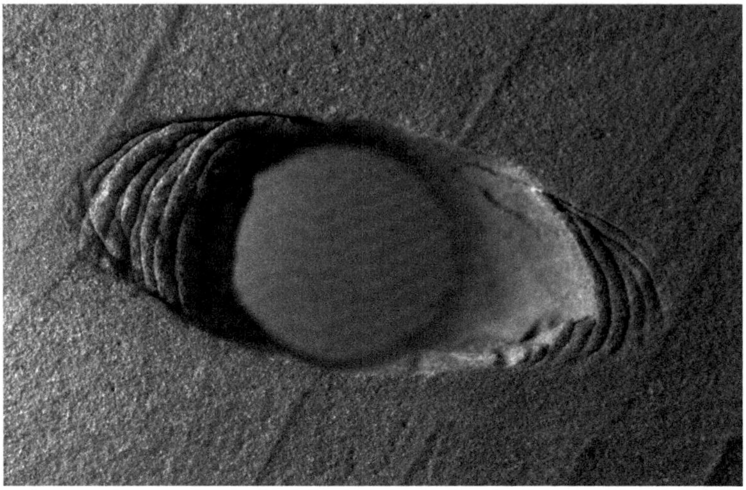

Figure 4.22 Cross-section of a cylindrical cavity after failure.

see Eq. (4.107). In contrast, the anisotropic solution gives a 4-fold symmetry, in which the maximum tangential stress is distinctly higher, about 81.6 MPa.

As we allow σ_v and σ_H to differ, the isotropic solution shows the 2-fold symmetry illustrated in Fig. 4.8, whereas the full 4-fold symmetry is broken in the anisotropic case. As the differences between the stresses increase, the stress anisotropy dominates increasingly more over the elastic anisotropy.

4.8 Beyond failure initiation

Borehole failure is commonly observed as breakouts in the borehole wall. Fig. 4.22 shows a typical example of a failure mode denoted "borehole elongation" due to the extension of the borehole cross-section in one direction. The stress conditions causing this cavity to fail corresponded to case a in Table 4.1. The diagram in Fig. 4.13 is split into various sectors by the dashed border lines where two principal stresses are equal. Each sector is associated with a failure condition in Table 4.1. A typical mode of rupture is associated with each sector. These are described in Table 4.2 and Figs. 4.23–4.25 for the most relevant cases.

The orientation of breakouts due to failure of type a are related to the orientation of the horizontal stresses, as discussed in Section 4.5.3. This

Table 4.2 Rupture modes for the failure of boreholes (Maury and Sauzay, 1987).

Case	Rupture mode	Figure
Shear, a	Rupture occurs in shear surfaces intersecting parallel to the axial stress	Fig. 4.23
Shear, b	Rupture occurs in fragments of toroidal shape	Fig. 4.24
Shear, c	Rupture occurs in multishear surfaces intersecting parallel to the radius	Fig. 4.25
Tensile	Hydraulic fracturing; isolated fractures parallel to the borehole axis	Fig. 11.1

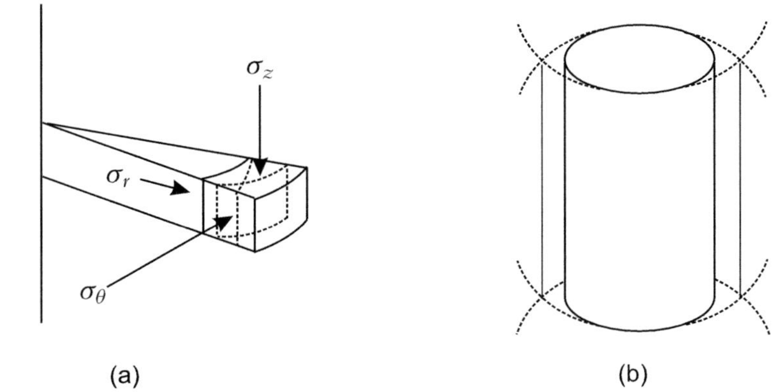

Figure 4.23 Failure planes caused by tangential (σ_1') and radial (σ_3') stresses. (a) Stresses on an element at the wellbore wall. (b) Breakouts.

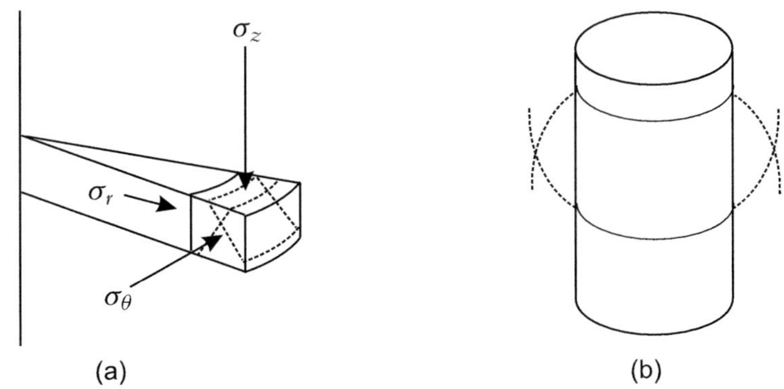

Figure 4.24 Failure planes generated by vertical (σ_1') and radial (σ_3') stresses. Breakouts will have toroidal shapes. (a) Stresses on an element at the wellbore wall. (b) Breakouts.

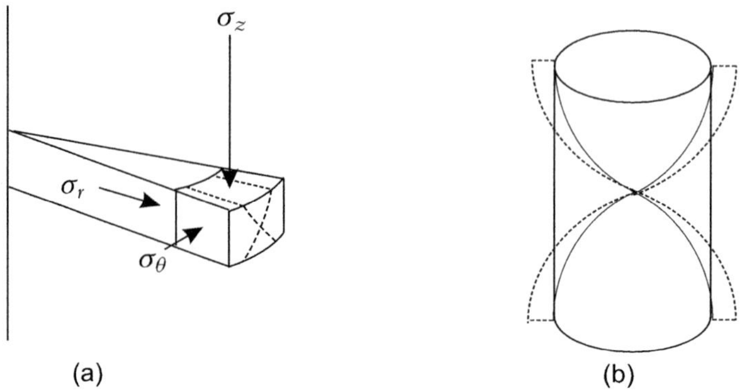

Figure 4.25 Failure planes generated by vertical (σ_1') and tangential (σ_3') stresses. Radial fractures may develop following a helical path along the well. (a) Stresses on an element at the wellbore wall. (b) Breakouts.

connection is used in the field for determining the direction of the stresses; see further discussion in Section 8.3.2.

Anisotropy in the material parameters of the rock also affects the orientation of breakouts. When the horizontal stress is isotropic, the breakouts tend to be oriented normal to the joints of bedding in the material (Kaiser et al., 1985). Note, however, that breakouts of the form shown in Fig. 4.22 may occur even if both the stress field and the material parameters are isotropic.

The criteria for the initiation of borehole failure discussed in Section 4.5 only define the initiation of borehole failure and do not reveal the consequences of failure initiation, which is of course of equal interest.

An idea of the development of borehole failure can be obtained by tracing a potential failure plane from the borehole wall into the formation. Here we choose the simplest case of a linearly elastic formation and isotropic formation stresses and assume that $p_w < \sigma_h$. The orientation of a potential failure plane at a distance r from the borehole centre is given by the orientation of the largest and smallest principal stresses (σ_θ and σ_r, respectively) and the failure angle β (which is related to the coefficient of internal friction; see Section 2.3.1). The situation is shown schematically in Fig. 4.26. The plane proceeds in the direction $(\pi/2 - \beta)$ relative to the direction of σ_θ. According to Fig. 4.26, we have

$$\frac{\Delta r}{r\Delta\theta} \to \frac{\mathrm{d}r}{r\mathrm{d}\theta} = \tan\left(\frac{\pi}{2} - \beta\right) \tag{4.139}$$

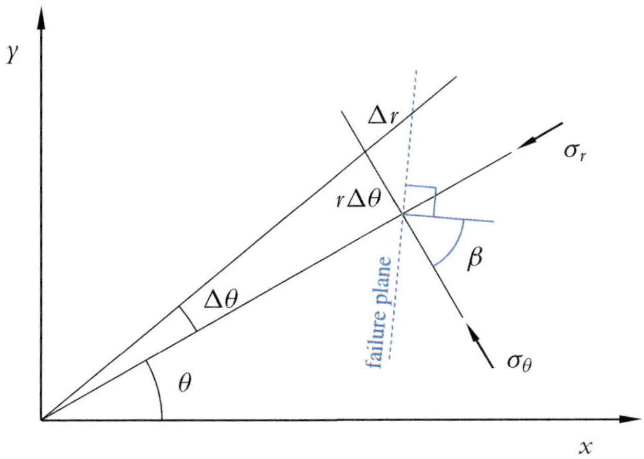

Figure 4.26 Sketch of the orientation of the failure plane at the borehole wall. β is the failure angle (Eq. (2.13)).

Eq. (4.139) is a differential equation with the solutions

$$r = C_1 e^{\pm b\theta} \tag{4.140}$$

where b is given by

$$b = \tan\left(\frac{\pi}{2} - \beta\right) = \sqrt{\mu^2 + 1} - \mu \tag{4.141}$$

and C_1 is a constant. (The coefficient of internal friction μ was defined in Eq. (2.6).) The $\pm$-sign in Eq. (4.140) arises because we have an alternate failure plane, which is the mirror image about the σ_r-direction of the one indicated in the figure.

Fig. 4.27 shows a set of curves given by Eq. (4.140). There are clear similarities in the crack patterns of Figs. 4.22 and 4.27. In Fig. 4.27 the number of failure planes and the positions of the failure plane initiations (determined by the constant C_1) were deliberately chosen such as to make the figure resemble the observation. In a more extended analysis, Zoback et al. (1985) connect these parameters to horizontal stress anisotropy. They also suggest that the formation of such breakouts takes place in successive stages, where new generations of failure planes develop outside the initial ones.

This suggestion is founded on an important aspect of the problem: the formation of a failure plane causes a redistribution of the stresses around

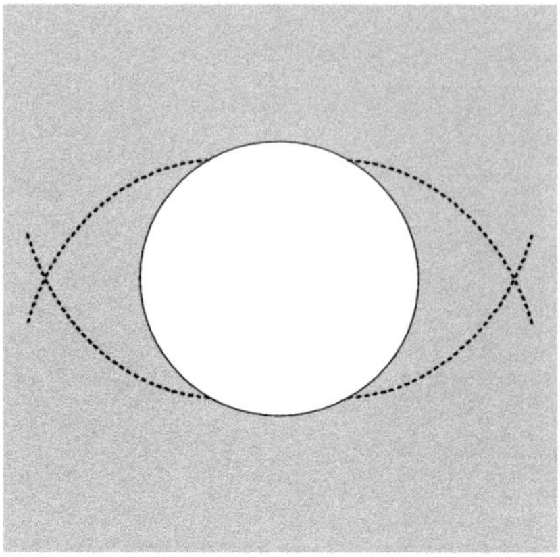

Figure 4.27 Potential failure planes surrounding a borehole.

the borehole. Thus the conditions for failure of the still intact parts of the borehole wall are being altered. Unfortunately, this alteration also implies a reduction in the symmetry of the problem, so that calculation of the stresses becomes more complicated.

Redistribution of stresses and a reduction of stress state symmetry may even occur prior to the formation of shear failure planes if the symmetric deformation becomes unstable relative to a surface buckling (Vardoulakis et al., 1988). Due to the reduced symmetry, the complexity of the problem rises quickly, and numerical techniques are normally needed for further modelling.

The failure process also involves opening of cracks oriented parallel to the major principal stress (or, more precisely, oriented with their normals parallel to the minor principal stress), as discussed in Section 6.4.4. In a situation corresponding to case a, Table 4.1, these cracks will be oriented parallel to the borehole wall. The opening and growth of these cracks is by some authors considered as the fundamental fracture mechanism for breakout formation in boreholes (Zheng et al., 1988; Ewy and Cook, 1989).

Theoretical simulations based on this consideration also resulted in elongated boreholes similar to Fig. 4.22. The simulations further gave the important result that the elongated borehole becomes stable when it has reached a specific shape. This shape depends on the actual combination of stresses and strength, and also on the stress path.

In these simulations, a breakout is initiated as a slab of rock covering a certain angle of the borehole wall is spalled off. This causes further stress concentrations in front of the spalled region, and further spalling takes place within this sector of the borehole. Thus, once a breakout is initiated within a sector, it will not become wider, but will deepen until it reaches its stable shape.

4.8.1 A simple plasticity model

Stabilisation of the borehole after failure initiation can also be described analytically, given some simplifying assumptions. Here we will discuss a simple model, in which the rock is assumed to behave according to the linear elastic/ideally plastic model described in Section 2.8.1 (Fig. 2.25). Rock failure in this model is associated with the transition from elastic to plastic behaviour. This transition does not involve the development of failure planes, and hence the symmetry of the stress state is preserved after failure.

Consider the situation discussed in Section 4.5.1, with a low well pressure. If the stress state at the borehole wall fulfils the failure criterion, then there will be a zone surrounding the borehole where the rock acts as a plastic material. Outside this region the material remains elastic. In the outer elastic region the general solutions (4.35) and (4.36) are still valid (although the coefficients of integration will change). In the inner "plastic zone" the equation of equilibrium (4.28) is valid, but the principal stresses must also satisfy a yield criterion.

4.8.1.1 The Tresca criterion

To illustrate the main principles with as simple mathematics as possible, we first use the Tresca criterion (2.5). We assume that σ_θ and σ_r will be the largest and smallest principal stresses in the plastic zone. The Tresca criterion then requires

$$\sigma_\theta - \sigma_r = C_0 \tag{4.142}$$

in the plastic zone. Introducing Eq. (4.142) into Eq. (4.28), we have

$$\frac{d\sigma_r}{dr} = \frac{C_0}{r} \tag{4.143}$$

The solution of Eq. (4.143) taking into account the boundary condition $\sigma_r(R_w) = p_w$ is

$$\sigma_r = p_w + C_0 \ln \frac{r}{R_w} \tag{4.144}$$

The tangential stress then follows from the failure criterion (Eq. (4.142)):

$$\sigma_\theta = p_w + C_0\left(\ln\frac{r}{R_w} + 1\right) \tag{4.145}$$

Eqs. (4.144) and (4.145) apply in the plastic zone $R_w < r < R_p$, where R_p is the radius of the plastic zone.

Outside the plastic zone $r > R_p$, the elastic solutions given by Eqs. (4.35) and (4.36) apply; C_1' is determined by the requirement that the stresses should be equal to σ_h as r approaches infinity, whereas C_2' is determined by requiring that $\sigma_\theta - \sigma_r$ fulfils the failure criterion (4.142) at $r = R_p$. The result is

$$\sigma_r = \sigma_h - \frac{1}{2}C_0\left(\frac{R_p}{r}\right)^2 \tag{4.146}$$

$$\sigma_\theta = \sigma_h + \frac{1}{2}C_0\left(\frac{R_p}{r}\right)^2 \tag{4.147}$$

The plastic zone radius is determined from the requirement that the radial stress should be continuous at $r = R_p$. From Eqs. (4.144) and (4.146) we then find

$$R_p = R_w e^{\frac{\sigma_h - p_w}{C_0} - \frac{1}{2}} \tag{4.148}$$

The stress distribution around the borehole is shown in Fig. 4.28. Although this solution has been obtained through a series of simplifying assumptions, it illustrates the general features of the plastic zone concept. It will, however, become clear below that the Tresca-based solution above is not even approximately accurate in frictional materials.

4.8.1.2 The Mohr–Coulomb criterion

The calculation steps are essentially the same for the Mohr–Coulomb criterion, although the formulas become more complex. As above, we assume that σ_r is the minimum stress and σ_θ is the maximum stress. We write the Mohr–Coulomb criterion (2.22) in the elastic zone in the form

$$\sigma_\theta' = C_0 + k\sigma_r' \tag{4.149}$$

where $k = \tan^2\beta$ and β is the failure angle.

For the plastified zone, we allow the possibility of altered parameters

$$\sigma_\theta' = C_0' + k'\sigma_r' \tag{4.150}$$

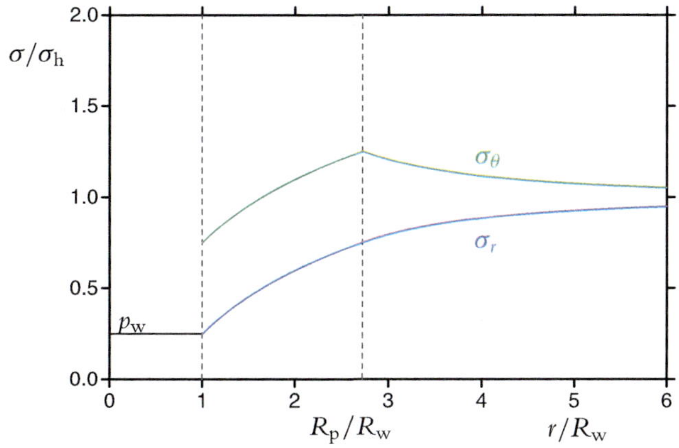

Figure 4.28 Stresses around a borehole with a plastic zone according to the Tresca model. Parameters are $C_0 = 0.5\sigma_h$ and $p_w = 0.25\sigma_h$.

This may be written as

$$\sigma_r - \sigma_\theta = \sigma_r' - \sigma_\theta' = \sigma_r'(1 - k') - C_0' \tag{4.151}$$

Inserting into the force balance equation (4.28), we obtain

$$\frac{\partial \sigma_r}{\partial r} + \frac{\sigma_r'(1 - k')}{r} = \frac{C_0'}{r} \tag{4.152}$$

If we now assume constant pore pressure $p_f = p_{fo}$, then we may replace σ_r with σ_r' in the first term. Solving Eq. (4.152) using the boundary condition $\sigma_r'(R_w) = p_w - p_{fo}$ gives the solution within the plastified region

$$\sigma_r' = \left(p_w - p_{fo} + \frac{C_0'}{k' - 1}\right)\left(\frac{r}{R_w}\right)^{k'-1} - \frac{C_0'}{k' - 1} \tag{4.153}$$

The tangential stress follows from the failure criterion (4.150):

$$\sigma_\theta' = C_0' + k'\sigma_r' = k'\left(p_w - p_{fo} + \frac{C_0'}{k' - 1}\right)\left(\frac{r}{R_w}\right)^{k'-1} - \frac{C_0'}{k' - 1} \tag{4.154}$$

From the requirement that $\sigma_\theta = \sigma_h$ as r becomes very large it follows that the elastic zone stresses (4.35) and (4.36) can be written

$$\sigma_r = \sigma_h + C_2' \frac{1}{r^2} \tag{4.155}$$

$$\sigma_\theta = \sigma_\mathrm{h} - C_2' \frac{1}{r^2} \tag{4.156}$$

Adding these equations gives the well-known relation

$$\sigma_\theta = 2\sigma_\mathrm{h} - \sigma_r \tag{4.157}$$

At the radius of the plastic zone the intact material yield criterion must be fulfilled, and hence by inserting Eq. (4.157) into Eq. (4.149) we have

$$\sigma_r'(R_\mathrm{p}) = \frac{2\sigma_\mathrm{h}' - C_0}{k+1} \tag{4.158}$$

Equating this expression to the inner zone expression for σ_r', Eq. (4.153), gives an equation to determine the plastic zone radius,

$$\frac{R_\mathrm{p}}{R_\mathrm{w}} = \left[\frac{2\sigma_\mathrm{h}' - C_0 + (k+1)\frac{C_0'}{k'-1}}{\left(p_\mathrm{w} - p_\mathrm{fo} + \frac{C_0'}{k'-1}\right)(k+1)} \right]^{\frac{1}{k'-1}} \tag{4.159}$$

The coefficient C_2' is determined from Eq. (4.155):

$$C_2' = -R_\mathrm{p}^2 \left(\sigma_\mathrm{h} - \sigma_r(R_\mathrm{p})\right) = -R_\mathrm{p}^2 \left(\sigma_\mathrm{h}' - \sigma_r'(R_\mathrm{p})\right) \tag{4.160}$$

which using Eq. (4.158) gives

$$C_2' = -R_\mathrm{w}^2 \frac{(k-1)\sigma_\mathrm{h}' + C_0}{k+1} \left[\frac{2\sigma_\mathrm{h}' - C_0 + (k+1)\frac{C_0'}{k'-1}}{\left(p_\mathrm{w} - p_\mathrm{fo} + \frac{C_0'}{k'-1}\right)(k+1)} \right]^{\frac{2}{k'-1}} \tag{4.161}$$

In terms of the *attraction*, $\mathcal{A} = C_0/(k-1)$ (see Eq. (2.9)), Eq. (4.159) may be written as

$$\frac{R_\mathrm{p}}{R_\mathrm{w}} = \left[\frac{2\sigma_\mathrm{h}' + (1-k)\mathcal{A} + (k+1)\mathcal{A}'}{\left(p_\mathrm{w} - p_\mathrm{fo} + \mathcal{A}'\right)(k+1)} \right]^{\frac{1}{k'-1}} \tag{4.162}$$

If $\mathcal{A}' = \mathcal{A}$, then this simplifies to

$$\frac{R_\mathrm{p}}{R_\mathrm{w}} = \left[\frac{2(\sigma_\mathrm{h}' + \mathcal{A})}{\left(p_\mathrm{w} - p_\mathrm{fo} + \mathcal{A}\right)(k+1)} \right]^{\frac{1}{k'-1}} \tag{4.163}$$

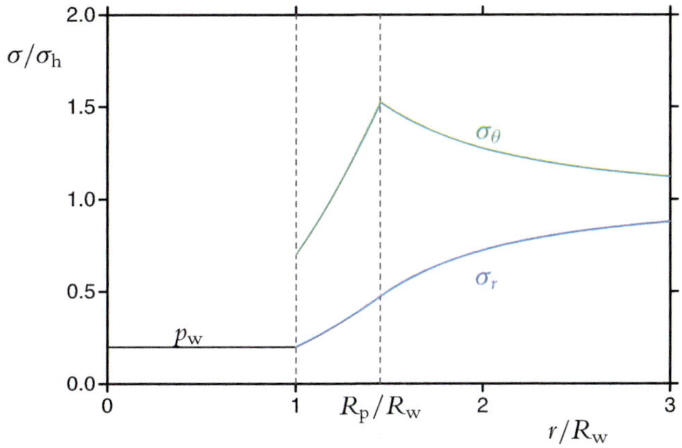

Figure 4.29 Stresses around a borehole with a plastic zone, assuming the same failure criterion in the intact and plastic zones. Parameters are $C_0 = 0.1\sigma_h$, $p_w = 0.2\sigma_h$ and $\varphi = 30°$.

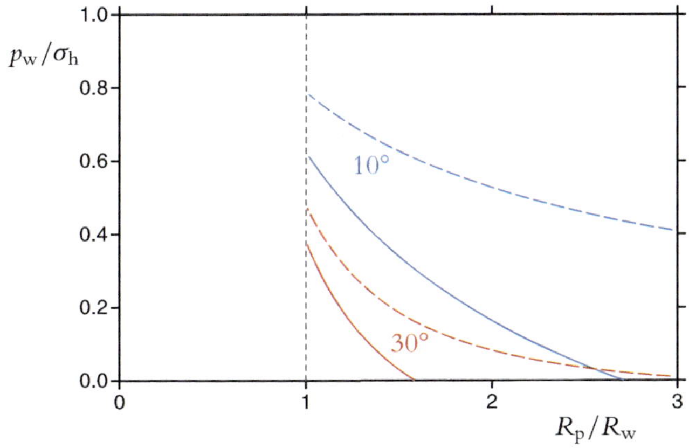

Figure 4.30 Radius of the plastic zone versus well pressure for friction angles 10° and 30°. Full lines correspond to $C_0 = 0.5\sigma_h$, whereas dashed lines are for $C_0 = 0.1\sigma_h$.

The stress distribution around the borehole, assuming a friction angle of 30°, is shown in Fig. 4.29. Note that the extent of the plastic zone is considerably smaller than for the Tresca case. Fig. 4.30 illustrates how the radius of the plastic zone depends on the well pressure. Observe that the extent of the plastic zone is reduced as the friction angle increases.

The axial stress σ_z is more tricky to determine. More sophisticated models indicate that the plastic region consists of two zones; in the inner zone the axial stress is equal to the tangential stress, whereas in the outer zone, all three principal stresses may differ.

The models presented above are simplifications, since both are based on ideal plasticity and do not take properly into account how the plastic strain changes the yield criterion. More realistic models need to incorporate a description of plastic flow and hardening/softening, and necessarily become significantly more complicated than the simple models sketched here. Although the Mohr–Coulomb-based equations above may be used for quick overview calculations, more advanced models, normally implemented in finite element codes, are necessary for precise calculations.

4.8.1.3 Deformation and plastic strain

The deformation and plastic strain may be computed by the general methods outlined in Section 2.8.1. To simplify the equations, we restrict ourselves to the Tresca criterion (2.5). We rewrite the criterion in the general form of Eq. (2.63),

$$f(\sigma_1', \sigma_2', \sigma_3') = f(\sigma_r', \sigma_\theta') = \sigma_\theta' - \sigma_r' - C_0 = 0 \tag{4.164}$$

and find the plastic strains according to the associated flow rule (2.68):

$$\varepsilon_r^{\mathrm{p}} = \mathrm{d}\lambda_{\mathrm{p}} \frac{\partial f}{\partial \sigma_r} = -\mathrm{d}\lambda_{\mathrm{p}} \tag{4.165}$$

$$\varepsilon_\theta^{\mathrm{p}} = \mathrm{d}\lambda_{\mathrm{p}} \frac{\partial f}{\partial \sigma_\theta} = \mathrm{d}\lambda_{\mathrm{p}} \tag{4.166}$$

$$\varepsilon_z^{\mathrm{p}} = \mathrm{d}\lambda_{\mathrm{p}} \frac{\partial f}{\partial \sigma_z} = 0 \tag{4.167}$$

The total strains are given by the deformations (Eqs. (4.13)–(4.15)), whereas the plastic strains can be found from Eq. (2.62). Inserting the resulting expressions into Eq. (4.19), we find

$$\sigma_r' = (\lambda_{\mathrm{fr}} + 2G_{\mathrm{fr}}) \left(\frac{\partial u}{\partial r} + \mathrm{d}\lambda_{\mathrm{p}} \right) + \lambda_{\mathrm{fr}} \left(\frac{u}{r} - \mathrm{d}\lambda_{\mathrm{p}} \right) \tag{4.168}$$

Proceeding in the same way for σ_θ' and using the yield criterion (4.164) give

$$\mathrm{d}\lambda_{\mathrm{p}} = \frac{1}{2}\frac{u}{r} - \frac{1}{2}\frac{\partial u}{\partial r} - \frac{C_0}{4G_{\mathrm{fr}}} \tag{4.169}$$

Inserting the expression for $d\lambda_p$ into Eq. (4.168) and using the force balance equation (4.28) then give the displacement equation

$$\frac{d^2u}{dr^2}+\frac{1}{r}\frac{du}{dr}-\frac{u}{r^2}-\frac{C_0}{\lambda_{fr}+G_{fr}}\frac{1}{r}+\frac{\alpha}{\lambda_{fr}+G_{fr}}\frac{dp_f}{dr}=0 \tag{4.170}$$

In the following, we restrict ourselves to the case where p_f is constant or where the derivative of p_f with respect to r is proportional to $1/r$. The latter case corresponds to stationary radial flow; see Eq. (4.64).

The displacement equation can then be rewritten as

$$\frac{d^2u}{dr^2}+\frac{1}{r}\frac{du}{dr}-\frac{u}{r^2}=\frac{P}{r} \tag{4.171}$$

where (using Eq. (4.64))

$$P=\frac{C_0}{\lambda_{fr}+G_{fr}}-\frac{\alpha}{\lambda_{fr}+G_{fr}}\frac{p_{fo}-p_w}{\ln(R_e/R_w)} \tag{4.172}$$

The second term is replaced by zero for the constant p_f case.

The general solution of Eq. (4.171) is

$$u=C_1r+\frac{C_2}{r}+\frac{P}{2}r\ln r \tag{4.173}$$

Entering this solution into Eq. (4.169) gives

$$d\lambda_p=\frac{C_2}{r^2}-\frac{P}{4}-\frac{C_0}{4G_{fr}} \tag{4.174}$$

The integration constant C_2 is now determined by requiring that $d\lambda_p$ should be zero at $r=R_p$, giving

$$d\lambda_p=\frac{1}{4}\left(\frac{C_0}{G_{fr}}+P\right)\left[\left(\frac{R_p}{r}\right)^2-1\right] \tag{4.175}$$

We have used the Tresca criterion, because it allows the main points to be demonstrated with simple mathematics. Using a more realistic criterion, like the Mohr–Coulomb criterion, increases significantly the complexity of the calculations. See Risnes et al. (1982).

Plastic borehole strain may exceed the range where infinitesimal strain is an acceptable approximation (see Section 1.8.3). This implies that borehole deformations derived under the assumption of infinitesimal strain (as

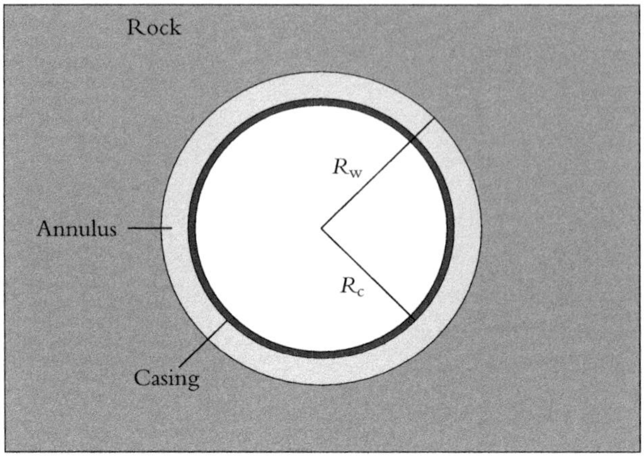

Figure 4.31 Cased hole.

above) tend to be overestimated and may need to be corrected. Vrakas and Anagnostou (2015) showed that the radial displacement u_{fs} derived from finite strain analysis can be expressed as a function of the displacement u_{is} derived under the assumption of infinitesimal strain:

$$\frac{u_{fs}}{R_w} = 1 - \frac{1}{\left[1 + (\kappa + 1)\frac{u_{is}}{R_w}\right]^{\frac{1}{\kappa+1}}} \tag{4.176}$$

The parameter κ is given as $\kappa = (1 + \sin\Psi)/(1 - \sin\Psi)$. Note that $\kappa = 1$ for incompressible plastic flow. The difference between u_{fs} and u_{is} is practically negligible for $u_{is}/R_w < 0.1$ but grows steadily for higher values of u_{is}/R_w.

4.9 Cased borehole

Consider a situation where the hole is equipped with a stiff cylinder (*casing*) that restricts the inwards movement of the rock (Fig. 4.31). After contact between the rock and the casing has been established, the stress distribution around the hole is controlled by the casing radius R_c rather than by the well pressure p_w.

First, we consider the simple case with isotropic rock, isotropic in situ stresses and constant pore pressure. We also assume plane strain conditions. The displacement of the borehole wall due to drillout is given by Eq. (4.47):

$$u_d(R_w) = \frac{R_w}{2G}(\sigma_h - p_w) \tag{4.177}$$

where R_w is the borehole radius, σ_h is the in situ stress, and G is the shear modulus. Assuming that the casing is much stiffer than the rock, the largest possible displacement of the borehole wall is $u_d(R_w) = R_w - R_c$. This restriction is enforced by a stress σ_c exerted by the casing in the radial direction, which takes over the role as the internal pressure acting on the borehole wall when contact between rock and casing has been established. Replacing p_w with σ_c and $u_d(R_w)$ with $R_w - R_c$ in Eq. (4.177) and solving for σ_c, we find

$$\sigma_c = \sigma_h - 2G\frac{R_w - R_c}{R_w} \tag{4.178}$$

By definition σ_c is a non-negative number (unless the rock is somehow glued to the casing). Moreover, σ_c is the total stress, and hence it has to exceed the pore pressure p_f for the rock to be in contact with the casing. If Eq. (4.178) yields a negative value for $\sigma_c - p_f$, then it simply means that the rock is not in contact with the casing and the following analysis does not apply.

The stresses around an open hole are given by Eqs. (4.43) and (4.44). By replacing p_w with σ_c (as expressed by Eq. (4.178)) in these equations and compensating for the movement of the borehole wall, we obtain expressions for the radial and tangential stresses around a cased hole:

$$\sigma_r = \sigma_h - 2G\frac{R_w - R_c}{R_w}\left(\frac{R_c}{r}\right)^2 \tag{4.179}$$

$$\sigma_\theta = \sigma_h + 2G\frac{R_w - R_c}{R_w}\left(\frac{R_c}{r}\right)^2 \tag{4.180}$$

Note that there are no shear stresses to be accounted for in this coordinate system, since we have rotational symmetry around the borehole.

Next, consider a vertical hole in a formation with anisotropic horizontal stresses. The stresses and displacements of the rock around the hole can be analysed by application of the complex variable method (see e.g. Jaeger et al., 2007 for a description of the method). We assume, as above, that the rock is in contact with the casing so that the displacement of the borehole wall is $u_d(R_w) = R_w - R_c$ all around, the casing is much stiffer than the rock, and we have plane strain conditions. We also assume that there is no shear force transmitted between the rock and the casing. A complete fulfilment of this assumption requires that there is no friction between the casing and the rock; however, the shear force at the rock-casing contact is likely to be small anyway since most of the tangential movement of the borehole wall that might give rise to such a force will occur before the rock hits the casing. Given these assumptions, we find that the stresses in the rock around the casing are

$$\sigma_r = \frac{1}{2}(\sigma_\mathrm{H} + \sigma_\mathrm{h}) - 2G\frac{R_\mathrm{w} - R_\mathrm{c}}{R_\mathrm{w}}\left(\frac{R_\mathrm{c}}{r}\right)^2 + \frac{1}{2}(\sigma_\mathrm{H} - \sigma_\mathrm{h})\left[1 - \frac{2}{5 - 6\nu_\mathrm{fr}}\left(\frac{R_\mathrm{c}}{r}\right)^2 + 6\frac{1 - \nu_\mathrm{fr}}{5 - 6\nu_\mathrm{fr}}\left(\frac{R_\mathrm{c}}{r}\right)^4\right]\cos 2\theta \tag{4.181}$$

$$\sigma_\theta = \frac{1}{2}(\sigma_\mathrm{H} + \sigma_\mathrm{h}) + 2G\frac{R_\mathrm{w} - R_\mathrm{c}}{R_\mathrm{w}}\left(\frac{R_\mathrm{c}}{r}\right)^2 - \frac{1}{2}(\sigma_\mathrm{H} - \sigma_\mathrm{h})\left[1 + 6\frac{1 - \nu_\mathrm{fr}}{5 - 6\nu_\mathrm{fr}}\left(\frac{R_\mathrm{c}}{r}\right)^4\right]\cos 2\theta \tag{4.182}$$

$$\tau_{r\theta} = -\frac{1}{2}(\sigma_\mathrm{H} - \sigma_\mathrm{h})\left[1 + \frac{1}{5 - 6\nu_\mathrm{fr}}\left(\frac{R_\mathrm{c}}{r}\right)^2 - 6\frac{1 - \nu_\mathrm{fr}}{5 - 6\nu_\mathrm{fr}}\left(\frac{R_\mathrm{c}}{r}\right)^4\right]\sin 2\theta \tag{4.183}$$

As before, θ is measured relative to the direction of the major horizontal stress.

The normal stress acting across the interface between rock and casing σ_c is given as σ_r (Eq. (4.181)) with $r = R_\mathrm{c}$:

$$\sigma_\mathrm{c} = \frac{1}{2}(\sigma_\mathrm{H} + \sigma_\mathrm{h}) - 2G\frac{R_\mathrm{w} - R_\mathrm{c}}{R_\mathrm{w}} + \frac{3}{2}(\sigma_\mathrm{H} - \sigma_\mathrm{h})\frac{3 - 4\nu_\mathrm{fr}}{5 - 6\nu_\mathrm{fr}}\cos 2\theta \tag{4.184}$$

where now σ_c is a function of the angle θ and has its largest value in the direction of σ_H. Again, a negative value for $\sigma_\mathrm{c} - p_\mathrm{f}$ implies that the rock is not in contact with the casing and that the conditions for this analysis are not fulfilled. By simple substitution corresponding expressions can be established for the rock around horizontal holes parallel to σ_H or σ_h.

The shear stiffness of rocks is usually much higher than the in situ stresses, typically two orders of magnitude or more. According to Eqs. (4.178) and (4.184), full contact between casing and an elastic rock may only happen if the rock is exceptionally soft or the gap $R_\mathrm{w} - R_\mathrm{c}$ is very small. Instead, the rock forms a stress arch that deflects the stresses around the hole and restricts shrinkage of the hole. Firm and full contact between rock and casing may however occur also for stiffer rocks if the in situ stresses are sufficiently high and the rock continues to deform as a plastic material after peak stress is reached. A region of plastified material will then develop around the hole, as described in Section 4.8.1, and this plastified rock may eventually establish contact with the casing.

Following a similar procedure as described above, it is possible to estimate the stress σ_c on the casing for an elastic–plastic model, as described in Section 4.8.1. No explicit expression for σ_c can be obtained in this case. However, using the model based on the Tresca criterion (Sections 4.8.1.1 and 4.8.1.3), we may establish an approximate expression

$$\sigma_c = \sigma_h - \sqrt{C_0 G \frac{R_w - R_c}{R_w}} \tag{4.185}$$

This expression is only relevant if $C_0 R_w < 4G(R_w - R_c)$; otherwise, no plastic zone will be developed around the hole. As before, contact between rock and casing requires that $\sigma_c - p_{fo}$ is non-negative.

Eq. (4.185) shows that σ_c decreases with increasing C_0. Numerical studies of a model based on the Mohr–Coulomb criterion (Section 4.8.1.2) show that σ_c also decreases with friction angle as well as the post peak parameters $\mathcal{A}'$ and k'. This follows from the fact that the more shear stress the rock can take before it reaches its peak stress level (controlled by C_0 and the friction angle), and the more shear stress the plastified zone can take (controlled by $\mathcal{A}'$ and k'), the less of the in situ stress will be transferred to the casing.

4.10 Spherical coordinates

In petroleum related rock mechanics, stress and strain solutions in spherical coordinates are of much less importance than solutions in cylindrical coordinates. Although the cylindrical geometry is often a good approximation for a borehole, spherical cavities are to a less extent a good approximation to real cases.

Still, solutions in spherical coordinates have been employed as a crude approximation to producing cavities (see Section 10.2.3) and are also useful in some methods for the estimation of surface subsidence (see Chapter 12).

Therefore here we give a short presentation of some basic equations.

4.10.1 Basic equations

We use the standard convention, denote the spherical coordinates by (r, θ, ϕ) and limit ourselves to a situation with full spherical symmetry.

Then the only deformation is in the radial direction, and the non-zero strains are given by (see Appendix D.1.2 for complete expressions for strain in spherical coordinates)

$$\varepsilon_r = \frac{\partial u}{\partial r} \tag{4.186}$$

$$\varepsilon_\phi = \varepsilon_\theta = \frac{u}{r} \tag{4.187}$$

The force equilibrium equation for the r-direction, corresponding to Eq. (4.28), becomes

$$\frac{\mathrm{d}\sigma_r}{\mathrm{d}r} + \frac{1}{r}(2\sigma_r - \sigma_\theta - \sigma_\phi) = 0 \tag{4.188}$$

whereas, due to the symmetry assumed, the force equilibrium equation for the θ-direction requires

$$\sigma_\theta = \sigma_\phi \tag{4.189}$$

Eq. (4.188) is then simplified to

$$\frac{\mathrm{d}\sigma_r}{\mathrm{d}r} + \frac{2}{r}(\sigma_r - \sigma_\theta) = 0 \tag{4.190}$$

Hooke's law is reduced to (using $\varepsilon_\phi = \varepsilon_\theta$)

$$\sigma_r - \alpha p_\mathrm{f} = (\lambda_\mathrm{fr} + 2G_\mathrm{fr})\varepsilon_r + 2\lambda_\mathrm{fr}\varepsilon_\theta \tag{4.191}$$

$$\sigma_\theta - \alpha p_\mathrm{f} = \lambda_\mathrm{fr}\varepsilon_r + 2(\lambda_\mathrm{fr} + G_\mathrm{fr})\varepsilon_\theta \tag{4.192}$$

Inserting the expressions for the strains, Eqs. (4.186) and (4.187), into Hooke's law and using the force equilibrium equation (4.190) give the displacement equation

$$\frac{\mathrm{d}^2 u}{\mathrm{d}r^2} + \frac{2}{r}\frac{\mathrm{d}u}{\mathrm{d}r} - \frac{2}{r^2}u + \frac{\alpha}{\lambda_\mathrm{fr} + 2G_\mathrm{fr}}\frac{\mathrm{d}p_\mathrm{f}}{\mathrm{d}r} = 0 \tag{4.193}$$

4.10.2 Stress distribution around a spherical cavity with no fluid flow

With no fluid flow the pore pressure will remain constant, equal to p_fo, and the displacement equation becomes

$$\frac{\mathrm{d}^2 u}{\mathrm{d}r^2} + \frac{2}{r}\frac{\mathrm{d}u}{\mathrm{d}r} - \frac{2}{r^2}u = \frac{\mathrm{d}}{\mathrm{d}r}\left(\frac{\mathrm{d}u}{\mathrm{d}r} + 2\frac{u}{r}\right) = \frac{\mathrm{d}}{\mathrm{d}r}\left(\frac{1}{r^2}\frac{\mathrm{d}}{\mathrm{d}r}\left(r^2 u\right)\right) = 0 \tag{4.194}$$

The general solution is

$$u = C_1 r + \frac{C_2}{r^2} \tag{4.195}$$

where the integration constants are to be determined from the boundary conditions

$$\sigma_r = p_w \quad \text{for} \quad r = R_c \tag{4.196}$$

$$\sigma_r = \sigma_o \quad \text{as} \quad r \to \infty \tag{4.197}$$

where R_c is the radius of the spherical cavity.

Proceeding as in Section 4.2.2, using Hooke's law (Eqs. (4.191)–(4.192)), the stress distribution around a spherical cavity is found to be given by

$$\sigma_r = \sigma_o - (\sigma_o - p_w)\left(\frac{R_c}{r}\right)^3 \tag{4.198}$$

$$\sigma_\theta = \sigma_o + \frac{1}{2}(\sigma_o - p_w)\left(\frac{R_c}{r}\right)^3 \tag{4.199}$$

These stress solutions are sketched in Fig. 4.32. The thin dashed lines show the corresponding distribution for a cylindrical hole; see Fig. 4.3. If we assume σ_o to be equal to σ_h, then the stress difference between the tangential and radial stresses at the cavity wall will be $2(\sigma_h - p_w)$ in the cylindrical hole case but only $(3/2)(\sigma_h - p_w)$ in the spherical cavity case. It follows that a

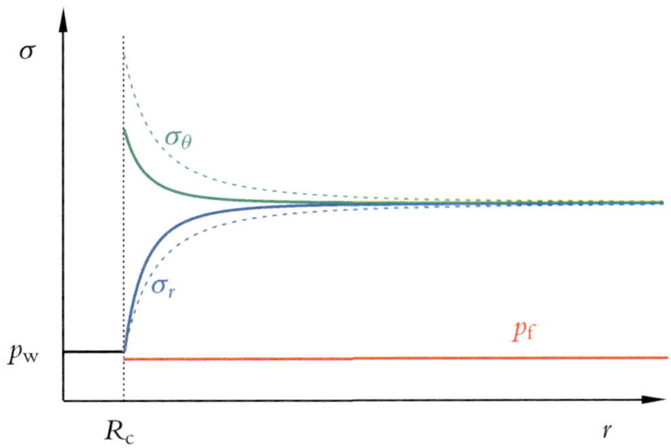

Figure 4.32 Stress distribution around a spherical cavity with no fluid flow. The thin dashed lines show the corresponding distribution for a cylinder; see Fig. 4.3.

spherical cavity tends to be more stable than an open hole with respect to shear failure.

4.10.3 Stress distribution with fluid flow

Darcy's law (1.308) for symmetric spherical flow into a cavity can be written

$$\frac{Q}{4\pi r^2} = \frac{k}{\eta_{\mathrm{f}}} \frac{\mathrm{d}p_{\mathrm{f}}}{\mathrm{d}r} \tag{4.200}$$

where $4\pi r^2$ is the area of a sphere of radius r. All the other parameters are the same as in Eq. (4.63). If we assume that the flow rate is independent of r, then the fluid pressure distribution is given by

$$p_{\mathrm{f}} = p_{\mathrm{fo}} - (p_{\mathrm{fo}} - p_{\mathrm{w}})\frac{R_{\mathrm{c}}}{r} \tag{4.201}$$

Solving the displacement equation, the steady-state solutions for fluid flow into the cavity can then be written as

$$\sigma_r = \sigma_{\mathrm{o}} - [\sigma_{\mathrm{o}} - p_{\mathrm{w}} - 2\eta(p_{\mathrm{fo}} - p_{\mathrm{w}})]\left(\frac{R_{\mathrm{c}}}{r}\right)^3 - 2\eta(p_{\mathrm{fo}} - p_{\mathrm{w}})\frac{R_{\mathrm{c}}}{r} \tag{4.202}$$

$$\sigma_\theta = \sigma_{\mathrm{o}} + \frac{1}{2}[\sigma_{\mathrm{o}} - p_{\mathrm{w}} - 2\eta(p_{\mathrm{fo}} - p_{\mathrm{w}})]\left(\frac{R_{\mathrm{c}}}{r}\right)^3 - \eta(p_{\mathrm{fo}} - p_{\mathrm{w}})\frac{R_{\mathrm{c}}}{r} \tag{4.203}$$

At the cavity wall, we have

$$\sigma_r = p_{\mathrm{w}} \tag{4.204}$$

$$\sigma_\theta = \sigma_{\mathrm{o}} + \frac{1}{2}(\sigma_{\mathrm{o}} - p_{\mathrm{w}}) - 2\eta(p_{\mathrm{fo}} - p_{\mathrm{w}}) \tag{4.205}$$

As for the cylindrical hole, the stresses at the free surface are the same as in the no-flow case, except that the tangential stress has an additional term due to pressure drawdown. This term is exactly the same as derived for the cylindrical vertical well.

References

Aadnøy, B.S., Angell-Olsen, F., 1995. Some effects of ellipticity on the fracturing and collapse behavior of a borehole. Int. J. Rock Mech. Min. Sci. Geomech. Abstr. 32 (6), 621–627. https://doi.org/10.1016/0148-9062(95)00016-A.

Amadei, B., 1983. Rock Anisotropy and the Theory of Stress Measurements. Springer, Berlin.

Asaka, M., Holt, R.M., 2020. Anisotropic wellbore stability analysis: impact on failure prediction. Rock Mech. Rock Eng. https://doi.org/10.1007/s00603-020-02283-0.

Bradley, W.B., 1979. Failure of inclined boreholes. J. Energy Resour. Technol. 101, 232–239. https://doi.org/10.1115/1.3446925.

Charlez, P.A., 1991. Rock Mechanics. Vol. I. Theoretical Fundamentals. Éditions Technip, Paris.

Charlez, P.A., 1997. Rock Mechanics. Vol. 2. Petroleum Applications. Éditions Technip, Paris.

Detournay, E., Cheng, A.H.-D., 1988. Poroelastic response of a borehole in a non-hydrostatic stress field. Int. J. Rock Mech. Min. Sci. Geomech. Abstr. 25, 171–182. https://doi.org/10.1016/0148-9062(88)92299-1.

Ewy, R.T., Cook, N.G.W., 1989. Fracture processes around highly stressed boreholes. In: Proc. 12th Annual Energy-Sources Technol. Conf. & Exhibition, Drilling Symp. ASME, New York, pp. 63–70.

Fairhurst, C., 1968. Methods of determining in-situ rock stresses at great depth. Technical report. Missouri River Division, U.S. Army Corps of Engineers. School of Mineral and Metallurgical Engineering, U. of Minnesota.

Fang, X., 2018. A revisit to the Lekhnitskii–Amadei solution for borehole stress calculation in tilted transversely isotropic media. Int. J. Rock Mech. Min. Sci. 104, 113–118. https://doi.org/10.1016/0148-9062(93)90073-M.

Gaede, O., 2011. Anisotropic damage evolution around boreholes in chemically reactive rock formations. PhD thesis. The University of Western Australia.

Guenot, A., 1987. Contraintes et ruptures autour des forages pétroliers. In: Proc. 6th Int. Conf. on Rock Mechanics, pp. 109–118.

Haimson, B., Fairhurst, C., 1967. Initiation and extension of hydraulic fractures in rocks. Soc. Pet. Eng. J. 7, 310–318. https://doi.org/10.2118/1710-PA.

Hiramatsu, Y., Oka, Y., 1962. Stress around a shaft or level excavated in ground with a three-dimensional stress state. Mem. Fac. Eng. Kyotu Univ. 24, 56–76.

Hiramatsu, Y., Oka, Y., 1968. Determination of the stress in rock unaffected by boreholes or drifts, from measured strains or deformations. Int. J. Rock Mech. Min. Sci. 5, 337–353. https://doi.org/10.1016/0148-9062(68)90005-3.

Jaeger, J.C., Cook, N.G.W., 1979. Fundamentals of Rock Mechanics, third edition. Chapman and Hall, London.

Jaeger, J.C., Cook, N.G.W., Zimmerman, R.W., 2007. Fundamentals of Rock Mechanics, fourth edition. Blackwell Publishing.

Jia, Q., Schmitt, D.R., 2014. Effects of formation anisotropy on borehole stress concentrations: implications to drilling induced tensile fractures. ARMA14-7462. In: 48th US Rock Mechanics/Geomechanics Symposium. Minneapolis, MN, USA.

Kaiser, P.K., Guenot, A., Morgenstern, N.R., 1985. Deformation of small tunnels. IV. Behaviour during failure. Int. J. Rock Mech. Min. Sci. Geomech. Abstr. 22, 141–152. https://doi.org/10.1016/0148-9062(85)93228-0.

Karpfinger, F., Prioul, R., Gaede, O., Jocker, J., 2011. Revisiting borehole stresses in anisotropic elastic media: comparison of analytical versus numerical solutions. ARMA11-273. In: 45th US Rock Mechanics/Geomechanics Symposium. San Francisco, CA, USA.

Kirsch, G., 1898. Die Theorie der Elastizität und die Bedürfnisse der Festigkeitslehre. Z. Ver. Dtsch. Ing. 42 (29), 797–807.

Lekhnitskii, S.G., 1963. Theory of Elasticity of an Anisotropic Elastic Body. Holden Day Inc., San Francisco.

Maury, V.M., Sauzay, J.-M., 1987. Borehole instability: case histories, rock mechanics approach, and results. SPE/IADC16051. In: SPE/IADC Drilling Conference. New Orleans, LA, USA.

Ong, S.H., Rogiers, J.-C., 1993. Influence of anisotropies in borehole stability. Int. J. Rock Mech. Min. Sci. Geomech. Abstr. 30 (7), 1069–1075. https://doi.org/10.1016/0148-9062(93)90073-M.

Peška, P., Zoback, M.D., 1995. Compressive and tensile failure of inclined well bores and determination of in situ stress and rock strength. J. Geophys. Res. 100 (B7), 12791–12811. https://doi.org/10.1029/95JB00319.

Pollard, D., 1973. Equations for stress and displacement fields around pressurized elliptical holes in elastic solids. Math. Geol. 5 (1), 11–25. https://doi.org/10.1007/BF02114084.

Raaen, A.M., Larsen, I., Fjær, E., Holt, R.M., 2019. Pore pressure response in rock: implications of tensorial Skempton *B* in an anisotropic formation. ARMA19-1918. In: 53rd US Rock Mechanics/Geomechanics Symposium. New York, NY, USA.

Risnes, R., Bratli, R., Horsrud, P., 1982. Sand stresses around a borehole. Soc. Pet. Eng. J. 22 (6), 883–898. https://doi.org/10.2118/9650-PA.

Santarelli, F.J., 1987. Theoretical and experimental investigation of the stability of the axisymmetric wellbore. PhD thesis. University of London.

Santarelli, F.J., Brown, E.T., Maury, V., 1986. Analysis of borehole stresses using pressure-dependent linear elasticity. Int. J. Rock Mech. Min. Sci. Geomech. Abstr. 23, 445–449. https://doi.org/10.1016/0148-9062(86)92310-7.

Setiawan, N.B., Zimmerman, R.W., 2018. Wellbore breakout prediction in transversely isotropic rocks using true-triaxial failure criteria. Int. J. Rock Mech. Min. Sci. 112 (7), 313–322. https://doi.org/10.1016/j.ijrmms.2018.10.033.

Vardoulakis, I., Sulem, J., Guenot, A., 1988. Borehole instabilities as bifurcation phenomena. Int. J. Rock Mech. Min. Sci. Geomech. Abstr. 25, 159–170. https://doi.org/10.1016/0148-9062(88)92298-X.

Vrakas, A., Anagnostou, G., 2015. A simple equation for obtaining finite strain solutions from small strain analyses of tunnels with very large convergences. Géotechnique 65 (11), 936–944. https://doi.org/10.1680/jgeot.15.P.036.

Zheng, Z., Cook, N.G.W., Myer, L.R., 1988. Borehole breakout and stress measurements. In: Cundall, P.A., Sterling, R.L., Starfield, A.M. (Eds.), Key Questions in Rock Mechanics. Balkema, Rotterdam, pp. 471–479.

Zoback, M.D., Moos, D., Mastin, L., Anderson, R.N., 1985. Well bore breakouts and in situ stress. J. Geophys. Res. 90 (B7), 5523–5555. https://doi.org/10.1029/JB090iB07p05523.

CHAPTER 5

Elastic wave propagation in rocks

Elastic waves are mechanical disturbances that propagate through a material. Such waves are able to travel over very long distances through the Earth and thus bring us information about parts of the formations that are otherwise inaccessible. In seismic surveys, elastic waves are generated at the surface, and the echoes from various layers in the formations below are used to map the structure of the underground. Earthquakes are generated by abrupt failure processes and propagate as elastic waves.

Elastic waves in air and water are usually called acoustic waves or sound waves. These terms are often used about elastic waves in rocks too. Elastic waves in rocks propagate with a velocity that is given by elastic stiffnesses and the density of the rock. These parameters depend on other parameters such as porosity. Thus elastic waves also provide a method by which specific formation parameters can be estimated in the field.

As we proceed, we will see that this evaluation is not straightforward, for example, the stiffness probed by an elastic wave is generally different from the static stiffness. However, the fact that elastic waves are mechanical disturbances means that there has to be a fundamental connection between rock acoustics and rock mechanics.

In this chapter, we first present the fundamentals of elastic wave propagation in porous media (Sections 5.1–5.5). Section 5.6 addresses the main focus of this chapter, the connection between rock acoustics and rock mechanics. Finally, we give some background to understand sonic wave propagation in boreholes and seismic waves through the subsurface (Sections 5.7–5.9).

5.1 The wave equation

When an elastic wave passes through a material, each part of the material is forced into oscillating motion. Consider a volume element within the material. The element is pushed by the adjacent element behind it and accelerates. The movement causes the volume element to push on the next element in front of it. This way the movement is propagating from volume element to volume element through the material as a wave.

Developments in Petroleum Science, Volume 72
ISSN 0376-7361, https://doi.org/10.1016/B978-0-12-822195-2.00014-0

Wave propagation is described mathematically by the wave equation. This equation follows directly from Newton's second law of motion: Force equals mass times acceleration. The force (per unit volume) acting on a volume element due to the stress the adjacent elements exert was derived in Chapter 1. The force F_x in the x-direction is given as (see Eq. (1.14))

$$F_x = \frac{\partial \sigma_x}{\partial x} + \frac{\partial \tau_{xy}}{\partial y} + \frac{\partial \tau_{xz}}{\partial z} \tag{5.1}$$

For simplicity, we will first consider a plane, unidirectional mode of deformation in the x-direction. We will further assume that the material is a fluid, so that no shear forces can exist, that is, $\tau_{xy} = \tau_{xz} = 0$. The mass (per unit volume) is given as the density ρ of the material, and the acceleration is given as the second derivative (with respect to time t) of the displacement u of the volume element. Newton's second law of motion thus requires that

$$\rho \frac{\partial^2 u}{\partial t^2} = \frac{\partial \sigma_x}{\partial x} \tag{5.2}$$

The stress σ_x can be expressed in terms of the displacement u according to Hooke's law (Eqs. (1.96)–(1.101)), which under the conditions assumed here (plane wave in the x-direction, which implies that $\partial/\partial y = \partial/\partial z = 0$, and fluid, which implies that $G = 0$) becomes

$$\sigma_x = \lambda \frac{\partial u}{\partial x} \tag{5.3}$$

Note that for a fluid, the Lamé modulus λ is equal to the bulk modulus. Combining Eqs. (5.2) and (5.3), we obtain

$$\rho \frac{\partial^2 u}{\partial t^2} = \lambda \frac{\partial^2 u}{\partial x^2} \tag{5.4}$$

This is the wave equation for a plane wave propagating in the x-direction in a fluid. The solution of the wave equation has the form

$$u = u^{\mathrm{o}} \mathrm{e}^{\mathrm{j}(\omega t - qx)} \tag{5.5}$$

where j represents the imaginary unit. The real part of this expression describes a periodic displacement in the x-direction with amplitude u^{o}. The complex formulation has been chosen here because it is mathematically simpler to work with.

The periodic displacement is a wave that propagates in the x-direction with angular frequency ω and wavenumber q. The angular frequency ω is related to the frequency f as

$$\omega = 2\pi f \tag{5.6}$$

The wave described by Eq. (5.5) extends from $-\infty$ to $+\infty$ both in space and time. A displacement pulse, which is limited in space and time, is described as a sum of waves like Eq. (5.5), each with different angular frequency and different amplitude, such that the net displacement vanishes for large values of $|x|$ and $|t|$.

The displacement $u(x, t)$ represents the displacement of a volume element at position x and time t as the wave passes by. In this context the volume element is considered as a particle attached with "springs" to its neighbouring elements. The displacement is therefore often called particle displacement. The displacement wave described by Eq. (5.5) can also be considered as a pressure wave, through the relation between displacement and stress given by Eq. (5.3).

Any two points x_1 and x_2 that are separated by a distance such that $qx_2 = qx_1 + 2\pi$ will at all times have equal displacement according to Eq. (5.5). The distance between two such points, $x_2 - x_1$, is called the wavelength (λ_w). The wavelength is thus related to the wavenumber as

$$q = \frac{2\pi}{\lambda_w} \tag{5.7}$$

The term $(\omega t - qx)$ in the exponent of expression (5.5) is called the *phase* of the wave. If we follow the wave locked on to a fixed displacement like a surfer on a sea wave, then we move with a velocity $v = x/t$ such that the phase $(\omega t - qx)$ remains constant. This velocity is called the *phase velocity* and is related to the angular frequency, the wavenumber and the wavelength as

$$v = \frac{\omega}{q} = \lambda_w f \tag{5.8}$$

Inserting Eq. (5.5) into Eq. (5.4) and solving for ω/q, we find that the phase velocity is given by the elastic stiffness and the density as

$$v = \sqrt{\frac{\lambda}{\rho}} \tag{5.9}$$

Thus we have derived the general result that the phase velocity of an elastic wave is proportional to the square root of an elastic modulus and inversely proportional to the square root of the density. This result provides our basis for using elastic waves to explore mechanical properties of rocks.

The phase velocity is the velocity at which the phase of an elastic wave travels. Elastic waves in fluids and gases are usually called sound waves or acoustic waves, and the phase velocity is usually called the *sound velocity*. For fresh water, the elastic modulus $\lambda = K_f \approx 2.25$ GPa, and the density $\rho \approx 1.0$ g/cm^3 (where both depend on temperature and pressure), and hence the sound velocity is 1500 m/s according to Eq. (5.9). For dry air under normal atmospheric conditions, $\lambda \approx 0.14$ MPa (adiabatic value; see Section 1.5), and $\rho \approx 0.0012$ g/cm^3, and thus the sound velocity is 343 m/s.

As shown above, the definition of the phase velocity is associated with a wave, which extends to infinity in space and time. In practice, we usually work with acoustic signals in the form of pulses of finite duration. The velocity of a pulse or, equivalently, the velocity of the wave energy is called the *group velocity* and is given as

$$v_{\text{group}} = \frac{\partial \omega}{\partial q} \tag{5.10}$$

By comparing Eqs. (5.8) and (5.10) we see that if the phase velocity varies with the frequency—a phenomenon called dispersion—the group and phase velocities will be different.

5.2 P- and S-waves

We now proceed to a more general case and allow the material to possess a shear stiffness so that shear forces may exist, that is, we will look at elastic wave propagation in an isotropic solid. Starting with Newton's second law of motion, and the general expression (5.1) for the force, we find the wave equation for displacement u in the x-direction:

$$\rho \frac{\partial^2 u}{\partial t^2} = \frac{\partial \sigma_x}{\partial x} + \frac{\partial \tau_{xy}}{\partial y} + \frac{\partial \tau_{xz}}{\partial z} \tag{5.11}$$

Introducing Hooke's law (Eqs. (1.96)–(1.101)), we find

$$\rho \frac{\partial^2 u}{\partial t^2} = (\lambda + G)\left(\frac{\partial^2 u}{\partial x^2} + \frac{\partial^2 v}{\partial x \partial y} + \frac{\partial^2 w}{\partial x \partial z}\right) + G\left(\frac{\partial^2 u}{\partial x^2} + \frac{\partial^2 u}{\partial y^2} + \frac{\partial^2 u}{\partial z^2}\right) \tag{5.12}$$

A general solution of this equation is

$$u = u^{\circ} e^{j(\omega t - q\hat{l}\cdot\vec{r})} \tag{5.13}$$

The vector $\vec{r}$ has the components x, y and z, whereas $\hat{l}$ is a unit vector parallel to the direction of wave propagation. The components l_x, l_y and l_z will be the direction cosines (see Eqs. (1.27)–(1.29)) of the propagation vector. Equations similar to Eq. (5.12) can be derived for particle displacement in the y- and z-directions. In general, these are coupled equations in u, v and w; however, there are two relatively straightforward solutions, which can be derived from Eq. (5.12). First, we search for a solution where the wave propagation is parallel to the displacement u (i.e. $l_x = 1$, $l_y = l_z = v = w = 0$):

$$u = u^{\circ} e^{j(\omega t - qx)} \tag{5.14}$$

Such a wave is called a *longitudinal* wave or, alternatively, a *compressional* wave, as it involves a periodic compression of the material. It is also often termed a *primary* wave, a name that originates from studies of earthquakes.

Since we are now considering a plane wave propagating parallel to the x-axis, we have $\partial/\partial y = \partial/\partial z = 0$, and Eq. (5.12) reduces to

$$\rho\omega^2 = q^2(\lambda + 2G) \tag{5.15}$$

The phase velocity is

$$v_{\mathrm{p}} = \frac{\omega}{q} = \sqrt{\frac{\lambda + 2G}{\rho}} \tag{5.16}$$

The subscript p denotes that this is a primary wave.

Eq. (5.12) also has other solutions. For instance, it is possible to establish a mode of motion in which the particles are displaced only in the x-direction (i.e. $u \neq 0$, $v = w = 0$), whereas the wave is propagating as a plane wave in the y-direction (which implies $\partial/\partial x = \partial/\partial z = 0$). Then the solution is

$$u = u^{\circ} e^{j(\omega t - qy)} \tag{5.17}$$

Inserting this into Eq. (5.12), we find

$$\rho\omega^2 = q^2 G \tag{5.18}$$

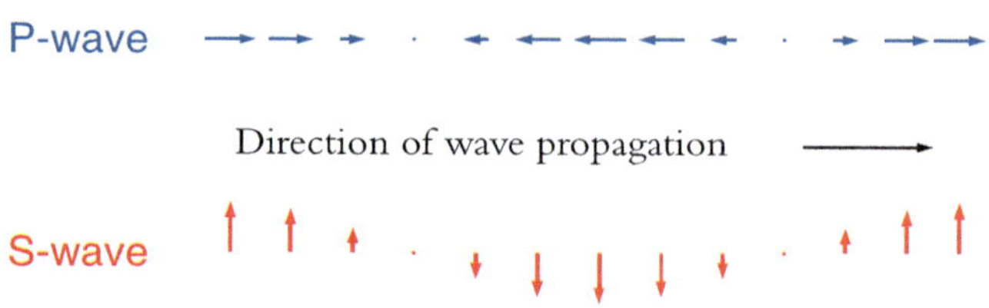

Figure 5.1 Particle motion in P-wave and S-wave propagation.

The phase velocity is

$$v_s = \frac{\omega}{q} = \sqrt{\frac{G}{\rho}} \tag{5.19}$$

The subscript s denotes that this is the velocity of a *secondary* wave. Such a wave is also called a *transversal* wave or, alternatively, a *shear* wave, since it involves a periodic shearing of the material. The motion of the volume elements (or "particles") in a material during propagation of compressional and shear waves is illustrated in Fig. 5.1. Comparing Eqs. (5.16) and (5.19) and recalling that $\nu > -1$ (see Section 1.4), we see that v_p is always larger than v_s in an isotropic linearly elastic solid.

The two solutions we found to Eq. (5.12), the compressional wave and the shear wave, are the only types of waves that exist for an isotropic homogeneous solid. For convenience, these are often denoted P-wave and S-wave, representing the primary and secondary waves, respectively. Also in this area, there is a divergence in the use of names: In physics and rock mechanics, "acoustic waves" and "elastic waves" mean both P- and S-waves. In seismics, "acoustic waves" mean only P-waves, whereas "elastic waves" mean both P- and S-waves.

The direction of particle motion is called the *polarisation* of the wave. For any given direction of wave propagation, there only exists one type of waves with polarisation parallel to the direction of propagation (the P-wave) and one type of waves with polarisation normal to the direction of propagation (the S-wave). (Strictly speaking, *two* independent S-waves exist for a given direction of propagation, since the polarisation can be in any direction in the plane normal to the direction of propagation. In an isotropic solid, however, these S-waves are equivalent in all respects except for the polarisation.)

By inverting Eqs. (5.16) and (5.19) we can express the elastic coefficients in terms of the phase velocities:

$$G = \rho v_s^2 \tag{5.20}$$

$$\lambda = \rho v_p^2 - 2\rho v_s^2 \tag{5.21}$$

Using Table 1.1, we find the corresponding expressions for the bulk modulus, Young's modulus and Poisson's ratio:

$$K = \rho v_p^2 - \frac{4}{3}\rho v_s^2 \tag{5.22}$$

$$E = \rho v_s^2 \frac{3v_p^2 - 4v_s^2}{v_p^2 - v_s^2} \tag{5.23}$$

$$\nu = \frac{v_p^2 - 2v_s^2}{2(v_p^2 - v_s^2)} \tag{5.24}$$

Further relations can be established by use of Table 1.1. Note that

$$\lambda + 2G = H = K + \frac{4}{3}G = \rho v_p^2 \tag{5.25}$$

where H is the uniaxial compaction modulus (also referred to as plane wave modulus, or P-wave modulus; see Section 1.3). Eqs. (5.20)–(5.25) were derived for a linearly elastic, isotropic and homogeneous material, and show how values for elastic moduli can be obtained from measurements of acoustic velocities in such a material. Rocks are usually neither isotropic nor homogeneous, nor are they linearly elastic, and we will see later (Section 5.6) that this complicates the relations between static elastic moduli and acoustic velocities.

5.3 Elastic waves in porous materials

The elastic response of a porous material may be significantly affected by the presence of a pore fluid. In poorly consolidated sediments, the P-wave velocity for a water saturated material can be several times larger than it is when the material is dry. Qualitatively, we can explain this as a consequence of the added resistance against compression provided by the pore fluid. For more competent materials or materials under high confining pressure, the frame stiffness is higher, and the contribution of the pore fluid is relatively less, and hence the effect of saturation is much smaller. As saturation also implies an increase in density, it may even result in a reduction of the P-wave velocity for a stiff rock. Saturation is assumed not to affect the shear modulus for a permeable rock (see Section 1.6.2), and hence it usually results in a small decrease in velocity for S-waves.

Clearly, the impact of the pore fluid on the velocities is strongly related to the porosity of the rock. A simple assumption, which has been frequently used for interpretation of well logs, is that the interval transit time Δt_c (which is the inverse of the P-wave velocity) is related linearly to the porosity according to the *time average equation* (Wyllie et al., 1958):

$$\Delta t_{\mathrm{c}} = \frac{1}{v_{\mathrm{p}}} = \frac{\phi}{v_{\mathrm{fluid}}} + \frac{1-\phi}{v_{\mathrm{solid}}} \tag{5.26}$$

where v_{fluid} and v_{solid} denote the P-wave velocities of the pore fluid and the solid grain material, respectively. The equation has an empirical basis (it is theoretically valid in the high-frequency limit for a material consisting of layers of fluid and non-porous solid), and it works with reasonable accuracy for several practical purposes. If the lithology (and hence v_{solid}) is known, then it can be used to estimate the porosity, or, alternatively, it can be used for identification of the lithology if the porosity can be estimated by other means. The equation is not applicable for predicting the effect of saturation and cannot it be used for S-waves.

5.3.1 Biot's theory of elastic wave propagation

In Chapter 1, we applied Biot's theory of poroelasticity to describe the effect of the pore fluid on the elastic properties of porous materials. The same formalism may also be used to describe elastic wave propagation in saturated rocks (Biot, 1962; Stoll, 1974). The procedure for establishing the wave equation and deriving the velocities is analogous to that of Sections 5.1 and 5.2; however, now we have to account for the pore pressure p_{f} in addition to the external stress. Also, we have to take into account that the displacement of the fluid ($u_{\mathrm{f}}^{\mathrm{fr}}$) may be different from the displacement of the solid material ($u_{\mathrm{s}}^{\mathrm{fr}}$).

Considering longitudinal motion parallel to the x-axis and following the same approach as in Section 5.1, we find that Newton's second law of motion requires that

$$\phi \rho_{\mathrm{f}} \frac{\partial^2 u_{\mathrm{f}}^{\mathrm{fr}}}{\partial t^2} + (1-\phi)\rho_{\mathrm{s}} \frac{\partial^2 u_{\mathrm{s}}^{\mathrm{fr}}}{\partial t^2} = \frac{\partial \sigma_x}{\partial x} \tag{5.27}$$

We now differentiate with respect to x and introduce the strain parameters ε_x, ζ (from Eq. (1.147)) and H_u (from Eq. (1.316)). As we consider only wave motion in the x-direction, we have $\partial/\partial y = \partial/\partial z = 0$. Expressing the stress σ_x in terms of the strains according to Eq. (1.150), we find that Eq. (5.27) may be written as

$$\frac{\partial^2}{\partial t^2}(-\rho_f\zeta + \rho\varepsilon_x) = \frac{\partial^2}{\partial x^2}(H_u\varepsilon_x - C\zeta) \tag{5.28}$$

Here ρ defined as

$$\rho = \phi\rho_f + (1-\phi)\rho_s \tag{5.29}$$

is the bulk density of the porous medium. Eq. (5.28) is a wave equation, similar to Eq. (5.4), although here we use strain rather than displacement as the variable. However, Eq. (5.28) has two variables, ζ and ε_x, and hence we need one more equation. Turning to the movement of the fluid, we find according to Eqs. (1.308) and (1.309) (Darcy's law) that the net effective force (per unit area) acting on the fluid part of a volume element is

$$\vec{\nabla} p_f - \frac{\eta_f}{k}\phi\left(\frac{\partial \vec{u}_f^{fr}}{\partial t} - \frac{\partial \vec{u}_s^{fr}}{\partial t}\right) \tag{5.30}$$

Considering only wave motion in the x-direction and denoting the x-components of $\vec{u}_f^{fr}$ and $\vec{u}_s^{fr}$ as u_f^{fr} and u_s^{fr}, respectively, Newton's second law of motion gives

$$\frac{\partial^2}{\partial t^2}(\rho_f u_f^{fr}) = \frac{\partial p_f}{\partial x} - \frac{\eta_f}{k}\phi\left(\frac{\partial u_f^{fr}}{\partial t} - \frac{\partial u_s^{fr}}{\partial t}\right) \tag{5.31}$$

We now differentiate this equation with respect to x, use Eq. (1.156) to eliminate p_f and Eq. (1.150) to introduce ζ, and find the second wave equation:

$$\frac{\partial^2}{\partial t^2}\left(\rho_f\varepsilon_x - \frac{\rho_f}{\phi}\zeta\right) = \frac{\partial^2}{\partial x^2}(C\varepsilon_x - M\zeta) + \frac{\eta_f}{k}\frac{\partial \zeta}{\partial t} \tag{5.32}$$

The two wave equations (5.28) and (5.32) have the solutions

$$\varepsilon_x = \varepsilon_x^{\text{o}} e^{j(\omega t - qx)} \tag{5.33}$$

$$\zeta = \zeta^{\text{o}} e^{j(\omega t - qx)} \tag{5.34}$$

Two modifications were introduced by Biot:

1. The pore fluid does not flow strictly in the direction of the pore pressure gradient, but is constrained to follow the pore channel network. To compensate for this, the second term on the left-hand side of Eq. (5.32) is multiplied by a parameter T, called the *tortuosity*. The tortuosity is a measure of the twistedness of the pore channels. Given a porous sample with thickness L, the pore fluid has to travel a distance $L_e > L$ to penetrate the sample. The tortuosity is defined as $T = (L_e/L)^2$. Thus $T \geq 1$. For sandstone, T is typically in the range 2–3.
2. The fluid flow at higher frequencies is not laminar. This is compensated for by modifying the viscosity $\eta_f \to \eta_f F(\kappa)$, where

$$\kappa = a_p \sqrt{\frac{\omega \rho_f}{\eta_f}} \tag{5.35}$$

   and a_p is equal to the radius of an average pore; $F(\kappa) = 1$ at $\omega = 0$, whereas $F(\kappa) \propto \sqrt{\omega}$ as $\omega \to \infty$. A specific expression for $F(\kappa)$ is given in Appendix D.2.1.

Introducing expressions (5.33) and (5.34) into the wave equations (5.28) and (5.32), including Biot's modifications above, and making use of the relation $v_p = \omega/q$, we obtain two equations for the amplitudes ε_x^o and ζ^o:

$$(H_u - \rho v_p^2)\varepsilon_x^o - (C - \rho_f v_p^2)\zeta^o = 0 \tag{5.36}$$

$$(C - \rho_f v_p^2)\varepsilon_x^o - (M - \frac{\rho_f T}{\phi} v_p^2 + j \frac{v_p^2}{\omega} \frac{\eta_f F(\kappa)}{k})\zeta^o = 0 \tag{5.37}$$

This homogeneous system of equations has non-trivial solutions if the determinant vanishes (see Appendix C.2.11), that is, if

$$\begin{vmatrix} (H_u - \rho v_p^2) & -(C - \rho_f v_p^2) \\ (C - \rho_f v_p^2) & -(M - \frac{\rho_f T}{\phi} v_p^2 + j \frac{v_p^2}{\omega} \frac{\eta_f F(\kappa)}{k}) \end{vmatrix} = 0 \tag{5.38}$$

Eq. (5.38) is an equation for v_p. Since it specifies how the velocity changes with frequency, it is a so-called *dispersion relation*. Attenuation is also implicitly included through the imaginary parts of the solution (see Section 5.4). In the low-frequency limit the term including ω^{-1} goes to infinity, and the only solution to Eq. (5.38) is

$$v_p(\omega \to 0) = \sqrt{\frac{H_u}{\rho}} = \sqrt{\frac{K_u + \frac{4}{3} G_{fr}}{\phi \rho_f + (1 - \phi)\rho_s}} \tag{5.39}$$

Here we have introduced expression (5.29) for the density ρ, whereas the modulus $H_u = \lambda_u + 2G_{fr} = K_u + \frac{4}{3}G_{fr}$, and K_u is given by Eq. (1.171).

Introducing the velocity given by Eq. (5.39) into Eq. (5.36), we find that $\zeta^o = 0$ in the low-frequency limit. Thus the solid and fluid are moving in phase (that is, together) during the oscillatory movements caused by the elastic wave.

We can establish an equation similar to Eq. (5.38) for the shear wave velocity:

$$\begin{vmatrix} (G_{fr} - \rho v_s^2) & \rho_f v_s^2 \\ -\rho_f v_s^2 & -(-\frac{\rho_f T}{\phi} v_s^2 + j\frac{v_s^2}{\omega}\frac{\eta_f F(\kappa)}{k}) \end{vmatrix} = 0 \tag{5.40}$$

In the low-frequency limit the solution of this equation is

$$v_s(\omega \to 0) = \sqrt{\frac{G_{fr}}{\rho}} = \sqrt{\frac{G_{fr}}{\phi\rho_f + (1-\phi)\rho_s}} \tag{5.41}$$

Eqs. (5.39) and (5.41) describe the effect of saturation on elastic wave velocities in the low-frequency limit.

To illustrate the predictions of the equations, let us look at an example. In a water-saturated sandstone, we have $K_f \approx 2.2$ GPa, $\rho_f \approx 1.0$ g/cm^3 and $\rho_s \approx 2.7$ g/cm^3, whereas for a dry rock, we may assume that $K_f = \rho_f = 0$. The frame moduli K_{fr} and G_{fr} are not specified by the Biot theory; however, these will vary with the porosity and the degree of cementation (see Chapter 6). Various empirical relations between the frame moduli and porosity have been presented. For this example, we choose to use the critical porosity relations. Fig. 5.2 illustrates how fluid saturation affects the elastic velocities in the low-frequency limit for sandstones with various porosities according to Eqs. (5.39), (5.41), (6.16) and (6.17). We can see that saturation will lead to an increase in v_p for soft (high-porosity) rocks, while it may lead to a slight decrease in v_p for stiff rocks. We also see that saturation always leads to a reduction in v_s. The effect of saturation depends on the ratio between the frame stiffness and the fluid modulus, and an increase in v_p may thus be seen also for low-porosity rocks if the frame stiffness is sufficiently low.

The frequency dependence of v_p and v_s is controlled by the term $j\frac{v_p^2}{\omega}\frac{\eta_f F(\kappa)}{k}$ in Eqs. (5.38) and (5.40). In the high-frequency limit, this term vanishes.

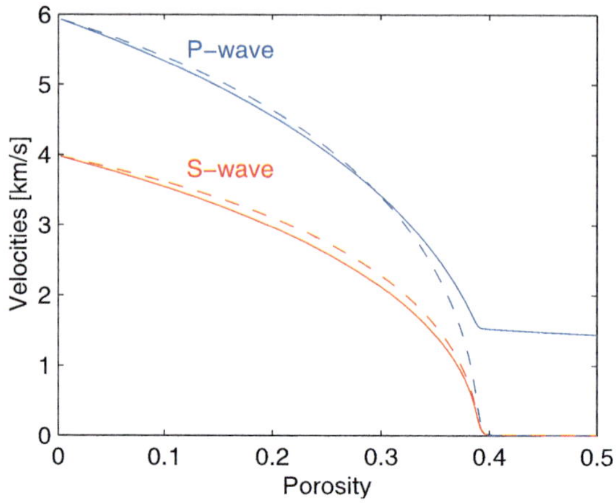

Figure 5.2 P- and S-wave velocities versus porosity for dry and saturated rocks, based on Biot's theory in the low-frequency limit. The frame moduli are given by the critical porosity relations (6.16)–(6.17) with a critical porosity of 0.39. Dashed lines: dry rock; solid lines: saturated rock.

For the shear wave, this gives the solution

$$v_s(\omega \to \infty) = \sqrt{\frac{G_{fr}}{\rho - \frac{\phi \rho_f}{T}}} = \sqrt{\frac{G_{fr}}{(1 - \frac{1}{T})\phi \rho_f + (1 - \phi)\rho_s}} \tag{5.42}$$

By comparing Eq. (5.41) and (5.42) we see that the shear wave velocity increases with frequency. The effect is quite small, typically less than 3 % from the low-frequency limit to the high-frequency limit for tortuosity $T = 3$. The velocity increase is a result of the coupling to the pore fluid motion. It may be shown that $\zeta^{\circ} \neq 0$ in the high-frequency limit, and thus the pore fluid moves out of phase with the solid framework. The actual mass involved in the wave motion is therefore reduced, and the velocity increases, as Eq. (5.42) shows. This flow of fluid relative to the solid framework is often called "Biot flow".

For the P-wave, the solution is more complicated for higher frequencies. First of all, Eq. (5.38) is a second-degree equation in v_p^2 for $\omega > 0$, and thus it has *two* solutions. The largest root is solution (5.39) in the low-frequency limit. This is the "normal" P-wave observed in seismic surveys and well logging. The expression for this solution in the high-frequency limit is somewhat complicated; however, it can be shown that there is only a marginal increase (typically, less than 1 %) in velocity from the low-frequency limit to the high-frequency limit in most cases. However, for very soft rocks (in the suspension limit), the dispersion can be significantly higher. In general, the dispersion is larger for higher porosity and lower tortuosity.

The other solution is a so-called "slow wave", which is highly attenuated. At low frequencies the wave does not exist as a propagating wave; however, it is rather a diffusive mode approaching ordinary (Darcy) flow through the porous rock in the limit as $\omega \to 0$. Although the slow wave is usually not observed as a propagating wave, its existence has been proven in laboratory experiments (first time by Plona, 1980), which provides a justification of the basis of the Biot theory.

5.3.2 Dispersion due to local flow

In reality, the actual P-wave dispersion is higher than predicted by Biot's theory. It is assumed that this is related to thin cracks in the solid framework (e.g. at grain contacts). At low frequencies the fluid pressure within these thin cracks will be equal to the pore pressure outside the cracks, and the cracks will thus be a part of the pore space in Biot's model. At high frequencies, the fluid pressure within the thin cracks will not be able to follow the rapid oscillations of the pore pressure outside. The cracks will then effectively be sealed off and no longer be a part of the pore space in Biot's model (which requires that the entire pore space is connected). As thin sealed fluid-filled cracks are almost incompliant (see Section 6.4), the effective stiffness of the solid framework will thus increase at higher frequencies.

The thin cracks tend to be closed at high external stresses. Thus we may expect to find a relation between the bulk and shear moduli of the framework in the high-frequency limit and the effective bulk modulus of the dry rock at high external pressure (Mavko and Jizba, 1991). The dispersion effect caused by the local flow is usually comparable to or larger than the effect predicted by the Biot theory alone.

5.4 Attenuation

The waves described in Sections 5.1 and 5.2 propagate with plane wavefronts without any loss of energy, so that the amplitude remains constant. In reality, the amplitude of an elastic wave that travels through a rock will be reduced as the wave propagates. We say that the wave is *attenuated*. This is partly due to geometrical spreading caused by a non-planar wavefront, partly due to scattering of wave energy in other directions, and partly due to transformation of wave energy into other forms of energy (absorption).

The amplitude reduction is a certain fraction of the amplitude itself, and hence the amplitude will fall off exponentially with travelled distance. For a plane wave propagating in the positive x-direction, the amplitude will thus fall off as

$$u = u^{\mathrm{o}} \mathrm{e}^{-\alpha x} \tag{5.43}$$

where α is a measure of the attenuation. The unit for α is inverse length m^{-1}, but it is often referred to in decibel per metre [dB/m]:

$$\alpha[\mathrm{dB/m}] \equiv (20 \log_{10} \mathrm{e})\alpha[\mathrm{m}^{-1}] \approx 8.686\,\alpha[\mathrm{m}^{-1}] \tag{5.44}$$

The quality factor Q (often called the Q-factor for short) is another parameter frequently used as a measure for attenuation. The Q-factor is inversely proportional to α, and hence a high Q-factor implies low attenuation and vice versa. It is defined as

$$Q = \frac{\omega}{2\alpha v} \tag{5.45}$$

Another measure for attenuation is the loss tangent

$$\delta = \frac{\pi}{Q} \tag{5.46}$$

Whereas α represents the loss per unit length, the loss tangent is a measure for the loss per wavelength. Measured Q-factors vary strongly with conditions and types of rock. Fig. 5.3 shows typical laboratory data (ultrasonic frequencies) for sandstone. Reliable data for attenuation are usually hard to obtain, and the amount of data reported in the literature is limited.

Geometrical spreading is associated with the *radius of curvature* for the wavefront. For a wave propagating spherically from a point source, the radius of curvature is equal to the radius (r) of the sphere that the wavefront constitutes at the time of investigation. The amplitude then falls off inversely proportional to r. For a plane wave, the radius of curvature is infinite. If the radius of curvature R is so large that it can be considered as a constant over the distance we consider, the amplitude falls off with travelled distance x as

$$u = u^{\mathrm{o}} \mathrm{e}^{-x/R} \tag{5.47}$$

Thus geometrical spreading causes an apparent attenuation. Note that the radius of curvature for a wavefront may be altered when the wave crosses

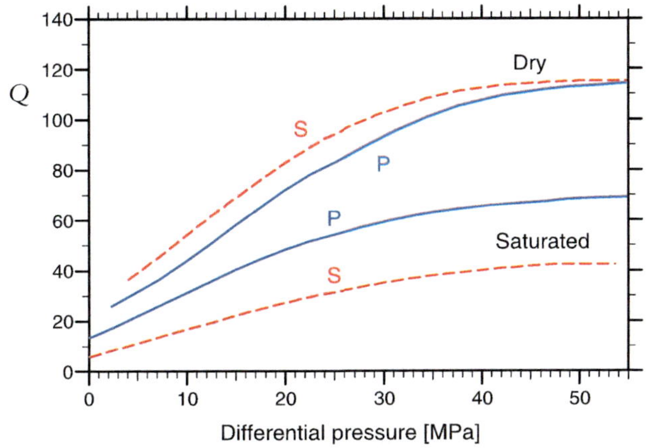

Figure 5.3 Q-factor for P- and S-waves vs. external pressure in dry and fluid-saturated sandstone, based on ultrasonic measurements. (After Johnston et al., 1979.)

an interface and enters into a different material where the phase velocity is different. This is a consequence of refraction, which will be discussed in Section 5.7. The radius of curvature will also be affected by anisotropy. Under special conditions (usually caused by reflection at a curved interface), the radius of curvature may become negative. This corresponds to a spherical wave travelling inwards, towards the sphere centre. As Eq. (5.47) shows, the amplitude will then increase as the wave propagates.

Geometrical spreading is mostly independent of the material the wave propagates through. More interesting to us is attenuation which depends explicitly on the rock properties. Several mechanisms are of importance, and which ones are most important depend on the conditions: lithology, saturation, frequency, strain amplitude and so on. The most important mechanisms are:

1. *Solid friction* (Walsh, 1966; Johnston et al., 1979). Elastic waves induce displacements between grains or crack faces. Due to friction, this implies that a part of the mechanical energy of the acoustic wave is absorbed and transformed into heat, and the wave is attenuated. Absorption due to friction gives a Q-factor independent of frequency, that is,

$$Q_{SF} = \text{const.}; \quad \alpha_{SF} \propto \omega \tag{5.48}$$

Since the static friction increases with increasing normal stress, the solid friction losses decrease with increasing external stress on a rock. Solid friction has traditionally been considered to be the dominating attenuation mechanism in rocks; however, recent studies indicate that fluid mobility is a more significant mechanism (Batzle et al., 2006).

2. *Local (squirt) flow* (Mavko and Nur, 1979; Murphy et al., 1984; Jones, 1986; Mavko and Jizba, 1991). These losses only occur in partially or fully saturated rocks. The periodic contraction and expansion in a rock when an acoustic wave passes by also involves contraction and expansion of the pore space. Due to local variations in shape, the volumetric strain of the pore space differs considerably over short distances. For instance, a flat crack may suffer large volumetric deformations, whereas a spherical pore close by only experiences minor volumetric strains. This gives rise to local variations in pore pressure, which will make the pore fluid flow back and forth following the oscillating deformations. The energy required for the fluid flow oscillations are taken from the acoustic wave, and thus elastic energy is absorbed and transformed into other forms of energy, and the wave is attenuated.

   The attenuation effect is most pronounced in a frequency range where the period of the oscillations matches the characteristic relaxation time for the local fluid flow. At lower frequencies the fluid has sufficient time to flow so that the pore pressure is almost equalised at any time. At high frequencies, the oscillations are so quick that the fluid has no time to move. Deformations of the most compliant parts of the pore space will then be counteracted by local pore pressure build-ups. The rock will effectively be stiffer, so that the velocity of the elastic wave will be higher, as described in Section 5.3.2. Fig. 5.4 shows schematically how attenuation and velocity change with frequency due to local flow at a grain to grain contact. The transition frequency is inversely proportional to the fluid viscosity. For water-saturated rocks, the transition frequency is typically a few kHz.

3. *Macroscopic (Biot) flow* Biot (1956a,b, 1962). An elastic wave may also induce fluid flow on a macroscopic scale, as described in Section 5.3.1. Like the local (squirt) flow, the fluid flow oscillations will suck energy out of the acoustic wave, which is thus attenuated. There will also be a velocity dispersion and a peak in attenuation associated with the macroscopic flow. An expression for the attenuation can be derived from the imaginary part of the solution of Eq. (5.38) and the relation outlined below (Eq. (5.50)). The transition frequency is proportional to the fluid viscosity and inversely proportional to the permeability of the rock. At low frequencies, $\alpha \propto \omega^2$, whereas $\alpha \propto \sqrt{\omega}$ at high frequencies.

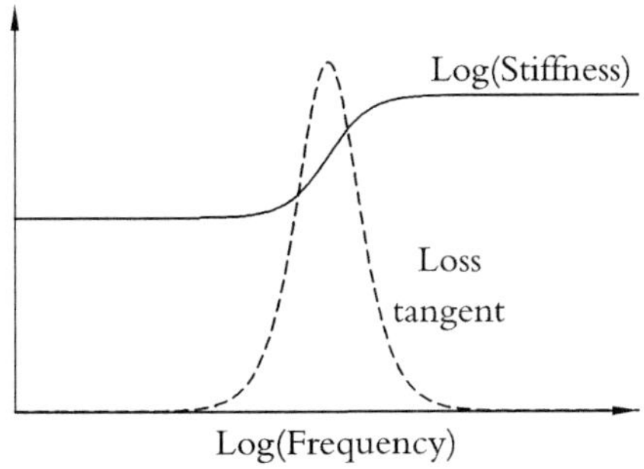

Figure 5.4 Sketch of stiffness (solid line) and Loss tangent (dashed line) versus frequency for a liquid saturated grain to grain contact. (After Murphy et al., 1984.)

For water-saturated rocks, the transition frequency is typically about 100 kHz. The effect is however rather small: The peak attenuation has a minimum Q-value of about 100.

4. *Scattering*. Heterogeneities in velocity or density within a rock result in Rayleigh scattering when an elastic wave passes by. The scattered energy is not absorbed and transformed into other forms of energy; however, the primary pulse loses energy and is thus attenuated. The effect is stronger the larger the impedance contrast of the heterogeneity is (see Section 5.7), and thus pores are strong scatterers, also for P-waves if the pores are empty (that is, the material is dry). Cracks and fractures may also be strong scatterers, unless both the direction of propagation and the polarisation are parallel to the crack/fracture. Scattering is strong when the wavelength λ_w is comparable or smaller than the size a of the heterogeneity. For longer wavelengths, attenuation falls off quickly with decreasing frequency:

$$\alpha_{\text{scatt.}} \propto \omega / Q_{\text{scatt.}} \propto a^3 \omega^4 \quad \text{for} \quad \lambda_w > a \tag{5.49}$$

In sandstones, scattering on grains and pores is dominating in the MHz range and effectively prohibits wave propagation over distances longer than a few centimetres at these (and higher) frequencies. At lower frequencies, scattering on grains and pores can be neglected; however, scattering on cracks and fractures may still be significant. Scattering loss usually decreases with increasing external stress, since confinement tends to close cracks and fractures in the rock. Scattering also reduces the velocity. This is further discussed in Chapter 6.

Mechanisms 2 and 3 are sometimes jointly referred to as wave-induced fluid flow (WIFF).

Comparing Eqs. (5.43) and (5.5), we find that the attenuation coefficient α can be associated with the imaginary part of the wavenumber q, that is,

$$\alpha = -\mathrm{Im}[q] \tag{5.50}$$

According to the wave equation, the wavenumber is complex if the corresponding elastic modulus is complex. Thus the phase velocity is related to the real part of the elastic modulus and attenuation to the imaginary part. Denoting by C the relevant elastic modulus (which may e.g. be $(\lambda + 2G)$ as in Eq. (5.15), or G as in Eq. (5.18), etc.), the general relations are

$$\rho v^2 \approx \mathrm{Re}[C] \tag{5.51}$$

$$\frac{1}{Q} \approx \frac{\mathrm{Im}[C]}{\mathrm{Re}[C]} \tag{5.52}$$

provided that the attenuation is sufficiently small, so that $Q^{-2} \ll 1$. Note that this link between the imaginary part of the elastic modulus and the attenuation is not valid for apparent attenuation due to geometrical spreading.

The real and imaginary parts of the elastic constants are related to each other according to the physical principle called causality (the principle that a reaction can never occur before its cause). A specific result of this is that within the frames of linearity, the velocity can only be independent of frequency if the corresponding attenuation is zero, that is, no velocity dispersion implies no attenuation. Fig. 5.4 shows a typical relation between attenuation and velocity dispersion in accordance with causality.

Finally, consider a simple mechanical system consisting of a spring in parallel with a viscous dash pot element coupled in series to another spring (Fig. 1.30, except for the right-hand dash pot element). This could represent a rock that responds to stress changes with some degree of transient creep. An elastic modulus C of this rock can be expressed as (see Eqs. (1.94), (1.338), (1.339) and (5.33))

$$C = \frac{\sigma}{\varepsilon} = \left(\frac{1}{\kappa_1 + \mathrm{j}\omega\chi} + \frac{1}{\kappa_2}\right)^{-1} = \kappa_2 \frac{\kappa_1(\kappa_1 + \kappa_2) + \omega^2\chi^2 + \mathrm{j}\omega\chi\kappa_2}{(\kappa_1 + \kappa_2)^2 + \omega^2\chi^2} \tag{5.53}$$

Thus the elastic modulus of this rock is complex, and the elastic wave corresponding to C will be attenuated. For sufficiently low frequencies, such that $\omega < (\kappa_1 + \kappa_2)/\chi$, the imaginary part of C will be proportional to ω, and, consequently, $1/Q \propto \omega$ according to Eq. (5.52). For higher frequencies, we find that $1/Q \propto 1/\omega$. On the other hand, the real part of C is seen to approach $\kappa_1\kappa_2/(\kappa_1 + \kappa_2)$ as $\omega \to 0$, whereas it approaches κ_2 as $\omega \to \infty$. Thus the model describes dispersion and a corresponding peak in attenuation similar to Fig. 5.4. This example also illustrates the relationship between creep and attenuation.

5.5 Anisotropy

Most sedimentary rocks are to some extent anisotropic. This was discussed in Section 1.7 regarding directional dependence of rock stiffness. This directional dependence does of course also affect the acoustic properties of the rock. To account for this, we need to establish a generalised version of the wave equation and its solutions. We will here present an outline of the procedure and the results. For a detailed discussion, see, for instance, Auld (1990).

5.5.1 The Christoffel equation

As described in Section 5.1, the wave equation can be established by looking at the stability of a volume element. For this general case, we choose not to simplify in terms of longitudinal or transversal wave motion from the start. Using the generalised form for the stress (1.236), we see that Newton's second law requires

$$\rho \frac{\partial^2 u_i}{\partial t^2} = \sum_j \frac{\partial \sigma_{ij}}{\partial x_j} \tag{5.54}$$

We assume that the solutions are of the type (Eq. (5.13))

$$u_i = u_i^{\rm o} \mathrm{e}^{\mathrm{j}(\omega t - q \sum_k l_k x_k)} \tag{5.55}$$

Combining Eqs. (5.54) and (5.55) with (1.236) and using expressions (1.77) and (5.8), we find (after some algebra) the so-called Christoffel equation

$$\sum_{j,l} (C_{ijkl} l_j l_l - \rho v^2 \delta_{ik}) u_k^{\rm o} = 0 \tag{5.56}$$

where δ_{ik} is Kronecker's δ (see Appendix C.12.2). Eq. (5.56) is in fact three equations with three unknowns $u_1^{\rm o}$, $u_2^{\rm o}$ and $u_3^{\rm o}$. The equations can be written in matrix form as

$$\mathbf{M}^{\mathrm{C}}\begin{pmatrix} u_1^0 \\ u_2^0 \\ u_3^0 \end{pmatrix} = 0 \tag{5.57}$$

where $\mathbf{M}^{\mathrm{C}}$ is called the Christoffel matrix. The elements of the matrix are defined as

$$\mathrm{M}_{ik}^{\mathrm{C}} = \sum_{j,l} C_{ijkl} l_j l_l - \rho v^2 \delta_{ik} \tag{5.58}$$

Non-trivial solutions of Eq. (5.57) exist if the determinant of the Christoffel matrix is zero, that is,

$$\left| \sum_{j,l} C_{ijkl} l_j l_l - \rho v^2 \delta_{ik} \right| = 0 \tag{5.59}$$

This is a compact way of writing a rather lengthy equation. The expanded expression for the determinant of a 3 × 3 matrix is shown in Appendix C.2.10. It turns out that Eq. (5.59) is a cubic equation for v^2. Thus we will in general expect to find three independent solutions for the wave velocity. One of these represents a P-wave (or, in the most general case, a *quasi*-P-wave, meaning a wave with polarisation nearly parallel to the direction of propagation). The other two represent S-waves with mutually orthogonal polarisations (or *quasi*-S-waves, meaning waves with polarisations nearly normal to the direction of propagation).

To demonstrate how velocities and polarisations can be found from these equations, we will consider the special case of transversal isotropy (see Section 1.7.2). Switching to the Voigt notation (see Appendix C.10) and choosing the z-axis as the unique axis, the stiffness matrix is given by expression (1.264). The Christoffel equation (5.56) then becomes, in expanded form,

$$\begin{pmatrix} C_{11}l_1^2 + C_{66}l_2^2 + C_{44}l_3^2 - \rho v^2 & (C_{11} - C_{66})\, l_1 l_2 & (C_{13} + C_{44})\, l_1 l_3 \\ (C_{11} - C_{66})\, l_1 l_2 & C_{66}l_1^2 + C_{11}l_2^2 + C_{44}l_3^2 - \rho v^2 & (C_{13} + C_{44})\, l_2 l_3 \\ (C_{13} + C_{44})\, l_1 l_3 & (C_{13} + C_{44})\, l_2 l_3 & C_{44}\left(l_1^2 + l_2^2\right) + C_{33}l_3^2 - \rho v^2 \end{pmatrix} \begin{pmatrix} u_1^0 \\ u_2^0 \\ u_3^0 \end{pmatrix} = 0 \tag{5.60}$$

The direction cosines l_1, l_2, and l_3 defines the direction of wave propagation, and the particle displacement vector $\vec{u}^{\mathrm{o}}$ defines the polarisation. The next step is to choose a direction of wave propagation that we want to study. For simplicity, we first choose the x-direction, for which $l_1 = 1$ and $l_2 = l_3 = 0$. The Christoffel equation (5.60) then simplifies to

$$\begin{pmatrix} C_{11} - \rho v^2 & 0 & 0 \\ 0 & C_{66} - \rho v^2 & 0 \\ 0 & 0 & C_{44} - \rho v^2 \end{pmatrix} \begin{pmatrix} u_1^o \\ u_2^o \\ u_3^o \end{pmatrix} = 0 \tag{5.61}$$

and the requirement that the determinant of this matrix will be zero becomes

$$(C_{11} - \rho v^2)(C_{66} - \rho v^2)(C_{44} - \rho v^2) = 0 \tag{5.62}$$

This is a cubic equation for v^2, as it should be, and hence it has three solutions. One solution is $v^2 = C_{11}/\rho$. Introducing this result for v into Eq. (5.61) shows that we must have $u_2^o = u_3^o = 0$ with $u_1^o \neq 0$ as the only non-trivial solution for the particle displacement. Thus, for this particular solution, both the direction of wave propagation and the polarisation are in the x-direction. This wave is therefore a P-wave.

Another solution of Eq. (5.62) is $v^2 = C_{66}/\rho$. Introducing this result into Eq. (5.61), we find $u_1^o = u_3^o = 0$ with $u_2^o \neq 0$ as the only non-trivial solution. For this solution, the polarisation is thus in the y-direction, which is normal to the direction of wave propagation, and hence this is an S-wave. The last of the three solutions is $v^2 = C_{44}/\rho$. Eq. (5.61) tells us that this is also an S-wave, polarised in the z-direction. Thus there are three different waves that can travel in the x-direction in this medium, one P-wave and two S-waves.

Considering wave propagation in the z-direction, the Christoffel equation becomes

$$\begin{pmatrix} C_{44} - \rho v^2 & 0 & 0 \\ 0 & C_{44} - \rho v^2 & 0 \\ 0 & 0 & C_{33} - \rho v^2 \end{pmatrix} \begin{pmatrix} u_1^o \\ u_2^o \\ u_3^o \end{pmatrix} = 0 \tag{5.63}$$

Following the same procedure as above, we find one P-wave solution with $v^2 = C_{33}/\rho$ and two S-wave solutions, both with $v^2 = C_{44}/\rho$. Introducing the S-wave solution for v, Eq. (5.63) only tells us that u_3^o has to be equal to 0, whereas both u_1^o and u_2^o may be non-zero. Thus, for an S-wave travelling parallel to the unique axis, the velocity is independent of the polarisation.

For wave propagation in a general direction, the off-diagonal terms of the Christoffel matrix no longer vanish and the solutions become more complicated mathematically. Typical for the transversely isotropic medium is that for any direction of wave propagation, there exists one pure S-wave with polarisation normal to the unique axis plus one P-wave and another S-wave. Except for wave propagation parallel or normal to the unique axis, the last two waves will not have polarisation exactly parallel or normal to the direction of propagation, and hence they will in general be a quasi-P-wave and a quasi-S-wave, respectively.

A full treatment of wave propagation in anisotropic media is beyond the scope of this book. We refer the interested reader to for instance Auld (1990).

5.5.2 Weak anisotropy

Thomsen (1986) simplified the description of anisotropy considering transversely isotropic rocks with weak anisotropy. His description is relevant for oilfield rocks and the data available in field situations, and has become widely applied. Thomsen introduced a new set of parameters:

$$\varepsilon_{\mathrm{Th}} = \frac{C_{11} - C_{33}}{2C_{33}} \tag{5.64}$$

$$\gamma_{\mathrm{Th}} = \frac{C_{66} - C_{44}}{2C_{44}} \tag{5.65}$$

$$\delta_{\mathrm{Th}} = \frac{(C_{13} + C_{44})^2 - (C_{33} - C_{44})^2}{2C_{33}(C_{33} - C_{44})} \tag{5.66}$$

The wave velocities in a general direction in a transversely isotropic material can be derived from Eq. (5.59). When linearised in terms of the Thomsen parameters (5.64)–(5.66), the expressions for the velocities take relatively simple forms:

$$v_{\mathrm{p}}(\theta) \approx \alpha_{\mathrm{Th}} \left(1 + \delta_{\mathrm{Th}} \sin^2\theta \cos^2\theta + \varepsilon_{\mathrm{Th}} \sin^4\theta\right) \tag{5.67}$$

$$v_{\mathrm{sv}}(\theta) \approx \beta_{\mathrm{Th}} \left(1 + \frac{\alpha_{\mathrm{Th}}^2}{\beta_{\mathrm{Th}}^2}(\varepsilon_{\mathrm{Th}} - \delta_{\mathrm{Th}}) \sin^2\theta \cos^2\theta\right) \tag{5.68}$$

$$v_{\mathrm{sh}}(\theta) \approx \beta_{\mathrm{Th}} \left(1 + \gamma_{\mathrm{Th}} \sin^2\theta\right) \tag{5.69}$$

where $\alpha_{\mathrm{Th}} = \sqrt{C_{33}/\rho}$, $\beta_{\mathrm{Th}} = \sqrt{C_{44}/\rho}$, θ is the angle between the unique axis and the direction of wave propagation, v_{sh} is the velocity of the S-wave polarised normal to the unique axis, v_{sv} is the velocity of the other S-wave (see Section 5.5.1).

$\varepsilon_{\mathrm{Th}}$ can be interpreted as the P-wave anisotropy, and γ_{Th} as the S-wave anisotropy. However, for P-waves propagating at small angles to the unique axis, the anisotropy will be dominated by δ_{Th}. The particular case where $\delta_{\text{Th}} = \varepsilon_{\text{Th}}$ is known as "elliptical anisotropy".

It can be shown (Tsvankin, 1997; Mensch and Rasolofosaon, 1997) that by making the Thomsen parameters depend on the azimuth these equations can also be applied for materials with even lower symmetry than transversely isotropy, provided that the anisotropy is weak.

5.6 Rock mechanics and rock acoustics

At the beginning of this chapter, we argued that elastic waves may provide means to estimate mechanical parameters of rocks. We will here take a closer look at the links between the parameters describing elastic waves and the parameters describing rock mechanical properties.

5.6.1 Static and dynamic moduli

Sound velocities depend explicitly on elastic moduli. This was shown in Section 5.2 (Eqs. (5.20)–(5.24)). These relations imply that we, for instance, should obtain the same value for E if we measure the acoustic velocities and the density and make use of Eq. (5.23), or we measure stress and strain in a uniaxial compression test and make use of Eq. (1.94). With knowledge about the P- and S-wave velocities and the density, for instance, from well logs or seismics, it should thus in principle be a simple job to obtain the elastic moduli even if we have no possibility to perform rock mechanical tests.

In reality, this is not quite so simple. There is a wide range of experimental evidence showing that the elastic moduli obtained from stress and strain measurements in a rock mechanical test ("static moduli") differ significantly from those obtained from acoustic velocities and density ("dynamic moduli"). Normally, the dynamic moduli are larger than the corresponding static ones. The difference is largest for weak rocks and is reduced with increasing confinement (King, 1970). In the relatively well-cemented Berea sandstone, the difference at a stress level corresponding to depth in the kilometre range is 20–30 %. In a weak sandstone, however, the difference can be an order of magnitude or more, depending on the stress state. Fig. 5.5 shows an example how the static and dynamic moduli change with the stress state in a relatively weak rock. These measurements were performed on a dry sample.

The pore fluid is a potential cause for the difference between static and dynamic moduli in rocks. In a velocity measurement the deformation of

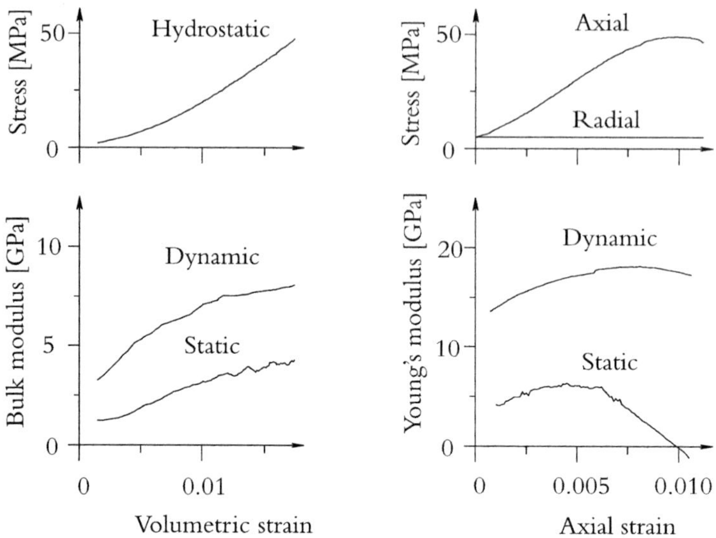

Figure 5.5 Static and dynamic bulk moduli as measured during a hydrostatic test (left) and static and dynamic Young's moduli as measured during a triaxial test (right) on dry Red Wildmoor sandstone.

the rock is undrained, which implies that the pore fluid contributes to the velocity (see Section 5.3.1). Thus the moduli derived from velocity measurements on a saturated rock are likely to be higher than the corresponding static moduli measured in a drained test. However, this effect does not explain the large differences between static and dynamic moduli observed in soft dry rocks, and we have to look for other causes.

As discussed above, velocity dispersion in fluid-saturated rocks typically amount to a few percent from seismic to ultrasonic frequencies. Thus this is not a likely mechanism to explain significant differences between static and dynamic moduli.

In a velocity measurement the strain rate varies from about $1\ s^{-1}$ (at ultrasonic frequencies) to about $10^{-4}\ s^{-1}$ (at seismic frequencies), whereas the strain amplitude is typically 10^{-6}–10^{-7}. For a measurement of static moduli, the strain rate is $10^{-2}\ s^{-1}$ or lower, whereas the strain amplitude is typically 10^{-2}–10^{-3}. Thus the major difference between static and dynamic measurements is—despite the name—the strain amplitude, not the strain rate. Static moduli, measured as slopes of stress–strain curves, differ from small strain amplitude dynamic (elastic) moduli because of plasticity or nonlinear effects.

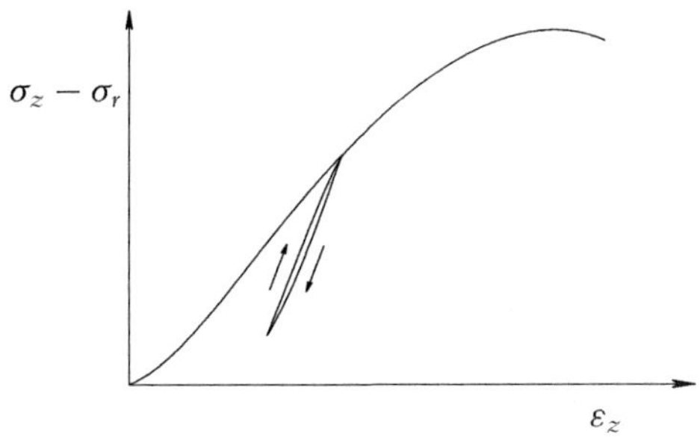

Figure 5.6 Unloading-reloading cycle in a rock mechanical test.

Unloading-reloading cycles in a static test (Fig. 5.6) are somewhat intermediate between a normal static test and a velocity measurement, as the material will experience similar stress cycles (with low amplitudes) when an acoustic wave passes by. In a set of measurements on dry Castlegate sandstone, Plona and Cook (1995) demonstrated that the elastic modulus derived from the slope of the stress–strain curve during such cycles approaches the corresponding elastic modulus derived from velocity measurements as the amplitude of the cycles approaches zero. This suggests that non-elastic deformation builds up gradually during unloading (and reloading). Fjær et al. (2013) showed that the non-elastic compliance, defined as the difference between static and elastic (dynamic) compliance, seems to increase linearly with the stress reduction during unloading, starting at zero at the turning point of the stress path. This observation makes it possible (through extrapolation) to eliminate strain amplitude effects from static measurements at single points on a stress path, so that static and dynamic measurements can be combined for estimation of dispersion. Other studies (Martin and Haupt, 1994; Tutuncu et al., 1998; Lozovyi et al., 2017) have shown that the difference between static and dynamic stiffness starts to grow when the strain amplitude exceeds $5 \cdot 10^{-6}$. Thus non-elastic contributions triggered by the strain amplitude contribute to the difference between dynamic moduli and the static moduli also during unloading-reloading cycles. During initial loading, the non-elastic contributions are usually much larger (cf. the difference between normally consolidated and overconsolidated states; Section 2.8.2), with correspondingly larger difference between static and dynamic moduli (Fig. 5.6).

To further understand the origin of the static-dynamic discrepancy, it is interesting to note that the static and dynamic moduli are equal for a homogeneous elastic material like steel (Ledbetter, 1993). Thus the physical origin of this discrepancy is likely to be related to the heterogeneous microstructure of the rocks. It is furthermore to be expected that much of the effect originates at the grain contacts, since the stress concentrations in the contact areas may exceed the elasticity limit of the material even at low external stresses.

A quantitative model for the relations between static and dynamic moduli was formulated by Fjær (1999), based on observations on weak sandstones:

$$K_{\text{stat}} = \frac{K_{\text{dyn}}}{1 + (P_z + 2P_r)K_{\text{dyn}}} \tag{5.70}$$

$$E_{\text{stat}} = \frac{E_{\text{dyn}}}{1 + P_z E_{\text{dyn}}}(1 - F) \tag{5.71}$$

where K_{dyn} and E_{dyn} are the bulk modulus and Young's modulus, respectively, obtained from velocity measurements, whereas K_{stat} and E_{stat} are the corresponding moduli obtained from the slope of the relevant stress–strain curve during initial loading. The expressions are related to a standard triaxial test, and the subscripts z and r refer to the axial and radial directions of the test geometry. The difference between the static and dynamic moduli is contained in the parameters P_i and F. Analyses of tests like that shown in Fig. 5.5 have shown that P_i is largest at lower stress levels, whereas F increases with increasing shear deformation. (Note that for $F = 1$, we have $E_{\text{stat}} = 0$ according to Eq. (5.71). This represents the peak point in the stress–strain curve.)

It is suggested that P_i is predominantly a measure of a process that involves crushing of asperities at the grain contacts, whereas F is associated with friction controlled sliding along contact points or closed microcracks in the material during shear loading. Thus the discrepancy between static and dynamic moduli in weak rocks is interpreted as being caused by a series of local failure processes on a microscopic scale occurring during the entire loading sequence.

Note that there may also be other reasons why observations of static and dynamic moduli show different values (Fjær, 2019). As static and dynamic measurements often involve different rock volumes, spatial fluctuations in rock properties may induce differences in the observed static and dynamic moduli. Such fluctuations may also produce dispersion effects, as shown in Section 6.1. Anisotropy is furthermore a complicating element when comparing static and dynamic moduli, as the static and dynamic measurements usually yield different components—or combinations of components—of the stiffness matrix. Thus estimation of static moduli from velocity measurements by application of Eqs. (5.22)–(5.24) under the assumption of isotropy may produce large errors.

5.6.2 Stress state and stress history

Sound velocities in rocks also depend on the state of stress. To some extent, this effect can be ascribed to stress-induced changes in porosity. However, this is not sufficient to explain the stress sensitivity of velocities in most rocks. Fig. 5.7 shows the typical behaviour for a sandstone. The behaviour can be understood in terms of microcracks (much smaller than the wavelength) that are opened or closed by the action of the stress. An open crack strongly reduces the velocity of a wave if the crack is oriented normal to the direction of propagation or polarisation of the wave, whereas its effect on the velocity is only marginal otherwise (see Section 6.4).

During hydrostatic loading, the velocities increase uniformly, with a gradually diminishing rate. This is to be expected due to closure of cracks, which make the rock stiffer, or similar processes, which broaden or multiply grain contacts. These effects are discussed in detail in Sections 6.3 and 6.4. During uniaxial loading, the velocities of waves with polarisation and/or direction of propagation parallel to the increasing load are seen to increase initially due to the same type of processes that cause increases in velocities during hydrostatic loading. Upon further loading, the velocities of waves with polarisation and/or direction of propagation parallel to the minimum principal stress are seen to be reduced. This is associated with the formation of tensile cracks (often illustrated as "wing cracks" formed during frictional sliding of closed cracks, as shown in Section 6.4.4, Fig. 6.11). Thus stress anisotropy induces acoustic anisotropy. The relevance of interpreting stress-induced velocity changes in sandstones in terms of changes in the occurrence of discontinuities, such as cracks, has been nicely demonstrated by Sayers (2002). By combining models for the impact of cracks on rock stiffness (Section 6.4) with a suitable model for the stress and strain dependence of crack density (Sayers, 2006; Fjær, 2006) models for stress-induced acoustic anisotropy and stress-dependent velocities may be established.

The impact of microcracks on acoustic wave propagation is also reflected in the attenuation, as the scattering losses due to such cracks can be significant. Thus attenuation is typically reduced with increasing confining pressure, as can be seen from Fig. 5.3.

Using observed acoustic anisotropy as a measure for stress anisotropy is however not straightforward. This is partly due to the fact that also other

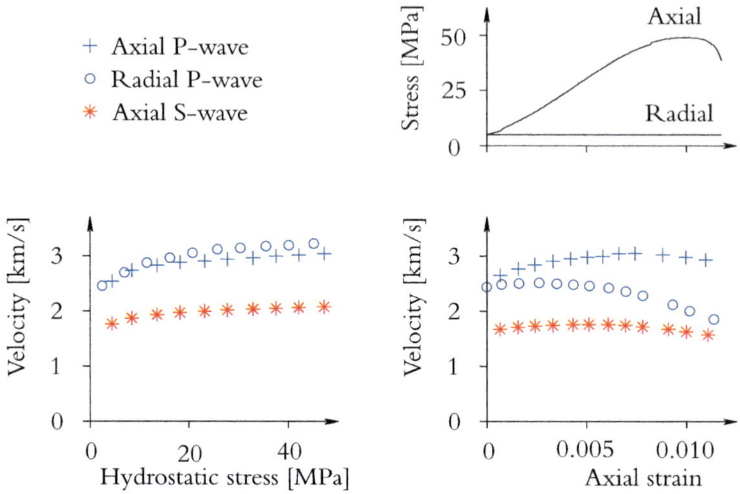

Figure 5.7 Ultrasonic velocities as functions of stress state as measured on dry Red Wildmoor sandstone.

effects, like fine layering or grain orientation, can cause anisotropy and partly to the fact that rocks are usually formed and cemented under stress. Stress-induced anisotropy in uncemented sediments is a consequence of static loading making the grain contacts stiffer (see Section 6.3). Different stresses in different directions thus induce different stiffness in different directions, which results in acoustic anisotropy. The cementation process provides added stiffness to all grain contacts, and so the relative stiffening is largest for the weakly loaded ones. The characteristic signature of the stress state is thus masked. Subsequent stress-induced anisotropy is characteristic for the stress *changes* occurring after cementation, rather than the current stress state. Thus the stress history of a rock may have a significant impact on the stress dependency of the velocities.

For a core taken from a deep well, the major recent stress history is the unloading process taking place during and after coring. The cement that was formed free from effective stress at the grain contacts will be subjected to tensile stress as the grain contacts deform when the load disappear. This will generate cracks that reduce the elastic stiffness and thus also the acoustic velocities. When sufficient stress is applied to the core in a laboratory test, these cracks will close, and the velocities will increase correspondingly. A significant part of the stress dependency of acoustic velocities observed in laboratory tests may therefore be related to core damage effects and not be representative for the in situ behaviour. This has been demonstrated

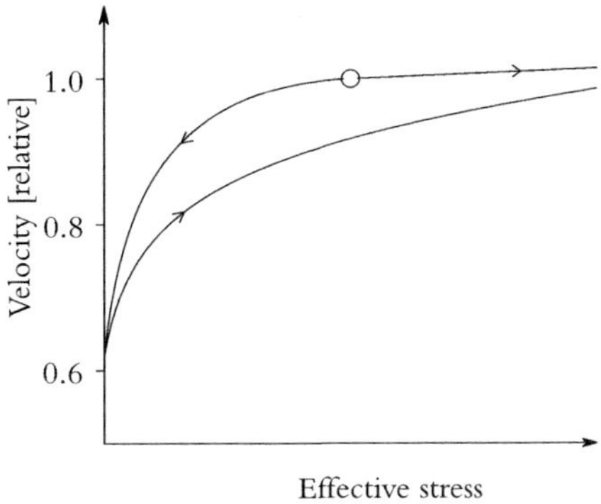

Figure 5.8 Schematic illustration of velocity versus stress for a rock, following two different stress paths from the cementation state (marked with a circle). The upper right curve could represent in situ loading (caused by depletion), the upper left curve in situ unloading (caused by injection or coring), and the lower curve a laboratory test on a core plug.

through tests on synthetic sandstones cemented under stress (Nes et al., 2002). The effect is illustrated in Fig. 5.8. Note that the velocity is quite insensitive to small changes in the stress while the rock is still in the vicinity of its cementation state and that it drops significantly upon extensive unloading. Upon reloading, as for a core tested in the laboratory, the velocity increases with a rate that is lower than the decreasing rate at the end of the unloading path, yet significantly higher than the initial ("in situ") rate in the vicinity of the cementation stress (see also Chapter 7).

It was shown in Section 1.6.4 that the deformation of a porous linear elastic rock is controlled entirely by an effective stress (Eq. (1.187)) that accounts for the combined action of the external stress and the pore pressure. It is tempting to assume that this principle is also valid for acoustic velocities; however, this is not the case in general. Often, but not always, the velocities may depend on the effective stress

$$\sigma' = \sigma - np_f \tag{5.72}$$

within a limited range of stress and pore pressure. The coefficient n depends on the type of wave and the type of fluid, and may be larger as well as smaller than 1. In some cases, this principle is not applicable at all. One simple example is a linearly elastic material, for which the frame moduli are stress independent (as will be the case e.g. for a well-cemented rock not suffering from core damage). Since the P-wave velocity also depends on the bulk modulus and density of the fluid, which in turn depend on the fluid pressure, it is clear that the effects of changes in σ and p_f are fundamentally different, and Eq. (5.72) is not valid in this case.

5.6.3 Additional effects

Elastic wave propagation in rocks is also affected by other parameters, which we will only briefly mention here.

5.6.3.1 Temperature

There is normally a slight reduction in velocities with increasing temperature. This effect is usually less than 5 % for a 100 °C increase in temperature (Bourbie et al., 1987; Christensen, 1982). The effect may be significantly larger if one or more of the rock constituents undergoes a phase transition within the actual temperature range, for instance, if the pore fluid is freezing or melting. Attenuation is also reduced with increasing temperature. This effect appears to be somewhat larger than the temperature effect on the velocities (Jones and Nur, 1983).

5.6.3.2 Partial saturation

A main objective of seismic exploration is to distinguish between oil or gas and water as pore fluids. In time-lapse (4D) seismic (see Section 12.4.5) the aim is to see saturation changes. Normally, the target rocks in such studies are partially saturated. For example, in CO_2 storage, it is of interest to quantify the amount of CO_2 accumulated in a storage reservoir, which initially may have a certain water saturation, and to detect any possible leakage.

Partial saturation may have a significant effect on both velocities and attenuation, making seismic and sonic logs relevant tools for characterisation. Consider a rock saturated with water and gas. At low frequencies the pore fluid may be considered as a suspension of gas bubbles in a liquid (at least if the water saturation S_w is larger than about 20 %). We may then assume that the gas pressure follows that of the water at any time, and then the effective fluid bulk modulus K_f is given by an equation similar to Eq. (1.137):

$$\frac{1}{K_f} = \frac{S_w}{K_w} + \frac{1 - S_w}{K_g} \tag{5.73}$$

where K_w and K_g are the bulk moduli of water and gas, respectively. Typically, $K_g \ll K_w$, and we thus have that $K_f \ll K_w$ unless S_w is very close to

1. The bulk density

$$\rho = \rho_s(1 - \phi) + \phi\left[S_w\rho_w + (1 - S_w)\rho_g\right] \quad (5.74)$$

is much less sensitive to S_w. The P-wave velocity of the rock may be calculated from the Biot–Gassmann equation using the suspension as the effective pore fluid (Gassmann–Wood approximation). Hence, the P-wave velocity (Eq. (5.39)) drops dramatically when S_w falls below 1 and increases slowly as S_w continues to decrease (Fig. 5.9). The S-wave velocity depends only on ρ in addition to G_{fr} (which is not affected by the degree of saturation, at least not for permeable materials at low frequencies), and hence the S-wave velocity is only marginally dependent on the saturation. For very low saturation, capillary forces or altered surface energy (see Section 2.6.2) may provide an additional stiffness that makes the velocities increase.

Note that gas bubbles in water also give rise to dispersion and attenuation in water itself (Andersen and Hampton, 1980) due to bubble resonance at frequencies above 30 kHz. Tisato et al. (2015) found experimentally CO_2 bubbles in brine to disappear above 2 MPa. Thus the effect of gas bubbles may not affect deeper subsurface monitoring. Gas will however still be present, but with much higher bulk modulus and density than at the surface (the bulk modulus of gas will be similar to the pore pressure and thus substantially lower than the bulk modulus of water).

For all systems of immiscible fluids, the distribution of fluids in the pore space becomes significant at high frequencies. For instance, water trapped in thin cracks may effectively respond to compression with a stiffness close to K_w, and the velocities will be correspondingly higher (see also Section 6.4). The distribution of water and gas is a result of the wetting properties of the rock and the way the actual saturation was reached. Consequently, there is no one-to-one relationship between P-wave velocity and water saturation (Endres and Knight, 1989), but rather a range of possible values for the velocity at each saturation level, as indicated by the grey area in Fig. 5.9. Laboratory experiments (Knight and Nolen-Hoeksema, 1990; Cadoret et al., 1995) showed that when the water saturation is increased by imbibition (the wetting phase is increasing), the saturation distribution becomes homogeneous on the pore scale ("fine scale mixing"), and the P-wave velocity follows the trend expected from the Gassmann–Wood approximation, incorporating Eqs. (5.73) and (5.74).

Drainage (the wetting phase is decreasing), on the other hand, was observed in the laboratory experiments to lead to a mixture of water and

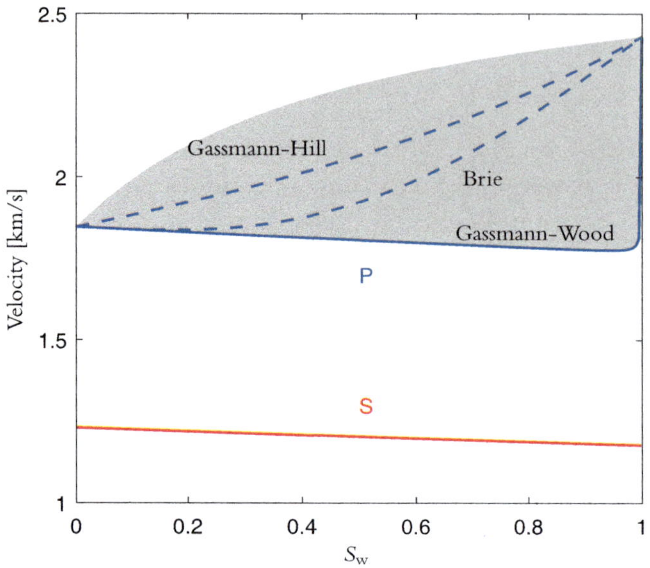

Figure 5.9 P- and S-wave velocities versus saturation in a gas/water saturated rock, as predicted by Eqs. (5.73)–(5.74), (5.39), (5.41) and (1.171) (solid lines), for a dataset representing a relatively weak rock at low pressures and low frequency. The dashed lines show the P-wave velocity for the Gassmann–Hill model (Eq. (5.76)) and the Brie model (Eq. (5.77) with $e = 3$), respectively. The grey area indicates the range of possible values for the P-wave velocity at higher frequencies for various fluid distributions.

air-saturated pores. In general, there may be different regions fully saturated by different saturating fluids, a situation often referred to as patchy saturation.

If the patch size is larger than the diffusion length, then pore pressure equilibrium will not be established. The fully saturated areas will behave as saturated undrained rock if the pore pressure diffusion length is less than the typical size l of the fully saturated areas, that is, if

$$\sqrt{\frac{D_h}{f}} < l \tag{5.75}$$

where D_h is the pore pressure diffusion constant (see Eq. (1.321)), and f is the frequency. Following Hill (1963), if the shear modulus can be considered homogeneous, then the overall bulk modulus and hence also the P-wave modulus depend only on concentrations of the different patches and not on patch shape. Incorporating Hill's solution into the Gassmann theory, which assumes shear modulus independent of saturation, the following equation (Gassmann–Hill) may be used to represent the P-wave modulus for patchy saturation:

$$H_{\mathrm{GH}} = \left[\sum_{i=1}^{n} \frac{S_i}{H(K_{\mathrm{f}i})} \right]^{-1} \tag{5.76}$$

where $K_{\mathrm{f}i}$ and S_i represent the bulk modulus and average saturation of each different pore fluid. Fig. 5.9 shows how the P-wave velocity changes with saturation for such a model. Note that we may use the Hashin–Shtrikman equations (6.14) and (6.15) to describe the P-wave modulus of the rock assuming that the patches are spherical. The result is identical to the Gassmann–Hill prediction regardless which of the fluids is inside the patches, in agreement with Hill's statement that the patch shape is irrelevant.

For water and gas saturated sandstone with 25% porosity and 1 Darcy permeability, the characteristic transition frequency based on Eq. (5.75) is in the seismic frequency range for $l \simeq 1$ m, in the sonic logging (kHz) range for $l \simeq 10$ cm, and in the ultrasonic (MHz) range for $l \simeq 1$ cm. For all cases, the patch size is assumed to be significantly smaller than the wavelength λ_{w} and significantly larger than the pore or grain size.

The transition from low frequencies to high frequencies has been the topic of a large number of publications. Müller et al. (2010) gave an overview of several models and the physics behind them. An underlying basis is provided by Biot's theory, with conversion of energy from fast to slow P-waves at heterogeneities. This gives rise to dispersion and attenuation caused by wave-induced fluid flow (WIFF). Whereas the early models were based on regular arrays of gas saturated pockets in water saturated rock (White, 1975; Dutta and Odé, 1979), newer models address patchiness within a random continuum and also heterogeneities in permeability and stiffness as well as in fluid saturation (Pride et al., 2004). All these models produce characteristic dispersion curves and attenuation behaviour similar to those illustrated in Fig. 5.4, but with transitional behaviour depending on the details of the models.

It is obvious that the original ambition of quantifying the amount of gas (e.g. CO_2) from velocity measurements is difficult in the case of fine-scale mixing when one essentially only can distinguish full saturation from partial saturation. In principle, changes in S-wave velocity due to density changes could be picked up, but these would normally be small. If the system is in a state of patchiness, then the detailed behaviour of velocity vs. saturation depends on frequency and further on patch sizes and their geometrical distribution. These are in general unknown, although we may envisage that future modelling may enable such predictions. In lack of this knowledge, a practical approach is required. As an example, an empirical relationship for the effective fluid bulk modulus given by Brie et al. (1995) is often used:

$$K_{\mathrm{f,eff}} = (K_{\mathrm{w}} - K_{\mathrm{g}})S_{\mathrm{w}}^{e} + K_{\mathrm{g}} \tag{5.77}$$

This equation was established from sonic log interpretation of gas-bearing sands. The equation reproduces the Voigt (isostrain; see Section 6.2) average when $e = 1$ and approximates the Reuss (isostress) average as e increases (with best-fitting e depending on the ratio of the bulk moduli). The exponent e can be regarded as a fitting parameter characterising the patchiness seen by the propagating waves, and values of around 3 are often used in practical situations.

A similar relationship to Eq. (5.77) can be obtained by considering the dynamic equilibrium of the two-phase system in the pore space without assuming patchy saturation: Papageorgiou and Chapman (2017) found that such a relationship would occur also at low frequencies with the ratio between the fluid bulk moduli as a controlling parameter, whereas Rozhko (2021), focusing on the dynamical hysteresis of the contact line between the two phases, showed that this may give rise to strain amplitude-dependent attenuation and velocities. Experimental verification is challenging, since the predictions cover large ranges of length scales and frequencies that cannot be fully addressed in laboratory studies.

Good knowledge of fluid properties is essential in predicting the impact of fluid and partial saturation on velocities. Properties under subsurface pressure and temperature conditions may differ significantly from those at surface conditions. Note however that oil under in situ conditions may contain significant amounts of dissolved gas, which reduce the density and the bulk modulus of the oil and hence reduce the contrast between the oil and gas. Similarly, brine also contains dissolved air at high pressures. The effects of partial saturation are significantly less for an oil–water system compared to the air–water system, since the differences in bulk modulus and density are much less for water and oil than for water and gas. Nevertheless, partly because of improved acquisition and processing technologies, it has become possible to distinguish between oil and water saturation in, for example, time-lapse seismic.

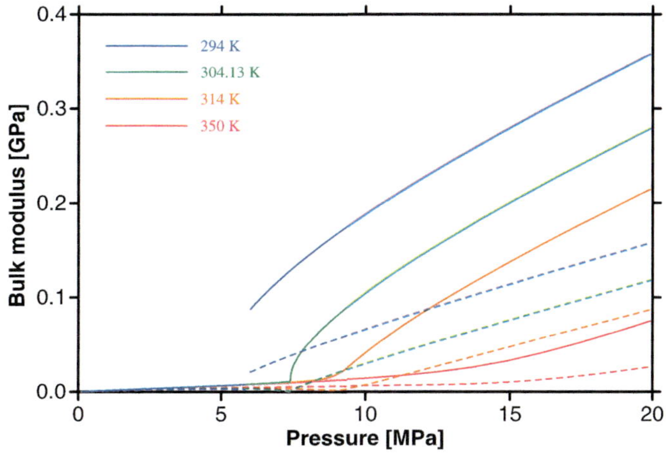

Figure 5.10 Bulk modulus of CO_2 as a function of pressure for a few temperatures (Lemmon et al., 2021). Note the strong dependence on both pressure and temperature, especially around the critical point (304.1282 K, 7.3773 MPa). Also note the significant difference between adiabatic (full lines) and isothermal (dashed lines) moduli.

Table A.6 shows some typical values for pore fluid properties, for a couple of pressure/temperature combinations corresponding roughly to about 1500-metres depth and 3000-metres depth, respectively. Note however that there are significant variations in the properties of oil and gas, depending on their chemical composition and variations in pressure and temperature conditions at a given depth. Also, the properties of brine vary somewhat with salinity and amount of dissolved gas.

CO_2 is a gas at room conditions but reaches a supercritical state at 7.4 MPa pressure and 31 °C, above which there are not distinct liquid and gas phases. At 20 °C, CO_2 condenses to a liquid at about 5.7 MPa. The bulk modulus of CO_2 changes significantly with conditions, especially near the critical point, as illustrated in Fig. 5.10.

5.6.3.3 Chemical effects

The minerals of the rock framework may react chemically with the pore fluid. In particular, chalk and clay minerals become soft or even dissolve in water (if the water is not in chemical equilibrium with the minerals). This implies that fluid substitution may actually change the framework moduli (K_{fr} and G_{fr}), due to chemical effects. Hence the elastic wave velocities and the static elastic moduli may be strongly sensitive to the type of saturating fluid (see also Section 2.6.3).

5.7 Reflections and refractions

When an elastic wave hits a boundary of the medium it is travelling through, the wave may be reflected, like light in a mirror, or refracted, like light at a water surface, or converted into other types of elastic waves. Such boundaries or interfaces between different parts of a medium are important for most aspects of rock acoustics. In particular, reflections at interfaces are the foundation of surface seismics, whereas sonic logging tools rely on refractions for obtaining a wave path through the rock along the hole.

Consider first the simple situation where a plane P-wave like (5.14) approaches a boundary normal to the direction of propagation. The boundary conditions are

- the displacement normal to the boundary is continuous at the boundary, and
- the stress normal to the boundary is continuous at the boundary.

To fulfil these requirements when the wave hits the boundary, two new waves are created at the boundary, one reflected wave and one transmitted wave. We will now see how these physical continuity requirements can be used to identify the amplitude and phase of both the reflected wave and the transmitted wave, relative to the amplitude and phase of the initial wave. We denote by subscript 1 the parameters of the medium through which the wave is initially travelling and by subscript 2 the parameters of the medium on the other side of the boundary. The three waves interacting at the boundary are then

$u^{\mathrm{i}} = u^{\mathrm{o,i}} \mathrm{e}^{\mathrm{j}(\omega t - q_1 x)}$ initial wave,
$u^{\mathrm{r}} = u^{\mathrm{o,r}} \mathrm{e}^{\mathrm{j}(\omega t + q_1 x)}$ reflected wave,
$u^{\mathrm{t}} = u^{\mathrm{o,t}} \mathrm{e}^{\mathrm{j}(\omega t - q_2 x)}$ transmitted wave.

The initial wave is travelling in medium 1 towards the boundary, the reflected wave is travelling in medium 1 away from the boundary, and the transmitted wave is travelling in medium 2 away from the boundary. Note that the frequency ω is the same in both media. The two physical continuity requirements can now be expressed as

$$u^{\mathrm{i}} + u^{\mathrm{r}} = u^{\mathrm{t}} \tag{5.78}$$

$$(\lambda_1 + 2G_1)\left(\frac{\partial u^{\mathrm{i}}}{\partial x} + \frac{\partial u^{\mathrm{r}}}{\partial x}\right) = (\lambda_2 + 2G_2)\frac{\partial u^{\mathrm{t}}}{\partial x} \tag{5.79}$$

Positioning for simplicity the x-axis so that the interface is located at $x = 0$, introducing the expressions for u^{i}, u^{r} and u^{t}, and dividing by $\mathrm{e}^{\mathrm{j}\omega t}$, we find

$$u^{o,i} + u^{o,r} = u^{o,t} \tag{5.80}$$

$$(\lambda_1 + 2G_1)(-jq_1 u^{o,i} + jq_1 u^{o,r}) = (\lambda_2 + 2G_2)(-jq_2 u^{o,t}) \tag{5.81}$$

These are two equations with two unknowns, namely $u^{o,r}$ and $u^{o,t}$. The use of complex numbers implies that the equations give us both the amplitude and phase of the reflected and transmitted wave. Rather than presenting these expressions explicitly, we now introduce the *reflection coefficient* r_{pp}, which expresses the stress amplitude of the reflected wave, and the *transmission coefficient* t_{pp}, which expresses the stress amplitude of the transmitted wave, both relative to the stress amplitude of the initial wave.

Expressed in terms of particle displacement amplitudes, these coefficients are defined as

$$r_{pp} = -\frac{u^{o,r}}{u^{o,i}} \tag{5.82}$$

$$t_{pp} = \frac{\rho_2 v_{p2}}{\rho_1 v_{p1}} \frac{u^{o,t}}{u^{o,i}} \tag{5.83}$$

Expressions for r_{pp} and t_{pp} can be found from Eqs. (5.80) and (5.81). Introducing further Eqs. (5.8) and (5.16), the expressions take the following form:

$$r_{pp} = \frac{\rho_2 v_{p2} - \rho_1 v_{p1}}{\rho_2 v_{p2} + \rho_1 v_{p1}} \tag{5.84}$$

$$t_{pp} = \frac{2\rho_2 v_{p2}}{\rho_2 v_{p2} + \rho_1 v_{p1}} \tag{5.85}$$

The product ρv_p is called the *acoustic impedance* of the medium. Thus the reflection and transmission coefficients depend on the impedance contrast between the two media. Note that r_{pp} and t_{pp} may be defined in different ways, depending on the choice of coordinate system and wave describing parameters. The definition used here implies that a compression is reflected as a compression if $r_{pp} > 0$, whereas for $r_{pp} < 0$, the reflection involves phase inversion.

The amplitude of the reflected wave depends on the density and velocity of the second medium, although the incoming and reflected waves have only propagated through medium 1. This may be used in analyses of seismic data, as we will see in Section 5.9.

This example was particularly simple, since the displacement and stress only had one component, normal to the interface between the two media.

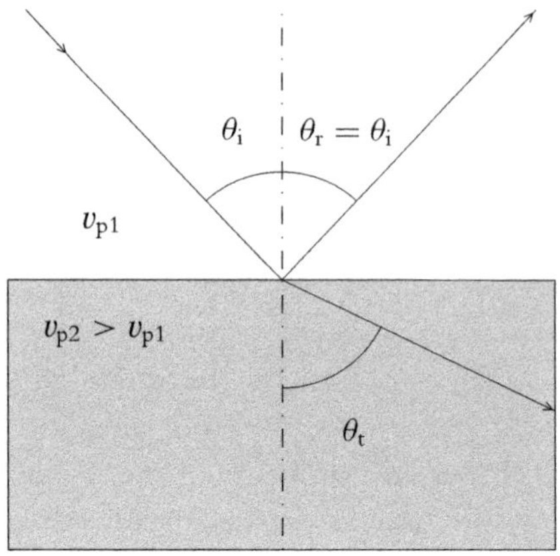

Figure 5.11 Reflection and refraction of an acoustic ray at the interface between two media.

Consider next a P-wave that hits an interface at an angle of incidence θ_i relative to the normal of the interface. It is now convenient to consider the wave as a ray, as we here wish to focus on the direction of propagation. A part of the wave will be reflected at an angle $\theta_r = \theta_i$ (see Fig. 5.11). Another part of the wave will be transmitted at an angle θ_t. The relation between the angle of incidence θ_i and the transmission angle θ_t is given by Snell's law

$$\frac{\sin\theta_t}{v_{p2}} = \frac{\sin\theta_i}{v_{p1}} \tag{5.86}$$

Thus the transmitted wave will proceed in a direction somewhat different from the incoming wave, provided that $v_{p2} \neq v_{p1}$. This effect is called *refraction*.

Note that we may have a situation where $(v_{p2}/v_{p1})\sin\theta_i > 1$, which would imply that $\sin\theta_t > 1$ according to Eq. (5.86). This is mathematically and physically unacceptable, and hence θ_t is not defined in this situation. This implies that the refracted wave does not exist, and we have *total reflection* at the interface. Thus the condition for total reflection of the P-wave is

$$\sin\theta_i > \sin\theta_{cr} \equiv \frac{v_{p1}}{v_{p2}} \tag{5.87}$$

The angle θ_{cr} is called the critical angle for total reflection.

P- and S-waves are different modes of propagating elastic energy and are not coupled in an ordinary medium due to symmetry. When a wave

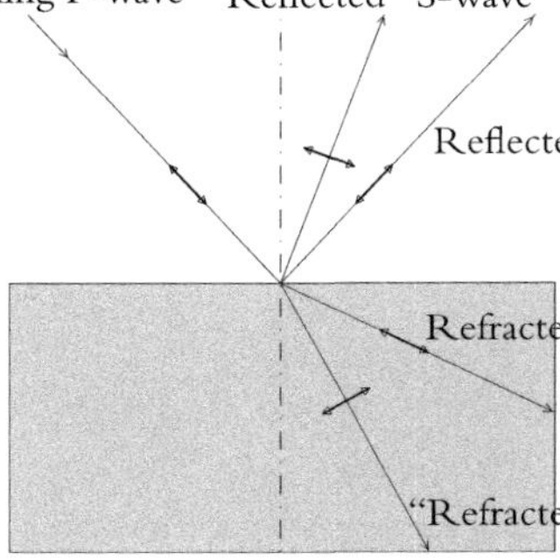

Figure 5.12 Polarisation (indicated by double arrows) of reflected, refracted and converted waves at an interface due to an incoming P-wave.

hits an interface at a skew angle relative to the direction of propagation, the symmetry is broken, and the waves may become coupled at the interface. In our example here the incoming P-wave has a particle motion, which is not fully orthogonal to the particle motion of an S-wave "reflected" from the interface at an angle θ_{rs} (Fig. 5.12), with polarisation in the same plane as the incoming wave and the normal to the interface. Thus such an S-wave may be generated at the interface. The angle θ_{rs} is given by the equation

$$\frac{\sin\theta_{rs}}{v_{s1}} = \frac{\sin\theta_i}{v_{p1}} \tag{5.88}$$

The incoming P-wave will also be coupled to a "refracted" S-wave, propagating in medium 2 at an angle θ_{ts} given by

$$\frac{\sin\theta_{ts}}{v_{s2}} = \frac{\sin\theta_i}{v_{p1}} \tag{5.89}$$

This wave also has its polarisation in the same plane as the incoming wave and the normal to the interface. The two S-waves are said to be *converted* waves, since they originated from a wave of different type. Similarly, if the incoming wave is an S-wave with polarisation in this plane, then two converted P-waves may be generated at the interface in addition to the reflected and refracted S-waves. Note that an S-wave with polarisation parallel to the interface is not coupled to any P-wave at the interface, and hence no such wave is created from an incoming P-wave, nor will such a wave generate converted P-waves.

The degree of coupling between the incoming wave and the reflected, transmitted and converted waves will vary with the angle θ_i. The expres-

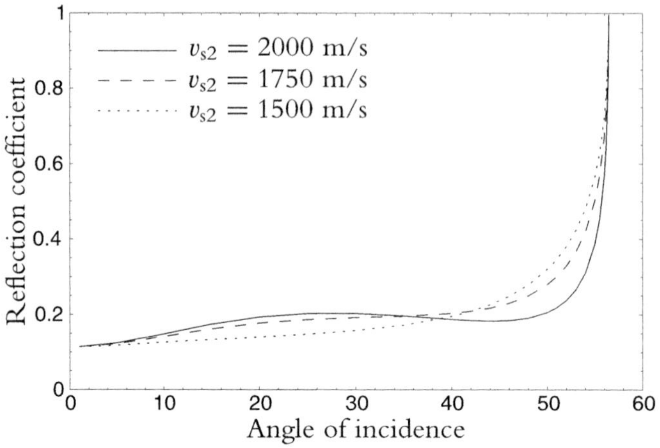

Figure 5.13 Reflection coefficient for a P-wave reflected at an interface between two media. Parameter values used in this example: $v_{p1} = 2500\,\mathrm{m/s}$, $v_{s1} = 1450\,\mathrm{m/s}$, $\rho_1 = 2.1\,\mathrm{g/cm^3}$, $v_{p2} = 3000\,\mathrm{m/s}$, $\rho_2 = 2.2\,\mathrm{g/cm^3}$.

sions for the reflection, transmission and conversion coefficients are rather complicated and are therefore given in Appendix D.2.2.

Fig. 5.13 shows an example where the reflection coefficient for the incoming P-wave has been calculated for various angles of incidence. The figure shows that the reflection coefficient increases abruptly just below the critical angle (Eq. (5.87)) for total reflection. The reflection coefficient is furthermore seen to depend on the shear wave velocity v_{s2} of the second medium.

5.7.1 Interface waves

In addition to the P- and S-waves, which can propagate through a material, there also exist some elastic waves that only propagate along the interface between two media. One familiar example is sea waves, which only exist at the surface of the water. An interface wave along the surface of a solid material is called a *Rayleigh wave*. This is the wave normally observed as a result of an earthquake. Since the wave is confined to the surface, it has a lower loss due to geometrical spreading than a body wave and therefore may propagate over larger distances. The velocity of a Rayleigh wave (v_R) is given by the shear wave velocity and Poisson's ratio of the solid (see e.g. Viktorov, 1979):

$$v_R \approx \frac{0.87 + 1.12\nu}{1 + \nu} v_s \tag{5.90}$$

Waves along a planar solid-fluid interface are called *Scholte waves*, whereas waves along a solid–solid interface are called *Stoneley waves*. There is a tradition within the petroleum industry, however, that both types of waves are called Stoneley waves. Such interface waves can be excited at the sea floor when a seismic wave passes through or at the borehole wall during sonic logging operations and thus can be observable during normal field data acquisition.

The direction of propagation of any interface wave is obviously along the interface. For the Rayleigh wave, the particle motion is an elliptical movement in the plane normal to the interface and parallel to the direction of propagation. The amplitude of this motion falls off exponentially into the solid material. Characteristic features of interface waves in borehole geometry are discussed in Section 5.8.

5.8 Borehole acoustics

Sonic well logging is an important application of elastic waves in the petroleum industry. The purpose of such logging is to measure the sonic velocities of the formation surrounding the well. In this section, we give a quick introduction and refer to dedicated literature such as Paillet and Cheng (1991) for details and deeper understanding. (See also Franco et al., 2006; Haldorsen et al., 2006.)

The basic elements of a sonic logging tool are shown in Fig. 5.14. The tool consists of a transmitter and a receiver separated by a distance L. When the transmitter emits a pulse, it generates a P-wave that travels through the mud and hits the borehole wall at different angles along the hole, where it may be reflected, refracted or converted into different modes. Eventually, a series of pulses arrive at the receiver: a refracted P-wave, a converted S-wave, and a set of borehole eigenmodes.

Fig. 5.15 illustrates refractions in the borehole geometry. Rays that hit the borehole wall at a sub-critical angle are refracted into the formation, and the energy can never return to the borehole. However, the ray that hits the wall at an angle such that the transmission angle $\theta_t = 90°$ will propagate parallel to the wall. This wave will continuously radiate energy into the borehole. We say that this wave is critically refracted. Some of this energy will reach the receiver. Since the critically refracted wave continuously loses energy, it will be attenuated even in the absence of absorption.

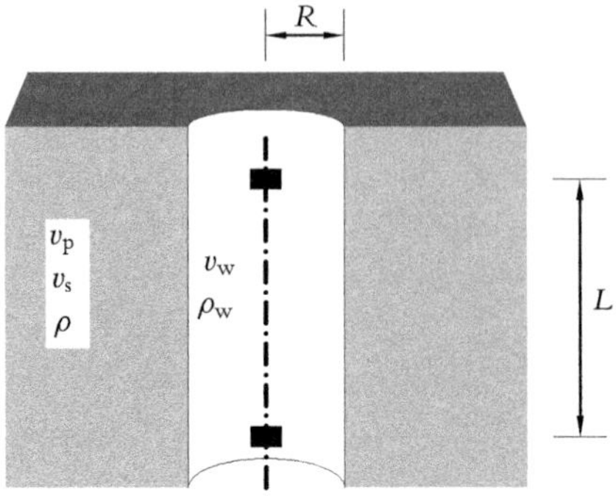

Figure 5.14 Schematic illustration of an acoustic logging tool with one transmitter and one receiver, situated in a borehole with radius R.

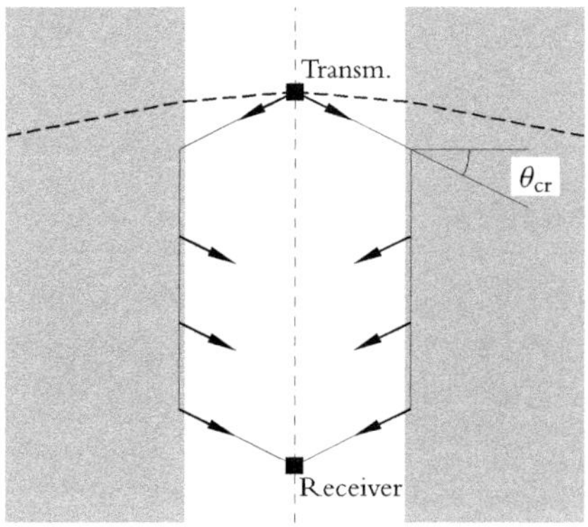

Figure 5.15 Refractions of waves transmitted from an acoustic logging tool. The dashed line represents a ray incident at sub-critical angle. The solid line is a critical refraction that continuously emits acoustic energy into the borehole.

Using the notation of Figs. 5.14 and 5.15, we can find an expression for the expected arrival time of the critical refraction as a function of transmitter–receiver separation. The length travelled in the fluid is

$$L_w = 2\frac{R}{\cos\theta_{cr}} \tag{5.91}$$

whereas the length travelled in the formation is

$$L' = L - 2R\tan\theta_{cr} \tag{5.92}$$

Introducing $\sin\theta_{cr} = v_w/v$ (see Eq. (5.87)), where v is the velocity of the refracted wave, and v_w is the velocity of the borehole fluid, we find the

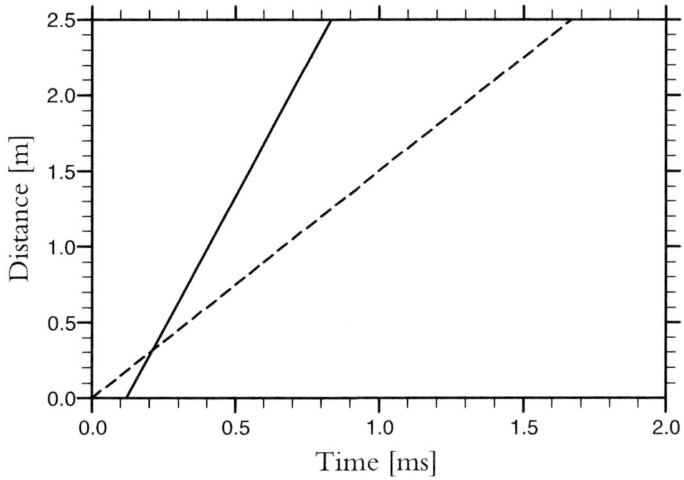

Figure 5.16 Arrival time of the critical refraction as a function of transmitter/receiver separation (solid line). Also shown is the arrival time of a direct wave with the velocity of the fluid (dashed line). The following parameters were used: $v = 3500\,\mathrm{m/s}$, $v_\mathrm{W} = 1500\,\mathrm{m/s}$, $R = 10\,\mathrm{cm}$.

arrival time

$$t = \frac{L_\mathrm{W}}{v_\mathrm{W}} + \frac{L'}{v} = \frac{L}{v} + 2R\left(\frac{1}{v_\mathrm{W}\sqrt{1-\left(\frac{v_\mathrm{W}}{v}\right)^2}} - \frac{v_\mathrm{W}}{v^2\sqrt{1-\left(\frac{v_\mathrm{W}}{v}\right)^2}}\right)$$

$$= \frac{L}{v} + 2R\sqrt{\frac{1}{v_\mathrm{W}^2} - \frac{1}{v^2}} \qquad (5.93)$$

Thus the arrival time is a constant plus a term equal to the separation between the transmitter and the receiver divided by the velocity of the formation. This means that the difference in arrival time between two receivers at different separations from the transmitter directly measures the velocity of the formation. Fig. 5.16 illustrates Eq. (5.93) and compares it with a direct wave travelling with the velocity of the borehole fluid. As can be seen, the refracted wave will be the first arrival as long as the source–receiver separation is above a limiting minimum.

In the above discussion, we have not specified whether v is the shear or compressional velocity of the formation. The arguments work equally well for both cases, which means that we may have both a critically refracted compressional wave and a critically refracted (converted) shear wave. This conclusion is, of course, subject to the condition that the formation velocities are higher than the fluid velocity. For the compressional wave, this will nearly always be fulfilled, but for the shear wave, this is not so. Indeed, formations are characterised as *slow* or *soft* when the shear wave velocity is slower than the fluid velocity, and as *fast* or *hard* otherwise.

5.8.1 Borehole modes

In addition to the refractions, the full wave train also consists of borehole modes. These are modes that are localised to the borehole and do not radiate into the formation (at least for some parameter ranges). This means that they are often the dominating components of the full wave train as far as amplitude is concerned. The borehole modes also contain information about the formation parameters, albeit in a more complicated manner than the refractions.

The most important eigenmode is the Stoneley wave, which is an interface wave. (The name "Stoneley wave" actually deviates from the strict definition of the type of interface wave we consider here; see page 299.) Its velocity is always smaller than that of the fluid in the borehole. The wave has relatively low dispersion. In the low-frequency limit the wave is known as the *tube wave*, and its velocity is

$$v_{\mathrm{St}} = \frac{v_{\mathrm{w}}}{\sqrt{1 + \frac{\rho_{\mathrm{w}} v_{\mathrm{w}}^2}{\rho v_{\mathrm{s}}^2}}} \tag{5.94}$$

where v_{w} and ρ_{w} are the velocity and density, respectively, of the fluid in the borehole, whereas v_{s} and ρ are the shear velocity and density, respectively, of the formation. This relation is plotted in Fig. 5.17 for two different density ratios. Note that for sufficiently small shear velocities, the Stoneley wave becomes attenuated in the low-frequency limit. This is due to the fact that it will radiate energy in the form of ordinary shear waves when v_{St} is higher than the shear velocity of the solid.

Eq. (5.94) indicates the possibility of using the Stoneley velocity to determine the shear wave velocity of a formation. Quite conveniently, the sensitivity to the shear wave velocity is highest when this velocity is small and no shear refraction exists. Some remarks of caution are however necessary. The high Stoneley wave attenuation in slow formations, as mentioned above and illustrated in Fig. 5.17, is the main limiting factor. Further, Eq. (5.94) is a low-frequency approximation and for practical applications, should be replaced by a numerical inversion in the frequency band of the

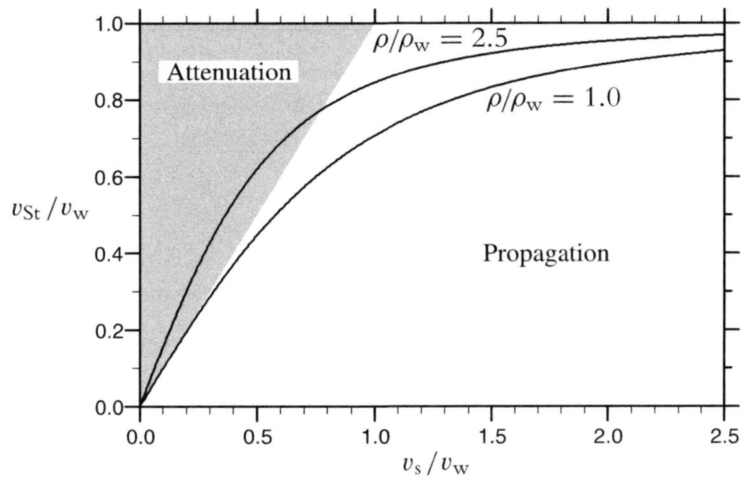

Figure 5.17 The low-frequency limit of the Stoneley velocity as a function of shear velocity for $\rho/\rho_W = 1.0$ and $\rho/\rho_W = 2.5$.

source. More important, the Stoneley wave velocity is quite sensitive to the formation permeability, especially at low frequencies (see e.g. Schmitt et al., 1988). Finally, the Stoneley wave and the refracted (converted) S-wave are sensitive to different components of the stiffness tensor, which may be significant in anisotropic formations. In principle, this offers the possibility of assessing the acoustic anisotropy of a fast formation, but in practice, this is very difficult since the Stoneley velocity has a low sensitivity to the shear velocity when the shear velocity is high.

The *shear eigenmodes* or *pseudo-Rayleigh* waves have phase velocities that are higher than the fluid velocity and slower than the shear velocity of the formation. Consequently, they only exist as undamped modes in fast formations. The pseudo-Rayleigh modes have a cut-off frequency that increases with mode number (in principle, infinitely many pseudo-Rayleigh modes exist). The cut-off frequency can be roughly estimated by requiring that an integer number of half wavelengths should fit into the borehole. For a borehole with diameter of 20 cm, this means that the order of magnitude of the cut-off frequency for the lowest mode is 10 kHz. Note that the presence of the logging tool in the borehole reduces the effective radius of the hole.

Fig. 5.18 shows examples of dispersion relations for the Stoneley mode and the lowest pseudo-Rayleigh modes. Fig. 5.19 shows examples of (computed) full wave trains in a fast formation for three different source spectra.

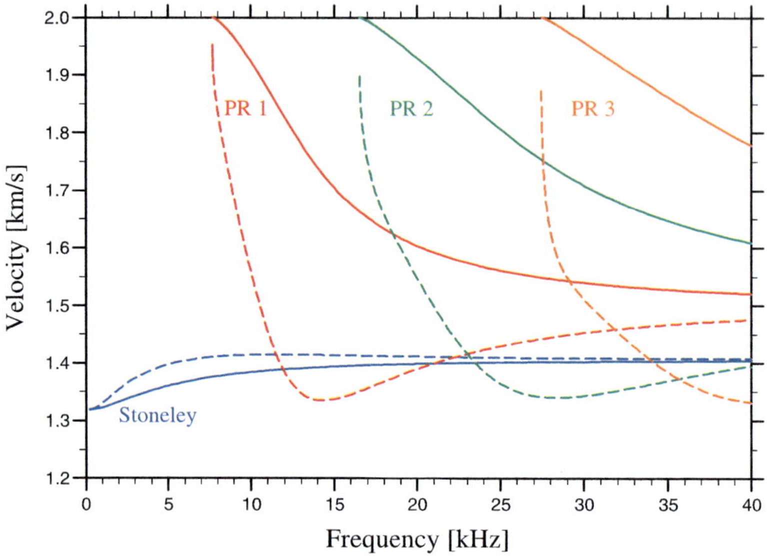

Figure 5.18 Example of dispersion of the Stoneley mode and the three lowest pseudo-Rayleigh modes in a fast formation. Solid lines are phase velocity, broken lines are group velocity. $v_P = 3500\,\mathrm{m/s}$, $v_S = 2000\,\mathrm{m/s}$, $v_W = 1500\,\mathrm{m/s}$, $\rho = 2.3\,\mathrm{g/cm^3}$, $\rho_W = 1.2\,\mathrm{g/cm^3}$, $R = 10\,\mathrm{cm}$.

All wave trains are shown in the same scale. In the lower left the early part of the wave trains are shown amplified by a factor 10. The figure illustrates that the full waveform depends strongly on the source spectrum. The arrows indicate the expected arrivals of the P- and S-refractions and the Stoneley wave based on ray theory.

Considering the high-frequency source in the lower frame, the wave train is dominated by the P-refraction and the pseudo-Rayleigh waves. No Stoneley wave is seen. In the middle frame the onset of the P- and S-refractions coincide well with the expectations from ray theory. The Stoneley overlays the pseudo-Rayleigh waves and is seen as a marked change in amplitude.

For the low-frequency case in the upper frame, there are no pseudo-Rayleigh waves, and the waveform is dominated by the Stoneley wave. Note that the Stoneley wave breaks before the time expected from the low-frequency velocity. This is due to dispersion. It is not possible to determine clear P- and S-refractions in this case.

The wave trains in a slow formation, as Fig. 5.20 show examples of, are considerably simpler due to the absence of the shear refraction and the

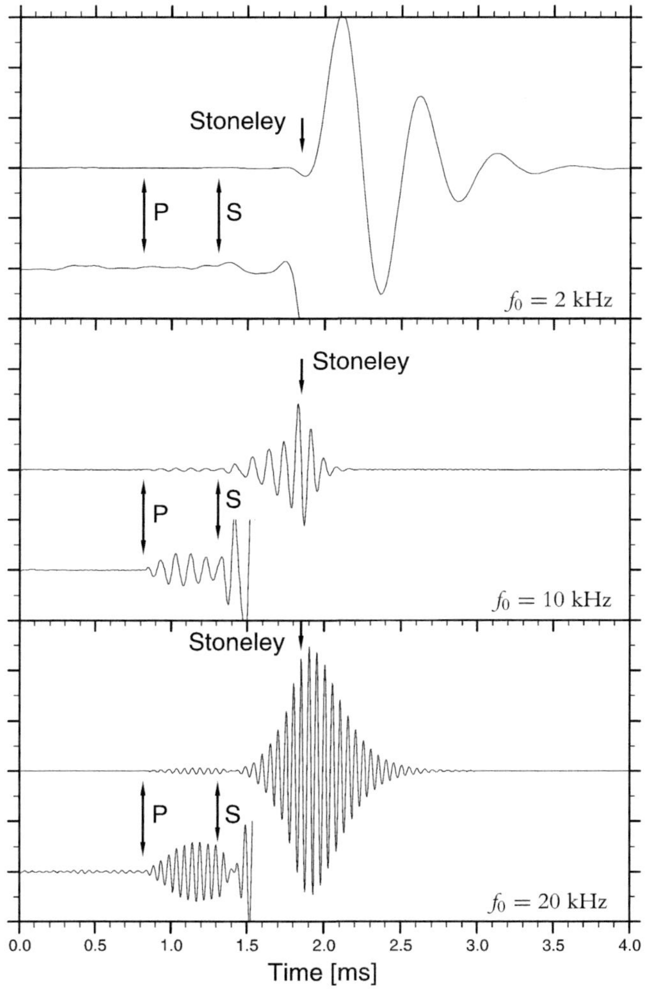

Figure 5.19 Example of (computed) full acoustic wave trains in a fast formation. Same parameters as in Fig. 5.18. $L = 2.44$ m. Source centre frequencies are 20 kHz, 10 kHz and 2 kHz for the lower, middle and upper frames, respectively. The early parts of the wave trains are also shown with 10 times amplification. The amplitude scale is the same in all three frames.

pseudo-Rayleigh modes. In certain cases, *damped* pseudo-Rayleigh modes may however be seen in slow formations.

In the high-frequency case, in the lower frame of Fig. 5.20, the only clear feature is the P-refraction. Looking at the middle frame of Fig. 5.20, we find two clear events: the P-refraction and the Stoneley wave. Note the

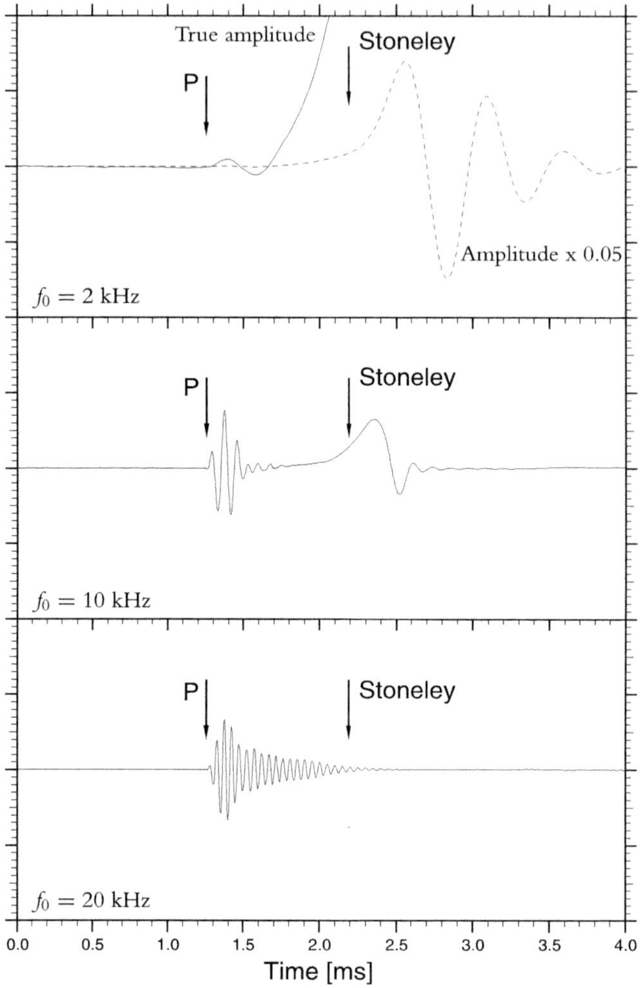

Figure 5.20 Example of (computed) full acoustic wave trains in a slow formation. $v_P = 2100\,\text{m/s}$, $v_S = 1200\,\text{m/s}$, $v_W = 1500\,\text{m/s}$, $\rho = 2.1\,\text{g/cm}^3$, $\rho_W = 1.2\,\text{g/cm}^3$, $L = 2.44\,\text{m}$, $R = 10\,\text{cm}$. Source centre frequencies are 20 kHz, 10 kHz and 2 kHz for the lower, middle and upper frames, respectively. The amplitude scale is the same in all three frames. The low-frequency wave train is also shown with the amplitude reduced by a factor 20 (dashed line).

relatively low frequency of the Stoneley wave. This is due to the fact that the Stoneley wave is excited most strongly at low frequencies, so that it favours the low frequencies of the excitation pulse. It is also evident that the early part of the Stoneley wave is lower in frequency than the later part. This is due to the dispersion of the Stoneley mode in slow formations, where the low-frequency parts propagate fastest. (In a fast formation, however, the high-frequency parts are the fastest.) The slow onset of the Stoneley mode in a slow formation of course makes picking of the arrival time rather difficult for real data, where noise is a complicating factor. This is another reason why the application of Eq. (5.94) for estimation of v_S is difficult.

For the low-frequency source in the upper frame, the amplitude of the Stoneley wave is strongly increased due to higher contents of low frequencies in the source. Note that the frequency of the Stoneley wave is more or less the same as in the middle frame. The P-refraction breaks where it is predicted from the ray theory.

In fast formations, we may in principle get the shear wave velocity from the shear refractions, but in practice the shear wave may be embedded in the ringing of the compressional wave and noise, such that its extraction is at least non-trivial. In slow formations, we may obtain the shear velocity from the Stoneley wave. However, as discussed, this method is subject to several restrictions making it a far from ideal method.

For estimation of mechanical properties, we are most interested in the shear velocity in slow formations, since a low shear velocity indicates low strength and hence potential stability problems. Dipole and quadrupole acoustic logging tools, which enable direct measurements of the shear wave velocity in both slow and fast formations, are very useful tools for this purpose.

The basic concept of the "Hula" log was suggested by White (1967). A dipole transmitter can in principle be pictured as two close monopole transmitters driven with opposite phase. The result may be thought of as a force oriented normal to the borehole wall, which induces a flexural motion of the borehole (see Fig. 5.21).

At sufficiently low frequencies, such that the wavelength is much longer than the borehole diameter, the flexural mode is little affected by the borehole and propagates with the shear velocity of the formation. At higher frequencies, the propagation velocity becomes lower. In fact, the dispersion of the flexural mode is quite similar to the Stoneley mode, with the exception that it is less affected by the permeability of the formation. An example of the flexural mode dispersion is shown in Fig. 5.22. A quadrupole tool excites a mode called the *screw* mode, which also propagates at the shear velocity at low frequencies and has a similar dispersion.

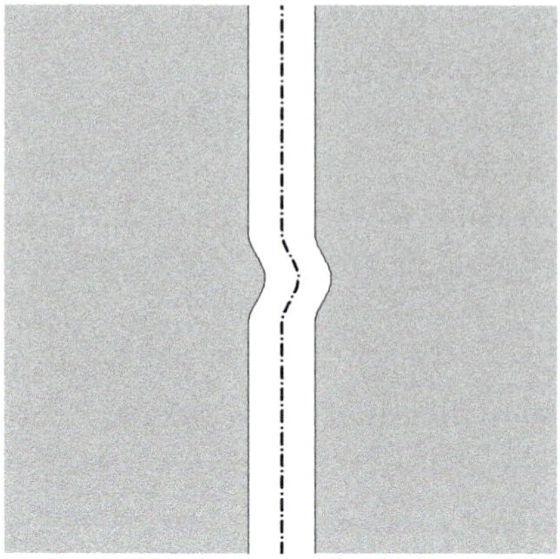

Figure 5.21 Schematic illustration of flexural borehole motion induced by a dipole transducer.

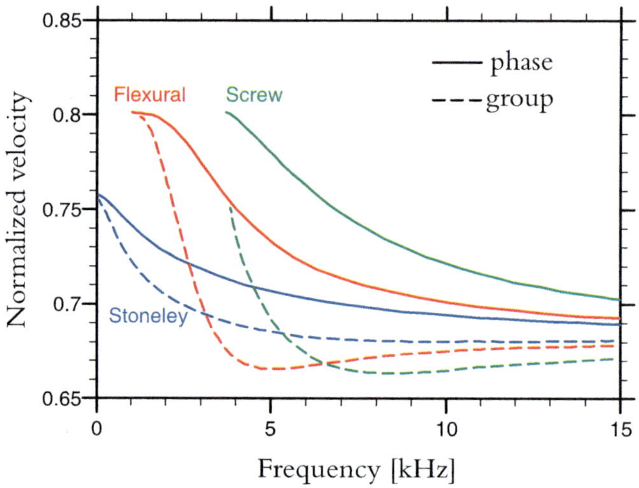

Figure 5.22 Dispersion of the flexural mode and the screw mode, compared to the Stoneley wave in a slow formation. Velocities are normalised to the borehole fluid velocity. (After Schmitt, 1988).

5.8.2 Borehole alteration

We have seen that the full waveform is quite complicated even in the simple model used above. Here we will discuss briefly a slightly more complicated situation, in which there is a concentric altered zone around the wellbore. The alteration may be due to mud filtrate invasion or stress-induced damage

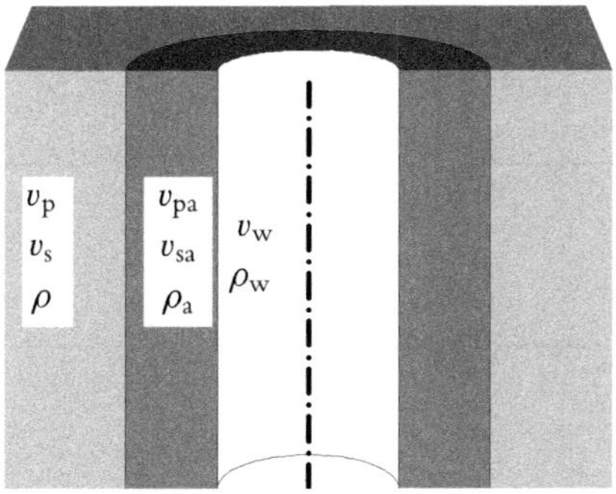

Figure 5.23 Principle sketch of an axisymmetric altered zone. The index a denotes altered parameter values.

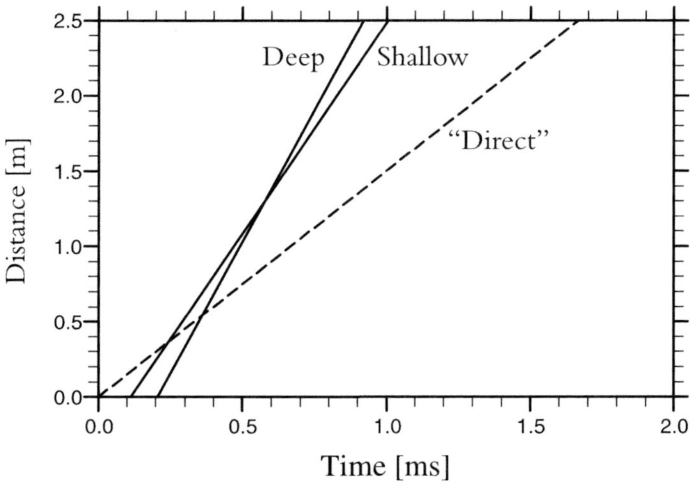

Figure 5.24 Arrival times for the model of Fig. 5.23. Virgin formation as in Fig. 5.16. The altered velocity is reduced by 20%. The altered zone thickness is 20 cm.

(see Section 4.5). We assume that the zone is homogeneous and isotropic. Fig. 5.23 sketches the situation.

From the ray-theory point of view, it is now clear that we may have refractions both from the borehole wall and from the interface between the two formation zones (assuming that the velocity of the altered zone is lower than that of the virgin formation). Fig. 5.24 shows the expected arrival times as a function of distance. The first refracted arrival will be the shallow refraction for short transmitter receiver distances and the deep refraction for

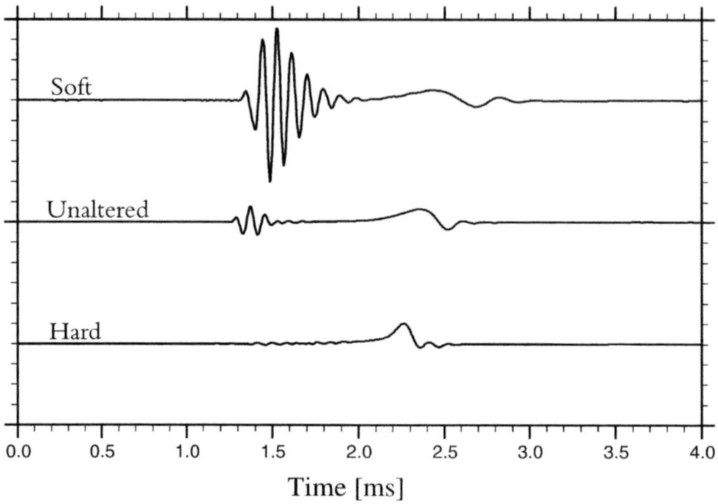

Figure 5.25 Synthetic wave trains in an altered slow formation. Soft means that the altered layer is 10 % slower than the virgin zone, and hard means that it is 10 % faster. The unaltered case is the same as in the middle frame of Fig. 5.20. The altered zone thickness is 10 cm.

longer separations. This effect has been used by Hornby and Chang (1985), who—using data from several runs of an experimental logging tool with variable transmitter receiver spacing—computed the altered zone thickness and velocity as functions of depth. Such calculations should, of course, be considered with some caution, since the model used is very simple. The assumption of an axisymmetric altered zone is dubious, and the abrupt transition from altered to unaltered formations is only a first approximation.

Fig. 5.25 shows examples of synthetic wave trains in an altered soft formation. The altered zone velocities are 10 % higher/lower than those of the virgin formation. It is clear from the figure that alteration affects significantly the character of the wave train. The amplitude of the P-refraction is changed considerably, and the dispersion characteristics of the Stoneley wave are also changed. This is due to the fact that the Stoneley wave samples some average of the altered and virgin shear velocities at low frequencies but sees only the altered zone at high frequencies.

5.9 Seismics

A seismic survey is the only tool available for mapping and characterisation of large subsurface structures. The major objectives for surface seismic

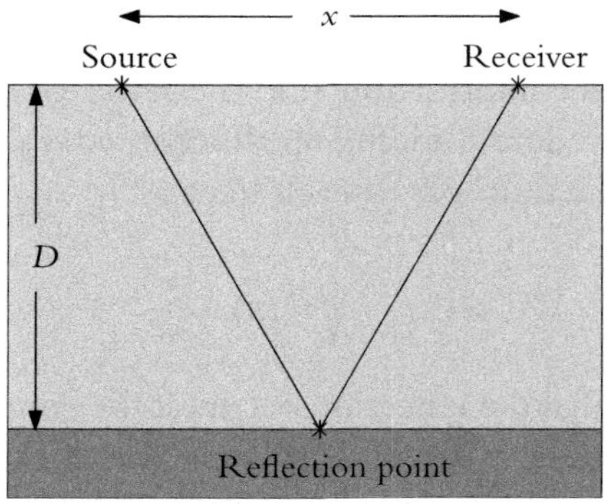

Figure 5.26 A seismic shot.

surveys are to identify and localise the subsurface structures. From a rock mechanical point of view, the fact that P- and S-wave velocities, density and to some extent anisotropy may be determined from seismic data provides a basis for large-scale applications.

The basic element in a seismic survey is a seismic shot, as shown in Fig. 5.26. The distance x between the source and the receiver is called the *offset*. When a shot is fired by the source, a semi-spherical wave is propagating down through the rock masses. At each interface between the various layers in the formation, a part of the wave is reflected and propagates back towards the surface, and is eventually detected at the receiver. The elapsed time t_{T} from the shot is fired until the reflection is detected at the receiver is called the *two-way traveltime* (TWT).

Consider a situation shown in Fig. 5.26, where the wave propagates with velocity v through a uniform formation until it is reflected at a plane, horizontal interface at depth D. By a simple geometric analysis we find that the TWT for a given offset can be expressed by the hyperbolic equation

$$t_{\mathrm{T}}^2(x) = t_{\mathrm{T}}^2(0) + \frac{x^2}{v^2} \tag{5.95}$$

where $t_{\mathrm{T}}(0) = 2(D/v)$.

Consider now a set of shots with different offsets around the same mid-point. The corresponding values of TWT and offset should then fall on a straight line in a plot of $t_{\mathrm{T}}^2(x)$ versus x^2. From this plot it is possible to identify both $t_{\mathrm{T}}(0)$ (from the intercept at $x = 0$) and v (from the slope). Thus it is possible to identify both the depth D and the velocity v from the seismic data.

A collection of shots around a common midpoint is called a CMP *gather*. To enhance the useful information (i.e. the reflected pulse) and suppress noise, a standard procedure is adding up all recorded traces in a CMP gather after having scaled the time axis of each trace as

$$t_T(x) \to \sqrt{t_T^2(x) - \frac{x^2}{v^2}} \tag{5.96}$$

This scaling ensures that the reflection occurs at the same place for all traces. Such a sum of traces is called a CMP *stack*. Note that stacking disturbs information about amplitude and phase in the signals.

If the formation consists of several layers, then the wave is refracted at each interface both before it reaches the actual interface we consider and on its way back to the surface. The expression for TWT is then rather complicated; however, in the hyperbolic approximation, we have (Dix, 1955)

$$t_T^2(x) \approx t_T^2(0) + \frac{x^2}{v_{RMS}^2} \tag{5.97}$$

with the so-called "root-mean-square velocity" v_{RMS} given as

$$v_{RMS}^2 = \frac{2}{t_T(0)} \sum_{j=1}^{N} L_j v_j \tag{5.98}$$

where L_j and v_j are the thickness and velocity of layer j, respectively, and N is the number of layers, whereas the TWT at zero offset is given as

$$t_T(0) = 2 \sum_{j=1}^{N} \frac{L_j}{v_j} \tag{5.99}$$

As we are able to obtain both v_{RMS} and $t_T(0)$ from the CMP gather for the reflection from each layer, we have $2N$ equations needed to determine the $2N$ unknowns L_j, v_j. This procedure, called Dix inversion, allows us to determine both velocity and thickness of each layer.

An alternative form of the traveltime equation (5.97) is found by Taylor expansion of the traveltime in x:

$$t_T(x) = t_T(0) + \frac{x^2}{2t_T(0)v_{NMO}^2} + \cdots \tag{5.100}$$

The function $\Delta t_T(x) = t_T(x) - t_T(0)$ is called the *normal moveout* (NMO), and the velocity v_{NMO} is called the *normal moveout velocity*. In general, v_{NMO}

differs from v_{RMS}; however, for plane, horizontal and homogeneous layers, we have $v_{NMO} = v_{RMS}$.

In practice the processing of seismic data is complicated by several factors, such as dipping layers (which imply that the reflection point differs from the common midpoint), lateral velocity variations and anisotropy (which both imply that the velocity is different for different offsets) and spurious reflections originating from refracted waves, converted waves, interface waves and multiple reflections. Data derived from seismic processing may thus suffer from large uncertainties.

We have shown earlier that the amplitude of a reflection generated at an interface depends on the acoustic impedance. Since the acoustic impedance depends on both velocity and density, the amplitude of a seismic reflection also carries information about the rock density. In general, the reflection coefficient depends on the angle of incidence, the P-wave velocities and densities of both layers, as well as the shear wave velocity v_{s2} of the lower layer (see example in Fig. 5.13). Thus the variation in amplitude versus offset (corresponding to the angle of incidence) for a P-wave reflected at the interface between two layers carries information about both the P- and S-wave velocities of the layer below the interface. In marine seismic surveys, both the source and the receivers are floating in water, and hence all information to be extracted about the formations below the sea floor has to be carried by P-waves. Application of the method outlined above is a possible way to extract also shear wave velocities from marine seismic data. The method is known as AVO (Amplitude Versus Offset). In land or ocean bottom-based seismic surveys, where both P- and S-waves are generated and detected, an extended method called multicomponent AVO may be applied to improve the confidence of the estimated shear wave velocity. This method also includes analyses of amplitudes of reflected S-waves by a similar set of equations as those described in Appendix D.2.2.

A velocity derived from a seismic data set is a kind of "average" velocity over a distance given roughly by the wavelength. If the rock is heterogeneous on a scale shorter than this distance, then the observed velocity will not be fully representative for the velocity on a shorter length scale. Examples of how the "average" velocity relates to the short-scale velocities and the geometrical structure of the heterogeneities are given in Section 6.1.

The resolution of seismic data—that is, the typical size of the smallest objects that can be seen as reflections—is usually assumed to be about a quarter of the wavelength. As a seismic wave contains a band of frequencies (typically 10–100 Hz), the resolution limit is usually defined as the wavelength of the component that has the highest frequency. The frequency content in a seismic signal is mainly given by the attenuation (which is a constant-Q type with Q in the order of 30–100 for underground formations), the distance travelled by the wave and the type of wave. For a P-wave reflected at about 1000-metres depth, the typical wavelength is about 100 m, whereas the resolution limit is about 20 m, depending of course on the local conditions. The loss of the highest frequencies with increasing travel time (which is a consequence of constant-Q attenuation) implies that the resolution is poorer for deeper reflections. For a P-wave reflected at about 4000-metres depth, the typical wavelength is about 200 m, and the resolution limit is about 50 m. Improved recording technology may improve the resolution somewhat, maybe as much as a factor of 2. Still the resolution is quite poor, and this limits the applicability of seismic data for determination of formation properties. Further improvements in the resolution require different processing techniques, which may be more sensitive to noise.

5.10 Acoustic emission

Non-elastic deformation, as described in Sections 1.8 and 2.8, is usually linked to localised failure events. Such events may be abrupt and induce transient elastic waves radiating from the failure point. This is called acoustic emission, abbreviated AE. The magnitude of such failure events can vary over many scales, and the magnitude of the acoustic emissions varies correspondingly. Earthquakes are essentially large-scale acoustic emissions. On field scale, AE activity is known as microseismicity.

AE activity can be monitored by simple listening devices, both in the laboratory and in the field. Monitoring of AE activity may provide information about onset and intensity of a fracturing process. It is known from earthquake monitoring that the position of single events can be identified with reasonable precision using a network of suitably distributed listening devices combined with proper processing algorithms. Downscaled versions of this technology are also used in oilfield monitoring and laboratory tests to reveal how a fracturing process develops in space and time.

Smaller events are less dramatic than larger events but tend to occur more frequently. This is expressed by a well-known relationship between magnitude and occurrence of events (Gutenberg and Richter, 1944),

$$\log N = a - bM \tag{5.101}$$

where $N = N(M)$ is the cumulative number of events with magnitude greater than M, whereas a and b are constants. These constants may, for instance, be determined through observations of earthquake activity, thus making it possible to estimate the probability of future earthquakes exceeding a given magnitude. Moreover, the b-value tends to drop significantly when a rock volume is close to global failure. Monitoring of this parameter over time may therefore serve as a failure warning.

A rock may display significant AE activity during initial loading, whereas it is more or less silent during unloading and reloading. This behaviour is a consequence of the fact that acoustic emissions originate from failure processes, which are by nature irreversible. Hence AE activity is not repeatable. When a reloading path reaches the previous peak stress level and proceeds beyond, the rock is experiencing stress states it has not experienced before, and new failure processes are triggered, resulting in a large increase in AE activity. This effect is known as the *Kaiser effect*. It implies that the rock has a memory about the largest stress it has experienced. For core material retrieved from deep wells, the in situ stress state is usually the most extreme stress it has experienced, and hence there is a potential for estimating in situ stresses by searching for the Kaiser effect during laboratory tests on the core materials (Holcomb, 1993; Pestman et al., 2001) (see also Section 7.4.3.2). However, the effect is rather diffuse unless the stress path in between the first and second visits to the peak stress state is relatively simple and only traversed back and forth. This requires, among others, that the orientation and relative magnitude of the stress components are known in advance and can be recreated in the laboratory. In practice the stress path that the core has experienced from coring to the re-establishment of in situ stresses in the laboratory is not sufficiently simple to make this an accurate tool (Lavrov, 2003). The "stress memory" of the rock may also be partly erased due to variations in humidity and temperature.

Several different mechanisms may generate AE activity (Michlmayr et al., 2012), some of which, like initiation and growth of cracks, are usually more interesting than others. Through analyses of recordings from single events (known as *moment tensor inversion*) it is possible to identify the moment tensor that describes the strength of the source and the force components that generate the radiation of acoustic waves. This makes it possible to identify the type of failure that created the event. Note that the level of detectable AE activity may differ significantly between different rock types. Absence of AE activity does not necessarily imply absence of failure processes.

References

Andersen, A.L., Hampton, L., 1980. Acoustics of gas-bearing sediments I. Background. J. Acoust. Soc. Am. 67, 1865–1889. https://doi.org/10.1121/1.384453.
Auld, B.A., 1990. Acoustic Fields and Waves in Solids, vol. I, second edition. Robert E. Krieger Publishing Company.
Batzle, M., Han, D.-H., Hofmann, R., 2006. Fluid mobility and frequency-dependent seismic velocity — direct measurements. Geophysics 71 (1), N1–N9. https://doi.org/10.1190/1.2159053.
Biot, M.A., 1956a. Theory of propagation of elastic waves in a fluid-saturated porous solid. I. Low-frequency range. J. Acoust. Soc. Am. 28, 168–178. https://doi.org/10.1121/1.1908239.
Biot, M.A., 1956b. Theory of propagation of elastic waves in a fluid-saturated porous solid. II. Higher frequency range. J. Acoust. Soc. Am. 28, 179–191. https://doi.org/10.1121/1.1908241.
Biot, M.A., 1962. Mechanics of deformation and acoustic propagation in porous media. J. Appl. Phys. 33, 1482–1498. https://doi.org/10.1063/1.1728759.
Bourbie, T., Coussy, O., Zinszner, B., 1987. Acoustics of Porous Media. IFP Publications. Éditions Technip.
Brie, A., Pampuri, F., Marsala, A.F., Meazza, O., 1995. Shear sonic interpretation in gas-bearing sands. SPE30595. In: SPE Annual Technical Conference and Exhibition. Dallas, TX, USA.
Cadoret, T., Marion, D., Zinszner, B., 1995. Influence of frequency and fluid distribution on elastic wave velocities in partially saturated limestones. J. Geophys. Res. 100 (B6), 9789–9803. https://doi.org/10.1029/95JB00757.
Christensen, N.I., 1982. Seismic velocities. In: Carmichael, R.S. (Ed.), Handbook of Physical Properties of Rocks, vol. II. CRC Press, pp. 2–228.
Dix, C.H., 1955. Seismic velocities from surface measurements. Geophysics 20 (1), 68–86. https://doi.org/10.1190/1.1438126.
Dutta, N.C., Odé, H., 1979. Attenuation and dispersion of compressional waves in fluid-filled porous rocks with partial gas saturation (White model). Part I. Biot theory. Geophysics 44 (11), 1777–1788. https://doi.org/10.1190/1.1440938.
Endres, A.L., Knight, R., 1989. The effect of microscopic fluid distribution on elastic wave velocities. Log Analyst Nov–Dec, 437–445.
Fjær, E., 1999. Static and dynamic moduli of weak sandstones. ARMA-99-0675. In: Proc. 37th U.S. Rock Mech. Symp. Vail, CO, USA.
Fjær, E., 2006. Modeling the stress dependence of elastic wave velocities in soft rocks. ARMA06-1070. In: 41st U.S. Symposium on Rock Mechanics. Golden, CO, USA.
Fjær, E., 2019. Relations between static and dynamic moduli of sedimentary rocks. Geophys. Prospect. 67 (1), 128–139. https://doi.org/10.1111/1365-2478.12711.
Fjær, E., Stroisz, A.M., Holt, R.M., 2013. Elastic dispersion derived from a combination of static and dynamic measurements. Rock Mech. Rock Eng. 46, 611–618. https://doi.org/10.1007/s00603-013-0385-8.
Franco, J.L.A., Ortiz, M.A.M., De, G.S., Renlie, L., Williams, S., 2006. Sonic investigations in and around the borehole. Oilfield Rev. Spring 2006, 14–31.
Gutenberg, B., Richter, C.F., 1944. Frequency of earthquakes in California. Bull. Seismol. Soc. Am. 34, 185–188.
Haldorsen, J.B.U., Johnson, D.L., Plona, T., Sinha, B., Valero, H.-P., Winkler, K., 2006. Borehole acoustic waves. Oilfield Rev. Spring 2006, 34–43.
Hill, R., 1963. Elastic properties of reinforced solids: some theoretical principles. J. Mech. Phys. Solids 11 (5), 357–372. https://doi.org/10.1016/0022-5096(63)90036-X.
Holcomb, D.J., 1993. Observations of the Kaiser effect under multiaxial stress states: implications for its use in determining in situ stress. Geophys. Res. Lett. 20 (19), 2119–2122. https://doi.org/10.1029/93GL01270.

Hornby, B.E., Chang, S.K., 1985. A case study of shale and sandstone alteration using a digital sonic tool. In: SPWLA 26th Ann. Logging Symp. Society of Petrophysicists and Well-Log Analysts.

Johnston, D.H., Toksöz, M.N., Timur, A., 1979. Attenuation of seismic waves in dry and saturated rocks: II. Mechanisms. Geophysics 44 (4), 691–711. https://doi.org/10.1190/1.1440970.

Jones, T.D., 1986. Pore fluids and frequency-dependent wave propagation in rocks. Geophysics 51, 1939–1953. https://doi.org/10.1190/1.1442050.

Jones, T.D., Nur, A., 1983. Velocity and attenuation in sandstone at elevated temperatures and pressures. Geophys. Res. Lett. 10 (2), 140–143. https://doi.org/10.1029/GL010i002p00140.

King, M.S., 1970. Static and dynamic elastic moduli of rocks under pressure. In: Proc. 11th US Symp. on Rock Mechanics, pp. 329–351.

Knight, R., Nolen-Hoeksema, R., 1990. A laboratory study of the dependence of elastic wave velocities on pore scale fluid distribution. Geophys. Res. Lett. 17 (10), 1529–1532. https://doi.org/10.1029/GL017i010p01529.

Lavrov, A., 2003. The Kaiser effect in rocks: principles and stress estimation techniques. Int. J. Rock Mech. Min. Sci. 40 (2), 151–171. https://doi.org/10.1016/S1365-1609(02)00138-7.

Ledbetter, H., 1993. Dynamic vs. static Young's moduli: a case study. Mater. Sci. Eng. A 165 (1), L9–L10. https://doi.org/10.1016/0921-5093(93)90634-Q.

Lemmon, E.W., McLinden, M.O., Friend, D.G., 2021. Thermophysical properties of fluid systems. In: Linstrom, P.J., Mallard, W.G. (Eds.), NIST Chemistry WebBook, NIST Standard Reference Database Number 69. National Institute of Standards and Technology, Gaithersburg MD, 20899.

Lozovyi, S., Sirevaag, T., Szewczyk, D., Bauer, A., Fjær, E., 2017. Non-elastic effects in static and dynamic rock stiffness. ARMA17-293. In: 51st US Rock Mechanics/Geomechanics Symposium. San Francisco, CA, USA.

Martin III, R.J., Haupt, R.W., 1994. Static and dynamic elastic moduli on granite: the effect of strain amplitude. In: Nelson, P.P., Lauback, S.E. (Eds.), Rock Mechanics. Balkema, Rotterdam, pp. 473–480.

Mavko, G., Jizba, D., 1991. Estimating grain-scale fluid effects on velocity dispersion in rocks. Geophysics 56 (12), 1940–1949. https://doi.org/10.1190/1.1443005.

Mavko, G.M., Nur, A., 1979. Wave attenuation in partially saturated rocks. Geophysics 44 (2), 161–178. https://doi.org/10.1190/1.1440958.

Mensch, T., Rasolofosaon, P., 1997. Elastic-wave velocities in elastic media of arbitrary symmetry—generalization of Thomsen's parameters ε, δ and γ. Geophys. J. Int. 128 (1), 43–64. https://doi.org/10.1111/j.1365-246X.1997.tb04070.x.

Michlmayr, G., Cohen, D., Or, D., 2012. Sources and characteristics of acoustic emissions from mechanically stressed geologic granular media — a review. Earth-Sci. Rev. 112 (3–4), 97–114. https://doi.org/10.1016/j.earscirev.2012.02.009.

Murphy III, W.F., Winkler, K.W., Kleinberg, R.L., 1984. Contact microphysics and viscous relaxation in sandstones. In: Johnson, D.L., Sen, P.N. (Eds.), Physics and Chemistry of Porous Media. Am. Inst. Phys., pp. 176–190.

Müller, T.M., Gurevich, B., Lebedev, M., 2010. Seismic wave attenuation and dispersion resulting from wave-induced flow in porous rocks—a review. Geophysics 75 (5), A147–A164. https://doi.org/10.1190/1.3463417.

Nes, O.-M., Holt, R.M., Fjær, E., 2002. The reliability of core data as input to seismic reservoir monitoring studies. SPE Reserv. Eval. Eng. 5 (1), 79–86. https://doi.org/10.2118/76641-PA.

Paillet, F.L., Cheng, C.H., 1991. Acoustic Waves in Boreholes. CRC Press. ISBN 978-0849388903.

Papageorgiou, G., Chapman, M., 2017. Wave-propagation in rocks saturated by two immiscible fluids. Geophys. J. Int. 209, 1761–1767. https://doi.org/10.1093/gji/ggx128.
Pestman, B.J., Kenter, C.J., van Munster, J.G., 2001. Core-based determination of in situ stress magnitudes. In: Proceedings of the DC Rocks. Washington, DC, USA.
Plona, T.J., 1980. Observation of a second bulk compressional wave in a porous medium at ultrasonic frequencies. Appl. Phys. Lett. 36 (4), 259–261. https://doi.org/10.1063/1.91445.
Plona, T.J., Cook, J.M., 1995. Effects of stress cycles on static and dynamic Young's moduli in Castlegate sandstone. In: Proc. of the 35th U.S. Rock Mech. Symp. Balkema, pp. 155–160.
Pride, S.R., Berryman, J.G., Harris, J.M., 2004. Seismic attenuation due to wave induced flow. J. Geophys. Res. 109 (B01201), 1–19. https://doi.org/10.1029/2003JB002639.
Rozhko, A.Y., 2021. On the spectral changes of seismic wave energy by a partially saturated crack due to the hysteresis of liquid bridges phenomenon. Geophysics 86, MR133–MR147. https://doi.org/10.1190/geo2020-0685.1.
Sayers, C.M., 2002. Stress-dependent elastic anisotropy of sandstones. Geophys. Prospect. 50 (1), 85–95. https://doi.org/10.1046/j.1365-2478.2002.00289.x.
Sayers, C.M., 2006. Sensitivity of time-lapse seismic to reservoir stress path. Geophys. Prospect. 54, 369–380. https://doi.org/10.1111/j.1365-2478.2006.00539.x.
Schmitt, D.P., 1988. Shear wave logging in elastic formations. J. Acoust. Soc. Am. 84 (6), 2215–2229. https://doi.org/10.1121/1.397015.
Schmitt, D.P., Zhu, Y., Cheng, C.H., 1988. Shear wave logging in semi-infinite saturated porous formations. J. Acoust. Soc. Am. 84 (6), 2230–2244. https://doi.org/10.1121/1.397016.
Stoll, R.D., 1974. Acoustic waves in saturated sediments. In: Hampton, L. (Ed.), Physics of Sound in Marine Sediments. Plenum, New York.
Thomsen, L., 1986. Weak elastic anisotropy. Geophysics 51 (10), 1954–1966. https://doi.org/10.1190/1.1442051.
Tisato, N., Quintal, B., Chapman, S., Podladchikov, Y., Burg, J.-P., 2015. Bubbles attenuate elastic waves at seismic frequencies: first experimental evidence. Geophys. Res. Lett. 42, 3880–3887. https://doi.org/10.1002/2015GL063538.
Tsvankin, I., 1997. Anisotropic parameters and P-wave velocity for orthorhombic media. Geophysics 62, 1292–1309. https://doi.org/10.1190/1.1444231.
Tutuncu, A.N., Podio, A.L., Gregory, A.R., Sharma, M.M., 1998. Nonlinear viscoelastic behavior of sedimentary rocks, Part I: effect of frequency and strain amplitude. Geophysics 63, 184–194. https://doi.org/10.1190/1.1444311.
Viktorov, I.A., 1979. Types of acoustic surface waves in solids (review). Sov. Phys. Acoust. 25, 1–9.
Walsh, J.B., 1966. Seismic wave attenuation in rock due to friction. J. Geophys. Res. 71 (10), 2591–2599. https://doi.org/10.1029/JZ071i010p02591.
White, J.E., 1975. Computed seismic speeds and attenuation in rocks with partial gas saturation. Geophysics 40 (2), 224–232. https://doi.org/10.1190/1.1440520.
White, J.E., 1967. The Hula log: a proposed acoustic tool. In: SPWLA 8th Ann. Logging Symp. Trans., pp. I1–I25.
Wyllie, M.R.J., Gregory, A.R., Gardner, G.H.F., 1958. An experimental investigation of factors affecting elastic wave velocities in porous media. Geophysics 23 (3), 459–493. https://doi.org/10.1190/1.1438493.

CHAPTER 6

Rock models

In the previous chapters, we discussed the mechanical behaviour of rocks in terms of continuum mechanics, that is, we treated the rocks as homogeneous materials. However, sedimentary rocks are made up of small particles and are largely heterogeneous materials on a length scale comparable to the particle size (Fig. 1.19). Furthermore, inhomogeneous distribution of particle types and particle sizes in the form of layers or clusters, as well as fractures, produce heterogeneities on larger length scales. The mechanical properties of the rock differ largely from one part to another if measured on a length scale, which is small or comparable to the size of the heterogeneities. The continuum approach can be applied with confidence only as an average and on a length scale, which is large compared to the size of the heterogeneities.

For many practical applications, the continuum approach is valid. However, it is intuitively clear that the mechanical properties of the rock even on a large scale must depend in some way on the microscopical nature of the rock. We have already touched this problem when we introduced the poroelastic formalism (Section 1.6). The theory of poroelasticity is based on the assumption that the rock consists of both solid and fluid parts, which are separated on a microscopic scale but coexist on a macroscopic scale. One of the statements of this theory is that the rock framework has its own moduli, separate from those of the solid material and the pore fluid.

The poroelastic formalism of Biot introduced earlier (Chapter 1) is capable of predicting the elastic properties of a porous material, provided that parameters like the frame bulk K_{fr} and the shear G_{fr} moduli are known. These are macroscopic parameters describing the properties of the rock on a length scale much larger than the grains and pores. The Biot theory is called a macroscopic theory since it only deals with macroscopic quantities. It is an ultimate goal to be able to express the macroscopic parameters like K_{fr} and G_{fr} in terms of the properties of the constituents the rock is made of and the microstructure of the rock fabric. Having obtained values for the macroscopic parameters from the microscopic theories, we may consider these parameters to represent a material that is homogeneous on a macroscopic scale. This virtual homogeneous material is commonly called an *effective medium*.

Developments in Petroleum Science, Volume 72
ISSN 0376-7361, https://doi.org/10.1016/B978-0-12-822195-2.00015-2

The elastic properties of a composite material like a sedimentary rock depend primarily on the following three factors:

1. the relative amount of each component present
2. the elastic properties of each component
3. the geometrical distribution of each component.

It is not possible to obtain accurate data for all three factors for a given rock, and hence any model of the rock will necessarily be based on simplifying assumptions. A model is therefore valid as a representation for the real rock only to the extent that the features governing the observed behaviour of the rock are accounted for in the model. This also implies that a model may be a good representation for the rock in a given situation, but it is not at all describing the actual behaviour of the rock in other situations.

A large number of rock models have been derived over the years, describing various properties of various rocks under various conditions. A comprehensive collection of such models was given by Mavko et al. (1998). Here we will first take a look at a model for a simple heterogeneous material to illustrate how the physical properties may depend on the length scale on which the material is being observed (Section 6.1). Some models for porous rocks involve only porosity, with no explicit reference to the shape of the pore space. Such models are presented in Section 6.2. Other models are based on the forming process of sedimentary rocks and describe essentially packs of grains. Such models are described in Section 6.3. Models describing rocks with cracks, including anisotropy and crack interactions and crack evolution under stress, are described in Section 6.4. In some cases the presence of multiple solid phases or solid pore fill may affect rock properties (Section 6.5). Fractured rocks are described in Section 6.6. Finally, a brief description of commonly applied numerical finite element analysis is given in Section 6.7.

6.1 Layered media

Consider a material made up of parallel isotropic layers. This is one of the simplest examples of a non-homogeneous material. We will use this as an example to see how a material can be considered as heterogeneous on one scale and effectively homogeneous on another.

Given a sample of a layered material, we denote the thickness of layer number i as L_i, so that the total thickness L of the sample is the sum of all L_i. If an external load σ_z is applied in the direction normal to the layers, then all layers have to carry the same load. The strain of layer i is $\varepsilon_{z,i} = \sigma_z / E_i$, where E_i denotes Young's modulus of layer i. The sample will thus deform according to

$$\varepsilon_z = -\frac{1}{L}\sum_i \Delta L_i = \frac{1}{L}\sum_i \varepsilon_{z,i} L_i = \frac{1}{L}\sum_i \frac{\sigma_z}{E_i} L_i = \sigma_z \frac{1}{L}\sum_i \frac{L_i}{E_i} \equiv \sigma_z \left\langle \frac{1}{E} \right\rangle \tag{6.1}$$

We use $\langle x \rangle$ to denote the average of the quantity x weighted by the volumetric proportion of each layer (often called *Backus average*). The effective Young's modulus E^* of the layered material is according to Eqs. (1.94) and (6.1) given as

$$E^* = \left\langle \frac{1}{E} \right\rangle^{-1} \tag{6.2}$$

Note that the layered material has transversely isotropic symmetry, so that the effective Young's modulus will vary with orientation of the applied stress. The total effective stiffness matrix of this layered material is given as (Backus, 1962; Helbig, 1994)

$$C_{11} = C_{22} = \left\langle \frac{4G(\lambda + G)}{\lambda + 2G} \right\rangle + \left\langle \frac{1}{\lambda + 2G} \right\rangle^{-1} \left\langle \frac{\lambda}{\lambda + 2G} \right\rangle^2 \tag{6.3}$$

$$C_{33} = \left\langle \frac{1}{\lambda + 2G} \right\rangle^{-1} \tag{6.4}$$

$$C_{12} = \left\langle \frac{2G\lambda}{\lambda + 2G} \right\rangle + \left\langle \frac{1}{\lambda + 2G} \right\rangle^{-1} \left\langle \frac{\lambda}{\lambda + 2G} \right\rangle^2 \tag{6.5}$$

$$C_{13} = C_{23} = \left\langle \frac{1}{\lambda + 2G} \right\rangle^{-1} \left\langle \frac{\lambda}{\lambda + 2G} \right\rangle \tag{6.6}$$

$$C_{44} = C_{55} = \left\langle \frac{1}{G} \right\rangle^{-1} \tag{6.7}$$

$$C_{66} = \langle G \rangle = \frac{1}{2}(C_{11} - C_{12}) \tag{6.8}$$

For an elastic wave with sufficiently low frequency, such that the wavelength λ_w is much larger than the thickness of any of the layers, the layered material will effectively appear as a homogeneous and anisotropic material with stiffness given by Eqs. (6.3)–(6.8). The velocity of a P-wave travelling normal to the layers is then, according to Eq. (6.4) and Section 5.5, given as

$$v_{\mathrm{p}}(\lambda_{\mathrm{w}} \gg L_i) = \sqrt{\frac{C_{33}}{\rho^*}} = \sqrt{\frac{1}{\langle\rho\rangle}\left\langle\frac{1}{\lambda+2G}\right\rangle^{-1}} \tag{6.9}$$

where ρ^* is the average density, $\rho^* = \langle\rho\rangle$.

For an elastic wave with high frequency, such that the wavelength is much shorter than the thickness of the layers, the material will appear as a sequence of layers with different velocities. The time Δt needed for a P-wave to travel through the sample (normal to the layers) is given as the sum of the time needed to travel through each of the layers, that is,

$$\Delta t = \frac{L}{v_{\mathrm{p}}(\lambda_{\mathrm{w}} \ll L_i)} = \sum_i \frac{L_i}{v_{\mathrm{p}}} = \sum_i L_i \sqrt{\frac{\rho_i}{\lambda_i + 2G_i}} \tag{6.10}$$

where v_{p} is the velocity of layer i. The average velocity is given as (see Eq. (5.26))

$$v_{\mathrm{p}}(\lambda_{\mathrm{w}} \ll L_i) = \left\langle\sqrt{\frac{\rho}{\lambda+2G}}\right\rangle^{-1} \tag{6.11}$$

The difference between Eqs. (6.11) and (6.9) represents the different properties displayed by a heterogeneous material when it is studied on two different length scales. Eq. (6.11) represents a situation where the relevant length scale of the observation (i.e. the wavelength of the elastic wave) is much smaller than the length scale of the heterogeneities (i.e. the layer thickness), and hence the material appears as heterogeneous. Eq. (6.9) represents on the other hand a situation where the length scale of the observation is much larger than the length scale of the heterogeneities, and the material appears effectively as homogeneous.

The transition between these two cases was nicely demonstrated in a laboratory experiment by Marion and Coudin (1992), where acoustic waves with different wavelengths were transmitted through a stack of plates of alternating steel and plexiglas. A simulation of a similar experiment, using a discrete element code, is shown in Fig. 6.1. Note that the velocity of the wave is significantly lower when the thickness of the layers is much smaller than the wavelength, as compared to the situation where the layer thickness is larger than the wavelength. Note also the strong attenuation for the intermediate cases, indicating that the wave is almost unable to propagate under such conditions. Hovem (1995) showed by analytical modelling that wave propagation is dispersive with no loss for frequencies below a

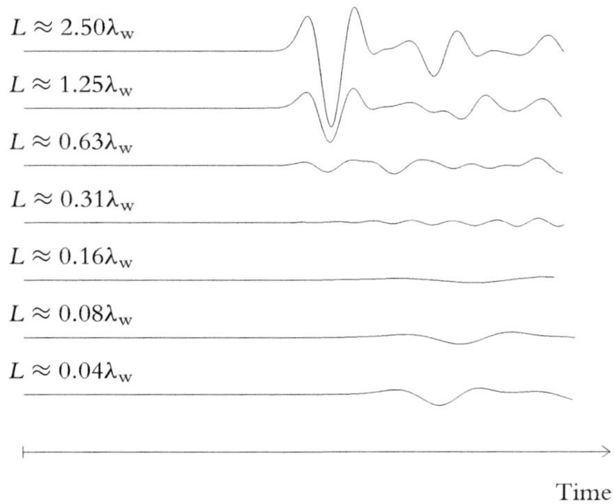

Figure 6.1 Recorded signals of P-waves transmitted through seven different samples, as simulated by a discrete element code. The excitation is a one-cycle sine pulse. Each sample is a stack of hard and soft layers of equal thickness. Only the layer thickness L is different for each sample. The hard-layer:soft-layer ratios for stiffness and density are 20:1 and 8:1, respectively.

critical limit, whereas waves with higher frequencies are evanescent, suffering loss at each interface. The critical frequency limit depends on the periodicity of the layered material and the impedance contrast between the layers.

This example demonstrates clearly that the properties of a heterogeneous material may be very different when observed on different length scales. An effective medium representing the material has the properties of the material when it is observed on a length scale much larger than that of the heterogeneities.

6.2 Models involving porosity only

For a simple porous rock consisting only of one type of solid material and saturated with only one type of fluid, the relative amount of the components is specified by the porosity ϕ, whereas the elastic properties of the components are given by the bulk modulus of the fluid K_f, and the bulk (K_s) and shear (G_s) moduli of the grain material. Many available expressions for the elastic properties of rocks are based only on these parameters, as details about the geometrical distribution of each component are normally not known.

Precise estimates for the elastic moduli of a rock cannot be obtained without the vital information about the geometrical distribution of each component; however, we may identify some upper and lower limits that the moduli are unlikely to exceed. Such limits—or *bounds* as they are usually called—can be established by considering extreme geometrical distributions of the constituents.

Assuming a distribution such that both the grain material and the fluid experiences the same strain due to an external hydrostatic stress, the undrained bulk modulus K_V of the rock is given as

$$K_V = \phi K_f + (1 - \phi)K_s \tag{6.12}$$

which is an upper bound for the undrained bulk modulus of a real rock and is called the Voigt bound. We may also refer to this as an isostrain average. Similarly, if we assume that the stress is the same in both the grains and the fluid, then we find that the undrained bulk modulus K_R of the rock is given by (see Eq. (1.137))

$$\frac{1}{K_R} = \frac{\phi}{K_f} + \frac{1 - \phi}{K_s} \tag{6.13}$$

It is a lower bound for the undrained bulk modulus and is called the Reuss (isostress) bound. Similar bounds can be found for the shear modulus simply by replacing the bulk moduli in Eqs. (6.12) and (6.13) by the corresponding shear moduli. Note that the Reuss bound for the shear modulus $G_R = 0$ for any $\phi > 0$ (since $G_f = 0$), whereas the Voigt bound $G_V > 0$ for any $\phi < 1$.

Note that the frame modulus $K_{fr} = 0$ in the Reuss limit and $K_{fr} = K_s(1 - \phi)$ in the Voigt limit.

Physically, the Reuss bound represents a suspension, for example, solid particles in a liquid. It may also describe gas bubbles in a liquid (in which case the bulk modulus K_g of the gas has to replace K_s in Eq. (6.13)). The Voigt bound represents an alloy of constituents and is most appropriate for a rock at low porosity.

The Voigt and Reuss bounds are very wide and of limited value for modelling rock properties. A narrower range is given by the Hashin–Shtrikman bounds. In a general form, these are given as (Hashin and Shtrikman, 1963)

$$K^{HS\pm} = K_1 + \frac{v_2}{\frac{1}{K_2 - K_1} + \frac{v_1}{K_1 + \frac{4}{3}G_1}} \tag{6.14}$$

$$G^{\mathrm{HS}\pm} = G_1 + \frac{v_2}{\frac{1}{G_2 - G_1} + \frac{2v_1(K_1 + 2G_1)}{5G_1(K_1 + \frac{4}{3}G_1)}} \tag{6.15}$$

The subscripts $_1$ and $_2$ denote components 1 and 2, respectively, whereas v represents the volume fraction of the actual component. The upper ($K^{\mathrm{HS}+}$, $G^{\mathrm{HS}+}$) and lower ($K^{\mathrm{HS}-}$, $G^{\mathrm{HS}-}$) bounds are found by interchanging materials 1 and 2 (the upper bound is found when material 1 is the stiffest one). Physically, the Hashin–Shtrikman bounds represent the moduli of an assembly of spheres of material 2, each of which is surrounded by a shell of material 1. Note that for the simple grain-fluid model of a rock described above, the lower Hashin–Shtrikman bound is identical to the Reuss bound for both the bulk and the shear moduli. This is also the exact solution for a suspension of grains in a fluid (see Section 1.6). Note also that the upper and lower bounds are identical if $G_1 = G_2$, that is, the two extreme geometries give the same result, which shows that the geometrical distribution of the components is irrelevant for this particular case (Hill, 1963).

The arithmetic average of the Voigt and Reuss bounds is sometimes used to obtain estimates of the moduli, rather than just the allowable range. These averages, $K = (K_{\mathrm{V}} + K_{\mathrm{R}})/2$ and similar for the shear modulus, are usually called the Voigt–Reuss–Hill average moduli (Hill, 1952). The Voigt–Reuss–Hill average moduli do not represent a well-defined model of the rock microstructure. They may rather be thought of as describing a specific mixture of the extreme structures represented by the Voigt and Reuss bounds.

An alternative approach would be to look for the most probable microstructure in a porous rock and try to estimate the moduli for such a structure. Considering, for instance, a collection of sand grains, it is clear that there exists a maximum porosity, above which the grains will no longer be in contact with each other and the frame moduli will vanish. For an ordered simple cubic packing of equally sized spheres, this critical porosity ϕ_c is 0.476, for a random packing, it is about 0.36, whereas for a typical clean sand, it is about 0.40. Based on this simple argument, combined with the expectation that the moduli should be equal to the moduli of the grain material if the porosity is zero, we find as the simplest possible expression

$$K_{\mathrm{fr}} = K_{\mathrm{s}}\left(1 - \frac{\phi}{\phi_c}\right) \tag{6.16}$$

$$G_{\mathrm{fr}} = G_{\mathrm{s}}\left(1 - \frac{\phi}{\phi_c}\right) \tag{6.17}$$

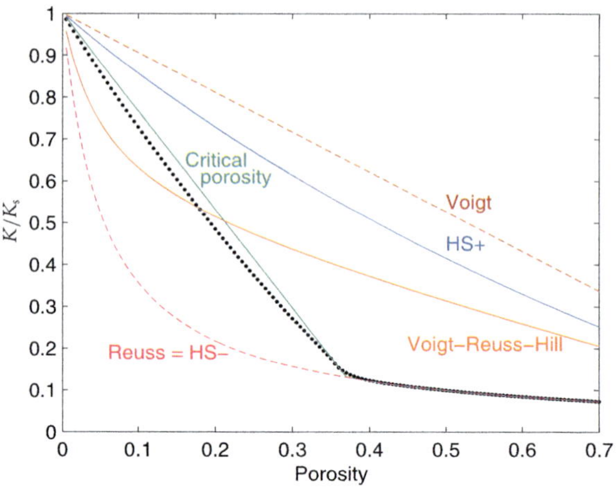

Figure 6.2 Undrained bulk modulus versus porosity for water saturated rocks according to various models. The dotted line is based on the empirical curve of Murphy et al. (1993) for clean sandstones. The fluid contribution has been accounted for by use of the Biot–Gassmann equation (1.171) for both the empirical curve and the critical porosity curve.

The critical porosity concept (Nur et al., 1991, 1995, 1998) is not really a micromechanical model, and the critical porosity is an empirical constant, which differs from one rock type to another; however, the concept provides an explanation for the characteristic behaviour of the frame moduli of sedimentary rocks. Fig. 6.2 shows the undrained bulk modulus versus porosity for a rock, as given by Eq. (6.16) with $\phi_c = 0.39$. The figure also shows the Voigt, Reuss and Hashin–Shtrikman bounds for a dry rock and the Voigt–Reuss–Hill average. For comparison, the empirical results of Murphy et al. (1993) for clean sands are also included.

The critical porosity concept also provides some understanding of the properties of sand–clay mixtures (Marion and Nur, 1991). Small amounts of clay in a sandstone usually appear as pore fill, and the mechanical properties of the rock are mainly determined by the framework of sand grains. The stiffness of the rock is however still largely determined by the sand grain framework, as the sand grains are stiffer than the clay. On the other hand, clay mixed with only a small amount of sand will behave as a kind of suspension, where the mechanical properties will be dominated by those of the clay. The transition between the two types of behaviour will occur when the relative volume of sand is about $1 - \phi_c$, that is, when the sand volume is just large enough for the grains to make a continuous framework.

6.3 Grain pack models

Uncemented sands (and clays) are primarily found at shallow depths. Overpressure can contribute to delay the cementation process (see Section 3.3.3) also at greater depths. Such sands have high porosity and represent possible hydrocarbon reservoirs. Obviously, soil mechanics models are applicable for these sediments, whereas granular media models are ideal for addressing their elastic behaviour from a microscopic perspective. Sedimentary rocks are compacted and to some extent cemented. Models describing the elastic properties of grain packs may also have applications for such materials.

Consider two spherical grains that are pressed together with a normal force F acting along the line through the centres of the spheres (Fig. 6.3). Due to the force, the grains will be deformed in the vicinity of the contact spot. The contact area will be a circle with radius b, and the displacement between the centres of the grains is denoted s.

We will first follow argumentation by de Gennes (1996), who demonstrated in an elegant manner that the grain contact exhibits nonlinear elasticity. Consider first contact stress between two grains, which is proportional to the contact strain b/a and an apparent stiffness parameter of the grain material (for now denoted as M_s). From geometrical considerations it follows that b/a is proportional to s/b, so that

$$\sigma_{\text{contact}} \propto \frac{F}{b^2} \propto M_s \frac{b}{a} \propto M_s \frac{s}{b} \tag{6.18}$$

$$F \propto M_s s b \propto M_s s^{\frac{3}{2}} a^{\frac{1}{2}} \propto M_s a^2 \left(\frac{s}{a}\right)^{\frac{3}{2}} \tag{6.19}$$

The macroscopic stress of a column of spheres is the force divided by the grain area, and the macroscopic strain is given by s/a. Hence the macroscopic stress and strain are related as

$$\sigma \propto \frac{F}{a^2} \propto M_s \varepsilon^{\frac{3}{2}} \tag{6.20}$$

$$\frac{\mathrm{d}\sigma}{\mathrm{d}\varepsilon} \propto \sigma^{\frac{1}{3}} \tag{6.21}$$

We see that the macroscopic stress is proportional to macroscopic strain raised to the power 3/2, from which the incremental stiffness is found to increase with stress as $\sigma^{1/3}$. Thus we have a nonlinear elastic system, where the nonlinearity is rooted in the geometry. The proportionality constant M_s is simply the plane strain modulus $E_s' = E_s/(1 - \nu_s^2)$. Here we use Young's

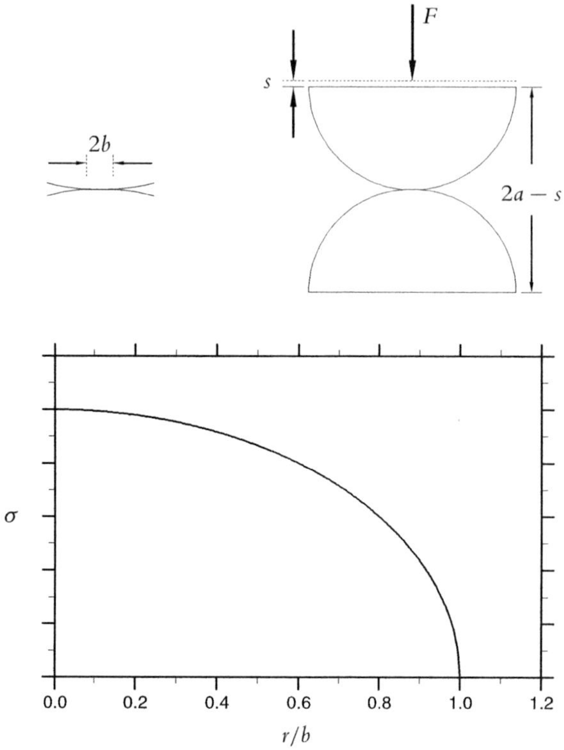

Figure 6.3 Normal stress distribution at the grain contact in a normally loaded contact between two equally sized, spherical grains. (After White, 1983).

modulus and Poisson's ratio (the resulting equations are often expressed by different combinations of elastic moduli from Table 1.1):

$$b = \left[\frac{3(1 - \nu_s^2)aF}{4E_s} \right]^{1/3} \tag{6.22}$$

This equation was originally derived by Hertz (1882) assuming linearly elastic grain material and that the contact radius is small ($b \ll a$); ν_s and E_s are Poisson's ratio and Young's modulus of the grain material, respectively. The distance between the centres of the grains is reduced with s due to the external force:

$$s = \left[\frac{9(1 - \nu_s^2)^2 F^2}{2E_s^2 a} \right]^{1/3} \tag{6.23}$$

The normal stress σ within the grain contact area is given as

$$\sigma = \frac{3F}{2\pi b^2}\left(1 - \frac{r^2}{b^2}\right)^{1/2} \tag{6.24}$$

where r is the lateral distance from the centre line between the two spheres. The force constant D_n relating the longitudinal deformation Δs to an increment ΔF can be derived from Eqs. (6.22) and (6.23) using the relationship between E_s and the shear modulus G_s (Table 1.1):

$$D_n = \frac{\Delta F}{\Delta s} = \left[\frac{3E_s^2 aF}{4(1-\nu_s^2)^2}\right]^{1/3} = \frac{2G_s b}{1-\nu_s} \tag{6.25}$$

The response to an incremental shear force $\Delta F'$ over a grain contact that is already under the same external load F as above can be found by a similar consideration of the forces in the grain contact area (see Fig. 6.4). A shear force constant

$$D_t = \frac{\Delta F'}{\Delta s'} = \frac{\left[6E_s^2 aF(1-\nu_s^2)^2\right]^{1/3}}{(2-\nu_s)(1+\nu_s)} = \frac{4G_s b}{2-\nu_s} \tag{6.26}$$

relates the shear force between the grains to the lateral displacement ($\Delta F'$ and $\Delta s'$ defined in Fig. 6.4). This theory was first developed by Mindlin (1949), so the grain pack theory as it stands is usually called the Hertz–Mindlin theory.

The relationship between the force F and the external pressure σ_p causing the force, as well as the relationship between the displacement s and macroscopic strain ε_{vol}, depend on the packing of the grains.

For a random packing of identical spheres, $\varepsilon_{vol} = 3s/(2a)$, whereas

$$F = \frac{4\pi a^2 \sigma_p}{N_c(1-\phi)} \tag{6.27}$$

where N_c is the average number of contacts per sphere, also called the coordination number. Note that the porosity is given essentially by the packing and is not a free variable in these equations. The bulk modulus for the random packing of identical spherical grains under isotropic external stress is thus given as

$$K = \frac{\partial \sigma_p}{\partial \varepsilon_{vol}} = \frac{N_c(1-\phi)}{6\pi a} D_n = \left[\frac{N_c^2(1-\phi)^2 E_s^2 \sigma_p}{72\pi^2(1-\nu_s^2)^2}\right]^{1/3} \tag{6.28}$$

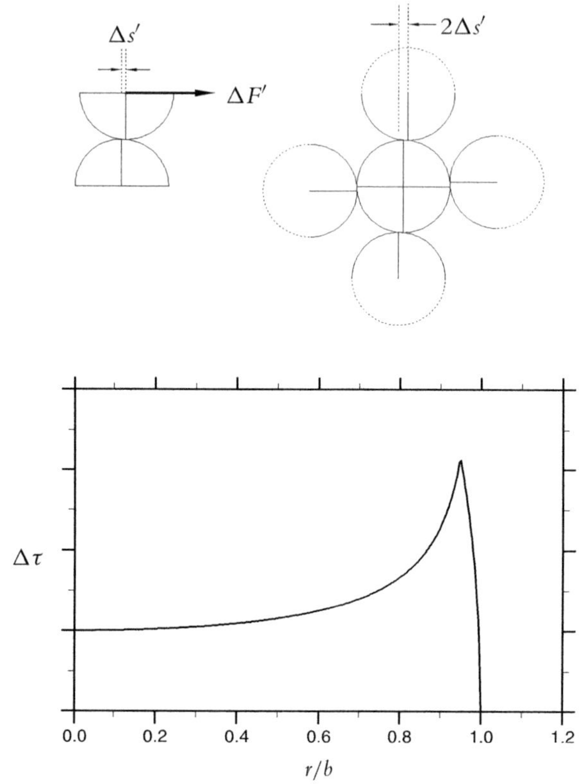

Figure 6.4 Tangential stress distribution at the grain contact in a contact between two equally sized, spherical grains, which is initially hydrostatically loaded, followed by an incremental shear load. (After White, 1983).

The shear modulus G of a random pack of identical spherical grains is given as (Digby, 1981)

$$G = \frac{N_c(1-\phi)}{10\pi a}\left(D_n + \frac{3}{2}D_t\right) = \frac{5-4\nu_s}{10(2-\nu_s)}\left[\frac{3N_c^2(1-\phi)^2E_s^2\sigma_p}{\pi^2(1-\nu_s^2)^2}\right]^{1/3} \tag{6.29}$$

These equations are valid when there is no slip between the spheres. Walton (1987) also considered possible slip at the contact surfaces, which implies that the friction coefficient at the contact area has an impact on the elastic moduli. For perfectly smooth spheres (the friction coefficient is zero; perfect slip), still under isotropic stress, Walton found

$$K = \frac{1}{6}\left[\frac{3N_c^2(1-\phi)^2\sigma_p}{\pi^4 B^2}\right]^{1/3} \tag{6.30}$$

$$G = \frac{1}{10}\left[\frac{3N_c^2(1-\phi)^2\sigma_p}{\pi^4 B^2}\right]^{1/3} = \frac{3}{5}K \tag{6.31}$$

where

$$B = \frac{1-\nu_s}{2\pi G_s} \tag{6.32}$$

To calculate stiffnesses from these equations, the main input parameters are the elastic stiffnesses of the grain material, the applied (isotropic) stress and the characteristics of the grain packing expressed by the porosity and the coordination number. The underlying assumptions in Walton's work are that the porosity and coordination number are constant, that is, no new contacts are created, and no contacts are lost as a result of stress changes. Further, all contacts are assumed to carry the same load. The minimum theoretical coordination number in a random packing of equally sized spheres is 6 (Makse et al., 2004), and the fixed porosity is 36–37.5 % (not well defined; see Torquato et al. 2000).

Eqs. (6.28) and (6.29) can be used to estimate how acoustic velocities depend on an external hydrostatic stress σ_p for a dry granular medium. Since both K and G depend on $\sigma_p^{1/3}$, we immediately see that both v_p and v_s depend on the stress as $\sigma_p^{1/6}$. Since there is no change in the number of grain contacts, there is no grain rearrangement. Thus porosity (hence also density) is assumed constant.

Using the porosity and coordination number as given above along with tabulated elastic moduli of quartz the no-slip limit tends to overestimate the velocities of unconsolidated natural sand, in particular, at low stress. There may be several reasons for this, for instance, that the sand may be close to the slip limit, that the effective (load bearing) coordination number may be lower than the theoretical value and that sand grains because of their shape and possible internal flaws are more compliant than solid quartz.

Experiments on loose sands (Domenico, 1977; Bachrach et al., 1998; Han and Batzle, 2006; Holt et al., 2007) have shown v_p and v_s depending on the stress as σ_p^q, where q is in the range from 1/4 to 1/5, usually with slightly higher q for S- than for P-waves. Furthermore, the classical Hertz–Mindlin model also underestimates the ratio between v_p and v_s and predicts that it is stress independent.

There are several possible reasons for these discrepancies. The analytical grain pack models assume that the load is carried equally by all grain contacts, whereas in reality it is carried by localised force chains throughout the grain assembly. Further, the Hertz–Mindlin approach does not account properly for shear relaxation and associated partial slip leading to grain rotation and rearrangement, as was shown by Makse et al. (2004) using a discrete particle numerical model. This explains the discrepancy in the ratio v_p over v_s as pointed to above. The stress dependence can be mimicked by introducing an effective stress-dependent coordination number and may also be captured by considering a binary mixing of slip and no-slip contacts (Bachrach and Avseth, 2008), where the fraction of no-slip contacts increases with stress (Duffaut et al., 2010; Torset and Holt, 2021).

Grains in natural sediments are not perfect solid spheres. It was shown by de Gennes (1996) that a thin soft outer shell (e.g. an oxide layer) on spherical grains could modify the exponent q from 1/6 to 1/4. When adapting the model to realistic grain shapes, the surface roughness of the grains may also lead to grain contact stiffnesses described by an exponent different from 1/6 (as would follow from Eqs. (6.25) and (6.26)). An example of such behaviour is the "bed of nails" model by Carlson and Gangi (1985), which is described in Section 6.6.1. Crushing of asperities during loading may also alter the exponent, and obviously the Hertz–Mindlin theory is no longer valid at high stress levels, when plastification of the grain contacts may occur.

The role of intergranular friction was addressed in detail by Mindlin and Deresiewicz (1953), who analysed the hysteretic behaviour of a single contact between two spheres under different loading paths. Their work demonstrates the impact of the loading history on each individual contact. Such behaviour is obviously difficult to incorporate in any analytical (and even numerical) model for a granular medium. Stoll (1989) built on this and derived expressions for loading and unloading of a regular packing of spheres in uniaxial strain conditions. Nihei et al. (2000) argued that the strains associated with wave propagation are too small to trigger frictional sliding, which enabled them to distinguish between static and dynamic moduli (Section 5.6.1). It is interesting to note that static and dynamic moduli in unconsolidated samples of sand or glass beads are close to equal during isotropic compression, whereas they become very different when shear stresses trigger frictional sliding for anisotropic stress paths (Holt et al., 2013).

The studies above demonstrate that wave velocities experience stress-induced anisotropy. In his original paper, Walton (1987) considered uniaxial strain in addition to isotropic loading. However, since grain rotations are not accounted for, the radial stress required to keep zero radial strain is strongly underestimated, leading to overestimated P- and S-wave anisotropies. Torset and Holt (2021), based on previous work by Bandyopadhyay (2009), generalised Walton's theory to arbitrary transversely isotropic stress paths, providing the slip and no-slip bounds as well as pointing to a feasible way of interpolating between them through a stress-dependent binary mixing factor. Using stresses as boundary conditions enables quantitative fit to laboratory experiments, which could not be done with the original formulation.

The Hertz–Mindlin contact theory has also been derived for various ordered packings of grains (see e.g. Wang and Nur, 1992, for a review). The assumption of equally sized grains may be relaxed, firstly by adjusting the coordination number (see Mavko et al., 1998 for details), and second, by permitting variable porosity. Dvorkin and Nur (1996) combined the no-slip limit (Eqs. (6.28) and (6.29)) with the Hashin–Shtrikman lower bound to model stress sensitivity of unconsolidated sand from a North Sea reservoir. Fluid saturation may be incorporated through the isotropic or anisotropic Biot–Gassmann models. Cemented rather than pressure-generated grain contacts can be incorporated (Digby, 1981; Schwartz, 1984; Dvorkin et al., 1991; Elata and Dvorkin, 1996). The resulting elastic properties depend on the mechanical stiffnesses of the grain material and of the cement, and on the spatial distribution of the cement (at grain contacts or as a grain-coating material). For cemented contacts, the force constants D_n and D_t are no longer given by the Hertz–Mindlin theory, and the stress dependency of acoustic waves tends to be much less pronounced than for uncemented granular media.

The general case of variously sized cemented particles in a random dense pack has not been solved analytically yet. However, such systems can be studied numerically. Discrete particle modelling has developed rapidly over the last few years, with the continuous development of faster and more powerful computers. One such approach is the PFC (Particle Flow Code[1]), which is based on the pioneering work of Cundall and Strack (1979).

In PFC, which may be formulated both in 2D and 3D, circular or spherical particles of arbitrary size distribution are packed into a granular assembly under a specific applied stress. Each disk or sphere is treated as a rigid body, with translational and rotational degrees of freedom. When two particles come in contact, forces are generated as a result of relative (normal and shear) displacements at the contact. Using a so-called soft contact

[1] Trademark of Itasca Consulting Group, Minneapolis, USA.

approach, the particles are permitted numerically to overlap each other, and the degree of overlap controls the contact forces through a contact law. The contact law may be Hertzian, linear or user-defined. By applying boundaries representative for the problem to be studied (e.g. flat walls to mimic pistons in case a laboratory test is simulated) the stress conditions can be changed and the contact forces and associated particle movements calculated, until equilibrium is reached and a new load increment can be applied. This procedure is cycled over and over again until a numerical experiment is completed.

The particle contacts may be unbonded or bonded. When simulated cement bonds are inserted, they have bond stiffnesses that are in general different from those of the uncemented contacts. In addition, the bonds have shear and tensile strength. When a shear or tensile failure criterion is exceeded locally, the bond fails, and it then retains the properties of the uncemented contact. With the features described above, the discrete model allows calculation of the complete stress–strain behaviour of an uncemented or cemented granular medium (Potyondy and Cundall, 2004). It thus serves as a tool for studying rock failure mechanics on the particle scale.

Several features may be added to allow direct comparison between laboratory measurements and discrete element computations, such as fluid coupling to simulate stress effects on permeability. For instance, wave propagation may be simulated directly in PFC, since the model is inherently dynamic (Li and Holt, 2002). Also, acoustic emissions may be simulated by calculating the energy associated with bond ruptures (Hazzard and Young, 2000).

Discrete particle modelling may be used to gain understanding and insight, in particular, in comparison with laboratory experiments. It may also be used to solve engineering problems, in competition with continuum methods. The main benefit (Cundall, 2001) is that no pre-assumed constitutive material law is needed: complex behaviour, including development of localised fractures and cracks is a natural output of the modelling.

From a fundamental viewpoint, we may ask to what extent the assumptions behind the simple effective medium theories sketched above are valid. As an illustration of that, a distribution of contact forces in a 2D network is shown in Fig. 6.5. The distribution is clearly heterogeneous, with a small number of contacts carrying a large part of the load (Radjai et al., 1996; Williams and Rege, 1997). This allows, amongst others, studies of various stress geometries, stress history and degrees of cementation. Note that the network of force lines is heterogeneous on a larger scale than the grain size.

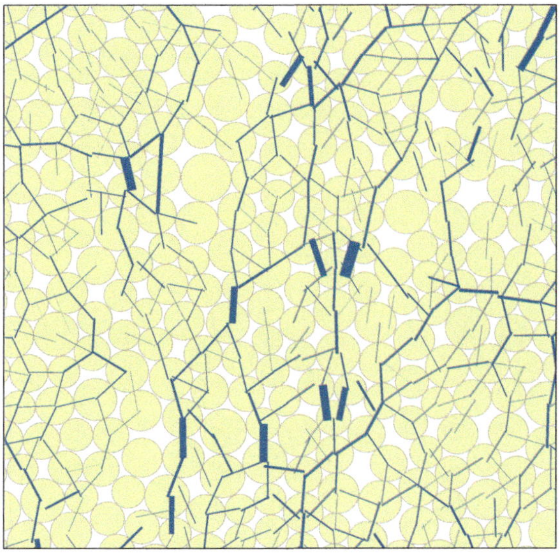

Figure 6.5 Distribution of contact forces in a 2D specimen created by PFC2D, loaded in the vertical direction. The width of the lines is proportional to the force magnitude, and the direction indicates the orientation. The figure shows only a section of the specimen.

The numerical calculations also allow for "snap-shots" of grain contact failures at given points in time, as shown in Fig. 6.6. Such figures may help to better understand deformation mechanisms and failure processes in sedimentary rocks.

6.4 Models for cracks and other inclusions

The grain pack models discussed above describe the porous material as a collection of grains with varying degree of contact between the grains. This may seem to be a logical path to follow when studying sedimentary rocks, since they are indeed formed this way. However, we may also consider a porous material as a solid with lots of holes in it, like a "Swiss cheese". For some purposes, it appears that this approach may be equally relevant, especially for well-cemented rocks. More generally, we talk about such models as inclusion models, as they consider the effect of including an object in a host material.

To establish a model for such a material, we need descriptions for how holes of various shapes affect the elastic properties of the solid. In a static

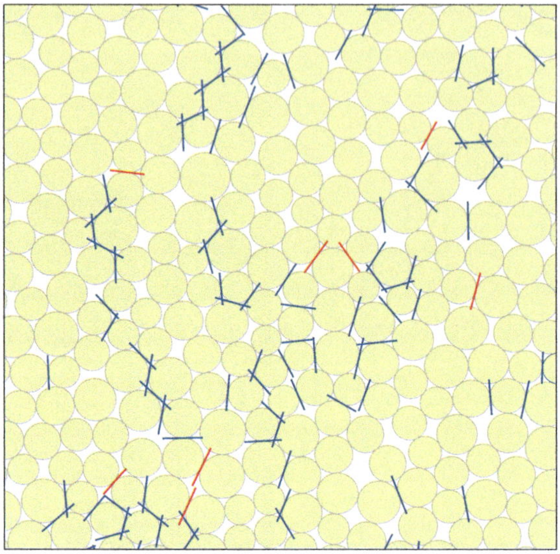

Figure 6.6 Grain contact failures at a given stress level in the 2D specimen shown in Fig. 6.5. The blue lines show failures that are predominantly tensile, whereas the red lines show shear failures.

approach (Eshelby, 1957; Walsh, 1965a,b), it is assumed that a static external load is imposed on a sample consisting of a host material with embedded inclusions. The elastic properties of the effective medium are obtained by requiring that the effective medium will suffer the same deformation as the real sample when subjected to the same external load. In a dynamic approach (Kuster and Toksöz, 1974), it is assumed that a region consisting of inclusions embedded in a host material is embedded in the effective medium. The elastic properties of the effective medium are defined such that the scattering of elastic waves from the embedded region is eliminated. The wavelength of the elastic wave is assumed to be much larger than the size of the region.

The two approaches converge when the concentration of inclusions is low. In Sections 6.4.1 and 6.4.2, we will take a look at models valid for low concentrations of inclusions. Models intended to account for higher concentrations of inclusions are discussed in Section 6.4.3.

6.4.1 Linear, isotropic models

Assuming that the concentration of inclusions is so low that interactions can be neglected, we may express the impact of the inclusion on the elastic stiffness of the material as a perturbation proportional to the concentration of the inclusions. Assuming also that the host material is isotropic, the effective bulk modulus K^* and the effective shear modulus G^* of a material containing inclusions can be written as

$$K^* = K_s(1 - Q_K \xi_{\text{incl}}) \tag{6.33}$$

$$G^* = G_s(1 - Q_G \xi_{\text{incl}}) \tag{6.34}$$

where K_s and G_s are the bulk modulus and shear modulus, respectively, of the solid material without inclusions, ξ_{incl} is an appropriate measure of the concentration of the inclusions, depending on their shape (as will be seen below), and Q_K and Q_G are measures of the impact these inclusions have on the respective moduli of the material. Q_K and Q_G are functions of the moduli of the host material and the material within the inclusions. Other moduli can be represented by similar expressions.

Consider first the case where the inclusions are spherical fluid-filled holes. The porosity ϕ, which is here given as the sum of the volumes of the holes relative to the total volume, is a suitable measure for the concentration of inclusions, and thus $\xi_{\text{incl}} = \phi$. The expressions for the effective moduli can, for instance, be derived from Eqs. (6.14) and (6.15) by defining $K^{\text{HS+}} = K^*$, $G^{\text{HS+}} = G^*$, $v_2 = 1 - v_1 = \phi$, $K_1 = K_s$, $K_2 = K_f$, $G_1 = G_s$, $G_2 = 0$ (the upper Hashin–Shtrikman bound then represents a solid surrounding a spherical fluid-filled inclusion) and linearising the expressions in ϕ. The results are (see also Mackenzie, 1950; Eshelby, 1957; Walsh, 1965a)

$$K^* = K_s \left(1 - \frac{3(1-\nu_s)(1-\frac{K_f}{K_s})}{2(1-2\nu_s) + (1+\nu_s)\frac{K_f}{K_s}} \phi \right) \tag{6.35}$$

$$G^* = G_s \left(1 - 15 \frac{1-\nu_s}{7-5\nu_s} \phi \right) \tag{6.36}$$

where ν_s is Poisson's ratio of the solid material, without inclusions. By comparing Eq. (6.35) to Eq. (1.171) we find (after some algebra) that K^* is in agreement with Biot's theory (Section 1.6.2). Note also that G^* does not depend on K_f, and thus G^* is the same whether the material is dry or saturated, again in agreement with Biot's theory. This is not the case in general for inclusion models, as we will see below.

We will now consider the case where the inclusions are thin flat cracks. For such inclusions, it turns out that the most suitable measure for the concentration of inclusions is the *crack density* ξ defined as

$$\xi = \frac{2n}{\pi}\left\langle\frac{A^2}{P}\right\rangle \tag{6.37}$$

where A is the area, P the perimeter of a crack, and n is the number of cracks per unit volume. The brackets $\langle\rangle$ denote the average over all cracks. For circular cracks with radius a, Eq. (6.37) becomes

$$\xi = n\langle a^3\rangle \tag{6.38}$$

Assuming that the cracks are randomly oriented, we have (Walsh, 1965a,b; Garbin and Knopoff, 1973, 1975; Budiansky and O'Connell, 1976; see also Watt et al., 1976)

$$K^* = K_s\left(1 - \frac{16}{9}\frac{1-\nu_s^2}{1-2\nu_s}D_c\xi\right) \tag{6.39}$$

$$G^* = G_s\left(1 - \frac{32}{45}(1-\nu_s)\left[D_c + \frac{3}{2-\nu_s}\right]\xi\right) \tag{6.40}$$

The drainage parameter D_c is given by

$$\frac{1}{D_c} = 1 + \frac{4}{3\pi\gamma}\frac{1-\nu_s^2}{1-2\nu_s}\frac{K_f}{K_s} \tag{6.41}$$

where γ is the aspect (thickness-to-diameter) ratio of the cracks. Note that for a dry material, for which $K_f = 0$ and consequently $D_c = 1$, the only crack-related parameter affecting the effective moduli is the crack density ξ. For very thin fluid-filled cracks, for which $\gamma \ll K_f/K_s$, we have that $D_c \to 0$, which implies that the cracks are almost incompliant. Again, only the crack density affects the effective moduli, and details regarding the shape of thin cracks have no impact on the effective moduli (Budiansky and O'Connell, 1976; Mavko and Nur, 1978).

Eq. (6.40) states that G^* is affected by saturation, and hence it is clear that these equations are not consistent with Biot's theory (Section 1.6.2). The reason for this discrepancy is that some of the cracks will necessarily have an orientation such that they will be compressed (or opened up) due to an applied shear stress. This change in volume of the crack implies that the fluid occupying the crack will be compressed (or expanded) to an extent that is partly given by K_f. Biot's theory, which has no assumptions on the shape of the pore space and hence should be valid also for thin flat (crack-shaped) pores, is based on the fundamental assumption that the material is permeable. Hence the compression of one crack due to an applied shear stress will only lead to flow of fluid from this crack into another crack whose orientation implies that it expands due to the same stress state. Thus there is no local build-up (or reduction) of fluid pressure, and the fluid bulk modulus does not affect the effective shear modulus.

It is possible to make an inclusion model consistent with Biot's theory by defining that G^* is not affected by saturation. This is a way of implementing hydraulic connections between the elements of the pore space (i.e. making the material permeable; see Thomsen (1985)). Alternatively, we may use the inclusion model only for the material in a dry state (for which $D_c = 1$), thus modelling the frame moduli, and add the effects of saturation by Biot's theory. The actual situation we are trying to model determines which alternative is the most representative. For instance, for high-frequency elastic waves (see Chapter 5), the fluid may not have time to flow into and out of the cracks, and hence the cracks will effectively be closed, and expressions like Eqs. (6.39)–(6.41) may give a good representation of the moduli. At low frequencies, the fluid pressure will have time to equalise, and a better representation of the moduli is found by using the inclusion model to estimate the frame moduli and the Biot model to add the fluid effects.

6.4.2 Anisotropic models

Cracks in rocks are not always randomly oriented. More typically, cracks grow as a result of anisotropic changes in the stress state, which leads to a preferred crack orientation with respect to the stress state (see also Section 1.8). The elastic constants of a material with parallel cracks can be expressed, similar to Eqs. (6.33) and (6.34), as

$$C^*_{ij} = C^{\text{o}}_{ij}(1 - Q_{ij}\xi) \tag{6.42}$$

where C^{o} is the stiffness matrix of the host material (without cracks), and Q_{ij} is the impact of the cracks on C^*_{ij}. Note that the Einstein summing convention (see Appendix C.12.1) is not applied here. If the cracks are oriented normal to the z-axis, then the coefficients Q_{ij} are given as the respective components of the matrix Q (see e.g. Hudson, 1981):

$$Q = \frac{16}{3}\begin{bmatrix} \frac{\nu_s^2}{1-2\nu_s}D_c & \frac{\nu_s(1-\nu_s)}{1-2\nu_s}D_c & \frac{(1-\nu_s)^2}{1-2\nu_s}D_c & 0 & 0 & 0 \\ \frac{\nu_s(1-\nu_s)}{1-2\nu_s}D_c & \frac{\nu_s^2}{1-2\nu_s}D_c & \frac{(1-\nu_s)^2}{1-2\nu_s}D_c & 0 & 0 & 0 \\ \frac{(1-\nu_s)^2}{1-2\nu_s}D_c & \frac{(1-\nu_s)^2}{1-2\nu_s}D_c & \frac{(1-\nu_s)^2}{1-2\nu_s}D_c & 0 & 0 & 0 \\ 0 & 0 & 0 & \frac{1-\nu_s}{2-\nu_s} & 0 & 0 \\ 0 & 0 & 0 & 0 & \frac{1-\nu_s}{2-\nu_s} & 0 \\ 0 & 0 & 0 & 0 & 0 & 0 \end{bmatrix} \tag{6.43}$$

Eqs. (6.42) and (6.43) show that $C_{33}^* < C_{11}^*$, and hence a rock containing a set of parallel cracks is softer when loaded perpendicular to the cracks than parallel to them. Thus the parallel cracks induce elastic anisotropy in the rock. Stress dependence of acoustic velocities in sedimentary rocks and stress-induced acoustic anisotropy can to a large extent be described in terms of cracks. This was also discussed in Chapter 5. This is a consequence of the fact that shear stresses tend to create cracks with a preferred orientation.

The impact of a crack set with a different orientation can be found by proper rotation of the matrix Q. For a rock that has several crack sets, with densities $\xi^{(1)}, \xi^{(2)}, \ldots$ and corresponding matrices $Q^{(1)}$, $Q^{(2)}, \ldots$, respectively, the effective elastic stiffness matrix is given as

$$C_{ij}^* = C_{ij}^{\text{o}}\left(1 - Q_{ij}^{(1)}\xi^{(1)} - Q_{ij}^{(2)}\xi^{(2)}\ldots\right) \tag{6.44}$$

assuming no mechanical interaction between the cracks.

We recall that $D_c \to 0$ for thin fluid-filled cracks (Section 6.4.1). This can be explained as follows: Even a small normal stress tends to impose a very large volumetric strain on an empty crack. If the crack is filled with a fluid, then the volumetric strain will immediately be counteracted by a large alteration of the fluid pressure. Consequently, the crack is almost non-deformable due to normal stresses.

These considerations imply that, according to Eqs. (6.43) and (5.59), cracks induce only a small P-wave anisotropy in a saturated rock. Thomsen (1995) pointed out that this model does not necessarily give a realistic description of a porous rock, since the pore fluid may flow between the cracks and pores that deform differently under stress. For a low-frequency wave, the fluid pressure in the cracks and pores will be nearly equalised at any time, which implies that the presence of the pore fluid cannot prevent the cracks from deforming, at least not to such an extent as for isolated cracks. Assuming perfect pore pressure equalisation, Thomsen derived a modified expression (D_{cp}) for the drainage parameter that accounts for the hydraulic interaction between flat cracks and spherical pores.

Assuming that $K_f \ll K_s$ so that higher-order terms of K_f/K_s can be neglected, and neglecting mechanical interaction between the inclusions, we find from Thomsen's work that the drainage parameter can be expressed as

$$\frac{1}{D_{cp}} = 1 + \left[\frac{3}{2}\frac{1-\nu_s}{1-2\nu_s}\frac{\phi_p}{\phi} + \frac{4}{3\pi\gamma}\frac{1-\nu_s^2}{1-2\nu_s}\left(1-\frac{\phi_p}{\phi}\right)\right]\frac{K_f}{K_s} \tag{6.45}$$

(Note that Eq. (6.45) is slightly different from Thomsen's definition of D_{cp} to keep expressions (6.45) and (6.41) consistent.) The porosity ϕ_p is due to the pores only. The porosity due to the cracks is given as

$$\phi - \phi_p = \frac{4}{3}\pi\gamma\xi \tag{6.46}$$

Hence $(1-(\phi_p/\phi))/\gamma$ does not depend on γ, and D_{cp} will not vanish regardless how thin the cracks are, provided that $\phi_p > 0$. The realism in Thomsen's model was confirmed in a model experiment by Rathore et al. (1995).

Expression (6.45) is valid for static or low-frequency oscillations of the external stress. At high frequencies the pore fluid will not have time to flow between the cracks and pores, and the cracks will effectively appear to be sealed. For this situation, Eq. (6.41) is the relevant expression for the drainage parameter.

For completeness, the direct impact of the pores should also be taken into account. Within the approximations used here, this can be done by adding an extra term to expression (6.42) for the effective elastic stiffness matrix:

$$C_{ij}^* = C_{ij}^o(1 - Q_{ij}\xi - Q_{ij}^P\phi_p) \tag{6.47}$$

The matrix Q_{ij}^P can be derived from Eqs. (6.35) and (6.36). Assuming that $K_f \ll K_s$, it can be expressed by

$$Q_{11}^P = Q_{22}^P = Q_{33}^P = \frac{1}{2}\frac{1+\nu_s}{1-2\nu_s}D_p + 10\frac{1-2\nu_s}{7-5\nu_s} \tag{6.48}$$

$$Q_{12}^P = Q_{13}^P = Q_{23}^P = \frac{1-\nu_s}{2\nu_s}\left(\frac{1+\nu_s}{1-2\nu_s}D_p - 10\frac{1-2\nu_s}{7-5\nu_s}\right) \tag{6.49}$$

$$Q_{44}^P = Q_{55}^P = Q_{66}^P = 15\frac{1-\nu_s}{7-5\nu_s} \tag{6.50}$$

At low frequencies, $D_p = D_{cp}$ (given by Eq. (6.45)), whereas at high frequencies where the pores and cracks are hydraulically isolated, D_p is given by

$$\frac{1}{D_p} = 1 + \frac{3}{2}\frac{1-\nu_s}{1-2\nu_s}\frac{K_f}{K_s} \tag{6.51}$$

Fluid flow between cracks and pores is a relaxation process driven by the difference in fluid pressure. Its impact on rock stiffness may be described by linear visco-elasticity (O'Connell and Budiansky, 1977). To describe the transition from low to high frequencies, we assign a characteristic relaxation time t^* to the process, such that for a time interval $t \ll t^*$, nearly no flow occurs, whereas for $t \gg t^*$, the process is completed, and pore pressure equalisation is established. Frequency dependence is introduced by making the drainage parameters dependent on the complex term $1/(1+j\omega t^*)$ while ensuring compliance with the limit values at high and low frequencies, where ω is the angular frequency (Eq. (5.6)). The frequency-dependent drainage parameter for cracks can then be expressed as

$$\frac{1}{D_c(\omega)} = 1 + \frac{4}{3\pi\gamma}\frac{1-\nu_s^2}{1-2\nu_s}\left[1 - \frac{1}{1+j\omega t^*}\frac{\phi_p}{\phi}\left(1 - \frac{9\pi\gamma}{8(1+\nu_s)}\right)\right]\frac{K_f}{K_s} \tag{6.52}$$

whereas the frequency-dependent drainage parameter for pores is given as

$$\frac{1}{D_p(\omega)} = 1 + \frac{3}{2}\frac{1-\nu_s}{1-2\nu_s}\left[1 - \frac{1}{1+j\omega t^*}\left(1 - \frac{\phi_p}{\phi}\right)\left(1 - \frac{8(1+\nu_s)}{9\pi\gamma}\right)\right]\frac{K_f}{K_s} \tag{6.53}$$

At high frequencies ($\omega \to \infty$) the real part of Eq. (6.52) approaches Eq. (6.41), whereas the real part of Eq. (6.53) approaches Eq. (6.51). At low frequencies ($\omega \to 0$) the real parts of Eqs. (6.52) and (6.53) are both approaching Eq. (6.45). The real parts enter the expressions for velocities, whereas the imaginary parts, which peak around $\omega = 1/t^*$, contribute to attenuation (cf. Eqs. (5.51) and (5.52)).

The equations above describe a material with spherical pores and a set of flat parallel cracks, where the hydraulic connection between the cracks and pores are accounted for. However, mechanical interactions between the inclusions have not been accounted for. Such mechanical interactions and the question whether it makes sense to include them or not are discussed in Section 6.4.3.

Sayers and Kachanov (1995) introduced a model where cracks are considered more generally as discontinuities, and their impact on the rock properties are described in terms of the displacement discontinuities they induce. In this model the compliance S^*_{ijkl} of a rock containing a set of cracks is expressed as

$$S^*_{ijkl} = S^{\mathrm{o}}_{ijkl} + \frac{1}{4}(\delta_{ik}\alpha_{jl} + \delta_{il}\alpha_{jk} + \delta_{jk}\alpha_{il} + \delta_{jl}\alpha_{ik}) + \beta_{ijkl} \tag{6.54}$$

where S^{o}_{ijkl} is the compliance of the rock without cracks, and the second-rank tensor α_{ij} and the fourth-rank tensor β_{ijkl} are defined as

$$\alpha_{ij} = \frac{1}{V}\sum_r B^r_{\mathrm{T}} n^r_i n^r_j S_r \tag{6.55}$$

$$\beta_{ijkl} = \frac{1}{V}\sum_r (B^r_{\mathrm{N}} - B^r_{\mathrm{T}}) n^r_i n^r_j n^r_k n^r_l S_r \tag{6.56}$$

where B^r_{N} and B^r_{T} are the normal and shear compliances of the rth discontinuity, n_i is the ith component of the normal to the discontinuity, S_r is the area of the discontinuity, and V is the rock volume. For open penny-shaped cracks, $B_{\mathrm{N}}/B_{\mathrm{T}} = 1 - \nu_{\mathrm{s}}/2$, which is close to 1, and hence β_{ijkl} is small and can be ignored, whereas it can be shown that α_{ij} is proportional to the crack density. In general, however, B_{N} and B_{T} may be considered as independent, and the model thus allows for a description of more general discontinuities than open cracks. The fact that this model describes the impact of the discontinuities as additions to the compliance, rather than deductions from the stiffness (as in Eq. (6.42)), is of minor importance as long as the concentration of discontinuities is small.

6.4.3 Models accounting for interactions

Eqs. (6.33) and (6.34) are based on the assumption that the concentration of inclusions is so small that interactions between the inclusions can be neglected. When several inclusions are present, they will interact in the sense that the impact of one inclusion on the material stiffness is affected by the presence of the others. This implies that Q_K and Q_G to some extent depend on ξ_{incl}. Alternatively, we can express this by adding higher-order terms in ξ_{incl} to Eqs. (6.33) and (6.34).

A number of models have been derived, where interactions between the inclusions are accounted for in one way or another to make the model valid also for high densities of inclusions. Several of these models are based on the concept of *self-consistency*. This implies that the inclusions are assumed to be surrounded by the effective medium, rather than the host material. Thus we can make Eqs. (6.33) and (6.34) self-consistent by redefining the coefficients of the impact terms ($K_s Q_K$ and $G_s Q_G$, respectively) by replacing the elastic moduli K_s and G_s of the solid host material with the resulting moduli K^* and G^* of the effective medium.

This approach was used by Budiansky and O'Connell (1976) when they introduced self-consistent expressions for the elastic moduli of a cracked material, except that they started by introducing the impact of the cracks as a perturbation to the compliance rather than the stiffness. That is, their initial equations were of the type

$$\frac{1}{K^*} = \frac{1}{K_s} + \frac{Q_K(\nu_s)}{K_s}\xi \tag{6.57}$$

$$\frac{1}{G^*} = \frac{1}{G_s} + \frac{Q_G(\nu_s)}{G_s}\xi \tag{6.58}$$

rather than Eqs. (6.33) and (6.34). For low concentrations of cracks, this makes no difference since the corresponding expressions converge for small values of ξ. The relevant expressions for Q_K and Q_G can be derived from Eqs. (6.39) and (6.40), respectively. Eqs. (6.57) and (6.58) are made self-consistent by making the impact terms functions of the effective medium parameters, that is,

$$\frac{1}{K^*} \rightarrow \frac{1}{K_s} + \frac{Q_K(\nu^*)}{K^*}\xi \tag{6.59}$$

$$\frac{1}{G^*} \rightarrow \frac{1}{G_s} + \frac{Q_G(\nu^*)}{G^*}\xi \tag{6.60}$$

Making in addition use of the relation between K, G and ν from Table 1.1, we find that for a dry rock ($D = 1$) containing isotropically distributed flat cracks with crack density ξ, the effective moduli K^*, G^* and ν^* may be expressed by the equations

$$K^* = K_s\left(1 - \frac{16}{9}\frac{1-\nu^{*2}}{1-2\nu^*}\xi\right) \tag{6.61}$$

$$G^* = G_s\left(1 - \frac{32}{45}\frac{(1-\nu^*)(5-\nu^*)}{2-\nu^*}\xi\right) \tag{6.62}$$

$$\xi = \frac{45}{16}\frac{(\nu_s-\nu^*)(2-\nu^*)}{(1-\nu^{*2})(10\nu_s-\nu^*-3\nu^*\nu_s)} \tag{6.63}$$

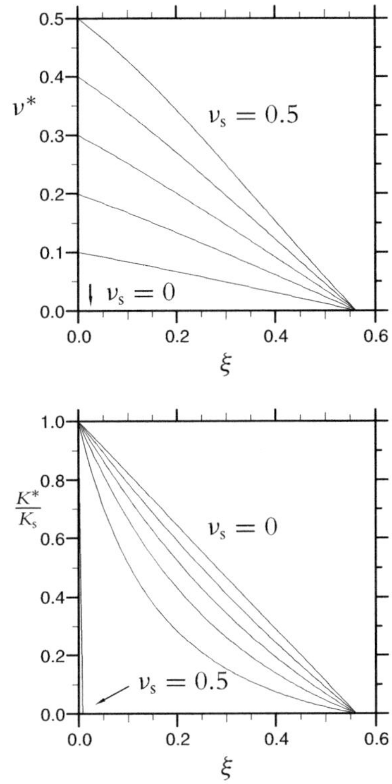

Figure 6.7 Effective elastic moduli for a material with flat dry cracks. (After Budiansky and O'Connell, 1976).

The effective Young's modulus may be derived by combination of these equations using suitable relations from Table 1.1. After linearisation in ξ, to make the expression consistent with the others, we find

$$E^* = E_s \left(1 - \frac{16}{45} \frac{(1 - \nu^{*2})(10 - 3\nu^*)}{2 - \nu^*} \xi \right) \tag{6.64}$$

The effective moduli derived from this model are shown in Fig. 6.7.

Berryman (1980) introduced self-consistency into the dynamic theory of Kuster and Toksöz (1974) and derived a set of implicit equations for the effective moduli. The equations are given in Appendix D.3.2. These solutions approach the linear solutions given in Section 6.4.1 for low concentrations of inclusions. Berryman also showed that if the material contains non-solid inclusions, then the effective shear modulus vanishes at a critical concentration of that inclusion. The critical concentration depends on the shape of the inclusions.

The concept of self-consistency is a simple and yet efficient way to account for interactions between inclusions. However, the approach does not give unique solutions. For instance, Berryman's model and the model of Budiansky and O'Connell, which express the impact of inclusions as a perturbation to the compliance, predict that the shear modulus vanishes when the concentration of non-solid inclusions reaches a critical limit, as shown above. On the other hand, a self-consistent version of Eqs. (6.33) and (6.34) predicts that the shear modulus never vanishes completely.

One attempt to get around this apparent problem was the Differential Effective Medium theory, initially introduced by Salganik (1973) and later studied in detail by Zimmerman (1991). This model considers the effect of introducing an infinitesimal amount of inclusions into a material. The equations describing the impact of the inclusions (like Eqs. (6.33) and (6.34)) can then be written in differential form, in which case it does not matter whether the inclusions are considered as perturbations to the compliance or the stiffness. The effective moduli for a finite concentration of inclusions can be found by integration.

The Differential Effective Medium model appears to eliminate the problem of non-uniqueness; however, this does not imply that this model is in general more correct than the others. Which of the models that gives the most accurate description of the interactions between inclusions in a material depends on the actual positions of the inclusions relative to each other. This is closely related to the statement we made at the beginning of this chapter, that the elastic properties of a composite material also depend on the geometrical distribution of each component.

None of the models described above include any explicit information about the positions of the inclusions relative to each other; however, the fact that the models predict different results implies that they represent different configurations. Unfortunately, most papers presenting such models do not specify which configuration the model represents. Thus it may be difficult to decide whether a given model is a suitable representation for a given material or not. However, we may get an indication by looking at an example that shows how different positions of inclusions affect the interactions between them.

Consider the simple two-dimensional sample shown in Fig. 6.8a. By applying a force on the sample (indicated by the arrows on the figure) and measuring the deformation we obtain a measure of the stiffness of the

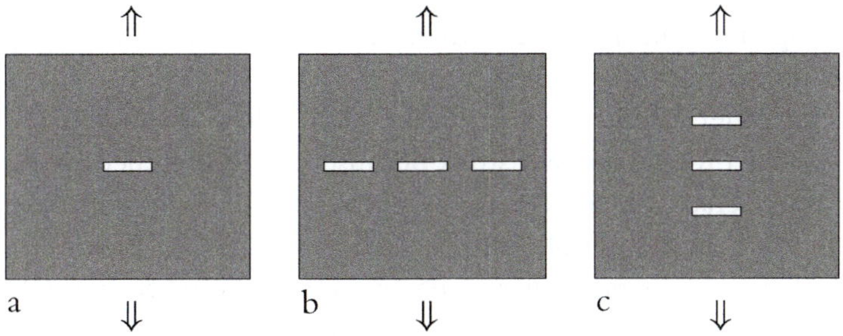

Figure 6.8 Two-dimensional material with cracks.

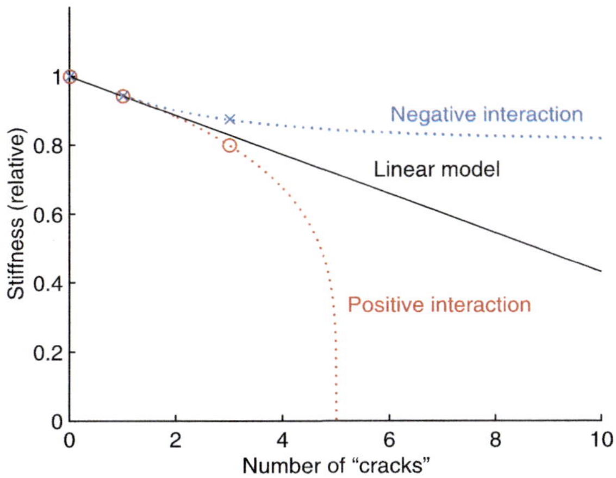

Figure 6.9 Stiffness of a two-dimensional sample (Fig. 6.8) versus number of cracks. "Positive interaction": the cracks are positioned so that they enhance the effect of each other, as in Fig. 6.8b. "Negative interaction": the cracks are positioned so that they reduce the effect of each other, as in Fig. 6.8c. The solid line shows the prediction of a linear (non-interacting) model. The markers show results obtained by finite element calculations for the geometries shown in Fig. 6.8.

material. The thin slit in the middle of the sample represents an inclusion similar to a crack. The presence of the crack reduces the stiffness of the sample, as shown in Fig. 6.9.

The presence of the crack in the sample also makes the stress distribution non-uniform. In the areas at the side of the crack the stress level will be larger than it would have been if the crack was not present. If new cracks are introduced in these areas, as shown in Fig. 6.8b, then the stiffness of the sample will be more reduced than a simple linear extrapolation of the impact of one crack would predict (positive interaction; see Fig. 6.9). If we keep adding cracks in these areas, then the sample will eventually split into two, and the stiffness will vanish.

On the other hand, the areas above and below the initial crack will be less exposed to the applied stress because of the presence of the crack. Adding more cracks in these areas as shown in Fig. 6.8c will only have a marginal effect on the sample stiffness, which will asymptotically approach the value for two separate pillars (negative interaction; see Fig. 6.9). Thus the reduction in stiffness due to the presence of a large number of cracks may be very different for different positions of the cracks relative to each other. In some cases the actual stiffness may be larger than predicted by a simple linear model, and in other cases, it may be less.

We may deduce from this example that a self-consistent model based on a perturbation of stiffness represents a material similar to that shown in Fig. 6.8c, where the inclusions are positioned so that they tend to shield each other from the external load. On the other hand, a self-consistent model based on a perturbation of compliance represents a material similar to that shown in Fig. 6.8b, where the inclusions are positioned in areas that are particularly exposed to the external load because of the presence of other inclusions.

To pick which model is best suited as a representation for a specific material, we thus need to know something about the distribution of inclusions in the material. Knowledge about the process that created the inclusions may be of some help. For instance, cracks induced by a tensile stress tend to appear in particularly exposed areas, like in Fig. 6.8b. A linear model like Eqs. (6.33) and (6.34) represents a kind of average over both types of geometries and may be a good starting point if we know nothing about the actual distribution of inclusions.

6.4.4 Crack development in stressed rocks

Changes in the stress state of a rock may cause cracks to initiate and grow, or maybe to close, depending on the orientation and sign of the principal stresses relative to the orientation of the cracks. Thus the number, size and predominant orientation of the cracks in a rock specimen largely reflects the stress state and stress history of the specimen.

A compressive stress acting in a direction more or less normal to a crack tends to close the crack. The closure process depends on the shape of the crack (Mavko and Nur, 1978). Mismatch between the crack faces implies that the closure process extends over a large range of stress values (see Section 6.6.1).

It is intuitively clear that a crack may grow if the crack faces are pulled apart with a sufficiently large force. Following Griffith (1921), we will consider crack growth induced by tensile stress in a planar (two-dimensional) system. The specimen under consideration is a plate of thickness t, and the crack is a flat, elliptical slit of length $2a$ penetrating the plate. A tensile stress $\sigma (< 0)$ acting normal to the crack causes the plate to deform (Fig. 6.10). The strain energy W_c due to the crack is defined as the strain energy of the specimen containing the crack minus the strain energy for a similar specimen without a crack, if it was exposed to the same stress. It is given as

$$W_c = \frac{\pi a^2 t}{E}\sigma^2 \tag{6.65}$$

where E is Young's modulus of the material the plate is made of. A free surface is supposed to possess a surface energy w_s per unit area (associated with broken atomic bonds), and thus the surface energy W_s of the crack is

$$W_s = 4w_s at \tag{6.66}$$

The surface energy required to increase the crack size by a small amount $2\Delta a$ is $(\partial W_s/\partial a)\Delta a$, whereas the strain energy available for this increase is $(\partial W_c/\partial a)\Delta a$. Thus the crack is unstable and will start to grow if

$$\frac{\partial W_c}{\partial a} \geq \frac{\partial W_s}{\partial a} \tag{6.67}$$

Combining Eqs. (6.65)–(6.67), we find that the crack will start to grow if the tensile stress exceeds the threshold value

$$T_0 = \sqrt{\frac{2w_s E}{\pi a}} \tag{6.68}$$

Crack growth implies that a increases, and consequently the threshold value decreases according to Eq. (6.68). The situation is thus unstable, and the crack will continue to grow until the specimen fails. Therefore T_0 represents the tensile strength of the specimen containing the crack. This demonstrates that pre-existing cracks have a pronounced effect on the tensile strength of a rock.

Crack growth is often analysed in terms of stress intensity factors, which define the stress state at the tip of a crack. For a crack experiencing tensile stress as described above (also called Mode I loading), the stress field in the vicinity of the crack tip can be expressed as

$$\sigma_{ij} = \frac{K_\mathrm{I}}{\sqrt{2\pi r}} f_{ij}^{(\mathrm{I})}(\theta) \tag{6.69}$$

where the coordinates r and θ are defined in Fig. 6.10, and $f_{ij}^{(\mathrm{I})}(\theta)$ are known functions of the angle θ (see e.g. Rice, 1968; Anderson, 2005; see also Appendix D.3.1). Notice that the stress diverges as $1/\sqrt{r}$ at the crack tip. This is due to the assumption of elasticity. In reality, stresses will be limited by plasticity and other effects. However, the elastic approach, called linear elastic fracture mechanics (LEFM), has proven to be a valuable starting point in material technology.

The factor K_I is called the stress intensity factor for Mode I loading and is a measure for the stress singularity at the crack tip. For a 2D crack of length $2a$, K_I may be calculated from the stress at the crack perimeter according to

$$K_\mathrm{I} = \frac{1}{\sqrt{\pi a}} \int_{-a}^{a} \sigma(y) \sqrt{\frac{a+y}{a-y}}\, \mathrm{d}y \tag{6.70}$$

For a crack with external loading, it is convenient to split the loading into two. From the far-field stresses the stresses at the crack position is computed in the absence of the crack. Then the crack problem is solved with the opposite stresses at its perimeter. The superposition of the two loadings then ensures the zero stress at the fracture. The singular term must clearly come from the second part.

For the case in the figure, we thus have $\sigma(y) = \sigma$, the integral evaluates to $a\pi$, and hence

$$K_\mathrm{I} = \sigma\sqrt{\pi a} \tag{6.71}$$

For a crack with only internal loading p_f, we replace σ with p_f.

For a penny-shaped crack (with radius a) in an infinite medium, we have

$$K_\mathrm{I} = \sigma\sqrt{\frac{4a}{\pi}} \tag{6.72}$$

If K_I exceeds a critical limit K_IC, called the *fracture toughness*, the crack will start to grow. Note that, as the crack grows, the stress intensity factor increases. Thus, unless loading is reduced, the growth will be unstable.

Crack growth may also occur under compressive stresses. If the stress state is anisotropic, then the stress conditions at the tip of a crack oriented

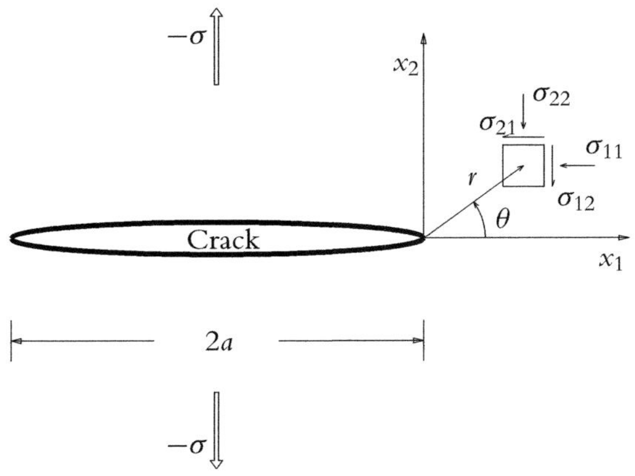

Figure 6.10 Crack tip coordinates.

at a skew angle relative to the largest principal stress may become tensile, and the crack will grow if this tensile stress exceeds a critical value. Based on such considerations, Griffith extended his theory to general stress conditions. The resulting failure criterion is given in Section 2.3.2. As the crack growth is assumed to be caused by local tensile stress, the tensile strength is a key parameter also for the compressive failure criterion.

The situation is illustrated in Fig. 6.11a (also discussed in Section 1.8). Even though the initial crack may be closed because of the compressive stress σ, the crack faces may start to slide relative to each other due to the shear stress τ if the static friction is overcome. The orientation of the local stresses at the tip of the crack is such that the crack growth will occur parallel to σ_1 and normal to σ_3, and thus the so-called *wing cracks* will develop.

The relation between crack growth and global failure of a sample is in general more complicated for compressive stresses. In unconfined situations ($\sigma_3 = 0$), failure typically occurs by axial splitting, starting with opening of axial cracks from the tip of the largest pre-existing crack or flaw in the specimen. Under confinement, like in a standard triaxial test, the wing cracks related to a suitably oriented crack will grow when the conditions are fulfilled; however, these cracks will not be able to grow critically and cause macroscopic failure on their own. Global failure is thus related to interaction and coalescence of several cracks forming an array along the plane, which eventually becomes the global failure plane.

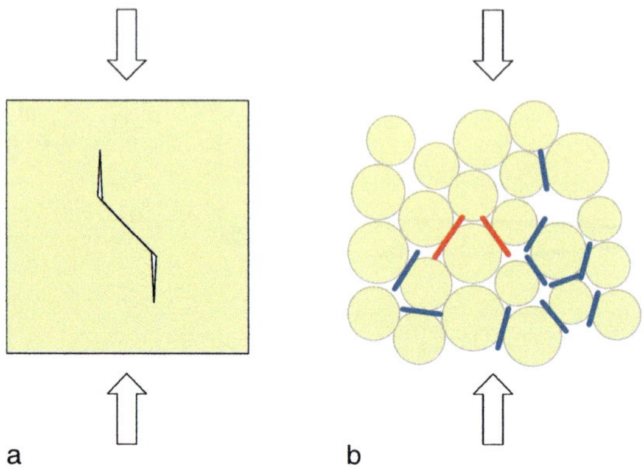

a b

Figure 6.11 a. Sliding grain boundary crack thought to be of central importance for rock failure under triaxial stress conditions. b. Local failure (shear: red lines; tensile: blue lines) caused by external loading on a 2D sample, as simulated by PFC. (This figure is a subset of Fig. 6.6.)

A mathematical model incorporating both shear failure and axial splitting was developed by Horii and Nemat-Nasser (1985). The input parameters to their model were the following:

1. The length and the orientation of the largest pre-existing crack
2. The cohesive yield stress and average friction coefficient of the crack interface
3. Fracture toughness for crack opening in the matrix material
4. The confining pressure

Clearly, such detailed knowledge of material properties may be hard to obtain in a general situation. The model is however useful in the sense that it can explain from microscopic principles the shape of the observed failure envelopes: the nonlinearity of the failure envelope, which is often observed in practice, is here explained by the transition from failure initiated at the largest pre-existing flaws (at low confining pressures) to failure initiated at critically oriented flaws (at high confining pressures). Horii and Nemat-Nasser obtained a good fit between model prediction and experimental observations for the variation of ultimate strength with confining pressure in Westerly Granite.

Cracks of the type shown in Fig. 6.11a are not normally observed in rocks. However, local damage induced by shear stress typically involves both shear and tensile failure, as illustrated on Fig. 6.11b. The construction shown in Fig. 6.11a represents a model that provides a coupling between shear and tensile failure, and is as such intuitively useful. The fact that the irregular microstructure of a granular material makes this coupling more complex is not surprising.

6.5 Multicomponent models

Most rock physics models treated so far consider porous materials with one solid phase and one fluid phase. Thermodynamically based macroscopic poroelasticity, such as in the Biot theory, treats porosity through one parameter only, with the underlying assumption of connected pore space (i.e. permeability). Effective medium theories, as seen in the above sections, may distinguish between different pore shapes, mostly by stiff (spherical) and soft (crack like) pores. Alternatively, spherical grains may define the solid phase with the elastic properties being controlled by grain contact stiffnesses. The microscopically defined porosity can then be treated as one parameter to obtain consistency with macroscopic poroelasticity.

Natural sedimentary rocks usually consist of several different solid minerals and sometimes of multi-phasic immiscible fluids. Homogenisation is possible, provided that we can define a representative elementary volume ("REV") with the same elastic properties as the whole material under consideration. We will here assume macroscopically homogeneous rock and leave the unresolved problem of scale dependence out of the discussion. However, a natural question arises: Since poroelasticity theory requires solid bulk modulus K_s, and effective medium theories normally also require solid shear modulus G_s, how do we estimate these parameters in case of several different minerals?

One obvious approach, if core material is available, is to measure the solid bulk modulus in unjacketed or incrementally unjacketed tests (keeping stress and pore pressure changes equal). A second alternative, knowing the mineral composition (from e.g. XRD analysis), is to use tabulated moduli and densities of minerals from handbooks and publications, and estimate solid moduli from upper and lower bounds. A simple choice is the Voigt and Reuss bounds with some kind of averaging (Voigt–Reuss–Hill) to define specific values. The Hashin–Shtrikman bounds (see Section 6.2) produce narrower bounds and can be generalised to multiple components through the so-called Walpole bounds (Walpole, 1966; Milton, 1981; Berryman, 1995).

Moduli calculated from such an approach are still approximate and may vary with the geometrical distribution of minerals. For instance, single crystals of rock-forming minerals are anisotropic, so the effective moduli will depend on their orientational distribution. Perfect alignment indicates that anisotropic solid moduli should be used. However, for simplicity, isotropic solid moduli are used in most practical applications, even if the rock frame model is anisotropic. The solid moduli for each mineral can then be derived from Voigt–Reuss–Hill- (see e.g. Antonangeli et al., 2005) or Hashin–Shtrikman-based averaging.

As an example, the symmetry group of quartz is trigonal (rhombohedral), with six independent elastic stiffnesses. Heyliger et al. (2003) measured these at room temperature in natural and artificially grown quartz crystals and found $C_{33}/C_{11} = 1.21$ and $C_{44}/C_{66} = 1.42$ for natural quartz. They estimated the corresponding bulk modulus to 37.7 GPa and shear modulus to 46.9 GPa for polycrystalline quartz. The corresponding values for laboratory grown quartz are very similar.

As pointed out in Section 1.6.2, the basic equations in poroelasticity (Biot, 1941; Rice and Cleary, 1976; Detournay and Cheng, 1993) allow that unjacketed pore compressibility $1/K_\phi$ (see Eq. (1.207)) is different from the solid grain compressibility $1/K_s$. Brown and Korringa (1975) formulated a generalised version (also allowing anisotropy) of the poroelastic theory with $K_\phi \neq K_s$. The total compressibility under constant stress difference ("unjacketed test") $1/K_M$ is given by

$$\frac{1}{K_M} = \frac{1}{V_{tot}}\left(\frac{\partial V_{tot}}{\partial p_f}\right)_{\sigma_d} = \frac{1-\phi}{K_s} + \frac{\phi}{K_\phi} \tag{6.73}$$

Note that $K_M = K_\phi = K_s$ in Gassmann's work. Introducing

$$\beta = \frac{K_s}{K_M} = 1 + \phi\left(\frac{K_s}{K_\phi} - 1\right) \tag{6.74}$$

alters the Biot parameters α and M as follows:

$$\alpha_{BK} = 1 - \beta\frac{K_{fr}}{K_s} \tag{6.75}$$

$$\frac{1}{M_{BK}} = \frac{\phi}{K_f} + \frac{1 - \beta^2\frac{K_{fr}}{K_s} - \phi}{K_s} \tag{6.76}$$

The relationship between undrained and drained bulk modulus following from the Brown–Korringa theory is

$$K_u = K_{fr} + \frac{K_f\left(1 - \beta\frac{K_{fr}}{K_s}\right)^2}{\phi + \frac{K_f}{K_s}\left(1 - \phi - \beta^2\frac{K_{fr}}{K_s}\right)} \tag{6.77}$$

Except for the appearance of the β, this relation is identical to the Biot–Gassmann equation.

The "compact form" of the Biot–Gassmann equation (1.174) becomes

$$\frac{1}{c_{\mathrm{u}} - c_{\mathrm{M}}} = \frac{1}{c_{\mathrm{fr}} - c_{\mathrm{M}}} + \frac{1}{\phi}\frac{1}{c_{\mathrm{f}} - c_{\phi}} \tag{6.78}$$

which is the original Brown–Korringa formulation.

The Brown–Korringa model is rarely applied, partly because measurements of K_ϕ and K_{s} are lacking. From the equations above it appears that unless the two are very different, the impact on computed undrained bulk modulus is small.

Mavko and Mukerji (2013) applied effective medium models to estimate K_ϕ for mixtures of stiff and soft grains by varying the aspect ratio of soft grains to mimic clay particles. From this they concluded that for modest contrast in bulk moduli (less than 50 %) K_{s} and K_ϕ were very similar and could well be represented by a Gassmann model with solid modulus computed by the Voig–Reuss–Hill average.

Amongst the few reported laboratory measurements, Tarokh et al. (2018) found K_ϕ of 22 GPa and K_{M} of 29 GPa for Berea sandstone, which would correspond to a difference of order one percent between the two estimates for undrained bulk modulus. Brown and Korringa introduced their theory with an assumption that it is valid also for monomineralic rocks. Tarokh et al. (2018) argue that the difference between the moduli they observed could be due to the presence of clay and unconnected pore space in Berea sandstone. In cleaner (more than 90 % quartz) Dunnville sandstone and in an artificial porous silica sample, their measurement did not reveal any difference between the unjacketed moduli.

Berryman (1992) showed that the assumptions above lead to modifications of effective stress coefficients, not only for Biot's α (as shown by Eq. (6.77)). For example, the porosity is now controlled by an effective stress coefficient

$$\chi = 1 - \phi\frac{c_\phi - c_{\mathrm{s}}}{c_{\mathrm{fr}}(\alpha_{\mathrm{BK}} - \phi)} = 1 - \frac{\beta - 1}{\alpha_{\mathrm{BK}} - \phi}\frac{K_{\mathrm{fr}}}{K_{\mathrm{s}}} \tag{6.79}$$

For the example above, χ for Berea sandstone becomes 0.96, whereas α is approximately 0.75. This difference from conventional Biot–Gassmann may be relevant in core-based estimates of in situ porosity, which plays an important role in assessment of in situ reserves. It also has impact on effective stress coefficients for other physical properties that depend on porosity, such as permeability and wave velocities.

Ciz and Shapiro (2007) developed the Brown–Korringa model further to describe a porous material where the pores are filled with solid. This could, for instance, represent heavy oil sand. In the limit when it becomes equivalent to the Biot–Gassmann equation, it is a good approximation for low contrast between pore filling and solid material simply to replace the fluid bulk modulus K_f by the bulk modulus of the pore filling solid (K_{pf}). A second equation has to be introduced for the shear modulus, having the same structure as the equation for bulk modulus, but with solid host and pore material represented by the corresponding shear moduli, that is,

$$G = G_{fr} + \frac{G_{pf}}{\phi} \frac{\left(1 - \frac{G_{fr}}{G_s}\right)^2}{1 + \frac{G_{pf}}{\phi G_s}\left(1 - \frac{G_{fr}}{G_s} - \phi\right)} \tag{6.80}$$

The terms corresponding to β in the Brown–Korringa equation may be included, one for the bulk and one for the shear modulus.

A special case arises in clay rich rocks like shales, where adsorbed water on mineral surfaces and interstitial bound water between clay mineral platelets may occupy a large portion of narrow (nanometre) pores. Molecular dynamics simulations (e.g. Skipper et al., 1991) have shown that such water has a crystal-like structure, whereas transverse atomic force measurements (Antognozzi et al., 2001) have verified that such water has a non-zero shear modulus and a strongly enhanced viscosity compared to bulk water. Bound water is likely to have a significant impact on the elastic and visco-elastic properties of clay rich shales. To illustrate this, Holt and Fjær (2003) (see also Holt and Kolstø, 2017) proposed a simple model where the Hashin–Shtrikman (or Voigt) upper bound was used to model the content of the pore space as an effective, rigid water phase consisting of free water embedded in a shell of bound water. The Hashin–Shtrikman (or Reuss) lower bound was used to model the elastic properties of the resulting clay material. Necessary input parameters were the amount of bound water, estimated from volume of swelling clay minerals and specific surface area, plus the assumed bound water bulk and shear moduli.

Sayers and den Boer (2018, 2020) used anisotropic extensions of Hashin–Shtrikman and modelled all five stiffnesses of a transversely isotropic material representing clay, for instance, inside a shale. Since clay minerals are too small to be measured directly, the solid minerals were characterised by their theoretical anisotropic stiffnesses or by experimental moduli for muscovite (which has a molecular structure identical to illite). By running a large number of realisations, allowing variation in bulk and shear modulus describing the interclay medium (containing bound and free water) and in volume fraction and aspect ratio of clay minerals, they arrived at elastic properties of a spatial domain of clay minerals. The result shows correlations between C_{ij}, which resemble those identified in the work of Horne (2013). The shear modulus of the interclay medium was roughly an order of magnitude lower than the bulk modulus of the free water. By averaging over an orientational distribution of domains and further embedding the clay into a rock physics model that also accounts for non-clay minerals, this may provide a useful tool for modelling shales, not at least predicting anisotropic properties that are not commonly obtained from field measurements.

6.6 Fractured rocks

Most rocks that we see in nature appear to be fractured masses rather than intact solid materials. We may also expect fracture systems to penetrate the formations in the underground. This may be the case on many length scales, from major faults on a scale larger than reservoirs to microfractures seen in a core plug. An example is the fractured chalk reservoir of the Ekofisk field in the North Sea. Here the fractures are responsible for the main part of the permeability. It is therefore of importance to be able to identify as early as possible the existence of such fracture systems in reservoirs and to find the directions of the fractures. It may also be important (with reference to e.g. the Ekofisk case and prediction of reservoir compaction during depletion) to know how much the fractures contribute to the overall mechanical behaviour of a rock mass. In this section, we will discuss how fractured (or jointed) rocks behave mechanically, starting with the mechanical behaviour of a single fracture (or joint). The terms *fracture* and *joint* will be used synonymously throughout this section, although there is a difference from the geological viewpoint (see Chapter 3). With the mathematical idealisations presented here, they can however be treated within the same framework.

6.6.1 Single fractures

Previously in this chapter, we considered the effect of cracks on rock mechanical and acoustical properties. By a crack we meant an open void of a given idealised shape (spheroid, penny-shaped), which acts as a local discontinuity. However, in practice, a crack in a porous rock does not have an ideal geometry. Fracture surfaces are rough, and they may touch each other at some points.

In general, the complete stress–strain behaviour of a fracture relates normal displacement δw^{F} to normal stress $\delta\sigma$ on the fracture and shear displacement δu^{F} to shear stress $\delta\tau$ along the fracture interface. We may define δw^{F} such that $\delta w^{\mathrm{F}} = w_{\mathrm{o}}^{\mathrm{F}} - w^{\mathrm{F}}$, where $w_{\mathrm{o}}^{\mathrm{F}}$ is the initial fracture opening, and w^{F} is the actual fracture opening. The fracture is closed when $\delta w^{\mathrm{F}} = w_{\mathrm{o}}^{\mathrm{F}} - w_{\mathrm{max}}^{\mathrm{F}}$, where $w_{\mathrm{max}}^{\mathrm{F}}$ is the fracture opening at maximum closure. The stress-displacement relation can be written as (see e.g. Bandis, 1990; Saeb and Amadei, 1992)

$$\begin{pmatrix} \delta\sigma \\ \delta\tau \end{pmatrix} = \begin{pmatrix} K_{\mathrm{nn}} & K_{\mathrm{nt}} \\ K_{\mathrm{tn}} & K_{\mathrm{tt}} \end{pmatrix} \begin{pmatrix} \delta w^{\mathrm{F}} \\ \delta u^{\mathrm{F}} \end{pmatrix} \tag{6.81}$$

where the indices in the fracture stiffness matrix (K_{ij}) denote n for normal and t for tangential. The observed deformational behaviour is highly nonlinear for increasing compressive load, whereas unloading causes large hysteresis and inelastic behaviour. Under tension, the associated fracture stiffness is close to zero. Under shearing, the stress–strain relation may exhibit "plastic" and "brittle" behaviours, depending on the fracture size. On an empirical basis, the following type of expression for K_{nn} has been proposed for joints (Goodman, 1974):

$$K_{\mathrm{nn}} = K_{\mathrm{nn}}^{\mathrm{o}} \left[\frac{K_{\mathrm{nn}}^{\mathrm{o}} w_{\mathrm{max}}^{\mathrm{F}} + \sigma}{K_{\mathrm{nn}}^{\mathrm{o}} w_{\mathrm{max}}^{\mathrm{F}}} \right]^2 \tag{6.82}$$

where $K_{\mathrm{nn}}^{\mathrm{o}}$ is the initial normal stiffness of the joint. The appearance of $K_{\mathrm{nn}}^{\mathrm{o}}$ and $w_{\mathrm{max}}^{\mathrm{F}}$ in Eq. (6.82) implies that stress history dependence may be accounted for in the joint stiffness.

The shear stress dependence of K_{tt} is described by a relationship of the type

$$K_{\mathrm{tt}} = K_{\mathrm{tt}}^{\mathrm{o}} \left[1 - \frac{\tau}{\tau_{\mathrm{s}}} \right]^2 \tag{6.83}$$

where $K_{\mathrm{tt}}^{\mathrm{o}}$ is the initial shear stiffness, and τ_{s} is the asymptotic value for τ at large shear displacements. The off-diagonal terms in the stiffness matrix (Eq. (6.81)) account for coupling between shear and normal behaviour due to dilatancy. A mathematical model incorporating these terms is presented by Saeb and Amadei (1992).

Given the fracture stiffness parameters (Eq. (6.81)), we may calculate the frequency-dependent time delay, and the sound transmission and reflection coefficients caused by a single fracture (Pyrak-Nolte et al., 1990). The calculations show an increased damping of high frequencies in a transmitted acoustic signal when the fracture stiffness is reduced. The effect of viscosity in a fluid filling the fracture was also analysed, showing increased high frequency damping with decreasing viscosity.

From a microscopic viewpoint, closure of a fracture can be described mathematically by parameters depending on the roughness characteristics of the fracture surfaces. Brown and Scholz (1986) derived a relationship where they accounted for the following fracture surface parameters: a distribution function for asperity heights, the number of local maxima per unit area, the effective radius of curvature of the asperities, the fracture width at zero stress, and a correction factor for tangential stress contributions. For a particular choice of the asperity height distribution function, the strain vs. stress relationship can be written

$$\delta\varepsilon^{\mathrm{F}} = A + B\ln\delta\sigma \tag{6.84}$$

where $\delta\varepsilon^{\mathrm{F}} = \delta w^{\mathrm{F}}/w^{\mathrm{F}}$ is the fracture strain, and A and B are functions of the fracture surface parameters described above. The joint closure is thus a highly nonlinear function of the normal stress.

Carlson and Gangi (1985) modelled a fracture using a "bed of nails" concept (see Fig. 6.12). A set of cylindrical rods of different lengths L_i are attached to the crack walls, simulating a distribution of asperities. The rods are assumed to act as springs. As the normal stress is increased across the fracture, the walls of the fracture move closer, and more and more rods come in contact with the opposite wall. Therefore the stiffness is increased as the strain increases, similarly to Eq. (6.84). For a throughgoing fracture, the fracture strain can be expressed as

$$\delta\varepsilon^{\mathrm{F}} = \left(\frac{\delta\sigma}{\sigma_{\mathrm{c}}}\right)^{m} \tag{6.85}$$

where σ_{c} is the stress required to close the crack. The exponent m (≤ 1) is given by the distribution of rod lengths. The fraction n of rods in contact is

$$n = (\delta\varepsilon^{\mathrm{F}})^{(1-m)/m} \tag{6.86}$$

where m is thought to be constant throughout a given experiment. The value $m = 1$ corresponds to a closed crack with perfectly smooth surfaces, whereas $m = 0$ corresponds to and open crack with smooth surfaces.

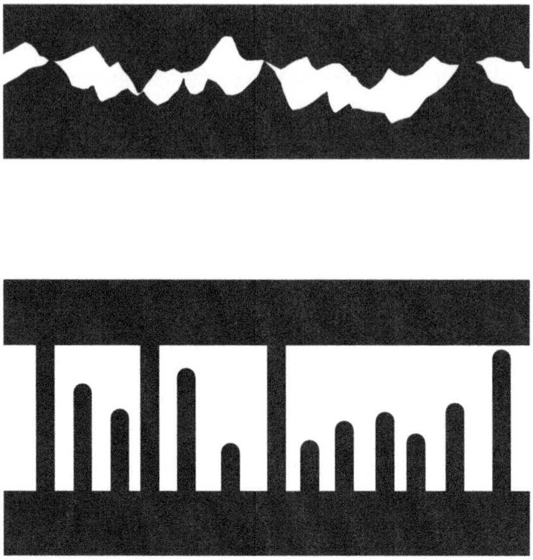

Figure 6.12 A natural fracture (above) and the mechanically and hydraulically equivalent "bed of nails" fracture model introduced by Carlson and Gangi (1985).

Carlson and Gangi modified expression (6.85) for the case of an isolated crack in a solid matrix, giving

$$\delta\varepsilon^{\mathrm{F}} = \left(\frac{\delta\sigma + \sigma_{\mathrm{o}}}{\sigma_c}\right)^m \tag{6.87}$$

where σ_o is equivalent to an initial stress (associated with the tensile strength of the material). The expressions can be used quantitatively to evaluate the stress dependence of P-wave velocities in cracked media, leading to a relationship of the following form, which is applicable for high-porosity sedimentary rocks:

$$v_{\mathrm{p}} = v_{\mathrm{p}}^{\mathrm{o}}\left(1 + \frac{\sigma}{\sigma_{\mathrm{o}}}\right)^{(1-m)/2} \tag{6.88}$$

In Section 6.3, we found for a model consisting of spheres in contact (where the contacts behave according to Hertzian theory) that the sound velocity would increase with stress according to a power law behaviour with exponent 1/6. This corresponds to $m = 2/3$ (and $\sigma_{\mathrm{o}} \to 0$) in Carlson and Gangi's theory. In many rocks, exponents $< 1/6$ have been observed, indicating that m is closer to 1 (Carlson and Gangi, 1985).

The shear strength of a joint or fracture is often expressed in terms of the Mohr–Coulomb criterion with zero cohesion (see Section 2.3.1):

$$\tau = \sigma \tan\varphi \tag{6.89}$$

Based on laboratory experiments with more than 200 artificial tensile fractures, Barton and Bandis (1990) found that the friction angle φ can be expressed as

$$\varphi = JRC \log\frac{JCS}{\sigma} + \varphi_r \tag{6.90}$$

where JRC is called the joint roughness coefficient, JCS is the wall strength identified as the unconfined compressive strength of the intact rock, and φ_r is the residual angle of friction ($\approx 30°$, see Chapter 2). The coefficient JRC can be found from simple shear tests and has typically a value around 20 for a rough joint without steps. It is a scale-dependent factor, and a scaling relation has been established (see Barton and Bandis, 1990), giving JRC as a function of roughness amplitude over the profile length.

6.6.2 Rocks with many fractures

A rock mass normally contains a system of fractures. Such a system is specified in terms of fracture orientation, fracture width, length and spacing. Whereas the orientation is mainly determined by the stress state at the time the fracture was formed, the remaining parameters are primarily given by the lithology. These parameters are necessary to quantify the flow of fluids in a fractured reservoir. In nature, there is of course a distribution of fracture parameters. Many investigators have claimed that fracture size distributions observed in outcrops tend to be scale invariant (fractal).

Several finite element and finite difference codes have been developed to analyse the mechanical behaviour of discontinuous media. Standard input parameters are a set of fracture stiffnesses and the elastic moduli of the intact material. We will not describe any of these methods here, but rather discuss one analytical model for the elastic behaviour of fractured rocks. This theory was developed by White (1983). Consider a cube (Fig. 6.13) consisting of fractured blocks. Each block is spot-welded to the neighbouring blocks (Fig. 6.14). The average dimensions of a block are L_x, L_y and L_z, and the numbers of contact points per unit area are N_x, N_y and N_z, respectively. The spot-welded contacts are assumed to be circles of average diameters D_x, D_y and D_z. The elastic properties of the fractured cube are derived based

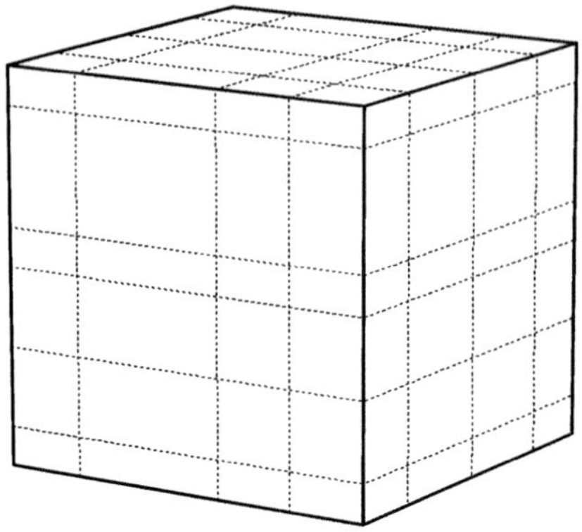

Figure 6.13 Elementary cube of fractured rock.

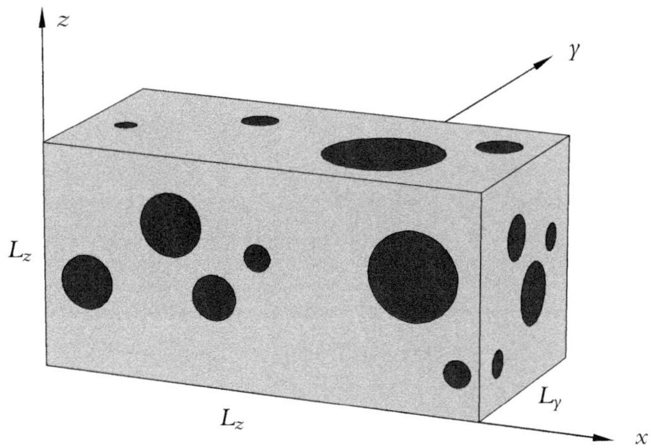

Figure 6.14 Average sized block of fractured rock and contact areas. (After White, 1983).

on the Hertz–Mindlin approach introduced in Section 6.3. For the effective contact radius b (Eq. (6.22)), we use the parameter $\bar{b}_x = (D_x N_x L_y L_z)/2$ for a surface normal to x and corresponding expressions for y and z. The compressive strain ε_x due to a force σ_x has two contributions:

Firstly, the deformation at the fracture planes normal to x causes the strain (see Eq. (6.25))

$$\varepsilon_x^{\rm f} = \frac{1-\nu_{\rm s}}{2G_{\rm s}\bar{b}_x}\frac{L_y L_z}{L_x}\sigma_x \tag{6.91}$$

Secondly, the deformation of the solid blocks gives the strain

$$\varepsilon_x^{\mathrm{s}} = \frac{1}{(\lambda_s + 2G_{\mathrm{s}})(1 - a_x)}\sigma_x \tag{6.92}$$

where a_x accounts for the effect of the fracture planes normal to y and z. The total strain is

$$\varepsilon_x = \varepsilon_x^{\mathrm{f}} + \varepsilon_x^{\mathrm{s}} = \frac{1}{C_{11}}\sigma_x \tag{6.93}$$

These equations allow us to find the elastic stiffness C_{11}. It is convenient to introduce the parameters R_x, R_y and R_z given by expressions of the form

$$R_x = N_x D_x L_x \tag{6.94}$$

The resulting expression for C_{11} is

$$C_{11} = \frac{\lambda_{\mathrm{s}} + 2G_{\mathrm{s}}}{\frac{1}{1-a_x} + \frac{2(1-\nu_{\mathrm{s}})^2}{1-2\nu_{\mathrm{s}}}\frac{1}{R_x}} \tag{6.95}$$

Similar expressions can be obtained for C_{22} and C_{33} with R_y, a_y and R_z, a_z, respectively, substituted for R_x, a_x. The derivation of a_x is rather tedious and is not included here. The result is

$$a_x = \frac{2\nu_{\mathrm{s}}^2}{1 - 2\nu_{\mathrm{s}}}\frac{4(1 - \nu_{\mathrm{s}}) + R_y + R_z}{\left(R_y + \frac{2(1-\nu_{\mathrm{s}})^2}{1-2\nu_{\mathrm{s}}}\right)\left(R_z + \frac{2(1-\nu_{\mathrm{s}})^2}{1-2\nu_{\mathrm{s}}}\right) - \frac{4\nu_{\mathrm{s}}^2(1-\nu_{\mathrm{s}})^2}{(1-2\nu_{\mathrm{s}})^2}} \tag{6.96}$$

The shear moduli are given by

$$C_{44} = \frac{G_{\mathrm{s}}}{1 + \frac{2-\nu_{\mathrm{s}}}{2}\left(\frac{1}{R_y} + \frac{1}{R_z}\right)} \tag{6.97}$$

and by similar expressions for C_{55} and C_{66}. Note that $R_x = R_y = R_z \to \infty$ means that the material effectively contains no fractures, that is, the elastic moduli are those of the solid. As a plausible example, consider the average distance between fractures $L_x = L_y = L_z = 1$ m with $N_x = N_y = 5\ \mathrm{m}^{-2}$ contacts per unit area and the average contact spot diameter of $D_x = D_y = 0.1$ m. Thus $R_x = R_y = 0.5$. Fig. 6.15 shows how the shear velocities associated with C_{44} (SV) and C_{66} (SH) vary as functions of R_z in this case (assuming that $\nu_{\mathrm{s}} = 0.25$). The figure also shows the case of a planar fracture system ($R_x = R_y \to \infty$). Both shear moduli are strongly reduced by the presence of fractures. This also underlines the existence of a size effect: if

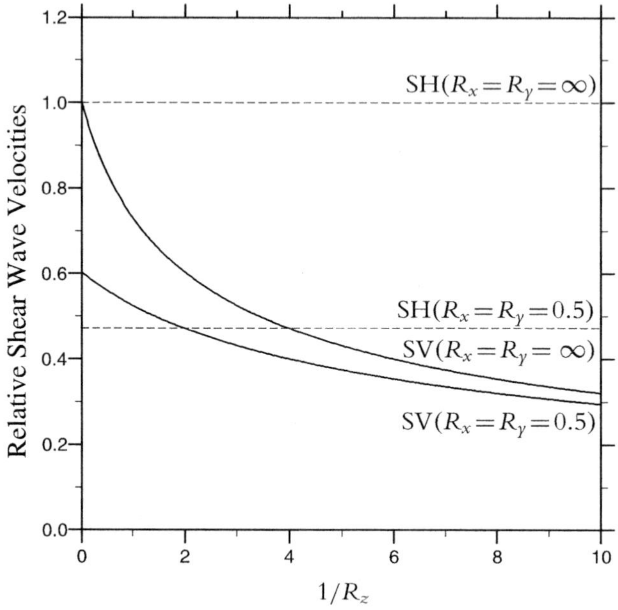

Figure 6.15 Shear wave velocities in a fractured rock relative to velocities in a solid block, calculated by White's theory. R_x, R_y and R_z are defined in the text.

a sample is smaller than the smallest fracture block, then we will measure the solid properties rather than the properties of the rock mass. White's model can also be used to reveal the effects of a directional distribution of fracture planes, which give rise to a significant anisotropy. Observations of seismic/VSP anisotropy can thus be used to characterise fractured reservoirs.

6.7 Finite element analysis

The rock models described previously in this chapter provide estimates for the elastic constants of an effective (homogeneous) medium, based on a description of the rock microstructure. These constants may be introduced in Hooke's law, which gives the relations between stresses and strains for the effective medium. However, analytical solutions of Hooke's law may be impractical or even inaccessible due to complicated geometry or spatial variations in the elastic constants. If plasticity or nonlinear effects are also introduced, analytical solutions are only accessible for exceptionally simple geometries. To solve more complex and practical problems, a set of numerical tools based on the finite element method (FEM) are commonly used.

The basic principle of FEM is the virtual division (discretisation) of the rock volume into subvolumes (called *elements*). The mechanical properties can vary from element to element; however, they are assumed to be uniform within each element. The complete assembly of elements, covering the entire sample under investigation, is called the *mesh*. Neighbouring elements are interacting with each other at the contact points (called *nodes*). Within each element, the stress and strain are assumed to vary in a predefined way, for instance, as a constant or as a linear or higher-order function. Based on the specified information about the properties of the elements, the FEM software establishes a *global stiffness matrix* $\mathbf{C}_g$, which controls displacements and forces (represented by the vectors $\vec{d}_g$ and $\vec{F}_g$, respectively) at the nodes. For a static analysis, the relation is

$$\mathbf{C}_g \vec{d}_g = \vec{F}_g \tag{6.98}$$

Next, Eq. (6.98) is solved for $\vec{d}_g$:

$$\vec{d}_g = \mathbf{C}_g^{-1} \vec{F}_g \tag{6.99}$$

in such a way that the initial conditions and boundary conditions are fulfilled.

For the user, one of the most important and time-consuming tasks is the division of the sample into elements. Given the restrictions on stress and strain within each element, it is clearly beneficial for accuracy to use a finely divided mesh. On the other hand, solution of Eq. (6.99) is very time consuming and may be not achievable if the number of elements is too high. Hence it is advisable to make the mesh as coarse as possible and only make a fine division in the areas where the stress and strain are expected to vary a lot over short distances.

The other important task for the user is evaluation of the results from the FEM calculations. FEM is a very useful tool, but it is essentially only an efficient way to solve a set of equations. It may easily produce meaningless results if some of the basic assumptions, discretisation of the mesh, input parameters or boundary conditions are inappropriate. FEM does not in any way free the user from the necessity to understand the physics of the problem. If you are not able to solve the problem without FEM—given enough time—you may not be ready to tackle it with the help from FEM either. Some FEM applications are equipped with automatic discretisation.

This may save a lot of efforts, but it also makes critical evaluation of the FEM results even more important.

Advanced FEM tools include a number of potentially relevant effects, like plasticity, creep, thermal diffusion and pore pressure diffusion as well as wave propagation and failure. Even the appearance and consequences of cracks may be taken into account, but only in such a way that location and orientation of the cracks are predefined. This cannot always be done in a realistic way, which has triggered the development of an extended finite element method, XFEM (Moës et al., 1999), which allows for cracks to be generated independently of the mesh. Hybrid tools that combine FEM with discrete particle modelling (Section 6.3) have also been developed to tackle the same problem (Munjiza et al., 1999; Klerck et al., 2004; Lisjak and Grasselli, 2014).

References

Anderson, T.L., 2005. Fracture Mechanics. Fundamentals and Applications, third edition. Taylor & Francis. (Note: Fourth edition 2017).

Antognozzi, M., Humphris, A.D.L., Miles, M.J., 2001. Observation of molecular layering in a confined water film and study of the layers viscoelastic properties. Appl. Phys. Lett. 78, 300–302. https://doi.org/10.1063/1.1339997.

Antonangeli, D., Krisch, M., Fiquet, G., Badro, J., Farber, D.L., Bossak, A., Merkel, S., 2005. Aggregate and single-crystalline elasticity of hcp cobalt at high pressure. Phys. Rev. B 72, 134303. https://doi.org/10.1103/PhysRevB.72.134303.

Bachrach, R., Avseth, P., 2008. Rock physics modeling of unconsolidated sands: accounting for nonuniform contacts and heterogeneous stress fields in the effective media approximation with applications to hydrocarbon exploration. Geophysics 73 (6), E197–E209. https://doi.org/10.1190/1.2985821.

Bachrach, R., Dvorkin, J., Nur, A., 1998. High-resolution shallow-seismic experiments in sand, Part II: velocities in shallow unconsolidated sand. Geophysics 63 (4), 1234–1240. https://doi.org/10.1190/1.1444424.

Backus, G.E., 1962. Long-wave elastic anisotropy produced by horizontal layering. J. Geophys. Res. 67 (11), 4427–4440. https://doi.org/10.1029/JZ067i011p04427.

Bandis, S.C., 1990. Mechanical properties of rock joints. In: Barton, N., Stephansson, O. (Eds.), Rock Joints. Balkema, Rotterdam, pp. 125–140.

Bandyopadhyay, K., 2009. Seismic anisotropy: geological causes and its implications to reservoir geophysics. PhD thesis. Stanford University.

Barton, N., Bandis, S.C., 1990. Review of predictive capabilities of JRC-JCS model in engineering practice. In: Barton, N., Stephansson, O. (Eds.), Rock Joints. Balkema, Rotterdam, pp. 603–610.

Berryman, J.G., 1980. Long-wavelength propagation in composite elastic media. II. Ellipsoidal inclusions. J. Acoust. Soc. Am. 68, 1820–1831. https://doi.org/10.1121/1.385172.

Berryman, J.G., 1992. Effective stress for transport properties of inhomogeneous porous rock. J. Geophys. Res. 97 (B12), 17409–17424. https://doi.org/10.1029/92JB01593.

Berryman, J.G., 1995. Mixture theories for rock properties. In: Ahrens, T.J. (Ed.), Rock Physics and Phase Relations: A Handbook of Physical Constants. AGU, pp. 205–228.

Biot, M.A., 1941. General theory of three-dimensional consolidation. J. Appl. Phys. 12, 155–164. https://doi.org/10.1063/1.1712886.
Brown, R.J.S., Korringa, J., 1975. On the dependence of the elastic properties of a porous rock on the compressibility of the pore fluid. Geophysics 40 (4), 608–616. https://doi.org/10.1190/1.1440551.
Brown, S.R., Scholz, C.H., 1986. Closure of rock joints. J. Geophys. Res. 90 (B5), 8675–8684. https://doi.org/10.1029/JB091iB05p04939.
Budiansky, B., O'Connell, R.J., 1976. Elastic moduli of a cracked solid. Int. J. Solids Struct. 12 (2), 81–97. https://doi.org/10.1016/0020-7683(76)90044-5.
Carlson, R.L., Gangi, A.F., 1985. Effect of cracks on the pressure dependence of P-wave velocities in crystalline rocks. J. Geophys. Res. 90 (B10), 8675–8684. https://doi.org/10.1029/JB090iB10p08675.
Ciz, R., Shapiro, S.A., 2007. Generalization of Gassmann equations for porous media saturated with a solid material. Geophysics 72, A75–A79. https://doi.org/10.1190/1.2772400.
Cundall, P.A., 2001. A discontinuous future for numerical modelling in geomechanics? Proc. Inst. Civ. Eng. – Geotec. Eng. 149 (1), 41–47. https://doi.org/10.1680/geng.2001.149.1.41.
Cundall, P.A., Strack, O.D.L., 1979. A discrete numerical model for granular assemblies. Géotechnique 29 (1), 47–65. https://doi.org/10.1680/geot.1979.29.1.47.
de Gennes, P.-G., 1996. Static compression of a granular medium: the "soft shell" model. Europhys. Lett. 35 (2), 145–149. https://doi.org/10.1209/epl/i1996-00546-1.
Detournay, E., Cheng, A.H.-D., 1993. Fundamentals of poroelasticity. In: Hudson, J.A. (Ed.), Comprehensive Rock Engineering, vol. 2. Pergamon Press, Oxford, pp. 113–171.
Digby, P.J., 1981. The effective elastic moduli of porous granular rocks. J. Appl. Mech. 48 (4), 803–808. https://doi.org/10.1115/1.3157738.
Domenico, S.N., 1977. Elastic properties of unconsolidated porous sand reservoirs. Geophysics 42 (7), 1339–1368. https://doi.org/10.1190/1.1440797.
Duffaut, K., Landrø, M., Sollie, R., 2010. Using Mindlin theory to model friction-dependent shear modulus in granular media. Geophysics 75 (3), E143–E152. https://doi.org/10.1190/1.3429998.
Dvorkin, J., Nur, A., 1996. Elasticity of high-porosity sandstones: theory for two North Sea data sets. Geophysics 61 (5), 1363–1370. https://doi.org/10.1190/1.1444059.
Dvorkin, J., Mavko, G., Nur, A., 1991. The effect of cementation on the elastic properties of granular materials. Mech. Mater. 12 (3–4), 207–217. https://doi.org/10.1016/0167-6636(91)90018-U.
Elata, D., Dvorkin, J., 1996. Pressure sensitivity of cemented granular materials. Mech. Mater. 23 (2), 147–154. https://doi.org/10.1016/0167-6636(96)00005-1.
Eshelby, J.D., 1957. The determination of the elastic field of an ellipsoidal inclusion, and related problems. Proc. R. Soc. Lond. Ser. A 241, 376–396. https://doi.org/10.1098/rspa.1957.0133.
Garbin, H.D., Knopoff, L., 1973. The compressional modulus of a material permeated by a random distribution of circular cracks. Q. Appl. Math. 30, 453–464. https://doi.org/10.1090/qam/99719.
Garbin, H.D., Knopoff, L., 1975. Elastic moduli of a medium with liquid-filled cracks. Q. Appl. Math. 33, 301–303. https://doi.org/10.1090/qam/99661.
Goodman, R.E., 1974. The mechanical properties of joints. In: Proc. 3rd Int. Congr. ISRM, vol. 1A. Denver Co., pp. 127–140.
Griffith, A.A., 1921. The phenomena of rupture and flow in solids. Phil. Trans. R. Soc. A 221, 163–198. https://doi.org/10.1098/rsta.1921.0006.
Han, D.-H., Batzle, M., 2006. Velocities of deepwater reservoir sands. Lead. Edge 25 (4), 460–466. https://doi.org/10.1190/1.2193222.

Hashin, Z., Shtrikman, S., 1963. A variational approach to the elastic behaviour of multiphase materials. J. Mech. Phys. Solids 11 (2), 127–140. https://doi.org/10.1016/0022-5096(63)90060-7.

Hazzard, J.F., Young, R.P., 2000. Simulating acoustic emissions in bonded-particle models of rock. Int. J. Rock Mech. Min. Sci. 37 (5), 867–872. https://doi.org/10.1016/S1365-1609(00)00017-4.

Helbig, K., 1994. Foundation of Anisotropy for Exploration Geophysicists. Pergamon, Oxford.

Hertz, H., 1882. Ueber die Berührung fester elastischer Körper. J. Reine Angew. Math. 92, 156–171. https://doi.org/10.1515/crll.1882.92.156.

Heyliger, P., Ledbetter, H., Kim, S., 2003. Elastic constants of natural quartz. J. Acoust. Soc. Am. 114 (2), 644–650. https://doi.org/10.1121/1.1593063.

Hill, R., 1952. The elastic behaviour of a crystalline aggregate. Proc. Phys. Soc. Lond. A 65 (5), 349–354. https://doi.org/10.1088/0370-1298/65/5/307.

Hill, R., 1963. Elastic properties of reinforced solids: some theoretical principles. J. Mech. Phys. Solids 11 (5), 357–372. https://doi.org/10.1016/0022-5096(63)90036-X.

Holt, R.M., Fjær, E., 2003. Wave velocities in shale – a rock physics model. In: Proc. 65th EAGE Conference & Exhibition. Stavanger, Norway.

Holt, R.M., Kolstø, M.I., 2017. How does water near clay mineral surfaces influence the rock physics of shales? Geophys. Prospect. 65 (6), 1615–1629. https://doi.org/10.1111/1365-2478.12503.

Holt, R.M., Li, L., Stenebråten, J.F., 2007. Compaction behavior of unbonded granular media: discrete particle vs. experimental vs. analytical modeling. ARMA-07-049. In: Rock Mechanics: Meeting Society's Challenges and Demands. Taylor & Francis, pp. 393–401.

Holt, R.M., Fjær, E., Bauer, A., 2013. Static and dynamic moduli—so equal, and yet so different. ARMA13-521. In: 47th U.S. Rock Mechanics/Geomechanics Symposium. San Francisco, CA, USA.

Horii, H., Nemat-Nasser, S., 1985. Compression-induced microcrack growth in brittle solids: axial splitting and shear failure. J. Geophys. Res. 90 (B4), 3105–3125. https://doi.org/10.1029/JB090iB04p03105.

Horne, S.A., 2013. A statistical review of mudrock elastic anisotropy. Geophys. Prospect. 61 (4), 817–826. https://doi.org/10.1111/1365-2478.12036.

Hovem, J.M., 1995. Acoustic waves in finely layered media. Geophysics 60 (4), 1217–1221. https://doi.org/10.1190/1.1443850.

Hudson, J.A., 1981. Wave speeds and attenuation of elastic waves in material containing cracks. Geophys. J. R. Astron. Soc. 64, 133–150. https://doi.org/10.1111/j.1365-246X.1981.tb02662.x.

Klerck, P.A., Sellers, E.J., Owen, D.R.J., 2004. Discrete fracture in quasi-brittle materials under compressive and tensile stress states. Comput. Methods Appl. Mech. Eng. 193, 3035–3056. https://doi.org/10.1016/j.cma.2003.10.015.

Kuster, G.T., Toksöz, M.N., 1974. Velocity and attenuation of seismic waves in two-phase media: Part 1. Theoretical formulations. Geophysics 39 (5), 587–606. https://doi.org/10.1190/1.1440450.

Li, L., Holt, R.M., 2002. Particle scale reservoir mechanics. Oil Gas Sci. Technol. – Rev. IFP 57 (5), 525–538. https://doi.org/10.2516/ogst:2002035.

Lisjak, A., Grasselli, G., 2014. A review of discrete modeling techniques for fracturing processes in discontinuous rock masses. J. Rock Mech. Geotech. Eng. 6 (4), 301–314. https://doi.org/10.1016/j.jrmge.2013.12.007.

Mackenzie, J.K., 1950. The elastic constants of a solid containing spherical holes. Proc. Phys. Soc. B 63 (1), 2–11. https://doi.org/10.1088/0370-1301/63/1/302.

Makse, H.A., Gland, N., Johnson, D.L., Schwartz, L., 2004. Granular packings: nonlinear elasticity, sound propagation, and collective relaxation dynamics. Phys. Rev. E 70, 061302. https://doi.org/10.1103/PhysRevE.70.061302.

Marion, D., Nur, A., 1991. Pore-filling material and its effect on velocity in rocks. Geophysics 56, 225–230. https://doi.org/10.1190/1.1443034.

Marion, D.P., Coudin, P., 1992. From ray to effective medium theories in stratified media: an experimental study. SEG Expand. Abstr. 11, 1341–1343. https://doi.org/10.1190/1.1821989.

Mavko, G., Mukerji, T., 2013. Estimating Brown–Korringa constants for fluid substitution in multimineralic rocks. Geophysics 78 (3), L27–L35. https://doi.org/10.1190/GEO20122-0056.1.

Mavko, G., Nur, A., 1978. The effect of nonelliptical cracks on the compressibility of rocks. J. Geophys. Res. 83 (B9), 4459–4468. https://doi.org/10.1029/JB083iB09p04459.

Mavko, G., Mukerji, T., Dvorkin, J., 1998. The Rock Physics Handbook. Cambridge University Press, New York. (Note: Third edition 2020).

Milton, G.W., 1981. Bounds on the electromagnetic, elastic and other properties of two-component composites. Phys. Rev. Lett. 46, 542–545. https://doi.org/10.1103/PhysRevLett.46.542.

Mindlin, R.D., 1949. Compliance of elastic bodies in contact. J. Appl. Mech. 16, 259–268.

Mindlin, R.D., Deresiewicz, H., 1953. Elastic spheres in contact under varying oblique forces. J. Appl. Mech. 20, 327–344.

Moës, N., Dolbow, J., Belytschko, T., 1999. A finite element method for crack growth without remeshing. Int. J. Numer. Methods Eng. 46 (1), 131–150. https://doi.org/10.1002/(SICI)1097-0207(19990910)46:1<131::AID-NME726>3.0.CO;2-J.

Munjiza, A., Andrews, K.R.F., White, J.K., 1999. Combined single and smeared crack model in combined finite-discrete element analysis. Int. J. Numer. Methods Eng. 44 (1), 41–57. https://doi.org/10.1002/(SICI)1097-0207(19990110)44:1<41::AID-NME487>3.0.CO;2-A.

Murphy, W., Reischer, A., Hsu, K., 1993. Modulus decomposition of compressional and shear velocities in sand bodies. Geophysics 58 (2), 227–239. https://doi.org/10.1190/1.1443408.

Nihei, K.T., Hilbert Jr., L.B., Cook, N.G.W., Nakagawa, S., Myer, L.R., 2000. Frictional effects on the volumetric strain of sandstone. Int. J. Rock Mech. Min. Sci. 37 (1), 121–132. https://doi.org/10.1016/S1365-1609(99)00098-2.

Nur, A., Marion, D., Yin, H., 1991. Wave velocities in sediments. In: Hovem, J.M., Richardson, M.D., Stoll, R.D. (Eds.), Shear Waves in Marine Sediments. Kluwer Academic Publishers, Dordrecht, Netherlands, pp. 131–140.

Nur, A., Mavko, G., Dvorkin, J., Gal, D., 1995. Critical porosity: the key to relating physical properties to porosity in rocks. SEG Expand. Abstr. 11, 878–880. https://doi.org/10.1190/1.1887540.

Nur, A., Mavko, G., Dvorkin, J., Galmudy, D., 1998. Critical porosity: the key to relating physical properties to porosity in rocks. Lead. Edge 17, 357–362. https://doi.org/10.1190/1.1437977.

O'Connell, R., Budiansky, B., 1977. Viscoelastic properties of fluid-saturated cracked solids. J. Geophys. Res. 82 (36), 5719–5735. https://doi.org/10.1029/JB082i036p05719.

Potyondy, D.O., Cundall, P.A., 2004. A bonded-particle model for rock. Int. J. Rock Mech. Min. Sci. 41 (8), 1329–1364. https://doi.org/10.1016/j.ijrmms.2004.09.011.

Pyrak-Nolte, L.J., Myer, L.R., Cook, N.G.W., 1990. Transmission of seismic waves across single natural fractures. J. Geophys. Res. 95 (B6), 8617–8638. https://doi.org/10.1029/JB095iB06p08617.

Radjai, F., Jean, M., Moreau, J-J., Roux, S., 1996. Force distributions in dense two-dimensional granular systems. Phys. Rev. Lett. 77, 274–277. https://doi.org/10.1103/PhysRevLett.77.274.

Rathore, J.S., Fjaer, E., Holt, R.M., Renlie , L., 1995. P- and S-wave anisotropy of a synthetic sandstone with controlled crack geometry. Geophys. Prospect. 43 (6), 711–728. https://doi.org/10.1111/j.1365-2478.1995.tb00276.x.

Rice, J.R., 1968. Mathematical analysis in the mechanics of fracture. In: Liebowitz, H. (Ed.), Fracture: An Advanced Treatise, vol. 2. Academic Press, New York, pp. 191–311.

Rice, J.R., Cleary, M.P., 1976. Some basic stress diffusion solutions for fluid-saturated elastic porous media with compressible constituents. Rev. Geophys. Space Phys. 14 (2), 222–241. https://doi.org/10.1029/RG014i002p00227.

Saeb, S., Amadei, B., 1992. Modelling rock joints under shear and normal loading. Int. J. Rock Mech. Min. Sci. Geomech. Abstr. 29 (3), 267–278. https://doi.org/10.1016/0148-9062(92)93660-C.

Salganik, R.L., 1973. Mechanics of bodies with many cracks. Mech. Solids 8 (4), 135–143.

Sayers, C.M., den Boer, L.D., 2018. The elastic properties of clay in shales. J. Geophys. Res. 123 (7), 5965–5974. https://doi.org/10.1029/2018JB015600.

Sayers, C.M., den Boer, L.D., 2020. Effect of variations in microstructure on clay elastic anisotropy. Geophysics 85 (2), MR73–MR82. https://doi.org/10.1190/GEO2019-0374.1.

Sayers, C.M., Kachanov, M., 1995. Microcrack-induced elastic wave anisotropy of brittle rocks. J. Geophys. Res. 100 (B3), 4149–4156. https://doi.org/10.1029/94JB03134.

Schwartz, L.M., 1984. Acoustic properties of porous systems: II. Microscopic description. In: Johnson, D.L., Sen, P.N. (Eds.), Physics and Chemistry of Porous Media. Am. Inst. Phys., New York, pp. 105–118.

Skipper, N.T., Refson, K., McConnell, J.D.C., 1991. Computer simulation of interlayer water in 2:1 clays. J. Chem. Phys. 94, 7434–7445. https://doi.org/10.1063/1.460175.

Stoll, R.D., 1989. Stress-induced anisotropy in sediment acoustics. J. Acoust. Soc. Am. 85 (2), 702–706. https://doi.org/10.1121/1.397596.

Tarokh, A., Detournay, E., Labuz, J., 2018. Direct measurement of the unjacketed pore modulus of porous solids. Proc. R. Soc. A 474. https://doi.org/10.1098/rspa.2018.0602.

Thomsen, L., 1985. Biot-consistent elastic moduli of porous rocks: low-frequency limit. Geophysics 50 (12), 2797–2807. https://doi.org/10.1190/1.1441900.

Thomsen, L., 1995. Elastic anisotropy due to aligned cracks in porous rock. Geophys. Prospect. 43 (6), 805–829. https://doi.org/10.1111/j.1365-2478.1995.tb00282.x.

Torquato, S., Truskett, T.M., Debenedetti, P.G., 2000. Is random close packing of spheres well defined? Phys. Rev. Lett. 84, 2064–2067. https://doi.org/10.1103/PhysRevLett.84.2064.

Torset, S., Holt, R.M., 2021. Stress path dependent effective medium model for granular media – comparison with experimental data. Geophysics 86 (1), EN27–V86. https://doi.org/10.1190/geo2020-0390.1.

Walpole, L.J., 1966. On bounds for the overall elastic moduli of inhomogeneous systems—I. J. Mech. Phys. Solids 14, 151–162. https://doi.org/10.1016/0022-5096(66)90035-4.

Walsh, J.B., 1965a. The effect of cracks on the compressibility of rock. J. Geophys. Res. 70 (2), 381–389. https://doi.org/10.1029/JZ070i002p00381.

Walsh, J.B., 1965b. The effect of cracks on the uniaxial elastic compression of rocks. J. Geophys. Res. 70 (2), 399–411. https://doi.org/10.1029/JZ070i002p00399.

Walton, K., 1987. The effective elastic moduli of a random packing of spheres. J. Mech. Phys. Solids 35 (2), 213–226. https://doi.org/10.1016/0022-5096(87)90036-6.

Wang, Z., Nur, A., 1992. Elastic wave velocities in porous media: a theoretical recipe. In: Wang, Z., Nur, A. (Eds.), Seismic and Acoustic Velocities in Reservoir Rocks, 2. In: SEG Geophysics Reprint Series, vol. 10. Society of Exploration Geophysicists, Tulsa, Oklahoma, pp. 1–35.

Watt, P., Davies, G.F., O'Connell, R., 1976. The elastic properties of composite materials. Rev. Geophys. Space Phys. 14 (4), 541–563. https://doi.org/10.1029/RG014i004p00541.

White, J.E., 1983. Underground Sound. Elsevier, New York.

Williams, J.R., Rege, N., 1997. Coherent vortex structures in deforming granular materials. Mech. Cohes.-Frict. Mater. 2, 223–236. https://doi.org/10.1002/(SICI)1099-1484(199707)2:3<223::AID-CFM30>3.0.CO;2-F.

Zimmerman, R.W., 1991. Compressibility of Sandstones. Elsevier Science Publishers B.V., Amsterdam.

CHAPTER 7

Mechanical properties and stress data from laboratory analysis

Evaluation of in situ rock mechanical behaviour requires relevant input data: What is the stress state and the pore pressure at the depth of interest? What are the elastic moduli and the strength parameters? Can we expect creep or other types of non-elastic behaviour to take place?

Our main data sources to answer these questions are core material and data recorded in the field (measurements while drilling, wireline logs, seismic data and various well tests). The logs provide data recorded continuously versus depth but do not measure directly the parameters that are needed for a rock mechanical analysis. For instance, the rock strength[1] cannot be measured from wireline logs. Strength estimates may, however, be given if a proper interpretation procedure is applied. Field measurements of rock mechanical parameters are discussed in Chapter 8.

In this chapter, we discuss rock mechanical core analysis. Cores provide a possibility for direct measurement of, for example, rock strength parameters and static elastic properties. We will start by looking at the rocks to be tested: the cores in the laboratory may not be fully representative of the formation we intend to study. They may also have been altered during coring and subsequent handling. Some of these problems cannot be solved, and others can be overcome by proper sample preparation methods, test procedures and correction procedures. On the other hand, if proper preparation procedures are not applied, then severe errors in the test results may be introduced. These factors will be discussed in more detail in Section 7.1. We will then discuss the laboratory equipment and measurement techniques (Section 7.2) applied in rock mechanical testing, including also acoustic measurements.

A rock mechanical core analysis has to be designed according to the purpose of the investigation. If the objective is to predict reservoir com-

[1] Throughout this book, we have avoided the term *formation strength* as a synonym for the mechanical strength of the rock/formation. The reason is that in the petroleum industry, formation strength is not clearly defined. It is frequently used to represent the limit against fracturing and losses but can also be seen to have a rather imprecise meaning related to failure processes or rock mechanical properties of the formation.

Developments in Petroleum Science, Volume 72
ISSN 0376-7361, https://doi.org/10.1016/B978-0-12-822195-2.00016-4

paction, then the tests and testing procedures may not be the same as, for instance, in a borehole stability study. Consideration also has to be given to the rock type which is going to be tested. Shales require both special preparation procedures and special test procedures. In Section 7.3, we describe relevant test procedures, which are applied in rock mechanical testing and give some recommendations for test interpretation.

Numerous attempts to determine in situ stresses from laboratory testing of core samples have been reported in the literature. The most relevant test methods are discussed in Section 7.4.

In some cases it may be useful to perform a set of simpler measurements, which can be used as indicators of rock strength and stiffness. This can provide information at less cost and material consumption. The latter is particularly relevant if the core is in a poor state and larger test plugs cannot be prepared. If performed at the rig-site, results may be available closer to real-time. Some of these so-called index tests are described in Section 7.5.

7.1 Core samples for rock mechanical laboratory analysis

7.1.1 Core representativeness and size effects

As we stated above, cores are our only possibility of obtaining direct measurements of rock strength and static mechanical parameters. Cores are, however, only available from discrete levels and may only be representative of the formations in or near the borehole itself. The limited availability of core material may cause large uncertainty in the use of results. Normally, cores are taken mainly or only in the reservoir, whereas the mechanical behaviour of the overlying strata may be of large importance for drilling problems, subsidence and repeated seismic interpretation during reservoir depletion.

A core plug used in standard mechanical testing has a diameter of 1″–2.5″. Only in special cases, larger (4″–5″) or smaller (0.1″–1″) samples will be used. This means that the tested cores may, in case of inhomogeneities, not be representative of the interval under investigation. For instance, in a fractured zone the rocks to be tested are normally taken from the intact parts. We have to recall that the elastic behaviour of the fractured rock mass can be very different from that of the intact rock (Chapter 6). This means that if the important properties are those of the rock mass, then the analysis will be biased. The samples should then, if possible, be taken so that fractures are contained in them. If, on the other hand, the data will be applied in, for example, sand prediction analysis, then the fractures may not be relevant, and the samples should be taken from the intact parts of the rock.

The discussion above shows that the size of the sample may be important. If the fractures occur on a scale of several metres, then it is very unlikely that tests on small samples will provide reliable data. Fractures and cracks may, however, occur on several length scales, down to and below the sample size. Statistically, a larger sample is expected to contain larger cracks than a small sample. Qualitatively, this would lead to an increased strength with decreasing sample size, since the strength decreases with the size of the largest crack. This hypothesis is often valid in experiments (see e.g. Vutukuri et al., 1974) but is not generally confirmed. In fact, experiments with sedimentary rocks sometimes may show an opposite trend (strength increases with sample size). Even though the size effect problem has been studied since the advent of material property testing in the 18th century (see the interesting historical overview in Hudson et al., 1972), it still remains unsolved.

Clearly, in granular rocks, there is a lower limit to the sample size given by the requirement that the sample should contain a large number of grains. In the procedures suggested by the ISRM (International Society for Rock Mechanics) (Bieniawski and Bernede, 1979), it is stated that the diameter of a specimen prepared for rock mechanical testing should be at least ten times the diameter of the largest grain in the rock.

7.1.2 Core alteration

A rock specimen is most likely altered when taken from in situ to laboratory conditions. In situ it may have been at a depth of one or more kilometres, at an ambient temperature of say 50–150 °C, and in chemical equilibrium with its own pore fluid. When cored, it is brought to atmospheric conditions in terms of stress, pore pressure and temperature. It is also in contact with some chemically active/inactive drilling mud, and a mud filtrate may penetrate the core. Thus the main core alteration mechanisms are stress release, pore pressure release (especially in low-permeability rocks), thermal effects and chemical effects due to fluid exposure. The cores may be further damaged by the action of the drill bit and by subsequent improper handling and storage (see Section 7.1.3).

Stress release probably has the most severe effect on rock mechanical properties. Fig. 7.1 schematically shows the effective stress field alterations experienced by a core during the coring process. This is under the assumption of a vertical well drilled in an initial stress field with a vertical stress

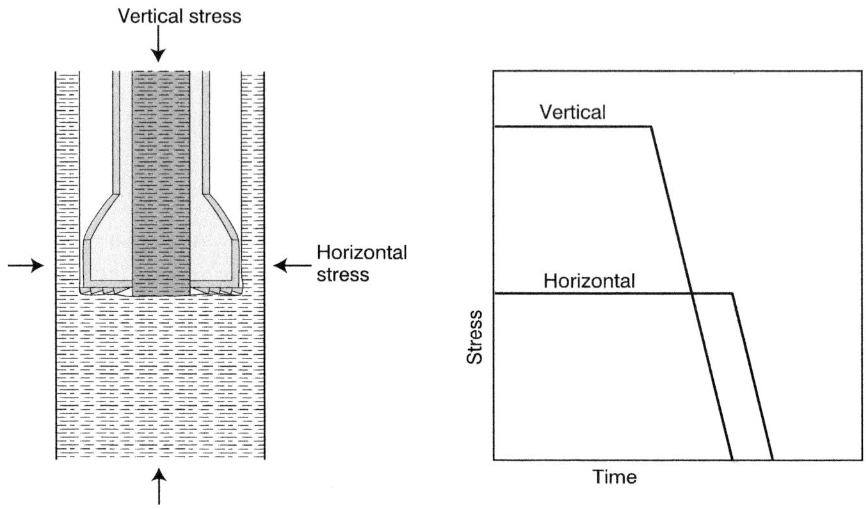

Figure 7.1 Schematic illustration of the effective stress field alteration experienced during coring of a vertical core in a normally stressed rock. (After Holt et al., 2000).

and isotropic horizontal stresses. As the core is drilled, the vertical stress is reduced first, whereas the horizontal stress is reduced as the core enters the core barrel. Thus there is a period where the horizontal stress may be significantly larger than the vertical stress. This may result in yielding and/or failure of the core. Thus the amount of core recovery and the size of the recovered pieces are qualitative measures of the rock strength. The weaker the rock, the less intact core material will be recovered. Macroscopic failure may also occur as tensile failure (core discing). This has been used to evaluate in particular the ratio between horizontal and vertical principal stress components (Maury et al., 1988).

Laboratory testing of rock alteration due to stress release has shown an effect on the mechanical properties of sandstone (Holt et al., 2000), resulting in a reduced uniaxial compressive strength, reduced stiffness, reduced acoustic velocities and increased stress-sensitivity.

Shale may have nanoDarcy or lower permeability (Section 3.4.3), which means that pore pressure equilibrium will not be maintained when the core is retrieved from depth. Thus, tensile failure is very likely to occur, either macroscopically or on a microscopic level. This process, as well as volumetric expansion during coring and retrieval, causes the shale core to be incompletely saturated when reaching the surface. As discussed further (see Section 7.1.4), this may affect shale properties significantly and needs to be accounted for in sample preparation and conditioning.

7.1.3 Core handling

We have seen above that cores may be disturbed by the coring process itself. Further treatment on the rig, during transport, during storage and prior to testing in the laboratory may also be critically important to the outcome of the rock mechanical analysis. There are two major problems which are relevant for sedimentary rocks. One is the fact that they can be mechanically weak and in some cases completely unconsolidated. The other is that they may be sensitive to the exposure to or loss of fluids.

For poorly consolidated *sandstones*, samples may not be kept intact unless they are frozen immediately after coring and kept so until mounted in the triaxial cell. Freezing should, however, be avoided if possible, since it may induce permanent changes in the rock. This is in particular the case if water is present. Since clay minerals contain bound water, freezing may cause disintegration and redistribution of these minerals. If the clay minerals act as cement, then this will clearly affect the mechanical properties.

For sandstones (in particular, clay-rich sandstones), the rock mechanical properties are also found to depend on moisture content. The general observation is that (see e.g. Colback and Wiid, 1965) uniaxial compressive strength decreases with increasing moisture content. Colback and Wiid relate this to a change in the surface energy.

It has been observed by the authors of this book that a sandstone stored in normal laboratory atmosphere for one year after sampling increased its strength by a factor of two due to drying. This was due to smectite in the sandstone, which contributed significantly to the cementation when dried.

Shales contain both bound water and free water and should therefore *never* be frozen. Shales are also extremely sensitive to contact with wetting fluids like water, or to loss of water from the pores. This means that the samples should be preserved as soon as the core is retrieved, so that the natural water content is not reduced. Dehydration may cause them to disintegrate completely within a matter of minutes. In other cases, depending on the microstructure and the shale composition, dehydration may lead to strengthening of the material.

Shale cores intended for rock mechanical tests should be sealed immediately after coring, for instance, by wrapping them in a plastic film with further protection like aluminium and wax on the outside. If this packing process cannot be performed immediately after the core has been landed on the rig, then it is still important to avoid water transport effects. One alternative is to cut the core and the core liner into appropriate lengths (typically, 1 m), wipe the mud from the core surface (especially if water-based mud has been used), fill the annulus between the core and the liner with an inert fluid like oil and then seal the liner. In all cases, it is essential to minimise the period when the core is exposed to air.

Shales should not be contacted with water or any other fluid that wets the shale. Even if the water is assumed to completely match the chemistry of the pore water, large capillary suction pressures may be induced because the shale core is no longer 100 % saturated when it is under atmospheric conditions. The capillary pressure is generated to counterbalance the surface tension that develops at the interface between the wetting (water) and non-wetting (gas) fluid.

The capillary pressure generated in a gas bubble as a rock sample is entirely surrounded by water is given by the Young–Laplace equation

$$p = \frac{2\gamma \cos\theta}{r} \tag{7.1}$$

where γ is the interfacial tension between the two phases, θ is the contact angle between the wetting phase and the solid, and r is the radius of the capillary tube, which in a porous medium would be characteristic of the pore size.

Consider a water/air interface in shale with a mean pore size (radius) of 20 nm (Horsrud et al., 1998). With an interfacial tension of 0.07 N/m and fully wetting conditions ($\theta = 0$), this gives a potential capillary pressure of 7 MPa. Such a large internal pressure may easily exceed the tensile strength of the shale and cause rupture and disintegration of the shale.

7.1.4 Preparation of test samples

Cores normally arrive at the laboratory at 4″–5″ in diameter and in lengths up to one metre. Core plugs then have to be drilled from these full-size core sections. When drilling the core plugs, it is important to be consistent with respect to drilling direction, since rock properties are usually anisotropic. This is often linked to the bedding direction. The strength has a minimum when the sample is oriented with its long axis at an oblique angle to the bedding (see Section 2.9). It may therefore be necessary to perform a series of measurements with specimens oriented with their axes parallel to, normal to, and at intermediate angles with respect to the bedding plane. The possibility of performing this type of investigation is in practice limited by the amount of core material available. In deviated wells with unoriented core, it is therefore necessary to make sure that the bedding direction can be established before core plugs are drilled.

It is recommended to avoid sampling the outer centimetre of the core due to potential alteration of this part of the core. This alteration may have been induced by the drilling mud and invasion of drilling mud filtrate.

Core drilling may be very difficult in weak rocks. Although not recommended, drilling and handling of unconsolidated sandstone may be impossible unless the sample is kept frozen. For a somewhat consolidated sandstone, pressurised air may be used for cooling and for removal of drill cuttings instead of water. If a drilling fluid is used, then it should be chemically compatible with the rock. When drilling chalk cores, it is customary to apply water saturated with the same chalk.

Shales and other strongly layered rocks may be difficult to drill normal to the bedding, since the bedding is often a weak plane. The tendency for core discing may be reduced by applying a small axial load on the sample (inside the coring bit) during drilling.

Shale samples should be drilled using an inert fluid like a mineral oil. For reasons already discussed, contact with water may induce capillary effects, even if the water is believed to have exactly the same chemical composition as the pore water. These capillary effects are due to incomplete saturation of the shale and the presence of air in the sample. It has been proposed to remedy this by resaturating the shale (e.g. Schmitt et al., 1994). This is not done by immersion, but in a controlled humidity atmosphere where water vapour condenses in the pores. To avoid trapped air, it is preferable to have a continuous gas phase before starting resaturation. This may require a first stage with drying of the sample. However, it has been strongly argued that such a process may change the structure and properties of the shale (Santos et al., 1996). Therefore attempts to resaturate shale should be avoided. The best approach is to make sure that the shale is well preserved, so that sufficient resaturation can be obtained by loading the sample prior to or during testing (Horsrud et al., 1998; Ewy, 2018). The water content of the rock should always be measured before rock mechanical tests are performed.

The significance of sample saturation is well known from soil mechanics. In rocks with small pores (shale, chalk) the presence of air may cause significant overestimation of the mechanical properties of the rock (Schmitt et al., 1994; Papamichos et al., 1997). As a first-order approximation, we may express this as a generalised effective stress

$$\sigma' = \sigma - p_{gas} + S_w(p_{gas} - p_{water}) \tag{7.2}$$

where S_w is the water saturation, and $p_{gas} - p_{water}$ is the pore-water capillary suction. The capillary suction effect may then be compared with the effect of a confining stress. See also Section 2.6.2.

Reservoir rocks (sandstone, chalk) may also be sensitive to the fluid in the pores. A dry rock is usually stronger than a fluid-saturated rock. When measuring petrophysical properties of sandstone, it is common procedure to clean the sample and then resaturate it with the relevant fluid(s). When doing rock mechanical testing of weak sandstone, this is not recommended, since the cleaning process may disturb the cementation of the rock. Weak sandstones should be well preserved and tested with the native fluid content.

Another important observation relates to the high-porosity North Sea chalks. Laboratory testing has shown that oil-saturated chalk samples are significantly stronger than water-saturated samples (Andersen, 1995). It has been suggested that this effect could be explained by capillary forces caused by the irreducible water still present in oil-saturated chalk. However, it has been shown that the chalk is strengthened (both in compression and tension) also when saturated with methanol, which is fully miscible with water (Risnes and Flaageng, 1999). Thus there must be other mechanisms present, which have not yet been explained.

The ISRM (International Society for Rock Mechanics) standards (Bieniawski and Bernede, 1979) require that specimens intended for standard rock mechanical tests are right circular cylinders with length-to-diameter ratio L/D between 2.5 and 3. The reason for this is that the sample should be sufficiently long to accommodate a shear plane penetrating through the side walls. If the sample is too short, then the shear plane will penetrate through the end faces, and an additional support will be provided. Fig. 7.2 illustrates this, and the expected effect of variations in L/D on strength. The chosen criterion therefore makes sure that the length-to-diameter ratio is such that the uniaxial compressive strength is rather insensitive to minor deviations in L/D. This type of relationship has been confirmed by experiments (see e.g. Vutukuri et al., 1974).

As mentioned above, the diameter should be at least ten times the diameter of the largest grain. The end surfaces should according to ISRM standards be flat to within ± 0.01 mm (which is unrealistic in coarse-grained, weak materials!), the parallelism better than $\Delta L/L = 0.001$, and the sides smooth and straight to within 0.3 mm over the full sample length. The sample diameter should be measured to the nearest 0.1 mm by averaging diameters measured at right angles to each other at the top, the middle and the bottom of the sample. The sample length should be determined to the

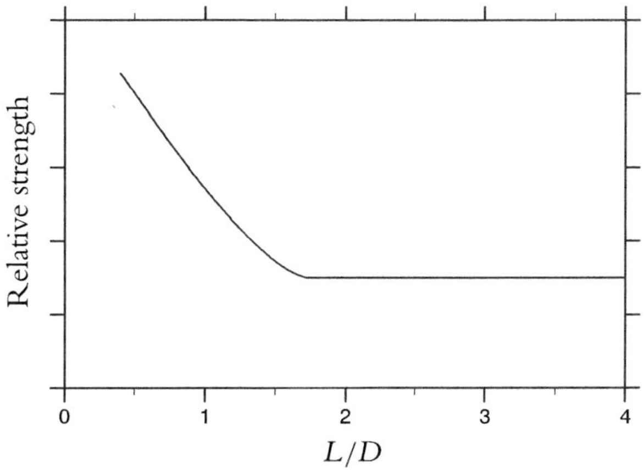

Figure 7.2 Sketch of the effect of length/diameter (L/D) ratio on the uniaxial compressive strength of rock.

nearest 1.0 mm, but a higher accuracy is requested if, for example, acoustic velocity measurements are performed during rock mechanical testing.

A geological description of the rock samples can be useful for understanding the mechanical response. This includes an overview of the mineralogy, with focus on how the rock is cemented. The grain size distribution should be obtained, either from sieve analysis or from thin section analysis. Photos of the samples should be made prior to and after testing.

Since the results are going to be used in downhole operations (drilling, completions, stimulation), it is important that the laboratory results can be compared with well log readings. An accurate measure of the core depth is therefore required. Abrupt lithology changes can easily be recognised in the recovered core and on logs. One possibility is to use a Gamma ray detection tool during coring and compare the signatures with readings of the Gamma ray logging tool. If anelastic strain recovery, differential strain curve analysis (see Section 7.4) or other techniques for determining the orientation of the in situ stress field are applied, then it is necessary also to have *oriented* cores. Cores may be oriented by knives inscribing the geographical direction on the core during coring or by using geological markers.

7.2 Laboratory equipment

A standard rock mechanical test laboratory for studies of sedimentary rocks includes (see Fig. 7.3) the following equipment:

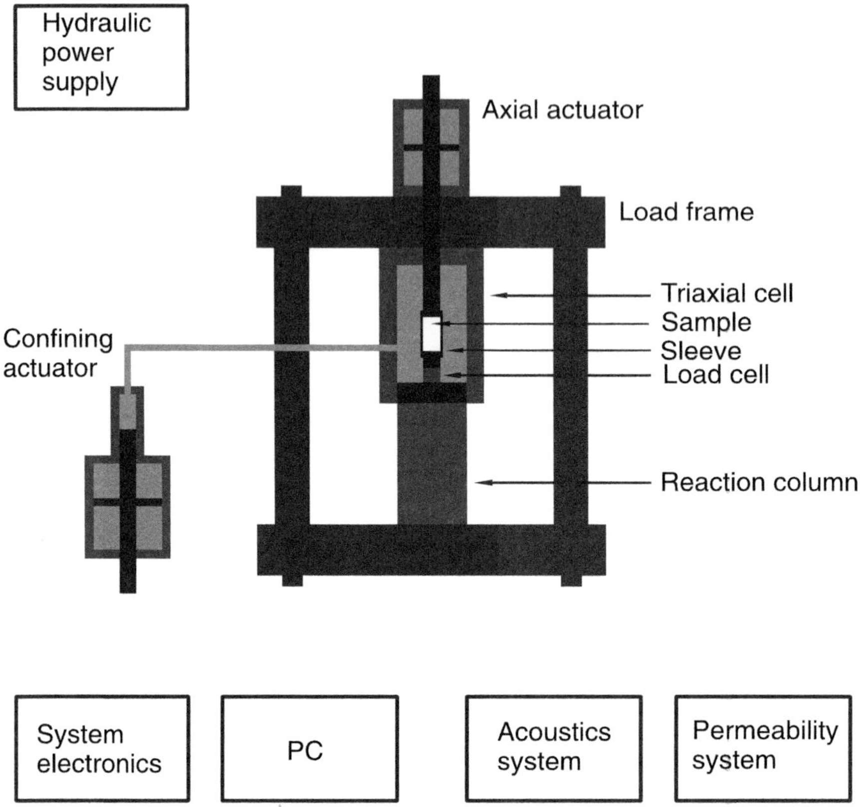

Figure 7.3 Schematic drawing of a rock mechanics laboratory system. On other systems the triaxial cell may, for example, be closed by bolts instead of a reaction column, and the axial actuator may be mounted on the lower platen.

- Load frame
- Triaxial cell
- Confining pressure system
- Pore pressure system
- Computer for control and data processing

7.2.1 The load frame

The load frame is a central element in the test equipment. Two features are essential for its operation in a rock mechanical laboratory, the frame stiffness and the ability to operate the system in a mode of constant displacement rate (Hudson et al., 1972).

Traditionally, the force was regarded as an independent variable in testing materials. If a brittle rock sample is loaded with a constant load increment per unit time, then the material will fail abruptly, and no controlled measurement can be achieved in the post-failure region. This is unsatisfactory for most rock testing, since the load bearing capacity of a rock after it has failed may be an important design parameter. For the post-failure behaviour to be properly recorded, the experiment has to be run in displacement control. This is achieved with most modern equipment, which is servo-hydraulically controlled. This means that a feedback control system is implemented: a detected difference between the recorded displacement and the programmed displacement can then be used to control a servo-valve, which induces further displacement of the loading pistons.

Studies of post-failure behaviour of brittle rocks require also that the testing apparatus is stiff compared to the maximum slope in the descending part of the stress–strain curve (Hudson et al., 1972). If not, a major part of the energy stored in the machine-specimen system will be released during the fracture process, causing the measured characteristics beyond failure to be an artifact induced by the test machine. However, by using a stiff and servo-hydraulically controlled test machine the experiment can be run at a constant deformation rate, provided that the response time of the feedback system is sufficiently fast. In this mode, post-failure behaviour can be studied in detail.

Whereas in situ changes have taken place on a geological time scale and changes during drilling or petroleum production occur within seconds, minutes or years, laboratory testing is normally performed with strain rates of 10^{-7}–$10^{-1}\,s^{-1}$. The choice of a particular strain rate depends on the nature of the rock and the problem that is actually studied. It is generally thought that the rock strength increases with increasing strain rate. Experiments with different rocks (Vutukuri et al., 1974) show significant increases in strength at strain rates above $\approx 10^{-2}\,s^{-1}$. There are at least two mechanisms that may cause such strain rate effects, namely creep and consolidation (see Section 2.10.1). In particular, if tests are performed with low-permeability rocks (e.g. shales) in which pore fluid drainage should be facilitated, then it is important that sufficient time is allowed for the pore fluid to escape during sample compression (see e.g. Swan et al., 1989). If we wish to quantify strain rate effects as part of a dedicated test programme, then at least three orders of magnitude in the strain rate should be spanned (Jones, 1988). This may mean that tests in extreme cases have to be run

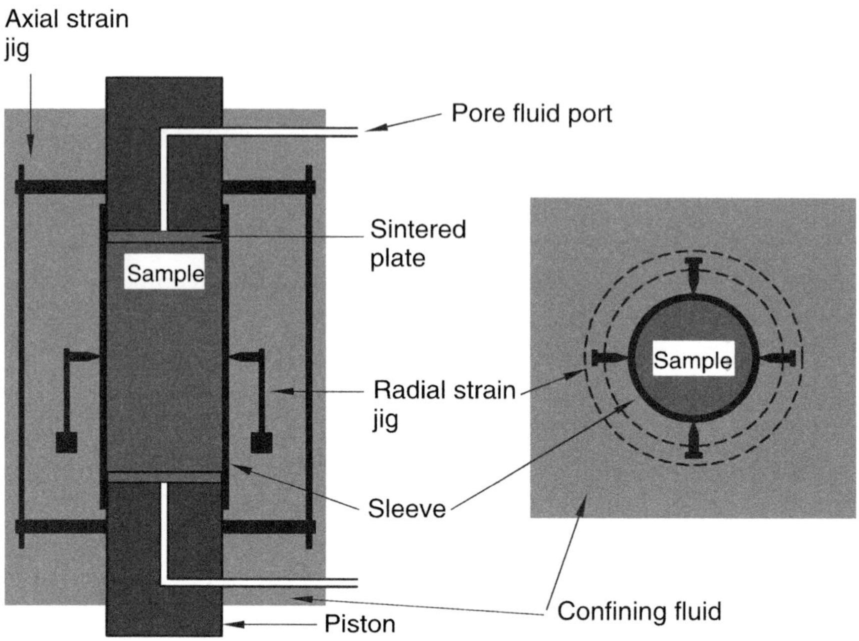

Figure 7.4 Principle sketch of the interior of a triaxial cell, showing the loading piston with fluid ports, optional sintered plates for fluid distribution, a radial strain jig (suspended from the specimen) measuring two orthogonal diameters and an axial strain jig measuring change in sample length between the pistons.

over periods of several weeks or even months, which imposes a strong requirement on system stability.

Further key parameters of the load frame are the maximum operating force in compression and extension, total stroke, free compression space and accuracy of the force or displacement control.

7.2.2 The triaxial cell

Another central element of the test system is the triaxial cell (Fig. 7.4). There are two types of triaxial cells used in rock testing (Jones, 1988). One is a cell where the sample is placed between two movable pistons (the "Hoek cell"; Hoek and Franklin, 1968). The other uses a fixed pedestal carrying the sample.

A triaxial cell (Fig. 7.4) contains the loading pistons, which are pressed against the sample. The sample is surrounded by a sleeve, which separates it from the confining fluid in the case of confined tests.

The pistons should be made of hard material. The surfaces should be ground and polished, and their flatness should be within 5 μm (Bieniawski and Bernede, 1979). The flatness of the pistons has an important effect on the test results. If the friction coefficient between the sample and the piston is large enough to prevent radial expansion of the sample at the end faces, then this is equivalent to providing the material with an additional support. This will manifest itself as a length:diameter effect, since the support of the end faces will give rise to a higher measured strength for a short than for a long sample.

In rock testing for petroleum applications, we may wish to have fluid inlet and outlet in the pistons to apply and measure pore pressure. This can be done by using sintered pistons or by introducing flow channels in the piston surfaces.

For acoustic measurements, the piston material should ideally have acoustic properties that provide an acoustic impedance-matching between the acoustic transducers and the rock material. The choice of steel as a piston material is in that case not ideal for studies of soft rock samples. Thus the requirements to piston material in terms of flatness and hardness may have to be relaxed to combine rock mechanical testing with other petrophysical measurements. When measuring the permeability during loading, it is recommended to have a system for distribution of the fluid flow at the end surfaces. This can be obtained by making grooves in the pistons or by inserting a sintered plate between the piston and the rock sample. This can, however, impose end effects (e.g. sedimentation of fines in the sintered plates). This is a potentially significant error source if the pressure drop along the core is measured on the outside of the sample.

The sleeve should be soft enough not to provide a significant support to the core. This is normally not a problem, but should be considered when testing very weak samples at low confining pressures. Sleeves are made from rubber, plastic, Teflon or other soft materials. Sleeve wall thicknesses are normally around a few millimetres. To improve the accuracy of radial deformation measurements (and also measurements of acoustic velocities for radially propagating waves), the sleeve should be made as thin as possible.

7.2.3 Measurements of stresses and strains

Stress measurements are performed using strain gauges in the load cell and fluid pressure transducers in the confining fluid in the triaxial cell. The load cell may be external or internal to the pressure vessel. The use of an internal load cell eliminates problems caused by loading ram friction but makes calibration more difficult (Jones, 1988). Pore pressures are measured with pressure transducers that are connected to the pistons. By monitoring the pore pressure drop along a specimen with transducers both in the top and bottom pistons, and in addition measuring the flow rate, it is possible to measure permeability as part of a rock mechanical test.

The radial deformation of the sample can be measured in several ways. One technique is to use LVDTs (Linear Variable Differential Transformers, which operate on a magnetic induction principle) pressed against the sleeve at various points circumferentially around the core. Another widely used method is to have strain gauges glued directly to the sample. The third technique is to use a cantilever system, where the displacement of the arms is measured by strain gauges. The fourth technique is to use a chain wrapped around the core plug, where the change in circumference is monitored with strain gauges. The latter method is accurate but does not have the potential to monitor anisotropic deformations. The other techniques are all essentially point measurements. The accuracy depends on calibration for sleeve deformation. The radial deformation sensors should be mounted on a jig that is not attached to the walls of the triaxial cell. Otherwise, the measurements must be calibrated for deformation of the entire cell.

Axial deformation can be monitored in several ways. Measurement of stroke (i.e. piston movement) is the most common method. To obtain the true sample deformation, this should be done between the pistons inside the triaxial cell and not between the loading platens on the outside. If measured outside the cell, corrections for piston deformation have to be made. Attempts have also been made to measure axial deformation by monitoring the relative displacement of two knife-edges pressed against the sleeve at different axial positions. This technique provides a strain measurement that eliminates end effects but will not cover the entire length of the sample.

Measuring devices mounted inside the triaxial cell must always be calibrated for pressure effects exerted by the confining fluid.

Volume changes may also be used to monitor deformations. The change in pore volume can be measured by monitoring the amount of pore fluid entering or leaving a fully saturated sample. The change in bulk volume may in principle be measured by monitoring the volume change of the confining fluid.

7.2.4 Acoustic measurements

The importance of acoustic measurements has been stated several times throughout this book. A major motivation for doing acoustic laboratory measurements is to support the interpretation of sonic log and seismic measurements. Acoustic laboratory measurements can provide correlations with rock strength and may help evaluate the difference between static and dynamic moduli (Section 5.6). They may also provide information about the in situ stress state.

The range of frequencies where acoustic measurements can be made is limited mainly by the sample and grain sizes. The attenuation of sound sets an upper limit, whereas the requirement that a sample should contain several wavelengths for a propagating wave to be set up imposes a lower limit to the frequency. With normal sample sizes (say 5–10 cm transmission length) the applicable frequencies will be in the 100 kHz–1 MHz range.

The ultrasonic sources and detectors used at these frequencies are piezoelectric (normally piezo-ceramic) transducer elements. They are deformed mechanically when an external voltage is applied to them, and vice versa: an electric field is generated when they are deformed by an external force. Differently manufactured piezo-elements are used for generation of P- and S-waves. By placing transducers at two opposing faces of a rock specimen the sound velocities are measured from transit times of pulsed wavelets through the sample. Measurements at various frequencies can be obtained by a spectral analysis of the waveforms (see Winkler and Plona, 1982), but this is only possible if the transducers are broadband in operation. The transducers can be embedded inside the pistons of a triaxial cell or in the confining fluid for radial transmission. Fig. 7.5 illustrates a set-up used for acoustic laboratory measurements and particularly shows how transducers can be integrated into a rock mechanical test system.

For practical purposes, the quality of acoustic signals is extremely sensitive to the acoustic coupling between the transducers and the sample. This coupling is normally obtained by a thin layer of grease, a gel or another viscous fluid placed between the face of the transducer and the surface of the specimen. It is important that the couplant layer is thin enough not to contribute significantly to the measured transit time.

With shear wave transducers, it is important to make sure that the polarisation of the transmitting and the receiving transducers point in the same direction. The polarisation will normally be marked on the transducers but should be checked by placing the transducers face to face. The maximum signal is then obtained when the polarisations are parallel, whereas the minimum signal is obtained when they are at 90° to each other. If frequencies lower than the 100 kHz–1 MHz range are required, then we may use the resonant bar technique. In this case, the resonance frequency of a vibrating

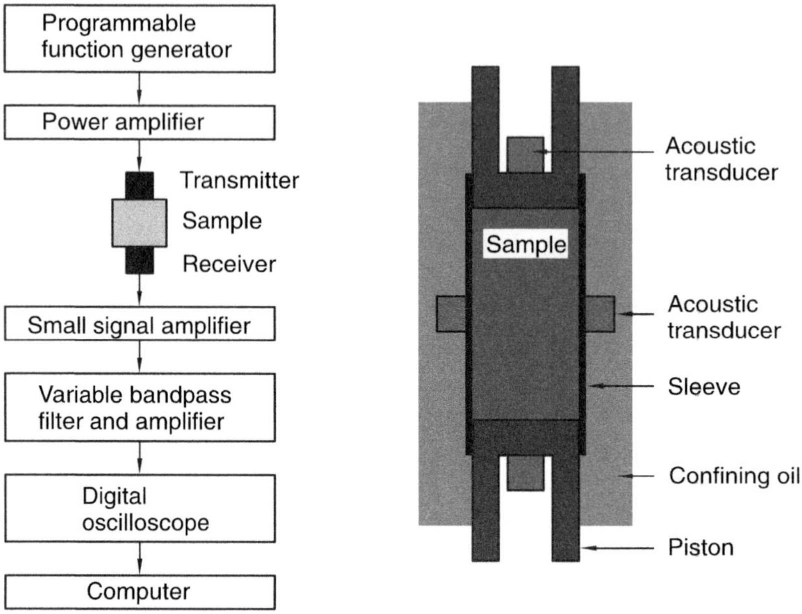

Figure 7.5 Ultrasonic test apparatus (left) and transducer positions inside the triaxial cell (right).

sample is determined. The resonance frequency is related directly to the wave velocity. Different frequencies can be obtained by using samples of various size.

In the seismic frequency range (10–100 Hz), velocities may be obtained from stiffnesses measured in very low strain ($<10^{-6}$) and stress cycles (Spencer, 1981; Batzle et al., 2006; Lozovyi and Bauer, 2019). By measuring strains in various directions and assuming isotropic and homogeneous specimens, both P- and S-wave moduli are estimated. Monitoring of phase shifts allows measurements of attenuation. These are however challenging experiments that cannot yet be considered as routine.

Acoustic measurements in the laboratory should be performed with rocks in unconfined and stressed conditions. As discussed later (Section 7.4), unloaded cores are thought to contain an orientational distribution of microcracks, reflecting the in situ stress field. The possibility of using acoustic velocity measurements on cores to evaluate the in situ stress field will be discussed further in Section 7.4.3.

Measurements of P- and S-wave velocities on core plugs at atmospheric conditions are sometimes performed and correlated with rock strength or elastic moduli measured in a pressure vessel or a triaxial cell. Such correlations are uncertain, since the two measurements refer to different stress conditions. Therefore, to obtain such correlations, acoustic velocities should be measured *during* the static tests. To compare with log readings, the acoustic velocities should be measured in the same direction, at the same stress state, and with the same saturating fluid as under downhole conditions. Even when this is done, experience shows that laboratory measured sound velocities can be different from those measured in the borehole. This may be due to a dispersion effect (Chapter 5), since the frequencies in laboratory studies normally are higher than those used in the field. It may also be due to core alteration, as discussed in Section 7.1.2.

7.3 Laboratory tests for rock mechanical property determination

In this section, we discuss procedures for performing the most common types of rock mechanical tests, including some comments about test interpretation.[2]

7.3.1 Stresses on cylindrical samples

With cylindrical samples and rotational symmetry of the test conditions, it is convenient to describe the tests in cylindrical coordinates r, θ and z. The coordinate axes are then aligned with the principal stress axes, and the corresponding stresses σ_r, σ_θ and σ_z are the principal stresses.

The confining pressure generates a rotationally symmetric and uniform stress field in the plane perpendicular to the sample axis, as long as the sample can be considered isotropic and homogeneous. (Note that the case of a hollow cylinder does not comply with these conditions. The material may be isotropic and homogeneous, but the sample is not). The radial and tangential stresses will be equal, and so will the radial and tangential strains:

$$\sigma_r = \sigma_\theta \tag{7.3}$$

$$\varepsilon_r = \varepsilon_\theta \tag{7.4}$$

[2] The International Society for Rock Mechanics and Rock Engineering (ISRM) has a comprehensive list of suggested methods (explanations of recommended procedures to follow in the various aspects of rock characterisation, testing and monitoring). Details are available on https://www.isrm.net/.

Thus there are only two independent stresses, the radial stress σ_r and the axial stress σ_z. The relations between stress increments and strain increments are given by Hooke's law:

$$E\Delta\varepsilon_r = (1-\nu)\Delta\sigma_r - \nu\Delta\sigma_z \tag{7.5}$$

$$E\Delta\varepsilon_z = \Delta\sigma_z - 2\nu\Delta\sigma_r \tag{7.6}$$

7.3.2 Drained and undrained test conditions

In drained tests the outlets through the pistons will be open, so that the pore fluid pressure can be kept at any prescribed value. Most often, the outlets are open to the atmosphere, so that the pore fluid pressure will be zero. However, it is sometimes of interest to run tests at actual reservoir conditions. The pore pressure must then be kept at the reservoir pressure level. Under drained conditions, the effective stress increments are equal to the total stress increments, and the strain increments are governed by the frame moduli. Then the relations between stress increments and strain increments can be written as

$$E_{\text{fr}}\Delta\varepsilon_r = (1-\nu_{\text{fr}})\Delta\sigma_r - \nu_{\text{fr}}\Delta\sigma_z \tag{7.7}$$

$$E_{\text{fr}}\Delta\varepsilon_z = \Delta\sigma_z - 2\nu_{\text{fr}}\Delta\sigma_r \tag{7.8}$$

In testing low-permeability materials, for instance chalk, a drained condition means that the test must be run sufficiently slow to avoid unacceptable pressure build-up when the sample deforms.

In undrained tests the outlets are in principle closed, so that the pore fluids contained in the sample cannot escape. If the pore fluid pressure is to be measured, then there must of course be hydraulic contact through one of the outlets, but the condition of no fluid flow must be ensured. Undrained conditions are commonly used when testing extremely low-permeability rocks like shale (Section 7.3.7).

For undrained tests, the relations between stress and strain increments are given by Eqs. (7.5) and (7.6) with E and ν representing undrained moduli. Eqs. (7.7) and (7.8) may also be used for analysis of an undrained test if the pore pressure increment Δp_{f} is measured and the Biot coefficient α is known.

From a more formal point of view, drained conditions mean that the pore pressure in the sample is an independent variable. On the other hand, in undrained conditions the pore fluid pressure will be a dependent variable.

7.3.3 Standard triaxial compression tests

Rocks at depth are exposed to an anisotropic stress field. During drilling or petroleum production, enhanced shear stresses occur near the borehole or the producing perforations (see Chapter 4). The lowest effective principal stress component is zero at the borehole or cavity wall if the wall is permeable and slightly higher if, for example, a mudcake blocks the fluid transport across the boundary (and $p_w > p_f$). The uniaxial compressive strength C_0 is the most important strength parameter to describe stability at the borehole wall. At some distance from the wall, however, the minimum principal stress increases. To evaluate borehole or cavity stability it is therefore of great value to know how the mechanical properties and the strength characteristics depend on external load conditions.

The most commonly adopted experimental procedure for this purpose is the so-called triaxial compression test. The axial stress σ_z and the radial confining pressure σ_r are kept equal and increased to a preset level, from which σ_z is increased further until failure has occurred, keeping σ_r constant. In this test, two of the principal stresses are always equal. This definition of a triaxial test (axial symmetry with confinement) is also the definition adopted by the International Society for Rock Mechanics and Rock Engineering. Note that the term *biaxial* testing in material testing is normally referring to the condition where two of the principal stresses act in the same plane and the third is zero. Tests where all the principal stresses have different magnitudes are commonly denoted *true triaxial* tests (see Section 7.3.10.3) or *polyaxial* tests. Note that the terminology used in the literature is not always consistent.

Results from a triaxial failure test are shown in Fig. 7.6. By performing a series of tests at different confining pressures it is possible to map the entire failure envelope (Section 2.7). If the aim of the test programme is a borehole stability or sand prediction analysis, then care should be taken that sufficient data are collected at low confining pressures, representing the stress state near the borehole. The maximum confining pressure should correspond to the minimum principal effective stress far from the borehole. For a producing reservoir, the corresponding value at maximum depletion should be selected.

The test can be run under drained or undrained conditions. Since both the elastic properties and the strength depend on the effective stress, drained tests provide a better experimental control because the pore pressure is kept constant at a known value. With extremely low-permeability shales (nano-Darcy range), however, undrained tests with pore pressure measurement

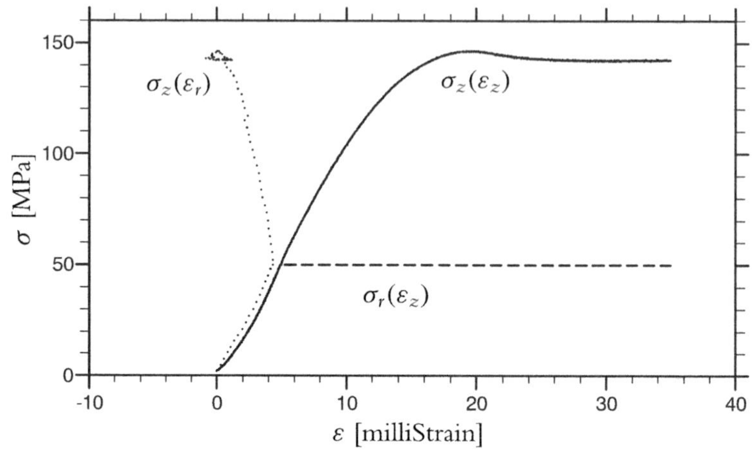

Figure 7.6 Triaxial test. Example of experimental result.

performed during the experiment may be advantageous, since undrained tests require shorter stabilisation times (see Section 7.3.7).

Beyond the brittle–ductile transition, it may be very difficult to define the failure point, which may be visible only as a minor change in the slope of the stress–strain curve. The post-failure behaviour is important, since it tells us the remnant strength of the rock, that is, how much load it can sustain after failure has occurred. It is therefore strongly advised that triaxial tests are run with the load frame in displacement control mode.

7.3.4 Interpretation of elastic moduli from triaxial tests

Fig. 7.6 illustrates a typical response of a *drained* triaxial test, including the hydrostatic loading phase. If we simplify the test response, assuming the stress–strain curves to be linear up to failure, then it is straightforward to interpret the elastic moduli from such a plot.

In the hydrostatic phase the stresses are equal (see also Section 7.3.1), so the *bulk modulus of the framework* is given as $K_{\text{fr}} = \Delta\sigma_z'/\Delta\varepsilon_{\text{vol}} = \Delta\sigma_z'/(\Delta\varepsilon_z + 2\Delta\varepsilon_r)$.

In the triaxial phase, $\Delta\sigma_r' = \Delta\sigma_\theta' = 0$. *Young's modulus* is then given by the slope of the axial stress–strain curve in the triaxial phase, that is, $E_{\text{fr}} = \Delta\sigma_z'/\Delta\varepsilon_z$, whereas *Poisson's ratio* is given by the ratio between the slopes of both the radial and the axial stress–strain curves in the triaxial phase, that is, $\nu_{\text{fr}} = -\Delta\varepsilon_r/\Delta\varepsilon_z$.

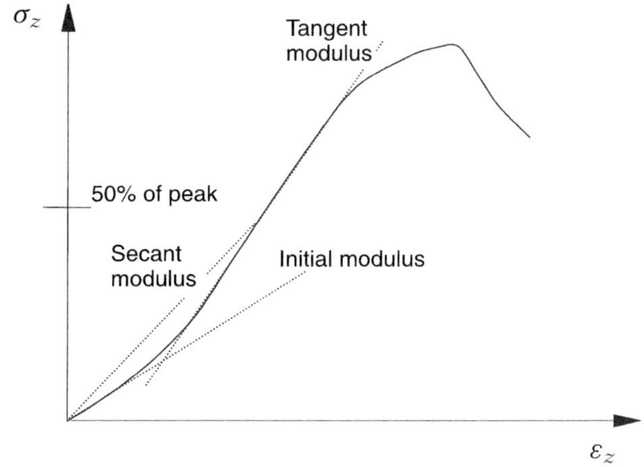

Figure 7.7 Illustration of different methods for calculating Young's modulus from axial stress–strain curves, using the slopes of the different lines.

The *shear modulus* can of course now be calculated directly from Young's modulus and Poisson's ratio (Table 1.1), given that the rock can be assumed to be isotropic. Alternatively, we have $G_{\text{fr}} = \frac{1}{2}\Delta\sigma_z'/(\Delta\varepsilon_z - \Delta\varepsilon_r)$.

It is important to realise that if the pore pressure is not kept constant during the test, then this interpretation requires that the Biot coefficient is known. In the general case, we should consider the full expression (Eqs. (1.150)–(1.156)) and then only eliminate parameters that are kept constant during the test phase of interest.

Another simplification we have made is that the elastic stress–strain response is linear. As illustrated by Fig. 7.6, this is generally not the case, neither in the hydrostatic phase nor in the triaxial phase. When presenting elastic moduli, it is therefore essential to define how the interpretation has been made. For the triaxial phase, the following alternatives are commonly used (see Fig. 7.7):

- Initial modulus, given as the initial slope of the stress–strain curve.
- Secant modulus, measured up to a fixed percentage of the peak stress.
- Tangent modulus, given at a specific percentage of the peak stress.
- Average modulus of a linear portion, given within a specific maximum and minimum stress level.

Out of these, the tangent modulus at 50% of the peak stress value is probably the most used method. However, if the curve is strongly non-linear, then complete information can only be given if the entire curve is presented. Comparing moduli from different sources, it is thus essential to ensure that the same interpretation method has been used. As illustrated by Fig. 7.7, the moduli can vary significantly with the method used for interpretation.

Such an interpretation is based on several assumptions, where the assumption of an elastic material has already been mentioned. Clearly, this assumption is violated if the stress–strain curve looks like Fig. 7.7. In such a case the rock is either nonlinearly elastic or elasto-plastic (or both). We should therefore refer to the measured quantities as static moduli and avoid the term "static elastic" moduli. To reduce the problem of nonlinearity and variation of the stress–strain slope, it is often recommended to perform unloading–reloading cycles at different stress levels and interpret the elastic moduli from the stress–strain response of these loading cycles. The stress levels should be chosen such that the sample is well within the elastic range. Both the stress level and the loading path should be chosen such that it is relevant for the use of the data.

Another assumption is that the sample is completely isotropic and homogeneous. Given perfect boundary conditions, this implies that the stress- and strain-fields are homogeneous throughout the sample and throughout the test sequence. Observations of experiments have shown that this is not always true. Barreling, buckling and shear-band formation are typical examples of loss of homogeneity during the test. This can occur significantly below the peak point of a load-displacement curve and can, for instance, be identified by loss of uniformity of radial strain measurements in different directions.

These phenomena have been the subject of extensive investigation using bifurcation theory, both from an experimental and a more theoretical approach. Examples with application to failure around boreholes are given by Santarelli and Guenot (1989) and Sulem and Vardoulakis (1989).

7.3.5 Uniaxial (unconfined) compression tests

In an unconfined compression test a sample is inserted into a load frame, and the axial load is increased with zero confining pressure. This test is hence a simplification of a standard triaxial compression test. By monitoring the axial stress plus the axial and the radial deformations we can under drained conditions measure (by definition; see Chapters 1 and 2):

- The uniaxial compressive strength (UCS) C_0 as the peak stress.
- Young's modulus E_{fr} (tangential modulus) as the slope of the axial stress versus axial strain curve.
- Poisson's ratio ν_{fr} as the ratio between radial and axial strains.

Most rocks exhibit brittle failure in unconfined failure tests, so it is straightforward to identify C_0. For interpretation of Young's modulus and Poisson's ratio, see Section 7.3.4.

If the rock is anisotropic, then Poisson's ratio and Young's modulus are not alone sufficient to describe the mechanical behaviour. In Section 1.7 the elastic stiffness tensor was defined. Eqs. (1.262)–(1.263) illustrate that the value of the measured Poisson's ratio depends both on the direction of the applied load and the direction of lateral strain measurement. Thus the orientation of the sample symmetry with respect to the directions of applied and measured stress and strains needs to be specified.

Theoretically, the uniaxial compressive strength depends on the largest crack or flaw present in the sample, as well as their orientation. Thus measurements of C_0 are expected to be very sensitive to sample heterogeneity and to cracks induced by coring and core treatment procedures. Large experimental uncertainties may therefore be expected, especially in weak rocks. Since the uniaxial compressive strength is a key parameter in many petroleum related rock mechanics applications, it may be worth considering other methods to obtain or estimate this parameter. One alternative may be to perform triaxial tests at low confining pressures, where some of the cracks not associated with the intrinsic rock behaviour are closed, and extrapolate the results towards zero confinement. Another alternative is a Constant Mean Stress (CMS) test; see also Section 7.3.9.

In low-permeability rocks like shale, unconfined tests are normally not run, since this test mode requires fully drained conditions, which are too time-consuming and difficult to obtain in low-permeability rocks.

7.3.6 Hydrostatic tests and determination of poroelastic parameters

Hydrostatic tests as such are performed mainly to determine the bulk modulus of the drained (framework) and undrained rock and of the solid grains. This test represents the first phase of a standard triaxial test. Normally, failure does not occur under hydrostatic loading, although pore collapse in chalk and grain crushing in sandstone may induce failure at high stress levels. In this section, we will mainly focus on elastic and poroelastic behaviour.

As said above, hydrostatic tests can be performed under drained or undrained conditions. The difference between moduli measured in drained and undrained conditions was outlined in Section 1.6. Here we repeat the most important results for laboratory testing.

In a drained hydrostatic experiment the slope of the stress vs. volumetric strain curve is the bulk frame modulus K_{fr} of the rock (see Section 7.3.4).

In an undrained hydrostatic experiment the pore pressure will change during the test, and the slope of the stress–strain curve will be equal to the undrained bulk modulus K_u of the rock, including the effect of the pore fluid (Eq. (1.171)).

In both cases we may often see nonlinear behaviour, that is, increasing bulk modulus with increasing stress. This may be caused by the granular nature of sand in case of poorly cemented sandstone and by the large deformability of clay in case of soft, normally consolidated shales. Part of the deformation may then also be non-elastic or plastic, which means that the slope of the stress–strain curve does not represent the elastic bulk modulus. To reduce the problem of nonlinearity and variation of the stress–strain slope, it is recommended to perform unloading–reloading cycles at different stress levels and interpret the elastic moduli from the stress–strain response of these loading cycles. The stress levels should be chosen such that the sample is well within the elastic range. This results in a much more consistent interpretation of the elastic moduli.

Hydrostatic tests are used to determine pore compressibility. The basic theory behind such measurements are given in Section 1.6.6. As was shown there, two different pore compressibilities c_p and c_{pp} may be obtained, depending on whether the tests are performed in drained conditions, measuring the effect on pore volume by changing the confining pressure c_p or with constant confining pressure, measuring the pore volume change versus change in pore pressure c_{pp}. In both cases, pore volume change needs to be measured by monitoring displaced fluid volumes. Pore compressibilities may be estimated, but with less accuracy, from independent measurements of porosity, bulk frame modulus and solid grain modulus (Eqs. (1.205) and (1.206)). For practical use to estimate production in a depleting reservoir, the impact of stress path has to be accounted for (see Section 12.4.1).

Hydrostatic experiments permit determination of poroelastic parameters. The increase of pore pressure during undrained loading in a poroelastic (Biot) medium is (see also Section 1.6.4)

$$p_f = (C/K_u)\sigma_p \tag{7.9}$$

The constant C is given in Eq. (1.170) and is seen to be equal to the product of the Biot coefficient α (Eq. (1.190)) and M (Eq. (1.169)). The ratio C/K_u is identical to Skempton's B-parameter (see Section 1.6.7).

For general estimates of undrained pore pressure response, accounting for anisotropic stress conditions, Skempton's A-parameter is also required (see Section 7.3.7).

For fluid and frame bulk moduli (K_f and K_{fr}) significantly smaller than the bulk modulus of the grain material (K_s), we find $C/K_u \to K_f/(\phi K_{fr} + K_f)$. It is seen that for a stiff framework, or a situation where gas is present in the pore fluid ($K_f \to 0$), the pore pressure development can be neglected. For a weak frame, however, the pore pressure may be almost as high as the external stress. One example of such behaviour is in soils. In shales, where pore fluid saturation may be hard to achieve, a measured pore pressure response during undrained loading is a useful indication that full saturation is actually reached. In case the rock also deforms plastically, like in normally consolidated clays or shales (Section 2.8.2), the pore pressure response may be larger than predicted by Eq. (7.9).

The Biot coefficient α is the most commonly used poroelastic parameter. It can be determined directly from the constitutive equations of Biot by performing an undrained constant volume (isochoric) test:

$$\alpha = \left(\frac{\Delta\sigma_p}{\Delta p_f}\right)_{\varepsilon_{vol}=0} \tag{7.10}$$

This follows from Eq. (1.197). A less direct approach is to base it on a combination of two stress paths, that is, a drained measurement (giving K_{fr}) and a measurement at constant net stress (constant difference between applied external stress σ_p and pore pressure p_f; supposed to provide K_s). The effective stress coefficient is then given by

$$\alpha = 1 - \frac{K_{fr}}{K_s} = 1 - \frac{\left(\frac{\Delta\varepsilon_{vol}}{\Delta p_f}\right)_{\Delta\sigma_p=\Delta p_f}}{\left(\frac{\Delta\varepsilon_{vol}}{\Delta\sigma_p}\right)_{\Delta p_f=\text{const.}}} \tag{7.11}$$

This expression can be generalised to determine effective stress coefficients for other physical properties than strain, for instance, wave velocities, permeability etc. Nermoen et al. (2013) made a comparison between these approaches, which in the case of their high-porosity chalk gave similar and reasonable results.

A constant net stress test is a generalised version of the "unjacketed" test in Section 1.6. In the constant net stress test, stress and pore pressure are in general different but are changed at equal rates. In an "unjacketed" test, that is, with the pore pressure equal to the confining pressure, the slope of the stress–strain curve gives the bulk modulus K_s of the solid particles ($1/K_s$ is often also termed "grain (or matrix) compressibility") (Eqs. (1.161) and (1.164)).

The unjacketed measurement can be performed on a core plug in direct contact with a fluid bath. The volume change of the fluid in the cell is measured as a function of pressure with and without the core present, and the results are subtracted to obtain the grain compressibility. Since the volume change is small, the results are very sensitive to changes in temperature or other factors that may influence the measurement of fluid compressibility.

In many cases, independent measurement of K_s is not available, and α is computed on the basis of tabulated values for mineral moduli. Since a rock may be composed of several minerals with different moduli, and since minerals mostly are anisotropic and do not have a well-defined bulk modulus, such a procedure may be not sufficiently accurate.

In case no static rock mechanics test is done, the frame modulus may in principle be derived from P- and S-wave velocities and density for an isotropic rock. The rock has then to be tested in non-saturated conditions, so that the velocities are not influenced by the pore fluid. Also, the value of K_s has to be estimated, based on mineralogy.

Since dynamic moduli in general exceed static moduli (Section 5.6.1), a value of α based on wave velocity measurements may not be comparable with one obtained from static measurements. However, when performing the static tests, we need to be aware that the poroelastic theory applied here assumes linear elasticity. Thus the most reliable data are obtained by performing stress cycles, and as said above, using the unloading (reducing effective stress) stages to determine the stiffnesses or the Biot coefficient when measured directly. We have to make sure that creep does not occur. The parameter values obtained will then be close to purely elastic, like what is obtained by use of wave velocities, except that the static test allows full determination of α and of K_{fr} and K_s.

If the rock is in a normal consolidation state in situ, that is, at its maximum burial depth, we may load it above in situ effective stress and use the loading part to estimate the effective stress coefficient. In this case, however, the influence of non-elasticity may become significant so that we have to be careful if the measured values are used within elastic theory.

One important application of poroelasticity is to use the Biot–Gassmann equation to estimate effects of fluid substitution on P-wave velocities. For this, in addition to bulk and frame shear modulus, Biot's α and the poroelastic modulus M (the storage modulus; see Section 1.6.5) are needed. The modulus M may be determined directly in isochoric (constant volume) tests, although this does not appear to be a commonly applied procedure. Eq. (1.200) shows that M can be measured from displaced fluid volume and pore pressure change in a test where the sample volume is kept constant. Recall that the definition of the ζ parameter in Eq. (1.199) is displaced fluid volume divided by the total sample volume (positive if fluid is produced) (see Eq. (1.141)).

7.3.7 Triaxial testing of shales

For testing shales it has been recommended to perform the so-called consolidated–undrained (CU) tests (Steiger and Leung, 1992; Nakken et al., 1989; Horsrud et al., 1998). The procedure is adopted from soil testing (e.g. Head, 1984). This test consists of three distinct phases:

1. Loading to the predetermined level of confining pressure and pore pressure.
2. Consolidation, maintaining a constant confining pressure and allowing drainage of the pore fluid against a constant pore pressure. A different approach to this phase (undrained) is presented by Giger et al. (2018).
3. Undrained axial loading under a constant axial displacement rate beyond failure of the sample.

This test sequence is illustrated schematically in Fig. 7.8. The triaxial phase is performed in undrained conditions to reduce the time required to complete a test. A fully drained test of a low-permeability shale may require weeks or months to complete. In the CU-test the pore pressure must be monitored throughout the test.

The time required to run such a test depends mainly on the permeability of the shale. Concepts from soil mechanics testing can be applied to assist in determining when consolidation is completed and also to determine the appropriate displacement rate in the undrained part of the test (Head, 1984). For instance, allowing drainage at the end surfaces and the sides of the core will reduce testing time considerably. As shown in Section 1.9, the consolidation response depends both on the permeability and elastic properties of the rock. It should thus be noted that in soil mechanics approaches, the "weak frame" assumption is implicit.

The displacement rate in the undrained phase is chosen such that pore pressure gradients within the sample are avoided. The total test time thus depends both on the permeability, the elastic properties, the sample dimensions and the drainage conditions around the sample. To reduce the

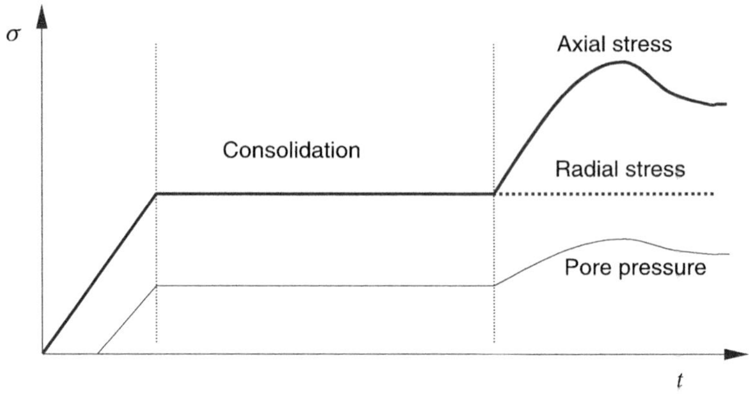

Figure 7.8 Schematic illustration of a consolidated–undrained test.

test time as much as possible, both axial and radial drainage should be implemented.

The results can be plotted in a (p', q) plot (see Section 2.7), which facilitates an interpretation of whether the shale behaves as normally consolidated or overconsolidated.

Examples from testing of a Tertiary shale are shown in Fig. 7.9 (from Horsrud et al., 1998). This shows three tests at different confining pressures that fail macroscopically by shear. The peak stress values fall on a straight line, which can thus be represented by a Mohr–Coulomb failure criterion.

The pore pressure response during the undrained triaxial test segment may be used to quantify Skempton's A. Following Eq. (1.215), the pore pressure change in this case (with constant confining pressure) is simply given by the product of A, B and the axial stress change. Undrained tests along any other non-isotropic stress path would also provide a measurement of A. Note that the parameter B_{11} should be derived from the measurements (Eq. (1.285)), allowing calculation of pore pressure changes in samples drilled at different angles to the symmetry axis (Holt et al., 2018).

7.3.8 Oedometer (K_o) test or uniaxial strain test (UST)

Laboratory experiments are often performed with rock specimens in uniaxial compression, that is, in a situation with no lateral deformation. This type of test is thought to simulate, for example, the compaction of a reservoir during depletion. For geotechnical applications, similar studies of soils are routinely performed to evaluate, for example, the stability of foundations. In that case the soil can be placed inside a solid, undeformable cylinder

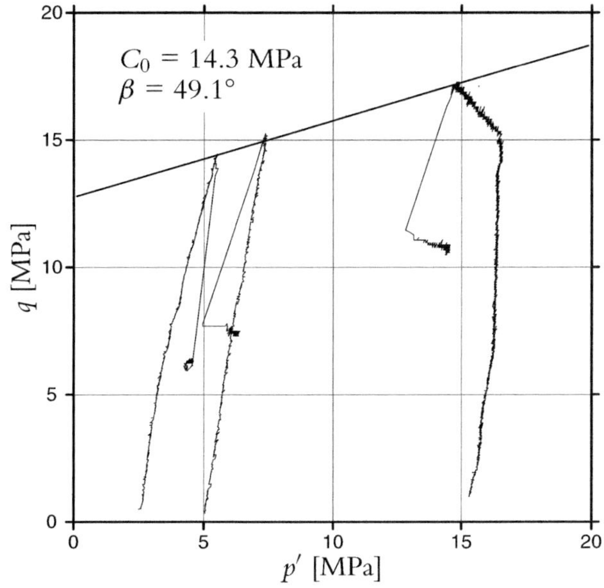

Figure 7.9 CU-tests of Tertiary shale. (After Horsrud et al., 1998).

and compressed vertically. With rocks, it is recommended to use a triaxial cell and adjust the confining pressure so that zero or negligible radial deformation is recorded as the axial load is increased. The normal way to apply the zero strain criterion is to use point measurements and diametrical strain recordings in one plane as the control parameters. This requires a homogeneous specimen, which is isotropic or transversely isotropic with the sample axis as the symmetry axis. This test is also commonly referred to as a uniaxial strain test (UST). Suggested methods for conducting this type of test has been published by ISRM (Dudley et al., 2016).

Theoretically, the stress–strain curve of an elastic isotropic rock in an oedometer test (drained) is given by Hooke's law (Eqs. (1.96)–(1.98)), keeping $\varepsilon_r = 0$. Written in terms of stress and strain increments, we have

$$\Delta\sigma_z' = (\lambda_{\text{fr}} + 2G_{\text{fr}})\Delta\varepsilon_z = \frac{E_{\text{fr}}(1-\nu_{\text{fr}})}{(1+\nu_{\text{fr}})(1-2\nu_{\text{fr}})}\Delta\varepsilon_z \tag{7.12}$$

and

$$\Delta\sigma_r' = \frac{\lambda_{\text{fr}}}{\lambda_{\text{fr}} + 2G_{\text{fr}}}\Delta\sigma_z' = \frac{\nu_{\text{fr}}}{1-\nu_{\text{fr}}}\Delta\sigma_z' \tag{7.13}$$

The modulus $d\sigma_z/d\varepsilon_z$ is called the oedometer modulus or uniaxial compaction modulus (H_{fr}) (see also Sections 1.3 and 12.2.1).

In the elastic limit, it is given by Eq. (7.12). The ratio between changes in the effective radial stress and effective vertical stress given by Eq. (7.13) represents a stress path coefficient for a reservoir that depletes under uniaxial strain conditions. Further discussion of reservoir stress path is found in Section 12.2.3.

The oedometer test is most commonly used in geotechnical engineering. The test is then performed by placing the sample (which normally has an L/D-ratio of $\approx 1/2$) in a steel ring. Drainage may be allowed both through top and bottom pistons or, alternatively, through only one of the pistons. Load is applied incrementally or in a continuous mode. Axial stress and strain vs. time plus pore pressure changes are the parameters measured throughout these tests.

It has been observed that cyclic loading of sandstones produce moduli that are very different from the first loading segment (see e.g. Mess, 1978). In a study of synthetic sandstones formed under stress (Holt et al., 2000) the modulus of the second and subsequent cycles was observed to be at least two times larger than the modulus of the first cycle. The initial modulus of the second cycle was found to be close to the initial modulus of the expected behaviour of a virgin rock, that is, a rock that had not been through the coring process.

In addition to the modulus given by Eq. (7.12), the oedometer tests can also provide information about the coefficient of consolidation D_h (see Section 1.9) if the test is run incrementally. Further, when specimens are brought beyond the elastic limit, the oedometer test provides a simple method of determining the preconsolidation stress. One way of obtaining this is to plot the oedometer modulus vs. effective vertical stress. A minimum in the modulus then corresponds to the preconsolidation stress (Janbu, 1985). Other ways of interpreting the preconsolidation stress from oedometer tests have also been proposed (see e.g. Lambe and Whitman, 1979).

7.3.9 Stress path tests

By the use of a triaxial test cell the tests may be predefined to follow a specific stress path other than those already discussed. This is mainly done for special purposes, for example, to mimic stress path conditions in a depleting reservoir, but can in principle follow a large variation of stress paths. In this section, we include two examples.

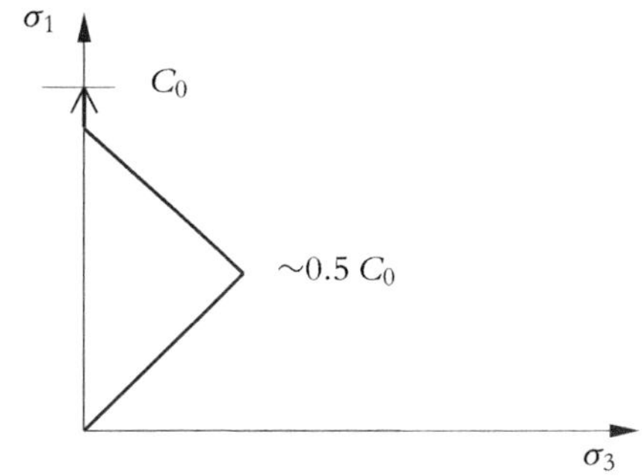

Figure 7.10 Principle sketch of stress path in CMS test.

7.3.9.1 Constant stress ratio tests

Constant stress ratio tests are compression tests where the ratio between the effective confining pressure and the effective axial stress

$$\kappa' = \frac{\Delta\sigma_r'}{\Delta\sigma_z'} \tag{7.14}$$

is kept constant; $\kappa' = 0$ corresponds to a drained triaxial compression test, whereas $\kappa' = 1$ represents a hydrostatic test.

The interest in this type of tests comes from the fact that stress measurements in compacting reservoirs have revealed that the boundary condition of no lateral movement is not always adequate (Section 12.2.3). A stress path with constant stress ratio may give a better description.

7.3.9.2 Constant mean stress (CMS) tests

The CMS test has been proposed as an alternative to provide more reliable estimates of the uniaxial compressive strength (UCS, C_0) of rocks (Kenter et al., 1997). The principle of the test is shown in Fig. 7.10. The rock sample is first loaded hydrostatically to a stress level close to 50% of the expected UCS. The sample is then loaded axially and unloaded radially until the radial stress is zero. Then the sample is loaded in a uniaxial stress mode until failure.

The idea behind this approach is that coring-induced microcracks are closed prior to failure, and thus more reliable and less scattered values of UCS are expected. This method requires that a rough estimate of the expected UCS can be made prior to testing. If failure occurs before the radial stress is completely unloaded, then the measured strength can be extrapolated towards unconfined conditions, or another test at a lower initial stress level can be run.

Note that the term "constant mean stress" is not entirely correct, since for a cylindrical sample, the mean stress is given as $(\sigma_z + 2\sigma_r)/3$, whereas here it is given as $(\sigma_z + \sigma_r)$, which is kept constant. The use of this term has been justified by the fact that the average of the largest and smallest principal stresses is kept constant.

7.3.10 Other triaxial failure tests

7.3.10.1 Extension tests

Triaxial tests can also be run in extension: after applying a hydrostatic confining pressure σ_r, just like in the standard compression tests, the axial load σ_z is decreased, keeping σ_r constant. If failure does not occur until σ_z reaches zero, then σ_r is increased. According to the Mohr–Coulomb criterion, which only depends on σ_1' and σ_3', the shear strength τ_c should be the same in extension and compression. However, several other criteria do also depend on the intermediate stress σ_2', which implies that they predict some difference in strength between extension and compression tests. It is also commonly observed that rocks are stiffer during unloading than loading, so that the elastic properties are also different between the two tests.

7.3.10.2 Multiple and continuous failure state triaxial tests

Determination of the failure envelope is a key task in the analysis of borehole stability and sand prediction. Normally, there are strong limitations on the amount of material available for rock mechanical tests. It may therefore be attractive to carry out tests where an entire failure envelope can be measured with just one sample. There are two slightly different methods for doing this, the multiple failure state test and the continuous failure state test (Kovari et al., 1983). In the first case (see Fig. 7.11) the test is started as a normal triaxial test from a given initial confining pressure. When failure is imminent (detected manually or automatically as a significant change in the slope of the stress–strain curve), the confining pressure is increased to a higher level. The axial stress is then increased further until failure is again imminent, and in this way the test continues at a multiple of confining stress levels. This test turns out to give quite reliable results with regards to failure stresses, at least in weak brittle rocks like sandstones (Holt and Fjær, 1991).

Reliable, real-time identification of proximity to failure is essential for this test. An alternative is to use the acoustic emission hit rate as an indicator. Changes in acoustic velocities (especially the radial velocity) measured

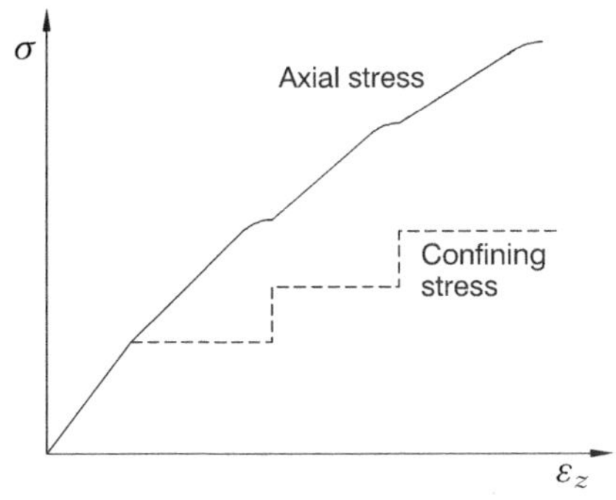

Figure 7.11 Sketch of the stress path followed in a multiple failure state test.

during the test may also be used as an indicator of failure (Holt and Fjær, 1991). Obviously, failure detection in ductile materials is more complicated. In addition, in hard rocks, failure in one of the steps may weaken the material considerably. Generally, the elastic moduli are influenced by damage caused in previous load steps, so this test is not recommended for evaluation of elastic properties.

The idea behind continuous failure state tests is similar to that of the multiple failure state tests. In the continuous failure state test (Kovari et al., 1983), both radial and axial stresses are increased to monitor the failure envelope continuously rather than stepwise.

7.3.10.3 True triaxial tests

To evaluate further the role of the intermediate principal stress on the failure envelope, it is possible to perform tests in test machines that can supply a true triaxial (polyaxial) stress state, that is, $\sigma_1 \neq \sigma_2 \neq \sigma_3$. These systems normally include three mutually perpendicular pairs of hydraulic jacks compressed against the faces of a cubically shaped sample. Other applications of polyaxial tests include testing of sand production, fracturing, perforating, injection and screen integrity. A limitation for the use of such tests within petroleum related rock mechanics is that the tests usually require larger specimens than can be obtained from field cores.

7.3.11 Hollow cylinder tests

Hollow cylinders can be used as scaled models to simulate the borehole situation in the laboratory (see e.g. van den Hoek et al., 1992). Besides that, by changing the external stresses and/or the internal (hole) pressure we can study a variety of load situations, including a true triaxial stress state at the borehole wall.

The stress state at the borehole wall of an externally loaded hollow cylinder is (follows from a calculation similar to that in Section 4.2)

$$\sigma_{zi} = \frac{F_z}{\pi(R_o^2 - R_i^2)} \tag{7.15}$$

$$\sigma_{\theta i} = \frac{2\sigma_{ro}R_o^2}{R_o^2 - R_i^2} \tag{7.16}$$

$$\sigma_{ri} = 0 \tag{7.17}$$

where F_z is the axial load, and σ_{ro} is the confining pressure. At the outer boundary of the sample, $\sigma_r = \sigma_{ro}$, whereas σ_z shows no radial variation. The ratio between the axial and radial external load determines which of the stresses σ_{zi} or $\sigma_{\theta i}$ is the maximum principal stress. From Eqs. (7.15) and (7.16) we find that σ_{zi} is larger than $\sigma_{\theta i}$ if $F_z > 2\pi\sigma_{ro}R_o^2$ and vice versa. Shear stresses will be largest at the borehole wall, and shear failure will occur when σ_{zi}, $\sigma_{\theta i}$ and σ_{ri} fulfil a shear failure criterion (Sections 4.5.1 and 4.5.2). By altering the ratio between σ_{zi} and $\sigma_{\theta i}$ it is possible to investigate the influence of the intermediate stress on the failure stress. In addition, it is of interest to investigate the failure mode (see Table 4.2 and Figs. 4.23–4.25). In both cases, also the ratio R_o/R_i can be used to tune the failure stress and failure mode. A convenient choice may be R_o/R_i = 5, and the same hydrostatic external pressure on the sample boundaries and on the loading pistons, that is, $\sigma_z = \sigma_{ro}R_o^2/(R_o^2 - R_i^2) = (25/24)\,\sigma_{ro}$. Failure will then be induced by the hoop stress $\sigma_{\theta i}$ on the borehole wall. According to Eq. (7.16), $\sigma_{\theta i} = (25/12)\,\sigma_{ro}$, that is, failure should occur when the confining pressure is 12/25 of the uniaxial rock strength. Experimental results, however, tend to show that failure occurs at larger stresses than predicted. It should be noted, however, that it may be difficult to determine the actual stress inducing borehole failure from the external stress–strain curve. Measurements inside the borehole are thus necessary. Such measurements include the use of cantilever systems with strain gauges monitoring borehole deformation.

An example of such a test is shown in Fig. 7.12, with internal monitoring of borehole wall deformation in two opposite directions. The test consists of three phases, first loading to a predetermined level of pore pressure and confining pressure, then a period of consolidation with constant

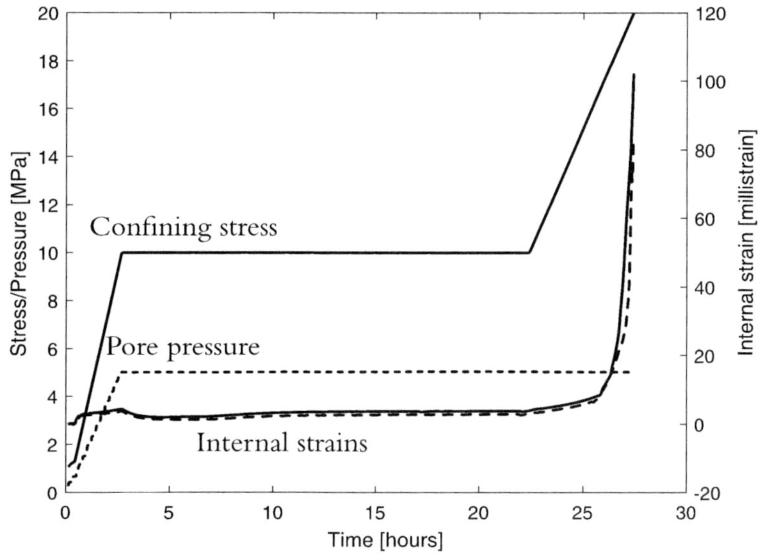

Figure 7.12 Hollow cylinder test of a shale sample. The sample is first loaded to predetermined levels of external stress (confining pressure) and pore pressure. The sample is then left to consolidate with constant boundary conditions (confining stress and pore pressure). In the final phase the confining pressure is increased, resulting in failure of the borehole wall. The failure process is monitored by measurement of the borehole diameter in two opposite directions (internal strains).

stress and pore pressure, and finally loading of the confining pressure until the borehole wall is failing. The failure of the borehole wall is clearly seen as a distinct change in slope of the internal strain. In general, however, borehole failure is not as well defined as the peak stress point in a triaxial test.

An alternative way of studying the stability of hollow cylinders is to apply a variable pressure in the hole while maintaining constant external conditions. For a permeable rock, a sleeve can be applied to simulate, for example, a mudcake. Borehole collapse is studied by reducing the internal pressure, keeping the confining pressure constant. In principle, hydraulic fracture generation can be simulated by increasing the borehole pressure.

The stability of cavities or perforation tunnels, especially in sandstone, can be studied by terminating the hole a representative distance into the sample (see e.g. Tronvoll et al., 1992). For simulation of the sand production potential, fluid flow should be included, and both confining pressure and flowrate should be varied during a test.

7.3.12 Measurements on small samples

The need for intact core material in laboratory testing in many cases represents a serious restriction. Either a core is simply not available, or the core may be in such a state that it is impossible to extract sufficiently large test samples. This has motivated the use of smaller samples (sub-cm) for testing. Especially, if drill cuttings can be used, then a new source of information is available without any additional cost.

Again, acoustic measurements are attractive because of the close relation to the mechanical response of the rock. Alternative techniques are available, both a sonic pulse technique (Santarelli et al., 1998) and a continuous wave technique (Nes et al., 1998a). The sonic pulse technique is in principle similar to the transmission technique described in Section 7.2.4. The continuous wave technique (CWT) is well known from solid state physics and fluid studies. This technique also requires a specimen with smooth and parallel end surfaces mounted between two acoustic transducers, one used for acoustic wave generation and the other for detection. This technique relies on the establishment of ultrasonic standing wave resonances as the excitation frequency is swept over a frequency range containing several resonance frequencies. A typical resonance spectrum is shown in Fig. 7.13. Resonances occur each time the sample length (thickness) equals an integer number of half wavelengths. The acoustic phase velocity is then given as

$$v = 2l\Delta f \tag{7.18}$$

where Δf is the frequency difference between resonance peaks, l is the sample thickness, and v is the phase velocity (P or S). CWT is particularly suited for measurements on fine-grained material like shale, where samples of sub-mm thickness may be used.

When used on core material, such measurements may represent a valuable supplement to more time-consuming and expensive tests, or an alternative when larger samples cannot be prepared. Efforts have also been made to simultaneously perform static measurements on such small samples (Nes et al., 1998b). However, questions related to sample size in relation to material heterogeneities should be considered.

These techniques may also be used on drill cuttings, and as such they have a potential for field use. This is discussed further in Chapter 8.

When drill cuttings are used, additional uncertainties are related to the quality and representativity of the cuttings, that is, stress unloading effects, drill bit effects, effects of the drilling fluid and so on.

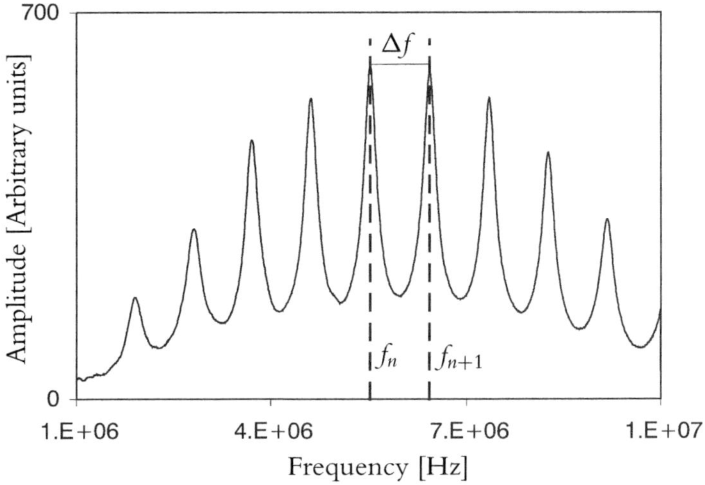

Figure 7.13 CWT test: example of recorded resonance spectrum.

7.4 Laboratory tests for stress determination

The stress release experienced by a rock sample during coring causes opening of microcracks by elastic and visco-elastic mechanisms. The orientational distribution of these cracks is thought to reflect the stress field experienced in situ. It is normally assumed that most cracks open so that the crack normals are parallel to the direction of maximum in situ stress, which is also the direction of maximum stress release. After the period of crack opening, the cracks remain open. The orientational distribution and the amount of cracks still contain information about the orientation and magnitudes of the previously experienced principal stress components. Clearly, influence may be expected from earlier loading–unloading cycles experienced by the rock throughout geological history. It is not clear to what extent the entire stress history influences the observations. It is however often assumed that the most recent and the most rapid unloading will dominate the behaviour (de Waal and Smits, 1988).

If the crack pattern contains information about the in situ stress field, then any technique by which the orientational distribution of microcracks can be measured may provide stress field data.

In this section, we discuss some methods based on core measurements. A requirement for such testing is of course that the core is *oriented*, that is, the geographical orientation of the core must be known (see Section 7.1.4). This is done by placing a mark, groove or line on the surface of the core and then determining the orientation with respect to geographic north. The described methods have mostly been used for determination of horizontal stress directions. The measurements thus have to be performed in the horizontal plane, either using a vertical core or mounting the measurement devices so that they measure in the true horizontal plane.

Attempts have also been made to estimate the stress magnitudes from these techniques, but the general experience is that the uncertainty of the results can be considerable.

7.4.1 Differential strain curve analysis

One way of obtaining this orientational information is to measure the anisotropic deformation of the core upon hydrostatic reloading. In the general case, this is done by machining a cube from the core and attaching strain gauges. The instrumented (and cleaned/dried) sample is subsequently cast in a compliable medium (e.g. silicone) to prevent confining fluid from entering the sample during testing. The result is a series of stress–strain curves for different directions. They will all exhibit initial nonlinearities related to crack closure. Most strain will occur in the direction of maximum in situ stress.

If one of the principal strain directions is known, then four independent strain measurements are required to determine the complete strain tensor. This will be the case if the vertical stress is a principal stress and the core is vertical. In that case, one measurement is made in the vertical direction and three in the transverse plane. In the general case, when a principal stress direction cannot be assumed to be known, it takes six independent measurements to determine the complete strain tensor. Note that if one principal direction is assumed vertical, then only three measurements in the horizontal plane are strictly required to determine the directions of the principal horizontal stresses. The horizontal measurements are made at 45° to each other, as illustrated in Fig. 7.14.

To see how the direction of principal stresses can be determined, we use Eq. (4.7):

$$\varepsilon_r = \varepsilon_x \cos^2\theta + \varepsilon_y \sin^2\theta + 2\Gamma_{xy} \sin\theta\cos\theta \tag{7.19}$$

Referring to Fig. 4.1, ε_x and ε_y are the strains in the x- and y-directions, and ε_r is the radial strain in a direction with an angle θ relative to the x-

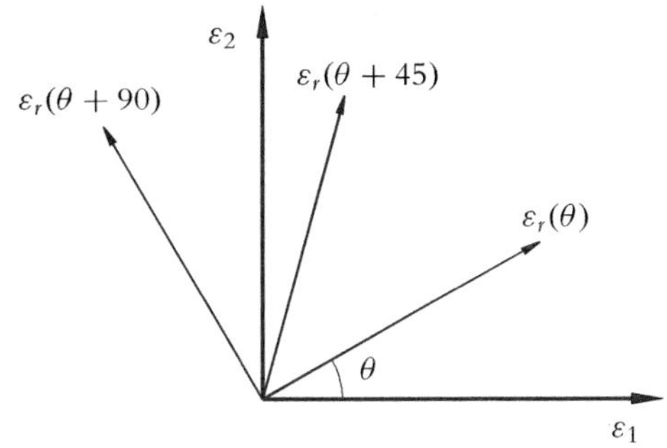

Figure 7.14 Geometry of measurement directions in an ASR experiment. Note that here we use ε_1 and ε_2 to represent the two principal directions in the horizontal plane.

axis. If we now assume that the *x*- and *y*-directions are the principal strain directions, then we have

$$\varepsilon_r = \varepsilon_1 \cos^2\theta + \varepsilon_2 \sin^2\theta \tag{7.20}$$

By measuring $\varepsilon_r(\theta)$, $\varepsilon_r(\theta + 45)$ and $\varepsilon_r(\theta + 90)$ we have three equations with three unknowns (ε_1, ε_2 and θ). This yields the following results:

$$\tan 2\theta = \frac{\varepsilon_r(\theta) + \varepsilon_r(\theta + 90) - 2\varepsilon_r(\theta + 45)}{\varepsilon_r(\theta) - \varepsilon_r(\theta + 90)} \tag{7.21}$$

$$\begin{aligned}\varepsilon_1 = {} & \frac{1}{2}\left[\varepsilon_r(\theta) + \varepsilon_r(\theta + 90)\right] \\ & + \sqrt{\frac{1}{2}\left[\varepsilon_r(\theta) - \varepsilon_r(\theta + 45)\right]^2 + \frac{1}{2}\left[\varepsilon_r(\theta + 90) - \varepsilon_r(\theta + 45)\right]^2}\end{aligned} \tag{7.22}$$

$$\begin{aligned}\varepsilon_2 = {} & \frac{1}{2}\left[\varepsilon_r(\theta) + \varepsilon_r(\theta + 90)\right] \\ & - \sqrt{\frac{1}{2}\left[\varepsilon_r(\theta) - \varepsilon_r(\theta + 45)\right]^2 + \frac{1}{2}\left[\varepsilon_r(\theta + 90) - \varepsilon_r(\theta + 45)\right]^2}\end{aligned} \tag{7.23}$$

By hypothesis the total strain associated with crack closure is proportional to the in situ stress. The constant of proportionality for stress magnitude estimates can be obtained from the vertical strain measurement, assuming that the effective vertical stress is known from, for example, the weight of the overburden and the pore pressure.

7.4.2 Anelastic strain recovery

This method is based on the anelastic behaviour (creep) occurring after unloading of a rock. In addition to the instantaneous elastic strain that takes place when the stress is relieved, an additional anelastic strain also develops over some time. For a core taken from an underground formation, the period of time over which most of the anelastic strain develops is typically 10–50 hours.

A core which is going to be used for ASR measurements should be instrumented as soon as possible after retrieval, since most of the anelastic strains will develop during the first few hours. It should be noted that the process of bringing the core out of the well takes several hours, so that the earliest possible start of the measurements is typically 3–5 hours after drilling-induced stress relief. Once the sample is instrumented, it is vital that it is kept in a constant temperature and saturated humidity atmosphere during the test.

The principle for determination of principal strain (and stress) directions is the same as that presented in Section 7.4.1.

Concerning the estimation of stress magnitudes, constitutive models that link the strains to stresses have to be established. Several methods based on visco-elasticity have been presented in the literature (Blanton, 1983; Warpinski and Teufel, 1989, and a general three-dimensional theory by Matsuki and Takeuchi, 1993). For an overview of these methods, see, for example, Amadei and Stephansson (1997).

There are, however, still questions which need to be resolved before reliable results from the ASR-technique can be expected. A study by Yassir et al. (1998) showed that for stress ratios[3] up to 1.5, the corresponding strain ratios were very similar, but for stress ratios above 2, the corresponding strain ratios showed more scatter. This study was done under carefully controlled laboratory conditions using a synthetic sandstone (see also Section 7.1.2), which was formed under stress and isotropically unloaded. Under these conditions, we will avoid the damage effects experienced during coring and core retrieval. Especially under conditions where the shear stress approaches the shear strength of the rock, we should expect this to affect the measured relaxation strains.

[3] The stress ratio is here defined as the ratio between the largest and smallest principal stresses.

7.4.3 Acoustic techniques

7.4.3.1 Differential wave velocity analysis

Since acoustic wave velocities and attenuations are extremely sensitive to cracks (see Chapters 5 and 6), the cracks will also contribute to acoustic anisotropy. This anisotropy is preserved after the core has entered the laboratory, and measurements of it may serve as additional information to evaluate in situ principal stress directions (and possibly also magnitudes). Engelder and Plumb (1984) demonstrated this effect for a number of rocks sampled at shallow depths. Experiments with reservoir sandstones show a clear minimum of sound velocity in a direction that corresponds to the in situ vertical, that is, the expected major principal stress direction (see e.g. Holt et al., 1989; Teufel, 1989).

A technique known as *differential wave velocity analysis* (DWVA) consists of hydrostatic loading of the core while sound velocity is measured in multiple directions. The directions with largest and smallest increase in velocity are interpreted as the directions of the largest and smallest principal in situ stresses, respectively.

The technique is based on the expectation of a relation between crack density and stress magnitude (Ren and Hudson, 1985) and the expectation that reloading the sample to its previous stress state will reduce the crack-induced anisotropy.

Care should however be taken before a directional velocity variation is interpreted as anisotropy. The variation may also be caused by inhomogeneities, since the wave paths will be slightly different. One way to check that anisotropy is really the source of variation is to measure S-wave velocities in the assumed principal symmetry directions with different polarisations, so that velocities of waves travelling exactly the same path but with different polarisations are measured. Any difference of these velocities cannot be due to heterogeneity and must therefore be caused by anisotropy.

7.4.3.2 Acoustic emission

Kaiser (1950) made the experimental observation that when loading a sample, the sample would start to produce acoustic emissions (AE) when the load exceeded the maximum stress level which the sample had experienced previously. This stress history or stress memory effect has been the basis for numerous attempts to use this phenomenon for determination of in situ stresses. Its application is, however, not straightforward. It is shown

(Holcomb, 1993) that for triaxial stress states, the maximum stress in a principal direction cannot be retrieved by uniaxial loading in that direction. The onset of acoustic emission in the loading direction is dependent also on the stress applied in the two other orthogonal principal directions.

Acoustic emissions are transient elastic waves generated from localised and sudden changes of the stresses in a sample. In a porous material like a rock, such sudden changes of local stresses can be the result of growth of microcracks and/or collapse of pores or grains. This is the basis for the concept of the damage surface of materials, analogous to a yield surface in the theory of plasticity (Section 2.8). A damage surface thus defines the limit in stress space where acoustic emissions are generated when passed (Holcomb, 1993).

It has been suggested that the rock will remember the critical stress state during coring and that the in situ horizontal stress state can be determined by mapping the 3D damage surface of a core (Kenter et al., 1998). This concept has been demonstrated experimentally on outcrop and field sandstone cores, and it has been supported by analytical and discrete particle modelling (Pestman et al., 2002). However, the effect can be rather diffuse for several reasons (see Section 5.10); hence precise, unambiguous results may not be expected.

7.5 Index tests and other characterisation tests

There is a number of test methods designed to measure parameters that are somehow relevant for rock mechanical analyses, even though they are not explicit measures of the basic rock mechanical properties. Several of these are simple and quick tests that can be applied to rocks without applying a confining pressure system. This includes measurements of index properties, which can be related in some way to strength or elastic parameters. We will describe tests that may give information about the tensile strength (point load strength and Brazilian test) and a hardness test, which has been found to give results that correlate with compressive strength. More recent test methods for correlation with rock mechanical properties based on indentation and scratching will be presented, and finally, some tests which may be particularly useful for studies of shales, will be discussed.

Most of the tests discussed below are described in more detail in Ulusay (2015).

7.5.1 Tensile strength indicators

Direct tensile strength measurements. By gluing both ends of a cylindrical rock specimen to metal front plates, which are fixed to the platens of the load frame, direct measurements of tensile strength can be obtained. The cement, of course, has to be more resistant to tensile stress than the specimen itself.

Point load strength. The point load strength index I_s is measured using spherically truncated, conical platens of a standard geometry (cone angle 60° and radius of curvature 5 mm at the tip); see Fig. 7.15a. The specimen may be in the form of a core (axial/radial), a cut block or an irregular lump. Hence, no special sample preparation is required. The index I_s is obtained as

$$I_s = \frac{F_c}{D_e^2} \tag{7.24}$$

where F_c is the load necessary to cause failure, D_e is the "equivalent core diameter", equal to the thickness in a diametrical test or $(4A/\pi)^{\frac{1}{2}}$ in the case of axial, block or lump tests with A the minimum cross sectional area of a plane through the platen contact points. Size corrections can be obtained by multiplying the right-hand side of Eq. (7.24) by a factor $(D_e/50)^{0.45}$ (D_e is measured in mm).

The point load strength index may be used as an indirect measurement of the uniaxial tensile strength T_0:

$$I_s = 0.80 T_0 \tag{7.25}$$

The Brazilian test.[4] This test is performed by applying a load with two platens diametrically compressed on a rock cylinder, which is normally shorter than or equal in thickness L to its diameter D (see Fig. 7.15b). Failure occurs by an extension fracture in or close to the loaded diametrical plane.

A plane stress elastic analysis shows that the stresses close to the centre of the sample are (Jaeger and Cook, 1979)

$$\sigma_h = -\frac{2F}{\pi DL} \tag{7.26}$$

$$\sigma_v = \frac{6F}{\pi DL} \tag{7.27}$$

[4] The test is also known as the *Lobo Carneiro test* in honour of the inventor.

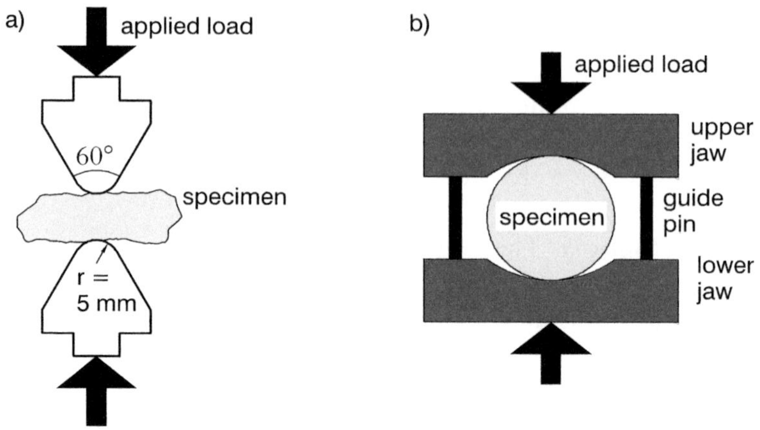

Figure 7.15 Simple sketches showing apparatus for a) point load strength test and b) Brazilian test.

The compressional stress is thus always about three times larger than the tensile stress, and since the ratio between the compressive strength and the tensile strength is normally larger than three, the sample will always fail in tension. By combining Eqs. (7.26) and (2.3) we find that the tensile strength (in MPa) is given by

$$T_0 = 0.636 \frac{F_c}{DL} \tag{7.28}$$

The peak load F_c is given in Newton and D and L must be given in millimetre. The Brazilian test tends to give results that are more reproducible than those obtained by point load strength measurements.

7.5.2 Hardness measurements

Brinell hardness. The Brinell hardness number N_{Br} follows from the ratio of applied load F on a spherical indenter to the indentation depth D:

$$N_{Br} = \frac{F}{2\pi rD} \approx \frac{F}{\pi r_s^2} \tag{7.29}$$

where r is the radius of the sphere, and r_s is the radius of the indentation surface area. The test is hence non-destructive. For metals and glasses, there is a theoretical relationship between N_{Br}, the yield stress σ_{yield} and Young's modulus E:

$$N_{Br} = \left(a + b \ln \frac{E}{\sigma_{yield}} \right) \sigma_{yield} \tag{7.30}$$

where a and b are constants, so that if the yield strength increases in proportion with E, then N_{Br} will be a measure of σ_{yield}.

Van der Vlis (1970) and later Geertsma (1985) studied the correlation between N_{Br}, dynamic elastic moduli and the compressive strength of rocks, and for sandstones in particular. Geertsma found a linear relationship between the Brinell hardness and the dynamic Young's modulus for well-consolidated sandstones:

$$E_{dyn} = 77.25 N_{Br} \tag{7.31}$$

He also presented an approximate result that the Brinell hardness equals the yield strength:

$$N_{Br} \approx \sigma_{yield} \tag{7.32}$$

Other hardness tests. There is a number of other tests whereby material hardness can be quantified. They have in common that they are either indentation tests (like the Brinell hardness) or rebound hardness tests (like the Schmidt hammer; for further details, see Ulusay, 2015).

7.5.3 Indentation and scratch tests

Indentation tests. The potential access to more information about rock mechanical properties from drill cuttings has motivated work to perform indentation tests on cuttings-size samples (Santarelli et al., 1998; Ringstad et al., 1998). These techniques are based on casting the samples in rapidly curing material (polyester, resin), which can act as a sample holder and also facilitate grinding and polishing of the sample surface. In these tests the entire load-displacement curve is normally recorded, and the slope of the linear regime is used to correlate with rock mechanical properties such as the uniaxial compressive strength.

Again, uncertainties related to sample size, material heterogeneity, fluid effects and so on need to be considered.

Scratch test. Another technique for determination of mechanical properties is the scratch test (Richard et al., 1998; Schei et al., 2000). The principle of this test method is to scratch the rock surface with a cutter while monitoring the applied force. The cutting depth is typically less than 1 mm. Testing of sandstone and carbonates was found to support a simple force cutting model in the ductile mode

$$F_t = EA \tag{7.33}$$

where F_t is the horizontal component of the cutting force averaged over a few cm of scratch length, E is the intrinsic specific energy, and A is the cross-sectional area of the cut.

A strong correlation between specific energy and the uniaxial compressive strength is well established for several rock types. The possibility to derive information about rock stiffness from scratch measurements has also been demonstrated (Schei et al., 2000). An obvious advantage of this method is the potential of continuous measurements along a core surface, providing continuous high-resolution data at low cost (Suarez-Rivera et al., 2002).

7.5.4 Punch tests

A punch test (Stenebråten et al., 2008) is a direct measurement of cohesion on a small rock sample, and as such it is strictly speaking not an index test. In practice, the test belongs to this group because of the simplicity of the test and the potential as well as the need for multiple tests due to the limited accuracy related to the small sample size. The test uses small disc-shaped samples of the same kind as the CWT test (Section 7.3.12) and may preferably be run subsequent to a CWT test on the same sample. During a punch test, a circular piston punches a hole through the sample, much like a paper puncher perforates a sheet of paper. The peak force F_p required to make the hole is recorded, and the cohesion S_0 can be estimated as

$$S_0 = \frac{F_p}{2\pi r_p l \cos\varphi} \tag{7.34}$$

where r_p is the radius of the piston, l is the sample thickness, and φ is the friction angle.

The tool is primarily intended for use on shales, for which the friction angle is so small that $\cos\varphi \approx 1$ may be an acceptable approximation.

Cohesion derived from punch tests tends to be somewhat higher than the values obtained from standard core plug tests at low confinement. This makes sense as the location of the failure plane is determined by the punching piston in a punch test. In a standard compression test, an infinite number of potential failure planes are available, and the sample will fail along the weakest one (see page 116). Hence the punch test seems to be less sensitive to small cracks than the standard low confinement test, despite the small sample size.

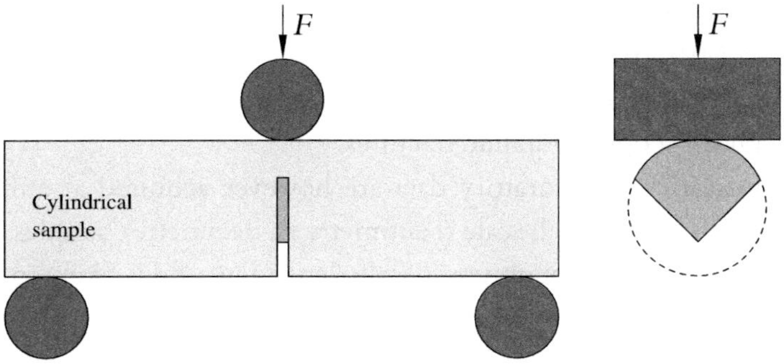

Figure 7.16 Principle sketch of chevron bend method for determination of fracture toughness. Left: side view. Right: cross section (normal to the sample axis) through the notch.

7.5.5 Fracture toughness

There are several different laboratory techniques available for measurement of fracture toughness K_{IC}. The International Society for Rock Mechanics (ISRM) suggests two methods for testing of rock, the chevron bend method and a short rod method (Ouchterlony, 1988). The chevron bend method uses a cylindrical specimen with a chevron notch (V-notch) machined into the middle of the specimen. The sample is loaded at three points by rollers placed as shown in Fig. 7.16. The main measurables are the crack opening displacement and the applied force. Also, the load point displacement of the sample is monitored.

In a perfectly brittle material the fracture toughness is given by the peak force F_{max} divided by $D^{1.5}$ (where D is the sample diameter) multiplied by a dimensionless geometrical factor (A_{min}):

$$K_{IC} = A_{min} \frac{F_{max}}{D^{1.5}} \tag{7.35}$$

With the failure load F_{max} given in kN and the diameter D in cm, the fracture toughness K_{IC} will be in MPa$\sqrt{m}$.

By performing repeated unload-reload cycles a plasticity parameter can be measured in addition to the fracture toughness, accounting for non-elastic behaviour. For further discussion of fracture toughness, see Sections 6.4 and 11.2.

Laboratory experiments with the technique described above and other related methods have provided a significant amount of fracture toughness data for rocks. Typical values for a variety of sedimentary rocks (Whittaker et al., 1992; Zhang, 2002; Beckers, 2004; Walle et al., 2017) range from 0.2 to 2.5 MPa$\sqrt{\text{m}}$, normally correlating well with tensile strength. These values imply that fracture toughness might play a minor role in propagation of long, macroscopic hydraulic fractures.

The majority of laboratory data are however acquired at atmospheric conditions and with small scale (centimetre to decimetre) samples. There is evidence that fracture toughness is scale dependent and is likely to be larger at the scale of the borehole than at the scale of a laboratory specimen. Further, as pointed out by Thiercelin (1987), experiments under confining pressure show a strong increase in fracture toughness. In the study by Funatsu et al. (2004), K_{IC} increased 5–10 times when raising the confining pressure from 0 to 9 MPa in a tuff and a sandstone. Finally, non-elastic behaviour plays a role. Fracture toughness is a concept from linear elastic fracture mechanics (LEFM), but since stresses diverge at the crack tip, non-elastic mechanisms come into play. The zone near the crack tip is a plastic zone, contributing to an apparent increase in fracture toughness. For a fracture generated by hydraulic fracturing, the pumped fluid will lag behind the rapidly propagating fracture, leading to a cohesive zone with high shear stresses around the crack tip. Papanastasiou (1999) calculated the impact of this on effective fracture toughness. He found that it depends on elastic parameters, friction angle, dilatancy angle, ratio between compressive and tensile strength, and ratio between maximum and minimum principal stress. He found that the effective fracture toughness would increase strongly (order of magnitude) by plasticity in the case of weak rocks in a strongly anisotropic stress field.

7.5.6 Specific shale characterisation tests

Some rocks, in particular shales or other clay rich materials, are sensitive to water. It is thus essential that characteristic parameters that can be related to the water sensitivity are measured. Such parameters include the water content, the cation exchange capacity (CEC), specific surface area and the mineralogy.

Water content is measured by weighing a sample before and after heat drying. The heating temperature should be kept lower than 105 °C to ensure that only free water is released. Above this temperature, crystalline water will start to be removed.

CEC can be obtained from several different methods, and since different methods can yield quite different results, we should take care when comparing CEC-values from different sources. A fairly rough and much used method is the so-called methylene blue test (see API, 1968), where a given amount of rock material is transformed into a slurry with de-ionised, distilled water. Strong sulphuric acid is then added to exchange all available cations in the interlayers of clay minerals with H^+ ions. After this, the slurry is diluted, and a solution of methylene blue ($C_{16}H_{18}N_3SCl{\cdot}3H_2O$) is added gradually. H^+ ions are then exchanged with NH_4^+, and the CEC is obtained from the amount of methylene blue required to form a blue corona when a suspended drop of the slurry is placed on a filter paper.

Specific surface area is obtained by adsorption of gas or liquid molecules of known size onto a surface of a known mass of the rock material (Gregg and Sing, 1967).

Specific surface area has also been used to correlate between wellsite measurements of the dielectric constant and the presence of hydratable clays (Leung and Steiger, 1992). A test kit for the dielectric constant measurement technique was developed using a small sample of cuttings.

The *mineralogy* of a rock may be determined from X-ray diffraction. Since the final numbers result from interpretation of diffractograms, this is not a completely objective method. Again, comparison of results from different sources should be done with great care. Note also that only crystalline minerals will appear on the diffractogram. X-ray amorphous minerals (e.g. certain opal minerals) will not be detected.

Measuring the *activity* (the relative humidity) of a shale sample is also frequently used as an indicator of whether the sample is well preserved and saturated or not. These measurements are commonly performed under atmospheric conditions. However, under atmospheric conditions, the curved interface between the pore water and the air will immediately reduce the measured activity below the true value (e.g. Carminati et al. (1997)). The amount of this reduction depends on, for example, the pore size distribution of the shale. The smaller the pores, the greater the error.

These characterisation tests give a general impression of the shale that can be used as indicators for how the shale will respond. They can also be used for comparison with other shales tested with the same methods. Provided that these results can be linked to more quantitative testing of rock properties, the characterisation tests can help categorise the shale in comparison with previously tested shales.

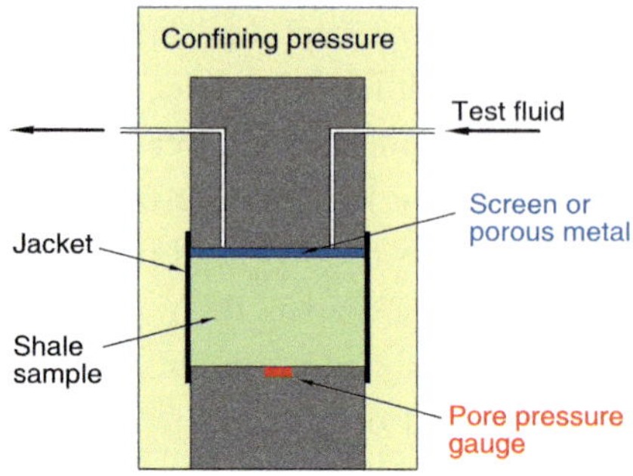

Figure 7.17 Schematic illustration of equipment for pressure transmission testing. After Ewy and Stankovic (2010).

7.5.7 Shale-fluid interaction tests

As already discussed, it is important that care is taken in sampling, preservation and handling of water-sensitive rocks like shale. This is particularly important when studying the response of these rocks to fluid contact. Tests to measure swelling and disintegration are commonly used by the drilling fluid industry for screening mud types and mud additives. It is critical that laboratory artifacts are avoided (e.g. due to capillary effects), as they may completely dominate the results (see also Section 7.1.4).

Another critical factor is the test conditions applied to the sample. Testing at atmospheric conditions can give results that are misleading with respect to downhole response under pressure and stress; see van Oort (2018). It is essential that the testing is planned and executed according to its purpose. Testing shale-fluid effects for wellbore stability application must be done under representative stress conditions on samples that have been properly preserved and are fully saturated when tested. The response of the wellbore wall downhole cannot be deduced from the appearance of the cuttings at the rig floor.

Some of the tests described previously in this chapter can also be used to study effects of different fluids (triaxial test, hollow cylinder test); see, for example, Fig. 9.9. Another test suitable for this purpose is the pressure transmission test (PTT) (van Oort, 1994; Ewy and Stankovic, 2010). A schematic illustration of such a test setup is shown in Fig. 7.17.

The principle of the test is to expose the sample to a test fluid on one side and measure the pore pressure response on the other side. When properly executed and analysed, the test can give information about osmotic effects, membrane efficiency, hydraulic diffusivity (permeability) and ion diffusivity; see also Section 9.4.

References

Amadei, B., Stephansson, O., 1997. Rock Stress and Its Measurement. Chapman & Hall, London.

Andersen, M.A., 1995. Petroleum Research in North Sea Chalk. RF — Rogaland Research, Stavanger, Norway.

API, 1968. Supplement 2 to API RP 13B (1st ed.) – Standard procedure for testing drilling fluids: Cation Exchange Capacity. American Petroleum Institute, New York.

Batzle, M., Han, D.-H., Hofmann, R., 2006. Fluid mobility and frequency-dependent seismic velocity — direct measurements. Geophysics 71 (1), N1–N9. https://doi.org/10.1190/1.2159053.

Beckers, T., 2004. Fracture toughness determination and micromechanisms of rock under mode I and mode II loading. PhD thesis. U Potsdam, Germany.

Bieniawski, Z.T., Bernede, M.J., 1979. Suggested methods for determining the uniaxial compressive strength and deformability of rock materials. Part 1. Int. J. Rock Mech. Min. Sci. Geomech. Abstr. 16 (2), 138–140. https://doi.org/10.1016/0148-9062(79)91451-7.

Blanton, T.L., 1983. The relation between recovery deformation and in-situ stress magnitudes. SPE/DOE11624. In: SPE/DOE Symp. on Low Permeability. Denver, Co, USA.

Carminati, S., Brignoli, M., Di Marco, A., Santarelli, F.J., 1997. The activity concept applied to shales: consequences for oil, tunnelling and civil engineering operations. Int. J. Rock Mech. Min. Sci. 34 (3–4), 038. https://doi.org/10.1016/S1365-1609(97)00069-5.

Colback, P.S.B., Wiid, B.L., 1965. The influence of moisture content on the compressive strength of rocks. In: Proc. 3rd Canadian Symp. Rock Mech. Toronto, Canada, pp. 65–83.

de Waal, J.A., Smits, R.M.M., 1988. Prediction of reservoir compaction and surface subsidence: field application of a new model. SPE Form. Eval. 3 (2), 347–356. https://doi.org/10.2118/14214-PA.

Dudley, J.W., Brignoli, M., Crawford, B.R., Ewy, R.T., Love, D.K., McLennan, J.D., Ramos, G.G., Shafer, J.L., Sharf-Aldin, M.H., Siebrits, E., Boyer, J., Chertov, M.A., 2016. ISRM suggested method for uniaxial-strain compressibility testing for reservoir geomechanics. Rock Mech. Rock Eng. 49, 4153–4178. https://doi.org/10.1007/s00603-016-1055-4.

Engelder, T., Plumb, R., 1984. Changes in in situ ultrasonic properties of rock on strain relaxation. Int. J. Rock Mech. Min. Sci. Geomech. Abstr. 21 (2), 75–82. https://doi.org/10.1016/0148-9062(84)91175-6.

Ewy, R.T., 2018. Practical approaches for addressing shale testing challenges associated with permeability, capillarity and brine interactions. Geomech. Energy Environ. 14, 3–15. https://doi.org/10.1016/j.gete.2018.01.001.

Ewy, R.T., Stankovic, R.J., 2010. Shale swelling, osmosis, and acoustic changes measured under simulated downhole conditions. SPE Drill. Completion 25, 177–186. https://doi.org/10.2118/78160-PA.

Funatsu, T., Seto, M., Shimada, H., Matsui, K., Kuruppu, M., 2004. Combined effects of increasing temperature and confining pressure on the fracture toughness of clay bearing rocks. Int. J. Rock Mech. Min. Sci. 41, 927–938. https://doi.org/10.1016/j.ijrmms.2004.02.008.

Geertsma, J., 1985. Some rock-mechanical aspects of oil and gas well completions. Soc. Pet. Eng. J. 25 (6), 848–856. https://doi.org/10.2118/8073-PA.

Giger, S.B., Ewy, R.T., Favero, V., Stankovic, R., Keller, L.M., 2018. Consolidated-undrained triaxial testing of Opalinus Clay: results and method validation. Geomech. Energy Environ. 14, 16–28. https://doi.org/10.1016/j.gete.2018.01.003.

Gregg, S.J., Sing, K.S.W., 1967. Adsorption, Surface Area and Porosity. Academic Press, New York.

Head, K.H., 1984. Manual of Soil Laboratory Testing, vols. 1–3. ELE International Ltd, London.

Hoek, E., Franklin, J.A., 1968. Simple triaxial cell for field or laboratory testing of rock. Trans. Inst. Min. Metall. 77A, 483–489.

Holcomb, D.J., 1993. Observations of the Kaiser effect under multiaxial stress states: implications for its use in determining in situ stress. Geophys. Res. Lett. 20 (19), 2119–2122. https://doi.org/10.1029/93GL01270.

Holt, R.M., Fjær, E., 1991. Validity of multiple failure state triaxial tests in sandstones. In: Proc. 7th ISRM Congress. Aachen, Germany.

Holt, R.M., Ingsøy, P., Mikkelsen, M., 1989. Rock mechanical analysis of North Sea reservoir formations. SPE Form. Eval. 4 (1), 33–37. https://doi.org/10.2118/16796-PA.

Holt, R.M., Brignoli, M., Kenter, C.J., 2000. Core quality: quantification of coring-induced rock alteration. Int. J. Rock Mech. Min. Sci. 37 (6), 889–907. https://doi.org/10.1016/S1365-1609(00)00009-5.

Holt, R.M., Bakk, A., Stenebråten, J.F., Bauer, A., Fjær, E., 2018. Skempton's A – a key to man-induced subsurface pore pressure changes. ARMA18-0949. In: 52nd US Rock Mechanics/Geomechanics Symposium. Seattle, WA, USA.

Horsrud, P., Sønstebø, E.F., Bøe, R., 1998. Mechanical and petrophysical properties of North Sea shales. Int. J. Rock Mech. Min. Sci. 35, 1009–1020. https://doi.org/10.1016/S0148-9062(98)00162-4.

Hudson, J.A., Crouch, S.L., Fairhurst, C., 1972. Soft, stiff and servocontrolled testing machines: a review with reference to rock failure. Eng. Geol. 6 (3), 155–189. https://doi.org/10.1016/0013-7952(72)90001-4.

Jaeger, J.C., Cook, N.G.W., 1979. Fundamentals of Rock Mechanics, third edition. Chapman and Hall, London.

Janbu, N., 1985. Soil models in offshore engineering. Géotechnique 35 (3), 239–281. https://doi.org/10.1680/geot.1985.35.3.241. 25th Rankine Lecture.

Jones, M., 1988. Determination of the Mechanical Properties of Reservoir Rocks Using the Triaxial Test. Experimental Guidelines. The Norwegian Petroleum Directorate, Stavanger.

Kaiser, J., 1950. Untersuchungen über das Auftreten von Geräuschen beim Zugversuch. PhD thesis. Technische Hochschule München.

Kenter, C.J., Brignoli, M., Holt, R.M., 1997. MS (Constant Mean Stress) vs. UCS (Unconfined Strength) tests: a tool to reduce core damage effects. Int. J. Rock Mech. Min. Sci. 34 (3–4), 129. https://doi.org/10.1016/S1365-1609(97)00139-1.

Kenter, C.J., Pestman, B.J., Van Munster, J.G., 1998. Determining in situ stress. Patent EPC 98202284.

Kovari, K., Tisa, A., Einstein, H.H., Franklin, J.A., 1983. Suggested methods for determining the strength of rock materials in triaxial compression: revised version. Int. J. Rock Mech. Min. Sci. Geomech. Abstr. 20 (6), 285–290. https://doi.org/10.1016/0148-9062(83)90598-3.

Lambe, T.W., Whitman, R.V., 1979. Soil Mechanics. John Wiley & Sons, New York.

Leung, P.K., Steiger, R.P., 1992. Dielectric constant measurements: a new, rapid method to characterize shale at the wellsite. IADC/SPE23887. In: Proc. 1992 IADC/SPE Drilling Conf. New Orleans, LA, USA, pp. 401–408.

Lozovyi, S., Bauer, A., 2019. Velocity dispersion in rocks: a laboratory technique for direct measurement of P-wave modulus at seismic frequencies. Rev. Sci. Instrum. 90, 1–15. https://doi.org/10.1063/1.5026969.

Matsuki, K., Takeuchi, K., 1993. Three-dimensional in situ stress determination by anelastic strain recovery of a rock core. Int. J. Rock Mech. Min. Sci. Geomech. Abstr. 30 (7), 1019–1022. https://doi.org/10.1016/0148-9062(93)90064-K.

Maury, V.M., Santarelli, F.J., Henry, J.P., 1988. Core discing: a review. In: Proc. SANGORM Symp.: Rock Mechanics in Africa, pp. 221–231.

Mess, K.W., 1978. On the interpretation of core compaction behaviour. In: Saxena, S.K. (Ed.), Evaluation and Prediction of Subsidence. American Society of Civil Engineers, New York, pp. 76–91.

Nakken, S.J., Christensen, T.L., Marsden, R., Holt, R.M., 1989. Mechanical behaviour of clays at high stress levels for wellbore stability applications. In: Maury, V., Fourmaintraux, D. (Eds.), Rock at Great Depth. Balkema, Rotterdam, pp. 141–148.

Nermoen, A., Korsnes, R., Christensen, H.F., Trads, N., Hiorth, A., Madland, M.V., 2013. Measuring the Biot stress coefficient and its implications on the effective stress estimate. ARMA13-282. In: 47th US Rock Mechanics/Geomechanics Symposium. San Francisco, CA, USA.

Nes, O.-M., Horsrud, P., Sønstebø, E.F., Holt, R.M., Ese, A.M., Økland, D., Kjørholt, H., 1998a. Rig site and laboratory use of CWT acoustic velocity measurements on cuttings. SPE Reserv. Eval. Eng. 1 (4), 282–287. https://doi.org/10.2118/50982-PA.

Nes, O.-M., Sønstebø, E.F., Horsrud, P., Holt, R.M., 1998b. Dynamic and static measurements on mm-size shale samples. SPE/ISRM47200. In: Proc. Eurock '98, vol. 2. Trondheim, Norway, pp. 23–32.

Ouchterlony, F. Co-ordinator. International Society for Rock Mechanics commission on testing methods. Suggested methods for determining the fracture toughness of rock. Int. J. Rock Mech. Min. Sci. Geomech. Abstr. 25 (2), 71–96. https://doi.org/10.1016/0148-9062(88)91871-2.

Papamichos, E., Brignoli, M., Santarelli, F.J., 1997. An experimental and theoretical study of a partially saturated collapsible rock. Mech. Cohes.-Frict. Mater. 2 (3), 251–278. https://doi.org/10.1002/(SICI)1099-1484(200001)5:1<1::AID-CFM74>3.0.CO;2-H.

Papanastasiou, P., 1999. The effective fracture toughness in hydraulic fracturing. Int. J. Fract. 96, 127–147. https://doi.org/10.1023/A:1018676212444.

Pestman, B.J., Holt, R.M., Kenter, C.J., van Munster, J.G., 2002. Field application of a novel core-based in-situ stress estimation technique. SPE/ISRM78158. In: Proc. Oil Rocks. Irving, TX, USA.

Ren, N.K., Hudson, P.J., 1985. Predicting the in situ state of stress using differential wave velocity analysis. In: Proc. 26th US Rock Mech. Symp., pp. 1235–1244.

Richard, T., Detournay, E., Drescher, A., Nicodeme, P., Fourmaintraux, D., 1998. The scratch test as a means to measure strength of sedimentary rocks. SPE/ISRM47196. In: Proc. Eurock '98, vol. 2. Trondheim, Norway, pp. 15–22.

Ringstad, C., Lofthus, E.B., Sønstebø, E.F., Fjær, E., 1998. Prediction of rock parameters from micro-indentation measurements: the effect of sample size. SPE/ISRM47313. In: Proc. Eurock '98, vol. 1. Trondheim, Norway, pp. 487–492.

Risnes, R., Flaageng, O., 1999. Mechanical properties of chalk with emphasis on chalk–fluid interactions and micromechanical aspects. Oil Gas Sci. Technol. – Rev. IFP 54 (6), 751–758. https://doi.org/10.2516/ogst:1999063.

Santarelli, F.J., Guenot, A., 1989. Experimental observations of bifurcation phenomena in rocks. In: Maury, V., Fourmaintraux, D. (Eds.), Rock at Great Depth. Balkema, Rotterdam, pp. 1563–1570.

Santarelli, F.J., Marsala, A.F., Brignoli, M., Rossi, E., Bona, N., 1998. Formation evaluation from logging on cuttings. SPE Reserv. Eval. Eng. 1 (3), 238–244. https://doi.org/10.2118/36851-PA.

Santos, H., Diek, A., Roegiers, J.C., Fontoura, S., 1996. Can shale swelling be (easily) controlled? In: Proc. Eurock '96. Torino, Italy, pp. 99–106.

Schei, G., Fjær, E., Detournay, E., Kenter, C.J., Fuh, G.-F., Zausa, F., 2000. The scratch test: an attractive technique for determining strength and elastic properties of sedimentary rocks. SPE63255. In: SPE Annual Technical Conf. and Exhibition. Dallas, TX, USA.

Schmitt, L., Forsans, T., Santarelli, F.J., 1994. Shale testing and capillary phenomena. Int. J. Rock Mech. Min. Sci. Geomech. Abstr. 31 (5), 411–427. https://doi.org/10.1016/0148-9062(94)90145-7.

Spencer Jr., J.W., 1981. Stress relaxations at low frequencies in fluid saturated rocks: attenuation and modulus dispersion. J. Geophys. Res. 86 (B3), 1803–1812. https://doi.org/10.1029/JB086iB03p01803.

Steiger, R.P., Leung, P.K., 1992. Quantitative determination of the mechanical properties of shale. SPE Drill. Eng. 7 (3), 181–185. https://doi.org/10.2118/18024-PA.

Stenebråten, J.F., Fjær, E., Haaland, S., Lavrov, A.V., Sønstebø, E.F., 2008. The shale puncher – a compact tool for fast testing of small shale samples. ARMA08-227. In: 42nd US Rock Mechanics Symposium and 2nd U.S.–Canada Rock Mechanics Symposium. San Francisco, CA, USA.

Suarez-Rivera, R., Stenebråten, J., Dagrain, F., 2002. Continuous scratch testing on core allows effective calibration of log-derived mechanical properties for use in sanding prediction evaluation. SPE/ISRM78157. In: Proc. Oil Rocks. Irving, TX, USA.

Sulem, J., Vardoulakis, I., 1989. Simplified bifurcation analysis of deep boreholes in rocks with microstructure. In: Maury, V., Fourmaintraux, D. (Eds.), Rock at Great Depth. Balkema, Rotterdam.

Swan, G., Cook, J., Bruce, S., Meehan, R., 1989. Strain rate effects in Kimmeridge Bay shale. Int. J. Rock Mech. Min. Sci. Geomech. Abstr. 26, 135–149. https://doi.org/10.1016/0148-9062(89)90002-8.

Teufel, L.W., 1989. Acoustic emissions during anelastic strain recovery from deep boreholes. In: Khair, A.W. (Ed.), Rock Mechanics as a Guide for Efficient Utilization of Natural Resources. Balkema, Rotterdam, pp. 269–276.

Thiercelin, M., 1987. Fracture toughness under confining pressure using the modified ring test. In: Proc. 28th U.S. Symp. on Rock Mechanics. Tucson, AZ, USA, pp. 149–156.

Tronvoll, J., Morita, N., Santarelli, F.J., 1992. Perforation cavity stability: comprehensive laboratory experiments and numerical analysis. SPE24799. In: SPE Annual Technical Conference and Exhibition. Washington, DC, USA.

Ulusay, R. (Ed.), 2015. The ISRM Suggested Methods for Rock Characterization, Testing and Monitoring: 2007–2014. Springer.

van den Hoek, P.J., Kooijman, A.P., Kenter, C.J., Khodaverdian, M., Hyland, C.R., McLennan, J.D., 1992. Size-dependency of hollow cylinder collapse strength. SPE24800. In: Proc. 67th Annual Technical Conference and Exhibition. Washington, DC, USA, pp. 351–360.

van der Vlis, A.C., 1970. Rock classification by a simple hardness test. In: Proc. 2nd Congress ISRM, pp. 23–30.

van Oort, E., 1994. A novel technique for the investigation of drilling fluid induced borehole instability in shales. SPE28064. In: SPE/ISRM Rock Mechanics in Petroleum Engineering Conference. Delft, Netherlands.

van Oort, E., 2018. How to test for compatibility between fluids and shales. IADC/SPE18963. In: IADC/SPE Drilling Conference and Exhibition. Fort Worth, TX, USA.

Vutukuri, V.S., Lama, R.D., Saluja, S.S., 1974. Mechanical Properties of Rocks, vols. I–IV. Trans Tech Publications, Clausthal.

Walle, L.E., Stroisz, A.M., Brevik, N.Ø., Jensen, S.S., Holt, R.M., 2017. Laboratory measurements of strength parameters for fracturing. ARMA17-724. In: 51st US Rock Mechanics/Geomechanics Symposium. San Francisco, CA, USA.

Warpinski, N.R., Teufel, L.W., 1989. A viscoelastic constitutive model for determining in situ stress magnitudes from anelastic strain recovery of core. SPE Prod. Eng. 4 (3), 272–289. https://doi.org/10.2118/15368-PA.

Whittaker, B.N., Singh, R.N., Sun, G., 1992. Rock Fracture Mechanics: Principles, Design and Applications. Elsevier.

Winkler, K.W., Plona, T.J., 1982. Technique for measuring ultrasonic velocity and attenuation spectra in rocks under pressure. J. Geophys. Res. 87 (B13), 10776–10780. https://doi.org/10.1029/JB087iB13p10776.

Yassir, N., Wang, D.-F., Enever, J.R., Davies, P.J., 1998. An experimental analysis of anelastic strain recovery of synthetic sandstone subjected to polyaxial stress. SPE/ISRM47238. In: Proc. Eurock '98, vol. 1. Trondheim, Norway, pp. 161–170.

Zhang, Z.X., 2002. An empirical relation between mode I fracture toughness and the tensile strength of rock. Int. J. Rock Mech. Min. Sci. 39 (3), 401–406. https://doi.org/10.1016/S1365-1609(02)00032-1.

CHAPTER 8

Mechanical properties and in situ stresses from field data

Knowledge of the mechanical properties and the in situ stresses of the subsurface formations is essential in connection with wellbore stability problems, fracturing operations, subsidence problems and sand production evaluation.

In this chapter, we will look at methods for estimating mechanical properties and in situ stresses from field tests and field data. The methods can be grouped into three main categories: measurements while drilling, wireline logging methods and well tests.

We start by recalling the mechanical parameters and in situ stresses of a formation.

Elastic parameters. For an isotropic medium, there are two independent elastic moduli. However, the elastic parameters vary with frequency and with stress level, so that even in the isotropic case the use of two parameters is clearly a simplification.

Strength parameters. We have seen that the strength of a material is dependent on the stress level and that failure criteria that describe actual data normally have at least 2–3 adjustable parameters.

In situ stresses. The in situ stresses are given by three principal stresses and three parameters giving the orientation of the principal stresses. The stress level determines whether a rock is critically loaded or not. Additionally, the in situ stresses influence both the elastic parameters and the strength parameters.

8.1 Estimation of elastic parameters

The most important method for estimation of elastic parameters is acoustic logging, in particular acoustic wireline logs and also acoustic logging while drilling.

Acoustic logging tools measure acoustic wave velocities, which, together with density information and the appropriate formulae discussed in Chapter 5, provide the dynamic elastic parameters (Eqs. (5.20)–(5.24)). We

Developments in Petroleum Science, Volume 72
ISSN 0376-7361, https://doi.org/10.1016/B978-0-12-822195-2.00017-6

repeat here the equations for dynamic Young's modulus E_{dyn} and dynamic Poisson's ratio ν_{dyn} for an isotropic material:

$$E_{\text{dyn}} = \frac{\rho v_s^2 \left(3v_p^2 - 4v_s^2\right)}{\left(v_p^2 - v_s^2\right)} \tag{8.1}$$

$$\nu_{\text{dyn}} = \frac{v_p^2 - 2v_s^2}{2\left(v_p^2 - v_s^2\right)} \tag{8.2}$$

Once the dynamic moduli are known, the problem is to relate them to the static parameters, which is what is needed for most applications. This problem is a difficult one, which has not been fully resolved.

The difference between the static Young's modulus and the dynamic Young's modulus may be quite significant, in particular at low stress levels. The dynamic modulus may be several times larger than the static modulus, for example, in a weak sandstone. The static-dynamic transition is complicated by the fact that it is not a constant shift, but varies independently with loading (Fjær, 1999, 2019). The main contribution to the difference is the strain amplitude, which is low in dynamic measurements and large in static measurements. In a porous rock, both pore fluid and cracks can give further contribution to this difference. Empirically based static-dynamic transition methods are thus of limited validity, and there is still also a lack of theoretically based methods. See also Section 5.6.1.

One step in the direction of a more fundamental approach was presented by Fjær (1999). This approach was developed primarily for weak sandstone, utilising a large number of carefully designed laboratory tests (mostly on North Sea cores) with simultaneous static and dynamic measurements. Further description of this quantitative model was given in Section 5.6.

The rest of this section is devoted to various types of acoustic measurements, primarily from downhole measurements, but also from rig measurements. The principles of acoustic wave propagation around boreholes were described in Section 5.8.

Indirect information which may be useful for evaluation of mechanical parameters, can be obtained from other logging tools, for example, the gamma density and neutron logs. These tools can be used to obtain lithological information.

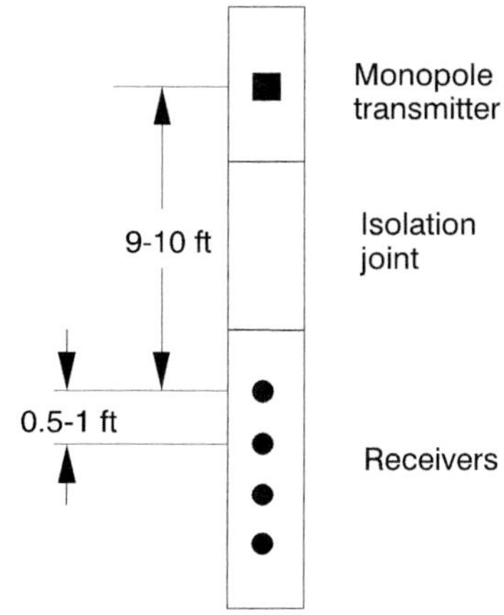

Figure 8.1 Principle sketch of a full waveform sonic tool. Typical distance from transmitter to first receiver is 10 ft (3 m), whereas the distance between each receiver is typically 0.5–1 ft (0.15–0.3 m).

8.1.1 Acoustic wireline logs

The primary output from acoustic logging tools is the velocities and waveforms of different waves, such as the compressional-, shear- and Stoneley waves. Note that acoustic logging tools may be used to generate information about a number of other characteristics of the formation, such as lithology identification, porosity estimates, gas detection, detection of natural fractures, qualitative evaluation of permeability variations and creation of synthetic seismograms for comparison with surface seismics.

See Section 5.8 for an introduction to borehole acoustics.

8.1.1.1 Full waveform sonic tools

A principle sketch of a full waveform sonic tool is shown in Fig. 8.1. This consists of an omnidirectional transmitter separated from an array of receivers.

The transmitter creates a pressure pulse in the borehole fluid. This pressure pulse generates compressional and shear waves in the formation, in addition to borehole waves like the Stoneley wave. These tools ideally provide compressional and shear velocities of the formation in addition to the Stoneley wave velocity. However, in slow formations (poorly consolidated) the shear wave velocity is slower than the fluid waves in the borehole. Consequently, the shear velocity cannot be measured directly.

From a mechanical properties point of view, we are most interested in the shear velocity in slow formations, since low wave velocities indicate low strength and hence potential stability problems. Direct measurements of both compressional and shear velocities also in slow formations are possible using dipole or multipole tools.

Typical centre frequency of the transmitter used in full waveform sonic tools is 15 kHz. However, to strongly excite the Stoneley mode, the source spectrum should include frequencies below 5 kHz. This can be obtained by increasing the bandwidth of the source and also by varying the excitation frequency of the source, depending on which waveform is requested.

8.1.1.2 Multipole sonic tools

The basic concept of this tool is the dipole source transducer. A dipole transducer can in principle be pictured as two closely spaced monopole transducers driven with opposite phases. This piston-like effect creates an increase in pressure on one side of the borehole and a decrease in pressure on the other side of the borehole. This force oriented normal to the borehole wall induces a flexural motion of the borehole (see Fig. 5.21).

The dipole source operates at low frequencies, typically with a centre frequency of 1–1.5 kHz. At low frequencies the flexural mode is very little affected by the borehole and travels at the same speed as the shear wave. Dipole tools can thus provide a direct measurement of the shear velocity even in slow formations. At higher frequencies the propagation velocity becomes smaller.

Dipole capabilities are often integrated in monopole tools, with the dipole transmitter close to the monopole source; see Fig. 8.1. By including two dipole transducers oriented perpendicular to each other and two receivers at each receiver station, both in-line and cross-line measurements can be performed. Cross-dipole measurements hence facilitate an evaluation of different types of anisotropy. It is a well-known observation that the acoustic velocity is faster along the rock bedding than normal to the bedding. This has commonly been observed when comparing P-wave velocities measured in vertical holes with measurements in highly deviated or horizontal holes in the same formation.

The crossed dipole sonic tools have a combination of 1–4 monopole sources and two orthogonally oriented dipole sources and can have up to about 100 receivers. The sources can sweep through a broad frequency band, from a few hundred Hz (e.g. 300 Hz) and up to approximately 10 kHz. The tool enables a fairly detailed characterisation of the elastic wave velocities in the vicinity of the borehole, as it is sensitive to axial, azimuthal and radial variations.

8.1.2 Acoustic logging while drilling

The tools configuration for acoustic logging while drilling is similar to that shown in Fig. 8.1, with a transmitter separated a few feet (typically 4–7 ft) from an array of receivers. The full waveforms are normally recorded in a downhole memory for later evaluation, but transit times can be transmitted uphole for real time use: lithology evaluation, porosity determination, pore pressure estimation and correlation of drill bit position to seismic maps.

Real-time access to good estimates of mechanical properties of the rock is obviously useful also for borehole stability evaluation while drilling. By logging while and after drilling we could possibly monitor changes in the rock response during the openhole period. This may give an early warning of possible deterioration of rock integrity with time, which eventually results in borehole stability problems. Consistent differences between acoustic velocities recorded while drilling and from wireline logging have been observed in shales (see e.g. Boonen et al., 1998).

8.1.3 Acoustic measurements on drill cuttings

Acoustic measurements on drill cuttings may represent a valuable source of information, which can supplement the logging techniques. Alternative techniques are available, both a sonic pulse technique (Santarelli et al., 1998) and a continuous wave technique (Nes et al., 1998). For further description, see also Section 7.3.12.

Drill cuttings represent a large source of information, which is available without any additional cost. If used on the rig, information will be available close to real time. There are, however, a number of limitations with these techniques, related primarily to the quality and representativity of the cuttings. How has stress unloading effects, drill bit effects and effects of the drilling fluid and so on changed the response compared to that of the virgin formation? Other uncertainties relate to possible length scale effects (typically mm-size for cuttings, whereas typically 30 cm for an acoustic log) and the accuracy in depth determination of cuttings origin. A bias in sample selection may also be introduced, promoting measurements on the more competent samples in the high-velocity range. The difference in test conditions (atmospheric for cuttings measurements compared to downhole conditions for logging tools) may also affect the result.

Despite all these possible uncertainties, reasonable agreement between logged values and values derived from cuttings measurements has been reported (Santarelli et al., 1998; Nes et al., 1998), especially with respect to depth trends. Cutting measurements thus represent a promising but not yet fully explored alternative and/or supplement to other measurements.

8.2 Estimation of strength parameters

There are no direct methods available for determination of the strength parameters in situ. Hence most of the efforts made to predict strength parameters have been based on empirical correlations, especially relations between acoustic velocities and strength parameters. In this section, we discuss both this and other relations, which are being used for estimation of strength parameters.

8.2.1 Log data (wireline and MWD)

Log data are attractive for prediction of rock mechanical properties since logging is one of the few downhole measurements available. As presented already, there is a direct relation between acoustic velocities and dynamic elastic properties. However, rock strength data cannot be inferred directly from logs. The traditional approach has therefore been more or less empirical, using both log derived acoustic velocities (especially the shear velocity) and log derived porosity. An obvious disadvantage of such an approach is the need for frequent calibration and checking of the validity of the empirical correlations when entering a new field or area.

The model briefly discussed in Section 8.1 (Fjær, 1999) can also be used for strength prediction (Raaen et al., 1996). This model also uses log data (acoustic velocities and density) and was developed primarily for prediction of weak sandstone strength. The fact that this model has a more theoretical basis makes it more robust, and it can thus be applied in new wells without extensive calibration. This therefore represents a step in the right direction. However, it is not trivial to establish methods that can predict the strength of different lithologies directly from logs, without any need of calibration.

The methods for predicting rock strength offered by the service companies are all subject to the limitations discussed above, being more or less empirical and requiring more or less calibration.

8.2.2 Drill cuttings measurements

Different types of measurements on drill cuttings have been discussed previously, both acoustic measurements (Sections 7.3.12 and 8.1.3) and indentation measurements (Section 7.5.3). To obtain strength values, correlations have to be established (Santarelli et al., 1998; Ringstad et al., 1998). Given the previously discussed uncertainties of using drill cuttings (Section 8.1.3), these methods must at present be considered quite coarse. However, they should be able to provide at least comparative evaluations, that is, a "weak" versus "strong" type of classification.

8.2.3 Empirical correlations

The lack of direct downhole strength measurements or other generally applicable theories for relating rock strength to measurable quantities has motivated considerable efforts to generate empirical correlations for rock strength prediction. Most of this has been based on log-measured or log-derived properties, such as acoustic velocities or porosities. Some examples of such work are Edlman et al. (1998); Farquhar et al. (1994); Mason (1987); Onyia (1988). See also Section 3.4 and Appendix A. Empirical correlations have been generated very much dependent on the need or the application. Production-related problems tend to focus on weak sandstones, whereas stability problems related to drilling deal with weak shales.

There are some general comments which should be made regarding empirical correlations:

- The correlations are normally based on one specific type of rock or lithology. A correlation developed in sandstone may not be valid in shale, that is, empirically based correlations are not universally applicable.
- Rock strength measurements may be very scarce for some lithologies, such as shales, because shales are normally not cored and tested. Shale-specific correlations are scarce (for exceptions, see Lashkaripour and Dusseault (1993); Horsrud (2001)).
- Lack of actual core material has promoted use of outcrop material, which may be not relevant for downhole sedimentary rocks (differences in sedimentation history, stress history, stress environment, diagenesis etc.).
- Laboratory test procedures or data interpretation procedures may vary, yielding results that are not fully comparable.
- Even if the correlation is based on real core material from one area, it may not be applicable to other geographical areas. It is always recom-

mended to check the validity of correlations when entering new areas and calibrate them if necessary.

- The published correlations are generally biased towards fairly competent rocks, thus reducing their accuracy with respect to weak rocks, which is the primary concern in all types of stability evaluations.

In general, we find that the rock strength increases when the rock is densified and cemented, that is, as depth, age and effective stresses increase. This is further reflected in normal trends of reduced porosity and increased acoustic velocity (Fig. 3.16).

Further discussion of mechanical properties of sedimentary rocks (sandstone, chalk and shale) is given in Section 3.4, including some examples of empirical correlations. A more extended list of correlations is included in Appendix A.

8.2.4 Drilling data

Intuitively, the response of the drill bit should reflect also the mechanical properties of the rock under true downhole conditions. Bit response is, however, also dependent on a number of other parameters. Traditionally, the motivation for rock-bit interaction models has been to improve bit selection and bit operation procedures. The process of cuttings generation and cuttings removal is fairly complicated, and most ROP (rate of penetration) models are therefore empirical or semi-empirical. Although ROP models are available for several bit types, most of the modelling efforts have been directed towards roller cone bits. A representative version of an ROP model is (Warren, 1987)

$$R = \left(\frac{aS^2D^3}{NW^2} + \frac{b}{ND} + \frac{cD\rho\mu}{F_m} \right)^{-1} \tag{8.3}$$

where R is the rate of penetration, W is the weight on bit, N is the bit rotary speed, S is the rock strength, D is the bit diameter, F_m is the modified jet impact force, ρ is the mud density, μ is the plastic viscosity, and a,b,c are dimensionless constants. This model is based on tests with low borehole pressure (low overbalance). Tests with increased overbalance reduced the ROP, but this appeared to be more complicated to integrate in a model.

Additional effects related to bit wear and chip hold down may also be included (e.g. Hareland and Hoberock, 1993).

Thus we can also solve Eq. (8.3) for the rock strength when the rate of penetration and the other parameters and functions are known (Hareland et al., 1996).

Determination of all the input parameters and model constants may not be trivial, and calibration against results from tests under controlled conditions or against previous bit runs in the same area may be required. Considerable uncertainty in quantitative predictions should therefore be expected. It is also obvious that such models are not easy to use in a predictive manner.

It is quite obvious that the drilling process is a response to the properties of the rock and its state of stress. Information about the rock mechanical properties should therefore be retrievable from drilling data, but continued efforts to understand and model the drilling process are required. Combining all measurements while drilling, both drilling data and log data, may eventually provide a real-time tool for follow-up of the drilling process.

8.3 Estimation of in situ stresses

In this section, we present available methods for estimating in situ stresses from field data. Methods based on core measurements were discussed in Chapter 7 and will not be repeated here. In this section we thus focus on data from well logs and well tests. For a more detailed description of all available stress measurement methods, see, for example, Amadei and Stephansson (1997) and Haimson and Cornet (2003). Note that not all these methods are suitable for petroleum applications.

The presentation given here is based on the common assumption of a vertical–horizontal stress field, that is, one principal stress is vertical, and the two others are horizontal. This is valid in tectonically inactive areas or areas that have been relaxed from previous tectonic activity. In active areas, this assumption may be not valid, and the stress field can be significantly distorted. See also Chapter 3.

There are thus three stress components to evaluate:

Overburden stress or vertical stress. This is determined with reasonable accuracy from integration of bulk density data; see Section 8.3.1.

Smallest horizontal stress. This parameter can be determined by direct measurement, because it is the smallest principal stress in the formation and is thus a controlling factor in the fracturing process (see Section 8.3.4). This property is essential as input to a number of evaluations and calculations. Some of these are:

- Borehole stability prognosis and well design.
- Sand prediction.
- Reservoir stress path due to depletion.

- Drilling window during depletion.
- Maximum safe injection pressure to determine seal integrity and avoid out-of-zone injection.
- Integrity at casing shoe, packer depth and other well barrier depths.
- Large-scale geomechanical modelling.

A sufficiently long fracture always grows in the direction of least resistance. The closure pressure therefore measures the smallest stress in the formation, even if the stress field is not oriented in a vertical–horizontal direction. The result will therefore also in such cases be a valid input to evaluations of barriers and integrity.

Largest horizontal stress. Unfortunately, it is not possible to measure this property. Some indirect approaches are available (Sections 8.3.3 and 8.3.5), but these suffer from large uncertainty. An evaluation of this stress component is still required, as it can have a severe impact on the resulting predictions; see, for example, Sections 4.3.3 and 4.5.3. A qualitative assessment can be done based on general knowledge of the structural geology, that is, whether this is a tectonically active or relaxed area (e.g. normal faulting or strike slip). Operational experience from the area can also be relevant, at least to determine the overall stress regime. Such experience can be wellbore stability during drilling for different well inclinations and azimuths, and use of oriented perforations for sand control; see Section 10.2.3. Finally, sensitivity analyses can be performed to evaluate the effect of uncertainty in the largest horizontal stress.

8.3.1 Overburden stress (density log)

The density tools are active gamma ray tools that use the Compton scattering of gamma rays to measure the electron density of the formation. By appropriate lithology corrections the electron density is converted to mass density with reasonable accuracy.

The density is useful for determining mechanical properties in two manners: first, the density is needed to convert from acoustic velocities to dynamic elastic moduli (see Section 8.1); second, the density integrated over the vertical depth of the well is usually considered to give a good estimate of the vertical stress, at least in areas of low tectonic activity. In such areas the vertical stress is also considered to be a principal stress. When the density log is available, the problem of determining the full in situ stress field is then reduced to determining the magnitude and orientation of the horizontal stresses.

However, the density log is rarely available in the first few hundred metres of a well. Then it is necessary to make estimates of the density to obtain the total vertical stress. In areas not subjected to tectonic activity, it is common to assume a density in the range 1.8–2.0 g/cm^3. This bulk density corresponds to a porosity in the range 50–38 % in a rock with a mineral density of 2.6 g/cm^3.

Offshore Mid-Norway, bulk densities above 2 g/cm^3 have been logged at very shallow depths. One plausible explanation to this unusually high density is that this area has also been exposed to ice loads. Thus the geological history should be taken into consideration when making estimates in non-logged intervals. Geotechnical data from site surveys may in some cases be available and provide additional information.

In offshore applications the total overburden stress also includes the water column; see also Section 9.7.2.

8.3.2 Horizontal stress directions (borehole logs)

Determination of horizontal stress directions is based on the possibility of failure at the borehole wall, which can be detected by borehole logging tools. To be detectable, the failures must occur in the period after drilling and prior to logging. In a vertical borehole that penetrates layers of significantly different horizontal stresses ($\sigma_H > \sigma_h$), two distinct failure modes can be detected, compressive and tensile failures. The directions of these two failure modes in an idealised situation are uniquely given by the directions of the two principal horizontal stresses, as illustrated in Fig. 8.2. Compressive failure or shear failure will be induced in the direction parallel with the smallest horizontal stress σ_h if the well pressure is low enough to induce shear failure. This is commonly referred to as *breakouts* in this context and will lead to an ovalisation of the borehole. Tensile failure will occur in the direction parallel with the largest horizontal stress σ_H if the well pressure is large enough to induce fracturing (drilling induced fracture, DIF). See also Chapter 4 for further discussion of stresses around boreholes.

Once a failure has occurred on the borehole wall, it is tempting to try to back-calculate also stress magnitudes, especially the magnitude of the largest horizontal stress by using elastic theory and appropriate failure criteria. However, a number of assumptions are required for such analysis, rendering the results uncertain. Such estimates can at best be considered as upper or lower bounds on the stress magnitudes (see also Section 8.3.5). The large amount of information that can be acquired by acoustic tools

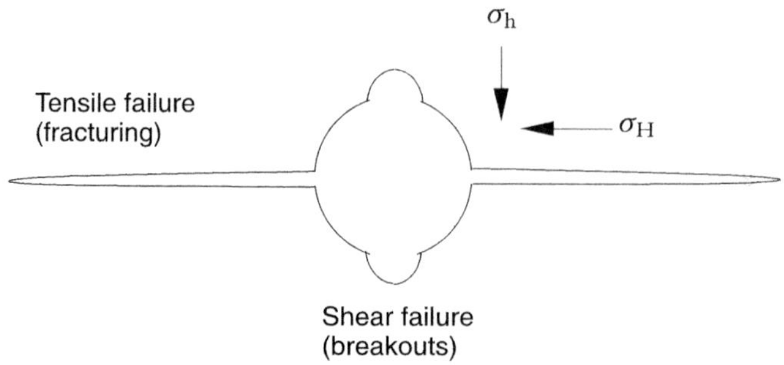

Figure 8.2 Illustration of directions for compressive and tensile failure around a vertical borehole. Since tensile failure occurs at high well pressures and compressive failure at low pressures, these failure modes are not normally observed at the same depth.

may eventually reduce some of the uncertainty and allow for more reliable estimation of the in situ stresses.

8.3.2.1 Caliper logs

The caliper log (four-arm) has commonly been used to estimate horizontal stress directions from breakout orientations. This tool provides two diameters of the borehole cross-section. To be able to identify stress-induced borehole breakouts, a set of identification criteria has to be implemented. We quote here the criteria published by Plumb and Hickman (1985):

1. The tool is rotating above and below a borehole breakout.
2. The rotation stops over the breakout zone.
3. The borehole elongation is clearly seen in the log. One pair of pads must show a relatively sharp ascent and descent of the borehole diameter.
4. The smaller of the caliper readings is close to bit size, or if the smaller caliper reading is greater than bit size, then it should exhibit less variation than the larger caliper.
5. The direction of elongation should not consistently coincide with the high side of the borehole when the hole deviates from vertical.

The last item in this list reflects the problem of differentiating between breakouts and keyseats.[1] This is a major problem, since it is not possible to distinguish between breakouts and keyseats when using four-arm caliper data. Generally, elongations are regarded as keyseats if their direction is

[1] A keyseat is an enlargement of the borehole in one preferred direction, normally created by the mechanical action of the drillstring on the borehole wall.

within ± 10°–15° of the hole azimuth. However, this can categorise a large portion of the elongations as keyseats, even at deviations less than 6° (Fejerskov and Bratli, 1998). This implies that either a large portion of the data points are wrongly rejected or much of the previous work on breakout analysis has to be rejected.

Six-arm caliper tools are also available, offering a potentially more accurate interpretation of borehole breakouts. However, analysis of six-arm caliper data is more complicated when the tool is off centre (Yang et al., 2018).

8.3.2.2 Image logs

Image logs include both electrical (resistivity) and acoustical borehole imaging logs. The electrical image log operates with a large number of electrodes in contact with the formation, usually distributed over several pads on independent arms (four or six). This shallow electrical investigation is well suited for investigation of fine structures like bedding planes, natural fractures and also drilling-induced fractures.

The acoustical imaging tool (often referred to as borehole televiewer, BHTV) is based on reflection of acoustic waves from the borehole wall, recording the travel time and amplitude of the reflected pulses. The pulses are generated by a rapidly rotating piezo-electric crystal, thus creating a helix-shape logging path with a short distance between each revolution. This tool is best suited for detection of borehole breakouts, as drilling-induced fractures do not create significant changes in borehole radius or reflectivity.

Examples of image logs are shown in Fig. 8.3. To the far left is an example of drilling-induced fractures in a vertical well, and in the middle an example of drilling-induced fractures in an inclined well. Note the en-echelon shape of these fractures. To the right is an example of breakouts from an acoustical imaging tool taken in a vertical well.

Since these tools provide a full image of the borehole wall, it is possible to distinguish between stress-induced breakouts and keyseats. Examples of the use of image logs for determination of horizontal stress directions are given by Brudy (1998).

If drilling-induced fractures are found at an inclination with respect to the borehole axis, then this implies that none of the principal stresses are parallel with the borehole axis. This can be the situation in an inclined borehole or in a vertical borehole, where the vertical stress is not a principal stress, as, for instance, close to faults (Brudy et al., 1997).

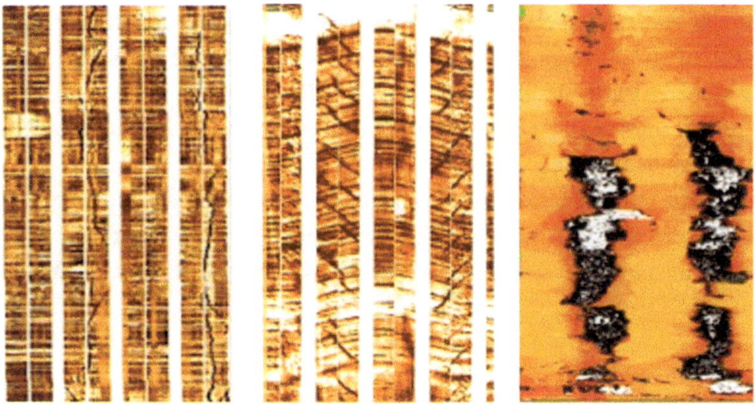

Figure 8.3 Examples of resistivity images (left and middle) showing drilling-induced fractures and acoustical image showing breakouts (right). The images to the left and right are from vertical wells. The image in the middle is from an inclined well, resulting in en-echelon type fractures. Dark regions represent low resistivity, for example, fluid filled fractures. Notice that these drilling-induced fractures are 180° apart at the borehole wall and are not interconnected around the borehole like traces of natural fractures or bedding planes. Courtesy of Equinor.

It has also been observed from image logs that in some cases, drilling-induced fractures and breakouts occur in the same interval. This phenomenon can in most cases be not explained by normal wellbore pressure variations in a circular borehole. AlBahrani and Morita (2020) showed that irregularities on the borehole wall can change the wellbore stability limits (shear failure and tensile failure) enough to make this happen. Any deviation from a circular hole can cause this to happen (e.g. breakouts, washouts, deformation due to non-symmetric stress). A discussion of elliptical holes is presented in Section 4.6.

In strike-slip stress regimes, drilling-induced fractures can occur at a well pressure lower than the minimum stress in the formation; see Eq. (4.128). In this case the fracture will be short, and loss of drilling fluid will not occur.

8.3.3 Horizontal stress magnitude (acoustic logs)

The simplest approach to estimating the minimum horizontal stress is the assumption of uniaxial compaction as given by Eq. (3.4) in combination with Poisson's ratio determined from shear and compressional sonic velocities as given by Eq. (8.2). In addition to sonic log data, estimates for the pore pressure and the vertical stress are then required.

This method involves numerous assumptions and simplifications. It is assumed that for millions of years, during sedimentation, burial, diagenesis, tectonic activity and so on, the rock has maintained elastic behaviour under uniaxial strain conditions. The dynamic Poisson's ratio measured today is assumed to contain all this information. Further uncertainty is introduced by the use of dynamic elastic data and the estimated values of the pore pressure and vertical stress. It is also evident that the small-scale depth variations seen in the sonic log are not realistically reflecting the variations of the minimum horizontal stress.

Despite all this, it seems that this approach can to some extent give realistic indications of overall trends and stress contrasts (Andrews and de Lesquen, 2019). The reasons are not obvious and the results maybe not universal, so direct measurements from fracture tests are strongly recommended.

Using the crossed-dipole option in acoustic logging tools (see Section 8.1.1), we may distinguish between lithological and stress-induced anisotropy (Plona et al., 2002; Sinha et al., 2006a, 2008). The flexural borehole mode is frequency dependent (dispersive), even in a homogeneous formation (see Fig. 5.22). The altered zone near the borehole alters the dispersive characteristics: The lower frequencies sense the formation 2–3 borehole diameters from the well, whereas higher frequencies detect the altered near-well zone (within 1 borehole radius), which in general will have lower velocities. Thus the deviation of observed dispersion from that of a homogeneous material is indicative of formation damage.

If the formation is transversely isotropic (TI) and the borehole is normal to the symmetry axis, then there will be two flexural modes associated with polarisation in the plane and out of the plane. Recorded waveforms can be processed to obtain these fast and slow flexural waves, which can be translated in fast and slow shear wave velocities. In case of intrinsic anisotropy the two dispersion curves will be parallel, and the low-frequency velocities will be most representative of the virgin formation. For stress-induced anisotropy, the two modes would however behave differently: Near a vertical well the tangential compressive stress peaks in the direction of the minimum horizontal stress, with possible associated damage and eventually breakouts originating in this direction. The associated flexural mode will hence be slowed down in the high-frequency range, whereas it will be the faster wave in the unaltered formation scanned by the low frequencies. Hence the dispersion curves of the two flexural modes will cross each other at some frequency, typically between 2 and 4 kHz. This cross-over is an attribute of stress-induced anisotropy. The data can hence be used to determine the in situ stress directions.

Estimation of stress magnitudes requires a nonlinear model that characterises the stress sensitivity of the rock stiffnesses (and thus velocities). An example of application is given by Sinha et al. (2006b). They used two flexural modes in a vertical well to derive C_{44} (slow) and C_{55} (fast), in addition to estimating C_{66} from the Stoneley mode in the monopole full waveform. The analysis was restricted to sand zones, where velocity anisotropy is assumed to be fully induced by stress anisotropy. The stress sensitivity of all stiffnesses was assumed to be given by one third-order elastic modulus, which was computed from observed velocity dispersions and the pre-knowledge of vertical stress, minimum horizontal stress and pore pressure. This was used to estimate the maximum horizontal stress versus depth in a reasonably homogeneous sand interval below the Norwegian Sea.

8.3.4 Horizontal stress magnitude (fracture tests)

The only fully reliable method for determination of the smallest horizontal stress σ_{h} is to fracture the formation and record the pressure at which the fracture closes. This requires that the fracture has penetrated far enough into the formation to feel only the resistance of the in situ horizontal stress. In a vertical well, this is achieved 2–3 wellbore diameters away from the borehole. In a deviated wellbore the fracture may have to travel farther away from the borehole due to the twisting of the fracture and the principal stresses close to a deviated borehole. Basic principles of the different phases of fracture initiation, growth and closure will be discussed in Chapter 11.

Determination of the largest horizontal stress σ_{H} is not trivial, and there is no straightforward method available for this. In an idealised linear elastic situation, the largest horizontal stress could be determined from a repeated fracture test. However, in practice, the fracture initiation pressure can vary considerably, rendering such an approach highly uncertain (see also Section 11.2). Normally, the fracture initiation pressure is lower than predicted from linear elastic theory. Investigation of Eq. (11.9) shows that if this is the case, then the largest horizontal stress is overestimated. This is probably the reason why field studies utilising such an approach consistently predict a strike-slip stress regime ($\sigma_{\mathrm{H}} > \sigma_{\mathrm{v}} > \sigma_{\mathrm{h}}$).

In this section, we focus on fracturing methods used for determination of the smallest horizontal stress.

8.3.4.1 Leak-off tests and extended leak-off tests

Leak-off tests (LOT) are performed during the drilling phase of a well in the formation immediately below each casing shoe. The purpose of this test is to verify the integrity at the casing shoe and to determine the maximum well pressure the new borehole section can sustain without fracturing and loss of drilling fluid. A traditional leak-off test is therefore not designed to determine the smallest horizontal stress, but to obtain a design value for the mud density in the next borehole section that is going to be drilled.

After a casing string has been cemented, the casing shoe is drilled out, and a few metres of new formation are penetrated. A leak-off test is then performed by pressurising this open-hole section. The principle of a leak-off test is shown in the first cycle in Fig. 8.4. The pressure in the hole is increased by pumping at a constant volume rate, typically 50–250 l/min. This yields a straight line of pressure versus volume (time) with the slope of the line given by the compressibility of the system (primarily the drilling fluid). The point where the pressure response starts to deviate from this straight line is defined as the leak-off point. This is in fact the point where a fracture is starting to initiate. Normally, a leak-off test is stopped shortly after this, even if the pressure continues to increase above the leak-off pressure.

The slope and shape of the pressure versus volume line can be affected by a number of operational aspects (equipment performance, cement channels, pump rate etc.). For a discussion of how these factors may affect LOT-interpretation, see, for example, Postler (1997).

As discussed in Chapter 11, this leak-off point or fracture initiation point is, however, not necessarily directly related to the smallest horizontal stress, and therefore there is no estimate of σ_h. It is also important to note that if the test is stopped shortly after the leak-off point, then the generated fracture is very short, and even if the shut-in phase is recorded (see Fig. 8.4), this shut-in pressure may significantly overestimate the smallest horizontal stress. Examples given by Raaen and Brudy (2001) and Raaen et al. (2006) demonstrate that using the leak-off pressure as an estimate of the smallest horizontal stress can cause significant errors. At the very best, the smallest horizontal stress will be a lower bound to a large population of leak-off pressures (Addis et al., 1998).

To make the leak-off test applicable also to stress determination, modifications have to be made. This has led to the so-called *extended leak-off* test (XLOT, ELOT) (see e.g. Kunze and Steiger, 1992). The main difference from the standard leak-off test is that pumping continues well beyond the

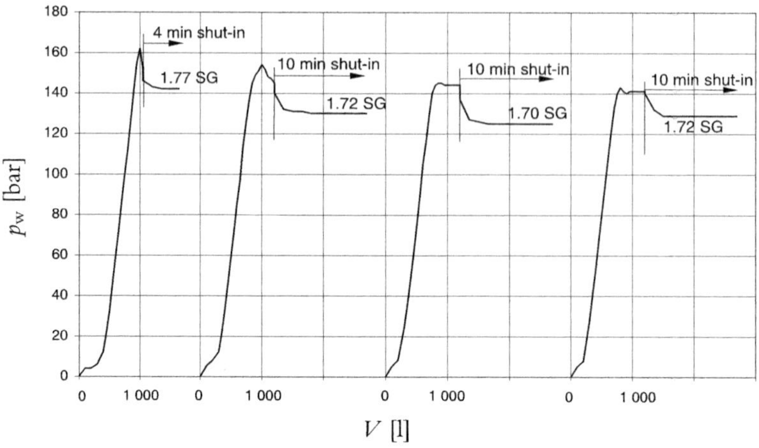

Figure 8.4 An example of a conventional leak-off test (first cycle) followed by three extended leak-off test cycles at 9 5/8″ casing shoe at 2550 m depth. The first cycle is shut in after 14 min of pumping at 75 l/min (1050 l pumped). The following cycles are shut in after 4.8 min of pumping at 250 l/min (1200 l pumped). The vertical lines indicate the shut-in point, after which the curves are functions of time, as indicated by the total shut-in period of each curve. Courtesy of Equinor.

leak-off point and also beyond the breakdown pressure. To get a reliable stress estimate, pumping should continue until stable fracture growth is obtained (see the third and fourth pump cycles of Fig. 8.4). After shut-in, the shut-in/decline phase should be recorded. As illustrated by Fig. 8.4, it is recommended to pump several cycles to obtain repeatable test results.

Since these casing shoe tests are normally performed in low-permeability formations like shale with drilling mud in the hole, very little leak-off from the fracture is expected. This implies that fracture closure may be slow in a shut-in/decline test, resulting in an almost flat pressure versus time response. The pressure recorded as the closure pressure is thus still an upper bound estimate of the smallest horizontal stress. Better estimates can then be obtained by including a flowback phase in the test.

A schematic illustration of an extended leak-off test with flowback is given in Fig. 8.5. This figure illustrates how we can maintain a record of the volumes during the different phases of the test. This can give information about when and whether the fracture has actually closed during the test.

The bold solid line represents the pressure-volume response during pump-in. V_{cin} represents the volume change due to compression of the fluid volume in the well, and V_{frac} represents the volume pumped into the fracture. This volume is spent creating fracture volume and also includes

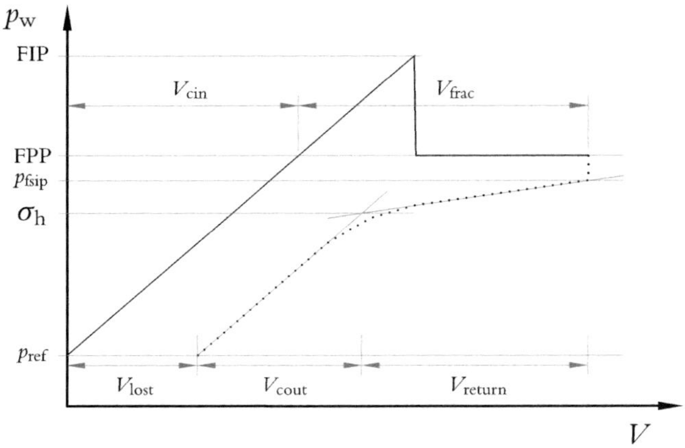

Figure 8.5 A schematic illustration of an extended leak-off test with a pump-in phase (bold solid line) and shut-in/decline and flowback phase (dotted line). Volumes during the different phases are indicated, and an estimation of the smallest horizontal stress from the flowback phase is included.

any volume lost to the formation. This volume may be used to check the length of the fracture and whether this is sufficient to penetrate the near-well region and into the virgin stress rock. The fracture initiation pressure (FIP) and fracture propagation pressure (FPP) are noted in the figure. When pumping stops and the well is shut-in, the pressure drops to the final shut-in pressure p_{fsip}. This drop in pressure results from removal of frictional losses, fluid loss to the formation and/or further growth of fracture volume.

The dotted line from p_{fsip} to p_{ref} represents the flowback phase. If the fracture has not closed during the shut-in phase, then monitoring the flowback phase can provide an estimate of the smallest horizontal stress as indicated in the figure. This occurs when the slope of the flowback curve changes, reflecting the large difference in compliance between the system with an open fracture and with a closed fracture. The last linear slope of the flowback curve should thus be close to the initial slope during pump-in, reflecting the compressibility of the fluid in the borehole. This volume is denoted as V_{cout}. If the entire flowback curve has the same slope as the initial pump-in phase, then the fracture has closed during the shut-in phase. V_{return} is thus the actual volume returned from the fracture, whilst V_{lost} is the volume lost to the formation during the test.

Fig. 8.5 is also idealised in the sense that σ_h is assumed to be equal to the fracture closure pressure (FCP). In practice, this is not always the case. Rather than a smooth closure of the fracture, the fracture can close near the wellbore. One consequence of this is that the closure pressure is higher than the minimum stress. Another consequence is that a significant volume of fluid will be left in the fracture, preventing the fracture from closing completely. This can result in a pressure rebound effect after flowback; see Section 11.5.2. A total return volume that is not much greater than the wellbore storage volume V_{cout} is another indication of this.

The extended leak-off test also provides information about fracture initiation and fracture propagation (see Fig. 8.5). This information is useful when assessing well integrity in an intact formation or a fractured formation. With access to a larger number of tests, it is possible to compare with theoretical predictions, to compare relations between the different parameters and how these are affected by test conditions, for example, the effect of water-based mud versus oil-based mud (Andrews et al., 2016).

A field example illustrating the principles outlined in Fig. 8.5 is shown in Fig. 8.6 (Raaen, 2006). Note that in the first cycle of this test (dotted line), no change in slope is seen before formation breakdown. In the second cycle (full line) the slope change takes place at a significantly lower pressure. Thus the fracture reopening pressure is considerably lower than the initial formation breakdown pressure.

The dashed lines show that the same closed-fracture slope is observed during pumping and flowback in both cycles. In cycle 2 the initial open fracture period during flowback is also well approximated by a straight line (dash-dot line).

In practice, it is convenient to perform the flowback with a fixed choke, which leads to a time-varying flowback rate. Since the flow rate through a choke is approximately proportional to the square root of the pressure drop, it is readily shown (Raaen et al., 2001) that a plot of the square root of pressure versus time should give a straight line if the system stiffness is constant. This is confirmed by Fig. 8.7, which shows a plot of data from cycle 2 of the test in Fig. 8.6. Both the initial open-fracture period and the closed fracture period are approximated well by the dashed straight lines, and the intercept is in accordance with that inferred from the pressure versus volume data.

For another field example, see Section 11.5, where also other alternatives for improving the interpretation of fracture tests are included.

To improve the quality of an extended leak-off test, there are thus several recommendations:

- Circulate mud to make sure that the mud column has uniform density and that there is no air in the system.

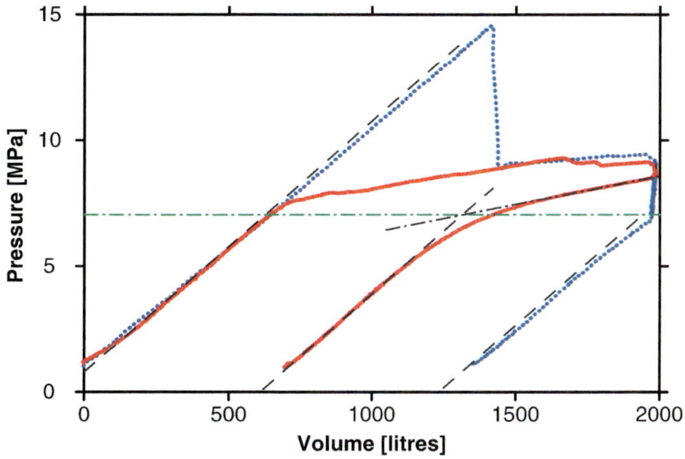

Figure 8.6 Pressure versus volume for a two-cycle flowback test with mud. The first cycle (dotted blue line) included a shut-in phase, whereas in the second cycle (full red line), flowback was started immediately after pumping. The horizontal dash-dot line shows the stress level inferred. Courtesy of Equinor.

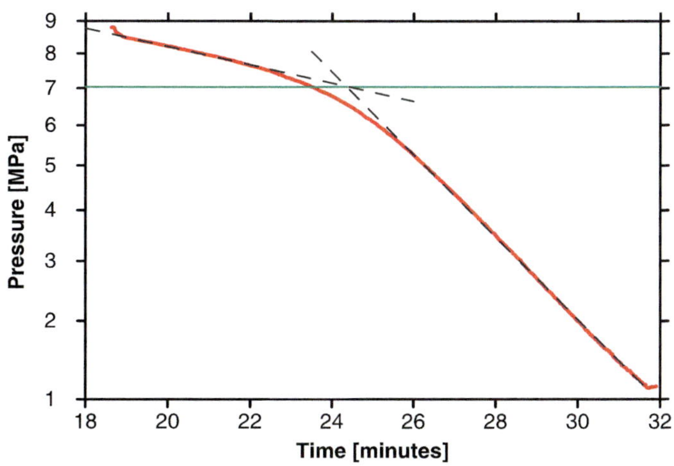

Figure 8.7 Square root of pressure versus time for the second cycle of the test in Fig. 8.6. Courtesy of Equinor.

- Calibrate the surface reading of the mud pressure or preferably use a downhole pressure gauge. Significant difference between surface pressure and downhole pressure can sometimes be seen at the start of flowback in deep water wells.

- Check that the slope prior to leak-off is in reasonable agreement with the expected fluid compliance. A significantly lower slope can indicate that there is a leakage in the system or poor cement integrity. A direct measurement of the compressibility is obtained by performing a casing test (i.e. pressurising the closed casing before drilling out the casing shoe).
- Make sure that the volume pumped into the fracture is large enough to create a fracture of sufficient length (typically 0.5–1 m^3). An estimate of the required volume can be obtained with information about the fracture width, the fracture length and the fracture height. The fracture width can be estimated from Eq. (11.30). The length of each fracture wing should be long enough to penetrate into virgin formation, for example, 10 wellbore diameters. Note that a longer fracture may be required in a deviated well. The total volume required for a test is thus given by the fracture volume in addition to the volume required to compress the fluid in the borehole prior to fracture initiation.
- Maintain a record of volumes pumped and bled back during flowback and record the volumes as functions of time during flowback.

The quality of the test should also be checked against possible operational problems such as cement channels. If the fluid can leak to shallower formations, then the test will not be relevant for the formation below the casing shoe.

The reluctance to perform extended leak-off tests is partly based on the extra time required to run this test compared to a conventional leak-off test. However, this may be the only possibility to obtain reliable stress data in the development of a field. Such stress data can be crucial in field development; see the introduction to Section 8.3.

In the drilling phase, an FIT or LOT is always required at the shoe. Then the extra time required to run a two-cycle extended leak-off test is typically only around 1.5 hours. In addition to stress data, the extended leak-off test provides information about fracture initiation pressure and fracture propagation pressure.

When using water-based mud, the fractures tend to heal and eventually restore the original leak-off pressure. When using oil-based mud, it is often claimed that healing does not occur. However, no published work on this is known to the authors. Such a difference can at least partly be attributed to differences in filtrate properties of the two mud systems. Water-based mud normally has higher spurt loss and filtrate loss. This creates a filter cake within the fracture, which seals the fracture tip, thus requiring a higher pressure to propagate the fracture (Onyia, 1991). This difference can be reduced by altering the filtrate properties of the oil-based mud (Morita et al., 1996). Note that this argumentation rests on the ability of the rock to accept filtrate, and hence this can not explain similar differences between the two mud systems in low-permeability rocks like shale.

In practice, it is often recommended not to perform extended leak-off tests if the margin between the leak-off pressure and the smallest horizontal stress is required for a successful operation. This may be critical in wells with a narrow margin between the lower and upper mud weight limits, such as exploration wells with large uncertainty in the pressure prognosis. An alternative then is—in case of a discovery—to run the test during P&A (plug and abandonment) of the well. The principle is to set cement plugs such that some metres of open hole is left between the casing shoe and top of a cement plug. Then the final cement plug is set after the XLOT has been run. It is important to look for such opportunities for stress measurements outside the drilling phase.

When several wells have been drilled in a field, it is not uncommon to run formation integrity tests (FIT) instead of leak-off tests. This test simply consists of increasing the well pressure up to a pre-defined level, which is considered sufficient for drilling of the next section. Pumping is then stopped, and a constant pressure in the shut-in phase is observed. Hence this test is normally stopped in the linear part before leak-off is reached. An FIT can therefore never be used for stress determination.

8.3.4.2 Mini-frac tests

As the name suggests, a mini-frac test is a fracture test where a relatively small volume is injected (typically around 10 m^3). Small in this sense relates to conventional fracture stimulation jobs, which commonly involve hundreds or even thousands of cubic metres. A mini-frac test is normally run prior to a fracture stimulation job, to obtain values for fracturing pressure, closure pressure, fluid loss parameters and so on, which are then used in the design of the fracturing treatment (see e.g. Tan et al., 1990). This implies that mini-frac tests are normally run in reservoir sections that require stimulation. Since the reservoir section may already be completed, it is not uncommon to run mini-frac tests in cased and perforated wells.

Again, it is recommended to use a downhole pressure gauge to improve the accuracy. A typical fluid used in a mini-frac test is 2% KCl brine. Gel may be added to reduce fluid loss and to make the test more similar to the main treatment.

8.3.4.3 Wireline tools

Wireline tools are also available for performing fracture tests in open holes (Kuhlman et al., 1993; Thiercelin et al., 1996; Desroches and Kurkjian, 1999; Franquet et al., 2018). The basic principle here is to isolate a small section of the hole, typically 1 m, by inflating two rubber packers against the formation. A relatively small volume (typically 5–30 l) of fluid is used during pumping at rates in the range from 0.1 l/min to 2 l/min. Due to the small volume, these tests are often referred to as *micro-frac* tests. The pressure is measured with a downhole gauge. As with other fracture tests, several cycles are usually performed.

When run in an open hole, good hole quality is essential, that is, a smooth hole surface and close to circular shape. Otherwise, it will be problematic to obtain good sealing by the packers. To provide sealing, the packer pressure should exceed the pressure in the isolated interval. This implies that under certain conditions, a fracture will be generated first behind the packer. Referring to Eq. (11.4), this depends on the response of the pore pressure around the borehole. In a permeable formation, it is possible to maintain a sealing pressure differential. However, in low-permeability formations, a fracture may be generated first behind the packer (*sleeve fracturing*). An alternative procedure in this situation is to do this deliberately and subsequently move the packers so that the generated fracture is within the isolated zone and then proceed as in a standard test (Desroches and Kurkjian, 1999).

The low volume implies that the fracture will be short. The fracture should therefore align with the stress field immediately upon initiation. It is therefore not recommended to run these tests in other than vertical or near-vertical wells.

Due to the low fracture volumes, these tests were initially run in formations with relatively low permeability to allow fracture closure detection during shut-in. Testing in sandstone with higher permeability (500–1000 mD), higher flow rate (7 l/min) and up to 60 litres per cycle was reported by Bautmans et al. (2013).

As discussed in Section 8.3.4, fracture closure will not occur during a reasonable shut-in period in rocks with very low permeability, such as shale. Controlled flowback of the fluid in the fracture is then required. An example of such an approach in micro-fracture testing is given by Prasad et al. (2020).

Small volumes, short fractures and effect of packers on the near-wellbore stresses are potential sources of uncertainty. It is not known how results from micro-frac tests compare with tests using larger volumes.

The azimuthal direction of the fracture and hence the direction of the horizontal stresses can also be estimated, provided that the test is performed in a vertical well. The method used to obtain the fracture direction varies somewhat with the type of tool, either using an imaging tool after fracturing or back calculating from measurements of borehole deformation during fracturing.

Such tools can of course also be used in cased and perforated holes. Problems with hole quality, packer sealing and sleeve fracturing are then no longer of any concern. The use is however then limited to producing zones in the reservoir.

8.3.4.4 Empirical relations

Breckels and van Eekelen (1982) developed empirical correlations for estimation of the smallest horizontal stress as functions of depth. These relationships were based on hydraulic fracture data from different regions around the world. The relationships for the US Gulf Coast are

$$\sigma_h = 0.0053D^{1.145} + 0.46\left(p_f - p_{fn}\right)\ (D < 3500\,\mathrm{m}) \tag{8.4}$$

$$\sigma_h = 0.0264D - 31.7 + 0.46\left(p_f - p_{fn}\right)\ (D > 3500\,\mathrm{m}) \tag{8.5}$$

where D is the depth in metres, p_f is the pore pressure in MPa, p_{fn} is the *normal* pore pressure, and σ_h is the total smallest horizontal stress in MPa.

The last term in these relations reflects abnormal pore pressures. The predicted horizontal stress will hence reflect changes in the pore pressure gradient.

Breckels and van Eekelen argue that these relationships can be used with a fair degree of confidence also in other tectonically relaxed areas of the world such as the North Sea. Note that these relations were developed at zero or shallow water depths. These authors' experience is that the relation for depths down to 3500 m gives fairly good estimates in most parts of the North Sea (down to about 2500–3000 m), even at water depths up to approximately 300 m. As the water depth increases, predictions at shallow formation depth should be avoided. Another experience is that the term reflecting abnormal pore pressure does not give reliable predictions in high-pressure environments. Thus these relations may provide reasonable estimates but should only be considered as a first estimate and should always be checked and/or calibrated against proper test data from each field. Having several stress measurements available, it is recommended to build local relations for each field.

8.3.5 Other methods

We will finish this section by briefly mentioning some other methods, which have not been discussed previously (in this chapter or in Chapter 3).

Inversion of leak-off test data has been suggested for back calculating the horizontal stresses (Aadnøy, 1990). This method uses data from a number of leak-off tests performed in boreholes of different inclination and azimuth. Assuming that the vertical stress and the pore pressure are known, the magnitude and direction of the horizontal stresses are back calculated from linear elastic theory.

Since the method is based on leak-off pressures, it suffers from the same basic weaknesses as the standard leak-off test, especially the large uncertainty of the leak-off pressure in relation to the fracture breakdown pressure as defined by elastic theory.

Another source of uncertainty is the potential spatial variation in the horizontal stresses in a field. When having to use test data from several wells, which can also cover a wide depth range, the tests may include compartments of significantly different stress regimes. Further discussion of these uncertainties can be found in Gjønnes et al. (1998).

Rather than trying to give precise numbers for the stresses, we can try to *constrain the stresses* (see e.g. Wiprut et al., 1997). This method uses observations of borehole failures (borehole breakouts, drilling-induced tensile fractures), primarily attempting to constrain the magnitude and direction of the largest horizontal stress. Especially when constraining the stress magnitude, it is essential to have information on both the smallest horizontal stress and rock strength properties.

This approach suffers from large uncertainty, not only due to uncertainty in the input parameters, but also due to the models used for back-calculation. The models will suffer from the same uncertainty as when used for forward prediction of shear failure and/or tensile failure. It has been shown that the result can in some cases be fundamentally wrong, as when predicting a strike-slip stress regime in what is actually a normal stress regime; see de Lesquen et al. (2020). Such errors can have serious consequences to further planning and execution. AlBahrani and Morita (2020) showed that irregularities on the borehole wall and deviation from a circular hole can change the wellbore stability limits (shear failure and tensile failure). This adds uncertainty to this approach.

Drilling data have also been proposed as a potential source for constraining stresses, especially the smallest horizontal stress (Hareland and Hoberock, 1993). The basis for this method is drill bit penetration models of the kind discussed in Section 8.2.4. This is used to predict the rock strength, which is further related to the ratio of vertical to horizontal stress in an iterative process.

The stress data predicted from borehole tests and borehole information are point data in a large rock mass. It is therefore often useful to put this into a large-scale framework of in situ stresses (see also Chapter 3).

8.3.6 Transferring stress data from offset wells

Data points from stress tests are often scarce and scattered both vertically and laterally. When making a prognosis for a new well in a field/area, it is then necessary to transfer available data points to the new location. There is no universal method available for this process, but it is necessary to take several factors into consideration, such as differences in total depth, water depth and air-gap.

The method presented here (Andrews and de Lesquen, 2019) is based on using the effective stress ratio K'. This method can in principle be used for several data types from stress testing, such as breakdown pressure, closure pressure, propagation pressure and leak-off pressure. The basic assumption is that the K'-value for a stress parameter is the same in both offset and target wells. Using the fracture closure pressure (FCP) as an example, the steps of the method are as follows:

1. Calculate the K'-value in the offset well at the test depth, $K' = (\text{FCP} - p_{\text{fo}})/(\sigma_{\text{v}} - p_{\text{fo}})$.
2. Determine the corresponding depth in the target well. This depth should be at the same relative position in the formation where the test was done.
3. Calculate the FCP-value in the target well at this depth using the K' from the offset well and the pore pressure and overburden gradient at the corresponding depth in the target well.

This approach accounts for the effects of depth and pore pressure. Another advantage is that the result is not dependent on the choice of reference depth (TVD RT or TVD MSL) as long as the choice is consistent. This is also the case for differences in water depth and air gap between the offset and target wells. Different units can also be used (absolute pressure or equivalent density), again as long as the choice is consistent.

The depth in the target well should be at the same relative position as in the offset well. The description above is based on using formation as the interval to consider. Other unit types may be used, but the depth in the target well should be in the same lithology as the offset well. The reason is that there may be stress contrasts between different lithologies (shale, sandstone, limestone). The pore pressure regime should also be similar at the two depths, so that a normally pressured interval does not end up in a high-pressure interval, and vice versa.

References

Aadnøy, B.S., 1990. Inversion technique to determine the in-situ stress field from fracturing data. J. Pet. Sci. Eng. 4, 127–141. https://doi.org/10.1016/0920-4105(90)90021-T.

Addis, M.A., Hanssen, T.H., Yassir, N., Willoughby, D.R., Enever, J., 1998. A comparison of leak-off and extended leak-off test data for stress estimation. SPE/ISRM47235. In: Proc. Eurock '98, vol. 1. Trondheim, Norway, pp. 131–140.

AlBahrani, H., Morita, N., 2020. Understanding the concurrence of drilling induced fractures and breakouts and its implications on drilling performance using 3D FEM and field imaging data. IPTC20171. In: International Petroleum Technology Conference. Dharhan, Saudi Arabia.

Amadei, B., Stephansson, O., 1997. Rock Stress and Its Measurement. Chapman & Hall, London.

Andrews, J.S., de Lesquen, C., 2019. Stress determination from logs. Why the simple uniaxial strain model is physically flawed but still gives relatively good matches to high quality stress measurements performed on several fields offshore Norway. ARMA-2019-0078. In: 53rd US Rock Mechanics/Geomechanics Symposium. New York, NY, USA.

Andrews, J.S., Fintland, T.G., Helstrup, O.A., Horsrud, P., Raaen, A.M., 2016. Use of unique database of good quality stress data to investigate theories of fracture initiation, fracture propagation and the stress state in the subsurface. ARMA16-887. In: 50th US Rock Mechanics/Geomechanics Symposium. Houston, TX, USA.

Bautmans, P., Helstrup, O.A., Constable, K.C., 2013. Wireline stress testing in reservoir sands. ARMA13-545. In: 47th US Rock Mechanics/Geomechanics Symposium. San Francisco, CA, USA.

Boonen, P., Bean, C., Tepper, R., Deady, R., 1998. Important implications from a comparison of LWD and wireline acoustic data from a Gulf of Mexico well. In: Trans. 39th SPWLA Annual Logging Symp. Keystone, CO, USA.

Breckels, I.M., van Eekelen, H.A.M., 1982. Relationship between horizontal stress and depth in sedimentary basins. J. Pet. Technol. 34 (9), 2191–2199. https://doi.org/10.2118/10336-PA.

Brudy, M., 1998. Determination of the state of stress by analysis of drilling-induced fractures – results from the northern North Sea. SPE/ISRM47236. In: Proc. Eurock '98, vol. 1. Trondheim, Norway, pp. 141–149.

Brudy, M., Zoback, M.D., Fuchs, K., Rummel, F., Baumgärtner, J., 1997. Estimation of the complete stress tensor to 8 km depth in the KTB scientific drill holes: implications for crustal strength. J. Geophys. Res. 102 (B8), 18453–18475. https://doi.org/10.1029/96JB02942.

de Lesquen, C., Andrews, J.S., Rasheva, S., 2020. Good quality LWD images in horizontal wells invalidate the strike-slip theory in the Northern North Sea. In: ISRM International Symposium. Eurock 2020 – Hard Rock Engineering. Trondheim, Norway.

Desroches, J., Kurkjian, A.L., 1999. Applications of wireline stress measurements. SPE Reserv. Eval. Eng. 2 (5), 451–461. https://doi.org/10.2118/58086-PA.

Edlman, K., Somerville, J.M., Smart, B.G.D., Hamilton, S.A., Crawford, B.R., 1998. Predicting rock mechanical properties from wireline porosities. SPE/ISRM 47344. In: Proc. Eurock '98, vol. 2. Trondheim, Norway, pp. 169–175.

Farquhar, R.A., Somerville, J.M., Smart, B.G.D., 1994. Porosity as a geomechanical indicator: an application of core and log data and rock mechanics. SPE28853. In: Proc. European Petr. Conf. London, pp. 481–489.

Fejerskov, M., Bratli, R.K., 1998. Can dipmeter logs be used to identify in-situ stress directions in the North Sea. SPE/ISRM47237. In: Proc. Eurock '98, vol. 1. Trondheim, Norway, pp. 151–160.

Fjær, E., 1999. Static and dynamic moduli of weak sandstones. ARMA-99-0675. In: Proc. 37th U.S. Rock Mech. Symp. Vail, CO, USA.

Fjær, E., 2019. Relations between static and dynamic moduli of sedimentary rocks. Geophys. Prospect. 67 (1), 128–139. https://doi.org/10.1111/1365-2478.12711.

Franquet, J., Shaver, M., Edwards, E., Al Neyadi, A., Noufal, A., Khairy, H., 2018. Microfrac in-situ stress measurements for drilling and completion design in Middle-Lower Cretaceous reservoirs in a giant field of Abu Dhabi, UAE offshore. SPE193004. In: Abu Dhabi International Petroleum Exhibition & Conference. Abu Dhabi, UAE.

Gjønnes, M., Cruz, A.M.G.L., Horsrud, P., Holt, R.M., 1998. Leak-off tests for horizontal stress determination? J. Pet. Sci. Eng. 20, 63–71. https://doi.org/10.1016/S0920-4105(97)00053-3.

Haimson, B.C., Cornet, F.H., 2003. ISRM suggested methods for rock stress estimation—Part 3: hydraulic fracturing (HF) and/or hydraulic testing of pre-existing fractures (HTPF). Int. J. Rock Mech. Min. Sci. 40, 1011–1020. https://doi.org/10.1016/j.ijrmms.2003.08.002.

Hareland, G., Hoberock, L.L., 1993. Use of drilling parameters to predict in-situ stress bounds. SPE/IADC25727. In: Proc. 1993 SPE/IADC Drilling Conf. Amsterdam, Netherlands, pp. 457–471.

Hareland, G., Bratli, R.K., Stene, F., Fagereng, S., Jorgensen, T., 1996. Safe mud weight window predictor – instantaneous, pre-planning and post analysis software. SPE36097. In: Proc. Fourth Latin American and Caribbean Petr. Eng. Conf. Port-of-Spain, Trinidad and Tobago, pp. 253–260.

Horsrud, P., 2001. Estimating mechanical properties of shale from empirical correlations. SPE Drill. Completion 16 (2), 68–73. https://doi.org/10.2118/56017-PA.

Kuhlman, R.D., Heemstra, T.R., Ray, T.G., Lin, P., Charlez, P.A., 1993. Field tests of downhole extensometer used to obtain formation in-situ stress data. SPE25905. In: Proc. SPE Rocky Mountain Regional/Low Permeability Res. Symp. Denver, CO, USA, pp. 625–634.

Kunze, K.R., Steiger, R.P., 1992. Accurate in-situ stress measurements during drilling operations. SPE24593. In: SPE 67th Ann. Techn. Conf. & Exh. Washington, DC, USA, pp. 491–499.

Lashkaripour, G.R., Dusseault, M.D., 1993. A statistical study of shale properties: relationships among principal shale properties. In: Li, K.S., Lo, S-C.R. (Eds.), Probabilistic Methods in Geotechnical Engineering. Balkema, pp. 195–200.

Mason, K.L., 1987. Three-cone bit selection with sonic logs. SPE Drill. Eng. 2, 135–142. https://doi.org/10.2118/13256-PA.

Morita, N., Black, A.D., Fuh, G.-F., 1996. Borehole breakdown pressure with drilling fluids – I. Empirical results. Int. J. Rock Mech. Min. Sci. Geomech. Abstr. 33, 39–51. https://doi.org/10.1016/0148-9062(95)00028-3.

Nes, O.-M., Horsrud, P., Sønstebø, E.F., Holt, R.M., Ese, A.M., Økland, D., Kjørholt, H., 1998. Rig site and laboratory use of CWT acoustic velocity measurements on cuttings. SPE Reserv. Eval. Eng. 1 (4), 282–287. https://doi.org/10.2118/50982-PA.

Onyia, E.C., 1988. Relationships between formation strength, drilling strength, and electric log properties. SPE18166. In: SPE 63rd Ann. Techn. Conf. & Exh. Houston, TX, USA, pp. 605–618.

Onyia, E.C., 1991. An analysis of experimental data on lost circulation problems while drilling with oil-base mud. SPE22581. In: SPE 66th Ann. Techn. Conf. & Exh. Dallas, TX, USA, pp. 425–436.

Plona, T.J., Kane, M.R., Sinha, B., Walsh, J., 2002. Evaluating stress-induced anisotropy and mechanical damage from cross-dipole sonic data using dispersion analysis. SPE/ISRM78233. In: SPE/ISRM Rock Mechanics Conference. 20–23 October, Irving, TX, USA.

Plumb, R.A., Hickman, S.H., 1985. Stress-induced borehole elongation: a comparison between the four-arm dipmeter and the borehole televiewer in the Auburn geothermal well. J. Geophys. Res. 90 (B7), 5513–5521. https://doi.org/10.1029/JB090iB07p05513.

Postler, D.P., 1997. Pressure integrity test interpretation. SPE/IADC37589. In: Proc. 1997 SPE/IADC Drilling Conf. Amsterdam, Netherlands, pp. 169–183.

Prasad, U., Franquet, J., Moronkeji, D., Gaz, I., Galiyeva, G., Bekkaliyev, Z., 2020. Controlled flowback technique during micro-fracturing testing for inducing fracture closure in non-permeable formation in the Caspian Sea. IPTC20077. In: International Petroleum Technology Conference. Dharhan, Saudi Arabia.

Raaen, A.M., 2006. The pump-in/flowback test improves routine minimum horizontal stress magnitude determination in deep wells. In: International Symposium on In-Situ Rock Stress. Trondheim, Norway.

Raaen, A.M., Brudy, M., 2001. Pump-in/flowback tests reduce the estimate of horizontal in-situ stress significantly. SPE71367. In: Proc. 2001 SPE Ann. Techn. Conf. & Exh. New Orleans, LA, USA.

Raaen, A.M., Hovem, K.A., Jøranson, H., Fjær, E., 1996. FORMEL: a step forward in strength logging. SPE36533. In: Proc. 1996 SPE Ann. Techn. Conf. & Exh. Denver, CO, USA, pp. 439–445.

Raaen, A.M., Skomedal, E., Kjørholt, H., Markestad, P., Økland, D., 2001. Stress determination from hydraulic fracturing tests: the system stiffness approach. Int. J. Rock Mech. Min. Sci. 38 (4), 529–541. https://doi.org/10.1016/S1365-1609(01)00020-X.

Raaen, A.M., Horsrud, P., Kjørholt, H., Økland, D., 2006. Improved routine estimation of the minimum horizontal stress component from extended leak-off tests. Int. J. Rock Mech. Min. Sci. 43, 37–48. https://doi.org/10.1016/j.ijrmms.2005.04.005.

Ringstad, C., Lofthus, E.B., Sønstebø, E.F., Fjær, E., 1998. Prediction of rock parameters from micro-indentation measurements: the effect of sample size. SPE/ISRM47313. In: Proc. Eurock '98, vol. 1. Trondheim, Norway, pp. 487–492.

Santarelli, F.J., Marsala, A.F., Brignoli, M., Rossi, E., Bona, N., 1998. Formation evaluation from logging on cuttings. SPE Reserv. Eval. Eng. 1 (3), 238–244. https://doi.org/10.2118/36851-PA.

Sinha, B.K., Vissapragada, B., Renlie, L., Tysse, S., 2006a. Radial profiling of the three formation shear moduli and its application to well completions. Geophysics 71 (6), E65–E77. https://doi.org/10.1190/1.2335879.

Sinha, B.K., Vissapragada, B., Renlie, L., Skomedal, E., 2006b. Horizontal stress magnitude estimation using three shear moduli – a Norwegian Sea case study. SPE103079. In: SPE Annual Technical Conference and Exhibition. San Antonio, TX, USA.

Sinha, B.K., Wang, J., Kisra, S., Li, J., Pistre, V., Bratton, T., Sanders, M., 2008. Estimation of formation stresses using borehole sonic data. In: SPWLA 49th Annual Logging Symposium. Edinburgh, Scotland.

Tan, H.C., McGowen, J.M., Soliman, M.Y., 1990. Field application of minifracture analysis to improve fracturing treatment design. SPE Prod. Eng. 5 (2), 125–132. https://doi.org/10.2118/17463-PA.

Thiercelin, M.J., Plumb, R.A., Desroches, J., Bixenman, P.W., Jonas, J.K., Davie, W.R., 1996. A new wireline tool for in-situ stress measurements. SPE Form. Eval. 11 (1), 19–25. https://doi.org/10.2118/25906-PA.

Warren, T.M., 1987. Penetration-rate performance of roller-cone bits. SPE Drill. Eng. 2 (1), 9–18. https://doi.org/10.2118/13259-PA.

Wiprut, D., Zoback, M., Hanssen, T.-H., Peška, P., 1997. Constraining the full stress tensor from observations of drilling-induced tensile fractures and leak-off tests: application to borehole stability and sand production on the Norwegian margin. Int. J. Rock Mech. Min. Sci. 34 (3–4). https://doi.org/10.1016/S1365-1609(97)00157-3.

Yang, Z., Fang, X., Huang, L., 2018. A robust approach for determining borehole shape from six-arm caliper logs. Geophysics 83, D203–D215. https://doi.org/10.1190/geo2017-0761.1.

CHAPTER 9

Stability during and after drilling

Borehole instabilities during drilling cause substantial problems in all areas of the world. A borehole stability problem is an example of what drillers refer to as a "tight hole" or "stuck pipe" incident. There are many possible reasons for being stuck, but in a majority of the field cases reported, the fundamental reason is mechanical collapse of the borehole (see e.g. Bol et al., 1994; Gazaniol et al., 1995). Borehole collapse is a term commonly used in the petroleum industry to represent an instability caused by insufficient mud weight, resulting in failure. Collapse can, however, also be used as a common term for all mechanisms causing rock fragments to fall off the wellbore wall and form an enlarged borehole. Shear failure due to insufficient mud weight is only one such mechanism. Other reasons for collapse of the wellbore wall are typically mechanical work from equipment in the hole (drill bit, BHA (bottom hole assembly), drill string, stabilisers etc.) and hydraulic work from circulation of the drilling fluid. The reason may also be combinations of these mechanisms. If a borehole is close to shear failure, then it is also more inclined to be influenced by mechanical and/or hydraulic forces. In this chapter, our focus is on collapse related to shear failure and tensile failure. It is, however, important to keep all these effects in mind when trying to diagnose wellbore instabilities.

Most instabilities of practical importance occur in shale or mudstone, predominantly in the overburden, but sometimes also within the reservoir. Often mechanical hole collapse is combined with a lack of hole cleaning ability. It is a common opinion in the industry that such stability problems amount to typically 5–10% of drilling costs in exploration and production, incorporating loss of time and sometimes also of equipment. These numbers imply a worldwide cost to the petroleum industry of hundreds of million dollars per year. Note that although this number appears to be relatively constant over time, both drilling areas and well trajectories have become more challenging and complex.

Borehole stability problems have been encountered for as long as wells have been drilled. Several new challenges have appeared over time, making the stability issue more difficult to handle, but also more important

Developments in Petroleum Science, Volume 72
ISSN 0376-7361, https://doi.org/10.1016/B978-0-12-822195-2.00018-8

to solve. For example, there has been an increasing demand by the industry for more sophisticated well trajectories. Highly deviated and horizontal wells are attractive, since a single production platform then can drain a larger area, which can reduce the number of platforms or subsea templates required to produce a given field. Stable drilling is however normally more difficult in deviated than in vertical boreholes. Other situations where borehole stability problems may be expected are during infill drilling in depleted reservoirs, when drilling in tectonically active areas and in deep and geologically complex surroundings. Deep water drilling is a special challenge. Due to high costs of drilling in such environments, additional time lost on borehole instability has an extra high price.

As mentioned above, borehole stability is an operational problem primarily in shale and claystone. Problems often occur in shales that are rich in swelling clay minerals and are often associated with high pore pressure. For a long period of time, the petroleum industry looked at borehole instability as being primarily caused by clay swelling, which can be treated by chemical additives (e.g. salt) to the drilling mud. The selection of mud weight has been governed by the pore pressure and the fracture gradient profiles: To prevent influx of fluids (in particular gas), it has been considered necessary to keep the mud weight above the pore pressure gradient. To prevent loss of mud into fractures ("lost circulation"), it has been found necessary to keep the mud weight below the fracture gradient. We will show in this chapter how the minimum required mud weight may be estimated from a rock mechanical perspective.

The fracture gradient is commonly referring to the pressure at which a fracture is induced in the formation. In the drilling community, it is implicitly understood that loss of drilling fluid can start to occur at this pressure. There are different methods available for determination of the fracture gradient, both practical (FIT, LOT, XLOT) and more theoretical. However, this pressure is not an inherent property of the formation and can vary significantly, even under seemingly comparable conditions (see Fig. 9.6). These issues will be discussed further in Section 9.2.3.

We will see how mud chemistry may affect the mechanical integrity of the formation near the well and hence the stability of the borehole. We will also consider the maximum allowed mud weight, associated with mud loss to new or existing fractures. This analysis leads to the "mud weight window", that is, the range of allowed mud weights associated with stable drilling. This is key input to well design.

Stability evaluation of a well represents a classical rock mechanical problem, the prediction of a rock response to mechanical loading. Some special circumstances make borehole stability evaluation particularly problematic:

- The drill bit may be thousands of metres away, and there are no methods available for visual observation of what is happening (contrary to e.g. tunnel drilling).
- There may be large variations in the formation stresses, for instance, when drilling through a depleted reservoir and non-depleted shale layers in the same well or when drilling through faults. Unfortunately, in situ stresses and in particular their variations are not measured systematically.
- There are large variations in formation properties. Coring costs are high, and only limited amounts of material are available for rock mechanics testing. Coring in layers above the reservoir is scarce.
- Many mechanisms contribute to the onset of borehole instability: mud chemistry, redistribution of stresses, temperature changes and so on.
- The operational condition of borehole instability does not necessarily coincide with the rock mechanical definition of borehole failure.

This illustrates that the reality cannot be exactly described by any model, no matter how complicated the model is. Based on some ideal assumptions, a stability analysis, however, can give some guiding limits, which then have to be related to practical conditions and experience.

All discussions and models used in this chapter are based on the assumption of a circular borehole. This assumption may be valid immediately after drilling, but caliper logs often show the appearance of an irregular borehole wall. AlBahrani and Morita (2020) showed that such changes in geometry affect the stability limits, both with respect to shear failure and tensile failure. In extreme cases, this can result in drilling induced fractures (DIF) and breakouts (BO) occurring in the same interval. This phenomenon in most cases cannot be explained by wellbore pressure variations alone, such as surge/swab and ECD (Equivalent Circulating Density) vs. ESD (Equivalent Static Density).

Effects on shear failure and tensile failure for the simplified case of an elliptical hole are discussed in Section 4.6.

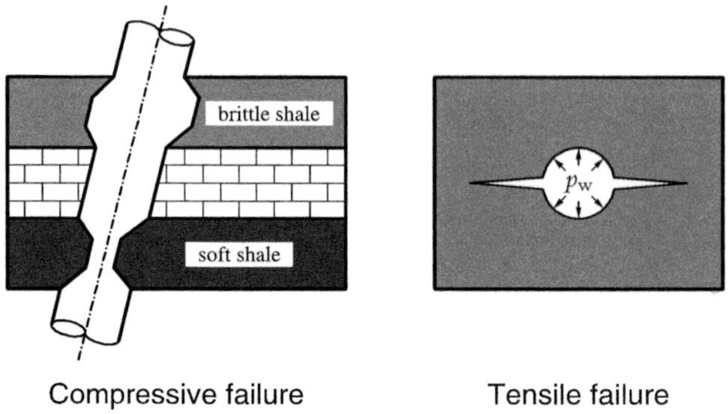

Figure 9.1 Stability problems during drilling (After Bradley, 1979).

9.1 Unstable boreholes: symptoms, reasons and consequences

In this section, we consider two main types of borehole instabilities, the so-called "tight hole" or "stuck pipe" incidents, which are time-consuming to solve and therefore expensive, and "lost circulation" or "mud loss" problems, which are potentially dangerous, thus representing a safety risk that has to be avoided. Fig. 9.1 illustrates in a schematic manner different instability problems that might occur.

9.1.1 Tight hole/stuck pipe

The main causes of tight hole/stuck pipe incidents are:

1. Hole collapse (rock mechanical failure)
2. Inappropriate hole cleaning
3. Differential sticking
4. Deviation from ideal borehole trajectory and shape

Cause 1 Hole collapse means that the formation near the borehole fails mechanically, most often by shear failure, but occasionally also by tensile failure. The results of such failures can be divided into two main types:

- Increased borehole size due to brittle failure and caving of the wellbore wall. If the cavings are not transported away, then this represents a potential source of a stuck pipe situation. This normally takes place in brittle rocks, but the borehole size may also increase by erosion (hydraulic or mechanical) in a weak rock. In driller's language, "sloughing shale" is often used (although not very well defined) to describe fragments or "spallings" generated from the borehole wall.

Although often thought to be of a chemical origin, this is first of all a mechanical problem, which to some extent may be influenced by shale-fluid interactions (see Section 9.4). The case of excessive hole enlargement is sometimes referred to as a "washout". A washout is primarily caused by erosion due to high mudflow intensity near the drill bit, or it may be related to softening of the formation by mechanical failure. Although a mechanical hole collapse leads to ovalisation of the hole (by formation of a breakout; see Chapter 4), this need not be the case with a washout.

- Reduced borehole diameter may occur in weak (plastic) shales, sandstones and salt. Some chalk formations can also show such behaviour. This phenomenon requires repeated reaming or may even result in a stuck drillpipe. The case of very soft (plastic) shale is sometimes referred to as "gumbo shale" (gumbo is a Creolian fish soup with rheology and appearance that may give associations with soft shale). Such shale is often sticky, contains considerable amounts of swelling minerals (smectite) and may cause problems like bit balling and solids accumulation.
- It has traditionally been thought that large hole diameter reductions might be caused by swelling clays. The potential chemical swelling of a shale in downhole stress conditions is however very limited, as was pointed out by Santarelli and Carminati (1995). Large hole deformation is thus a result of primarily plastic shale deformation (see also Section 9.10). The effect of drilling fluid chemistry on borehole stability is explained in more detail in Section 9.4.

Cause 2 Inappropriate hole cleaning means that drill cuttings or, in the case of a collapsing hole, cavings (rock fragments produced by formation failure) cannot be fully removed by the drilling fluid. Mechanisms 1 and 2 are therefore not independent, but often act together. Hole cleaning is less problematic in sand than in shale formations, since the drilling mud can more easily remove sand particles than shale cavings.

Cause 3 Differential sticking is the only likely reason for stuck pipe in a permeable reservoir rock. When differentially stuck, the drilling tool is clamped against the borehole wall as a result of a differential pressure between the well and the formation (Outmans, 1958). This differential pressure is caused by an overpressure in the borehole and maintained by a mud-cake. Since shales have extremely low permeabilities and mudcakes do not form on shales, this mechanism is not possible in shale zones.

Cause 4 Deviation from an ideal borehole trajectory and shape may result in a stuck pipe situation. Such deviations may be caused by a non-ideal hole shape. In deviated wells the lower side of the drilling tool may dig into the bottom of the hole and create what is known as a "keyseat". The tool may also be guided by washouts and breakouts. Furthermore, if the hole trajectory has a too sharp bend ("dog leg"), then the tool may be stuck. The problems described here are typical for deviated holes. Hole collapse in general occurs more easily in deviated holes as can be seen from the mechanical stability analysis in Chapter 4 and discussed further below. Hole cleaning is also more difficult in deviated holes, in particular, at intermediate angles in the range 40°–60°.

The main consequence of tight hole/stuck pipe situations is loss of time during drilling, since the remedial actions usually are reaming or sidetracking. Instabilities may also cause considerable problems to later operations in the borehole. It may become difficult to run wireline logs and, in particular, to interpret the logs, since log interpretation is usually based on the assumption of a gauge hole with known size. Irregular borehole shape also leads to large uncertainty in the required cement volume. Poor cementing of the casing can lead to problems for well integrity, perforation, sand control, production and stimulation. Finally, instabilities may trigger new instabilities.

Good well design is the key to stable drilling. When a well is designed, mud weight and composition, casing setting depths, and well trajectory (including deviation and azimuthal orientation) are selected. The well design serves many purposes: First of all, the target reservoir needs to be penetrated in a way that assures optimum drainage during production. Furthermore, the drilling speed must be satisfactory. A central part of well design is however to ensure safe and stable drilling. If an unexpected instability occurs during drilling, then the drilling fluid is more or less the only adjustable factor. It is important to underline that the solution to a borehole stability problem depends on the cause, which means there is a strong need for diagnostic tools. If, for example, the reason for the instability is shear failure, then the standard solution is to increase the mud weight (as we will see below). If, however, differential sticking is the cause of the problem, then the solution is usually to reduce the mud weight. Thus, if the diagnosis is wrong, then the selected solution may have a destabilising effect on the borehole.

Some elements of a diagnostic analysis are shown in Table 9.1. The table shows how field observations can be linked to the three major causes for

Table 9.1 Example of stuck pipe diagnostics inspired by Kenter (1995); Charlez and Onaisi (1998). Symbols: ∗ likely cause of stuck pipe; 0 indifferent; ÷ unlikely; ⊘ can not be cause of stuck pipe.

	Hole collapse	Inappropriate hole cleaning	Differential sticking
Drilling environment			
Shale	∗	∗	⊘
(Permeable) reservoir rock	÷	∗	∗
Observations during drilling			
Rotating before stuck	∗	0	÷
Moving up/down before stuck	∗	0	÷
Rotating after stuck	÷	÷	⊘
Circulating after stuck	÷	÷	∗
Excessive cuttings and cavings	∗	÷	÷
Observations after drilling			
Non-gauge hole diameter from caliper	∗	÷	⊘
Low-density/high-porosity/low-acoustic wave velocities	∗	÷	÷

stuck pipe. Note that it may be difficult to distinguish hole collapse from hole cleaning problems.

9.1.2 Lost circulation

Lost circulation means that a significant amount of drilling fluid is lost into the formation. This implies that a fracture has been created (see Chapter 11) or that mud is lost into an existing fracture. This is an operational problem, partly because the mud is expensive, and partly because there is a limit to the amount of mud available on the rig. The mud loss may also lead to a temporary pressure drop in the well, since a part of the mud column disappears into the formation. As a consequence, pore fluid may flow into the well from permeable layers. In the presence of gas, this may lead to a rapid increase in well pressure ("kick") and a high risk of a blowout. This is a potentially dangerous situation, which may result in loss of lives and equipment.

The main solution is to keep the mud weight sufficiently low that fluid loss does not occur, that is, below the limit for fracture initiation and growth in non-fractured formations and below the fracture reopening/propagation pressure in naturally fractured formations. If the margins are small enough, then the equivalent circulating density (ECD) may be sufficient to exceed the fracturing pressure. The ECD (or dynamic mud weight) equals the static mud weight (ESD) plus a term proportional to the pressure drop in the annulus.

This is an argument to keep the mud weight well below the critical limit. In some cases, fractures may initiate, but the stress conditions may not allow fracture growth. We will come back to this, but we will keep in mind that in a previously unfractured formation, fracture growth is necessary to lose significant amounts of drilling fluid. In cases where hole cleaning is not done properly, for instance, in relation to hole collapse, pack-off during backreaming may lead to lost circulation. This shows the need for good diagnostics: an apparent loss problem may have been initiated by hole collapse and enhanced by insufficient cleaning.

Mud loss prevention additives (lost circulation material, LCM) is commonly used to heal fractures created during loss situations. This treatment has a better effect in permeable rocks than in low-permeability rocks like shale, because leak-off from a fracture in shale is very slow. The requirement not to induce fluid loss situations sets an upper limit to the mud density with an associated strong impact on the entire well design.

Particles may also be used in a preventive manner by including a specific concentration of particles (e.g. graphite and calcium carbonate) in the drilling fluid while drilling. The purpose is to arrest drilling-induced fractures at an early stage, before significant losses occur. The particles are supposed to bridge off the fracture opening and reduce fluid loss to the fracture to the extent that the fracture pressure drops and fracture growth stops. Used successfully (Aston et al., 2004; Davison et al., 2004), the fracture pressure can be increased relative to drilling with a mud without particles. Such an approach may extend drilling in depleting reservoirs by maintaining a positive mud weight window. Other potential applications include deep water drilling, where the window between the pore pressure and the fracture gradient is often initially low, and drilling into cooled reservoir sections (due to cold water injection), where cooling has reduced the fracture pressure significantly.

Further discussion of the lost circulation problem can be found in Lavrov (2016). This book discusses mechanisms, diagnostics and suggestions for preventing and curing losses.

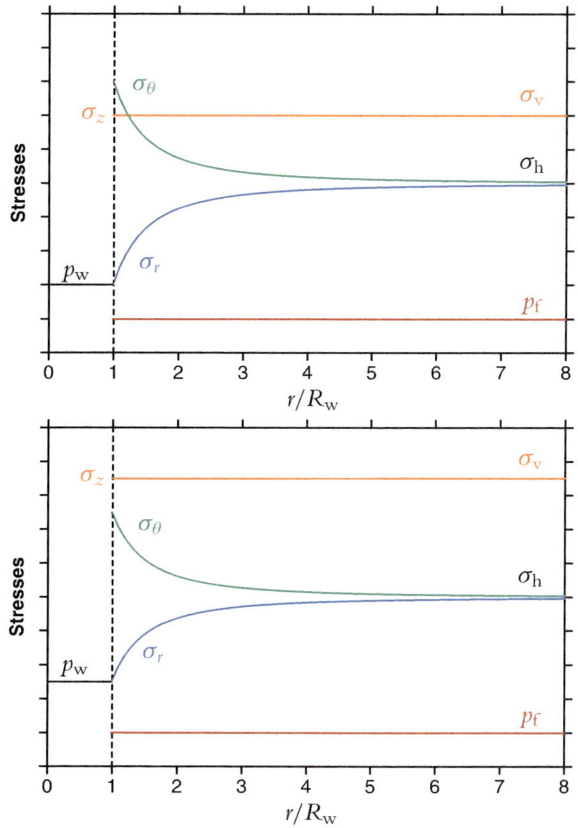

Figure 9.2 Stresses around a vertical borehole with an impermeable wall (elastic formation; isotropic horizontal far-field stresses). The upper part corresponds to the case where the tangential stress is the maximum principal stress at the borehole wall. In the lower part the vertical is the maximum principal stress.

9.2 Rock mechanics analysis of borehole stability

9.2.1 Introduction

Formations at a given depth in the Earth are exposed to vertical and (generally anisotropic) horizontal compressive stresses, as well as a pore pressure (Chapter 3). When a hole is drilled, the surrounding rock must carry the load previously carried by the removed rock. Chapter 4 presents the theoretical basis of stresses around boreholes. In a rock that behaves linearly elastic, this leads to a stress concentration near the well as shown in Fig. 9.2. If the rock is sufficiently weak, then this stress concentration can lead to failure of the borehole.

In drilling of a deep well, such as in petroleum fields, the well is filled with drilling fluid. The mud serves two main purposes. One is to prevent flow of pore fluid into the well and hole instabilities. The other purpose of the mud is to transport drill cuttings from the hole to the surface. The mud density ρ_w controls the pressure in the well:

$$p_w = \rho_w g D \tag{9.1}$$

where g is the acceleration due to gravity, and D is the vertical depth. In oil well drilling language, we usually refer to mud weight (in density units) and gradients of stress or pressure (also measured in equivalent density units) instead of pressures. Circulation of the mud implies that the effective (dynamic) mud pressure in the well is higher than the static pressure expressed by Eq. (9.1). As mentioned above, the dynamic well pressure is often referred to as an equivalent circulating density (ECD). The difference may be in the order of 5–10 %. Under normal drilling operations, the pressure will vary between the static and dynamic values (see Sections 9.3.1 and 9.7.3).

The mud will carry part of the stress concentration. We will now use rock mechanics to analyse how the mud weight controls the mechanical stability of the borehole, building mainly on the theory outlined in Chapter 4.

To begin with, we assume linearly elastic rock behaviour and look at the deformation of the hole resulting from drill-out. Eq. (4.47) tells us that a vertical borehole will contract proportionally to the difference between the total horizontal stress and the well pressure, and inversely proportional to the shear modulus of the formation:

$$\frac{\Delta R_w}{R_w} = \frac{|\sigma_h - p_w|}{2G_{fr}} \tag{9.2}$$

This equation produces a relatively small change in borehole radius: With a shear modulus G_{fr} of 1 GPa (representative of very soft shale), a difference of 5 MPa between horizontal stress and well pressure leads to a 0.5 mm shrinkage of a 10 cm radius borehole. This is practically insignificant for the stability of the hole, since such small displacements will not lead to a tight hole. An operational problem would only occur if the shear modulus of the formation becomes orders of magnitude lower. For shale at depth, this means that the shale will have to be taken beyond the elasticity limit and into a state of brittle failure or plastic deformation. For analysing borehole stability, it is therefore more fruitful to focus on criteria for borehole failure.

9.2.2 Lower mud weight limit

As was seen in Chapter 4, there are two basic types of borehole failure (illustrated in Fig. 9.1), shear (or compressive) failure and tensile failure. In the case of linear elasticity the maximum deviatoric stress occurs at the borehole wall, so this is where shear failure will occur. We will use this criterion for borehole collapse, which means that borehole failure is regarded as perfectly brittle (that is, a complete loss of load-bearing capacity once failure occurs). The shear failures most critical to the stability of a borehole are failures where either the tangential (hoop) stress or the axial stress along the borehole axis is the maximum, and the radial stress (= well pressure) is the minimum (these are referred to as cases a and b in Table 4.1). For the vertical hole, we can write the solutions for critical well pressure in a borehole with impermeable wall conditions as follows (see Eqs. (4.126) and (4.127)):

$$p_{\mathrm{w,min}}^{(a)} = \frac{3\sigma'_{\mathrm{H}} - \sigma'_{\mathrm{h}} - C_0}{\tan^2\beta + 1} + p_{\mathrm{fo}} \tag{9.3}$$

$$p_{\mathrm{w,min}}^{(b)} = \frac{\sigma'_{\mathrm{v}} + 2|\nu_{\mathrm{fr}}|(\sigma'_{\mathrm{H}} - \sigma'_{\mathrm{h}}) - C_0}{\tan^2\beta} + p_{\mathrm{fo}} \tag{9.4}$$

This analysis is based on the Mohr–Coulomb criterion for shear failure.

In shale, where the permeability is very low (nanoDarcy or so), an impermeable borehole wall corresponds to the situation during or shortly after drilling when no pore pressure penetration occurs. An impermeable wall may also be an appropriate boundary condition in a permeable formation when a perfect mud cake is formed at the borehole wall. A mud cake in a traditional meaning cannot be established on shale, although in case of a non-wetting drilling fluid or a fluid that contains molecular species (e.g. polymers) that block the shale surface, the impermeable wall approach may be valid also after initial drill-out.

We see from Eqs. (9.3) and (9.4) that the limiting well pressure may be larger or smaller than the in situ pore pressure p_{fo}, depending on in situ stresses and on the strength parameters of the rock. Only one of conditions (a) or (b) represents a valid solution to the hole stability problem, so when the limiting well pressure has been calculated, we need to check that the borehole wall stresses are in agreement with the assumed conditions. This solution will in practice be the one with the larger of the calculated pressures. These types of borehole failure result in the formation of a breakout (e.g. as shown in Figs. 4.22 and 9.4). As discussed in the preceding section,

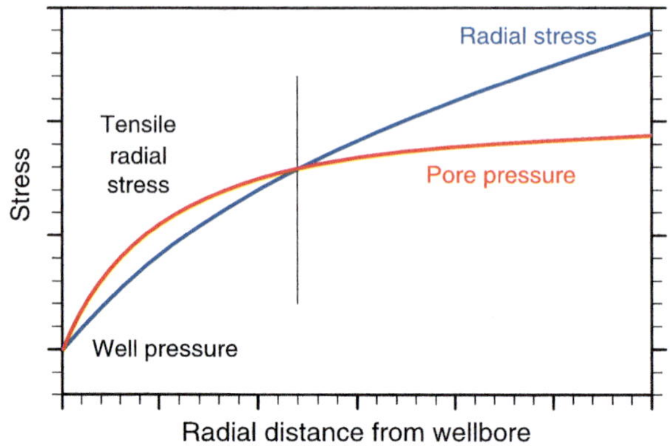

Figure 9.3 Zone of tensile radial stress caused by underbalanced drilling.

this may cause a stuck pipe situation. The obvious reason is that the mud weight is too low.

The lower allowed value of the mud weight is limited not only by the criterion for shear failure, but also by the condition for radial tensile failure (see Fig. 9.3), that is,

$$\sigma_r' = -T_0 \Longrightarrow p_{\mathrm{w,min}}^{\mathrm{rad.tension}} = p_{\mathrm{f}} - T_0 \tag{9.5}$$

Since the tensile strength of shale is low, this means that if the borehole pressure is lower than the pore pressure (underbalanced drilling), then there is a risk of tensile failure. This failure mode leads to blade-shaped cavings and may cause a tight hole. Since the new free surface generated will be exposed to exactly the same effect, this is a process that is not self-stabilising. Being in underbalance is also risky from the point of view of pressure control, since kicks may occur from permeable layers. In pure shale, there is however no risk of fluid influx to the well, since the shale permeability is extremely low (typically nanoDarcy; see Chapter 3). All in all, the risks of generating borehole instabilities by being in underbalance have to be weighted against the possible benefits of higher drilling rate and less formation damage by mud filtrate influx during drilling of producible reservoir zones.

In practice, the lower mud weight limit is associated with the maximum of the critical well pressures calculated from shear and tensile failure criteria in Eqs. (9.3)–(9.5) in addition to the pore pressure. If the well pressure has to be in overbalance (by regulations or to reduce the risk of kicks), then

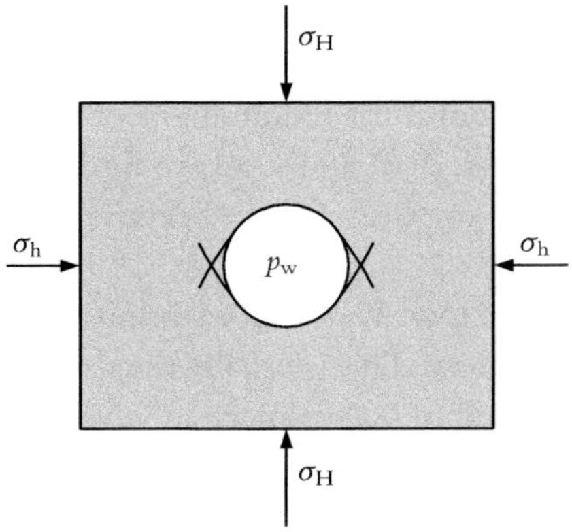

Figure 9.4 Illustration of compressive failure direction around a vertical borehole with unequal horizontal stresses.

the minimum required mud weight is only controlled by shear failure if the shear failure limit is above the pore pressure gradient.

The shear failures described by cases d–f in Table 4.1 all imply that the well pressure is higher than the axial stress along the borehole. For a vertical well, this means that the well pressure exceeds the vertical stress, which is not a common practical situation. In horizontal or deviated wells, such a high well pressure would always exceed the fracturing limit. In borehole stability analysis, these cases are therefore not considered.

The stability of boreholes drilled along horizontal principal stress directions can easily be analysed using the same mathematical framework as above. We have to remember that the far-field stresses controlling the tangential stress (and hence the shear failure criterion (a) at the borehole wall) are the stresses acting in the plane perpendicular to the borehole, so that the term $(3\sigma'_H - \sigma'_h)$ in Eqs. (9.3) and (9.4) in general has to be replaced by $(3\sigma'_{max\perp} - \sigma'_{min\perp})$, where $\sigma'_{max\perp}$ and $\sigma'_{min\perp}$ are the maximum and minimum effective principal stresses, respectively, in the plane perpendicular to the borehole axis:

$$p^{(a)}_{w,min} = \frac{3\sigma'_{max\perp} - \sigma'_{min\perp} - C_0}{\tan^2\beta + 1} + p_{fo} \tag{9.6}$$

$$p^{(b)}_{w,min} = \frac{\sigma'_{\parallel} + 2|\nu_{fr}|(\sigma'_{max\perp} - \sigma'_{min\perp}) - C_0}{\tan^2\beta} + p_{fo} \tag{9.7}$$

where $\sigma'_{\parallel}$ is the principal stress parallel to the direction of the hole.

In summary, the lower mud weight limit in the case of a linearly elastic formation undergoing perfectly brittle failure is found from Eqs. (9.3) and (9.4) and the possible additional requirement of overbalanced drilling. By further inspection of Eqs. (9.3) and (9.4) we find that the risk of borehole instability with respect to shear failure becomes high (high critical well pressure) if

- The rock strength is low. This is as we intuitively would expect.
- The failure angle is low. This is usually the case in shale, where β often is 50°–55°, compared to 60° or more in sandstones (see Chapter 3).
- The pore pressure is high. This is often the case in shales above the reservoir (see also Chapter 3), as well as in some reservoir sections prior to depletion.

These results are all in good agreement with operational experience, indicating the validity of a rock mechanics approach to handle borehole stability.

Furthermore, from Eqs. (9.3) and (9.4) and the discussion above we see that the lower mud weight limit increases with $(3\sigma'_{\max\perp} - \sigma'_{\min\perp})$. Thus shear failure is triggered by a high stress level and high stress anisotropy perpendicular to the well. However, a high stress level will reduce the risk of mud losses; high stress anisotropy may facilitate fracture initiation.

Given an elastic and isotropic rock, the most unfavourable direction for a borehole with respect to initiation of hole instability will be parallel to the intermediate principal stress. The most favourable direction for a borehole will normally be along the largest principal stress direction. This illustrates the importance of knowing in situ stresses and principal stress directions. Remark that Zheng (1998) found that if one is able to handle the initial formation of the instability, then a hole drilled along the intermediate stress may in fact be beneficial, since the breakout geometry may give a better post-failure stability. Effects of other borehole shapes are discussed in Section 4.6 and AlBahrani and Morita (2020).

When a deviated borehole is drilled along an arbitrary inclination (relative to vertical) and azimuthal direction in the horizontal plane, the stability analysis has to account for shear stresses at the borehole wall, since the tangential and axial stresses are not principal stresses in general. The principal stresses and their orientations have to be computed first and then inserted into the failure criteria. The equations for a generally deviated well are given in Section 4.3. The results depend on the stress regime and are difficult to generalise.

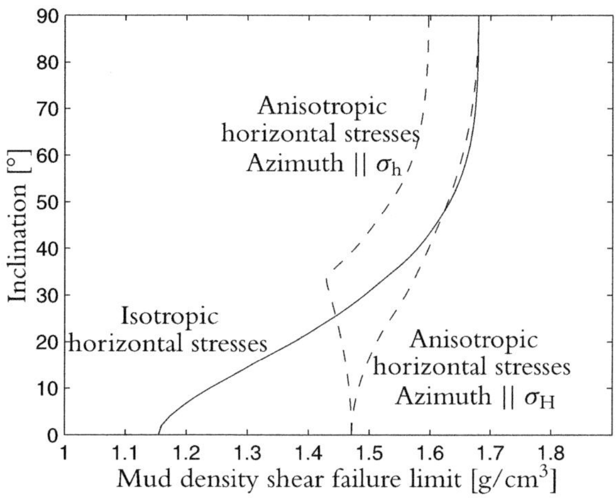

Figure 9.5 Illustration of stability analysis for a deviated wellbore at 1500 m depth with vertical stress 30 MPa, isotropic horizontal stresses 25 MPa and pore pressure 15.5 MPa. The unconfined strength is set to 10 MPa, the friction angle is 30°, and Poisson's ratio is 0.25. Also included is a case with anisotropic horizontal stresses, where all parameters are kept the same as above, except the maximum horizontal stress 28 MPa.

Fig. 9.5 illustrates a case with a normal in situ stress situation (the vertical stress is the major principal stress), showing the required minimum mud weight to prevent shear failure as a function of hole deviation. The maximum allowed mud weight may conservatively be taken as the minimum horizontal stress, in this case 25 MPa corresponding to a mud density of 1.70 g/cm^3. For isotropic horizontal stresses, the stable range is narrowed continuously as the inclination increases towards horizontal, in particular, from 30° until 60°. If the horizontal stresses are anisotropic, then the stability of a vertical hole becomes significantly reduced.

The critical mud weight for a deviated well depends both on inclination and azimuthal direction. For a horizontal well parallel to the minimum horizontal stress, the shear failure limit is improved when compared to the case with horizontal stress isotropy, because the stress anisotropy perpendicular to the hole is reduced.

When the horizontal stresses and the vertical stress become more equal (isotropic), the effect of inclination and azimuth will be reduced accordingly. A prime example is a homogeneous salt formation, where all three principal stresses are expected to be equal.

9.2.3 Upper mud weight limit

Although shear failure may also occur if the well pressure is high (e.g. case c in Table 4.1), this situation is not thought to give rise to significant drilling problems. In addition, the limit for this shear failure mode is normally very close to the limit for hydraulic fracturing by tensile failure (see Fig. 4.13). For the case of an impermeable borehole wall in a vertical well, fracture initiation is given by

$$p_{\mathrm{w,max}}^{\mathrm{frac}} = 3\sigma_{\mathrm{h}} - \sigma_{\mathrm{H}} - p_{\mathrm{fo}} + T_0 \tag{9.8}$$

provided that the minimum horizontal is the minimum principal rock stress. For fracture initiation, the term $(3\sigma_{\mathrm{h}} - \sigma_{\mathrm{H}})$ is in general written $(3\sigma_{\mathrm{min}\perp} - \sigma_{\mathrm{max}\perp})$. This is analogous to what is done in Chapters 10 and 11, to be used for calculation of critical drawdown and fracture initiation pressures. In case the vertical stress is the minimum, a fracture is initiated if $\sigma'_{\mathrm{v}} = -T_0$. In the following, we will not consider this possibility.

Most of the overburden consists of shale, where the assumption of a perfect filter cake is not valid. Traditionally, leak-off tests below the casing shoe have been used to establish a fracture gradient in the overburden (see Section 8.3.4 and Fig. 11.4). However, leak-off pressures have a tendency to have a very large spread, making the fracture gradient uncertain. Such variation is a consequence of natural variation in the conditions at the borehole wall. Even if tests are done at the same depth in the same formation, the leak-off and/or breakdown pressures can vary significantly, since this is not an inherent property of the formation. This is shown in Fig. 9.6, where by extended leak-off tests it is established that the minimum stress in the formation is the same, whereas leak-off and breakdown pressures vary significantly.

There are also more theoretical approaches available to estimate the fracture gradient. An overview of these (Matthews and Kelly, 1967; Daines, 1982; Eaton, 1969) can be found, for example, in Zhang and Yin (2017). These methods all somehow relate to in situ stresses and pore pressure and therefore do not capture the natural variability of the tests. The same uncertainty will therefore persist.

It should be noted that for a hydraulic fracture to create a drilling problem (i.e. for significant mud loss to occur), the fracture needs not only to be initiated, but also to propagate beyond the near well region. A hydraulic fracture will propagate if the pressure in the fracture exceeds the minimum principal stress, plus an additional term depending on the conditions for

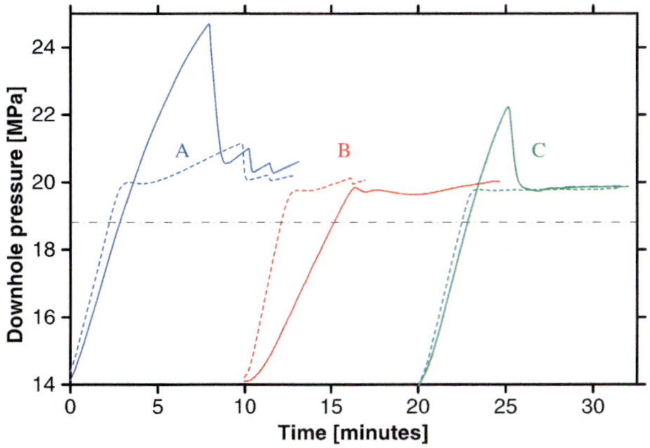

Figure 9.6 Pressure versus time for the pumping phase of two-cycle XLOTs at the same depth in three neighbouring vertical wells. (For clarity, the start times for tests B and C have been shifted by 10 and 20 min.). The first cycles are shown by solid lines, whereas the second cycles are shown by dashed lines. The fracture closure pressure (indicated by the dashed-dotted line) was determined to be the same in all wells, but the pump-in responses are strikingly different. Note that in test B the pumping rate in the second cycle was double that in the first cycle. After Raaen et al. (2006).

fracture growth at the tip, leakage from the fracture to the formation and so on. Often, mud loss into pre-existing fractures may be a problem. This occurs if the well pressure is high enough to allow reopening of such a fracture. Therefore in practice the well pressure should not exceed the fracture closure pressure (minimum principal, usually, horizontal, stress), plus an additional contribution Δp_{exp} to be quantified on the basis of experience, that is,

$$p_{\text{w,max}}^{\text{fracprop}} > \sigma_3 + \Delta p_{\text{exp}} \tag{9.9}$$

Such experience can be obtained from XLOTs, since they capture both the propagation pressure and the closure pressure. The propagation pressure will be less affected by variations from one test to another but can be influenced by changes in fluid properties and well azimuth and inclination (fracture tortuosity near the wellbore). A fracture gradient based on propagation pressure will hence show less variation than a fracture gradient based on leak-off tests, but will be on the conservative side in formations that are not naturally fractured.

The improved consistency given by the propagation pressure (and also the re-opening pressure) is clearly shown in Fig. 9.6.

There are at least two situations where the fracture initiation pressure predicted by elastic theory can be less than the minimum stress in the formation:

- For a vertical well in a strike-slip stress regime, see Eq. (9.8).
- For a horizontal well in a normal stress regime where σ_h is reduced due to significant depletion, see Section 11.3.

However, losses can only occur if the well pressure is increased to a level somewhat above σ_h.

The discussion so far has been based on the assumptions and solutions of the Kirsch equation for tensile failure. Experience has shown that this equation can both underestimate and overestimate the actual fracture initiation pressure (Andrews et al., 2016).

Sampath et al. (2018) present an overview of different analytical models for formation breakdown, based on tensile strength, energy release rate and stress intensity factor. Their conclusion is that the two latter methods often provide more realistic predictions. The disadvantage is however that the more complex models require input data that are seldom available in a practical situation. Therefore, at present, there is no unique equation available for precise prediction of the upper mud weight limit at which losses will occur.

So far no distinction has been made between the fracture initiation pressure (FIP) and the fracture breakdown pressure (FBP). Such a difference has been observed experimentally; see discussion in Section 11.2. The concept is that a fracture can initiate and grow significantly (a few cm) before the formation breaks down and unstable fracture growth occurs. This is explained by an internal filter cake in the fracture, and thus it implies that a permeable rock like a sandstone is required. Models that account for this difference between FIP and FBP are presented by Morita and Fuh (2012) and Zhang et al. (2018).

Further understanding of the lost circulation problem can be found in Lavrov (2016), both regarding mechanisms and diagnostics, and suggestions for preventing and curing losses.

9.3 Time-delayed borehole failure

9.3.1 Establishment of pore pressure equilibrium

In the field it is often observed that initial drilling may be stable, but that tight hole/stuck pipe situations occur during tripping or logging runs, typically several days after drill-out. The model presented in Section 9.2 does not account for time-dependent borehole stability. One limitation of the analysis above was the assumption of an impermeable borehole wall. As pointed out, this is only true if a perfect mud cake is established or if the rock is impermeable. Shale has a very low (nanoDarcy) but still finite permeability, which means that impermeable wall conditions are valid only during drilling. After drilling, the pore pressure close to the borehole wall will gradually approach the well pressure. Eventually, if the formation is exposed to the mud pressure for sufficiently long time, then steady-state pore pressure equilibrium is reached.

The situation may be different when drilling with an oil-based fluid with a capillary entry pressure. This is discussed in Section 9.4.

The stress field under permeable wall conditions is given in Chapter 4. Applying the Mohr–Coulomb failure criterion to a piece of rock at the borehole wall, the minimum required well pressure (analogous to Eqs. (9.3) and (9.4)) for a vertical borehole is given by one of cases (a) and (b) (see Eqs. (4.122) and (4.123)):

$$p_{\mathrm{w,min}}^{(a)} = \frac{3\sigma'_{\mathrm{H}} - \sigma'_{\mathrm{h}} - C_0}{2 - \alpha\frac{1-2\nu_{\mathrm{fr}}}{1-\nu_{\mathrm{fr}}}} + p_{\mathrm{fo}} \tag{9.10}$$

$$p_{\mathrm{w,min}}^{(b)} = \frac{\sigma'_{\mathrm{v}} + 2|\nu_{\mathrm{fr}}|(\sigma'_{\mathrm{H}} - \sigma'_{\mathrm{h}}) - C_0}{1 - \alpha\frac{1-2\nu_{\mathrm{fr}}}{1-\nu_{\mathrm{fr}}}} + p_{\mathrm{fo}} \tag{9.11}$$

When the hole is drilled along one of the horizontal principal stress directions, we have to replace the principal stresses with the appropriate ones, as was done in Eqs. (9.6) and (9.7).

The time τ_{D} it takes to reach pore pressure equilibrium after a hole is drilled in unbalanced conditions is based on consolidation theory (Section 1.9), which states that

$$\tau_{\mathrm{D}} \approx \frac{l_{\mathrm{D}}^2}{D_{\mathrm{h}}} \tag{9.12}$$

where l_{D} is the characteristic length, and D_{h} is the diffusion coefficient defined in Eq. (1.321), which is proportional to the permeability. Taking $l_{\mathrm{D}} \sim$ borehole radius $\sim$10 cm and the shale permeability $\sim$1 nanoDarcy, we find that the characteristic time typically is 5–10 days. This corresponds to the time period after which instabilities occur in practice; see, for example, van Oort et al. (2019).

If the well is kept in overbalance, then the pore pressure near the borehole wall will increase, resulting in reduced hole stability with time. This is

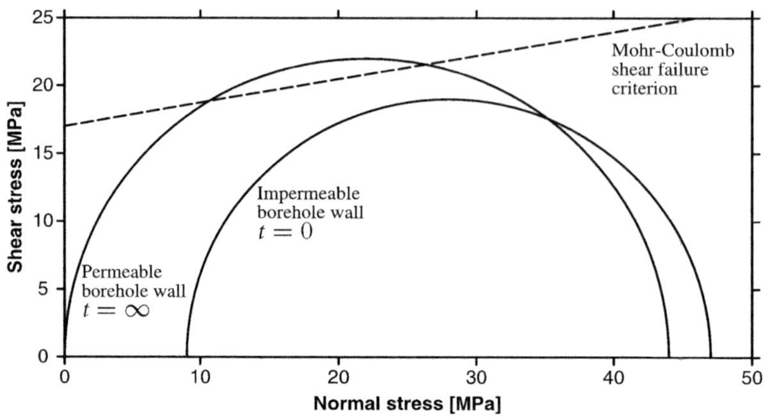

Figure 9.7 Illustration of how Mohr's circle representing the borehole stress field moves from time $t = 0$ (drillout) to time $t \rightarrow \infty$ (pore pressure equilibrium), creating a more unstable situation with respect to shear failure (hole collapse).

illustrated in Fig. 9.7, which shows the displacement of Mohr's circle in the direction of the shear failure criterion from short time ($t = 0$) to long time ($t = \infty$). The reason is that the effective confinement (minimum effective principle stress) approaches zero as pore pressure equilibrium is established. For the case described for a vertical well in Fig. 9.5, the minimum required well pressure increases with time to 26 MPa (corresponding to mud density 1.77 g/cm^3), which means that the horizontal stress is exceeded, and stable drilling no longer is possible. On the other hand, if the well is drilled in underbalance, then the pore pressure will decrease, and stability with respect to hole collapse will improve with time. The change in pore pressure as a function of time is illustrated in Fig. 4.9.

Two practical implications can be drawn from the above: The likelihood of stability problems (at least during overbalanced drilling) will be reduced the sooner one is able to leave a well section behind by setting casing. Furthermore, maintaining an impermeable boundary condition is a way to stabilise the hole over time. Solutions to this challenge are being sought in terms of oil-based or polymer-containing drilling fluids (see Section 9.4).

During the drilling process, the pressure in the wellbore is at circulating density (ECD) most of the time. It is only when the pumps are off (during pipe connections) that the pressure is at the static value (ESD). If the stability limit is somewhere between ECD and ESD, then the wellbore may stay stable during drilling since the ESD-intervals are

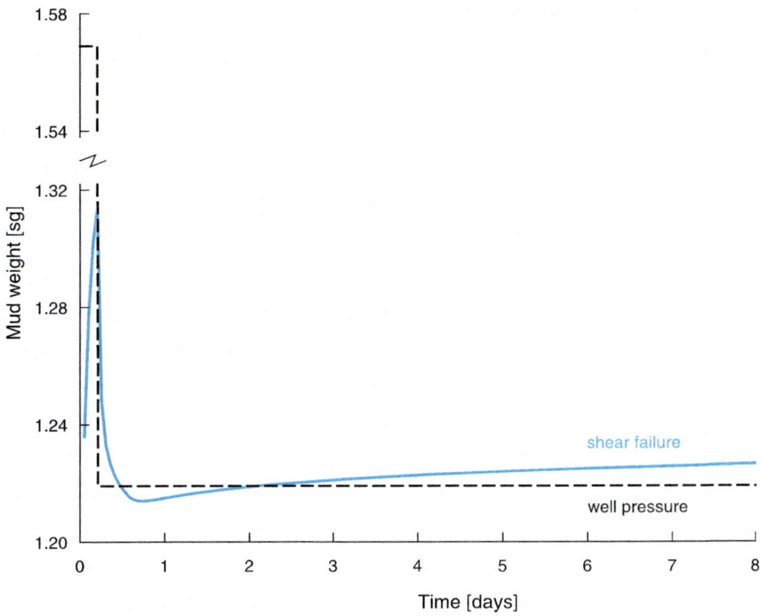

Figure 9.8 Illustration of change in well pressure compared to shear failure pressure as a function of time. The well pressure is reduced from the ECD-value to the static value. The difference is slightly exaggerated to make the effect clearer. It is assumed that drill-out takes place at $t = 0$. The initial increase in the shear failure curve, before the pumps are shut off, is mainly due to pore pressure penetration.

short. However, after having completed drilling of a section, the wellbore will experience the ESD-value for a longer period. This results in reduced wellbore pressure and time-dependent change of near-wellbore pore pressure, thus increasing the risk of instability. This instability can occur already when starting to trip out, or it can occur while out of the hole. These changes in well pressure and failure pressure are illustrated schematically as a function of time in Fig. 9.8. The latter event can result in problems re-entering the hole. If the well pressure is close to the stability limit, then creep can also be a contributing factor to potential long-term instability; see Section 9.10. A remedy to this situation can be to displace to a drilling fluid with higher density before pulling out (Nes et al., 2015).

Anisotropic horizontal stresses can also have a direct effect on the pore pressure. In Section 4.4.2, it was shown how drill-out in such a situation leads to a change in pore pressure that will dissipate with time.

9.3.2 Temperature effects

Consolidation (or establishment of pore pressure equilibrium), as described in the preceding section, is not the only possible explanation of time delayed borehole failures. The drilling fluid is usually (at $t = 0$) colder than the formation to be drilled, since it has surface temperature when circulation starts. The fluid and formation temperatures will gradually adjust depending on the circulation rate. After a stop in the circulation, the formation near the well will gradually heat up. Maury and Sauzay (1987) found that this could explain delayed failure. Shortly after drilling, the borehole is stable. However, as the temperature increases, the tangential and axial stresses at the borehole wall will both increase by the same amount.

The thermoelastic stresses induced at an impermeable borehole wall are given by Eqs. (4.71)–(4.73). If the borehole fluid is colder than the formation, then the tangential and axial stresses are reduced proportionally to the temperature difference between formation and borehole fluid. This acts as a strengthening of the borehole with respect to shear failure. Deliberate cooling of the mud can therefore be a practical approach to mitigate stability problems. This was applied with success in the field (Guenot and Santarelli, 1989; Maury and Guenot, 1995). The thermal effect is proportional to rock stiffness, which means that it is more significant in hard than soft rocks. It is also proportional to the thermal expansion coefficient.

There is not a lot of data on thermal expansion of sedimentary rocks, but typically α_T is around 10^{-5} K^{-1} (see Appendix A). This gives a typical thermal stress contribution of a few MPa, which reduces the lower mud weight limit by 5–10 % compared to a situation with no thermal effect. However, since cooling reduces the tangential stress, it not only reduces the risk of shear failure, but it also promotes fracturing and may hence be a destabilising factor with respect to lost circulation problems. The extent of the cold zone is however limited, and this may limit the growth of the fracture.

Cooling of a low permeability rock like shale will also influence the pore pressure due to a larger thermal expansion coefficient for the fluid than for the solid parts of the rock. Thus cooling reduces the pore pressure, which in general improves stability, whereas heating has the opposite effect. This is demonstrated in Section 1.9.2 and illustrated in Fig. 1.25. Rock properties (strength, stiffness) may be altered as a result of temperature changes: Normally, strength and stiffness will increase with decreasing temperature. Finally, mud properties are also temperature dependent: Cooling the mud will lead to a slight mud density increase due to thermal contraction, again resulting in improvement of stability with respect to shear failure.

Note that if one drills into a reservoir that is cooled by waterflooding, then the in situ stresses may have been altered in a way that makes mud loss and fracturing more likely than under initial normal temperature conditions. Field evidence for this phenomenon is given by Hettema et al. (2004).

9.3.3 Creep

When discussing time-dependent borehole stability, it is obvious that creep may cause a borehole to collapse with time. It is general knowledge that a material that is brought above its yield point may creep to failure. Translating to the borehole situation, this means that if the mud weight is kept marginally above the lower limit for long enough time, then the borehole may collapse as a result of creep. The risk of high (accelerating) creep rates increases with increasing temperature. Although truly constant stress conditions over time never occur during drilling, the intrinsic time-dependent properties of the formation may still have an influence on stability. In practice, this is difficult to distinguish from other time-dependent effects, like consolidation, thermal and chemical effects.

Fundamentals about rock creep can be found in Section 1.9.4, and a more elaborate discussion of shale creep in Section 9.10.

Creep as a phenomenon is also well known in salt; see Section 3.4.4 and Section 9.8.3.

9.4 Interaction between shale and drilling fluid

In the previous section, three sources of time-dependent borehole stability were considered: consolidation, mud cooling, and creep. In this section, we will discuss one more mechanism, chemical interaction between the drilling fluid and the formation. This is however not just a source of time-delayed hole collapse, but may be used to improve shale stability, in particular, in cases where instabilities occur in shales with high contents of swelling clay minerals.

It is often observed that oil-based mud gives better stability than water-based mud. To start out, let us consider the (somewhat academic) example of drilling with pure oil (or another fluid that is completely non-wetting to the shale). The capillary pressure for oil (or other non-wetting fluid) in contact with a water-wet shale is

$$p_{cp} = \frac{2\gamma}{r} \tag{9.13}$$

where γ is the surface tension, $\gamma_{oil-water} = 50 \cdot 10^{-3}$N/m, and r is the pore size, which in shale typically is 10 nm. From that we find the capillary entry pressure for pure oil to enter a water-saturated shale to be ~10 MPa. Thus an overbalance of 10 MPa is required for oil (or another non-wetting fluid with similar surface tension) to penetrate into intact shale. This means that for lower mud overbalance, the borehole wall will remain impermeable, and the $t = 0$ stability will prevail.

Oil-based muds usually give better hole stability than water-based muds, but the latter are preferred for environmental reasons. Oil-based muds are however not pure oils. They contain a water phase, and the chemistry of the water phase has an influence on the hole stability.

Various mud additives have been introduced in water-based systems to help maintain impermeable borehole wall conditions over time.

Such additives—polymers, asphaltenes, graphite and other products—are designed primarily to work by physically sealing the shale. Silicate has also been in use for some time for this purpose (van Oort et al., 1996). More recently, the use of nanoparticles has been introduced. It has however not been possible to find one single solution to this problem, nor to the problem of shale-inhibition in general. In practical use, it is not possible to avoid the fact that shales are very heterogeneous even on a small scale and that additives need to be tailored to each application (Gomez and Patel, 2013; Hoxha et al., 2019).

We will now discuss chemical mechanisms that may lead to interaction between shale and drilling fluid. Let us first consider osmosis. The basis for osmosis is the existence of a semi-permeable membrane that allows water molecules to pass but prevents ions to enter the shale. The membrane may be thought of in two different ways:

1. Oil-based mud may act as a semi-permeable membrane. Ions associated with salts are prevented to move between the water phase of the mud and the formation.
2. The shale is often regarded as a semi-permeable membrane when contacted by water-based mud. Because the clay minerals have surface charge, the movement of ions is thought to be hampered, giving rise to membrane properties.

The osmotic potential $\Delta\Pi$ (from e.g. Marine and Fritz, 1981) is

$$\Delta\Pi = \frac{RT}{V_w} \ln \frac{a_{w,df}}{a_{w,sh}} \tag{9.14}$$

where R is the molar gas constant ($8.31\,\text{J}\,\text{mol}^{-1}\text{K}^{-1}$), V_w is the molar volume of water (0.018 l/mol), $a_{w,df}$ is the chemical activity of water in the drilling fluid, and $a_{w,sh}$ is the chemical activity of pore water in the shale. The activity denotes the effective concentration of water in a solution, so that $a_w = 1$ for fresh water, whereas $a_w < 1$ for salt water. For ideal solutions, activity is directly given by the salt concentration. For non-ideal solutions with approximately 0.1 M salt concentration, the difference between activity and water concentration may be 10–50 %, depending on type of ion (Mahan, 1975).

Mody and Hale (1993) implemented osmotic theory into a rock mechanics model for borehole stability. They added the osmotic potential through a stress term equivalent to the poroelastic contribution seen in Eqs. (4.58)–(4.60):

$$\sigma_r = p_w \tag{9.15}$$

$$\sigma_\theta = 2\sigma_h - p_w + \alpha \frac{1 - 2\nu_{fr}}{1 - \nu_{fr}} \Delta\Pi \tag{9.16}$$

$$\sigma_z = \sigma_v + \alpha \frac{1 - 2\nu_{fr}}{1 - \nu_{fr}} \Delta\Pi \tag{9.17}$$

Adding salt to the drilling fluid so that $a_{w,df} < a_{w,sh}$ sets up an osmotic potential $\Delta\Pi < 0$, which will tend to drive water out of the shale and hence acts as an effective pore pressure reduction. This has an instantaneous stabilising effect on the borehole.

It is well established that ions do move through shale (see e.g. Ballard et al., 1994). This means that the osmotic membrane is leaky, which is handled by introducing a membrane efficiency $\sigma < 1$. The membrane efficiency reduces the osmotic potential

$$\Delta\Pi = \sigma \frac{RT}{V_w} \ln \frac{a_{w,df}}{a_{w,sh}} \tag{9.18}$$

The membrane efficiency depends on ionic diffusivity and will hence in general depend both on clay type and on the type of ions present in the drilling fluid (for more elaborate analysis, see Marine and Fritz, 1981; Bailey et al., 1991; Sherwood, 1994; Heidug and Wong, 1996; Lomba et al., 2000a; Heidug and Wong, 1996; Lomba et al., 2000b; Sherwood and Craster, 2000). Experiments (Chenevert, 1970; van Oort et al., 1996; Ewy and Stankovich, 2002; Schackelford et al., 2003) seem to indicate that for shales, σ is in the range 0.05–0.30. The osmotic potential may however still be several MPa, which means that reducing drilling fluid activity by adding salt may improve borehole stability significantly, at least during the initial phase. The effect will however, due to membrane leakiness, decay with time after drillout. In case of very low activity drilling fluid, we may expect damage as a result of reduced tangential and/or axial stresses at the borehole wall.

There is also experimental evidence that ionic exchange may occur within the shale when it is exposed to brine (Steiger, 1982; Denis et al., 1991; Horsrud et al., 1998, see Fig. 9.9). In particular, exposure of smectite-rich shale to KCl causes significant shrinkage (Fig. 9.9), even when the potassium concentration is so small that osmotic swelling would be expected. The rate of exchange depends on the ion concentration, whereas the maximum shrinkage depends on the amount of available sites in the shale, that is, on the diffusing ion type, and on the amount of swelling clay minerals and their exchange capacity. Physically, shrinkage happens because the hydrated potassium ion fits more easily into the clay mineral crystal lattice than the hydrated sodium ion, which is the dominant native ion in shales. The chemically induced shrinkage strain can be looked at in analogy with the thermal cooling effect. This means that it may be modelled similarly (see Eqs. (4.71)–(4.73)):

$$\sigma_r = p_{\rm w} \tag{9.19}$$

$$\sigma_\theta = 2\sigma_{\rm h} - p_{\rm w} - \frac{E_{\rm fr}}{1-\nu_{\rm fr}}\varepsilon^{\rm ch} \tag{9.20}$$

$$\sigma_z = \sigma_{\rm v} - \frac{E_{\rm fr}}{1-\nu_{\rm fr}}\varepsilon^{\rm ch} \tag{9.21}$$

Ionic exchange affects shale properties as well. Apparently, KCl exposure enhances plasticity, which is beneficial for borehole stability. The shrinkage effect leads to improved stability with respect to the lower mud weight limit, but if the chemical strain is high, then tensile or shear failure may be induced. There are indications from the literature that ionic exchange effects may also be important for ions other than potassium and that the type of anion also plays a role (Carminati et al., 1999; Sønstebø and Holt, 2001).

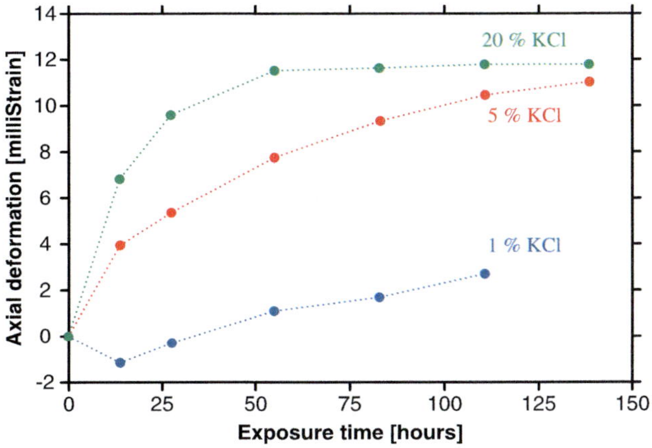

Figure 9.9 Effect of KCl brine exposure on the deformation of a shale sample under effective confining stress of 5 MPa at 80 °C (After Horsrud et al., 1998).

9.5 Borehole stability analysis for well design: incorporating effects of nonlinear elasticity, plasticity and rock anisotropy

In practical situations, boreholes are often found to be more stable than predicted by simple linear elastic theory and perfectly brittle failure mechanics analysis, as used in the previous sections. There may be a number of reasons for this, which we will discuss further below. Let us however keep in mind that a fundamental challenge in borehole stability analysis is to define failure in practical terms. The driller is only concerned with stability problems that interfere with the drilling operation. Rock fragments breaking loose from the borehole wall or a minor reduction in the wellbore diameter due to plastic flow are not necessarily equivalent to drilling problems. Guenot (1989) pointed out the need for a new "failure criterion", which can account for these practical aspects.

Such a criterion has to bridge field experience with fundamental rock mechanics knowledge. The industry learns from previous failures and may use their experience in future drilling operations. The approach may be purely empirical, but to have predictive power, it must at some stage line up with rock mechanics analysis. What we can do with rock mechanics as our main tool is to refine the theory to deal in a better way with the shortcomings identified in the field.

As said above, initiation of failure is not necessarily going to lead to a borehole stability problem. However; the brittle failure theory that we have used predicts instability as soon as the Mohr–Coulomb criterion (or any other criterion) is violated at the borehole wall. The calculations of minimum well pressure (minimum mud weight) thus implicitly assume that the rock loses its entire load bearing capacity once the shear failure limit is reached. The estimated minimum mud weight is therefore likely to be too low. A simple way to try to compensate for this is to increase the rock strength with a "fudge factor" > 1, which is calibrated on the basis of field experience, and then use the analytical approach presented in the preceding sections.

Alternatively, which is of course more satisfactory from the theoretical viewpoint, the rock mechanics model may be improved to describe the constitutive behaviour of the rock in a more correct manner. One obvious modification is to account for the post-failure behaviour of the rock. Strong rocks tend to be more brittle with little ability to deform and carry load after the peak stress has been passed. The brittleness tends to decrease with increasing confining pressure. Weaker rocks are normally more ductile, that is, they have larger ability to deform and carry load beyond their elastic limit. The brittle–ductile transition depends on the stress level. Reaching peak stress is of course more critical if the rock is brittle. If the rock is ductile, then the failure limit may be exceeded without dramatic operational consequences.

To take care of the factors discussed above, models incorporating plasticity and/or nonlinear elasticity must be applied. This represents no fundamental problem but makes the calculations more cumbersome, especially for deviated holes. When plasticity effects are taken into consideration, this gives rise to a so-called plastic zone near the borehole wall (as outlined in Chapter 4). This is a weakened zone but shields the rock beyond and to some extent improves stability. The stability limit is set either as an upper limit to the plastic deformation or to the extent of the plastic zone. This gives results that are numerically more realistic for the actual well failure situation than the simple elastic model. Such a failure criterion has however to be specified on the basis of experience, be it from field operations or controlled hollow cylinder failure tests performed in the laboratory.

Santarelli and Brown (1987) incorporated a pressure-dependent elastic modulus into the analytical model for borehole failure. Triaxial testing shows that the elastic modulus tends to increase with the confining pressure (see also Chapter 4). With this model, much of the discrepancy between a linear elastic model and results from hollow cylinder tests could be accounted for. It can also explain why failure can occur at some distance behind the wall and not at the wellbore wall surface, where the stress concentration should be maximum according to constant-modulus theory.

We have based our borehole stability analysis on the Mohr–Coulomb criterion for shear failure. This has many advantages, including the fact that analytical calculations are facilitated. However, the Mohr–Coulomb criterion does not account for the experimentally observed effect of the intermediate principal stress on rock strength, which makes the results often too conservative for practical use. The stress state in the vicinity of the borehole wall is true triaxial, and this can be accounted for by using failure criteria that depend on the intermediate principal stress, like the Drucker–Prager criterion; see Chapter 2 or the Mogi–Coulomb approach (Al-Ajmi and Zimmerman, 2006).

When more complex material models are used, more extensive rock characterisation is also required. This is often difficult due to lack of core material. In addition, the rock properties can change significantly over short distances. We should therefore in each case consider whether more complex models are worth using. This should not prevent the use of such models for studying phenomenological problems and for sensitivity analysis.

One aspect of borehole stability not considered in the analysis presented so far is the role of rock anisotropy. Shale is normally anisotropic due to textural alignment of clay minerals (Chapter 3) and therefore exhibits anisotropic strength, stiffness and permeability. Combining the weak plane model for rock strength (Section 2.9) with the borehole failure criterion, we may incorporate the effect of strength anisotropy in the modelling of hole collapse.

Fig. 9.10 shows an example of such a calculation using the same input parameters as in the isotropic stress case displayed in Fig. 9.5, except that the shale strength and friction angle both are reduced in the horizontal plane. The result is a significantly reduced stability with respect to shear failure for inclinations above 30°. For this particular case, stable drilling becomes impossible or more risky with further increase in hole angle, since the mud weight corresponding to the shear failure limit rapidly exceeds the minimum in situ stress. By adding plasticity the stability is significantly improved. We should however be aware that buckling of bedded shale when drilling along or near the symmetry plane (Økland and Cook, 1998) may be an additional cause of instability not caught by the analysis presented here.

Fig. 9.11 illustrates how different the failure process can be in an anisotropic shale with distinct bedding, depending on the angle of attack.

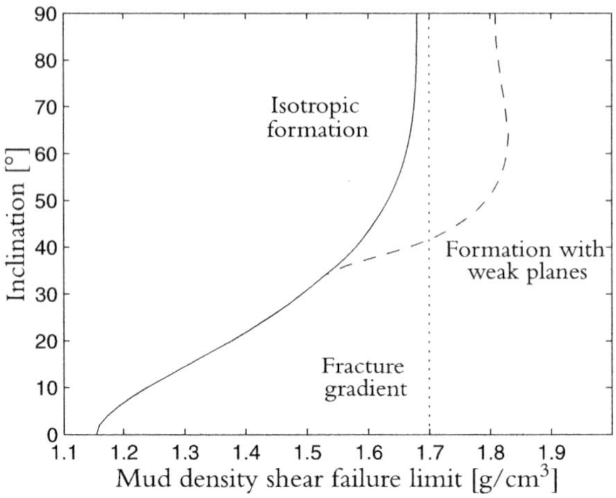

Figure 9.10 Illustration of stability analysis for a deviated wellbore at 1500 m depth. All input parameters are as in Fig. 9.5, with isotropic horizontal stresses 25 MPa. The figure shows the effect of horizontal weak planes with reduced strength ($C_0/2$) and reduced friction angle ($\varphi/2$).

The left figure shows a shale sample where the hole has been drilled normal to the bedding. The sample has been run to failure by increasing the confining pressure on the hollow cylinder. The result clearly illustrates the expected shape of the failure planes due to shear failure (see also Fig. 9.4). Similar response was reported by Meier et al. (2013) when testing a Lower Jurassic outcrop shale. The figure to the right shows another hollow cylinder sample from a different shale where the hole is drilled parallel with the bedding. This is similar to the result presented by Økland and Cook (1998). The failed zone grows normal to the bedding by a combination of shear failure at the "corners" and tensile failure of the bedding planes. It is evident from these pictures that diagnosing wellbore instability based only on shape and size of cavings is not straightforward, especially in wells inclined at intermediate angles.

Fig. 9.12 sketches the procedure of complete borehole stability analysis. The input data required are rock properties, earth stresses and pore pressure, plus the planned hole trajectory. For a simple analysis, only the parameters listed in the first row of boxes are required. For a more advanced level of sophistication, chemical, thermal, plastic, anisotropic and time-dependent features are added. In most cases the effects are simply added by superimposing poroelastic, thermoelastic and osmotic contributions to the borehole

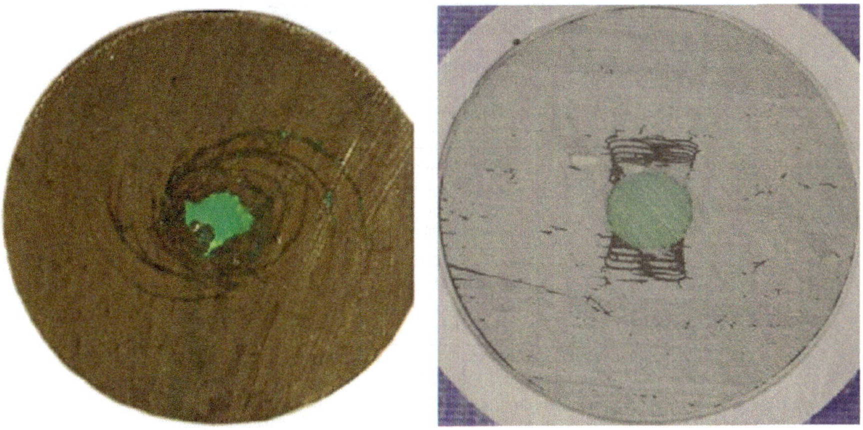

Figure 9.11 Shape of cavings for different hole orientations in anisotropic shale. The left picture is from a hollow cylinder test where the hole is drilled normal to the bedding, whereas the picture to the right is from a test where the hole is drilled parallel with the bedding. The rock sample to the left is from the North Sea, Sleipner Øst well 15/9-A-11, at 914.35 m MD in the Nordland Fm (courtesy of Equinor and Sleipner PL046 license partners). The rock sample to the right is from another Equinor-operated field (courtesy of Equinor and partners).

stresses. This may be satisfactory for most purposes, but implies that coupling between chemical and thermal processes are neglected.

The result of the analysis is the mud weight window, that is, the minimum well pressure required to prevent hole collapse (or fluid influx) and the maximum allowable well pressure to prevent loss of fluid to the formation by flow into existing or induced fractures. When these limits are known, the well may be designed.

Well design is of course more than a question of borehole stability. The purpose of the well is to reach a certain target to ensure optimum drainage of a reservoir or reservoir zone. Then the suggested hole trajectory required to reach the target has to be evaluated to see if it is drillable, and then the drilling process has to be optimised to reduce costs by drilling fast and by using as few casing strings as possible. At large depths, this is crucial: The number of casing strings has to be kept low, since the casing diameter decreases for each new string.

Fig. 9.13 illustrates the role of wellbore stability analysis in well design. Because it is not possible to drill the entire section shown with one mud weight, a casing has to be set to seal off the upper part of the section before continuing with an increased mud weight in the lower part. This illustrates

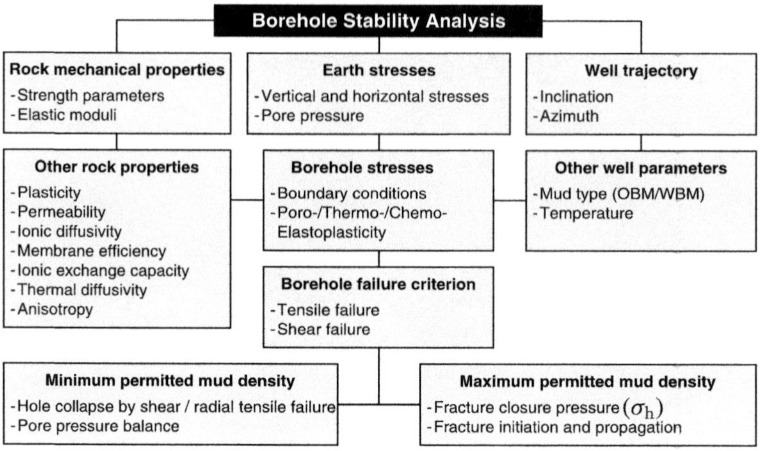

Figure 9.12 Schematic borehole stability analysis.

two main tools available to drill stable boreholes, the mud weight and the casing program.

9.6 Use of pressure gradients

9.6.1 Introduction

So far we have largely used pressure in our stability calculations. However, since this has to be related to a mud weight or mud density, it is customary to convert the different pressures at a given depth to a density value. This is often referred to as an equivalent mud weight or a pressure gradient. Note that the term gradient is not entirely correct, since it is not the slope of the pressure curve at a given depth, but rather a secant value from surface to the given depth, that is,

$$\rho = \frac{p}{gD} \tag{9.22}$$

where p is the pressure at depth D, g is the acceleration of gravity, and ρ is the equivalent mud weight or pressure gradient. Common units for ρ are g/cm^3 or SG (specific gravity relative to water).

In SI-units we have

$$\rho \text{ [SG]} = \frac{p \text{ [MPa]}}{0.00981D \text{ [m]}} \tag{9.23}$$

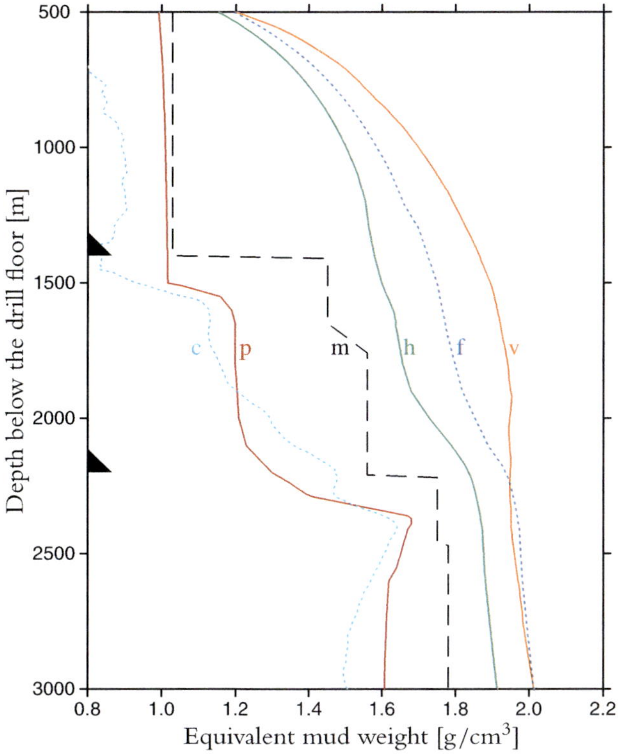

Figure 9.13 Example of a stability chart for a well from the Norwegian Continental Shelf. The full lines are, from left to right, the estimated pore pressure gradient (p), the minimum horizontal stress gradient (h) and the overburden stress gradient (v), respectively. The dotted lines are the estimated shear failure gradient (c) and fracturing gradient (f), whereas the dashed line is the planned mud weight gradient (m). Casing shoes are indicated by the black triangles. Observe how the mud weight is increased below each shoe. Further note that for this particular well, the shear failure gradient is larger than the pore pressure gradient for only a limited interval. This may be different in other wells. The airgap is 18 m, and the water depth is 370 m. Courtesy of Equinor.

Any relevant stress or pressure parameter, for example, pore pressure, horizontal stress, overburden stress and fracture pressure may be represented as a gradient, as exemplified in Fig. 9.13.

9.6.2 Depth reference and depth corrections

The use of Eq. (9.22) introduces the need of a depth reference level, that is, a surface reference level. Since the well pressure gradient is related to the mud column, it is natural to use the same depth when calculating the other pressure gradients. The surface reference level will then obviously be the drill floor or rotary table (RT), that is, the top of the mud column. When drilling an onshore well, it is then quite clear that D is essentially measured from the Earth surface. When drilling an offshore well, D is however made up of several depth elements:

$$D = D_{formation} + D_{water} + D_{airgap} \tag{9.24}$$

where D_{airgap} is the air gap or the height from the sea to the drill floor, D_{water} is the water depth, and $D_{formation}$ is the formation depth (from the sea floor to the depth of investigation).

When comparing gradients from different wells in a field, it is therefore essential to keep track of the reference depth. In the exploration phase the variation in D_{airgap} is normally small (typically less than 5 m), as these wells are normally drilled from semisubmersible rigs with D_{airgap} typically ranging from 23 to 28 m.

In production wells, D_{airgap} depends on how the field is developed. If a fixed platform is installed, then D_{airgap} may be significantly larger. Thus we need to be able to convert from one drill floor level to another. The pressure at the same depth underground must be the same, so that

$$\rho_1 g D_1 = \rho_2 g D_2 \tag{9.25}$$

where the subscripts represent different reference levels. This yields

$$\rho_2 = \rho_1 \frac{D_1}{D_2} \tag{9.26}$$

To avoid keeping track of different rig floor elevations, it is convenient to use a more general reference level for the different pressure gradients, for example, the mean sea level (MSL). This is a convenient approach when collecting and comparing data from different drilling rigs, data that will be the basis of a model that is going to be used for planning of production drilling from a platform.

Having established the different pressure gradients used as input for stability analysis, we must always refer back to the same reference level as the mud column (the rotary table). When planning a specific well, we must therefore remember to convert back to a depth that includes the air gap D_{airgap} of that specific well.

9.7 Beyond simple stability analysis

9.7.1 Field cases: the borehole stability problem in complex geology

In this section we present a few field cases from the literature, which illustrate how rock mechanics analysis can be applied to borehole stability problems.

9.7.1.1 Case 1: drilling-induced lateral shifts along pre-existing fractures (Meillon St. Faust Field, France)

The Meillon St. Faust is a gas field located in the southwest of France. For a more detailed description, the reader is referred to Maury and Zurdo (1996). Severe casing failures occurred in a fault zone above the reservoir during early gas production. These failures could be interpreted by migration of gas or water from abnormally pressured zones at depth through damaged cement and into uphole faults or bedding joints. This increased pore pressure led to reduced effective normal stress on the faults, which triggered shear displacement and thus casing collapse.

It is likely that pressure increase in pre-existing fractured rock mass may also cause stuck pipe situations during drilling by a similar mechanism (Santarelli et al., 1992). If the well is drilled with too high overbalance (like in high pressure cycles during surge-swab; see below), and the mud does not create an efficient seal towards the fracture, then the pressure may build up in the fractures, and the normal effective stress decreases, leading to shear displacements that may reach a few cm. There are cases where tight hole occurred with no indication of cavings and where the problem could not be solved by mud weight increase. Other symptoms of this failure mode is that tight hole problems occur at the same depth, together with problems to run in and pull out of hole, abnormal localised torques, deviation of trajectories, and possible drillstring failures. The remedies to solve problems of this kind are to keep the mud weight low (but still above the shear failure criterion for the intact rock) and to increase the sealing capacity of the mud. In the field case reported here, this was done by adding asphalt-type products associated with inert colloidal particles.

9.7.1.2 Case 2: drilling in a complex geological setting with high tectonic stresses (Cusiana Field, Colombia)

The Cusiana field is located in the foothills of the Andes Mountains in Colombia. Major borehole stability problems (amounting to several millions of dollars per well) have been encountered (a brief summary was published by Last et al., 1996; see also Willson et al., 1999). They report hole enlargement as the cause of the majority of problems, resulting in large amounts of cavings, hole cleaning problems and stuck pipe, poor cementing and necessary side-trackings. The most difficult formation was an alternating sand-shale sequence, where mud losses and tight hole occurred in the sandy units and hole enlargements in the shaly units. It was found that increasing the mud weight to limit hole enlargements did not work, because it led to significant mud losses.

A borehole stability task force established by the operator acquired a large amount of data. Stress measurements were done, downhole borehole images and 4-arm caliper data were collected, and shale core tests were performed in addition to characterisation and monitoring of cavings at surface and MWD surveys. An important result was the determination of the in situ stress state, which proved the maximum principal stress to be (sub-)horizontal and much larger than the vertical stress, and the minimum principal stress to be horizontal (parallel to the mountains) and much smaller than the vertical stress. In addition, the sand-shale sequence mentioned above turned out to be fractured throughout the entire field. It was concluded that borehole mechanical instabilities occurred as a result of shear failure in weak fractured siltstones and shales and that lost circulation occurred when the mud pressure exceeded the minimum (horizontal) stress. Further, mud invasion contributed to weakening the fractured rock (by the same mechanism as explained in the preceding paragraph). An important consequence of these mechanisms was that the permitted mud weight window was shrinking to zero, since the minimum mud weight in the shale sections exceeded the lost circulation threshold in the sands. The challenge was then not to find a way to drill a stable hole, but rather to find a way to drill an inherently unstable hole.

Compromises were made by using mud weights that were too high for the sands and too low for the shales and then trying to manage the instabilities by efficient drilling practices. This included improvement of mud sealing capacity and optimisation of hole cleaning, and having lost circulation pills available in case of loss in the sands. The bit was allowed to seek its own trajectory where possible, and drilling was performed as smooth and fast (minimising reaming, back-reaming and vibrations, increasing bit life) as possible. In addition, hole conditions (including caving rates) were carefully monitored.

9.7.1.3 Case 3: the Heidrun Field, Norwegian Sea

The Heidrun field offshore Mid-Norway experienced persistent hole stability problems in Tertiary shale formations. A high KCl content drilling fluid was used to handle possible reactive clay formations above the reservoir. The majority of the wells are extended reach wells with inclination angles in the range 55°–70°. The problems manifested themselves mainly as cavings, pack-offs and lost circulation, and were handled by increasing the mud weight. However, in several cases, this did not reduce the caving production problem. The details of this case are reported by Stjern et al. (2000).

The operator expected the drilling problems to occur mainly in a low-density and high-porosity zone. Extensive data collection was performed, including logging (6-arm caliper, sonic and density, temperature), shale coring and laboratory testing, and sampling and measurements of sonic velocities on cuttings. In addition, the stress field was evaluated using extended leak-off tests.

As a result of the systematic data collection, it was found that the problem zone was not the originally assumed low-density zone, but rather a normal density but fractured zone. This led to the conclusion, like in the preceding cases, that the mud density could be reduced. At the same time the salt concentration in the mud was reduced, which apparently also improved stability. It was also concluded that hole cleaning had to be optimised by using low-viscosity mud and provide close to turbulent flow. The authors point out that the borehole stability study performed has given a cost reduction of close to 2.5 million dollars per well in pure saving on flat time during drilling.

9.7.2 Drilling below deep water

When drilling below deep water, the water depth contributes to reducing both the shear failure and the fracturing gradients. The well pressure is given by Eq. (9.1), where the total depth includes water depth from the top of the mud column (neglecting the air gap). This means that the critical mud density $\rho_{\text{w,crit}}$ for shear failure and fracturing will be given by the corresponding critical mud densities in absence of water ($\rho^0_{\text{w,crit}}$) and by the water depth (D_{water}) and the depth measured from the sea floor ($D_{\text{formation}}$) as follows:

$$\rho_{\text{w,crit}} = \frac{\rho^0_{\text{w,crit}} D_{\text{formation}} + \rho_{\text{water}} D_{\text{water}}}{D_{\text{formation}} + D_{\text{water}}} \tag{9.27}$$

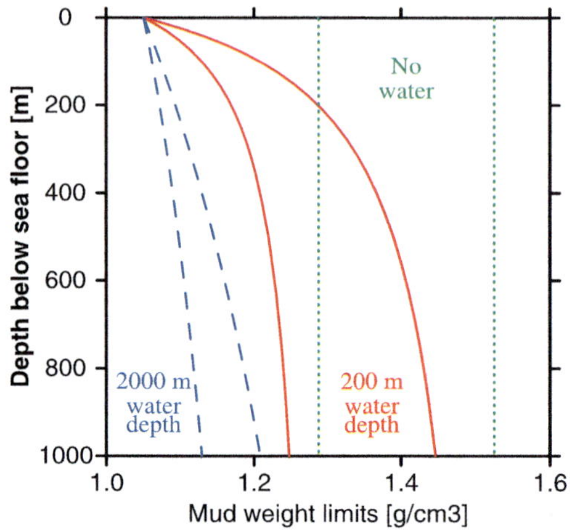

Figure 9.14 Illustration of the effect of water depth on the mud weight window. Note the narrow window near the sea-floor in case of deep water.

As can be seen from this equation and from Fig. 9.14, the main effect of the water is shrinking the mud weight window at shallow depth within the formation. At the sea floor ($D_{\text{formation}} = 0$), there is no mud weight window, since both the upper and lower limits are given by the water density (effective stresses = 0). At shallow depth below the sea floor the effect of water depth still dominates, so that both the upper and lower mud weight limits are lower than they would have been if drilling took place onshore or in shallow water. The consequences of instability in deep water are however much more severe.

A potential solution to the drilling problems is the so-called dual gradient or dual density drilling. The basic idea is to use only seawater above the sea floor and the heavier mud from the seafloor and down into the formations. This is achieved by using a separate mud lift system to take care of the return mud and cuttings from the borehole. Another useful approach is to apply managed pressure drilling (MPD) to reduce the ECD-effect. MPD is a closed-loop circulation system where the drilling fluid density is supplemented by a surface backpressure. This backpressure can be adjusted much faster in response to downhole conditions compared with changing mud weights conventionally.

In some areas, a special drilling problem known as shallow water flow (Alberty et al., 1999; Furlow, 1998) has occurred, with high resulting costs. This problem occurs in uncemented and overpressured sands, so the main cause of the sand collapse is an underbalanced situation. This problem is as such more related to sand production (see Chapter 10).

9.7.3 Surge and swab effects

When the drill string is pulled out or run into the hole, this will cause cyclic loading of the rock near the borehole. The string acts as a piston in the hole because the mud cannot flow without restriction, and hence the well pressure changes. How large this effect will be depends on the tripping speed and the mud viscosity. Bourgoyne et al. (1986) have presented equations needed to calculate this effect.

We may foresee the following effects of cyclic loading on boreholes in shales:

1. The failure stress is directly exceeded on a low pressure cycle. The duration and magnitude of the low pressure cycle will influence the severity of the failure. This could apply both to pressure cycles causing failure in underbalance or shear failure in overbalance.
2. Even when the well pressure oscillates within the failure limits, the delayed pore pressure response could lead to an increased failure tendency from cycle to cycle. Although initially within its failure limits, the borehole may fail due to pore pressure increase near the borehole wall.
3. Stress cycling may also lead to rock fatigue, depending on the amplitude and the number of cycles.

Some of these issues were discussed in more detail in Section 9.3.1.

9.7.4 Hole cleaning

As mentioned in Section 9.1, hole cleaning and borehole stability problems may often be difficult to distinguish from each other. Hole cleaning problems may be caused by borehole instabilities, leading to production of noticeable amounts of cavings. Formation fragments (both cuttings and cavings) will tend to accumulate in hole enlargements, because of reduced flow rate in such places. These fragments may fall or migrate back into the hole, especially when pumps are stopped. This will cause restrictions in the mud flow and may also stick the drill string if the fragments are large enough. The depth where the string gets stuck may thus be in gauge sections below the unstable zone, and not at the depth of failure. We have seen above that although increasing the mud weight appears to be the most obvious solution, this may in practice not always be recommendable.

High rheology mud (high viscosity, high yield point (YP)), together with high annular velocities (AV), has in general been accepted as the correct philosophy for cleaning of near-vertical wells. Cleaning high angle wells is, however, not straightforward, and the same philosophy does not readily apply to this situation. Shaw and Sutherland (1988) report the use of low-viscosity sweeps (using seawater) together with frequent wiper trips of varying length as an appropriate approach in deviated wells. Seawater goes into turbulence at lower velocity than normal mud, thus picking up cuttings at lower annular velocity. This also minimises hole washouts in weaker formations. In large diameter holes, larger drill pipe has been used, allowing higher flow rate with the same injection pressure. Of course, a problem here may be that low mud weight is detrimental to borehole stability.

Both experimental studies and field experience indicate (e.g. Sewell and Billingsley, 2002) that hole cleaning problems are most severe around intermediate inclination angles (40°–60°). It is important to minimise caving production due to borehole instability. If that problem can be overcome, hole cleaning can be satisfactorily obtained by using appropriate operational procedures. Essentials here are: mud rheology, casing programme, string rotation, flow rate and general drilling practices. Use of low/high-viscosity pills should be considered with respect to the effect on the equivalent circulation density (ECD), the effects on surge/swab pressures and eventually the effect on borehole stability.

9.7.5 Amount and quality of input data

Wellbore stability prediction is currently a routine exercise in planning of wells. As we have seen, there are models available for most relevant processes (e.g. plasticity, time, chemistry, temperature, anisotropy). Numerical modelling gives the possibility to include geometrical effects on a field scale, together with effects of production, both in the reservoir and in the overburden (e.g. Kristiansen and Flatebø, 2010; DeGennaro et al., 2017). However, almost all examples show that in practical use, it is difficult to acquire all the data needed to perform an analysis with sophisticated models. This is probably the reason why most analyses are still based on linear elastic theory. The user should however be aware of the limitations given by the elastic analysis and be able to evaluate the importance of these in relation to the rock type, its deformational behaviour and so on. Given the limitations of the model and the available data, one possible approach is to calibrate the model with field results.

The input data required for borehole stability analysis are shown in Fig. 9.12. These include Earth stresses, pore pressure and rock properties. When a well is drilled in an unexplored area, there is virtually no information available that may allow estimation of the required parameters. From regional geological knowledge, seismic data and experienced trends of variation in rock parameters and stresses with depth, it may still be possible to establish a first guess of the stability limits. When a few wells are drilled, additional sources of information also become available, and we may try to design a field scale mechanical Earth model. Clearly, the more data are collected, the better will be the quality of the stability prediction. Data collection is however time consuming and expensive, and the possible gain is often not considered during field planning. Ideally, cores should be taken, and the failure envelope determined experimentally. For shales, this requires tedious tests, like the CU triaxial tests described in Chapter 7. Furthermore, extended leak-off tests (Chapters 8 and 11) should be performed so that the horizontal stress field (including orientation) is known. Finally, pore pressure needs to be estimated, which is very difficult in shale sections. We have to rely on identification of abnormal pressures by studying deviations from expected trends in MWD or log data. The accuracy of the existing pore pressure prediction procedures may be questioned. High pore pressures ahead of the bit may in principle be identified from seismic or sonic while drilling measurements.

If less than required data are available, then borehole stability analysis needs to be based on educated parameter guessing. For this purpose, systematic analysis of available core and log data may be used to establish correlations between mechanical properties and more easily measurable index properties. For instance, shale strength shows good correlation to porosity, to sonic wave velocities and to hardness by indentation (Horsrud, 2001; see also Chapter 3). To obtain these parameters, we may use drill cuttings or caving fragments for more rapid—possibly rig-site—measurements (Santarelli et al., 1998; Nes et al., 1998) and estimation of shale strength. Held together with drilling parameters and pressure while drilling (PWD) measurements, and fed into an appropriate model, better estimates of the mud weight limits for borehole stability may be obtained in real time or quasi-real time. The mud weight is however only one piece in the puzzle required for full automation of the drilling process and the choice of drilling parameters.

Another possible approach when data are missing is calibrating against experience. The obvious choice is to use incidents with instability when calibrating. However, as we have seen, correct diagnosis of the root cause of the problem is not always straightforward. An alternative or supplement is to also use cases where problems have not occurred and where the prediction of a stable borehole may have been challenged. A certain combination of, for example, well inclination and mud weight can then—through back-calculation—provide a calibration factor larger than one to the rock strength. This calibration factor can then be applied to later predictions in the same formation. By systematically challenging the stability limits the predictions can be moved towards the actual stability limit. This is especially applicable to development drilling of a field.

9.8 Stability issues in different lithologies

Most instabilities occur in shale, but other lithologies can also be problematic under certain conditions.

9.8.1 Sandstone

Experience has shown that even very weak sandstone can be drilled successfully without indications of shear failure. There are several possible reasons for this:

- When drilled in overbalance, a filter cake can be established that effectively creates support to the formation. This is illustrated in Fig. 9.15 with two Mohr's circles, one without overbalance representing the situation in shale (no filter cake) and one with 5 MPa overbalance representing the situation in sandstone (with filter cake). This example is for a vertical well with equal horizontal stresses. The support from the drilling fluid thus increases the radial stress and at the same time reduces the tangential stress with the same amount. The total effect is a stress situation that is moved away from failure. Even if the sandstone is weaker, it will be stable according to the figure, whereas the shale is cutting the failure line, risking shear failure to occur.
- The friction angle is often higher in sandstone than in weak shale. As illustrated in Fig. 9.15, this also brings the stress situation further away from the failure line.
- Chemical effects, temperature effects and time effects are not present in sandstone in the same way as in shale.

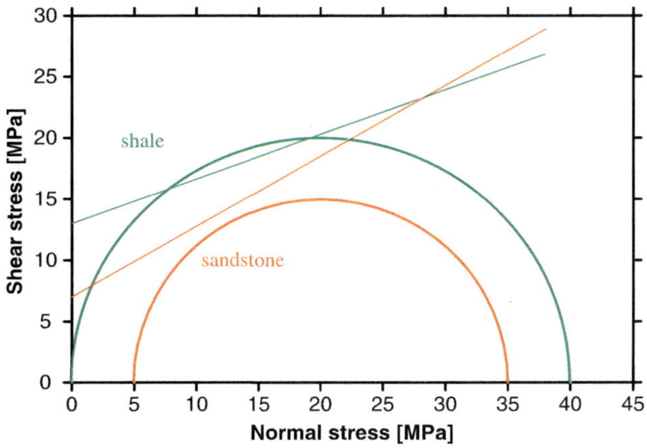

Figure 9.15 Illustration of drilling fluid support at the wellbore wall. No overbalance is representing the situation in a shale (largest Mohr's circle), whereas 5 MPa overbalance represents the situation in sandstone with a perfect filter cake. Two straight lines illustrate the failure lines in sandstone and shale.

9.8.2 Carbonate rock

Many carbonate rocks have a double porosity system, a low-permeability matrix together with natural fractures, vugs and caves (also known as karst) (Chilingarian et al., 1992). These carbonates are usually quite stiff and strong, so instability due to shear failure is rarely a problem.[1] However, the porosity in fractures and vugs represent potential pathways for loss of drilling fluid (see e.g. Tangen et al., 2019). Such losses are hence primarily a result of overbalance, that is, the difference between mud weight and pore pressure. Then drilling-induced fractures due to tensile failure are not the root cause of the problem.

Since the openings representing the secondary porosity can be up to tens of centimetres in size, such losses can be severe and difficult to control by conventional methods. Some losses may be unavoidable, but measures can be taken to minimise the losses: low overbalance, low ECD (e.g. using Managed Pressure Drilling (MPD) methods) and addition of LCM (Lost Circulation Material) to the drilling fluid.

[1] In this discussion of carbonates the primarily soft chalk deposits of Western Europe are hence not included (see Section 3.4.2).

9.8.3 Salt

Drilling through intervals of salt is most often the result of exploring reservoirs beneath salt. Such reservoirs are found in several regions of the world, for example, in the Gulf of Mexico, offshore Brazil and offshore West Africa. The salt overlying these reservoirs can vary substantially in thickness, from a few hundred metres to several thousand metres (2000–3000 m). The most common constituent of these salt layers is rock salt (halite) (Section 3.4.4).

Rock salt (halite) and other evaporitic salt minerals (e.g. sylvite, carnallite, tachyhydrite) have certain characteristics that affect the drilling process. The porosity can be considered so low that the pore pressure is zero. Another special feature is the significant ability of the salt to creep (viscous flow). The creep rate depends strongly on the temperature (Section 3.4.4).

One consequence of creep is that the stresses in the salt have evolved to be isotropic and equal to the vertical stress. This is beneficial in the sense that inclination in principle has no effect on the choice of mud weight. The mud weight is determined primarily by the creep process. Mud weights of up to 80–95 % of the vertical stress are commonly used to ensure a hole without restrictions until the casing has been set. Since the creep rate is different for different salts (Costa et al., 2010), it is important to map the entire salt sequence, as this rarely consists of halite only.

In addition to other evaporitic minerals, the salt sequence can also contain inclusions of other lithologies (sandstone, shale, tar). This can introduce problems like instabilities, losses, kicks, stuck pipe and so on (Chatar et al., 2010).

A generalised plot of pore pressure and stresses down and into a pre-salt or sub-salt reservoir is shown in Fig. 9.16. Pre-salt means that the section is stratigraphically older than the salt, whereas sub-salt means that the section is structurally below the salt but can represent a sediment that is younger than the salt.

It is obvious from this figure that the transition from the salt to the reservoir can be challenging. The salt requires a high density of the drilling fluid, whereas the same density in the reservoir may result in losses.

It is also interesting to note the change in slope for the overburden gradient through the salt. This is due to the lower density of the salt, typically $2.16\,\mathrm{g/cm^3}$ for halite.

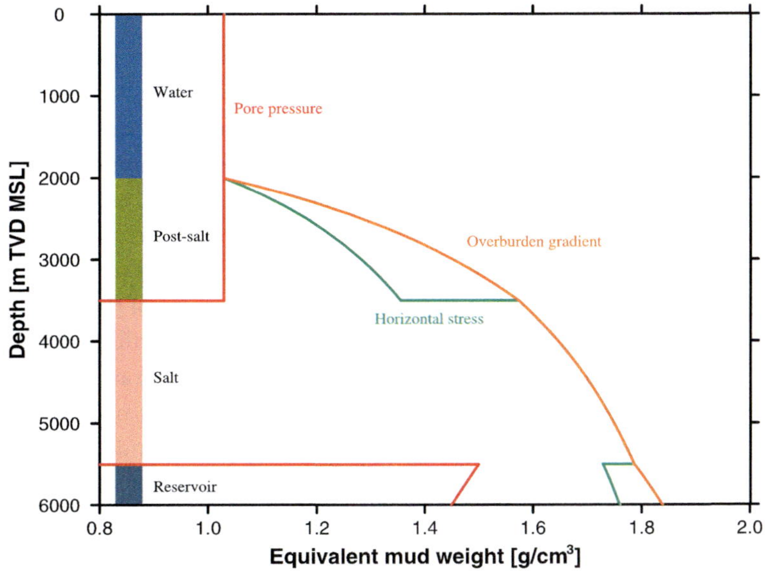

Figure 9.16 A generalised example showing pore pressure and stresses in a sequence of post-salt (shale), salt and pre(sub)-salt (reservoir).

Several potential hazards can thus be present: uncertainty in depth, uncertainty in pore pressure beneath the salt, and rubble zones.[2]

The salt body has a stress regime which can be very different from surrounding formations. The formation in the vicinity of the salt is therefore affected, especially with respect to direction of the principal stresses. Principal stresses tend be aligned normal to and parallel with the boundary surface. It is therefore recommended to penetrate the surface at right angle to the surface, to reduce the risk of instability in the formation beneath or outside the salt. Penetrating the surface at an angle is analogous to drilling an inclined well in a normal stress regime.

It is important that the drilling fluid used in the salt prevents or minimises leaching of the salt and consequently washouts in the salt. Whether the drilling fluid is water-based or oil-based, the salinity of the water phase should be at or near saturation (Dusseault et al., 2004).

[2] Rubble zone is a sediment just beneath the salt that has been affected by the movement of the salt, typically resulting in fracturing and/or crushed rock material. The rubble zone is thus potentially unstable, and loss of drilling fluid can also occur.

9.8.4 Coal

Coal is drilled for gas production (coal-bed methane) but can also be encountered as sequences within other rocks, for example, a sandstone reservoir. Coal is characterised by a system of cleats and fractures, where these occur mostly perpendicular to each other and also perpendicular to the bedding plane. Coal is in itself not especially weak, but the fracture system can significantly affect the stability. A standard wellbore stability prediction is still required to account for basic shear failure stability. However, as with other fractured rocks (Section 9.7.1), continued increase of the mud weight can have an adverse effect on stability due to invasion of drilling fluid filtrate in the fractures. It is important to avoid pressurising the fracture system, as this will reduce the effective stress and increase the risk of mobilisation.

The effect of the fracture system must therefore be handled more operationally than by modelling alone. As with drilling of all potentially weak rocks, it is important to reduce the impact of the drilling process itself, both mechanical and hydraulic impact. An example of such an approach (low-energy drilling operations, LEDO) applied to paleosols is presented by Cook et al. (2014). Another remedial measure is obviously to prevent drilling fluid from entering the fractures. This can be done by adding particles (LCM) to the drilling fluid.

9.9 Drilling in depleted reservoirs

A particularly difficult situation is infill drilling in depleted reservoirs, where the pore pressure may still be the initial one in shale zones or in the cap rock right above. This is due to the very low permeability of shales, which means that the time required to establish pore pressure equilibrium in a shale zone of tens of metres thickness may be several years and easily may exceed reservoir lifetime. In the permeable reservoir zones the reduced pore pressure is established almost instantaneously (except for possible undepleted pockets). Thus the high pore pressure in the shale requires that the mud weight is kept above the initial shear failure limit to avoid shale instabilities. In neighbouring reservoir sand zones, this mud weight may be too high, since the reduced pore pressure also has caused reduced horizontal stresses (reduced fracture gradient). Thus drilling with initial mud weight may lead to mud losses. Normally, we do not know while drilling which zones are depleted and which zones are at original pore pressure: This requires a coupling to reservoir management and reservoir monitoring by, for example, time-lapse (4D) seismics.

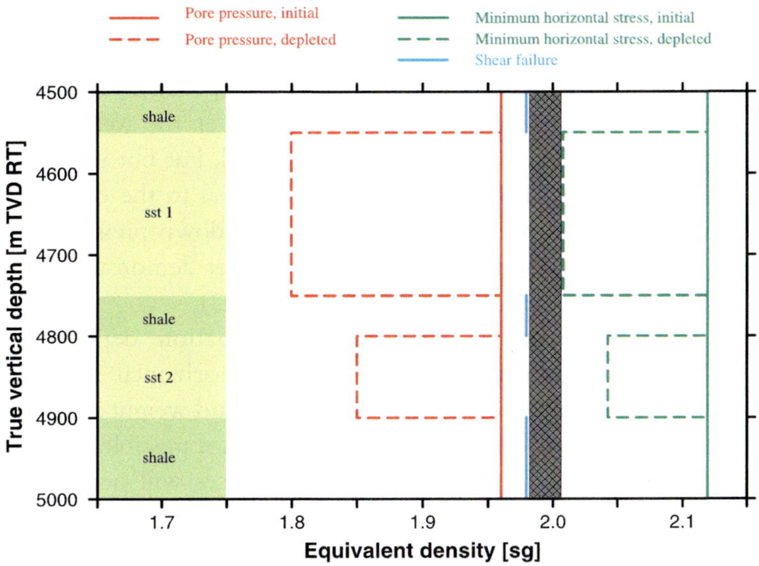

Figure 9.17 Illustration of depletion effects in a reservoir section with both sandstone and shale intervals. The hatched area represents the available mud weight range.

Unless maintained by injection, the reservoir pressure will thus decrease as a result of production. This situation is illustrated schematically in Fig. 9.17, with two intervals of sandstone with different depletion. The hatched rectangle now represents the mud weight range that is available for maintaining a stable borehole. Further depletion will eventually lead to a potential conflict where the hatched rectangle disappears. The minimum horizontal stress continues to be reduced, requiring a reduction of the mud weight to avoid fracturing and losses. However, further reduction of the mud weight will eventually result in borehole instability and/or an influx. Thus there is apparently a theoretical limit to the depletion at which further drilling into the reservoir is impossible.

In the case illustrated in Fig. 9.17 the upper mud weight limit is given by the minimum horizontal stress. That is not necessarily the case in practice, since a loss situation has several levels of varying risk depending on the mud weight (Section 9.2.3 and Chapter 11). To initiate losses, the fracture breakdown pressure (FBP) has to be exceeded. This can be significantly higher than the minimum horizontal stress. Estimates can be obtained using the linear elastic approach presented in Chapter 11. However, as pointed out in Section 9.2, exact values are difficult to predict using this and other analytical equations.

An important boundary condition is that a proper filter cake is created, so that there is no pressure communication between the wellbore and the pore space. This is achievable in a permeable sand, but not in a shale with very low permeability. FBP is directly proportional to the effective stress, so increasing the pore pressure reduces the breakdown pressure. The importance of the fluid and fluid properties is further demonstrated through the distinction between FIP and FBP (Section 11.2).

If a fracture is created, then the loss situation depends on the ESD/ECD-values compared to the minimum horizontal stress and the fracture propagation pressure (FPP). If the static mud weight is less than the minimum horizontal stress, then static losses are not possible, but dynamic losses can occur if ECD exceeds FPP. Static losses will occur if ESD is higher than FPP (see also discussion in Section 9.1.2).

The real situation is however more complex:

- The reservoir section contains different lithologies including also potential weaknesses in lithology boundaries.
- The depletion in each reservoir zone is not known exactly. There is always a possibility that zones with initial reservoir pressure are penetrated. Therefore the static mud weight has to balance the initial reservoir pressure.
- The stress path is in general only an estimate from modelling. It is not constant, but tends to vary both with time and space (see also Section 12.2.3). It is recommended to do stress measurements when possible to calibrate the models and predictions.

If a situation with total mud loss occurs, then the consequences can be severe. Unless the losses can be controlled, the mud support in the hole is reduced significantly. This can result in collapse of shale in the cap rock and/or intra-reservoir shale, depending on what is exposed in the section. Another possible consequence is crossflow between reservoir intervals. This occurs if the mud pressure has dropped below the pore pressure in a reservoir zone with high pressure, allowing fluid to flow from this zone along the wellbore and into a zone with lower pressure. This is also known as an underground blowout. It can be both difficult and costly to control an underground blowout, and it may be necessary to drill a separate well to kill and control the blowout.

Experience shows that reservoirs at initially high pressure can be drilled successfully in depleted conditions with as much as 500–600 bar overbalance (Fambon and Joffroy, 2008; Ay et al., 2016; DeGennaro et al., 2017). This requires careful planning and execution applying the correct measures for risk reduction. Some key issues are listed below.

- Geomechanical modelling is recommended for prediction of deformations and stress changes within and outside the reservoir along the planned well path. It should be noted that the effects can vary in the field, making it necessary to investigate several different well paths.
- Careful design of the drilling fluid and especially the particle content has proven to be an essential success factor for drilling of depleted reservoirs.
- For the execution phase, managed pressure drilling (MPD) and the use of drill-in liner are available technologies that should be considered.
- A pilot hole penetrating the depleted reservoir should be evaluated. A primary purpose is to measure the pore pressure. If possible, a stress test to establish the minimum formation stress and fracturing pressure should be included.

Depletion of several hundred bars is also known to result in structural changes in the reservoir and its surroundings, especially in the close overburden. Stress alterations in the overburden, also known as stress arching (see Section 12.3.2), can cause re-activation of faults, collapse of casing and loss of production (Section 12.5).

In a real situation the transitions between different lithologies will not be as sharp as in Fig. 9.17. When the depletion becomes significant—and takes place over some time—the pore pressure in the shale will also be affected, reducing the pore pressure a few metres into the shale. The pore pressure can also change as a result of changes in the mean stress under undrained conditions. This is described in Section 1.6.7 introducing the Skempton coefficients A and B. A pore pressure reduction is most likely, but an increase may occur (Holt et al., 2018).

9.10 Shale as a barrier

After drilling, deep wells are usually lined with a steel tube (*casing*) to prevent borehole instabilities from disturbing well operations. To facilitate the installation procedure, the radius of the casing R_c is smaller than the borehole radius R_w, typically, $R_c/R_w \approx 0.75$–0.8 (Fig. 4.31). The annulus between the casing and the rock is filled with cement for some distance above the casing shoe; however, long sections of the annulus may be left open after completion. Open sections are potential passageways for migration of hydrocarbons, which may eventually reach the surface and cause leakage from the well. To prevent such leakage, it is necessary to establish throughgoing barriers in some sealing formations along the well. This is particularly important prior to well abandonment, as a leakage need not appear immediately, but rather reach the surface at a later stage and go on unnoticed from an abandoned well.

Establishment of a throughgoing barrier implies that also the annulus has to be properly sealed off in the barrier sections, which should be at least 30–50 m long according to regulations. The standard method to achieve this is to fill the annulus with cement, either through perforations in the casing or after removing the casing through a milling process. This is an expensive and time-consuming process. However, in some shale formations the annulus apparently closes by itself after completion (Williams et al., 2009; Kristiansen et al., 2018), forming what is known as a *shale barrier*.

The existence of a shale barrier is usually revealed by cement bond logs, which are designed to identify the state of matter (solid, liquid or gas) of the material behind the casing. Pressure integrity tests or pressure communication tests are used to verify the sealing efficiency of the barrier. Note that verification of a hydraulic barrier outside the casing is challenging, whether it is a shale barrier or a cement barrier. Interpretation of cement bond logs can be ambiguous, whereas the resolution of pressure integrity and communication tests is relatively poor.

Shale is in most ways the ideal barrier material. It has the same consistency and largely the same properties as the material that covered the whole rock section prior to drilling and made it a sealing formation. It exists in endless amounts right next to the well, and a strong force—the in situ stress—is available to move it into position. Annulus closure may occur if the difference between in situ stress $\sigma_{\rm h}$ and annulus pressure $p_{\rm w}$ is sufficiently large, and either the shear stiffness or the shear strength of the rock is sufficiently low. This can be seen from Eq. (4.178), and the requirement that the stress on the casing $\sigma_{\rm c}$ must exceed the pore pressure, as discussed in Section 4.9. In weak formations, a significant overbalance may be needed to keep the hole open until the casing is in place.

After the casing has been installed, the annulus is sealed off from the well by a cement plug, and the annulus pressure will move towards a final state in equilibrium with the surroundings. The borehole wall is not protected by a mud cake in shale formations, hence the final state for the annulus pressure is likely to be the in situ pore pressure $p_{\rm fo}$. Thus the overbalance needed to keep the hole open during drilling disappears after completion, and the difference between in situ stress and annulus pressure may become sufficiently large to close the annulus. From these considerations and a combination of Eqs. (4.178) and (4.185) we find that closure of the annulus may happen if

$$\sigma_\mathrm{h} - p_\mathrm{fo} > \min\left[2G\left(1 - \frac{R_\mathrm{c}}{R_\mathrm{w}}\right), \sqrt{C_0 G\left(1 - \frac{R_\mathrm{c}}{R_\mathrm{w}}\right)}\right] \tag{9.28}$$

The first case of min[..., ...] represents a situation where the shear stiffness G of the shale is so low that the annulus closes while the rock around the hole remains in a pre-peak condition.

As a typical example, we may choose $\sigma_\mathrm{h} - p_\mathrm{fo} = 10\,\mathrm{MPa}$ and $R_\mathrm{c}/R_\mathrm{w} = 0.75$ and find that annulus closure may happen if $G < 0.02\,\mathrm{GPa}$. This is much lower than the *elastic* shear stiffness of even the softest shales. However, the static stiffness that involves a combination of elastic and plastic deformation (Section 5.6.1) may be sufficiently low for some shales.

It is important to keep in mind that annulus closure is a necessary, but not sufficient, requirement for the formation of a shale barrier. The second case of min[..., ...] in Eq. (9.28) represents a situation where a part of the shale around the hole has to be deformed beyond peak stress to close the annulus. This enhances the probability for formation of fractures during the annulus closure process, so that the barrier may not be sealing properly even if the annulus is closed. If the shale suffers brittle failure, then the material filling the annulus will be heavily fractured and may remain highly permeable unless it is strongly compacted or sealed off in other ways (Kristiansen et al., 2018; Holt et al., 2020). The margin with which the inequality of (9.28) is fulfilled represents the net stress on the casing and hence the compactive stress acting on the material filling the annulus. The larger this margin is, the more likely it is that the barrier is sealing. (Note that Eq. (9.28) is a simplified expression—see Section 4.9—and that friction angle as well as post-peak strength parameters also have an impact on the conditions for annulus closure.)

Both the shear stiffness G and the compressive strength C_0 tend to decrease with time due to creep (Sections 1.9.4 and 2.10.1), and hence the conditions for the formation and sealing efficiency of a shale barrier as expressed by Eq. (9.28) tend to change favourably with time. A shale barrier is not exposed to deterioration either, as it is in chemical equilibrium with its surroundings, in contrast to the alternative, a cement barrier. Note however that the creep process is largely reduced when the shale establishes contact with the casing (Xie et al., 2019).

Eq. (9.28) is based on the assumption of isotropic in situ stress, and involves σ_h as the only stress parameter. The situation is more complex when the in situ stresses are anisotropic, as indicated by Eq. (4.184). The driving force in the formation of a barrier is the compressive stress normal to the borehole axis, and the lowest value this stress has along the perimeter of the hole has a critical impact on the sealing efficiency of the whole barrier. For an inclined well oriented parallel to the minimum horizontal stress σ_h, the lowest value of the compressive stress normal to the borehole axis is equal to σ_h for zero inclination and increases with inclination until it becomes equal to the largest horizontal stress σ_H (assuming that $\sigma_v > \sigma_H > \sigma_h$). This effect is reduced for other well azimuths, but in general, it tends to enhance the probability for shale barrier formation for inclined wells.

Whereas the mechanical properties of the shale control the conditions for annulus closure, low permeability of highly deformed shale is essential for the sealing efficiency of the barrier. Laboratory measurements on quartz/clay mixtures (Mondol et al., 2008) show that permeability increases rapidly when the quartz content exceeds about 50 per cent, and hence rocks with higher quartz content are unlikely to form sealing barriers. These measurements also show that smectite reduces permeability much more than kaolinite, and hence high content of smectite is favourable for the formation of a sealing barrier. Note that mud additives may tend to clog defects in the shale barrier structure and thus have a positive impact on the sealing efficiency.

The conditions for formation of a shale barrier as expressed by Eq. (9.28) are based on the assumption that annulus pressure eventually equalises with the in situ pore pressure. This is not always the case. If the annulus contains oil based mud and is sealed off in both ends, then a permanent overpressure may be established in the annulus, counteracting annulus closure. On the other hand, annulus closure may be induced by bringing the rock close to the borehole wall temporarily into an elevated shear stress state. Methods to achieve this may, for instance, be (Fjær and Larsen, 2018; van Oort et al., 2020; Xie et al., 2020):

- reduced annulus pressure, which increases shear stress; cf. Eqs. (4.43)–(4.44)
- heating, which increases shear stress and also increases pore pressure; cf. Section 9.3.2
- exposure to fresh water, which induces swelling; cf. Section 9.4

These methods may induce plastic deformations that remain after the temporary action is terminated and the conditions approach long-term equilibrium. A common feature with these methods is that the achieved effect is partly relaxed after the action has been terminated, and hence the sealing efficiency of the barrier may for a while appear to be better than its long-term efficiency. On the other hand, in the context of radioactive waste storage, several laboratory and field studies have addressed fractured core samples and excavation damaged and disturbed zones around underground openings (Bock et al. (2010), and references therein). Various experiments demonstrate that fractures in clay and claystones may be self-sealing (and possibly self-healing) over variable periods of time, up to a few years, and that water plays an essential role in this process.

References

Al-Ajmi, A.M., Zimmerman, R.W., 2006. Stability analysis of vertical boreholes using the Mogi–Coulomb failure criterion. Int. J. Rock Mech. Min. Sci. 43 (8), 1200–1211. https://doi.org/10.1016/j.ijrmms.2006.04.001.

AlBahrani, H., Morita, N., 2020. Understanding the concurrence of drilling induced fractures and breakouts and its implications on drilling performance using 3D FEM and field imaging data. IPTC20171. In: International Petroleum Technology Conference. Dharhan, Saudi Arabia.

Alberty, M.W., Hafle, M.E., Minge, J.C., Byrd, T.M., 1999. Mechanisms of shallow waterflows and drilling practices for intervention. SPE Drill. Completion 14 (2). https://doi.org/10.2118/56868-PA.

Andrews, J.S., Fintland, T.G., Helstrup, O.A., Horsrud, P., Raaen, A.M., 2016. Use of unique database of good quality stress data to investigate theories of fracture initiation, fracture propagation and the stress state in the subsurface. ARMA16-887. In: 50th US Rock Mechanics/Geomechanics Symposium. Houston, TX, USA.

Aston, M.S., Alberty, M.W., McLean, M.R., de Jong, H.J., Amagost, K., 2004. Drilling fluids for wellbore strengthening. SPE/IADC87130. In: IADC/SPE Drilling Conference. Dallas, TX, USA.

Ay, A., Sagberg, T., Gyland, K.R., Askø, A., Jordal, O.H., Døssland, Ø., 2016. Use of specially treated micronized barite fluid system with wellbore strengthening approach to drill extremely depleted reservoir – HPHT Kvitebjørn field, offshore Norway. SPE180055. In: SPE Bergen One Day Seminar. Bergen, Norway.

Bailey, L., Denis, J.H., Maitland, G.C., 1991. Drilling fluids and wellbore stability — current performance and future challenges. In: Ogden, P.H. (Ed.), Chemicals in the Oil Industry. Roy. Soc. of Chemistry, London, pp. 53–70.

Ballard, T.J., Beare, S.P., Lawless, T.A., 1994. Fundamentals of shale stabilisation: water transport through shales. SPE Form. Eval. 9, 129–134. https://doi.org/10.2118/24974-PA.

Bock, H., Dehandschutter, B., Martin, C.D., Mazurek, M., de Haller, A., Skoczylas, F., Davy, C., 2010. Self-sealing of fractures in argillaceous formations in the context of geological disposal of radioactive waste. OECD Report NEA 6184. ISBN 978-92-64-99095-1. https://www.oecd-nea.org/rwm/reports/2010/nea6184-self-sealing.pdf.

Bol, G.M., Wong, S.-W., Davidson, C.J., Woodland, D.C., 1994. Borehole stability in shales. SPE Drill. Completion 9, 87–94. https://doi.org/10.2118/24975-PA.

Bourgoyne Jr., A.T., Millheim, K.K., Chenevert, M.E., Young Jr., F.S., 1986. Applied Drilling Engineering. SPE Textbook Series, vol. 2.

Bradley, W.B., 1979. Failure of inclined boreholes. J. Energy Resour. Technol. 101, 232–239. https://doi.org/10.1115/1.3446925.
Carminati, S., Del Gaudio, L., Zausa, F., Brignoli, M., 1999. How do anions in water-based muds affect shale stability? SPE50712. In: SPE Int. Symp. Oilfield Chemistry. Houston, TX, USA.
Charlez, P.A., Onaisi, A., 1998. Three history cases of rock mechanics related stuck pipes while drilling extended reach wells in North Sea. SPE/ISRM47287. In: Proc. Eurock '98. Trondheim, Norway.
Chatar, C., Muhoan, S., Imler, M., 2010. Overcoming a difficult salt drilling environment in the Gulf of Mexico: a case study. IADC/SPE128192. In: IADC/SPE Drilling Conference and Exhibition. New Orleans, LA, USA.
Chenevert, M.E., 1970. Shale alteration by water adsorption. J. Pet. Technol. 22, 1141–1148. https://doi.org/10.2118/2401-PA.
Chilingarian, G.V., Mazzullo, S.J., Rieke, H.H., 1992. Carbonate Reservoir Characterization: A Geologic-Engineering Analysis, Part I. Elsevier.
Cook, M., Horstman, M., Irondelle, M., Hugall, J., 2014. Integration of real-time LWD spectroscopy data to optimise drilling through paleosols supports low energy drilling operations. SPE169193. In: SPE Bergen One Day Seminar. Bergen, Norway.
Costa, A.M., Poiate Jr., E., Amaral, C.S., Gonçalves, C.J.C., Falcão, J.L., 2010. Geomechanics applied to the well design through salt layers in Brazil: a history of success. ARMA10-239. In: 44th US Rock Mechanics Symposium. Salt Lake City, UT, USA.
Daines, S.R., 1982. The prediction of fracture pressures for wildcat wells. J. Pet. Technol. 34 (4), 863–872. https://doi.org/10.2118/9254-PA.
Davison, J.M., Leaper, R., Cauley, M.B., Bennett, B., Mackenzie, A., Higgins, C.J., Shuttleworth, N., Wilkinson, D., 2004. Extending the drilling operating window in Brent: solutions for infill drilling in depleting reservoirs. SPE/IADC87174. In: IADC/SPE Drilling Conference. Dallas, TX, USA.
DeGennaro, S., Taylor, B., Bevaart, M., van Bergen, P., Harris, T., Jones, D., Hodzic, M., Watson, J., 2017. A comprehensive 3D geomechanical model used to deliver safe HPHT wells in the challenging Shearwater field. ARMA17-735. In: 51st US Rock Mechanics/Geomechanics Symposium. San Francisco, CA, USA.
Denis, J.H., Keall, M.J., Hall, P.L., Meeten, G.H., 1991. Influence of potassium concentration on the swelling and compaction of mixed (Na,K) ion-exchanged montmorillonite. Clay Miner. 26, 255–268. https://doi.org/10.1180/claymin.1991.026.2.09.
Dusseault, M.B., Maury, V., Sanfilippo, F., Santarelli, F.J., 2004. Drilling through salt: constitutive behavior and drilling strategies. ARMA/NARMS04-608. In: 6th North America Rock Mechanics Symposium (NARMS). Houston, TX, USA.
Eaton, B.A., 1969. Fracture gradient prediction and its application in oilfield operations. J. Pet. Technol. 21 (10), 25–32. https://doi.org/10.2118/2163-PA.
Ewy, R.T., Stankovich, R.J., 2002. Shale-fluid interactions measured under simulated downhole conditions. SPE/ISRM78160. In: SPE/ISRM Rock Mechanics Conference. Irving, TX, USA.
Fambon, L., Joffroy, G., 2008. Successful development drilling of an HP/HT infill well in a highly depleted reservoir: case study. IADC/SPE112708. In: IADC/SPE Drilling Conference. Orlando, FL, USA.
Fjær, E., Larsen, I., 2018. Shale as a sealing barrier around deep wells. OMAE2018-78749. In: ASME 37th Int. Conf. on Ocean, Offshore and Arctic Eng. Madrid, Spain.
Furlow, W., 1998. Shallow water flows: how they develop; what to do about them. Offshore 58 (9), 70–71, 161.
Gazaniol, D., Forsans, T., Boisson, M.J.F., Piau, J-M., 1995. Wellbore failure mechanisms in shale: prediction and prevention. J. Pet. Technol. 47 (4), 589–595. https://doi.org/10.2118/28851-PA.

Gomez, S., Patel, A., 2013. Shale inhibition: what works? SPE164108. In: SPE Int. Symp. on Oilfield Chemistry. The Woodlands, TX, USA.

Guenot, A., 1989. Instability problems at great depth, drilling boreholes and wells. In: Maury, V., Formaintraux, D. (Eds.), Rock at Great Depth. Balkema, Rotterdam, pp. 1199–1208.

Guenot, A., Santarelli, F.J., 1989. Influence of mud temperature on deep borehole behaviour. In: Maury, V., Formaintraux, D. (Eds.), Rock at Great Depth. Balkema, Rotterdam, pp. 809–817.

Heidug, W.K., Wong, S.W., 1996. Hydration swelling of water-absorbing rocks: a constitutive model. Int. J. Numer. Anal. Methods Geomech. 20 (6), 403–430. https://doi.org/10.1002/(SICI)1096-9853(199606)20:6<403::AID-NAG832>3.0.CO;2-7.

Hettema, M., Bostrøm, B., Lund, T., 2004. Analysis of lost circulation during drilling in cooled formations. SPE90442. In: SPE Annual Technical Conference and Exhibition. Houston, TX, USA.

Holt, R.M., Bakk, A., Stenebråten, J.F., Bauer, A., Fjær, E., 2018. Skempton's A – a key to man-induced subsurface pore pressure changes. ARMA18-0949. In: 52nd US Rock Mechanics/Geomechanics Symposium. Seattle, WA, USA.

Holt, R.M., Larsen, I., Fjær, E., Stenebråten, J.F., 2020. Comparing mechanical and ultrasonic behaviour of a brittle and a ductile shale: relevance to prediction of borehole stability and verification of shale barriers. J. Pet. Sci. Eng. 187, 106746. https://doi.org/10.1016/j.petrol.2019.106746.

Horsrud, P., 2001. Estimating mechanical properties of shale from empirical correlations. SPE Drill. Completion 16 (2), 68–73. https://doi.org/10.2118/56017-PA.

Horsrud, P., Bostrøm, B., Sønstebø, E.F., Holt, R.M., 1998. Interaction between shale and water-based drilling fluids: laboratory exposure tests give new insight into mechanisms and field consequences of kcl contents. SPE48986. In: SPE Annual Technical Conference and Exhibition. New Orleans, Louisiana.

Hoxha, B.B., van Oort, E., Daigle, H., 2019. How do nanoparticles stabilize shale? SPE Drill. Completion 34, 143–158. https://doi.org/10.2118/184574-MS.

Kenter, C.J., 1995. Unpublished lecture notes. NTNU, Norway.

Kristiansen, T.G., Flatebø, R.E., 2010. Sixty days ahead of schedule: reducing drilling risk at Valhall using computational geomechanics. SPE Drill. Completion 25, 544–554. https://doi.org/10.2118/119509-PA.

Kristiansen, T.G., Dyngeland, T., Kinn, S., Flatebø, R., Aarseth, N.A., 2018. Activating shale for form well barriers: theory and field examples. SPE191607. In: SPE Ann. Techn. Conf. & Exh. Dallas, TX, USA.

Last, N.C., Plumb, R., Harkness, R., Charlez, P., Alsen, J., McLean. Brief, M., 1996. An integrated approach to wellbore instability in the Cusiana field. J. Pet. Technol. 48, 245–246.

Lavrov, A., 2016. Lost Circulation: Mechanisms and Solutions. Elsevier.

Lomba, R.F.T., Chenevert, M.E., Sharma, M.M., 2000a. The ion-selective membrane behavior of native shales. J. Pet. Sci. Eng. 25, 9–23. https://doi.org/10.1016/S0920-4105(99)00028-5.

Lomba, R.F.T., Chenevert, M.E., Sharma, M.M., 2000b. The role of osmotic effects in fluid flow through shales. J. Pet. Sci. Eng. 25, 25–35. https://doi.org/10.1016/S0920-4105(99)00029-7.

Mahan, B.H., 1975. University Chemistry, second edition. Addison-Wesley Publ. Co., Reading, Massachusetts.

Marine, I.W., Fritz, S.J., 1981. Osmotic model to explain anomalous hydraulic heads. Water Resour. Res. 17, 73–82. https://doi.org/10.1029/WR017i001p00073.

Matthews, W.R., Kelly, J., 1967. How to predict formation pressure and fracture gradient. Oil Gas J. 65 (8), 92–106.

Maury, V., Guenot, A., 1995. Practical advantages of mud cooling systems for drilling. SPE Drill. Completion 10 (1), 42–48. https://doi.org/10.2118/25732-PA.

Maury, V., Zurdo, C., 1996. Drilling-induced lateral shifts along pre-existing fractures: a common cause of drilling problems. SPE Drill. Completion 11, 17–23. https://doi.org/10.2118/27492-PA.

Maury, V.M., Sauzay, J.-M., 1987. Borehole instability: case histories, rock mechanics approach, and results. SPE/IADC16051. In: SPE/IADC Drilling Conference. New Orleans, LA, USA.

Meier, T., Rybacki, E., Reinicke, A., Dresen, G., 2013. Influence of borehole diameter on the formation of borehole breakouts in black shale. Int. J. Rock Mech. Min. Sci. 62, 74–85. https://doi.org/10.1016/j.ijrmms.2013.03.012.

Mody, F.K., Hale, A.H., 1993. Borehole stability model to couple the mechanics and chemistry of drilling fluid shale interaction. J. Pet. Technol. 45 (11), 1093–1101. https://doi.org/10.2118/25728-PA.

Mondol, N.H., Bjørlykke, K., Jahren, J., 2008. Experimental compaction of clays: relationship between permeability and petrophysical properties in mudstones. Pet. Geosci. 14 (4), 319–337. https://doi.org/10.1144/1354-079308-773.

Morita, N., Fuh, G.-F., 2012. Parametric analysis of wellbore strengthening methods from basic rock mechanics. SPE Drill. Completion 27 (2), 315–327. https://doi.org/10.2118/145765-PA.

Nes, O.-M., Horsrud, P., Sønstebø, E.F., Holt, R.M., Ese, A.M., Økland, D., Kjørholt, H., 1998. Rig site and laboratory use of CWT acoustic velocity measurements on cuttings. SPE Reserv. Eval. Eng. 1 (4), 282–287. https://doi.org/10.2118/50982-PA.

Nes, O.-M., Stenebråten, J., Fjær, E., 2015. Time dependent borehole stability: effect of static vs. dynamic mud weight. SPE/IADC173048. In: SPE/IADC Drilling Conference and Exhibition. London, United Kingdom.

Outmans, H.D., 1958. Mechanics of differential pressure sticking of drill collars. Petr. Trans. AIME 213 (1), 265–274. https://doi.org/10.2118/963-G.

Raaen, A.M., Horsrud, P., Kjørholt, H., Økland, D., 2006. Improved routine estimation of the minimum horizontal stress component from extended leak-off tests. Int. J. Rock Mech. Min. Sci. 43, 37–48. https://doi.org/10.1016/j.ijrmms.2005.04.005.

Sampath, K.H.S.M., Perera, M.S.A., Ranjith, P.G., 2018. Theoretical overview of hydraulic fracturing break-down pressure. J. Nat. Gas Sci. Eng. 58, 251–265. https://doi.org/10.1016/j.jngse.2018.08.012.

Santarelli, F.J., Brown, E.T., 1987. Performance of deep wellbores in rock with a confining pressure dependent elastic modulus. In: Proc. 6th ISRM Congress. Montreal, Canada, pp. 1217–1222.

Santarelli, F.J., Carminati, S., 1995. Do shales swell? A critical review of available evidence. SPE/IADC29421. In: SPE/IADC Drilling Conference. Amsterdam, Netherlands.

Santarelli, F.J., Dahen, D., Baroud, H., Sliman, K.B., 1992. Mechanisms of borehole instability in heavily fractured rock media. Int. J. Rock Mech. Min. Sci. Geomech. Abstr. 29 (5), 457–467. https://doi.org/10.1016/0148-9062(92)92630-U.

Santarelli, F.J., Marsala, A.F., Brignoli, M., Rossi, E., Bona, N., 1998. Formation evaluation from logging on cuttings. SPE Reserv. Eval. Eng. 1 (3), 238–244. https://doi.org/10.2118/36851-PA.

Schackelford, C.D., Malusius, M.A., Olsen, H.W., 2003. Clay membrane behaviour for geoenvironmental containment. In: Culligan, P.J. (Ed.), Soil Rock America, vol. I. VGE Verlag, Essen, pp. 767–774.

Sewell, M., Billingsley, J., 2002. An effective approach to keeping the hole clean in high-angle wells. World Oil 223 (10).

Shaw, M.N., Sutherland, W.G., 1988. Innovative approaches to drilling high-angle wells in Bass Strait. SPE/IADC17233. In: SPE/IADC Drill. Conf. Dallas, TX, USA.

Sherwood, J.D., 1994. A model of hindered solute transport in a poroelastic shale. Proc. R. Soc. Lond. Ser. A 445, 679–692. https://doi.org/10.1098/rspa.1994.0086.

Sherwood, J.D., Craster, B., 2000. Transport of water and ions through a clay membrane. J. Colloid Interface Sci. 230, 349–358. https://doi.org/10.1006/jcis.2000.7100.

Steiger, R.P., 1982. Fundamentals and use of potassium/polymer drilling fluids to minimize drilling and completion problems associated with hydratable clays. J. Pet. Technol. 34, 1661–1669. https://doi.org/10.2118/10100-PA.

Stjern, G., Horsrud, P., Agle, A., 2000. Improving drilling performance in troublesome clay formations in the Heidrun field. SPE/IADC59219. In: SPE/IADC Drill. Conf. New Orleans, LA, USA.

Sønstebø, E.F., Holt, R.M., 2001. Brine exposure effects on a tertiary North Sea shale. In: Proc. DCRocks. Balkema, Washington, DC, USA.

Tangen, G.I., Smaaskjaer, G., Bergseth, E., Clark, A., Fossli, B., Claudey, E., Qiang, Z., 2019. Experience from drilling a horizontal well in a naturally fractured and karstified carbonate reservoir in the Barents Sea using a CML MPD system. SPE/IADC194545. In: IADC/SPE Managed Pressure Drilling and Underbalanced Operations Conference and Exhibition. Amsterdam, Netherlands.

van Oort, E., Hale, A.H., Mody, F.K., Sanjit, R., 1996. Transport in shales and the design of improved water-based shale drilling fluids. SPE Drill. Completion 11 (3), 137–146. https://doi.org/10.2118/28309-PA.

van Oort, E., Pasturel, C., Bryla, J., Ditlevsen, F., 2019. Improved wellbore stability in North Sea Lark and Horda shales through shale/fluid-compatibility optimization. SPE Drill. Completion 34 (1), 46–59. https://doi.org/10.2118/184661-PA.

van Oort, E., Juenger, M., Aldin, M., Thombare, A., McDonald, M., 2020. Simplifying well abandonments using shale as a barrier. IADC/SPE199654. In: IADC/SPE Drilling Conf. & Exh. Galveston, TX, USA.

Williams, S., Carlsen, T., Constable, K., Guldahl, A., 2009. Identification and qualification of shale annular barriers using wireline logs during plug and abandonment operations. SPE/IADC119321. In: SPE/IADC Drill. Conf. & Exh. Amsterdam, Netherlands.

Willson, S., Last, N.C., Zoback, M.D., Moos, D., 1999. Drilling in South America: a wellbore stability approach for complex geologic conditions. SPE53940. In: Latin American and Caribbean Petroleum Engineering Conference. Caracas, Venezuela.

Xie, A., Bauer, X., Stenebråten, J.F., Bakheim, S., Lavrov, A., Fjær, E., Kristiansen, T.G., 2020. Can heating induce borehole closure? Rock Mech. Rock Eng. 53, 5715–5744. https://doi.org/10.1007/s00603-020-02238-5.

Xie, X., Fjær, E., Detournay, E., 2019. Time-dependent closure of a borehole in a viscoplastic rock. Geomech. Energy Environ. 19, 100115. https://doi.org/10.1016/j.gete.2019.02.001. pp. 1–14.

Zhang, J., Yin, S.-X., 2017. Fracture gradient prediction: an overview and an improved method. Pet. Geosci. 14 (4), 720–730. https://doi.org/10.1007/s12182-017-0182-1.

Zhang, Y., Zhang, J., Yuan, B., Yin, S., 2018. In-situ stresses controlling hydraulic fracture propagation and fracture breakdown pressure. J. Pet. Sci. Eng. 164, 164–173. https://doi.org/10.1016/j.petrol.2018.01.050.

Zheng, Z., 1998. Integrated borehole stability analysis — against tradition. SPE/ISRM47282. In: Proc. Eurock '98. Trondheim, Norway.

Økland, D., Cook, J.M., 1998. Bedding-related borehole instability in high-angle wells. SPE/ISRM47285 In: Proc. Eurock '98. Trondheim, Norway.

CHAPTER 10

Solids production

When hydrocarbons are produced from a reservoir, solid particles sometimes follow the reservoir fluid into the well. This unintended byproduct of the hydrocarbon production is called *solids production*.

The amount of solids produced can vary from a few grams per cubic metre of reservoir fluid, which usually represents a minor problem, to catastrophic amounts, possibly leading to complete filling of the borehole. It is estimated that seventy percent of the world's hydrocarbon reserves are contained in reservoirs where solids production is likely to become a problem at some point in time. The problem is most pronounced in sand reservoirs, and hence sand production has attracted most attention. However, it may also be a problem in chalk and coal reservoirs.

There is a large variation in the way solids production has been handled, from not considering it as a problem at all to not accepting anything but insignificant amounts of solids. The latter attitude has been spurred by the observation that sand production may result in serious accidents. However, methods to control solids production involves additional costs and usually results in reduced production, so that in later years a more relaxed tolerance for solids production is sometimes considered. When dealing with heavy oil, solids production is even deliberately provoked in some cases.

We will first take a look at some operational aspects related to solids production. Mechanisms that may cause sand production are then described, followed by discussions on how to predict sand production. Chalk production is discussed more briefly, which reflects that the public literature is rather limited on this subject.

10.1 Operational aspects of solids production

10.1.1 Consequences of solids production

Solids production causes a series of problems to the petroleum production. The main types of problems are:

1. Erosion of the production equipment. This is primarily a safety problem, as well as an economical one. It is also a typical sand production problem due to the abrasiveness of the quartz grains. The erosion

Developments in Petroleum Science, Volume 72
ISSN 0376-7361, https://doi.org/10.1016/B978-0-12-822195-2.00019-X

caused by a given amount of sand depends strongly on the velocity of the fluid in the pipeline; typically, the amount of eroded material is proportional to $(\text{velocity})^n$, where n is a number between 2 and 3 (Haugen et al., 1995). Consequently, the amount of acceptable sand production is much smaller for a gas well—where the fluid velocity is high—than for an oil well.

2. Instability of the production cavities and the wellbore itself, which may in extreme cases result in a complete filling of the borehole ("sand up") so that the well has to be abandoned.
3. The necessity to handle large amounts of polluted solids at the rig.

On the other hand, solids production sometimes has the positive effect that it enhances the productivity of the well (Geilikman et al., 1994). In a production technique known as CHOPS (cold heavy oil production with sand), solids production is actually encouraged, since the heavy oil is otherwise very hard to produce (Geilikman and Dusseault, 1997; Dusseault and El-Sayed, 2000).

10.1.2 Well completion and solids control

After a well has been drilled, it is prepared for hydrocarbon production. This process is called *completion*. Completion methods that do not involve specific equipment to prevent or reduce the consequences of solids production are called *natural* (or barefoot) completions. Instead, the potential for solids production is carefully evaluated prior to completion, and care is taken to stay within the limits of solids free production. Natural completion is cheap and attractive since it does not impair the productivity of the well. However, the required restrictions on the production to avoid solids may prevent economical production. The use of special equipment to prevent solids from being released from the formation or to prevent produced solids from following the hydrocarbon flow to the surface is called *active solids control*.

Well completion can be done in many different ways. Here we will restrict our attention to the most common ones.

The hole may be left open, without casing and cement, which allows the hydrocarbons to flow freely into the well through the wellbore wall. The inflow area is equal to the circumference of the borehole times the length of the open hole section within the reservoir. This simple completion technique requires the formation to be relatively strong and stable. If not, some support for the borehole wall can be obtained by placing gravel in the hole (gravel packing). The hydrocarbons are then produced through

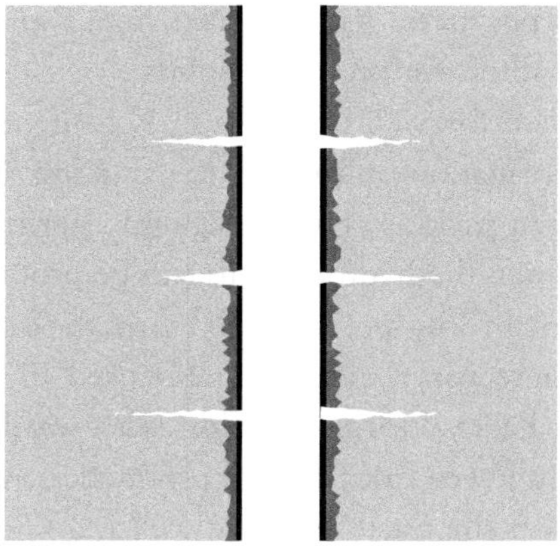

Figure 10.1 Principle sketch of perforations through the casing and the cement.

the gravel, which may also act as a filter for produced solids. Screens, which are finely meshed metal nets placed in the open hole, may alternatively be used as filters. The use of screens or gravel packs increases the completion costs and reduces the productivity of the well. Clogging of these filters may also be a problem. Other methods for solids control include chemical consolidation, where the formation is strengthened by a resin injected into the rock.

In weak formations, casing is usually set and cemented to the formation to stabilise the borehole. The casing is then perforated in the reservoir zones, so that the hydrocarbons can flow into the well (Fig. 10.1). Each perforation is created by a charge fired through the casing into the formation. This typically generates a cylindrical hole, 1–2 cm in diameter and 20–50 cm long. The size and shape of the perforations may however vary considerably with the type of charge used, the formation properties and the well pressure relative to the pore pressure in the formation at the time the perforation is created.

It is normally recommended to perforate in underbalance (i.e. the well pressure is lower than the pore pressure, typically, a few MPa) to clean up the perforations and to avoid too large reduction of the near-well permeability. Too large underbalance during the perforation operation may however result in a rapid growth of the perforation cavities. The inflow area for a perforated zone is equal to the surface area (circumference times length) of each perforation times the number of perforations, which is usually in the order of 10–40 per metre. Gravel packs may also be used in cased and perforated holes, as filters for produced solids.

A third alternative for well completion is to generate a fracture from the well into the formation and to fill the fracture with proppants. This technique, called frac packing, creates a highly permeable slit in the formation, through which the hydrocarbons can be produced. The technique is particularly useful in low-permeability formations, as the effective inflow area can become very large. It is also used in weak formations, as the large inflow area allows for production with smaller pressure gradients and correspondingly lower risk for sand production. Clogging of the perforations that connect the fracture to the well is a known problem, which may possibly be linked to sand produced within the fracture. However, no studies on this effect have so far emerged in the literature. Frac packing is extensively used in some regions (like the Gulf of Mexico), whereas the use is limited in other regions, partly due to the costs involved and partly due to the risk and uncertainty involved in the fracturing operation.

10.2 Sand

Observations of sand production can be classified into three types:

1. *Transient sand production*, where a burst of sand is followed by a continuous production of sand with declining rate under constant conditions. Transient sand production is frequently observed right after a perforation job, after a change in the production conditions (usually a reduction in the well pressure) and after water breakthrough.
2. *Continuous sand production*, where sand is continuously produced at a relatively constant rate.
3. *Catastrophic sand production*, where sand is produced at such a high rate that the well is choked.

Continuous sand production, even only a few grams per m^3, may add up to amounts in the order of hundreds of kilos per metre of the well after several years of production. The removal of such amounts of sand obviously changes the size and shape of the producing cavities. For instance, perforations that are initially formed more or less as separate cylindrical holes (Fig. 10.1), may eventually merge and form larger cavities behind the casing.

10.2.1 Necessary and sufficient conditions for sand production

It is generally accepted that sand production cannot occur from intact rock, even if it is poorly consolidated, since a flowing fluid does not normally provide sufficient force to pull sand grains out of an intact rock. This is illustrated by a model example in the next section. Sand production may thus occur only if the rock in the vicinity of the producing cavity is unconsolidated or has been damaged. Such damage may be caused by the stress concentrations that occur around open holes in a stressed rock, as described in Chapter 4. Onset of sand production is therefore closely related to stress-induced damage of the rock. Criteria for failure of the rock around production cavities are discussed in Section 10.2.3.

Although unconsolidated or damaged rock is a necessary condition for sand production, it is not necessarily a sufficient condition. Post failure stabilisation of a production cavity may occur after some initial sand production due to, for instance, a change in the cavity shape or increased permeability in the damaged region. It has also been shown that a damaged rock may sometimes form stable sand arches that allow for sand-free production at significantly higher drawdowns than the rock failure conditions indicate. Criteria describing the stability of sand arches are described in Section 10.2.4. When the conditions exceed the stability limit for the sand arches, sand will definitively be produced into the well.

Monitoring of sand production is usually based on measurements or observations at the surface. However, the produced sand will only reach the surface if the well flow is sufficiently strong to carry the sand grains all the way up through the well. Thus sand transport in the well is also a significant part of the sand production problem; however, that part will only be discussed briefly here (Section 10.2.6).

Prevention of sand production problems is most effective if it is done or planned during well completion. Thus it is essential to be able to predict possible sanding problems before the well completion starts. To an increasing extent, this also involves prediction of the rate of produced sand in the cases where sand-free production is unachievable or economically unfavourable. A model describing sand rate is described in Section 10.2.5, whereas sand prediction is discussed in Section 10.2.7.

10.2.2 Forces on a sand grain

Consider a sand grain of diameter d_g squeezed in between its neighbouring grains at the wall of a cylindrical cavity, as illustrated schematically in

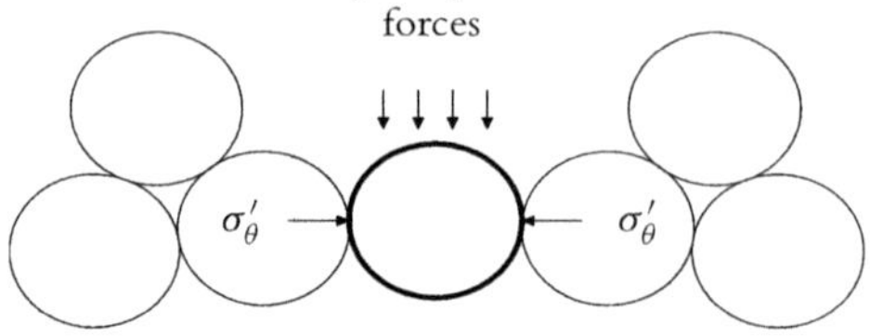

Figure 10.2 Sand grain at the wall of a cylindrical cavity.

Fig. 10.2. The force F_r needed to remove the grain can be estimated as the sum of the shear forces needed to induce shear failure in the four contact planes at the side of the grain (given by e.g. the Mohr–Coulomb criterion) plus the force needed to induce tensile failure in the contact plane behind the grain (given by the tensile failure criterion). On average, this is given as

$$F_r = \pi \left(\frac{d_g}{2} \right)^2 [4S_0 + \mu(2\sigma'_z + 2\sigma'_\theta) + T_0] \tag{10.1}$$

where T_0 and S_0 are the tensile strength and the cohesion, respectively, μ is the coefficient of internal friction (see Sections 2.2 and 2.3), and σ'_z and σ'_θ are the effective axial and tangential stresses, respectively, at the cavity wall.

The hydrodynamic forces pulling on the grain are caused by the fluid flowing through the rock. An estimate of the forces can be obtained as follows: The force F acting on a volume element of the rock due to a fluid flowing through it is $F = -A\Delta p_f$, where A is the cross-sectional area through which the fluid is flowing, Δx is the length of the volume element, and Δp_f is the pore pressure drop over Δx. From Darcy's law (Eq. (1.308)) we derive that $F = -A\Delta p_f = (\eta_f/k)Q\Delta x$. An expression for the permeability k of a porous rock is given by the Kozeny–Carman equation (Carman, 1956; see also Dullien, 1992); an alternative version of this equation is given in Eq. (12.64):

$$k = \frac{1}{180} \frac{\phi^3}{(1-\phi)^2} d_g^2 \tag{10.2}$$

The number N of grains in this volume element is roughly given as the volume of solid material in the element, which is $(1-\phi)A\Delta x$, divided by the volume of one grain, which is $(1/6)\pi d_g^3$. From these considerations we may deduce that the average hydrodynamic force F_h acting on one grain

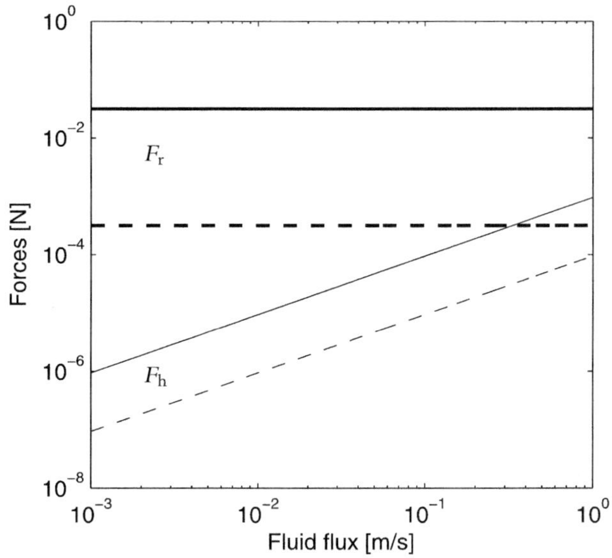

Figure 10.3 Hydrodynamic force F_h acting on a sand grain, compared to the forces F_r needed to remove the grain. Estimated from Eqs. (10.1) and (10.3); $T_0 = 0$, $\mu = 0.5$, $S_0 = 1$ MPa, $\sigma_r' = \sigma_\theta' = 0$, $\eta_f = 1$ cP, $\phi = 0.2$. Solid lines: $d_g = 100$ μm; dashed lines: $d_g = 10$ μm.

within the volume element is

$$F_h = \frac{F}{N} = 30\pi\,\eta_f \frac{1-\phi}{\phi^3}\frac{Q}{A}d_g \tag{10.3}$$

(As the grain we consider is placed at the surface of the rock, we might alternatively consider the hydrodynamic forces acting on a free particle. The result, as shown by Charlez (1997), is somewhat different; however, the following conclusion is the same. See also Vardoulakis (2004)).

Fig. 10.3 compares the forces on a grain in a very weak rock ($S_0 = 1$ MPa). It is assumed that the grain is only attached by cohesional forces (which is probably the case for some of the grains). The figure shows that the hydrodynamic forces F_h increase with the fluid flux Q/A; however, for the typical range shown here, they remain for the most part several orders of magnitude smaller than the force F_r needed to pull the grain away. Thus the hydrodynamic forces are not able to destroy a competent rock and induce failure on their own. The major role for the hydrodynamic forces is to remove grains from unconsolidated or damaged rock and to transport loose material into the well.

10.2.3 Critical drawdown for cylindrical cavities

As the fluid flow is normally not able to destroy the rock and induce sand production, a necessary condition for sand production is that the rock is unconsolidated or has been damaged by some other mechanism. Such damage is usually caused by the effective stresses in the vicinity of the producing cavity. These depend on several factors, in particular, the far-field in situ stresses, the pore pressure and the geometry of the producing cavity. Since the stress field around a cavity is not homogeneous, the effective stresses in the vicinity of the cavity may also depend on the rock properties.

In this section, we will take a look at the criteria for sand failure around a producing opening, caused by the in situ stresses. We restrict our attention to cylindrical holes, which may be seen as a representative for a perforation or an open hole (i.e. a well with no casing), and to (semi)spherical holes, which may represent a perforation after some sand has been produced. A semispherical hole may also represent the tip of a cylindrical perforation.

When working with production, it is often convenient to use *drawdown* rather than well pressure to describe the pressure conditions in the well. Drawdown $\mathcal{D}$ is defined as the difference between the pore pressure p_{fo} far from the well and the well pressure p_{w}, that is,

$$\mathcal{D} = p_{\mathrm{fo}} - p_{\mathrm{w}} \tag{10.4}$$

When the well pressure has been reduced so much that the well starts to produce sand, we have reached the *critical drawdown* $\mathcal{D}_{\mathrm{c}}$ for sand production in the well. Some expressions for the critical drawdown are derived and discussed in this section.

10.2.3.1 Shear failure

The condition for shear failure at the wall of a cylindrical hole can be derived from expressions for the stresses at the cavity wall combined with a failure criterion, as shown in Chapter 4.

Consider first the simple case where the stress σ_{h}' is isotropic. During production, the pore pressure at the cavity wall is $p_{\mathrm{f}}(R_{\mathrm{c}}) = p_{\mathrm{w}}$, the smallest principal stress is $\sigma_r(R_{\mathrm{c}}) = p_{\mathrm{w}}$, and the largest principal stress is (see Eq. (4.120))

$$\sigma_\theta(R_{\mathrm{c}}) = 2\sigma_{\mathrm{h}} - p_{\mathrm{w}} - \frac{1 - 2\nu_{\mathrm{fr}}}{1 - \nu_{\mathrm{fr}}}\alpha(p_{\mathrm{fo}} - p_{\mathrm{f}}(R_{\mathrm{c}})) \tag{10.5}$$

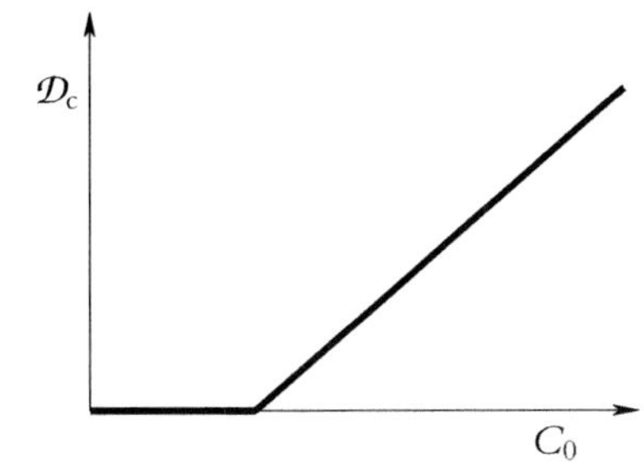

Figure 10.4 Characteristic relationship between critical drawdown $\mathcal{D}_c$ and uniaxial compressive strength C_0 for a cylindrical cavity.

Failure according to the Mohr–Coulomb failure criterion (2.22) requires that

$$\sigma_\theta(R_c) - p_f(R_c) = C_0 + (\sigma_r(R_c) - p_f(R_c)) \tan^2 \beta \tag{10.6}$$

Solution of these equations in terms of p_w gives us the lower well pressure limit $p_{w,min}$ where failure is initiated.

Critical drawdown $\mathcal{D}_c$ for the cavity is defined as

$$\mathcal{D}_c = p_{fo} - p_{w,min} \tag{10.7}$$

Assuming that the rock is sufficiently soft so that the Biot coefficient $\alpha \approx 1$, the solution of Eqs. (10.5) and (10.6) can be expressed as

$$\mathcal{D}_c = (1 - \nu_{fr})(C_0 - 2\sigma'_h) \tag{10.8}$$

where we have introduced $\sigma'_h = \sigma_h - p_{fo}$ as the effective stress far from the well. Note that Eq. (10.8) is essentially the same as Eq. (4.122), only that the failure condition is expressed in terms of $\mathcal{D}_c$ rather than $p_{w,min}$.

Fig. 10.4 graphically illustrates the shear failure criterion (10.8). Although it is based on a very simplified model, the picture is qualitatively in agreement with observations. It shows that a minimum rock strength is required to prevent failure of the rock during production. It also shows that the critical drawdown depends on the effective stress far from the well in such a way that $\mathcal{D}_c$ is reduced when the reservoir is depleted, that is, when p_{fo} is reduced. This implies that the probability for sand production increases as the reservoir is being produced under constant drawdown.

Consider next a situation where the principal in situ stresses are all different. The stability of the producing cavity (whether it is an open hole or a perforation) now depends on the orientation of the cavity relative to the in situ stresses. The stresses at the cavity wall for a non-flow situation can be obtained from Eqs. (4.78)–(4.90), whereas Eq. (4.59) tells us to subtract $(p_{fo} - p_w)((1 - 2\nu_{fr})/(1 - \nu_{fr}))\alpha$ from the tangential stress to account for the pore pressure gradient. Introducing these expressions into a failure criterion, for instance, the Mohr–Coulomb criterion (2.22), assuming that $\alpha \approx 1$ and following the same procedure as described in Chapter 4, we obtain the critical drawdown for any orientation of the cavity.

In general, the result cannot be expressed in an explicit form. However, if the cavity is parallel to one of the principal in situ stress directions, the result may be written as

$$\mathcal{D}_c = (1 - \nu_{fr})(C_0 - 2\sigma'_{is}) \tag{10.9}$$

The effective in situ stress σ'_{is} is given by

$$2\sigma'_{is} = 3\sigma'_{max\perp} - \sigma'_{min\perp} \tag{10.10}$$

where $\sigma'_{max\perp}$ and $\sigma'_{min\perp}$ are the maximum and minimum effective principal stresses, respectively, in the plane perpendicular to the cavity axis (see also Section 9.2).

For completeness, we should also consider the possibility that the axial stress is the largest principal stress at the critical point on the cavity wall (Case b in Table 4.1). This gives the criterion

$$\mathcal{D}_c = \frac{1 - \nu_{fr}}{\nu_{fr}}(C_0 - \sigma'_{is,b}) \tag{10.11}$$

where

$$\sigma'_{is,b} = \sigma'_{\parallel} + 2\nu_{fr}(\sigma'_{max\perp} - \sigma'_{min\perp}) \tag{10.12}$$

and $\sigma'_{\parallel}$ is the principal stress parallel to the cavity axis (see Section 9.2). The criterion that gives the lowest critical drawdown, either Eq. (10.9) or Eq. (10.11), is the relevant one to use at any time. Note however that criterion (10.11) is unlikely to be the relevant one except for rather special stress conditions and that this criterion has no relevance if $\nu_{fr} \leq 0$ (except for the situation that $C_0 < \sigma'_{is,b}$, in which case the cavity is unstable at zero drawdown).

As a rule of thumb, we can say that stress anisotropy in the plane normal to the cavity axis reduces the critical drawdown. Consider as an example a

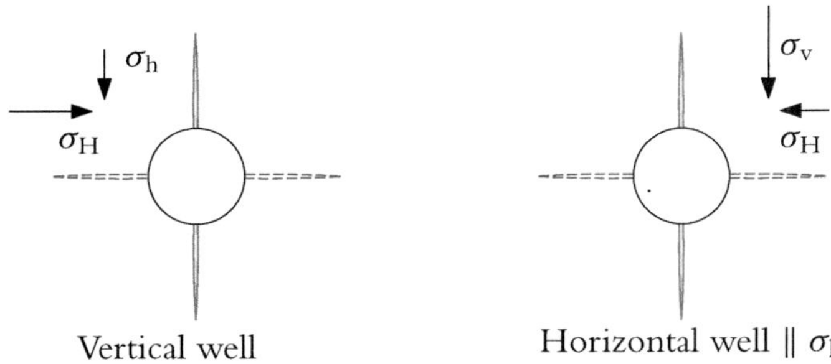

Figure 10.5 Examples of preferred orientation for perforations in a formation where $\sigma_v > \sigma_H > \sigma_h$. Perforations with highest critical drawdown are drawn as solid blue lines, and perforations with lowest critical drawdown are dashed red lines.

vertical well in a formation where $\sigma_v > \sigma_H > \sigma_h$. For a cavity parallel to σ_H (inclination $i = 90°$, azimuth $a = 0$), we have according to Eq. (10.9)

$$\mathcal{D}_c = (1 - \nu_{fr})(C_0 - 3\sigma_v' + \sigma_h') \tag{10.13}$$

whereas for a cavity parallel to σ_h ($i = 90°$, $a = 90°$), we have

$$\mathcal{D}_c = (1 - \nu_{fr})(C_0 - 3\sigma_v' + \sigma_H') \tag{10.14}$$

We see that $\mathcal{D}_c$ is largest for the cavity parallel to σ_h, which has the smallest stress anisotropy in the plane normal to the cavity axis. Thus from a vertical well it is preferable to orient the perforations parallel to σ_h to obtain the largest critical drawdown. Furthermore, Eqs. (10.9) and (10.10) tell us that from a horizontal well parallel to σ_h it is preferable to perforate vertically, provided that $\sigma_v > \sigma_H$. The examples are plotted in Fig. 10.5.

A complicating factor considering the stability of perforations is that the presence of the well itself disturbs the stress field at the perforation. Note in particular that the perforations we identified as the most preferable (to obtain largest critical drawdown for the vertical well; see Fig. 10.5) are positioned in the parts of the borehole wall, where we may expect to find breakouts around the well if the conditions for such failure have been fulfilled at some stage (see Section 4.5.3). If we have reasons to believe that such breakouts have been formed, then the perforations should preferably be placed outside the breakout sectors.

In addition, the presence of other nearby cavities may affect the stress field to some extent and reduce the stability of a perforation. A minimum distance between perforations is therefore recommended.

In Chapter 4, it was shown that plastic deformation of the rock tends to stabilise a hole, so that it can take larger loads than the purely elastic solutions predict. To see how this affects the critical drawdown, we assume that the material behaves according to a linear elastic/perfectly plastic model as described in Chapter 4 and that failure occurs when the plastic strain λ reaches a critical limit λ_c. Although this model is obviously a simplification of the true rock behaviour, it captures fairly well the characteristics of soft rocks. Based on expression (4.175) for λ and the assumption that λ_c is so small that we can ignore higher-order terms, we find that the criterion for failure in an isotropic stress situation becomes

$$\mathcal{D}_{\mathrm{c}} = \frac{1}{2}(C_0 - 2\sigma'_{\mathrm{h}}) + \lambda_c \Sigma_c \tag{10.15}$$

where

$$\Sigma_{\mathrm{c}} = \frac{4C_0(\lambda + G)}{2C_0\frac{\lambda+G}{G} - \frac{C_0 - 2\sigma'_{\mathrm{h}}}{\ln\frac{R_{\mathrm{e}}}{R_{\mathrm{w}}}}} \approx 2G \tag{10.16}$$

Note that Eq. (10.15) does not become exactly equal to Eq. (10.8) when $\lambda_c = 0$. This is because Eq. (10.8) is based on the Mohr–Coulomb criterion, whereas Eq. (10.15) is based on the Tresca criterion.

We may combine the results obtained above into a generalised equation for critical drawdown

$$\mathcal{D}_{\mathrm{c}} = A_1(C_0 - 2\sigma'_{is}) + A_2 \tag{10.17}$$

where A_1 is a non-dimensional parameter depending on Poisson's ratio and the failure criterion, and A_2 is a parameter with the dimension of stress, depending on the plasticity of the rock. In practice, A_1 and A_2 may be empirical constants. It has been shown (van den Hoek et al., 2000a,b,c) that A_2 also depends on the size of the cavity. Due to the increasing difficulty in rotating sand grains at the cavity wall for larger grain diameter-to-cavity diameter ratios, the smaller cavities can take larger plastic strains before failing, and therefore A_2 is larger for the smaller cavities.

An alternative approach is to apply strength data from thick-walled cylinder tests (TWC) rather than the unconfined compressive strength C_0 in the critical drawdown criterion. The philosophy of this approach is that nonlinear elasticity and plasticity are implicitly included in the TWC results (Veeken et al., 1991). The critical drawdown can still be computed from

Eq. (10.17) with the modification

$$C_0 + \frac{A_2}{A_1} \rightarrow sC_{\mathrm{TWC}} \tag{10.18}$$

where C_{TWC} is the collapse pressure of the thick-walled cylinder. By introducing Eq. (10.18) into Eq. (10.17) we obtain an expression for the critical drawdown of a cylindrical cavity in terms of thick-walled cylinder strength:

$$\mathcal{D}_{\mathrm{c}} = A_1(sC_{\mathrm{TWC}} - 2\sigma'_{\mathrm{is}}) \tag{10.19}$$

The constant s accounts for the effects of size and the ratio of the outer and inner diameters of the thick-walled cylinder, as the TWC tests conducted in the laboratory do not exactly match the field conditions on these points. Willson et al. (2002) proposed to use $s = 3.1$ when C_{TWC} is derived from tests on standard $1\frac{1}{2}''$ outer diameter, $\frac{1}{2}''$ inner diameter and $3''$ long TWC samples.

Although perforations are most conveniently made as cylindrical holes, this is not the most stable shape for a producing cavity. In Chapter 4 the stresses at the wall of a spherical cavity were given. From these results, in combination with the Mohr–Coulomb failure criterion, we find that the critical drawdown for a semi-spherical cavity in an isotropic stress field is

$$\mathcal{D}_{\mathrm{c}} = \frac{1 - \nu_{\mathrm{fr}}}{1 + \nu_{\mathrm{fr}}}(2C_0 - 3\sigma'_{\mathrm{h}}) \tag{10.20}$$

This expression corresponds to Eq. (10.8) for a cylindrical cavity. Both expressions are illustrated graphically in Fig. 10.6. We see that the spherical cavity is more stable than the cylindrical cavity, which tells us that the cylindrical shape typically created by the perforation charge need not be the most stable one. Thus it is possible for a cavity that starts to produce sand at a given drawdown, to obtain a more stable shape and eventually stop producing sand at this drawdown level. Fig. 10.7 shows an axial cross-section through a sand producing cavity, as observed after a laboratory test. The shape and size of the cavity has been affected by the sand production.

A change in shape of the cavities is one possible explanation for a typical observation of transient sand production, where sand production is initiated when the drawdown is increased and subsequently stops when the drawdown has been kept constant for a while.

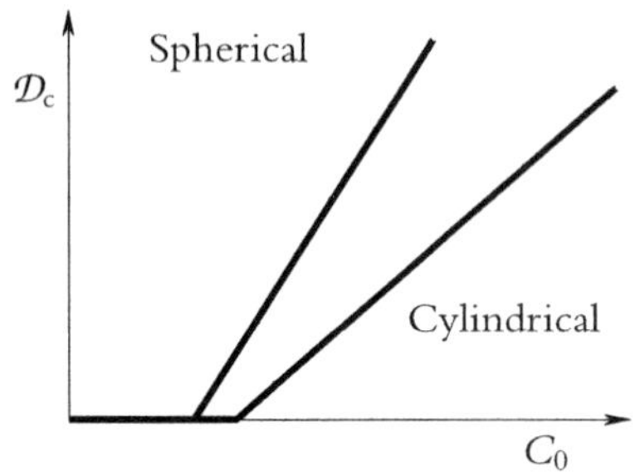

Figure 10.6 Critical drawdown versus rock strength for cylindrical and spherical cavities.

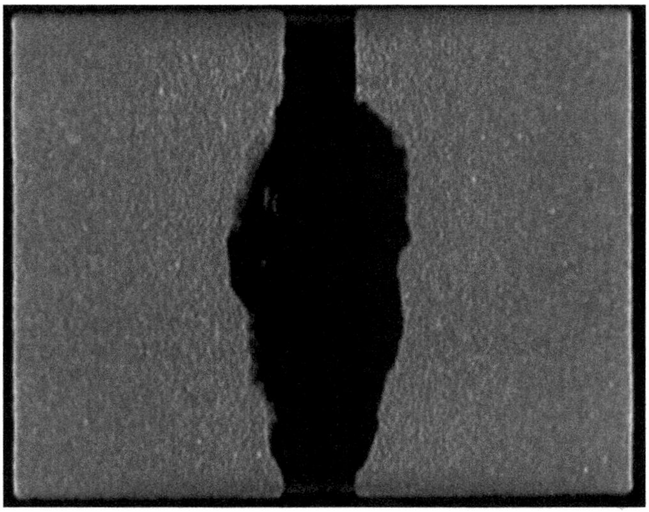

Figure 10.7 CT scan from a laboratory test on an outcrop sandstone, showing a cross-section through a producing cavity after some sand production. The cavity was initially cylindrical. Fluid and sand has been produced through the opening at the lower end. Courtesy of SINTEF.

10.2.3.2 Tensile failure

Another possible failure mode is tensile failure. At the cavity wall the radial stress and the pore pressure are both equal to the well pressure, which implies that the effective radial stress is zero at the cavity wall. If the pore pressure gradient is larger than the radial stress gradient at the cavity wall, then the effective radial stress becomes negative, and the condition for tensile failure may be fulfilled at some point inside the wall. Thus a minimum criterion for tensile failure to occur is that the pore pressure gradient is larger than the radial stress gradient at the cavity wall, that is,

$$\left.\frac{\partial p_{\mathrm{f}}}{\partial r}\right|_{r=R_{\mathrm{c}}} > \left.\frac{\partial \sigma_r}{\partial r}\right|_{r=R_{\mathrm{c}}} \tag{10.21}$$

If the tensile strength is larger than zero, then this criterion may not be sufficient for tensile failure to occur.

The *normalised drawdown pressure gradient* (Morita et al., 1989b,a) is defined as

$$g_{pn} = R_{\mathrm{c}} \left.\frac{\partial p_{\mathrm{f}}}{\partial r}\right|_{r=R_{\mathrm{c}}} \tag{10.22}$$

where R_{c} is the radius of the cavity. The *critical drawdown pressure gradient* g_{pn}^{c} is the largest possible normalised drawdown pressure gradient without sand failure, and thus Eq. (10.21) expresses that

$$g_{pn}^{\mathrm{c}} = R_{\mathrm{c}} \left.\frac{\partial \sigma_r}{\partial r}\right|_{r=R_{\mathrm{c}}} \tag{10.23}$$

Again, we consider the simple case of a cylindrical cavity in a formation where the stress is isotropic. The radial stress is given by Eq. (4.55). Taking the derivative with respect to r, we find that the critical drawdown pressure gradient is given as

$$g_{pn}^{\mathrm{c}} = 2\left[\sigma_{\mathrm{h}} - p_{\mathrm{w}} - (p_{\mathrm{fo}} - p_{\mathrm{w}})\alpha \frac{1 - 2\nu_{\mathrm{fr}}}{2(1 - \nu_{\mathrm{fr}})}\right] \tag{10.24}$$

By combining Eqs. (4.64) and (10.22) we may find an expression for g_{pn} during stable production. The results shows that it is highly unlikely that g_{pn} will reach the critical limit (10.23).

Van den Hoek et al. (2000a) concluded, based on modelling accounting for strain localisation and grain rotations, that tensile failure may only occur in small holes like perforations, not in open holes. The reason for this is that shear failure will always precede tensile failure for a large cavity like an open hole. Due to the size effect, a small cavity like a perforation has a much higher threshold for shear failure, and hence tensile failure may occur first; however, even for small cavities it will only just precede shear failure. Van den Hoek et al. (2000a) found support for their conclusion in laboratory experiments showing no correlation between the (non-)occurrence of sand production and the flow rate (which controls the pore pressure gradient).

However, in a transient period when p_w is reduced, the pore pressure gradient at the cavity wall can be much larger. During start-up of the production, the well pressure is lowered, and the pore pressure gradient at the cavity wall will be significantly larger for a period until stable conditions are established. Thus there may be a critical rate for well pressure reduction to prevent sand production in the start-up period (Santarelli, 1994).

To find an expression for the critical well pressure reduction rate, we need to consider the pore pressure distribution around the cavity, which is given by the pressure diffusion equation

$$\frac{\partial p_f}{\partial t} = D_h \nabla^2 p_f \tag{10.25}$$

where D_h is the diffusion constant (see Eq. (1.320)). The boundary conditions are given by p_w and p_{fo}. A complete solution of this equation is somewhat complicated. However, for the situation where the fluid pressure at the cavity wall is changed abruptly from p_{fo} to p_w at time $t=0$, a rough approximation for the transient time range $0 < t < R_c^2/D_h$ is given by

$$p_f(r,t) \approx p_w + (p_{fo} - p_w)(1 - e^{-x}), \quad t \geq 0 \tag{10.26}$$

where

$$x = \frac{r - R_c}{\sqrt{\pi D_h t}} \tag{10.27}$$

Consider a situation where the well pressure is reduced from $p_w = p_{fo}$ to the production level p_{pr} at a constant rate $r_o = -\mathrm{d}p_w/\mathrm{d}t$ over a time $t_o = (p_{fo} - p_{pr})/r_o$. The pore pressure gradient at the cavity wall at time t_o (which is the time when the gradient is largest) is then given as

$$\frac{g_{pn}(t_o)}{R_c} = \left[\frac{\partial p_f}{\partial r}\right]_{\substack{r=R_c\\ t=t_o}} = \int_0^{t_o} \frac{\partial}{\partial r}\left[\frac{dp_f}{dp_w}\right]\frac{dp_w}{dt}\,\mathrm{d}t \approx 2\frac{p_{fo} - p_{pr}}{\sqrt{\pi D_h t_o}} \tag{10.28}$$

Recalling that failure occurs if $g_{pn} > g_{pn}^c$, we find that tensile failure may be triggered during start-up of the production if the well pressure is reduced so quickly that

$$t_o < \frac{4R_c^2}{\pi D_h}\left[\frac{p_{fo} - p_{pr}}{g_{pn}^c}\right]^2 \approx \frac{4R_c^2 \eta_f}{\pi k(K_{fr} + \frac{4}{3}G_{fr})}\left[\frac{p_{fo} - p_{pr}}{g_{pn}^c}\right]^2 \tag{10.29}$$

Here we have introduced the approximate expression (1.325) for the diffusion constant D_h. Note that the critical value for t_o is proportional to

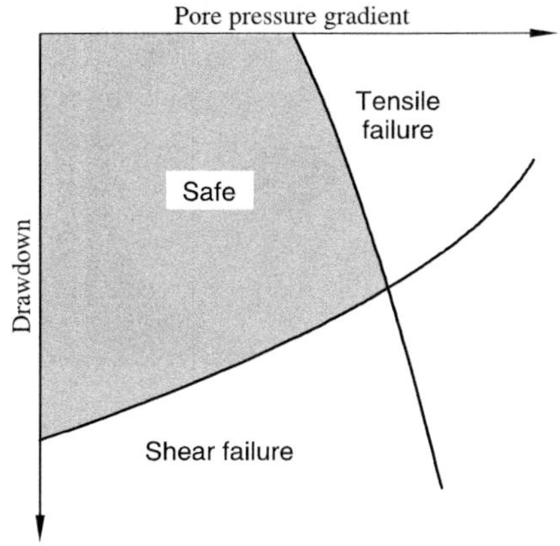

Figure 10.8 Stability diagram for production cavities.

the inverse of the permeability, and hence the lower the permeability is, the longer the start-up period should be to avoid stability problems. The critical value for t_o is also proportional to R_c^2, and hence this problem is of greater concern for open holes than for perforations.

We now have two criteria, one related to drawdown (Eqs. (10.8)–(10.15)) and one related to the pore pressure gradient, which limit the range for sand free production. The conditions are illustrated graphically in Fig. 10.8. This figure is an elementary version of a sand production stability diagram (Morita et al., 1989b,a).

Laboratory tests of drawdown induced sand production on outcrop and reservoir sandstones (Papamichos et al., 2008) revealed that cylindrical cavities may fail in three distinct patterns, depending on the rock type. The rocks were classified as Class A – brittle, Class B – ductile, and Class C – compactive. The corresponding failure patterns are shown in Fig. 10.9.

The failure mechanism for Class A rocks involves the formation of a slitlike failure zone that grows from the tip of the initial shear failure zone. The growth of such a failure zone may be explained as follows: At the tip of the slit the material fails in compression (see Chapter 2). If the material is contractant, which may be the case for high-porosity rocks, then the failed material takes up less space than it did before failure. The failed material is therefore only loosely attached and may easily be washed away by the flowing fluid. When the failed material is washed away, the condition for such failure is maintained at the tip of the slit. The slit may therefore grow rapidly, resulting in significant sand production. The formation of such

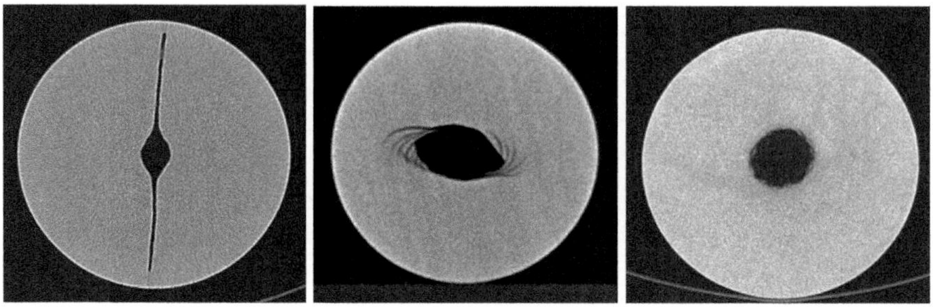

Figure 10.9 Characteristic failure patterns for sand producing cavities. Left: Class A; middle: Class B; right: Class C. Courtesy of SINTEF.

slitlike failure zones has been associated with compaction bands, which are zones of localised compactive failure (Haimson and Kovacich, 2003).

For Class B, the characteristic breakouts are formed in an initial sand production process. After that, the process slows down unless the conditions changes. The breakouts grow with increasing effective in situ stress. Class C rocks develop a uniform failure zone around the cavity, so that sand production results in a uniform enlargement of the cavity. The sand rate increases with increasing effective in situ stress.

The formation of a stable, yet continuously growing "wormhole", may take place under special conditions, for instance, at the interface between soft sand and a shale layer (van den Hoek et al., 2000c). This mechanism implies improved productivity and reduced sand production; however, it can only take place in the presence of a stronger, supporting rock.

10.2.4 Stability and collapse of sand arches

Observations have shown that it is possible to maintain stable sand arches while producing from unconsolidated sand (Hall and Harrisberger, 1970; Bratli and Risnes, 1981). Sand arches may be formed when damaged rock or loose sand is pushed towards a small opening like the perforated holes in the casing. Highly damaged rock, which is still able to form a stable structure around a cavity, possibly assisted by capillary forces, also represents a sand arch.

However, eventually, even the sand arches may fail. Sand production in the form of collapse of thin inner shells around cavities in wet, unconsolidated sand was observed by Bratli and Risnes (1981). They concluded that the sand production was initiated when the criterion for tensile failure (10.21) was fulfilled at the cavity wall. We will here outline the deriva-

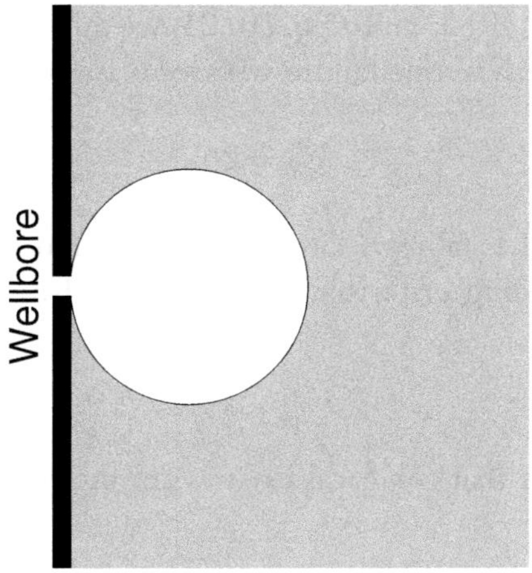

Figure 10.10 Schematic illustration of a spherical production cavity.

tion of Bratli and Risnes for a spherical cavity. The model they studied is shown schematically in Fig. 10.10.

Although the partial saturation gives the sand a finite cohesion (see Section 2.6.2), the material surrounding the cavity can be considered to be in a plastified state.

The radial stress gradient can be derived from one of the equations of equilibrium in spherical coordinates (Eq. (4.190)) giving

$$\frac{\partial \sigma_r}{\partial r} = 2\frac{\sigma_\theta - \sigma_r}{r} \tag{10.30}$$

We further assume that the yield criterion is defined by the Mohr–Coulomb criterion, which implies that the radial and tangential stresses are related by (Eq. (2.22))

$$\sigma_\theta - p_\mathrm{f} = C_0 + (\sigma_r - p_\mathrm{f}) \tan^2 \beta \tag{10.31}$$

At the cavity wall, $p_\mathrm{f} = p_\mathrm{w} = \sigma_r$, and thus, according to Eq. (10.31),

$$\sigma_\theta - \sigma_r = C_0 \tag{10.32}$$

Combining Eqs. (10.30) and (10.32), at the cavity wall, we have

$$\left.\frac{\partial \sigma_r}{\partial r}\right|_{r=R_\mathrm{c}} = 2\frac{C_0}{R_\mathrm{c}} \tag{10.33}$$

By introducing Eq. (10.33) into Eq. (10.23) we find that the critical pressure gradient g_{pn}^{c} at which tensile failure will occur is given as

$$g_{pn}^{c} = 2C_0 \tag{10.34}$$

This is essentially the stability criterion of Bratli and Risnes (1981).

The corresponding criterion for a cylindrical cavity is (Risnes et al., 1982)

$$g_{pn}^{c} = C_0 \tag{10.35}$$

which again shows that spherical cavities are more stable than cylindrical ones.

Eq. (10.34) is often expressed in terms of drawdown rather than flow rate. This is a valid approach in a situation with stable production, as drawdown and pore pressure gradient are then closely related. By combining Eq. (4.201) with Eqs. (10.22) and (10.4) we find that Eq. (10.34) can be expressed as

$$\mathcal{D}_c = 2C_0 = \frac{4S_0 \cos\varphi}{1 - \sin\varphi} \tag{10.36}$$

This is the critical drawdown for tensile failure of a spherical sand arch under stable production. The right part of Eq. (10.36) is derived from the relation between cohesion, friction angle and unconfined strength (Eq. (2.21)), and has been included since this form of the criterion is frequently used.

Note the line of arguments used in the derivation of Eq. (10.34) (and Eq. (10.36)): First, the material near the cavity is assumed to be in a plastified state due to shear failure. Next, sand production is provoked by tensile failure of the plastified material at the cavity wall.

Thus Eq. (10.36) describes the stability limit for a spherical sand arch formed by plastified rock. Beyond this limit, sand *will* be produced. Note that Eq. (10.20), which predicts a lower critical drawdown than Eq. (10.36), only describes the onset of shear failure for a spherical cavity surrounded by intact rock. Beyond that limit, sand *may* be produced from a cavity of this shape, since rock failure is a necessary but not necessarily sufficient requirement for sand production.

For a gas producing cavity, the pore pressure distribution is somewhat different since gas is a highly compressible fluid. This affects the pore pressure gradient at the cavity wall, and hence it also affects the critical drawdown. Weingarten and Perkins (1995) derived a generalised version of

Eq. (10.36), which includes the effect of fluid compressibility. The generalised equation can only be expressed in implicit form:

$$\frac{1}{m+1}\left[\mathcal{D}_c - p_{fo}\left(1-\left(\frac{p_{fo}}{p_{fo}-\mathcal{D}_c}\right)^m\right)\right]=2C_0 \qquad (10.37)$$

where the exponent m specifies the relation between the fluid density and the fluid pressure:

$$\rho_f=\gamma p_f^m \qquad (10.38)$$

The gas density coefficient γ is a constant. For an incompressible fluid, which we here can assume for water or oil, $m=0$, and Eq. (10.37) reduces to Eq. (10.36). For an ideal gas, $m=1$, and Eq. (10.37) becomes

$$\mathcal{D}_c=2C_0+p_{fo}-\sqrt{4C_0^2+p_{fo}^2} \qquad (10.39)$$

We notice that fluid compressibility reduces the critical drawdown.

The expressions above are based on the assumption that the fluid flow is following Darcy's law (1.308). This assumption does not hold for flow into high-rate gas wells due to the high flow rates. Ong et al. (2000) derived an expression corresponding to Eqs. (10.36) and (10.37) for non-Darcy flow. The somewhat complicated expression is shown in Appendix D.4.1. It is found that non-Darcy flow reduces $\mathcal{D}_c$ significantly, and hence it can be overly optimistic to neglect the influence of this effect in high-rate gas wells.

Note that the in situ stresses do not appear at all in the stability criteria for sand arches (Eqs. (10.34) and following). This is due to the plastified state of the material, which relaxes the stresses near the cavity wall. Although the in situ stresses are still important for inducing plastification, they obviously play a less important role for the onset of sand production according to these criteria than shear failure criteria like Eq. (10.17) indicate. For wells in HPHT reservoirs, this difference could be significant, as the conventional models based on shear failure alone tend to be overly conservative in their estimates of the onset of sanding under such conditions (Vaziri et al., 2002).

10.2.4.1 The impact of water

The sand in a hydrocarbon reservoir will normally be partly water saturated and waterwetting, whereas the remaining pore space is filled with hydrocarbons. This implies that the sand will have an additional cohesion due to capillary forces, as described in Chapter 2. Thus even unconsolidated sand reservoirs may have sufficient strength to build stable arches and resist the drag forces of the flowing fluid.

Water breakthrough into a well changes the water saturation of the rock near the well. This tends to reduce the capillary forces and thereby also the strength of the partially water-saturated reservoir sand (Section 2.6.2). Chemical effects may also induce a strength reduction in sandstone when oil or gas is replaced with water (Section 2.6.3). Reduced strength implies a reduction in critical drawdown. Hence a possible consequence of water breakthrough is that the conditions for sand failure may change from not fulfilled to fulfilled. This may explain the common observation that sand production accompanies water breakthrough (Morita et al., 1989a).

However, it has been claimed (Skjærstein et al., 1997) that sand production following water breakthrough is typically a temporary problem and that the well may even produce sand free at a larger drawdown some time after water breakthrough than it did before. Observations described by Willson et al. (2002) indicate that sand production increases with water cut up to about 50 % and declines towards the initial (no water) level when the water cut approaches 100 %. Possible explanations of such observations could be that the water breakthrough only releases loose sand that has been trapped in the cavities by capillary forces or that the cavities may obtain a more stable shape after the production of some sand. (Note that a spherical cavity tends to be more stable than a cylindrical one; see Section 4.10.2.)

Changes in fluid drag forces during a water breakthrough event are also considered as a potential explanation for these observations. These drag forces increase initially with increasing water cut as the relative permeability of hydrocarbon decreases. The relative permeability of water increases on the other hand with increasing water cut, and at some point, this effect dominates the drag forces so that a further increase in water cut leads to a decrease in the drag forces. This temporary increase in the drag forces may induce a temporary increase in sand production as the rate of produced sand is assumed to be proportional to these forces (Section 10.2.5).

10.2.5 Rate of produced sand

Even when the conditions for sand production are fulfilled, the sand may be produced at such a low rate that it is still acceptable from an operational point of view. It is therefore of interest to be able to predict not only the conditions for onset of sand production, but also the rate of sand production given that it occurs.

Based on a qualitative analysis, Morita (1994) stated that the rate of sand production depends on three factors: (1) how much the well pressure is reduced below the critical sand production pressure, (2) fluid flow rate and viscosity, and (3) cementation.

Since then, various numerical models have been developed for the prediction of sand production rate. The topic is still under development, however. Here we will here present a simplified analytical model (Fjær et al., 2004) for continuous sand production, based on a numerical model by Papamichos et al. (2001). The model is describing sand production from the rock around a cavity that has been brought beyond the onset of sand production.

Consider a volume element of the rock in the vicinity of the cavity. If sand is being produced from this element, then the porosity increases. Mathematically, this relation is described as

$$\frac{\partial m_{\mathrm{sand}}}{\partial t} = \rho_{\mathrm{s}} \frac{\partial \phi}{\partial t} \tag{10.40}$$

where m_{sand} represents the (cumulative) mass of the sand being produced per unit volume of rock, and ρ_{s} is the density of the sand grains.

We now assume that the erosion is driven by the flowing fluid in such a way that the rate of produced sand $\partial m_{\mathrm{sand}}/\partial t$ is proportional to the hydrodynamic forces (Eq. (10.3)) acting on the grains. This implies that the rate of produced sand is proportional to $\eta_{\mathrm{f}}(1-\phi)\phi^{-3}q_{\mathrm{fl}}$, where $q_{\mathrm{fl}} = Q/A$ is the fluid flux. We may also assume that a minimum force is required to uphold the erosion process, and hence no sand is produced if the fluid flux is lower than some critical limit $q_{\mathrm{fl}}^{\mathrm{cr}}$. Thus we have established a general equation for sand erosion:

$$\frac{\partial m_{\mathrm{sand}}}{\partial t} = \lambda_{\mathrm{sand}} \eta_{\mathrm{f}} \frac{1-\phi}{\phi^3} \left(q_{\mathrm{fl}} - q_{\mathrm{fl}}^{\mathrm{cr}}\right) \tag{10.41}$$

The *sand production coefficient* λ_{sand} has the dimension of s/m^3. The combination of Eqs. (10.40) and (10.41) gives us a differential equation for the porosity of the eroding rock. The solution is a little complicated, so we settle here for the approximate solution

$$\phi \approx \phi_o \left[1 + 4\frac{\lambda_{\mathrm{sand}} \eta_{\mathrm{f}}}{\rho_{\mathrm{s}} \phi_o^4} \left(q_{\mathrm{fl}} - q_{\mathrm{fl}}^{\mathrm{cr}}\right)(t - t_o)\right]^{\frac{1}{4}} \tag{10.42}$$

given the boundary condition $\phi = \phi_o$ (initial porosity) at time $t = t_o$. (Note that this approximation does not obey the restriction $\phi \leq 1$ for sufficiently large values of $t - t_o$; however, the process will be truncated before this becomes a problem, as we will see below.)

Next, we need to identify the volume of the rock that is in a state where it will produce sand. Considering a cylindrical cavity, we assume that the rock can be described by a simple linear elastic/perfectly plastic model as described in Sections 4.8.1 and 10.2.3. The plastic strain at a point r from the cavity centre can be expressed as (Eq. (4.175))

$$\lambda_p \propto \left(\frac{R_p}{r}\right)^2 - 1 \tag{10.43}$$

The radius R_p of the plastic zone is given by Eq. (4.148) and can be expressed as

$$R_p \propto R_c e^{\frac{\mathcal{D}}{C_0}} \tag{10.44}$$

using Eq. (10.4).

Sand is produced if a part of the rock has been deformed beyond its critical plastic strain limit λ_c (see Section 10.2.3.1). This is the case for a zone of the rock inside a critical radius R_{cr}. The situation is illustrated in Fig. 10.11.

Onset of sand production occurs when $\mathcal{D} = \mathcal{D}_c$. At this point, we also have $R_{cr} = R_c$. We may deduce from Eq. (10.43) that $R_{cr} \propto R_p$, and hence we find

$$R_{cr} = R_c e^{\frac{\mathcal{D}-\mathcal{D}_c}{C_0}} \approx R_c \left(1 + \frac{\mathcal{D} - \mathcal{D}_c}{C_0}\right) \tag{10.45}$$

The approximation is valid if the drawdown exceeds the critical value by a relatively small amount such that $\mathcal{D} - \mathcal{D}_c \ll C_0$. The volume of the sand producing zone V_{sp} can now be expressed as

$$V_{sp} = \pi L(R_{cr}^2 - R_c^2) \approx 2\pi L R_c^2 \frac{\mathcal{D} - \mathcal{D}_c}{C_0} \tag{10.46}$$

where L is the length of the cavity.

The erosional process that gradually increases the porosity of the sand producing zone will also reduce its ability to carry load. We will here account for this erosional-mechanical coupling by simply assuming that the load carrying ability of the zone remains constant until the porosity has reached a critical value ϕ_{cr}. At that point the zone collapses, and the remaining solids in the zone are produced in one burst.

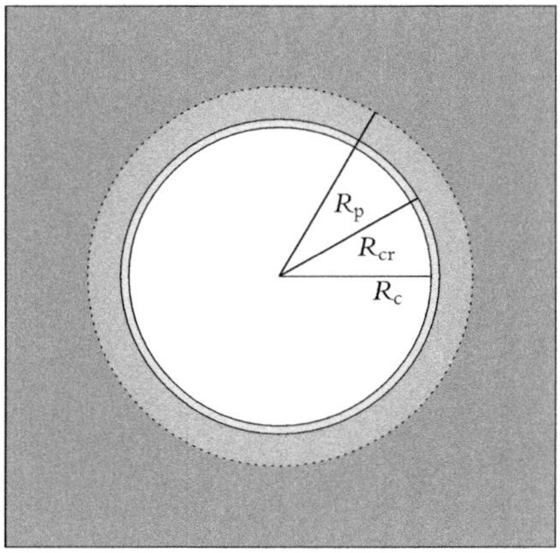

Figure 10.11 Schematic illustration of the plastified zone and the sand producing zone around a cavity.

The time Δt_{cr}, from the erosional process starts until the zone collapses (when $\phi = \phi_{cr}$), can be found from Eq. (10.42) and may be written as

$$\Delta t_{cr} = \tau_s \left(\frac{Q}{Aq_{fl}^{cr}} - 1 \right)^{-1} \tag{10.47}$$

The inflow area $A = 2\pi LR_c$ is the surface area of the cavity, and $Q = Aq_{fl}$ is the fluid flow rate for the cavity. The characteristic time constant τ_s for the erosion process is defined as

$$\tau_s = \frac{\rho_s \left(\phi_{cr}^4 - \phi_o^4 \right)}{4\lambda_{sand} \eta_f q_{fl}^{cr}} \tag{10.48}$$

The collapse of the sand producing zone implies that the radius of the cavity increases from R_c to $(1 + a)R_c$, where a can be found from Eq. (10.45):

$$a = \frac{R_{cr}}{R_c} - 1 \approx \frac{\mathcal{D} - \mathcal{D}_c}{C_0} \tag{10.49}$$

The stresses around the cavity are now redistributed, and the situation is the same as it was at $t = t_o$ (Fig. 10.11), except that the radii have been scaled:

$$R_c \to (1 + a)R_c; \quad R_{cr} \to (1 + a)R_{cr}; \quad R_p \to (1 + a)R_p \tag{10.50}$$

Thus the erosion process starts on a new cycle.

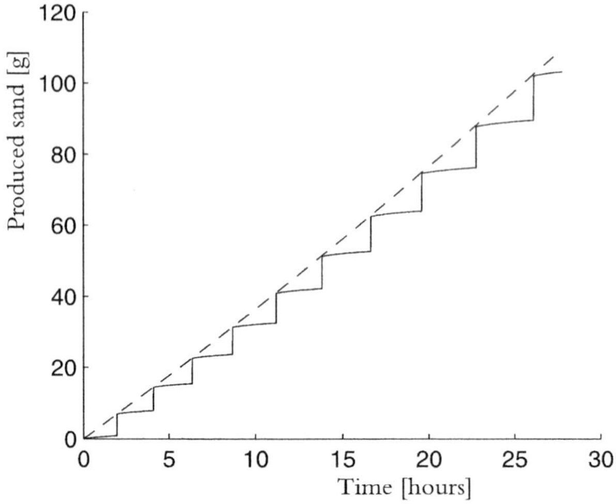

Figure 10.12 Predicted cumulative sand production from a perforation. Parameter values: $Q = 3.77$ l/min, $(\mathcal{D} - \mathcal{D}_c)/C_0 = 0.04$, $\lambda_{\text{sand}} = 200$ s/m^3, $\phi_o = 0.2$, $\phi_{\text{cr}} = 0.3$, $\rho_s = 2.65$ g/cm^3, $R_c = 5$ mm, $L = 0.5$ m, $\eta_f = 1$ cP, $q_{\text{fl}}^{\text{cr}} = 0.001$ m/s. The dashed line is an approximate average with slope given by Eq. (10.51).

The predictions of this model in terms of continuous erosion (Eqs. (10.40) and (10.42)) and the repeated collapses of the sand producing zone (Eqs. (10.46) and (10.47)) are illustrated in Fig. 10.12. We see that the predicted sand production is a cyclic process, where a period of relatively low sand production due to erosion is followed by a burst of sand caused by the collapse of the sand producing zone.

The average sand production rate $\dot{M}_{\text{sand}}$ can be estimated as the amount of sand in the sand producing zone ($= V_{\text{sp}}\rho_s(1 - \phi_o)$) divided by the time Δt_{cr} between each collapse. From Eqs. (10.46) and (10.47) we find that the average sand production rate is given as

$$\dot{M}_{\text{sand}} = M_{\text{sand}} \frac{2}{\tau_s} \frac{\mathcal{D} - \mathcal{D}_c}{C_0} \left(\frac{Q}{A q_{\text{fl}}^{\text{cr}}} - 1 \right) \tag{10.51}$$

where

$$M_{\text{sand}} = \pi L R_c^2 \rho_s (1 - \phi_o) \tag{10.52}$$

is the cumulative amount of sand produced from the cavity plus the amount of sand initially removed to create the cavity.

The rate $\dot{M}_{\text{sand}}$ has the dimension of kg/s and corresponds to the slope of the dashed curve shown in Fig. 10.12. This simplified model reproduces several of the features of sand production observed in laboratory tests (Papamichos et al., 2001):

- Sand is typically produced in bursts—only on a longer time scale the sand production appears as continuous.
- The average sand production rate increases with increasing flow rate.
- The average sand production rate increases with increasing drawdown above the critical drawdown limit.

We also notice that the model largely confirms the predictions of Morita (1994) (listed at the beginning of this section) as $\dot{M}_{\text{sand}}$ explicitly depends on (1) $\mathcal{D} - \mathcal{D}_c$, (2) Q and η_f (through τ_s), and (3) C_0.

Note that the sand production rate changes with time, since R_c increases as the cavity wall is eroded away. We may estimate the long-term development of $\dot{M}_{\text{sand}}$ by establishing an expression for the development of R_c.

We assume that the drawdown $\mathcal{D}$ is kept constant. Note that the surface area $A = 2\pi L R_c$ also increases with time. On the other hand, Q which is proportional to A/R_c (see Eqs. (4.63) and (4.64)) is approximately constant as $\mathcal{D}$ remains constant. By taking the change in R_c during one erosion cycle, which is $\Delta R_c = aR_c$, and dividing by the time Δt_{cr} between each collapse, we find that the average rate of change for R_c can be expressed as

$$\frac{dR_c}{dt} = \frac{\Delta R_c}{\Delta t_{cr}} = \frac{aR_c}{\tau_s}\left(\frac{Q}{Aq_{fl}^{cr}} - 1\right) = \frac{a}{\tau_s}[(b+1)R_{co} - R_c] \tag{10.53}$$

where R_{co} is the cavity radius at the onset of sand production, and the dimensionless parameter b is defined as

$$b = \frac{Q}{A_o q_{fl}^{cr}} - 1 \tag{10.54}$$

where $A_o = 2\pi L R_{co}$ is the surface area of the cavity at the onset of sand production. The parameter b is a constant as long as Q is constant, whereas a will change with time if $\mathcal{D}_c$ changes, for instance, as a result of depletion.

Integration of Eq. (10.53) gives us an expression for R_c as a function of time

$$R_c = R_{co}\left[1 + b(1 - e^{-t_a/\tau_s})\right] \tag{10.55}$$

where

$$t_a = \int_0^t a(t)\,dt \tag{10.56}$$

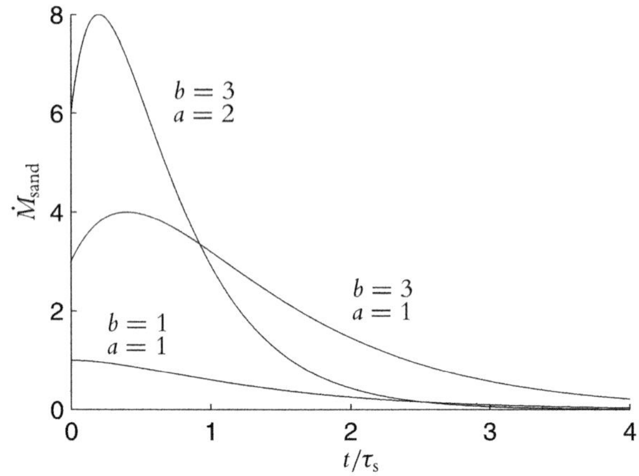

Figure 10.13 Predicted sand rate as given by Eq. (10.57), for different values of a and b. It is assumed that a does not change with time.

Introducing expression (10.55) into Eq. (10.51), we find that the sand rate as a function of time can be expressed as

$$\dot{M}_{\text{sand}} = M^{\text{o}}_{\text{sand}} \frac{2ab}{\tau_s} \left[1 + b(1 - e^{-t_a/\tau_s})\right] e^{-t_a/\tau_s} \tag{10.57}$$

where

$$M^{\text{o}}_{\text{sand}} = M_{\text{sand}}(t = 0) = \pi L R^2_{\text{co}} \rho_s (1 - \phi_o) \tag{10.58}$$

The time development of the sand rate $\dot{M}_{\text{sand}}$ is illustrated in Fig. 10.13. Here it is assumed that a is constant, so that $t_a = at$. Depending on the time scale and the amount of sand produced, the model predicts that sand production will appear either as a time limited event (transient sand production), as a stable process with a relatively constant sand production rate (continuous sand production) or as an accelerating process reaching disastrous proportions (catastrophic sand production). Note that a, which depends on the drawdown, affects the time scale, whereas b, which depends on the flow rate, affects the total amount of sand produced.

The cumulative amount of sand produced up to a given point in time can be found by integration of Eq. (10.57). The total amount of sand ($M^{\text{T}}_{\text{sand}}$) produced during one event, that is, the cumulative amount of sand produced from time $t = 0$ to some time $t \gg \tau_s/a$, where the sand rate is practically zero, is found to be

$$M_{\text{sand}}^{\text{T}} = M_{\text{sand}}^{\text{o}} b[b+2] = M_{\text{sand}}^{\text{o}} \left[\left(\frac{Q}{A_o q_{\text{fl}}^{\text{cr}}} \right)^2 - 1 \right] \tag{10.59}$$

Hence the total amount of produced sand depends only on the flow rate. Recall however that $\mathcal{D} > \mathcal{D}_c$ is still a necessary requirement for sand to be produced at all. Note in particular that $\mathcal{D}_c$ may increase with time as a result of changing shape of the cavity or some other form of post failure stabilisation. This slows down the erosion process and reduces the sand production rate. If $\mathcal{D}_c$ becomes equal to $\mathcal{D}$ at some point in time, then the sand production stops.

There is no known procedure for independent measurement of τ_s (or of λ_{sand} and ϕ_{cr}), and it is not clear to what extent these parameters depend on, for instance, cementation. For practical applications, the model therefore needs calibration, either from field observations or from sand production tests in the laboratory.

In its basic form, as presented here, the model describes a Class C failure pattern (Fig. 10.9). With some adjustments, the model may also be adapted to Class A and Class B cases (Fjær and Papamichos, 2008).

10.2.6 Sand transport

Sand production in the reservoir zone does not necessarily imply that the sand ultimately appears at the surface. For this to happen, the fluid flow must be strong enough to wash the sand up the entire well. Otherwise, the sand will remain in the well and possibly will be washed out at a later time if the flow rate is increased. This effect complicates the analysis of sand production in a well: no sand at the surface does not necessarily imply that no sand is being produced from the reservoir. Similarly, if sand appears at the surface, then it may be difficult to tell whether this sand has just been produced from the reservoir under the current conditions or was produced earlier under different conditions.

Sand transport by a flowing fluid is a rather complex problem to describe in general (see e.g. Gerhart et al. (1992) for an introduction to fluid mechanics), and we will here only take a brief look at some basic effects. Consider a spherical grain with diameter d_g and density ρ_s. Gravity and buoyancy produces a net downward pointing force F_g on this particle:

$$F_g = \frac{\pi}{6} d_g^3 (\rho_s - \rho_f) g \tag{10.60}$$

where g is the acceleration of gravity. In a stationary fluid, this force causes the grain to sink. As the grain is moving in the fluid, the motion is counteracted by the viscous drag force

$$F_{\mathrm{d}} = \frac{\pi}{8} d_{\mathrm{g}}^2 \rho_{\mathrm{f}} U^2 C_{\mathrm{dr}} \tag{10.61}$$

where U is the velocity of the grain, and C_{dr} is the dimensionless drag coefficient. The drag coefficient is a (rather complicated) function of the so-called Reynolds number defined as

$$Re \equiv d_{\mathrm{g}} \rho_{\mathrm{f}} (U/\eta_{\mathrm{f}}) \tag{10.62}$$

where η_{f} is the fluid viscosity. For values of $Re < 10^5$ (typically, for a non-smooth sphere), we may use the approximate expression

$$C_{\mathrm{dr}} \approx (24/Re)\sqrt{1 + 0.2Re + 0.0003Re^2} \tag{10.63}$$

For values of $Re > 10^5$, C_{dr} drops to about 0.1.

Eventually, the velocity of the grain will reach a value U_{t} (called the terminal settling velocity), where $F_{\mathrm{d}} = F_{\mathrm{g}}$, so that no net force is acting on it, and the velocity remains constant. By equating Eqs. (10.60) and (10.61) we find the terminal settling velocity

$$U_{\mathrm{t}} = 2\sqrt{\frac{d_{\mathrm{g}} g (\rho_{\mathrm{s}} - \rho_{\mathrm{f}})}{3 \rho_{\mathrm{f}} C_{\mathrm{dr}}}} \tag{10.64}$$

The fluid's ability to move the sand is closely linked to the ratio of the velocity of the fluid to the terminal settling velocity. As an approximation, we may say that the terminal settling velocity indicates the flow velocity required to move the sand grain along with the flowing fluid. If the terminal settling velocity is low, then the grains are easily carried away by the fluid, whereas a high terminal settling velocity requires a higher flow velocity to move the grains.

For sufficiently small grains, $Re < 1$, and the drag coefficient can be approximated by $C_{\mathrm{dr}} \approx 24/Re$. This gives

$$U_{\mathrm{t}} \approx \frac{d_{\mathrm{g}}^2 g}{18 \eta_{\mathrm{f}}} (\rho_{\mathrm{s}} - \rho_{\mathrm{f}}) \tag{10.65}$$

This expression is often referred to as Stokes' law. The general relation between terminal settling velocity, grain diameter and fluid viscosity, based on Eqs. (10.64) and (10.63), is illustrated graphically in Fig. 10.14.

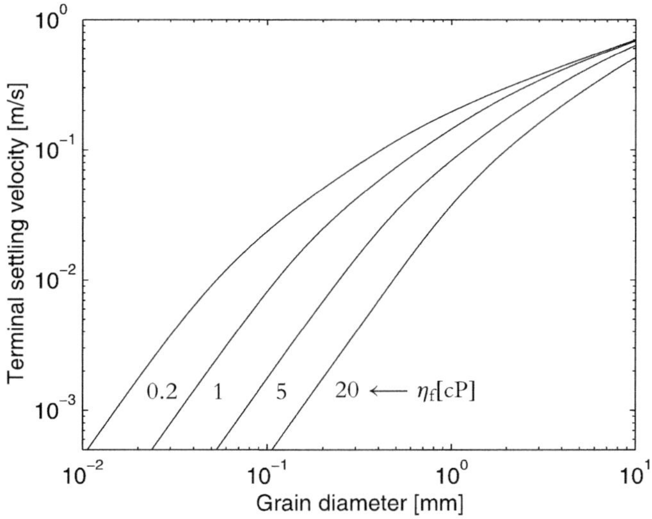

Figure 10.14 Terminal settling velocity as a function of grain diameter, derived from Eqs. (10.64) and (10.63), for four different values of the fluid viscosity.

These considerations tell us that larger grains have a much higher terminal settling velocity than smaller grains, and hence the larger grains are more likely to be trapped in the well. It also shows that low-viscosity fluids require a much higher flow velocity to be able to carry the same amount of sand as compared to high-viscosity fluids. As the flow velocity may vary along the well, the conditions for sand transport may also vary. Note that sand may be trapped at the lower end of the production zone due to the low fluid velocity down there. Also note that sand that has been trapped in the well while the fluid velocity was relatively low may be mobilised and washed to the surface if the fluid velocity is increased.

These results clearly show that there is not necessarily a one-to-one relationship between the amount of sand observed at the surface and the sand production in the reservoir. This effect obviously makes it difficult to evaluate the validity of sand prediction and complicates sand management. It also illustrates the need for reliable downhole sand detectors.

10.2.7 Sand prediction

Prediction of possible sand production is an important part of the development phase for a sandstone reservoir. Accurate predictions for where and under which conditions sand production will occur assist in the design of an optimal development plan and completion strategy for the wells in the reservoir.

Reliable sand prediction requires good, relevant models for sand production and a complete set of input data for the models. It is a general opinion that the models for sand production available today are able to predict the onset of sanding with reasonable accuracy, whereas models for predicting the amount of produced sand are still under development.

The relatively simple analytical models for onset of sand production described in Section 10.2.3 show that two groups of data are required for sand prediction, rock properties (primarily strength) and formation conditions (in situ stresses and pore pressure). However, a complete foot-by-foot coverage of such data is rarely available, and simpler tools for sand prediction are therefore frequently used.

The simplest possible tool is a criterion stating that sand control is not needed below a certain depth. This critical depth is established from field experience and varies from one region to another. The criterion may have a physical bearing in that rock strength usually increases with depth. Other tools relate the critical drawdown directly to the sonic transit time (compressional, shear or both), which can be argued for as there is some correlation between the sonic transit time and rock strength (see Section 8.2.3).

Models based on expressions like Eq. (10.17) may be used once a suitable calibration has been established for the actual formation, and some estimate of C_0 (alternatively C_{TWC}; cf. Eq. (10.18)) is available.

Some companies use finite element modelling assuming linear elastic behaviour to obtain a better description of the stress state around the cavities. The predictions from such modelling may be calibrated with strength data from TWC tests on material from the actual formation. This technique is based on the philosophy that nonlinear elasticity and plasticity are implicitly included in the TWC results and that linear elasticity may therefore be assumed for the finite element modelling of the stress state around the perforation.

Even more advanced numerical models are based on elasto-plastic modelling, like the highly recognised sand prediction model of Morita (Morita et al., 1989b,a). The kinematic hardening model describing the rock (see Raaen, 1998) is quite extensive in this case and requires several core tests for calibration.

For prediction of sand volume, additional parameters are needed, and the challenge to obtain relevant data is even larger. The model described in Section 10.2.5 requires, for instance, the sand production coefficient λ_{sand}, a parameter that so far can only be derived from sand production tests (Papamichos et al., 2001). Also, the flow rate per cavity (Q) is seen to be of large importance for the rate of sand production, and hence rock permeability and fluid viscosity are implicitly involved and need to be evaluated.

The choice of parameters is somewhat different among other models for prediction of sand volume, like the semi-empirical model of Willson et al. (2002) and the coupled numerical model of Chin and Ramos (2002), which links the amount of produced sand directly to the plastic deformation of the rock. However, the basic elements of the models appear to be the same.

10.3 Chalk

Solids production can be a serious problem also in chalk reservoirs. The mechanisms causing chalk production have however not been as well described as those causing sand production. Prediction and control of chalk production can therefore be more problematic.

In some cases, chalk is produced slowly at a relatively constant rate. Massive chalk production sometimes occur as sudden, rather unexpected events. In such situations, large amounts of chalk enter into the well, occasionally enough to kill the well. Although the mechanisms causing these events are not fully understood, it is known that sudden changes in the well conditions, like quick closure or opening of the fluid flow, may trigger such instabilities.

We may expect that the mechanisms believed to cause sand production also could cause chalk production under corresponding conditions. A couple of effects related to the nature of chalk versus that of sandstone make the situation somewhat different however. First, the matrix permeability in chalk is usually quite low, typically of the order 1 milliDarcy. This may imply that the conditions for tensile failure are more likely to be fulfilled for chalk. However, reservoir chalk is usually extensively fractured. This enhances the effective permeability while at the same time it implies that the fluid pressure distribution may be rather complicated on a small scale. It may therefore be questioned whether expressions like Eqs. (10.24) and (10.29) are valid for chalk. Second, the collapse pressure for high-porosity chalk is quite low so that this type of failure has to be considered in addition to shear failure and tensile failure. Collapse of the rock surrounding a producing cavity may have a devastating effect on the productivity of the cavity, even if it does not result in solids production.

When a rock collapses, the pore volume is largely reduced (see Section 2.4). This induces an abrupt increase in pore pressure, which takes some time to dissipate in a low-permeability material as chalk. The result may be a significant pore pressure build-up, leading to tensile effective mean stress and even liquefaction conditions. This process may explain observed chalk production in the field and in laboratory tests, where the chalk is seen to flow like toothpaste (Raaen and Renlie, 1990; Andersen, 1995; Kristiansen and Meling, 1996).

Tensile failure may occur if the smallest effective stress $\sigma'_{\min\perp}$ becomes negative. In the vicinity of a cavity, this may, for instance, occur if the pore pressure exceeds the radial stress, as described in Sections 10.2.3 and 10.2.4. The requirement for liquefaction to occur is that the mean effective stress $\bar{\sigma}'$ becomes negative (Section 2.8.4), which is less likely but more dramatic if it occurs.

Chalk is sensitive to the composition of the pore fluid and may exhibit a strong water-weakening effect (Section 2.6). Papamichos (1998) points out that chalk has a relatively low cohesion, often provided primarily by capillary forces, which exist only when the water saturation is low. Collapse, with subsequent liquefaction and extensive chalk production, may thus be induced by water flooding. Numerical modelling shows that this condition may extend deeply into the formation around a producer at water breakthrough.

References

Andersen, M.A., 1995. Petroleum Research in North Sea Chalk. RF — Rogaland Research, Stavanger, Norway.

Bratli, R.K., Risnes, R., 1981. Stability and failure of sand arches. Soc. Pet. Eng. J. 21 (2), 236–248. https://doi.org/10.2118/8427-PA.

Carman, P.C., 1956. Flow of Gases Through Porous Media. Butterworths, London.

Charlez, P.A., 1997. Rock Mechanics. Volume 2. Petroleum Applications. Éditions Technip, Paris.

Chin, L.Y., Ramos, G.G., 2002. Predicting volumetric sand production in weak reservoirs. SPE/ISRM78169. In: SPE/ISRM Rock Mechanics Conference. Irving, TX, USA.

Dullien, F.A.L., 1992. Porous Media. Fluid Transport and Pore Structure, second edition. Academic Press, New York.

Dusseault, M.B., El-Sayed, S., 2000. Heavy-oil production enhancement by encouraging sand production. SPE59276. In: SPE/DOE Improved Oil Recovery Symposium. Tulsa, OK, USA.

Fjær, E., Papamichos, E., 2008. The variable rate of sand production captured by an analytical model. In: Proc. of the 1st Southern Hemisphere International Rock Mechanics Symposium (SHIRMS 2008). Perth, Australia.

Fjær, E., Cerasi, P., Li, L., Papamichos, E., 2004. Modeling the rate of sand production. ARMA/NARMS04-588. In: Gulf Rocks. Houston, TX, USA.

Geilikman, M.B., Dusseault, M.B., 1997. Fluid-rate enhancement from massive sand production in heavy oil reservoirs. J. Pet. Sci. Eng. 17 (1–2), 5–18. https://doi.org/10.1016/S0920-4105(96)00052-6.

Geilikman, M.B., Dusseault, M.B., Dullien, F.A., 1994. Sand production as a viscoplastic granular flow. SPE27343. In: SPE International Symposium on Formation Damage Control. Lafayette, LA, USA.

Gerhart, P.M., Gross, R.J., Hochstein, J.I., 1992. Fundamentals of Fluid Mechanics. Addison-Wesley Publishing Company.

Haimson, B., Kovacich, J., 2003. Borehole instability in high-porosity Berea sandstone and factors affecting dimensions and shape of fracture-like breakouts. Eng. Geol. 69 (3–4), 219–231. https://doi.org/10.1016/S0013-7952(02)00283-1.

Hall, C.D., Harrisberger, W.H., 1970. Stability of sand arches: a key to sand control. J. Pet. Technol. 22 (7), 821–829. https://doi.org/10.2118/2399-PA.

Haugen, K., Kvernvold, O., Ronold, A., Sandberg, R., 1995. Sand erosion of wear-resistant materials: erosion in choke valves. Wear 186–187, 179–188. https://doi.org/10.1016/0043-1648(95)07158-X.

Kristiansen, T.G., Meling, S., 1996. A production parameter analysis of chalk influxes in the Valhall field. In: Proc. 5th North Sea Chalk Symposium. Reims, France.

Morita, N., 1994. Field and laboratory verification of sand-production prediction models. SPE Drill. Completion 9 (4), 227–235. https://doi.org/10.2118/27341-PA.

Morita, N., Whitfill, D.L., Fedde, Ø.P., Løvik, T.H., 1989a. Parametric study of sand production prediction: analytical approach. SPE Prod. Eng. 4 (1), 25–33. https://doi.org/10.2118/16990-PA.

Morita, N., Whitfill, D.L., Massie, I., Knudsen, T.W., 1989b. Realistic sand-production prediction: numerical approach. SPE Prod. Eng. 4 (1), 15–24. https://doi.org/10.2118/16989-PA.

Ong, S., Ramos, G.G., Zheng, Z., 2000. Sand production prediction in high rate, perforated and open-hole gas wells. SPE58721. In: SPE International Symposium on Formation Damage Control. Lafayette, LA, USA.

Papamichos, E., 1998. Chalk production and effects of water weakening. Int. J. Rock Mech. Min. Sci. 35 (4–5), 529–530. https://doi.org/10.1016/S0148-9062(98)00126-0.

Papamichos, E., Vardoulakis, I., Tronvoll, J., Skjærstein, A., 2001. Volumetric sand production model and experiment. Int. J. Numer. Anal. Methods Geomech. 25 (8), 789–808. https://doi.org/10.1002/nag.154.

Papamichos, E., Stenebråten, J., Cerasi, P., Lavrov, A., Vardoulakis, I., Fuh, G.-F., Brignoli, M., de Castro Gonçalves, C.J., Havmøller, O., 2008. Rock type and hole failure pattern effects on sand production. ARMA08-217. In: 42nd US Rock Mechanics Symposium and 2nd Canada–US Rock Mechanics Symposium. San Francisco, CA, USA.

Raaen, A.M., 1998. Efficient determination of the parameters of an elastoplastic model. SPE/ISRM47362. In: Proc. Eurock '98. Trondheim, Norway, pp. 277–283.

Raaen, A.M., Renlie, L., 1990. Relations between acoustic and static mechanical properties of chalk. In: Proc. 3rd North Sea Chalk Symposium. Copenhagen, Denmark.

Risnes, R., Bratli, R., Horsrud, P., 1982. Sand stresses around a borehole. Soc. Pet. Eng. J. 22 (6), 883–898. https://doi.org/10.2118/9650-PA.

Santarelli, F.J., 1994. Similarities and differences between wellbore stability and sand production. In: Aasen, J.O., Berg, E., Buller, A.T., Hjelmeland, O., Holt, R.M., Kleppe, J., Torsæter, O. (Eds.), North Sea Oil and Gas Reservoirs — III. Kluwer Academic Publishers, pp. 331–338.

Skjærstein, A., Tronvoll, J., Santarelli, F.J., Jøranson, H., 1997. Effect of water breakthrough on sand production: experimental and field evidence. SPE38806. In: SPE Annual Technical Conference and Exhibition. San Antonio, TX, USA.

van den Hoek, P.J., Hertogh, G.M.M., Kooijman, A.P., de Bree, P., Kenter, C.J., Papamichos, E., 2000a. A new concept of sand production prediction: theory and laboratory experiments. SPE Drill. Completion 15 (4), 261–273. https://doi.org/10.2118/65756-PA.
van den Hoek, P.J., Kooijman, A.P., de Bree, P., Kenter, C.J., Zheng, Z., Khodaverdian, M., 2000b. Horizontal-wellbore stability and sand production in weakly consolidated sandstones. SPE Drill. Completion 15 (4), 274–283. https://doi.org/10.2118/65755-PA.
van den Hoek, P.J., Kooijman, A.P., de Bree, Ph., Kenter, C.J., Sellmeyer, H.J., Willson, S.M., 2000c. Mechanisms of downhole sand cavity re-stabilisation in weakly consolidated sandstones. SPE65183. In: SPE European Petroleum Conference. Paris, France.
Vardoulakis, I., 2004. Fluidisation in artesian flow conditions: hydromechanically stable granular media. Géotechnique 54 (2), 117–130. https://doi.org/10.1680/geot.2004.54.2.117.
Vaziri, H., Xiao, Y., Palmer, I., 2002. Assessment of several sand prediction models with particular reference to HPHT wells. SPE/ISRM78235. In: SPE/ISRM Rock Mechanics Conference. Irving, TX, USA.
Veeken, C.A.M., Davies, D.R., Kenter, C.J., Kooijman, A.P., 1991. Sand production prediction review: developing an integrated approach. SPE22792. In: SPE Annual Technical Conference and Exhibition. Dallas, TX, USA.
Weingarten, J.S., Perkins, T.K., 1995. Prediction of sand production in gas wells: methods and Gulf of Mexico studies. J. Pet. Technol. 47 (7), 596–600. https://doi.org/10.2118/24797-PA.
Willson, S.M., Moschovidis, Z.A., Cameron, J.R., Palmer, I.D., 2002. New model for predicting the rate of sand production. SPE/ISRM78168. In: SPE/ISRM Rock Mechanics Conference. Irving, TX, USA.

CHAPTER 11

Mechanics of hydraulic fracturing

Hydraulic fracturing in rocks takes place when the fluid pressure within the rock exceeds the smallest principal stress plus the tensile strength of the rock. This results in tensile failure or splitting. A hydraulic fracture may be initiated by natural geological processes in the Earth whereby the fluid pressure increases and/or the smallest principal stress decreases. Artificial or man-made hydraulic fractures in petroleum activities are normally initiated by increasing the fluid pressure in the borehole to the point where the smallest principal stress at the borehole becomes tensile. Continued pumping at an elevated pressure causes the formation to split, and the fracture grows in the direction of least resistance. Some distance away from the borehole the fracture will always propagate in the plane normal to the smallest principal stress in that specific formation.

As the least principal stress often is in a horizontal direction, the resulting fractures will normally be vertical. If we consider a vertical open hole, then the picture will be as indicated in Fig. 11.1. As the figure shows, two symmetric fracture wings develop perpendicularly to the least principal stress.

Hydraulic fracturing has been used commercially as a stimulation technique in the petroleum industry since the early 1950s. Such fracturing jobs are designed to stimulate production from reservoirs with low permeability. This often involves pumping large amounts of fluid and solids (proppants), thus creating long fractures filled with proppants. A massive hydraulic fracturing (MHF) job may exceed one thousand cubic metres of fluid and one million kilograms of proppant. The fracture thus creates a high-permeability flow channel towards the wellbore that has a large drainage area towards the low-permeability formation. If the fracture was not filled with solid material, it would close when the fluid pressure drops.

Unintentional fracturing may occur during drilling operations, often referred to as lost circulation; see also Section 9.1.2. This is a potentially hazardous situation, since the hydraulic support of the fluid column in the wellbore is reduced. The wellbore pressure may drop below the pore pressure in the formation, and if the formation is permeable, then formation fluids may start to flow into the wellbore in an uncontrolled manner. In the worst case, a blowout can be the result.

Developments in Petroleum Science, Volume 72
ISSN 0376-7361, https://doi.org/10.1016/B978-0-12-822195-2.00020-6

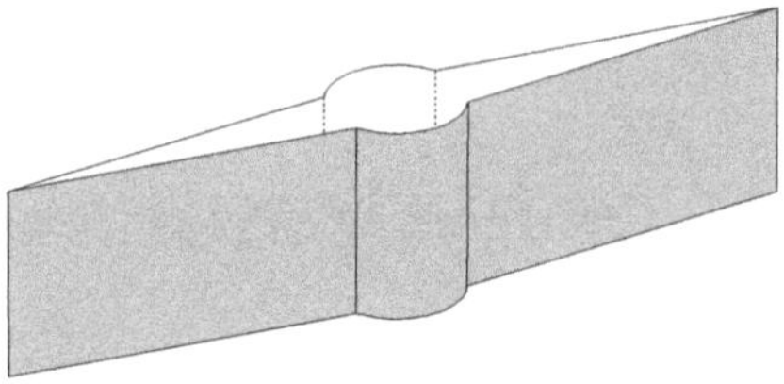

Figure 11.1 Vertical fracture around a vertical well.

Today, hydraulic fracturing has several applications:

- Unconventionals utilise multi-stage hydraulic fracturing from horizontal wells. As mentioned above, enabling economic production from low-permeability rock has been the main driver for hydraulic fracturing. Whereas several attempts to rubbilise hard rock have been made (by e.g. injection of liquid nitroglycerin in the late 1800s (Montgomery and Smith, 2010)); the first planned hydraulic fracturing operation took place in the Hugoton gas field in Kansas (USA) in 1947. Since then, millions of wells have been fractured, and the combination of horizontal drilling and staged fracturing has made way for the development of shale gas and shale oil (tight oil) resources in North America during the last decades.
- "Frac and pack" operations are used in weak sandstone reservoirs. These are relatively short and wide proppant-filled fractures, designed to penetrate near-well damaged zones and thus reduce the pressure drop close to the borehole and hence also the sanding potential of the well.
- Fractures can be created also during water injection in reservoirs. Injectivity may be reduced with time due to partial plugging of the formation around the injector to the point where the injection rates can only be maintained if the injection pressure is raised above the fracture pressure. Fracturing during injection can occur also at constant injection pressure. This occurs during water injection when the water is consistently colder than the formation. Cooling causes the formation to shrink, and thus the stresses are reduced below the initial fracturing pressure.
- Massive hydraulic fracturing jobs are normally preceded by a fracture test with a small injection volume (mini-frac) to determine the smallest stress in the formation and thus the required treating pressure and other design parameters of the fracturing job.

- In drilling operations, leak-off tests are performed to determine the fracture gradient of the formation below a casing shoe and thus the maximum allowable well pressure to prevent lost circulation. If a sufficiently large volume is pumped, then this will create a small fracture around the borehole.
- An extended leak-off test is essentially a standard leak-off test where pumping continues long enough to ensure that a fracture is created. By monitoring the shut-in phase and the flowback phase the smallest principal stress can be estimated.
- Hydraulic fracturing can be used for subsurface deposition of waste. In the petroleum industry, reinjection of contaminated drill cuttings is one example. In this application, it is important to have a permeable formation at or somewhere above the injection point, which can act as a recipient and pressure reducer. If this condition is not present, then the pressure will remain high due to very slow leak-off, and the fracture created during injection may eventually penetrate all the way to the surface.

In this chapter, we discuss basic processes related to hydraulic fracturing, whereas hydraulic fracturing applied as a technique for stress determination is discussed further in Chapter 8. Some aspects of hydraulic fracturing related to borehole stability (lost circulation) are treated in Chapter 9. Details concerning design, operation and interpretation of hydraulic fracturing for stimulation has been treated extensively in the literature and here will not be discussed further. The reference list includes some key references also in this field.

11.1 Conditions for tensile failure

From a macroscopic point of view, fracturing is related to tensile failure. A general background to tensile failure is given in Section 2.2. Man-made hydraulic fractures, which are the main topic of this chapter, are tensile failures initiated at the borehole wall. The mechanics of this process is described in Section 4.5, and the application to borehole stability analysis is given in Section 9.2.

Here we recall that the criterion for tensile failure initiation Eq. (2.3) is

$$\sigma_3' = -T_0 \tag{11.1}$$

This means that a hydraulic fracture is generated by increasing the fluid pressure and that the fracture opens up in the energetically most favourable direction, that is, normal to the minimum principal stress. This implies that the shape of the fracture eventually becomes more or less planar, depending on local heterogeneities. Further, the rock parameter controlling fracture initiation is the tensile strength T_0.

Consider a fracture that has been formed by the tensile failure mechanism but is closed by a decrease in pore pressure or an increase in the external stress. To reopen this fracture, the fluid pressure p_f in the fracture has to be increased so that

$$\sigma_3 - p_f < 0 \tag{11.2}$$

There are two points to make here: First, there is no tensile strength anymore, since the grain bonds were broken during the fracture initiation. Second, the fluid pressure inside the fracture is the relevant pressure. This can be seen by assuming that the two sides of the fracture are covered by an impermeable film and that a fluid of pressure p_f is injected between the films.

The condition for reopening or keeping a fracture open is given by Eq. (11.2). The internal pore pressure in the formation plays no role as long as it remains below σ_3 and p_f. Note that the criterion for closing a fracture is given by the expression in Eq. (11.2) becoming > 0.

The continued growth of the fracture depends on fracture tip processes incorporating fracture toughness and operational parameters related to the fracturing fluid, like pressure in the fracture, flow rate and fluid viscosity. This will be discussed further in the following sections.

11.2 Fracture initiation and formation breakdown

The basics of borehole failure (in shear and tension) were discussed in Section 4.5 and represent a starting point for further elaborations in this chapter.

To illustrate the concepts of fracture initiation and formation breakdown, let us consider the following idealised situation: A vertical borehole penetrates a stress field with isotropic horizontal stresses, that is, $\sigma_v > \sigma_H = \sigma_h$. The rock is assumed to be isotropic and homogeneous and obey Hooke's law of linear elastic behaviour. We further assume an *impermeable* borehole wall, implying that the pore pressure in the formation remains constant and unaffected by the wellbore pressure. In fracturing terminology, we may refer to this as a situation of no fluid invasion (non-penetrating

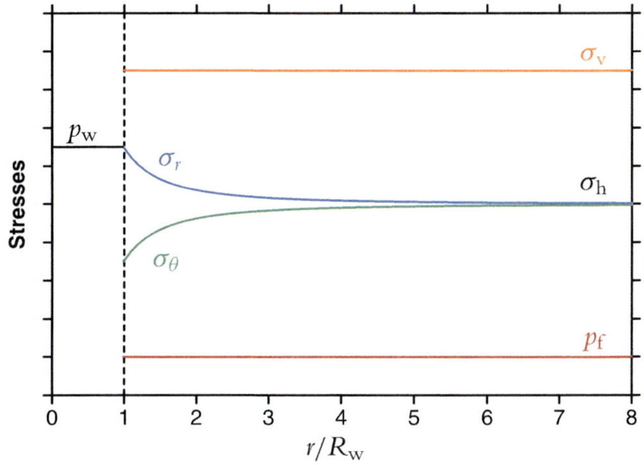

Figure 11.2 Stress distribution around an open vertical hole in an impermeable formation at high well pressure.

fluid). As the well pressure p_w is increased, the tangential stress σ_θ will be reduced correspondingly (Eq. (4.113)); see Fig. 11.2.

The condition for initiation of a hydraulic fracture is thus reached when

$$\sigma_\theta - p_f = -T_0 \tag{11.3}$$

or

$$p_{w,max}^{frac} = 2\sigma_h - p_f + T_0 \tag{11.4}$$

The pressure response in the borehole will then be as illustrated in Fig. 11.3.

The first linear part represents the elastic deformation of the system in and around the borehole, primarily compression of the fluid in the borehole. The peak represents the fracture initiation condition, that is, the creation of a vertical fracture on the borehole wall. The well pressure drops instantaneously at this point. This implies a situation of unstable fracture growth, whereby the volume of the fracture grows at a higher rate than that of fluid injection. Continued pumping will eventually result in stable fracture growth represented by the constant well pressure level. In this idealised case the points of fracture initiation and formation breakdown are thus identical.

The second curve in Fig. 11.3 shows the response that would occur if a second pressure cycle was run. Then the only resistance to fracture initiation and formation breakdown would be the stress concentration around the borehole. The tensile strength is now zero, since the fracture already

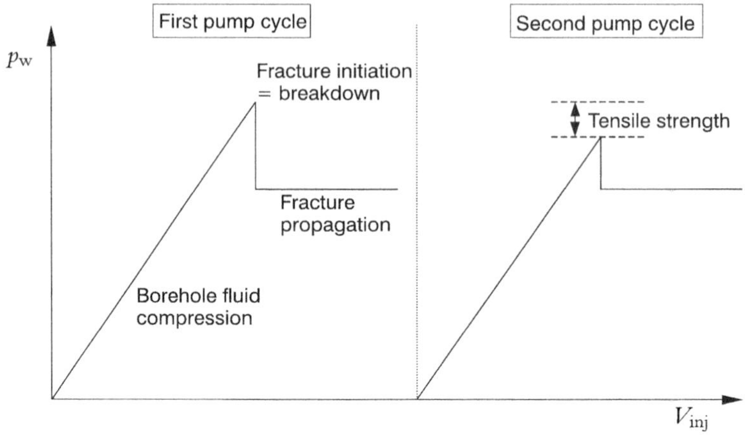

Figure 11.3 Idealised borehole pressure versus injected volume during hydraulic fracturing of a vertical wellbore. Two pressure cycles are included.

exists. The difference between the well pressures corresponding to the first and the second peak would thus ideally be a direct measure of the tensile strength of the formation. In practice, however, the presence of the fracture may make the effective stress concentration smaller in the repeat cycle than in the first, meaning that the difference is not only related to the tensile strength.

If the borehole is assumed to be fully *permeable*, so that fluid invasion occurs, and the pressurisation rate is slow enough to ensure steady-state conditions during pumping, then the tangential stress at the borehole wall is given by Eq. (4.59):

$$\sigma_\theta = 2\sigma_h - p_w - (p_{fo} - p_w)\frac{1 - 2\nu_{fr}}{1 - \nu_{fr}}\alpha \tag{11.5}$$

Inserting this expression for the tangential stress into Eq. (11.3) and accounting for the fact that the pore pressure near the borehole and the well pressure now are equal, the fracture initiation pressure becomes (Eq. (4.124))

$$p_{w,max}^{frac} = \frac{2\sigma_h - \alpha\frac{1-2\nu_{fr}}{1-\nu_{fr}}p_{fo} + T_0}{2 - \alpha\frac{1-2\nu_{fr}}{1-\nu_{fr}}} \tag{11.6}$$

If we assume that $\alpha = 1$, then the expression can be simplified to

$$p_{w,max}^{frac} = 2(1 - \nu_{fr})\sigma_h - (1 - 2\nu_{fr})p_{fo} + (1 - \nu_{fr})T_0 \tag{11.7}$$

Eq. (11.4) represents the upper limit for the fracture initiation pressure, and Eq. (11.6) represents the lower limit for the fracture initiation pressure. The upper limit is often referred to as the "fast" pressurisation limit, and the lower limit as the "slow" pressurisation limit (Detournay and Carbonell, 1997). In a real situation the fracture initiation pressure will usually be somewhere between these two limits, depending on several factors:

- There may be cracks and flaws at the borehole wall, so that the effective tensile strength of the rock is essentially zero, even if the tensile strength of the intact formation may be significant.
- There may be some time-dependent transfer of pressure between the wellbore and the formation. This will depend on the permeability of the formation, the ability of the fracture fluid to build a filter cake prior to fracture initiation and also the pump rate.

Considering the effect of pump rate on pore pressure evolution alone, this effect is only significant for very low rock permeabilities and/or if a low-permeability filter cake is established. Detournay and Cheng (1992) showed that for a fluid like water, the permeability of the rock would have to be lower than 1 μDarcy before any effects of rate would be experienced. Above this value the rock could be assumed to be fully permeable, and the lower limit of fracture initiation would apply (Eq. (11.6)).

In a real situation the time-pressure curves will not be as schematic as shown in Fig. 11.3. Fig. 11.4 illustrates more realistic curves. The curve to the left has a distinct breakdown or initiation pressure. This is commonly seen in an intact borehole with a straight line up to breakdown. Sometimes the curve starts to deviate from linear shortly before breakdown, thus representing a leak-off point; see also Section 8.3.4. The exact reasons for the appearance of a leak-off point are not known, but some probable causes may be listed randomly: filter cake efficiency, plastification, stress dependent elastic properties, and temperature effects. However, if a significant volume is pumped between leak-off and breakdown, then this can be an indication of a test of lower quality, for example, caused by upward leakage behind the casing. The figure to the right is the second cycle of the same test, illustrating the much lower initiation pressure experienced when the wellbore wall is already fractured. Then there is little difference between the pressure required to re-open the fracture and the pressure required to propagate the fracture. Similar response is seen when the formation is naturally fractured.

Let us assume that cracks or fractures exist at the borehole wall such that the inherent tensile strength of the rock is broken. We further assume

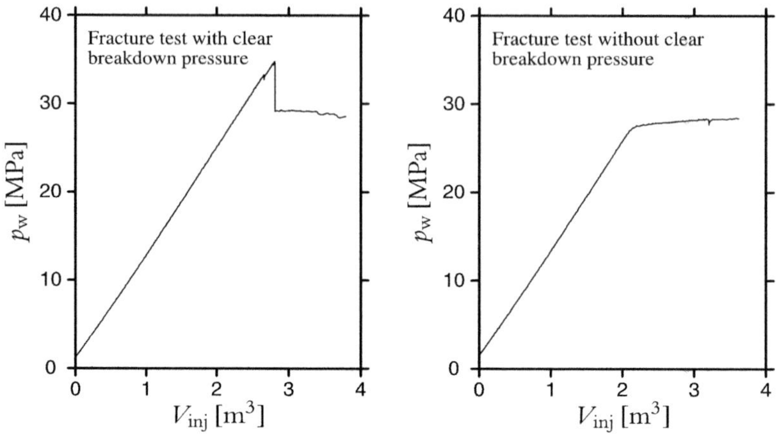

Figure 11.4 Topside pressure verses injected volume during hydraulic fracturing of a nearly vertical wellbore, illustrating an intact wellbore wall (left) and a pre-fractured wellbore wall (right). The figure to the left has a distinct breakdown pressure, whereas this is not the case in the figure to the right. The data are from two cycles of an XLOT with mud (After Gederaas and Raaen, 2009). Notice the "noise" before the breakdown in cycle 1. This is due to a short stop in pumping. The same applies to the "glitch" at about 3.2 m^3 in cycle 2.

that the tangential stress is transmitted across the fracture but that the well pressure can penetrate into the fracture without progressing further into the formation, which would require a sealing mud cake around the fracture. Such a scenario is possible when repressurising a previously fractured borehole in a sandstone. Referring to Eq. (11.4), this would give

$$p_{\mathrm{w,max}}^{\mathrm{frac}} = \sigma_{\mathrm{h}} \tag{11.8}$$

This would hence imply a fracture initiation pressure equal to the smallest horizontal stress. This situation is however not automatically detectable in the pressure-volume response, especially if the fractures are small, because there is still resistance to fracture growth at the tip. It is not until this fracture tip resistance is overcome and the fracture starts to grow that a deviation from linearity in the pressure-volume response can be observed. Deviation is most likely to be observed at a lower pressure in a shale than in a sandstone, as the discussion below explains. This difference between shale and sandstone is however independent of whether fractures already exist or not.

It has been observed experimentally that a fracture initiated on a borehole wall can grow significantly (several cm) before it becomes unstable

(Morita et al., 1996). This is explained by an internal filter cake in the fracture, which protects the fracture tip from the pressure in the rest of the fracture. This implies that a certain overpressure in the fracture is required before the pressure at the tip is large enough to cause breakdown and unstable growth of the fracture. This behaviour is only possible in a permeable rock like a sandstone and can thus explain why there is apparently less difference between fracture initiation and breakdown in a sandstone than a shale. For this type of behaviour, the slope of the curve will remain almost constant also after fracture initiation, until fluid is actually beginning to flow into the fracture. Another factor which can contribute to this apparent difference between sandstone and shale is the much larger permeability of a sandstone. In a sandstone a filter cake is built, which helps seal the wellbore wall, and there will be much less effect of borehole pressurisation on the pore pressure near the borehole.

In some cases a distinct breakdown pressure is not observed (see right-hand curve in Fig. 11.4). In that case the fracture growth is stable throughout.

If the horizontal stresses are different, then the stress distribution is modified. The basic formulae have been presented in Chapter 4 and will not be repeated here. If we assume σ_{h} to be the smallest principal stress and σ_{H} is the intermediate principal stress, then the fracture initiation pressure for the impermeable or fast pressurisation case in a vertical well is given as

$$p_{\mathrm{w,max}}^{\mathrm{frac}} = 3\sigma_{\mathrm{h}} - \sigma_{\mathrm{H}} - p_{\mathrm{f}} + T_0 \tag{11.9}$$

Note that this expression can be obtained from Eq. (11.4) by simply substituting $2\sigma_{\mathrm{h}}$ by $(3\sigma_{\mathrm{h}} - \sigma_{\mathrm{H}})$. Similarly, the same substitution can be made in Eq. (11.6) to get the expression for the lower limit of fracture initiation pressure. Since σ_{H} by definition is larger than σ_{h}, this also implies that in this case of anisotropic horizontal stresses the fracture initiation pressure is lower than if the horizontal stresses are equal. Note, as explained in Chapter 4, that fracture initiation occurs in the direction of minimum tangential stress at the borehole wall.

Consider Eq. (11.7) with $\nu \to 0.5$ and a negligible tensile strength. This yields

$$p_{\mathrm{w,max}}^{\mathrm{frac}} \approx \sigma_{\mathrm{h}} \tag{11.10}$$

Thus the fracture will initiate at a pressure equal to the pressure required to propagate the fracture (neglecting any other effects on the fracture propagation pressure). In this case, there will hence be no stress concentration effect around the borehole on the fracture initiation pressure.

If we again consider the same equation (11.7) but now with the substitution that reflects anisotropic horizontal stresses, then the result is

$$p_{\mathrm{w,max}}^{\mathrm{frac}} = \frac{3\sigma_{\mathrm{h}} - \sigma_{\mathrm{H}}}{2} \tag{11.11}$$

or

$$p_{\mathrm{w,max}}^{\mathrm{frac}} = \sigma_{\mathrm{h}} - \frac{\sigma_{\mathrm{H}} - \sigma_{\mathrm{h}}}{2} \tag{11.12}$$

This simply shows that in this special case the fracture will initiate at a pressure that is in fact lower than the smallest principal stress in the formation σ_{h}. Thus the fracture initiation pressure is lower than the propagation pressure, and an increase in the borehole pressure would be required to make the fracture propagate. Similar effects can be observed in deviated boreholes, also in formations with isotropic horizontal stresses, since the stress anisotropy then is induced by the difference between the vertical and horizontal stresses.

Again it should be emphasised that the above expressions are based on idealised conditions. The real fracture initiation pressure will depend on stress conditions, borehole direction and inclination, rock properties (e.g. tensile strength, permeability), borehole fluid properties and operational procedures.

The concept of fracture toughness was introduced in Section 6.4, where it was used to introduce a microscopic description of rock failure initiation. Fracture toughness expresses the resistance of a material to unstable crack growth, and it represents the consumption of surface energy in forming new surfaces (Ouchterlony, 1990). Fracture toughness can be seen as a key parameter in describing fracture growth after initiation, in the sense that it may be incorporated in numerical models to predict the extent of fracture growth. In mechanical engineering, fracture toughness measurements on solids provide valuable information about the risk for unwanted catastrophic failure.

Eq. (6.71) can be rearranged to estimate the *net fracture pressure* p_{net},

$$p_{\mathrm{net}} = p_{\mathrm{frac}} - \sigma_{\mathrm{h}} \tag{11.13}$$

required to make a hydraulic fracture grow in terms of fracture toughness K_{IC} and the initial fracture size (here taken as the fracture half-length L):

$$p_{net} \sim \frac{K_{IC}}{L} \tag{11.14}$$

where it is assumed that σ_h is the smallest principal stress. The proportionality factor depends on fracture geometry. This equation illustrates that the longer the initial fracture length is, the less significant is the impact of fracture toughness on crack growth. However, this is obviously also a question of the magnitude of K_{IC}; see Section 7.5.5, where laboratory techniques and observations are described.

11.3 Fracture orientation, growth and confinement

So far we have only considered a vertical hole for which a vertical fracture normal to the smallest horizontal stress will always be the result when $\sigma_v > \sigma_H > \sigma_h$. For deviated and horizontal holes the situation is much more complicated and thus difficult to generalise.

For illustration, let us consider a horizontal hole in a stress field where $\sigma_v > \sigma_H > \sigma_h$. In this situation, it is obvious that the fracture initiation pressure depends also on the azimuthal direction of the borehole. Consider two situations:

1. Borehole parallel with σ_H: $\quad p_{w,max}^{frac} = 3\sigma_h - \sigma_v - p_f + T_0 \quad$ (11.15)
2. Borehole parallel with σ_h: $\quad p_{w,max}^{frac} = 3\sigma_H - \sigma_v - p_f + T_0 \quad$ (11.16)

Obviously, the fracture initiation pressure is larger in case 2 than in case 1. The direction of the fracture relative to the borehole will also be different in the two cases. In case 1 the smallest in situ stress is normal to the borehole, and the fracture will be parallel with the borehole (Fig. 11.5a). In Case 2 the fracture will hence be normal to the borehole (Fig. 11.5b).

In cases where the borehole is not aligned with any of the principal stress directions the situation is further complicated. This will always be the situation in a deviated borehole in a vertical/horizontal stress field. The fracture will start to grow in the direction of least resistance and with some type of non-planar characteristics. However, this direction may not coincide with the direction of the smallest principal stress in situ. Hence the fracture may start to grow in one direction and then twist to eventually align itself normal to the smallest in situ principal stress.

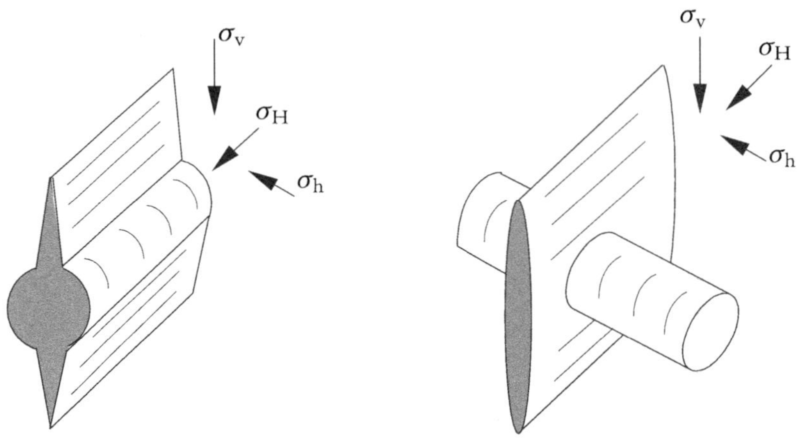

Figure 11.5 Illustration of different fracture directions, parallel to the borehole (left) and normal to the borehole (right).

In the normal stress regime assumed above, a fracture will always tend to initiate along the wellbore at the high and low sides of the wellbore, even in deviated wellbores drilled at an angle to the horizontal stresses. Abass et al. (1992) showed experimentally that even in a horizontal hole along the smallest horizontal stress, both a transverse fracture and a fracture along the wellbore was created, forming a T-shaped fracture. The longitudinal fracture would however disappear, resulting in propagation of only the transverse fracture. At intermediate angles, Abass et al. (1992) also found that multiple fractures could form at the wellbore.

In perforated wells the orientation and density of the perforations may complicate this picture further, resulting in both multiple fractures and non-planar geometries. Recommendations for perforation design to obtain successful hydraulic treatments have been given by e.g. El Rabaa (1989) and Abass et al. (1992). A theoretical study for prediction of fracture initiation (pressure, orientation) both in open and perforated holes has been presented by Hossain et al. (2000).

When one of the horizontal stresses is smallest, the generated fracture will always end up as a vertical fracture. We can of course also imagine the special situation where the vertical stress is smallest, and the generated fracture should then be horizontal. This may be the case in areas of extensive tectonic activity or in areas that have been uplifted without release of the horizontal stresses. As long as the rock is reasonably homogeneous and isotropic, a fracture will propagate normal to the least principal stress,

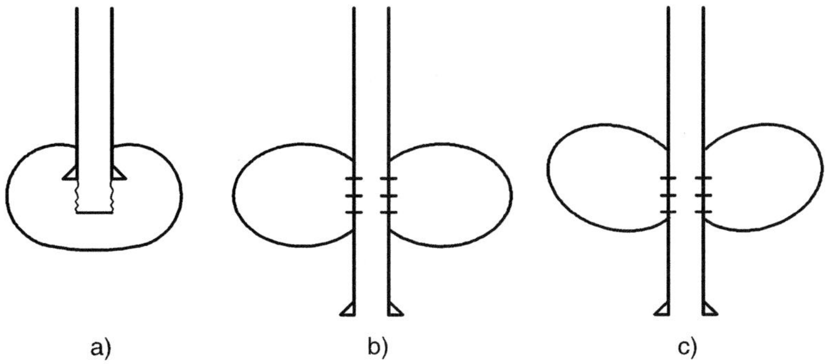

Figure 11.6 Well configuration effects on fracture growth: casing shoe (a); perforations (b); gravity (c).

whatever that direction may be. In areas of tectonic activity the principal stresses may be tilted from the vertical/horizontal direction. Then a fracture will also end up as tilted.

A fracture expanding from a point source in an isotropic rock in a homogeneous stress field will expand equally in all directions in the fracture plane, thus forming a circular or "penny-shaped" fracture. When a fracture is initiated in a well, the geometry may be modified by the well configuration, as indicated in Fig. 11.6. There is a tendency for the fractures to grow upwards as sketched in Fig. 11.6c, as the gravitational stresses are reduced in this direction.

Although the overburden stress normally increases monotonously with depth, the horizontal stresses may depend also on lithology or type of formation. An investigation by Warpinski et al. (1985) showed that in marine shale layers the horizontal stresses could approach lithostatic stress conditions, whereas sandstones showed significantly lower values of horizontal stress. Note however that lithological differences may occur simply because of the way a test is performed (see Section 11.5).

This contrast in horizontal stresses is important when considering confinement of fractures in layered formations. Differences in elastic properties and strength may also affect the propagation of a fracture from one layer to another, but contrasts in in situ stresses are more efficient.

Fracturing of a sandstone layer with over- and underlying shale formations can be illustrated as in Fig. 11.7. The difference in least horizontal principal stresses is indicated in part (a) of the figure. This difference may be large enough to prevent the fracture from propagating into the shale layers. This will result in a fracture with elliptical wings, as illustrated in part (b).

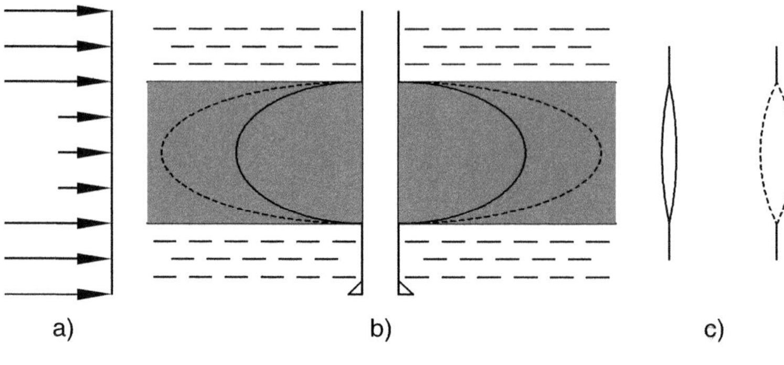

Figure 11.7 Confinement of a fracture between layers of higher stress.

To extend such a fracture, the walls must be forced outwards, increasing the width of the fracture (c). As the fracture is confined at the top and bottom, this will result in a more curved fracture wall, and a higher pressure in the fracture is needed to achieve this. A characteristic feature of this type of fracture is that the fracture propagation pressure will increase with fracture length. The pressure required to extend the fracture can thus conceptually be divided into three parts:

1. Pressure required to keep the fracture open towards the minimum stress.
2. Pressure required to flow fluid through the fracture.
3. Pressure required to overcome the resistance at the fracture tip and thus create new fracture volume.

Shortly,

$$p_{\mathrm{net}} = p(\sigma_{\mathrm{h}}) + p(\mathrm{flow}) + p(\mathrm{tip}) \tag{11.17}$$

The first term thus represents both the minimum in situ stress and the additional pressure required to maintain a fracture aperture. For small, narrow fractures with a low-viscosity fluid, the extension pressure will remain reasonably constant. For larger fractures using viscous fluids (or fluids with proppant), especially the fluid resistance will increase and thus also the pressure required to propagate the fracture.

For extension of fractures of circular shape, the extension pressure will remain relatively constant. On the other hand, for larger fracturing jobs, the fracture may also become confined in the vertical direction as illustrated in Fig. 11.7, and the extension pressure increases with fracture growth. The first term of Eq. (11.17) will then start to increase, since the fracture walls are bent outwards and the aperture thus increases. Eventually the extension

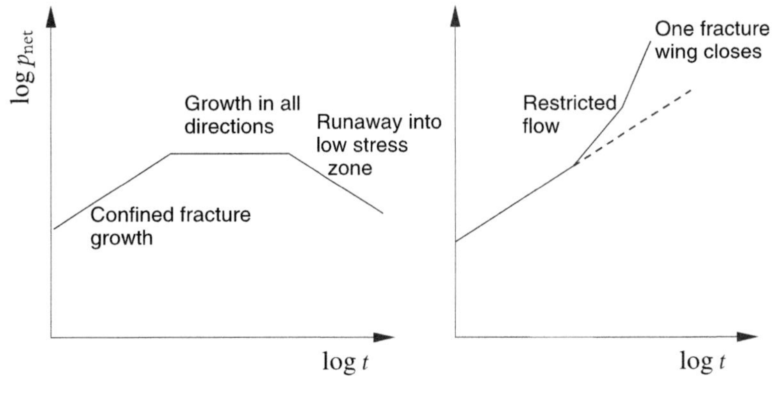

Figure 11.8 Log-log plot of net fracturing pressure versus time.

pressure becomes larger than the horizontal stress in the confining layers. Then the fracture will start to penetrate into these high-stress regions. Contrasts in in situ stresses thus represent barriers to fracture propagation but not in an absolute sense. In the case of alternate layers of shale and sand, a fracture induced in one layer may break through to another sand formation and extend there. This will result in a complicated fracture geometry.

A still more complicated geometry will result if hydraulic fractures are introduced in naturally fractured reservoirs (Warpinski and Teufel, 1987). This will result in combinations of opening of existing fractures linked with hydraulically induced fractures, and the resulting fracture pattern will be extremely difficult to predict.

It is quite obvious that the minimum in situ stress is an important design parameter for fracture stimulation jobs. Equally important is the understanding of the pressure response during fracturing. A comprehensive discussion of fracturing pressures has been given by Nolte (1988a,b). For confined fractures, it can be shown that there will be a power law relation between net fracturing pressure p_{net} and time when the injection rate is constant. The exponent will however be small, between 1/8 and 1/4. In a log-log plot of the net fracturing pressure versus time, the extension of such a fracture will plot as a straight line. This is illustrated in Fig. 11.8 (left) as the first part of the curve. The second part of the curve corresponds to an extension of a fracture that grows in all directions, also in height. This type of behaviour could however also be caused by increased fluid loss to the formation. This pressure level may therefore be regarded as the maximum pressure capacity of the formation. The third part of the curve would correspond to fracture extension into a low-stress zone, leading to a kind of runaway condition.

In Fig. 11.8 (right) the pressure curve indicates pressure increase above the normal fracture extension pressure. This is often observed in actual fracturing jobs when proppants are pumped into the fracture. An increasing pressure of approximately unit slope is characteristic of restricted flow into the fracture wings caused by proppant bridging. The effect could also be caused by excessive fluid viscosity. If one of the wings subsequently becomes blocked, then this will result in a doubling of the log-log slope.

11.3.1 Stress shadowing

When a fracture opens up, it compresses the surrounding formation, which results in an increase in the local stress. This makes it harder to create a new fracture in this region, and this effect is referred to as the *stress shadow* of the fracture.

Stress shadowing is of particular importance when creating multiple fractures from a horizontal well but was first discussed in the context of fractures from different wells by Warpinski and Branagan (1989).

Sneddon (1946) gave the formulas for the stress around a plane strain fracture with internal pressure. His Eq. (2·1·5) may be written as

$$\frac{1}{2}\left(\sigma_x+\sigma_y\right)=p_{\mathrm{w}}\frac{r}{\sqrt{r_1 r_2}}\cos\left(\theta-\frac{\theta_1+\theta_2}{2}\right)-p_{\mathrm{w}} \tag{11.18}$$

$$\frac{1}{2}\left(\sigma_y-\sigma_x\right)=p_{\mathrm{w}}\frac{r\sin\theta}{4H}\left(\frac{H^2}{r_1 r_2}\right)^{\frac{3}{2}}\sin\frac{3}{2}(\theta_1+\theta_2) \tag{11.19}$$

$$\tau_{xy}=p_{\mathrm{w}}\frac{r\sin\theta}{4H}\left(\frac{H^2}{r_1 r_2}\right)^{\frac{3}{2}}\cos\frac{3}{2}(\theta_1+\theta_2) \tag{11.20}$$

where we have written p_{w} for the fracture pressure, and the notation is clarified in Fig. 11.9.

Fig. 11.10 shows the radial and tangential stresses at the middle of the fracture ($x=0$). Note that at the fracture wall, both the radial and tangential stresses are compressive and equal to p_{w}. This is consistent with Fig. 4.18 since a narrow fracture is the limiting case of an elliptic borehole as $f\to 1$.

Fig. 11.10 shows that the governing factor for the stress shadow is the fracture height. At a distance of $y=H$, corresponding to the tip-to-tip height of the fracture, the stress shadow effect is reduced to less than 30 %. This means that to evaluate stress shadow effects, knowledge of factors that limit fracture height growth, such as stress variations, is required.

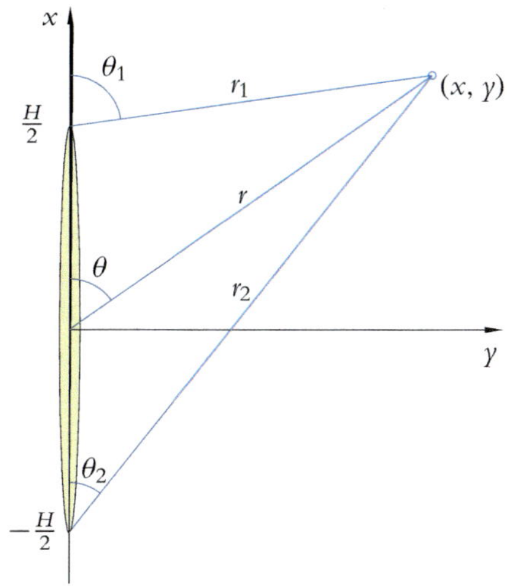

Figure 11.9 Sneddon's geometry. Notice the direction of the x- and y-axes.

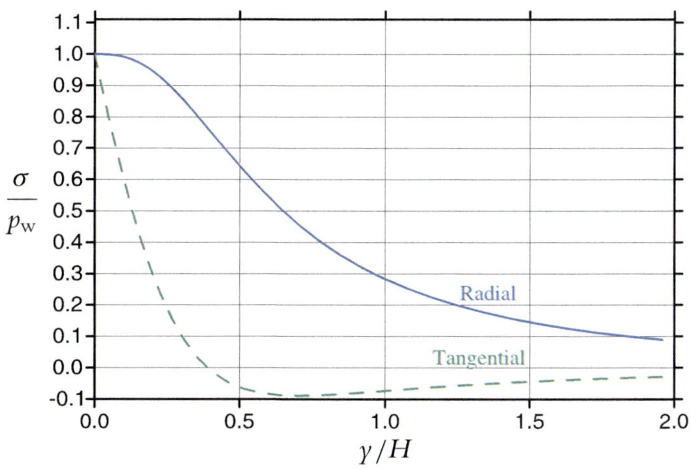

Figure 11.10 Radial and tangential stresses at the middle of the fracture ($x = 0$) as functions of distance.

At the fracture wall, τ_{xy} is zero. However, as we move away from the wall, τ_{xy} is non-zero except at $x = 0$. This means that the principal stresses are rotated within the stress shadow. Thus, if we initiate a fracture in the stress shadow of another fracture, then the new fracture will tend to rotate away from the first fracture.

Since the range of the stress shadow depends directly on the fracture height, even relatively close fractures will not see each other initially. However, as height grows, they will begin to influence each other. If the rock was perfectly homogeneous, we would be able to propagate all the fractures at the cost of an increased pumping pressure.

However, due to the inhomogeneity of rock, there will be small differences between neighbouring fractures, such that some grow more and quench their neighbours, limiting the number of effective fractures. For an example of simple modelling illustrating this, see, for example, Olson (2008).

11.3.2 The "wedge effect"

It is clear that to propagate a fracture in a homogeneous formation in a homogeneous stress field, the pressure in the fracture must be larger than the minimum stress. However, if there is a variation in stress, then it is possible for the fracture to propagate into a region with minimum stress higher than the pressure in the fracture. The mechanism is similar to what happens when slitting a piece of wood with a wedge, where the fracture propagates ahead of where the force is applied.

Adachi et al. (2010) discuss a three-layer case with a low-stress layer sandwiched between two layers with higher stress. It is assumed that the mechanical properties of all layers are the same and that there is full bonding between the layers. We here discuss a few of their basic results but recommend the paper for details and deeper understanding.

Fig. 11.11 shows how the pressure in the fracture is related to the height of the fracture for two different values of fracture toughness. For zero fracture toughness, the fracture starts to grow into the higher stress layers immediately. The height of the fracture is twice the thickness of the low-stress layer when the pressure has increased by 2/3 of stress contrast, that is, the fracture has penetrated significantly into the high-stress layer even though the pressure is lower than the stress there.

For a non-zero fracture toughness, there is some pressure increase before the penetration starts, but the fracture still penetrates into the high-stress layer at a pressure below its stress.

For $K_{IC} = 0$, the curve is monotonously increasing with height. However, for non-zero K_{IC}, it has a peak and falls off. As an example, for $\kappa = 2$, the maximum is for $H/h = 2 + \sqrt{5} \approx 4.24$ (beyond the scale of the figure). This means that unstable fracture growth is possible.

Fig. 11.12 illustrates the fracture shape for two different pressure levels.

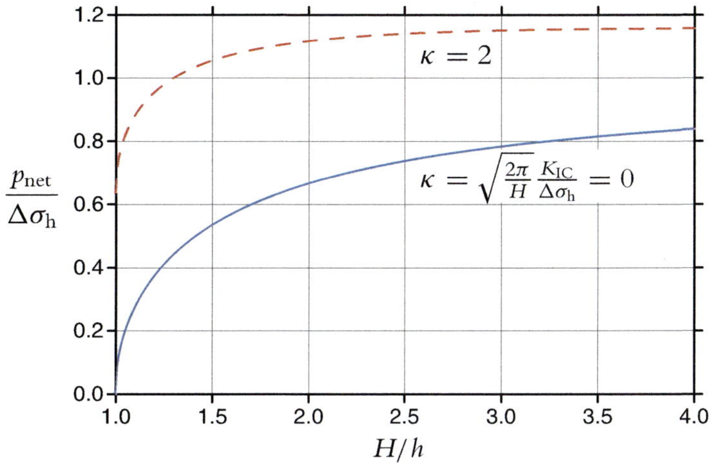

Figure 11.11 Scaled fracture pressure versus scaled fracture height. σ_h is the minimum stress in the low stress layer, while $\Delta\sigma_h$ is the stress contrast to the surrounding layers. κ is a scaled fracture toughness. $\kappa = 0$ is a definite lower bound, while $\kappa = 2$ is an *approximate* upper bound. h is the thickness of the low-stress layer, and H is the tip to tip height of the fracture (see Fig. 11.12).

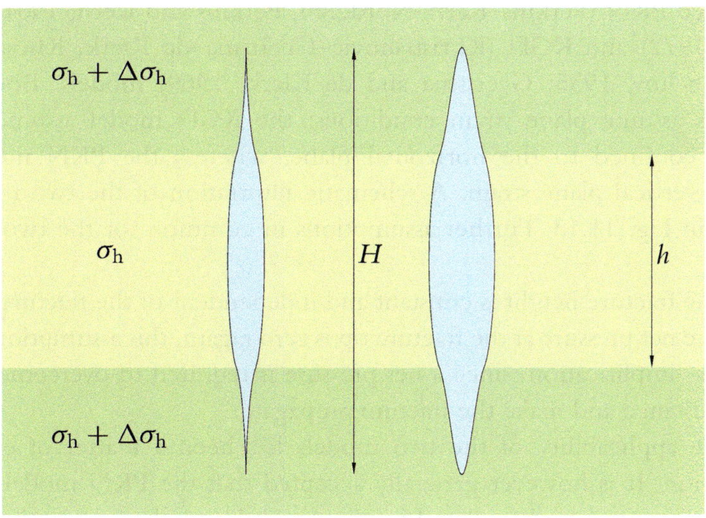

Figure 11.12 Illustration of fracture shapes, as computed from Adachi et al. (2010). To the left, $p_w = \sigma_h + (2/3)\Delta\sigma_h$ and $K_{IC} = 0$. To the right, $p_w = \sigma_h + \Delta\sigma_h$. This requires $\kappa > \sqrt{2\pi/3} \approx 1.48$, as the fracture would propagate farther for lower values. Fracture width has been strongly exaggerated.

11.4 Fracture size and shape

Modelling of fracture geometry (width, length, height) has been a topic of investigation for several decades, again primarily in the interest of hydraulic fracturing as a stimulation technique. It is beyond the scope of this book to present a detailed background for such calculations. The reader is referred to other published material for details on this topic (for a starting point, see e.g. Smith and Montgomery, 2015; Economides and Nolte, 2000; Valkó and Economides, 1995; Gidley et al., 1989).

In this section, we only briefly describe some basic principles of fracture geometry modelling and some analytical expressions which can be used to get estimates of fracture geometry. The motivation is that when performing stress tests in open boreholes, it is also necessary to know approximately what volume to pump (see Sections 11.5 and 8.3).

Classical analytical modelling of two-dimensional fractures starts with the solution of a line crack in an infinite elastic plane. Note that already at this point, there is a potentially serious deficiency in the model, as fracturing is a failure process that requires that the material is already outside the elastic regime.

The classical models for fracture geometry in two dimensions are the so-called PKN (Perkins–Kern–Nordgren, Perkins and Kern, 1961; Nordgren, 1972) and KGD (Khristianovic–Geertsma–de Klerk, Khristianovic and Zheltov, 1955; Geertsma and de Klerk, 1969) models. Both these models assume plane strain conditions: the KGD model assumes strain to be confined to the horizontal plane, whereas the PKN model assumes vertical plane strain. A schematic illustration of the two models is given in Fig. 11.13. Further assumptions in common for the two models are:

- The fracture height is constant and independent of the fracture length.
- The net pressure at the fracture tip is zero (again, this assumption clearly is a simplification, since a net pressure is required to overcome the tip resistance and make the fracture propagate).

The applicability of the two models has been a matter of extensive discussion. It is however generally accepted that the PKN model is most appropriate for length-to-height ratios much larger than one, whereas the KGD model is most appropriate for small length-to-height ratios (less than one). This implies that the PKN model is more used in conventional hydraulic fracture modelling, where the fracture is long compared to the fracture height. However, for open hole stress tests, the fracture is nor-

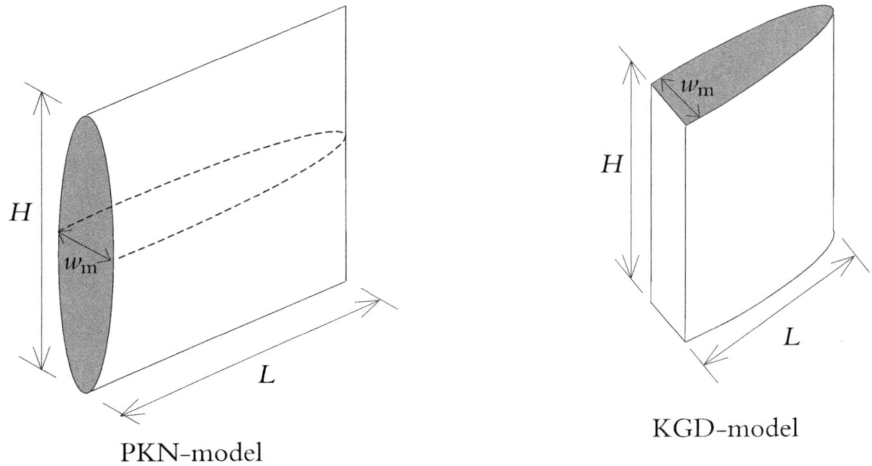

Figure 11.13 Illustration of fracture shapes for the PKN and KGD models.

mally short compared to the height, and hence the KGD model should give a better approximation.

Here we give an introduction to how analytical expressions for fracture length, width and internal fracture pressure can be obtained in the basic simple analytical models, based on Smith and Shlyapobersky (2000).

The starting point is mass balance: The injected volume of fluid V_{inj} is shared between fracture volume V_{frac} and the volume that leaks off into the formation V_{lo}:

$$V_{inj} = Q_{inj} t_{inj} = V_{frac} + V_{lo} \tag{11.21}$$

where Q_{inj} and t_{inj} are the pumping rate and time, respectively.

Injection rates in stimulation projects are usually thousands of litres per minute, and the duration of pumping is typically an hour or two.

Let us first assume that the fracturing fluid is the same as the reservoir fluid. From each small area of the fracture ΔA, the flow is essentially one-dimensional, normal to the fracture face. The pressure development may then be found by solving the one-dimensional version of Eq. (1.320). The result is

$$p(r, t) = \Delta p \,\mathrm{Erfc}\left(\frac{x}{2\sqrt{D_h t}}\right) \tag{11.22}$$

where Erfc is the complementary error function, and Δp is the difference between the fracture pressure and the far-field pressure.

The derivative of Erfc is

$$\frac{\mathrm{dErfc}(z)}{\mathrm{d}z} = -\frac{2}{\sqrt{\pi}}\mathrm{e}^{-z^2} \tag{11.23}$$

The flow rate at the fracture wall is then found using Darcy's law (Eq. (1.308))

$$\frac{Q(t)}{\Delta A} = \frac{k}{\eta_{\mathrm{f}}}\frac{\Delta p}{\sqrt{\pi D_{\mathrm{h}} t}} = C_{\mathrm{lo}}\frac{1}{\sqrt{t}} \tag{11.24}$$

This equation defining the *leak-off coefficient* or *fluid loss coefficient* C_{lo} is often referred to as Carter's law (Howard and Fast, 1957).

We see that in the approximation above, we have

$$C_{\mathrm{lo}} = \frac{k}{\eta_{\mathrm{f}}}\frac{\Delta p}{\sqrt{\pi D_{\mathrm{h}}}} \approx \sqrt{\frac{k\phi}{\pi\eta_{\mathrm{f}}K_{\mathrm{f}}}}\Delta p \tag{11.25}$$

where the right-hand expression is the "stiff frame limit", Eq. (1.324).

In real-world cases, we must take into account the fact that the fracturing fluid differs from the formation fluid and also that the fracturing fluid may build a filter cake. Further, we must take into account that different parts of the fracture are opened at different times, giving different starting times for the leak-off.

Then the volume that leaks off is as formulated by Smith and Shlyapobersky (2000):

$$V_{\mathrm{lo}} \simeq 4LHS_{\mathrm{p}} + 6C_{\mathrm{lo}}HL\sqrt{t_{\mathrm{inj}}} \tag{11.26}$$

where L is the fracture half length, and H is the fracture height. We have simplified and assumed that the total fracture height and length contribute to leak-off. Since a fracture may include several formations, this need not be the case. The factor 4 is needed to get the total fracture area (two fracture wings with two faces each), whereas 6 is an approximate number, which will vary a little based on fracture geometry (see e.g. Nolte, 1986).

The first term is called the *spurt loss* and represents the "instantaneous" loss of fluid. This is relevant for "wall-building" or filter-cake building fracturing fluids and represents the initial loss before the filter cake is effective. "Instantaneous" hence refers to the fact that this period is normally short relative to the total pumping time. The parameter S_{p} has the unit length and can be thought of as a depth of filtrate invasion before the filter cake is effective, and is typically a few centimetres.

The second term is due to the diffusional flow of fluid from the fracture, as described above. Typical range of values for C_{lo} are from 10^{-5} to 10^{-2} m/$\sqrt{\text{min}}$.

Combining Eqs. (11.21) and (11.26) yields the fracture length

$$L = \frac{Q_{inj} t_{inj}}{4LHS_p + 6C_{lo}H\sqrt{t_{inj}} + 2\overline{w}H} \tag{11.27}$$

where the fracture volume is written as the mean width $\overline{w}$ multiplied by the fracture height, and the factor 2 results from the two wings of the fracture. Note that Eq. (11.27) is not in general an explicit expression for L, since $\overline{w}$ depends on L.

If we neglect all fluid loss in Eq. (11.27) and assume the constant width $\overline{w}$, then we see that L increases linearly with time. In reality the width increases with length so that the length growth is a little slower, see below. On the other hand, if the leak-off term containing C_{lo} dominates, then the fracture length grows as the square root of time.

The width profile of a thin fracture in plane strain with tip-tip length H (see Fig. 11.9) and internal pressure $p_{net} + \sigma_h$ can be written (Sneddon and Elliot, 1946) as

$$w(x) = 4\frac{p_{net}}{E'}\sqrt{\left(\frac{H}{2}\right)^2 - x^2} \tag{11.28}$$

where E' is the plane strain modulus; see Eq. (1.112).

The maximum value of the fracture width w_{max} is at $x = 0$:

$$w_{max} = 2\frac{p_{net}}{E'}H \tag{11.29}$$

whereas the average width is

$$\overline{w} = \frac{\pi}{2}\frac{p_{net}}{E'}H \tag{11.30}$$

Considering that a fracture is a 2D object rather than a one-dimensional line crack, the expressions in Eqs. (11.29) and (11.30) have to be modified by replacing the fracture half length L with the minimum of the fracture height H and L. The constraints on fracture height were discussed in Section 11.3.

Since the net fracture pressure is normally at most a few MPa, whereas E' is in the GPa range, it is immediately clear from Eq. (11.30) that the fracture width is much smaller than its length. Assuming a fracture with minimum dimension (L or H) of 30 m, a plane strain modulus of 20 GPa and a net pressure of 2 MPa gives a fracture width of 6–8 mm. According to Smith and Shlyapobersky (2000), modelling of hydraulic fracture propagation in medium to low-permeability formations shows an average width of about 1/4″ (6.5 mm) within a variation of ±50 %. The fracture width generally decreases with increasing depth, since stiffness of a given formation tends to increase with burial depth (lithification).

Eq. (11.30) allows us to replace $\overline{w}$ in Eq. (11.27) with p_{net}, which means that we have to relate p_{net} to the other parameters. The net pressure depends on the flow resistance in the fracture (including that caused by leak-off to the surrounding formations) and mechanical breakdown processes at the fracture tip.

Here we will consider the PKN model, that is, a long fracture with constant height, and simplify by neglecting leak-off and tip effects. For laminar flow along an elliptical pipe with half axes a and b (see e.g. Landau and Lifshitz, 1987 and also Appendix D.6), the flow rate Q is given by

$$Q = \frac{\pi}{4\eta_{\text{f}}} \frac{a^3 b^3}{a^2 + b^2} \frac{\mathrm{d}p}{\mathrm{d}z} \tag{11.31}$$

We now use Eq. (11.31) as an approximation for the fracture (along which the cross-section varies slightly). We replace a and b with $H/2$ and $w_{\text{max}}(z)/2$, where z runs from 0 to L, and assume that $w_{\text{max}}(z) \ll H$. We also make the approximation that the assumption of no leak-off means that the flow rate is independent of z (that is, most of the flow goes to extending the fracture rather than widening the existing part). Darcy's law can then be written

$$\frac{1}{2} Q_{\text{inj}} = \frac{\pi}{64\eta_{\text{f}}} w_{\text{max}}(z)^3 H \frac{\mathrm{d}p_{\text{net}}(z)}{\mathrm{d}z} \tag{11.32}$$

where the factor 1/2 reflects the two equivalent wings of the fracture.

Next we use Eq. (11.29) to eliminate w_{max}. We obtain a simple differential equation $p_{\text{net}}^3 p_{\text{net}}' = \text{const}$ and using $p_{\text{net}}(L) = 0$, find

$$p_{\text{net}}(z) = \frac{2}{\pi^{\frac{1}{4}}} \frac{E'^{\frac{3}{4}}}{H} \left[\eta_{\text{f}} Q_{\text{inj}} (L - z)\right]^{\frac{1}{4}} \tag{11.33}$$

The pressure profile from Eq. (11.33) is shown in Fig. 11.14. The average fracture width follows from Eq. (11.30):

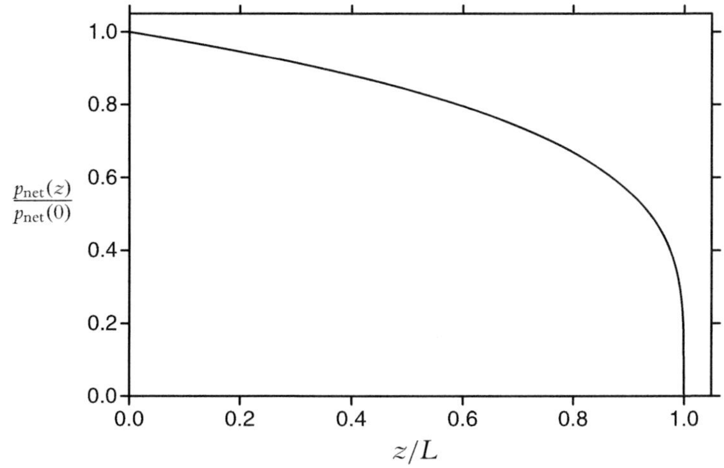

Figure 11.14 Pressure profile for a fracture without leak-off. The profile is only moderately changed for a fracture where leak-off dominates (see e.g. Nordgren, 1972), but clearly the length is reduced. (Note that the fracture width and the net pressure are proportional quantities according to Eq. (11.28).)

$$\overline{w}(z) = \frac{\pi^{\frac{3}{4}}}{E'^{\frac{1}{4}}} \left[\eta_{\mathrm{f}} Q_{\mathrm{inj}} (L - z)\right]^{\frac{1}{4}} \tag{11.34}$$

These equations should not be taken to literally, since they are non-rigorous solutions to a simplified problem, where we have assumed Newtonian flow and neglected leak-off and tip effects. They do however reveal some principles such as the width depends weakly on both rate, viscosity, fracture length and formation stiffness.

At first glance, it may be puzzling that the width is independent of H. This is because p_{net} is inversely proportional to H, since we are assuming the same rate independent of H.

The fracture volume (with no leak-off) may be found by integrating Eq. (11.34). We find

$$\frac{1}{2} V_{\text{frac}} = \frac{1}{2} Q_{\text{inj}} t_{\text{inj}} = H \int_0^L \overline{w}(z)\, \mathrm{d}z = \frac{4}{5} H \frac{\pi^{\frac{3}{4}}}{E'^{\frac{1}{4}}} (\eta_{\mathrm{f}} Q_{\mathrm{inj}})^{\frac{1}{4}} L^{\frac{5}{4}} \tag{11.35}$$

This allows us to infer

$$L = \left(\frac{5}{8}\right)^{\frac{4}{5}} \pi^{-\frac{3}{5}} \frac{Q_{\mathrm{inj}}^{\frac{3}{5}} E'^{\frac{1}{5}}}{\eta_{\mathrm{f}}^{\frac{1}{5}} H^{\frac{4}{5}}} t_{\mathrm{inj}}^{\frac{4}{5}} \tag{11.36}$$

and

$$p_{\mathrm{net}}(z=0) \propto \overline{w}(z=0) \propto t_{\mathrm{inj}}^{\frac{1}{5}} \tag{11.37}$$

Nordgren (1972) showed that accounting for the small flow rate variation along the fracture gives the same equations, but with a small change of the prefactors.

As shown above, in the leak-off dominated limit, we have

$$L \propto t_{\mathrm{inj}}^{\frac{1}{2}} \tag{11.38}$$

The net pressure and fracture width in this case develops as (see e.g. Nordgren, 1972)

$$p_{\mathrm{net}}(z=0) \propto \overline{w}(z=0) \propto t_{\mathrm{inj}}^{\frac{1}{8}} \tag{11.39}$$

Thus, for a constant height fracture, the net pressure increases with time (see Fig. 11.8), irrespective of the leak-off regime.

Accounting for the fracture tip pressure, Eq. (11.33) can be reformulated as

$$p_{\mathrm{net}} \propto \left[\frac{E'^3 \eta_{\mathrm{f}} Q_{\mathrm{inj}} L}{H^4} + p_{\mathrm{tip}}^4\right]^{\frac{1}{4}} \tag{11.40}$$

This is not a rigorous equation (as seen e.g. by the lack of a numerical prefactor in the first term), but a convenient way of obtaining linearity in p_{tip} when it dominates and linearity in the viscous pressure drop when it dominates.

Note that the fracturing fluid has been regarded as Newtonian, which is usually not the case in practical fracturing operations. The fluid contains proppants, in many cases, gels and other components, which may give it a reasonably high but rate-dependent viscosity (50–200 cP).

For the fracture to grow, the pressure at the tip has to overcome the fracture toughness. For a circular fracture geometry ("radial fracture"), this crack extension pressure is (see Eq. (6.72))

$$p_{\mathrm{tip}}^{\mathrm{cr}} = K_{\mathrm{IC}}\sqrt{\frac{\pi}{4L}} \tag{11.41}$$

This expression is an approximation, since it assumes constant pressure within a penny-shaped fracture. A main take-away is that it becomes smaller with fracture length, that is, the tip effect becomes less important as the

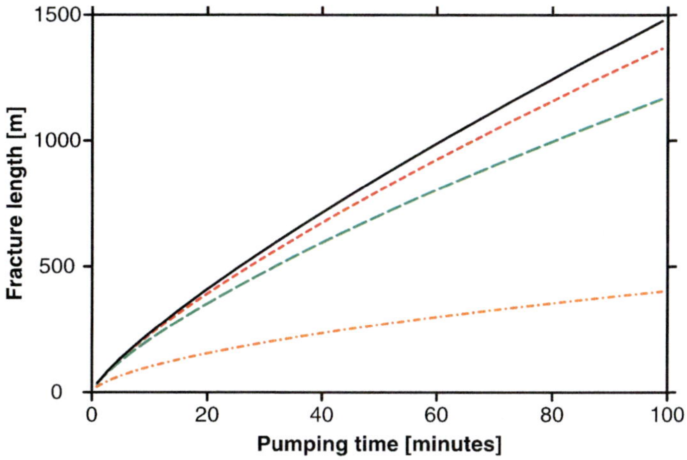

Figure 11.15 Fracture length vs. time of injection for a fictitious case with a 30 m high fracture, a constant injection rate of 10 m^3/min with a fluid that has a viscosity of 100 cP. The plane strain modulus E' is 20 GPa. In addition to a case with no fluid loss, three cases with leak-off coefficient $C_{lo} = 3 \cdot 10^{-5}$, 10^{-4} and $10^{-3}\,\text{m}/\sqrt{\text{min}}$ are shown.

fracture grows. For a non-circular fracture, the relevant dimension is the smallest, that is H for a PKN geometry.

As pointed out in Section 11.2, for any fracture that has reached a certain length (a few metres), this pressure is much smaller than commonly applied pressures as a result of pumping and viscous flow. Hence the fracture toughness derived from linear elastic fracture mechanics seems to play a minor role in propagation away from the borehole. However, accounting for nonlinear and non-elastic processes by an apparent or effective fracture toughness, this statement may be modified, but then the challenge is to determine the right apparent toughness for the field situation, which probably only can be done based on field experience.

Let us finally look at a simplified example shown in Fig. 11.15. We inject at 10 m^3/min in a fracture assumed to be 30 m in height. The no leak-off case and three cases with increasing leak-off are shown. The figure illustrates how the fracture growth is severely halted by leak-off.

This calculation demonstrates what can be done with a relatively basic analytical approach. It may serve as a guideline to qualitatively study the impact of some parameters but clearly is not an operational tool. We refer to the specialised literature mentioned earlier for in-depth treatment and many topics not addressed here.

See also Detournay (2004, 2016) for a discussion of the relative importance of fluid (viscous) and fracture tip (mechanical) processes.

11.5 Fracture closure

The fracture closure phase is of interest because it provides an opportunity to get an estimate of the smallest principal in situ stress σ_3. If the fracture has penetrated beyond the influence zone of the wellbore and into the virgin formation, then it is clear from the discussion in Section 11.1 that the fracture will close when the fluid pressure in the fracture equals the stress acting normal to the fracture σ_3. In most cases, it is assumed that this closure stress equals the smallest horizontal stress in the formation σ_{h}.

However, as we will see in the next two sections, interpretation of the minimum in situ stress from fracture closure is not straightforward. This uncertainty in interpretation of the closure pressure is due to the fact that fracture closure is not instantaneous. Fracture closure is gradual, from the moment of the first physical contact between the two fracture faces, until there is no further deformation of the fracture faces. Tests with downhole tiltmeters (Warpinski et al., 1998) have indeed shown that the fracture closure process is characterised by smoothly decreasing deformation, without evidence of well-defined fracture closure. The minimum in situ stress is of course passed at some point on the pressure decline curve, but it is not obvious that this point can be accurately determined by recording the pressure in the wellbore.

This can be illustrated further by considering the pressure versus volume during a flowback test in Fig. 11.16. The first linear part represents the stiffness of the open fracture. Then there is a gradual transition towards the final linear part, representing the stiffness of the borehole system, that is, after fracture closure. This transition period is significant, spanning more than 50 bar. Following the argumentation of Raaen et al. (2001), mechanical closure is associated with the pressure where the system stiffness starts to deviate from the first linear part. This change in system stiffness must be a result of reduced fracture size, that is, the pressure is equal to or less than the smallest principal stress in some part of the fracture. The transition period thus represents further fracture closure, and eventually there is no hydraulic communication between the fracture and the wellbore.

As mechanical closure is considered to represent the in situ stress, the interpretation of the smallest principal stress would be in the upper part of this range. However, there is still considerable uncertainty related to picking

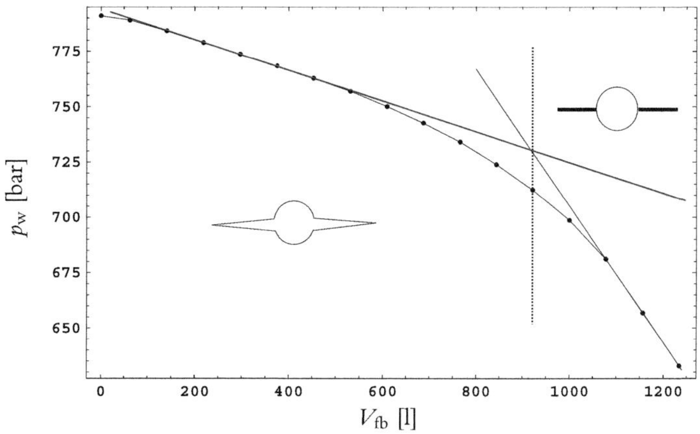

Figure 11.16 Illustration of well pressure versus volume during a flowback test. After (Raaen et al., 2001). An open fracture is indicated to represent the first linear part and a closed fracture to represent the final linear part.

of an exact value for the smallest principal stress, as the discussion in the next two sections will demonstrate.

11.5.1 Estimation of σ_3 from shut-in/decline tests

When the pumps are stopped and the fracture is shut-in, the pressure required for fracture opening and the pressure required for fluid flow in the fracture immediately drop to zero (Eq. (11.17)). This implies that the pressure in the fracture is equal to the smallest principal stress plus whatever additional pressure is left in the fracture. If the viscosity of the fracture fluid is low (e.g. water) and the fracture is small and unpropped, then the additional pressure in the fracture may be small. This implies that under these conditions, the pressure measured immediately after shut-in is a reasonable approximation to the smallest in situ principal stress σ_3. This is commonly referred to as the instantaneous shut-in pressure (ISIP).

This situation is illustrated in Fig. 11.17. Shown in this figure is also the stress where the surplus pressure has bled off and the fracture is actually closing. This is denoted as the closure pressure and is usually the best estimate of σ_3.

The instantaneous shut-in pressure may be significantly higher than the closure pressure. This depends on several factors such as fluid viscosity, leak-off into the formation and pump rates. The best estimate of σ_3 is then obtained by waiting until the fracture closes. The instantaneous shut-in

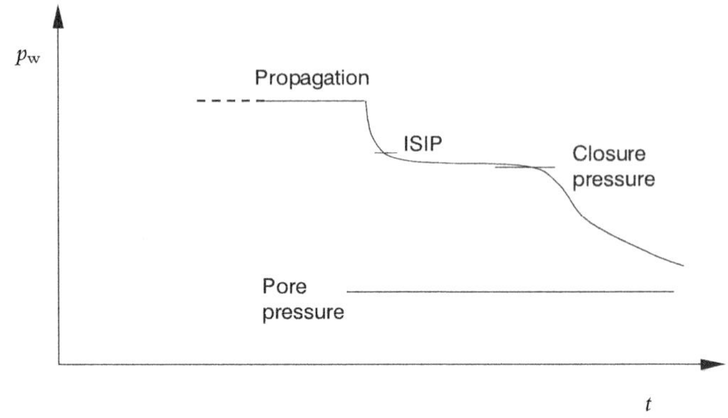

Figure 11.17 Illustration of well pressure response after shut-in.

pressure is thus always an upper bound for the σ_3 value, but the difference between the shut-in pressure and the value of σ_3 may be significant, and this difference will vary from one test to another. The instantaneous shut-in pressure may in some cases be difficult to identify on the decline curve, for instance, if the rock has very low permeability, and thus bleed-off from the fracture to obtain closure is extremely slow. Then it may also be impractical to wait for fracture closure.

Accurate identification of the closure pressure may not be easy, since fracture closure happens gradually rather than instantly. When a physical contact between the two fracture faces is established, the fluid in the fracture may still be free to flow. The two fracture faces may also have been distorted relative to each other. For the ideal case of instant closure, there should be a distinct change in the response, indicating the start of rapid pressure drop. This is caused by the drastic increase of the system stiffness as the fracture closes. The real response will depend on the nature of the fracture fluid, the permeability of the rock and the stiffness of the system.

To assist in the interpretation of fracture pressure response, a number of different plotting methods can be used (square root of time, log time, log pressure versus log time etc.). This originates from the similarity to pressure transient analysis in reservoir engineering. An example plotting decline pressure versus square root of shut-in time is shown in Fig. 11.18.

The curve approaches a linear trend as long as the fracture remains open. When the fracture closes, the pressure curve will depart from this linear trend. The change may be in either direction depending on fracture and formation properties. If the change is dominated by the increase in system

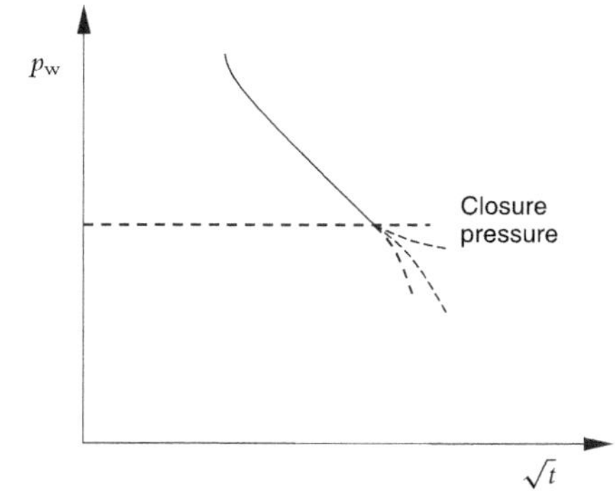

Figure 11.18 Pressure decline as a function of square root of time.

stiffness, then the slope increases. However, the slope may also decrease. This can occur if the fracture pinches off near the wellbore and the leak-off rate to the formation decreases. This implies that the closure pressure may be difficult to detect in certain cases.

A number of more or less elaborate methods have been proposed for interpretation of shut-in/decline data. See, for example, Guo et al. (1993) for an overview of some methods and Nolte (1989) for an in-depth treatment. Raaen et al. (2001) proposed a method applicable for tests with mud, where a pressure drop across a tight mud cake may dominate the leak-off signature.

In formations with very low permeability (e.g. shales), these classical methods for testing and interpretation may not be applicable. Leak-off from the fracture will then be too slow to close the fracture within reasonable time. This implies that the ISIP is poorly defined and/or the time to reach closure is too long. A pressure decrease may still be observed after shut-in, but this can also be a result of fracture extension due to energy stored in the fracture. To obtain more precise estimates of the smallest principal stress, both in shales and more permeable rocks, a pump-in/flowback test is recommended.

11.5.2 Estimation of σ_3 from flowback tests

In a pump-in/flowback test a volume of fluid is injected at fracturing rates. The well is then allowed to flow back until closure has been detected. A plot of pressure versus time will show an inflection point at the closure pressure. This is the upper inflection point shown in Fig. 11.19. The physical reason for this change is the increase in system stiffness as the fracture closes.

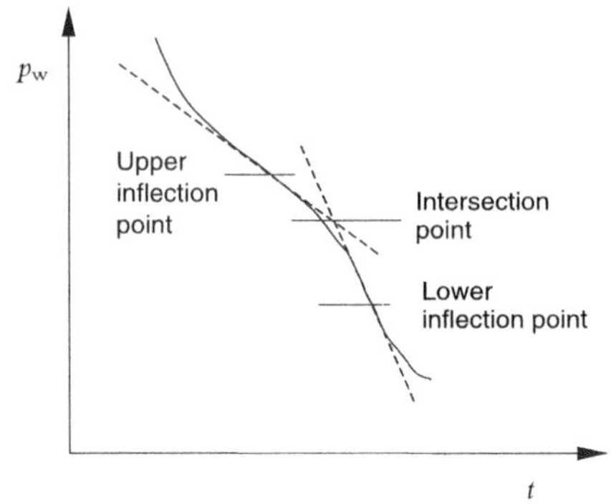

Figure 11.19 Well pressure response during flowback.

An alternative interpretation was presented by Shlyapobersky et al. (1988), utilising the lower inflection point (see Fig. 11.19). This is based on the assumption that mechanical closure occurs when the fracture storage equals zero.

Plahn et al. (1997) suggest to use the intersection point between two straight lines drawn through the upper and lower inflection points as illustrated in Fig. 11.19. They support this procedure with numerical modelling of the flowback and rebound process, using both KGD (Khristianovic–Geertsma–de Klerk) and PKN (Perkins–Kern–Nordgren) fracture models (Gidley et al., 1989). If the flowback test is idealised to an impermeable rock and inviscid fluid, then the closure pressure corresponds to the intersection of two uniquely defined lines. In a more realistic case (viscous fluid), a unique point that corresponds to the closure pressure cannot be found. However, they find that the intersection point is always closer to the closure pressure than either of the two inflection points.

More recently, an alternative approach was presented by Raaen et al. (2001, 2006). In this approach the flowed back volume is measured, and the change of system stiffness as the fracture closes is directly observed. See Section 8.3.4 for further discussion and a field example.

With reference to Fig. 11.19, it is evident that both the methods of Plahn et al. (1997) and Raaen et al. (2001) provide estimates of the closure pressure that are closer in numerical value to the upper inflection point.

It has traditionally been assumed that it is necessary to have a constant flowback rate in an appropriate range to be able to get a distinct and interpretable response. However, the data and interpretations provided by Raaen et al. are based on constant choke setting during flowback, and they also provide detailed recommendations for the design of such a job. Savitski and Dudley (2011) discussed a case where they used a start-stop procedure to prolong flowback time.

The pump-in/flowback test may be further extended to include recording of the pressure after the flowback phase. This phase is sometimes characterised by pressure build-up or rebound, which may be an indication of near-borehole fracture pinch-off (see e.g. de Bree and Walters, 1989).

When withdrawing fluid from the fracture during flowback, the fracture will tend to close near the wellbore, isolating the remaining part of the fracture, which is still open (Plahn et al., 1997), especially if flowback is rapid. After flowback, the fracture will continue to produce back to the wellbore, thus creating this pressure rebound effect. The appearance of a rebound effect has been used to confirm the existence of a fracture (Desroches and Kurkjian, 1999). It has also been proposed to use the maximum pressure during the rebound phase as an estimate of the closure stress, especially in impermeable or very low-permeability formations (Desroches and Kurkjian, 1999). Note, however, that this experience is based on testing with very small volumes, in packer isolated zones. Such an approach may not be applicable to testing where the entire hole volume is active.

11.6 Thermal effects on hydraulic fracturing

Thermal effects on hydraulic fracturing were briefly mentioned in the introduction to this chapter. Thermally induced fracturing is normally observed during water injection, especially when there is a large temperature difference between the (cold) injection water and the (hot) reservoir. Typical response is a sudden increase in injectivity after a significant period of stable injection (see e.g. Svendsen et al., 1991). This reflects that the reservoir rock has been gradually cooled during injection of the cold water. The reservoir rock shrinks due to cooling, and eventually the smallest in situ stress is reduced to a level below the bottom hole injection pressure. This results in the creation of a fracture, which provides a much larger contact area with the formation and hence a dramatic increase in injectivity. The length of the fracture is limited by the extent of the cooled zone. This process is normally referred to as thermally induced fracturing (TIF).

In the early days of cold water injection, TIF was an unexpected and in some cases unwanted effect. Today, however, this effect is taken into consideration in the planning process and is actively used. Thermal fracturing has to be considered with respect to well placement, pump capacities, fluid displacement pattern and other operational issues.

This process differs from conventional hydraulic fracturing jobs in several ways, primarily with respect to the viscosity of the fluid and the time scale of the process. Simultaneously, the fracturing pressure is also affected by changes in the pore pressure in the reservoir. Modelling injectivity is therefore a relatively complex matter, which should also include a number of other factors (damage due to particles in the water, reservoir heterogeneities, variation in fluid properties, well interaction etc.). Ideally, the modelling of this process requires a fully coupled fracture and reservoir simulation model (e.g. Clifford, 1989). Simplified models have however also been introduced, such as the partially de-coupled model of Settari and Warren (1994) and the analytical approach of Detienne et al. (1998).

For further details on injectivity modelling, the reader is referred to the above and other references. Here we restrict ourselves to a brief look at the purely mechanical aspect of this problem.

Perkins and Gonzalez (1984) gave analytical expressions for the stress change resulting from a cylindrical cooled zone of arbitrary height. In the limit of small diameter-to-height ratio ($d/h \to 0$), they found

$$\Delta\sigma_r(\Delta T) = \Delta\sigma_\theta(\Delta T) = \frac{E_{\text{fr}}}{2(1-\nu_{\text{fr}})}\alpha_{\text{T}}(T_{\text{cool}} - T_{\text{o}}) \tag{11.42}$$

where T_{o} represents the initial temperature of the formation, and T_{cool} is the temperature of the cooled region.

For the opposite limiting case ($d/h \to \infty$), which is applicable to the situation after some time when the cooled region has grown considerably into the reservoir, the result is

$$\Delta\sigma_r(\Delta T) = \Delta\sigma_\theta(\Delta T) = \frac{E_{\text{fr}}}{1-\nu_{\text{fr}}}\alpha_{\text{T}}(T_{\text{cool}} - T_{\text{o}}) \tag{11.43}$$

Note that for this case, the stress change is a factor 2 larger. For typical sandstone values ($E_{\text{fr}} = 5$ GPa, $\nu_{\text{fr}} = 0.3$, $\alpha_{\text{T}} = 1.5 \cdot 10^{-5}/°\text{C}$), Eq. (11.43) gives a stress change of 0.1 MPa/°C.

Perkins and Gonzalez (1985) considered thermal fracturing, where the presence of the fracture leads to a non-circular temperature distribution. They modelled this as an elliptical cooled zone and, based on numerical modelling, gave approximate analytical expressions for the stress change in cooled regions as functions of height and ellipticity.

As seen above, the stresses reduce considerably as a result of cooling, thereby facilitating fracture growth. In addition, depending on the aspect ratio of the ellipse, the stress field orientation may be altered: The reduction in the initially major horizontal stress will be larger than in the initially minor horizontal stress. Perkins and Gonzalez suggested that the directions of the principal stresses may flip, leading to possible generation of fractures orthogonal to the initial ones.

A pore pressure change also influences the reservoir stresses (see the discussion of the reservoir stress path in Section 12.2.3). From the correspondence between thermoelasticity and poroelasticity (see Section 1.6.9),

$$\Delta p_{\mathrm{f}} \leftrightarrow \Delta T \tag{11.44}$$

$$(1-2\nu_{\mathrm{fr}})\alpha \leftrightarrow E_{\mathrm{fr}}\alpha_{\mathrm{T}} \tag{11.45}$$

the general expressions of Perkins and Gonzalez (1985) may also be used for a homogeneous pressure change within an elliptical region.

Using Eqs. (11.44) and (11.45), and the numbers above, we see that for 5 °C cooling of the rock, the thermally induced stress reduction is compensated by approximately 1 MPa increase in the cooled zone pressure. This implies that for certain cases (low thermal expansion coefficient, low elastic stiffness), the pore pressure-induced stress may dominate over the thermally induced stress (Eltvik et al., 1992).

During injection of a hot or cold fluid, heat transport is primarily by convection. In addition, thermal conduction to/from over- and underlying formations and, to a smaller extent, thermal conduction in the direction of fluid movement, contribute.

If thermal conduction can be neglected, then a sharp temperature front results. Inside the front the temperature is that of the injection fluid, whereas outside, it is the original reservoir temperature. The position of the front can be estimated from energy balance considerations. For a radially symmetric injection pattern, we find

$$\frac{R_{\mathrm{cool}}}{R_{\mathrm{flood}}} = \sqrt{\frac{\phi\rho_{\mathrm{w}}C_{\mathrm{w}}}{(1-\phi)\rho_{\mathrm{s}}C_{\mathrm{s}} + \phi\rho_{\mathrm{w}}C_{\mathrm{w}}}} \tag{11.46}$$

where R_{cool} is the radius of the cooled zone, R_{flood} is the radius of the flooded zone, ρ_{w} and C_{w} are the density and specific heat capacity of the injection fluid, and ρ_{s} and C_{s} apply for the rock matrix. Inserting typical numbers, we find that thermal front moves at 30–50 % of the speed of the fluid front.

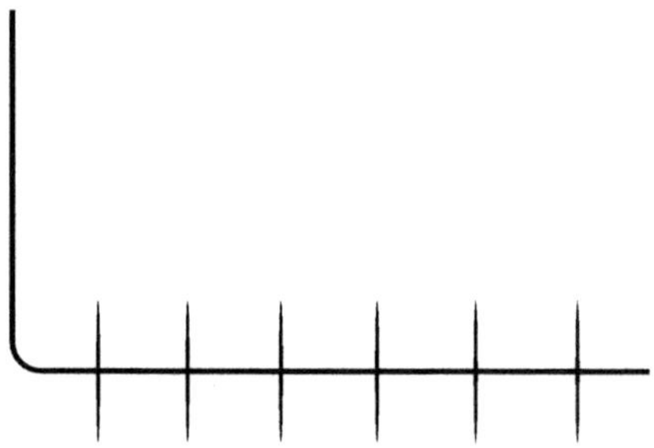

Figure 11.20 Sketch of vertical well that is turned horizontal and then fractured in multiple stages, a typical configuration in shale gas and shale oil exploitation.

Thermal conduction smoothens the temperature front, more strongly for thin beds. For some simple models taking thermal conduction into account, see Lauwerier (1955); Spillette (1965); Marx and Langenheim (1959); Platenkamp (1985).

11.7 Fracturing in unconventional reservoirs

The term "unconventional" is used for low-permeability hydrocarbon reservoirs that require special recovery operations to be produced economically. The term thus covers shale gas and shale oil (tight oil) in addition to tight gas and heavy oil sands, coalbed methane, gas hydrate deposits, and oil shale. Although all these resources have been known for decades, technological breakthroughs have been necessary to make them profitable. In the case of heavy oil, thermal techniques have been implemented, whereas for the shale reservoirs in particular, the game changer is the combination of horizontal wells and multi-stage hydraulic fracturing, as illustrated in Fig. 11.20.

In shale plays the reservoir is also the source rock. The formations vary in composition and are not always shale in the same sense as clay rich overburden shales in many conventional basins. They are rather carboniferous or siliceous in many cases with a reasonably low amount of clay minerals. What they have in common is essentially the low (nanoDarcy range) permeability. Hence fracturing is needed to enhance the flow performance. Often shale reservoirs contain natural fracture networks, so the purpose of stimulation is both to fracture the intact formation in an efficient way and to hook induced fractures to initial fractures to enhance and maintain connectivity.

Geomechanics can obviously play a role in all phases of the operations and may be part of the response to questions such as:

- Can optimum positions for fracturing be decided based on pre-existing knowledge about the local geology, the stress field and the rock properties?
- Can the fracture growth be predicted, that is, how far will fractures go, and is there a risk that they grow out of the reservoir zone?
- How can one monitor the fracture operation, both using microseismic events during fracturing or some kind of seismic imaging after it, to see where the fracture actually went?
- The fracturing process also includes transport of proppants. Can we be sure the fracture stays open after the pressure has gone back to normal and below during depletion?
- How can the amount of fracturing fluid used be limited and reduce the need for reinjection of water, which may be of limited access in the area where the operation takes place? Massive water injection is also known to be a source for induced earthquakes.

We will address some of these topics below based on the relevance of rock mechanics for hydraulic fracturing. For further readings, textbooks can be consulted, such as Zoback and Kohli (2019).

Ideally, decisions on where to place wells and fractures are taken on the basis of petrophysical reservoir characterisation. However, in unconventional reservoirs, production relies heavily on successful fracturing, so identifying sweet spots for fracture placement is of great interest.

For this purpose, the industry has searched for simple index parameters that can be easily determined using available information from logs or rock fragments (not necessarily core samples). The ideal index parameter should provide an attribute for fracturing the rock such that the stimulated reservoir volume (SRV) is maximised. The complex processes involved depend on the rock itself, on the in situ stress state, on the fracturing fluid and on operational constraints (pumping rate etc.). This calls for a more robust and science-based analysis than can be achieved with pragmatic index parameters.

The industry has to some extent adapted "brittleness" index as an attribute to cover these needs, since brittle behaviour is a pre-requisite for obtaining the requested enhanced permeability. The rock mechanics literature (Hucka and Das, 1974) provides several brittleness parameters, including the ratio between elastic strain and total strain prior to failure in a triaxial test, ratio between strain energies rather than strains, drop in shear stress to residual stress after peak (Bishop, 1967), or simply uses $\sin\varphi$ (φ is friction angle). Ingram and Urai (1999) suggested to relate brittleness directly to the overconsolidation ratio, taking stress history into account, which makes sense for soils and clay-rich materials. Several additional brittleness indices have been suggested as part of shale gas and shale oil developments. Jarvie et al. (2007) related brittleness to mineralogical content using the weight fraction of quartz as guidance, whereas several subsequent publications have tried to generalise this by adding weight fractions of brittle minerals and normalised to the total solid mass, a challenge being to define which minerals are brittle and which are not.

One simple approach that has become both popular and controversial was suggested by Rickman et al. (2008) and later reformulated by Jin et al. (2014). These relations express brittleness as a kind of weighted average of Young's modulus E and Poisson's ratio ν subtracted from an upper limit for the given formation. The argument is that a stiff formation (high E) tends to be brittle, whereas a ductile formation should have a high value of ν, approaching 1/2 for a fluid. Of course, this has nothing to do with the non-elastic nature of brittleness and, as such, lacks scientific basis. Holt et al. (2015) pointed out that this brittleness index will increase with increasing stress, whereas fundamentally, rocks go through a brittle to ductile transition with increasing stress (see Chapter 2).

Jin et al. (2014) argued that a high Young's modulus is beneficial for creating complex fracture networks, whereas the SRV increases with reducing elastic energy release rate. The energy release rate is proportional to K_{IC}^2/E' and thus increases with fracture toughness, but it decreases with increasing plane strain modulus. They found that Young's modulus above a certain value (typically between 10 and 15 GPa) reduces the positive effect on SRV. Thus a high E-modulus may be detrimental to the purpose of the fracturing operation.

Whereas all the index parameters above are somehow properties of the rock material, the previous sections in this chapter have shown that fracturing is a complex process, also strongly depending on the stress field where the fracture is forming and on the properties of the fracturing fluid. Papanastasiou and Atkinson (2015) suggested an alternative definition of brittleness, based on an elasto-plastic model for hydraulic fracturing (Papanastasiou, 1999). The resulting brittleness parameter is related entirely to the stress state and rock strength parameters. In the absence of stress anisotropy, no plasticity evolves, so we have perfect brittleness. The higher the stress anisotropy, the more planar and less complex the generated fracture trajectories. It also demonstrates that vertical fractures grow more significantly in the horizontal direction than vertically, which is beneficial for fracture containment.

Rheology and pumping rate of the fracturing fluid also affect fracture growth. Britt (2012) summarised fundamentals of fracture stimulation in low-permeability reservoirs and pointed out the need to integrate the entire fracturing operations within a geomechanical perspective. Gil et al. (2011) performed numerical (DEM) simulations of fracturing in a naturally fractured reservoir. They found that the amount of shear to tensile damage was much higher with use of low- (water) than high-viscosity (gel) fluid. They also found that shear displacement of the fracture surfaces was more prominent at low injection rates. Shearing combined with dilation would be a way of optimising the SRV.

A key to success is to generate multi-stage fractures that together generate the complex network with natural fractures and optimise the SRV. The multi-stage operation needs to consider the risk of stress shadowing, which may prevent neighbouring fractures to grow. It is important that the complexity of the fracture process is accounted for in modelling. This includes heterogeneities, shale anisotropy, plasticity and visco-elasticity.

11.8 Microseismic monitoring of fracturing

Microseismic monitoring of hydraulic fracturing operations (Warpinski, 2009; van der Baan et al., 2013) is commonly applied in shale oil and gas fields with the main aim of detecting, locating and characterising the stimulated rock volume (SRV). Similar technology is also used to monitor the effect of steam injection in heavy oil reservoirs and the efficiency of enhanced (or engineered) geothermal systems (EGS). Passive seismic listening is occasionally applied to surveillance of injection- or production-induced seismicity on reservoir scale (Maxwell and Urbanic, 2001; Maxwell et al., 2010) and to monitor stability of underground mines.

The principles behind microseismic monitoring are the same as for acoustic emissions in rock samples stressed in the laboratory (see Section 5.10) and as for conventional seismology (earthquake detection): Damage by a rock failure process generates elastic waves that can be picked up by receivers (piezoelectric transducers, geophones or accelerometers). In fact, a relatively unique trend appears to link the magnitude to the size of the fault responsible for the event and to the amount of slip generated (see Fig. 11.21).

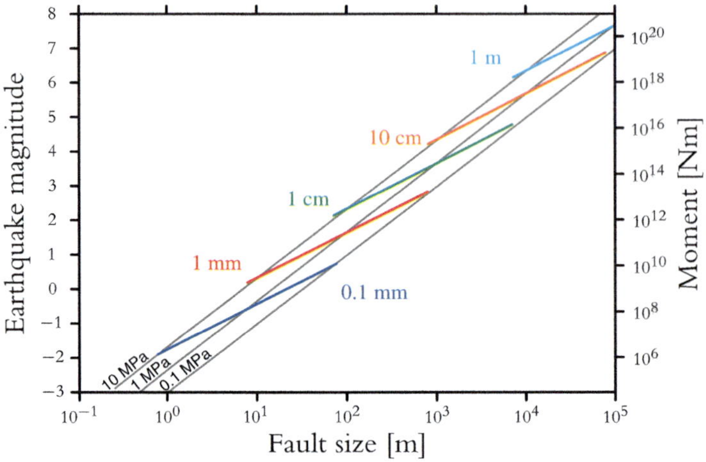

Figure 11.21 Earthquake magnitude and moment vs. fault size. Coloured line segments indicate amount of slip, and three lines denote associated stress drops (after Zoback and Gorelick, 2012).

Detection of microseismic events in hydraulic fracture monitoring is most often done with three-component geophones in one or two observation wells, and occasionally by surface sensors in addition or instead. Borehole data are recorded from a few Hz to the kHz range. The larger events may be detectable within a km or so from the source. Since data are desired in near real time for application in the ongoing completion, the focus is on processing data to obtain hypocentre locations and magnitudes. This requires relatively simple processing and short recording times. Location algorithms are based on an estimated underground velocity structure, which is optimised by a best fit between a forward model and the acquired data and can be supported by using sources in known positions for calibration. The nature of the source (shear or tensile) can be inferred from the recorded direction of particle motion, which points towards the source for P-waves and is orthogonal for S-waves. Sonic log data prior to fracturing may be used to guide the choices of velocities. Still, the velocity structure represents a main possible source of error. Most log data are from vertical wells, whereas the detected microseismic waves are moving close to horizontal. Shales are host rocks for unconventional reserves, so anisotropy may be significant and needs to be accounted for, at least in a simplified way by adjusting vertical log velocities to mimic assumed horizontal velocities. However, most location algorithms are limited to isotropic media.

Dispersion between seismic and sonic frequencies is also a potential source of error.

Observed microseismic magnitudes associated with hydraulic fracturing are typically between −4 and −1, which means (Fig. 11.21) that the associated slip is on sub-mm to mm scale. They also stem primarily from shear events and not from the tensile opening of the hydraulic fracture itself. If the cumulative energy put into the fracturing operation would be converted to one single seismic event, then the magnitude would be around 3 (Warpinski, 2009), but data show that only 10^{-9}–10^{-5} of this energy is converted to normally observed seismicity. Using broadband detection, Maxwell (2011a) showed considerable energy in the very low frequency range (10^{-4}–10^{-1} Hz) when monitoring a single fracturing operation over long time. Most of this is due to slow plastic or creep deformation associated with tensile opening and closure of the fracture, usually termed aseismic energy, since it is not seen in standard microseismic surveys. The aseismic response is larger in soft clay-rich shales than, for instance, in sandstones and is connected to the plasticity of the formation. Hence it is influenced both by rock properties and local in situ stress conditions plus the natural fracture network. In addition to using broadband receivers, long observation time (hours to days) is required to explore the low-frequency part of the spectrum. This is detrimental to the aim of near real-time monitoring. We have to be aware that in standard surveys, primarily brittle events associated with failure processes near the fracture tip are detected in addition to interactions between induced and natural fractures. The observed spatial distribution of seismic events (seismic cloud), including uncertainties due to erroneous velocity structure and noise, tends to overestimate the flow active volume (SRV) generated by the hydraulic fracture.

Several examples of located microseismic events as a result of multistage fracturing operations can be found in Zoback and Kohli (2019). Data normally show that hydraulic fractures are relatively well contained (see e.g. Maxwell, 2011b; Fisher and Warpinski, 2012). Fisher and Warpinski argue on the basis of thousands of monitored fracturing operations that vertical fracture growth is hindered by overlying formations and claim that "there is no method by which a fracture is going to propagate through the various rock layers and reach the surface". Possible leakage paths are mainly found in the vicinity of wells as a result of insufficient casing quality or bad cement (Dusseault and Jackson, 2014).

Long duration monitoring may be useful to quantify and better understand the noise disturbing microseismic surveys. In addition to environmental noise (from e.g. drilling or other ongoing processes in nearby wells), part of the noise may be structured, such as wellbore resonances and guided waves along the casing. Use of multiple observation wells along with waveform processing offers analysis of further information such as moment tensors and resonance frequencies, that is, identification of shear vs. tensile failure processes and ability to sort out waves that do not belong to the events of direct interest.

Microseismic analysis benefits from geomechanical modelling, preferably including sophisticated modelling of the fracturing process itself, such as by permitting non-planar fractures and incorporation of plasticity and time-dependent deformation. The microseismic data may also contain valuable information to be fed back to the geomechanical model. As an example, van der Baan et al. (2013) argued that the *b*-value in the Gutenberg–Richter relationship (Eq. (5.101)) provides information about the local stress regime. Grob and van der Baan (2011) used this to reveal change in stress regime during cyclic steam injection in a heavy oil reservoir.

References

Abass, H.H., Hedayati, S., Meadows, D.L., 1992. Nonplanar fracture propagation from a horizontal wellbore: experimental study. SPE Prod. Facil. 11 (3). https://doi.org/10.2118/24823-PA.

Adachi, J.I., Detournay, E., Peirce, A.P., 2010. Analysis of the classical pseudo-3D model for hydraulic fracture with equilibrium height growth across stress barriers. Int. J. Rock Mech. Min. Sci. 47 (4), 625–639. https://doi.org/10.1016/j.ijrmms.2010.03.008.

Bishop, A.W., 1967. Progressive failure with special reference to the mechanism causing it. In: Proc. Geotechnical Conference, vol. 2. Oslo, Norway, pp. 142–150.

Britt, L., 2012. Fracture stimulation fundamentals. J. Nat. Gas Sci. Eng. 8, 34–51. https://doi.org/10.1016/j.jngse.2012.06.006.

Clifford, P.J., 1989. Simulation of waterflood fracture growth with coupled fluid flow, temperature and rock elasticity. In: Proc. Second IMA/SPE European Conf. on the Mathematics of Oil Recovery. Cambridge, United Kingdom.

de Bree, P., Walters, J.V., 1989. Micro/minifrac test procedures and interpretation for in situ stress determination. Int. J. Rock Mech. Min. Sci. Geomech. Abstr. 26, 515–521. https://doi.org/10.1016/0148-9062(89)91429-0.

Desroches, J., Kurkjian, A.L., 1999. Applications of wireline stress measurements. SPE Reserv. Eval. Eng. 2 (5), 451–461. https://doi.org/10.2118/58086-PA.

Detienne, J.-L., Creusot, M., Kessler, N., Sahuquet, B., Bergerot, J.-L., 1998. Thermally induced fractures: a field proven analytical model. SPE Reserv. Eval. Eng. 1 (1), 30–35. https://doi.org/10.2118/30777-PA.

Detournay, E., 2004. Propagation regimes of fluid-driven fractures in impermeable rocks. Int. J. Geomech. 4, 35–45. https://doi.org/10.1061/(ASCE)1532-3641(2004)4:1(35).

Detournay, E., 2016. Mechanics of hydraulic fractures. Annu. Rev. Fluid Mech. 48, 311–339. https://doi.org/10.1146/annurev-fluid-010814-014736.

Detournay, E., Carbonell, R., 1997. Fracture mechanics analysis of the breakdown process in minifrac or leak-off tests. SPE Prod. Facil. 12 (3), 195–199. https://doi.org/10.2118/28076-PA.

Detournay, E., Cheng, A., 1992. Influence of pressurization rate on the magnitude of the breakdown pressure. In: Tillerson, J.R., Wawersik, W.R. (Eds.), Rock Mechanics. Balkema, Rotterdam, pp. 325–333.

Dusseault, M.B., Jackson, R., 2014. Seepage pathway assessment for natural gas to shallow groundwater during well stimulation, in production, and after abandonment. Environ. Geosci. 21, 107–126. https://doi.org/10.1306/eg.04231414004.

Economides, M.J., Nolte, K.G. (Eds.), 2000. Reservoir Stimulation, third edition. Wiley.

El Rabaa, W., 1989. Experimental study of hydraulic fracture geometry initiated from horizontal wells. SPE19720. In: SPE Ann. Tech. Conf. & Exh. San Antonio, TX, USA.

Eltvik, P., Skoglunn, T., Settari, A., 1992. Waterflood-induced fracturing: water injection above parting pressure at Valhall. SPE24912. In: SPE Ann. Tech. Conf. & Exh. Washington, DC, USA.

Fisher, K., Warpinski, N., 2012. Hydraulic-fracture-height growth: real data. SPE Prod. Oper. 27 (1), 8–19. https://doi.org/10.2118/145949-PA.

Gederaas, T.B., Raaen, A.M., 2009. Precise minimum horizontal stress determination from pump-in/flowback tests with drilling mud. ARMA09-88. In: 43rd US Rock Mechanics Symposium and 4th U.S.–Canada Rock Mechanics Symposium. Asheville, NC, USA.

Geertsma, J., de Klerk, F., 1969. A rapid method of predicting width and extent of hydraulically induced fractures. J. Pet. Technol. 21 (12), 1571–1581. https://doi.org/10.2118/2458-PA.

Gidley, J.L., Holditch, S.A., Nierode, D.E., Veatch Jr., R.W. (Eds.), 1989. Recent Advances in Hydraulic Fracturing. SPE Monograph Series, vol. 12. SPE, Richardson, TX.

Gil, I., Nagel, N., Sanchez-Nagel, M., Damjanac, B., 2011. The effect of operational parameters on hydraulic fracture propagation in naturally fractured reservoirs – getting control of the fracture optimization process. ARMA11-391. In: 45th US Rock Mechanics / Geomechanics Symposium. San Francisco, CA, USA.

Grob, M., van der Baan, M., 2011. Inferring in-situ stress changes by statistical analysis of microseismic event characteristics. Lead. Edge 30, 1296–1302. https://doi.org/10.1190/1.3663403.

Guo, F., Morgenstern, N.R., Scott, J.D., 1993. Interpretation of hydraulic fracturing pressure: a comparison of eight methods used to identify shut-in pressure. Int. J. Rock Mech. Min. Sci. Geomech. Abstr. 30, 627–631. https://doi.org/10.1016/0148-9062(93)91222-5.

Holt, R.M., Fjær, E., Stenebråten, J.F., Nes, O.-M., 2015. Brittleness of shales: relevance to borehole collapse and hydraulic fracturing. J. Pet. Sci. Eng. 131, 200–209. https://doi.org/10.1016/j.petrol.2015.04.006.

Hossain, M.M., Rahman, M.K., Rahman, S.S., 2000. Hydraulic fracture initiation and propagation: roles of wellbore trajectory, perforation and stress regimes. J. Pet. Sci. Eng. 27 (3–4), 129–149. https://doi.org/10.1016/S0920-4105(00)00056-5.

Howard, G.C., Fast, C.R., 1957. Optimum fluid characteristics for fracture extension? In: Drilling and Production Practice. API.

Hucka, V., Das, B., 1974. Brittleness determination of rocks by different methods. Int. J. Rock Mech. Min. Sci. 11, 389–392. https://doi.org/10.1016/0148-9062(74)91109-7.

Ingram, G.M., Urai, J.L., 1999. Top seal leakage through faults and fractures: the role of mudrock properties. In: Muds and Mudstones: Physical and Fluid Flow Properties. In: Geol. Soc. (London) Spec. Publ., vol. 158, pp. 125–135.

Jarvie, D.M., Hill, R.J., Ruble, T.E., Pollastro, R.M., 2007. Unconventional shale-gas systems: the Mississippian Barnett shale of north-central Texas as one model for thermogenic shale-gas assessment. AAPG Bull. 91, 475–499. https://doi.org/10.1306/12190606068.

Jin, X., Shah, S.N., Roegiers, J.-C., Zhang, B., 2014. Fracability evaluation in shale reservoirs – an integrated petrophysics and geomechanics approach. SPE168589. In: SPE Hydraulic Fracturing Technology Conference. The Woodlands, TX, USA, pp. 1–14.

Khristianovic, S.A., Zheltov, Y.P., 1955. Formation of vertical fractures by means of highly viscous fluids. In: Proc. 4th World Petroleum Congress. Rome, Italy, pp. 579–586.

Landau, L.D., Lifshitz, E.M., 1987. Fluid Mechanics, second edition. Pergamon Press.

Lauwerier, H.A., 1955. The transport of heat in an oil layer caused by the injection of fluid. Appl. Sci. Res. A 5, 145–150. https://doi.org/10.1007/BF03184614.

Marx, J.W., Langenheim, R.H., 1959. Reservoir heating by hot fluid injection. Pet. Trans., AIME 216 (1), 312–315. https://doi.org/10.2118/1266-G.

Maxwell, S.C., 2011a. What does microseismicity tell us about hydraulic fractures? In: SEG Technical Program Expanded Abstracts, pp. 1565–1569.

Maxwell, S.C., 2011b. Microseismic hydraulic fracture imaging: the path toward optimizing shale gas production. Lead. Edge 30, 340–346. https://doi.org/10.1190/1.3567266.

Maxwell, S.C., Urbanic, T.I., 2001. The role of passive microseismic monitoring in the instrumented oil field. Lead. Edge 20, 636–639. https://doi.org/10.1190/1.1439012.

Maxwell, S.C., Rutledge, J., Jones, R., Fehler, M., 2010. Petroleum reservoir characterization using downhole microseismic monitoring. Geophysics 75, 75. https://doi.org/10.1190/1.3477966.

Montgomery, C.T., Smith, M.B., 2010. Hydraulic fracturing: history of an enduring technology. J. Pet. Technol. 62 (12), 26–32. https://doi.org/10.2118/1210-0026-JPT.

Morita, N., Black, A.D., Fuh, G.-F., 1996. Borehole breakdown pressure with drilling fluids – I. Empirical results. Int. J. Rock Mech. Min. Sci. Geomech. Abstr. 33, 39–51. https://doi.org/10.1016/0148-9062(95)00028-3.

Nolte, K.G., 1986. A general analysis of fracturing pressure decline with application to three models. SPE Form. Eval. 1 (6), 571–583. https://doi.org/10.2118/12941-PA.

Nolte, K.G., 1988a. Principles for fracture design based on pressure analysis. SPE Prod. Eng. 3 (1), 22–30. https://doi.org/10.2118/10911-PA.

Nolte, K.G., 1988b. Application of fracture design based on pressure analysis. SPE Prod. Eng. 3 (1), 31–42. https://doi.org/10.2118/13393-PA.

Nolte, K.G., 1989. Fracturing-pressure analysis. In: Gidley, J.L., Holditch, S.A., Nierode, D.E., Veatch Jr., R.W. (Eds.), Recent Advances in Hydraulic Fracturing. SPE, pp. 297–318.

Nordgren, R.P., 1972. A propagation of a vertical hydraulic fracture. Soc. Pet. Eng. J. 12 (4), 306–314. https://doi.org/10.2118/3009-PA.

Olson, J.E., 2008. Multi-fracture propagation modeling: applications to hydraulic fracturing in shales and tight gas sands. ARMA08-327. In: 42nd US Rock Mechanics Symposium and 2nd U.S.–Canada Rock Mechanics Symposium. San Francisco, CA, USA.

Ouchterlony, F., 1990. Fracture toughness testing of rock with core based specimens. Eng. Fract. Mech. 35, 355–366. https://doi.org/10.1016/0013-7944(90)90214-2.

Papanastasiou, P., 1999. The effective fracture toughness in hydraulic fracturing. Int. J. Fract. 96, 127–147. https://doi.org/10.1023/A:1018676212444.

Papanastasiou, P., Atkinson, C., 2015. The brittleness index in hydraulic fracturing. ARMA15-489. In: 49th US Rock Mechanics/Geomechanics Symposium. San Francisco, CA, USA.

Perkins, T.K., Gonzalez, J.A., 1984. Changes in earth stresses around a wellbore caused by radially symmetrical pressure and temperature gradients. Soc. Pet. Eng. J. 24 (2), 129–140. https://doi.org/10.2118/10080-PA.

Perkins, T.K., Gonzalez, J.A., 1985. The effect of thermoelastic stresses on injection well fracturing. Soc. Pet. Eng. J. 25 (1), 78–88. https://doi.org/10.2118/11332-PA.

Perkins, T.K., Kern, L.R., 1961. Width of hydraulic fractures. J. Pet. Technol. 13 (9), 937–949. https://doi.org/10.2118/89-PA.

Plahn, S.V., Nolte, K.G., Miska, S., 1997. A quantitative investigation of the fracture pump-in/flowback test. SPE Prod. Facil. 12 (1), 20–27. https://doi.org/10.2118/30504-PA.

Platenkamp, R.J., 1985. Temperature distribution around water injectors: effects on injection performance. SPE13746. In: SPE Middle East Oil Technical Conference and Exhibition. Bahrain.

Raaen, A.M., Skomedal, E., Kjørholt, H., Markestad, P., Økland, D., 2001. Stress determination from hydraulic fracturing tests: the system stiffness approach. Int. J. Rock Mech. Min. Sci. 38 (4), 529–541. https://doi.org/10.1016/S1365-1609(01)00020-X.

Raaen, A.M., Horsrud, P., Kjørholt, H., Økland, D., 2006. Improved routine estimation of the minimum horizontal stress component from extended leak-off tests. Int. J. Rock Mech. Min. Sci. 43, 37–48. https://doi.org/10.1016/j.ijrmms.2005.04.005.

Rickman, R., Mullen, M., Petre, E., Grieser, B., Kundert, D., 2008. A practical use of shale petrophysics for stimulation design optimization: all shale plays are not clones of the Barnett shale. SPE115258. In: SPE Ann. Tech. Conf. Denver, CO, USA.

Savitski, A.A., Dudley, J.W., 2011. Revisiting microfrac in-situ stress measurement via flow back—a new protocol. SPE147248. In: SPE Annual Technical Conference and Exhibition. Denver, CO, USA.

Settari, A., Warren, G.M., 1994. Simulation and field analysis of waterflood induced fracturing. In: Proc. Eurock '94. Delft, Netherlands.

Shlyapobersky, J., Walhaug, W.W., Sheffield, R.E., Huckabee, P.T., 1988. Field determination of fracturing parameters for overpressure calibrated design of hydraulic fracturing. SPE18195. In: SPE Ann. Tech. Conf. & Exh. Houston, TX, USA.

Smith, M.B., Montgomery, C.T., 2015. Hydraulic Fracturing. CRC Press, Boca Raton.

Smith, M.B., Shlyapobersky, J.W., 2000. Hydraulic fracture mechanics. In: Economides, M.J., Nolte, K.G. (Eds.), Reservoir Stimulation, third edition. Wiley, pp. 5.1–5.28.

Sneddon, I.N., 1946. The distribution of stress in the neighbourhood of a crack in an elastic solid. Proc. R. Soc. Lond. A 187 (1009), 229–260. https://doi.org/10.1098/rspa.1946.0077.

Sneddon, I.N., Elliot, H.A., 1946. The opening of a Griffith crack under internal pressure. Q. Appl. Math. 4, 262–267. https://doi.org/10.1090/qam/17161.

Spillette, A.G., 1965. Heat transfer during hot fluid injection into an oil reservoir. J. Can. Pet. Technol. 4 (4), 213–218. https://doi.org/10.2118/65-04-06.

Svendsen, A.P., Wright, M.S., Clifford, P.J., Berry, P.J., 1991. Thermally induced fracturing of Ula water injectors. SPE Prod. Eng. 6 (4), 384–390. https://doi.org/10.2118/20898-PA.

Valkó, P., Economides, M.J., 1995. Hydraulic Fracture Mechanics. John Wiley & Sons Ltd, Chichester.

van der Baan, M., Eaton, D., Dusseault, M., 2013. Microseismic monitoring developments in hydraulic fracture stimulation. In: Effective and Sustainable Hydraulic Fracturing. IntechOpen.

Warpinski, N., 2009. Microseismic monitoring: inside and out. J. Pet. Technol. 61, 80–85. https://doi.org/10.2118/118537-JPT.

Warpinski, N.R., Branagan, P.T., 1989. Altered-stress fracturing. J. Pet. Technol. 41 (9), 990–997. https://doi.org/10.2118/17533-PA.

Warpinski, N.R., Teufel, L.W., 1987. Influence of geologic discontinuities on hydraulic fracture propagation. J. Pet. Technol. 39 (2), 209–220. https://doi.org/10.2118/13224-PA.

Warpinski, N.R., Branagan, P., Wilmer, R., 1985. In-situ stress measurements at U.S. DOE's multiwell experiment site, Mesaverde Group, Rifle, Colorado. J. Pet. Technol. 37 (3), 527–536. https://doi.org/10.2118/12142-PA.

Warpinski, N.R., Peterson, R.E., Branagan, P.T., Wolhart, S.L., 1998. In situ stress and moduli: comparison of values derived from multiple techniques. SPE49190. In: SPE Ann. Tech. Conf. & Exh. New Orleans, LA, USA.

Zoback, M.D., Gorelick, S.M., 2012. Earthquake triggering and large-scale geologic storage of carbon dioxide. Proc. Natl. Acad. Sci. USA 109 (26), 10164–10168. https://doi.org/10.1073/pnas.1202473109.

Zoback, M.D., Kohli, A.H., 2019. Unconventional Reservoir Geomechanics: Shale Gas, Tight Oil, and Induced Seismicity. Cambridge University Press.

CHAPTER 12

Reservoir geomechanics

The probably most apparent examples of rock mechanical effects on reservoir scale behaviour are reservoir compaction and surface subsidence associated with pore pressure depletion. Operational problems related to subsidence are environmental concerns, like risk of flooding in land operations or platform safety concerns in offshore production. A majority of compaction-induced problems are associated with casing collapse within or adjacent to the compacting reservoir. It is on the other hand well established that compaction may be a prominent drive mechanism in relatively soft oil reservoirs.

Pore pressure depletion leads to changes in effective and total earth stresses. These stress changes control not only compaction and subsidence, but may also lead to changes in the fluid flow performance of the reservoir. Permeability may change, and the directions of preferred flow may be altered. In some cases the stress changes within and near the reservoir triggers seismicity during reservoir depletion. The concepts derived here also apply to injection processes, either to enhance hydrocarbon recovery or to ensure safe and secure storage of CO_2. The dynamic nature of petroleum reservoirs affects the stability of wells to be drilled and produced during the lifetime of the reservoir. Last but not least, geomechanical alteration of the reservoir and its surroundings may be observed in time-lapse (4D) seismic surveys, enabling the efficiency of the production or storage process to be monitored. Reservoir geomechanics has become established as an essential part of reservoir management.

12.1 Compaction and subsidence

When oil or gas is produced from a reservoir in the underground, the fluid pressure will generally decline. Reduced pore pressure in the reservoir rock will increase the effective stress and thereby cause the rock itself to shrink, and thus the reservoir will compact. Reservoir compaction may then in turn cause subsidence at the surface as sketched in Fig. 12.1.

Most oil and gas reservoirs will experience only a small degree of compaction, and the corresponding subsidence at the surface will be negligible.

Developments in Petroleum Science, Volume 72
ISSN 0376-7361, https://doi.org/10.1016/B978-0-12-822195-2.00021-8

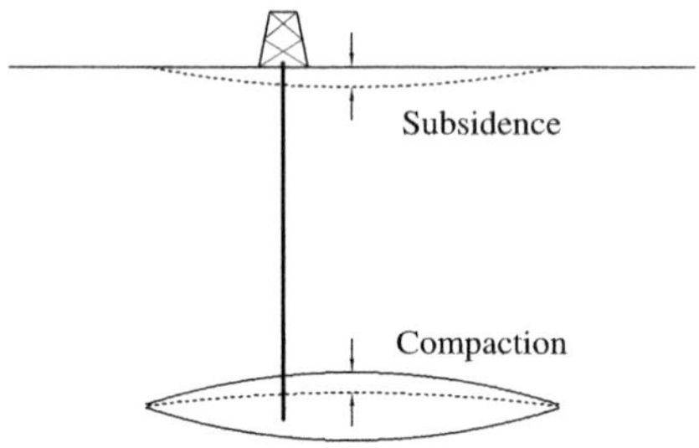

Figure 12.1 Compaction and subsidence.

To see a considerable degree of subsidence, one or several of the following conditions must be present:

- The reservoir pressure drop must be significant. Pressure maintenance by, for example, water flooding is likely to counteract compaction.
- The reservoir rock must be highly compressible. Compaction is more likely to be important in soft rocks.
- The reservoir must have a considerable thickness. However, the whole depleted zone must be considered. An adjacent aquifer will contribute to compaction, and so may part of the overburden if drained to the reservoir.
- For subsidence to occur, the reservoir compaction must be significant and, in addition, not shielded by the overburden rock. The degree of shielding depends on the depth and geometry of the reservoir and on the contrast in mechanical properties between the reservoir and its surroundings.

Considering the conditions above, we would expect only a few reservoirs to cause severe subsidence problems. This is also the case, but the frequent severe consequences of subsidence make it necessary for the industry to predict possible subsidence and compaction upfront, so that platforms and casings are designed properly, and to decide on an appropriate scheme for infill drilling, placement of injector wells, well completion strategy and so on.

The first report on ground subsidence is from the Goose Creek field in Texas and is dated back in the late 1910s. Later in the 1920s the Bolivar oil field in Venezuela showed the same behaviour. A well-known case is the Wilmington field in Long Beach, California, where subsidence of almost 9 metres was detected. More recent examples are the Ekofisk and Valhall reservoirs in the Norwegian sector of the North Sea and in the Groningen gas field onshore and offshore the Netherlands (Doornhof et al., 2006). The subsidence and induced seismicity in these fields are described further in Section 12.6. An extended review of subsidence and upheaval and other geomechanical impacts of the related processes of water withdrawal and injection is given by Gambolati and Teatini (2015).

In the following, we will present an overview of how we can model compaction and subsidence. The models presented here are largely analytical, which means that several simplifying assumptions need to be made. A key assumption is to consider the reservoir as a homogeneous medium. In reality the geology is complex, and we deal with the behaviour of a composite and inhomogeneous medium. The modelling needs to account for that: The reservoir needs to be divided into compartments with different properties and maybe with different depletion schemes. This requires numerical modelling and a detailed knowledge of the geology and associated rock properties. Often, however, simple models like those presented below may be sufficient to give an idea of the possible severity of compaction and subsidence in a given field case. Based on a simple analysis, we may then decide whether a further and more refined analysis is necessary.

12.2 Modelling of reservoir compaction

12.2.1 Uniaxial reservoir compaction

In this section, we calculate reservoir compaction in a simple basic case. We assume linear poroelasticity and consider a homogeneous reservoir consisting of isotropic rock. Then the deformation of the reservoir can be expressed by Hooke's law from Eqs. (1.105)–(1.107) in terms of changes in effective earth stresses, using the initial stress state (i.e. start of production) as reference for the strains:

$$E_{\mathrm{fr}}\varepsilon_{\mathrm{h}} = \Delta\sigma_{\mathrm{h}}' - \nu_{\mathrm{fr}}(\Delta\sigma_{\mathrm{H}}' + \Delta\sigma_{\mathrm{v}}') \tag{12.1}$$

$$E_{\mathrm{fr}}\varepsilon_{\mathrm{H}} = \Delta\sigma_{\mathrm{H}}' - \nu_{\mathrm{fr}}(\Delta\sigma_{\mathrm{h}}' + \Delta\sigma_{\mathrm{v}}') \tag{12.2}$$

$$E_{\mathrm{fr}}\varepsilon_{\mathrm{v}} = \Delta\sigma_{\mathrm{v}}' - \nu_{\mathrm{fr}}(\Delta\sigma_{\mathrm{H}}' + \Delta\sigma_{\mathrm{h}}') \tag{12.3}$$

The change in reservoir thickness Δh is given by the vertical strain ε_{v} and the reservoir thickness h:

$$\Delta h = -\varepsilon_{\mathrm{v}} h \tag{12.4}$$

Thus vertical compaction of the reservoir corresponds to a negative Δh.

To compute the compaction, we need to know how the stresses evolve. We will use the classical approach used within the industry and then, in a later section, discuss possible refinements.

Two assumptions are made: First, if the lateral extent of a reservoir is much larger than its thickness, then it is reasonable as a first assumption to neglect the lateral strain:

$$\varepsilon_{\mathrm{h}} = \varepsilon_{\mathrm{H}} = 0 \tag{12.5}$$

Thus the reservoir is assumed to compact only vertically, and we refer to this as uniaxial compaction. This is the same deformation mode as in an oedometer (or K_{o}) test as described in Section 7.3.8.

To maintain uniaxial vertical compaction during depletion, the effective horizontal stresses have to increase. Inserting Eq. (12.5) into Eqs. (12.1)–(12.2), we find

$$\Delta\sigma'_{\mathrm{h}} = \Delta\sigma'_{\mathrm{H}} = \frac{\nu_{\mathrm{fr}}}{1-\nu_{\mathrm{fr}}}\Delta\sigma'_{\mathrm{v}} \tag{12.6}$$

Secondly, we assume that the total vertical stress acting on the reservoir remains constant during production, which means that the full weight of the overburden is sensed by the reservoir at all times as the pore pressure is reduced ($\Delta\sigma_{\mathrm{v}} = 0$). This is valid for an infinitely wide and thin reservoir. Thus stress arching (or shielding of stress by the overburden) is neglected. This gives

$$\Delta\sigma'_{\mathrm{v}} = \Delta\sigma_{\mathrm{v}} - \alpha\Delta p_{\mathrm{f}} = -\alpha\Delta p_{\mathrm{f}} \tag{12.7}$$

We can then insert for all three principal stress changes in Eq. (12.3) and arrive at the following compaction formula:

$$\frac{\Delta h}{h} = \frac{1}{E_{\mathrm{fr}}}\frac{(1+\nu_{\mathrm{fr}})(1-2\nu_{\mathrm{fr}})}{1-\nu_{\mathrm{fr}}}\alpha\,\Delta p_{\mathrm{f}} \tag{12.8}$$

The compaction resulting from a certain pore pressure depletion can now be estimated from Eq. (12.8), provided that the elastic properties E_{fr} and ν_{fr}, the poroelastic coefficient α and the reservoir thickness h are known.

We may define a *compaction coefficient* or *uniaxial compressibility* c_{m} as

$$\frac{\Delta h}{h} = c_{\mathrm{m}}\,\alpha\,\Delta p_{\mathrm{f}} \tag{12.9}$$

The uniaxial compressibility relates to different frame moduli of the rock and is equal to the inverse of the uniaxial compaction modulus H_{fr}:

$$c_{\mathrm{m}} = \frac{1}{E_{\mathrm{fr}}} \frac{(1+\nu_{\mathrm{fr}})(1-2\nu_{\mathrm{fr}})}{1-\nu_{\mathrm{fr}}} = \frac{1}{3K_{\mathrm{fr}}} \frac{1+\nu_{\mathrm{fr}}}{1-\nu_{\mathrm{fr}}} = \frac{1}{H_{\mathrm{fr}}} = \frac{1}{\lambda_{\mathrm{fr}} + 2G_{\mathrm{fr}}} \tag{12.10}$$

Note that the uniaxial compaction modulus corresponds to the plane wave modulus (see Section 1.3). Since the modulus in Eq. (12.10) is a framework modulus, the P-wave modulus must be measured in dry conditions (or converted to its drained value through the Biot–Gassmann equation (1.171)) and Eq. (1.160). Furthermore, because dynamic moduli are often larger than static moduli (see Section 5.6), a direct estimate of compaction from sonic data is likely to provide a minimum estimate of the true compaction.

In addition to the underlying assumption of isotropic, linear poroelasticity, we need to be aware that pore pressure change is not uniform: Obviously, only the part of the reservoir where pore pressure is changing contributes to compaction and subsidence. Further, upscaling mechanical properties from laboratory or well to field scale is a not yet resolved challenge.

12.2.2 The depleting sphere

As another basic example, we will consider a depleting sphere with radius R_{o}, in which the pore pressure is reduced uniformly. This may also be seen as a limiting (although somewhat unrealistic) case of a depleting reservoir.

To compute the deformation and stresses, we need the equilibrium equation for spherically symmetrical displacement in *spherical* coordinates, as discussed in Section 4.10. The general expression for the radial displacement around a shrinking sphere is (Eq. (4.195))

$$u(r) = C_1 r + \frac{C_2}{r^2} \tag{12.11}$$

where the constants C_1 and C_2 are fixed by the boundary conditions.

The radial displacement should vanish at infinity and remain finite at $r = 0$, and both the radial displacement and the radial stress must be continuous at $r = R_0$, the border between the depleted region and the surroundings.

The result is that the radial stress change is constant within the depleted region and is given by

$$\Delta\sigma_r = \frac{2}{3} \frac{1-2\nu_{\mathrm{fr}}}{1-\nu_{\mathrm{fr}}} \alpha \Delta p_{\mathrm{f}} = \frac{4}{3} \eta \, \Delta p_{\mathrm{f}} \tag{12.12}$$

where η is the poroelastic stress coefficient defined in Eq. (4.54).

The volumetric strain in the depleted region is

$$\varepsilon_{\mathrm{vol}} = -\frac{\alpha \Delta p_{\mathrm{f}}}{\lambda_{\mathrm{fr}} + 2G_{\mathrm{fr}}} = -c_{\mathrm{m}} \alpha \Delta p_{\mathrm{f}} \tag{12.13}$$

Observe that depletion leads to a decrease in the total stress but an increase in the effective stress and hence a positive strain (compaction).

Eq. (12.13) may at first glance be slightly surprising, as we might be inclined to expect α/K_{fr} instead of αc_{m}. The reason why this is not so is the stress concentration that develops around the sphere, which shields it partly from the external stress and makes it appear stiffer than it would otherwise be.

12.2.3 Reservoir stress path

During uniaxial reservoir compaction (described in Section 12.2.1), the vertical stress remains constant, whereas the horizontal stresses change according to Eq. (12.6) during depletion. In general, the stress evolution during depletion (the reservoir stress path) may be defined by the stress path coefficients, as introduced by Hettema et al. (2000):

$$\gamma_{\mathrm{v}} = \frac{\Delta \sigma_{\mathrm{v}}}{\Delta p_{\mathrm{f}}} \tag{12.14}$$

$$\gamma_{\mathrm{H}} = \frac{\Delta \sigma_{\mathrm{H}}}{\Delta p_{\mathrm{f}}} \tag{12.15}$$

$$\gamma_{\mathrm{h}} = \frac{\Delta \sigma_{\mathrm{h}}}{\Delta p_{\mathrm{f}}} \tag{12.16}$$

The coefficient γ_{v} is called the arching coefficient and is zero when the vertical stress remains constant as the pore pressure is changed. The coefficients γ_{h} and γ_{H} describe the change in the horizontal stress field. Assuming full rotational symmetry in the horizontal plane, $\gamma_{\mathrm{h}} = \gamma_{\mathrm{H}}$. The coefficient γ_{h} is commonly termed the depletion coefficient. Recall that for a depleting reservoir, the pressure drop is negative, so that positive values of the stress path coefficients imply reductions in total vertical and horizontal stresses.

The stress path coefficients may also be similarly defined in terms of the effective stresses. It is straightforward to show that the relation between the stress path coefficients γ' for effective stresses and the coefficients defined above is

$$\gamma' = \gamma - \alpha \tag{12.17}$$

Since $\gamma < \alpha$ in practice, the γ' values will be negative. This simply implies that effective stresses increase as a result of depletion or decrease as a result of inflation.

We may also define another stress path coefficient κ' as the ratio between effective horizontal and vertical stress changes:

$$\kappa' = \frac{\Delta\sigma'_h}{\Delta\sigma'_v} \tag{12.18}$$

This parameter is convenient in laboratory tests when the individual values of γ_v and γ_h cannot be measured. The coefficient κ' expresses the ratio between horizontal and vertical effective stress changes, which can be obtained by controlling axial and radial effective stress changes in the laboratory (Eq. (7.14)). If this is done under uniaxial strain conditions, then κ' will be equal to the parameter K_o in Section 7.3.8. The following relation is seen to exist between γ_h, γ_v and κ':

$$\kappa' = \frac{1 - \dfrac{\gamma_h}{\alpha}}{1 - \dfrac{\gamma_v}{\alpha}} = \frac{\alpha - \gamma_h}{\alpha - \gamma_v} = \frac{\gamma'_h}{\gamma'_v} \tag{12.19}$$

For uniaxial compaction (no arching) as described in Section 12.2.1, the stress path coefficients are

$$\kappa' = \frac{\nu_{fr}}{1 - \nu_{fr}} \tag{12.20}$$

$$\gamma_h = \gamma_H = \alpha\frac{1 - 2\nu_{fr}}{1 - \nu_{fr}} \tag{12.21}$$

$$\gamma_v = 0 \tag{12.22}$$

This has more or less been the default assumption within the petroleum industry for decades. Through the analysis of compaction and subsidence in the Ekofisk field (Teufel et al., 1991), the relevance of the stress path concept was recognised. In that case, stress measurements during reservoir depletion showed that the effective horizontal stress increase was much lower than expected from laboratory data (κ' was close to 0.2 in the field, whereas a value of approximately 0.5 was measured on cores). This has had profound impact on the entire Ekofisk field development and has led to increased focus on the reservoir stress path for many other fields.

From Hooke's law (12.1)–(12.3) and Eqs. (12.4), (12.14) and (12.16) elastic reservoir compaction can now be calculated for a general stress path:

$$\frac{\Delta h}{h} = \frac{\alpha}{E_{\rm fr}}\left[\left(1-\frac{\gamma_{\rm v}}{\alpha}\right) - 2\nu_{\rm fr}\left(1-\frac{\gamma_{\rm h}}{\alpha}\right)\right]\Delta p_{\rm f} = -\frac{1}{E_{\rm fr}}\left[\gamma'_{\rm v} - 2\nu_{\rm fr}\gamma'_{\rm h}\right]\Delta p_{\rm f} \tag{12.23}$$

The values of $\gamma_{\rm v}$ and $\gamma_{\rm h}$ depend on the geometry of the depleting reservoir zone and also on the elastic contrast between the reservoir and its surroundings. For example, if the reservoir is soft, then stress will be concentrated in the stiffer surroundings, thus shielding the depleting zone.

The horizontal stress path (or depletion) coefficient $\gamma_{\rm h}$ depends on the boundary conditions at the sides of the reservoir. In Section 12.2.1, we assumed uniaxial compaction based on the argument that horizontal strain is negligible because the lateral extent of the reservoir is normally much larger than its thickness. This underlines the importance of the geometry of the depleting reservoir zone.

As an example of the effect of geometry, consider the analytical solution for the stress field within a depleting sphere, as given in Eq. (12.12). This shows that for a spherical reservoir embedded in an elastic medium with properties identical to those of the surroundings, the horizontal and vertical stress path coefficients are equal (in contrast to Eqs. (12.21) and (12.22))

$$\gamma_{\rm h} = \gamma_{\rm v} = \frac{2}{3}\alpha\frac{1-2\nu_{\rm fr}}{1-\nu_{\rm fr}} = \frac{4}{3}\eta \tag{12.24}$$

12.2.3.1 The ellipsoidal reservoir

Rudnicki (1999) modelled the stress path coefficients analytically for a more general case. Building on the theory of Eshelby (1957) (see also Chapter 6), the reservoir is considered as an ellipsoidal poroelastic inclusion embedded in an infinite solid medium. "Infinite" here means that the influence of the free surface is neglected, which limits the validity of the results to reservoirs at depths greater than their lateral extents. The depleting part of the reservoir is oriented with the short principal axis in the vertical direction. In the derivation the strain occurring in a depleting stress-free reservoir is calculated first. Then the stresses required to restore the original reservoir shape are calculated, and finally these stresses are added to the initial in situ stresses. Rudnicki's model also allows contrast in elastic properties between the reservoir and its surroundings. Note that here and in the following, we consider drainage to occur only within the reservoir, so the relevant elastic properties of the surrounding rock are undrained moduli.

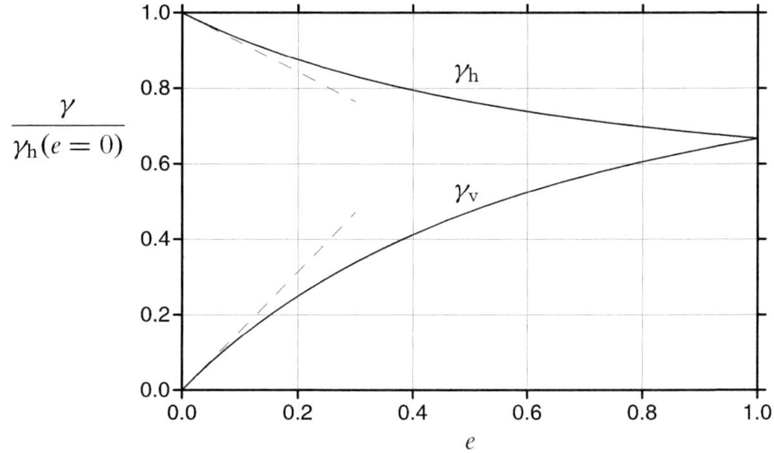

Figure 12.2 Normalised stress path coefficients versus the aspect ratio of the depleting reservoir zone, calculated from Rudnicki's model for the case of no elastic contrast. The stress path coefficients have been normalised by dividing by γ_h for uniaxial strain conditions; see Eq. (12.21). The dashed lines show the linearisation from Eqs. (12.27) and (12.28).

The resulting coefficients depend, as expected, on the thickness-to-diameter ratio (also called aspect ratio) e and on the elastic contrast between the reservoir and its surroundings. If we assume that all elastic moduli are the same within the reservoir and in the surrounding rock, then the coefficients can be calculated using the following expressions:

$$\gamma_h = \alpha \frac{1-2\nu_{fr}}{1-\nu_{fr}} \left[1 - \frac{e}{2\sqrt{(1-e^2)^3}} \left(\arccos e - e\sqrt{1-e^2} \right) \right] \tag{12.25}$$

$$\gamma_v = \alpha \frac{1-2\nu_{fr}}{1-\nu_{fr}} \frac{e}{\sqrt{(1-e^2)^3}} \left(\arccos e - e\sqrt{1-e^2} \right) \tag{12.26}$$

Fig. 12.2 shows γ_h and γ_v as functions of the aspect ratio. For flat reservoirs (small e), these expressions can be approximated (as shown previously by Segall and Fitzgerald (1998)):

$$\gamma_h = \alpha \frac{1-2\nu_{fr}}{1-\nu_{fr}} \left(1 - \frac{\pi}{4} e \right) \tag{12.27}$$

$$\gamma_v = \alpha \frac{1-2\nu_{fr}}{1-\nu_{fr}} \frac{\pi}{2} e \tag{12.28}$$

In the limit as $e \to 0$, the stress path coefficients are identical to the expressions found above (Eqs. (12.21) and (12.22)) for uniaxial compaction. The horizontal stress path coefficient decreases as the shape of the depleting zone is changed from infinitely flat towards spherical ($e \to 1$), whereas the vertical (arching) coefficient increases with increasing aspect ratio. In the limit as $e \to 1$ (spherical inclusion), the arching coefficients become equal, in agreement with Eq. (12.24).

Note that from Eqs. (12.25) and (12.26) it follows that

$$\gamma_v + 2\gamma_h = 2\alpha \frac{1 - 2\nu_{fr}}{1 - \nu_{fr}} \tag{12.29}$$

Thus, for an ellipsoidal reservoir, when the surroundings have the same elastic properties as the reservoir, the sum $\gamma_v + 2\gamma_h$ is independent of the aspect ratio e.

The volumetric compaction of the reservoir can be expressed in terms of the stress path coefficients by introducing the definitions into Hooke's law. The result is

$$\varepsilon_{vol} = -\left(\alpha - \frac{\gamma_v + 2\gamma_h}{3}\right)\frac{\Delta p_f}{K_{fr}} = \frac{\overline{\gamma'}}{K_{fr}}\Delta p_f \tag{12.30}$$

where $\overline{\gamma'} = (\gamma_v' + 2\gamma_h')/3$.

For an ellipsoidal reservoir with no contrast to the elastic properties of the surroundings, this means (combining Eqs. (12.10) and (12.29)) that the volumetric deformation of the depleting reservoir is always given by the uniaxial compaction modulus, that is,

$$\varepsilon_{vol} = -\alpha c_m \Delta p_f \tag{12.31}$$

This is a direct consequence of stress arching. It also implies that the pore compressibility is independent of the stress path (but it is different from the basic definition in Eq. (1.206)); see Section 12.4.1.

Comparing the vertical compaction of the reservoir calculated from Eq. (12.23) with the simple case of uniaxial compaction ($e = 0$, Eq. (12.8)), we find that the compaction is largest for uniaxial deformation. The physical explanation of this is that arching (for $e > 0$) will redistribute the stresses so that the reservoir is partly shielded from the overburden stress. When the reservoir approaches spherical shape ($e = 1$), the compaction drops to 1/3 of the uniaxial compaction case. This also follows directly from the volumetric strain being independent of the reservoir shape; see Eq. (12.31).

12.2.3.2 *Elastic contrast*

The effect of elastic contrast between the reservoir and the surroundings can be studied using the full equations in Rudnicki's article (1999) or based on finite element (FEM) simulations (Morita et al., 1989; Mulders, 2003). In all cases, it is found that arching is promoted if the stiffness of the depleting reservoir is significantly lower than the stiffness of the surrounding rock.

Morita et al. (1989) argued from FEM simulations with a disk-shaped reservoir that if the ratio between the shear modulus of the reservoir and the overburden rock is between 0.2 and 1.5, and if the aspect ratio of the reservoir is less than 0.05, then the resulting compaction is essentially uniaxial.

Similar conclusions can be drawn from Rudnicki's model, which shows an increase both in γ_v and to some extent in γ_h compared to the elastically matched case (Holt et al., 2004). For a finite aspect ratio, both stress path coefficients approach 1 in the limit of infinite elastic contrast. On the other hand, a stiff reservoir in soft surroundings will have lower stress path coefficients, approaching zero in the case of infinite contrast. Note that the value of $\gamma_v + 2\gamma_h$, which controls volumetric compaction (see Eq. (12.30)) and also pore compressibility (see Section 12.4.1), does not remain constant for all aspect ratios and decreases with increasing ratio between the stiffness of the reservoir and the surroundings.

12.2.3.3 *Non-ellipsoidal reservoirs*

Mulders (2003) compared numerical FEM simulations of stress path coefficients for a disk-shaped reservoir with those calculated by Rudnicki's model. The assumption of ellipsoidal geometry by Rudnicki causes the stress path coefficients to be homogeneous over the entire depleted zone.

For a disk-shaped reservoir, Mulders found that the stress path coefficients near the centre of the reservoir agree with the analytical approach. However, γ_v increases, and γ_h decreases towards the edge of the reservoir. From FEM simulations we may inspect the effect of reservoir depth and tilt with respect to the vertical. In particular, tilt is found to enhance arching.

Thermoelastic effects may be added to this model (Rudnicki, 1999; Segall and Fitzgerald, 1998). This has relevance for recovery of hydrocarbon reserves by injection of cold water or hot steam.

12.2.4 Beyond simple elastic theory

So far we have considered the reservoir (and its surroundings) to behave linearly elastic. A reservoir may however develop plastic deformation during production and possibly reach a state where faults are generated or pre-existing faults are triggered. Faulting may also occur outside the reservoir, in particular, near the edges, where the stress concentration resulting from stress arching in the surrounding rock is particularly large (see Section 12.3.2).

For a reservoir rock being intact (non-fractured and in an elastic state) before production starts, non-elastic processes may occur during depletion if the stress path brings the rock past yield and in the direction of its failure surface. Failure processes may take place from the grain scale to the scale of large heterogeneities, and the compaction behaviour gradually becomes plastic. Laboratory experiments, where synthetic sandstone was formed under stress (Holt et al., 2000), showed that after an initial phase with largely elastic deformation, permanent (plastic) deformation gradually takes over as the material compacts. In uniaxial strain conditions, this led to approximately a doubling of the compaction rate, and it also led to enhanced lateral stress development (corresponding to a reduction of the coefficient γ_h) after onset of plasticity.

Plastic behaviour also implies that if water or gas is injected to raise the pore pressure and counteract compaction, we may not expect to recover the compaction that has occurred, as explained by Santarelli et al. (1998). They observed however a decreasing trend in γ_h. Thermal cooling of the reservoir may explain this, as pointed to by Søreide et al. (2014), on the basis of laboratory experiments and field observations. Depending on the stress path and on the constitutive behaviour of the reservoir rock, we may also reach a stress state that allows development of localised deformation bands or activation of existing faults.

Fig. 12.3 illustrates schematically how the stress paths will evolve towards the failure surface in the case of a perfectly linear elastic material. We see that arching directs the path more in direction of the end cap, whereas lack of arching apparently may lead towards a shear failure (faulting) situation. The development of plasticity after the rock reaches its yield point may however redirect the stress path, as shown in Fig. 12.4. These examples are taken from laboratory tests with a high-porosity sandstone. They show that, depending on the initial stress state, a uniaxial compaction path may lead to various scenarios. In both cases the stress path bends over to become increasingly parallel to the shear failure part of the failure envelope. In one case (curve a; low initial shear stress) the stress path eventually reaches failure near the end cap, whereas in the other case (curve b; high initial shear stress), strain softening occurs after a ride along the failure envelope.

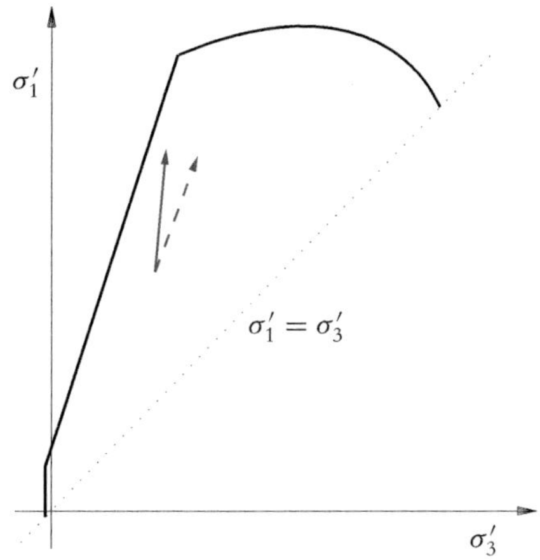

Figure 12.3 Influence of arching on the reservoir stress path. The solid arrow shows the stress path in the absence of arching, whereas the dashed arrow indicates how arching may redirect the stress path.

An initially faulted and fractured reservoir in a tectonically active environment may be assumed to be in a continuous failure state (Addis, 1997). Then a failure criterion (such as Mohr–Coulomb) should apply at all stages during depletion. In a normally faulted stress regime, where the vertical stress is the largest principal stress, this gives (see Eq. (3.5))

$$\Delta\sigma_{\mathrm{v}}' = \Delta\sigma_{\mathrm{h}}' \tan^2\beta \tag{12.32}$$

Thus the stress path coefficient is

$$\kappa' = \frac{1}{\tan^2\beta} = \frac{1-\sin\varphi}{1+\sin\varphi} \tag{12.33}$$

where β is the failure angle, and φ is the angle of internal friction (see Section 2.3.1). Provided that the total vertical stress remains constant (no arching), it follows from Eq. (12.19) that

$$\gamma_{\mathrm{h}} = \frac{\Delta\sigma_{\mathrm{h}}}{\Delta p_{\mathrm{f}}} = 1 - \frac{1}{\tan^2\beta} = \frac{2\sin\varphi}{1+\sin\varphi} \tag{12.34}$$

It is assumed that the appropriate effective stress constant is equal to 1, since this situation describes rock failure (Eq. (2.50)). This applies to critically oriented fractures. Since fractures and faults may have developed under different stress regimes, their orientations may vary. A generalised equation for

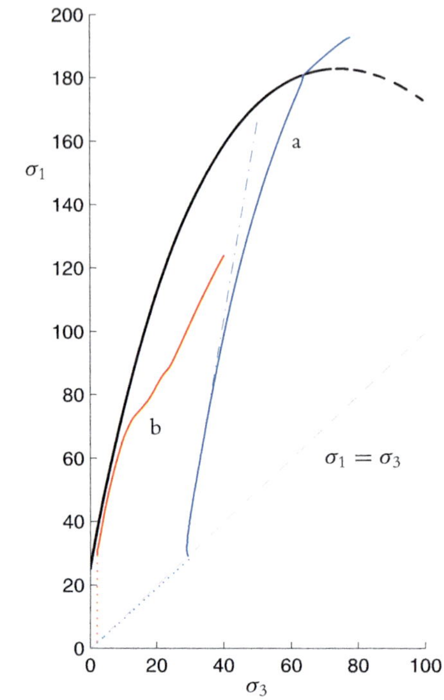

Figure 12.4 Stress paths (thin solid lines) followed by a 29% porosity sandstone under uniaxial compaction, starting from the state $\sigma_1 = \sigma_3 = 30$ MPa (path a) and from the state $\sigma_1 = 31$ MPa, $\sigma_3 = 2$ MPa (path b). The dash-dot line shows the theoretical stress path corresponding to path a assuming that the rock is linearly elastic. The failure envelope (thick line) is estimated on the basis of several other tests. Courtesy of SINTEF.

the depletion coefficient in Eq. (12.34) for fractures of arbitrary orientation (= maximum principal stress) can be found in Holt et al. (2016).

12.2.5 Time-delayed reservoir compaction

It is observed in several field cases that compaction and/or subsidence occurs with some delay with respect to change in pore pressure (Hettema et al., 2002). Time-dependent deformation (Section 1.9) is normally due to consolidation, that is, establishment of pore pressure equilibrium, or it is associated with creep.

In a homogeneous high-permeability reservoir, pore pressure equilibration is a reasonably rapid process, taking place within hours or days. Longer time constants must be associated with very low permeability rock, such as

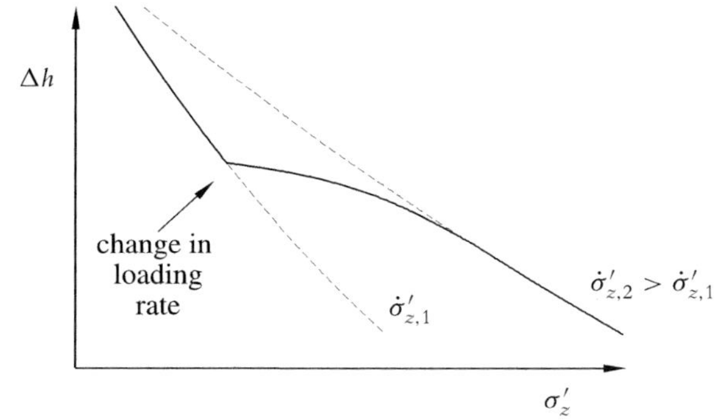

Figure 12.5 Effect of change in loading rate.

shale, which may be present as lenses or layers within the reservoir section. For instance, it would take about 100 years to reach pore pressure equilibrium within a 10 m thick shale layer with 1 nanoDarcy permeability. Thus shales continue to compact as their pore pressure approaches equilibrium with that of the reservoir rock.

Reservoir depletion occurs over a time scale of 50 years or so, which means that the deformation rate is considerably faster than during natural compaction on a geological time scale. Most geomaterials are visco-elastic (they creep), which means that an increased loading rate will cause an increased stiffness at a given stress level (see Fig. 12.5) (de Waal and Smits, 1988; see also Section 1.9.4). A slow initial transient compaction followed by an increased compaction rate is then expected in accordance with field observations. This model however also predicts that reservoir rocks would be more compliant than laboratory cores, because the loading rate (test duration: hours or days) employed in the laboratory is much higher than in the field. This appears to be in disagreement with most observations.

Delayed compaction and subsidence may also originate from the drainage pattern of the reservoir: If the pore pressure is reduced only within a zone of limited lateral extent, then stress arching is enhanced (as seen in Section 12.2.3), and compaction is reduced accordingly, compared to what it would be if the entire reservoir was drained uniformly. As the drainage radius increases, stress arching is reduced, and compaction accelerates and eventually reaches the level it would have for uniform drainage.

A transition from elastic to plastic behaviour may also be interpreted as a time-delayed compaction, because the compaction rate would increase with increasing pore pressure depletion. This is however not a time-delayed compaction effect as such (although plastic strain often involves creep).

Above we have listed mechanisms that may lead to time-delayed compaction and hence time-delayed subsidence with respect to pore pressure depletion. In addition, subsidence may be delayed with respect to compaction. To see this in field data, separate compaction and subsidence measurements are required. Such data are usually not available. Time-delayed stress alteration and creep in the cap rock may contribute to time-delayed subsidence (Christensen et al., 1988).

12.3 From compaction to subsidence

12.3.1 Geertsma's nucleus of strain model

To get a more precise picture of the subsidence, we will take a look at how the deformation of the compacting reservoir propagates through the overburden. A simple approach is the so-called nucleus of strain model presented by Geertsma (1973a,b). The idea is to calculate the subsidence resulting from the compaction of a small sphere and then to apply the superposition principle to calculate the total subsidence by adding the influence of many such spheres filling the entire reservoir.

Geertsma's model is limited to the case where there is no contrast in elastic properties between the reservoir and surroundings. As mentioned in Section 12.2.3, although the drained (frame) moduli are relevant for the reservoir, the surrounding rocks do not drain, at least not on the same time-scale, so undrained moduli are therefore relevant for these rocks. Thus a homogeneous subsurface does not simply imply the same rock properties everywhere.

Considering the region outside the depleting sphere (of radius R_0) and denoting by u_0 the radial displacement at the surface of the depleting sphere, the radial displacement at a distance r from the centre of the sphere is (see Eq. (12.11))

$$u(r) = u_0 \frac{R_0^2}{r^2} \tag{12.35}$$

where u_0 may be related to the volume change of the depleted region by (recall the sign convention, displacements point in the opposite direction of the coordinate axes; see e.g. Fig. 4.1)

$$\Delta V = -4\pi R_0^2 u_0 \tag{12.36}$$

Combined with Eq. (12.31) and the definition of volumetric strain, Eq. (1.79), this gives

$$u_0 = -\frac{c_m V \alpha \Delta p_f}{4\pi R_0^2} \tag{12.37}$$

and hence

$$u(r) = -\frac{c_m}{4\pi} V \alpha \Delta p_f \frac{1}{r^2} \tag{12.38}$$

which describes the displacement field outside a depleting sphere in an infinite medium.

Note that the magnitude of the displacement field is governed by the product $V\Delta p_f$. This means that we get the same displacement field outside R_0 for a sphere of half the volume if p_f is doubled. This is similar to gravitation theory, in which we get the same gravitational field outside the Earth if we assume all the mass to be situated at the centre of the Earth.

This principle, as we will see, is important for the Geertsma theory.

12.3.1.1 The effect of the free surface

Eq. (12.38) has one major shortcoming: it does not take the surface properly into account. At the surface the vertical stress must be zero. This problem was solved by Geertsma, who used known solutions from thermoelasticity (Mindlin and Cheng, 1950a,b; Sen, 1951) to arrive at the corrected expression

$$\vec{u} = \frac{c_m}{4\pi}\left(\frac{\vec{R}_1}{R_1^3} + (3-4\nu)\frac{\vec{R}_2}{R_2^3} - \frac{6z(z+D)\vec{R}_2}{R_2^5} + \frac{2\hat{z}}{R_2^3}\left[(3-4\nu)(z+D) - z\right]\right) V\alpha\Delta p_f \tag{12.39}$$

The geometry and symbols are defined in Fig. 12.6. The first term is recognised as that developed in Eq. (12.38), whereas the remaining terms are correction terms ensuring zero vertical stress at $z=0$. It is a straightforward but admittedly tedious job to compute the strain components from Eq. (12.39) and to insert them into Hooke's law to show that $\sigma_z = 0$ at $z = 0$.

It is easy vector algebra (observe from Fig. 12.6 that $\vec{R}_2 + 2D\hat{z} = \vec{R}_1$) to show that at the surface, $\vec{u}$ points in the $\vec{R}_1$ direction and that the displacement is given by

$$\vec{u} = \frac{c_m(1-\nu)}{\pi} V\alpha\Delta p_f \frac{\vec{R}_1}{R_1^3} \tag{12.40}$$

where R_1 is the distance from the nucleus to the surface point we consider.

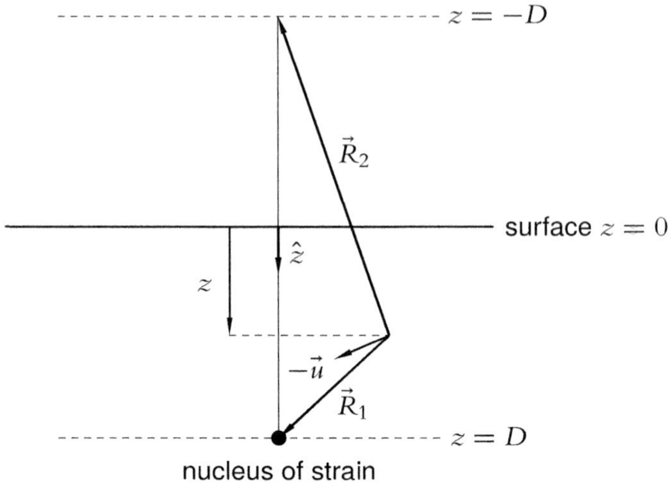

Figure 12.6 Geometry for the Geertsma solution. $\hat{z}$ is the unit vector in the z-direction, and $-\vec{u}$ indicates the direction of particle movement for subsidence. (Recall that due to the sign convention, the positive direction for displacement is opposite of the coordinate axes).

Thus the presence of the free surface enhances the displacement by typically a factor three!

12.3.1.2 The size of the subsidence bowl

From Eq. (12.40) we can compute the volume ΔV_{subs} of the subsidence bowl by integrating the vertical deformation over the entire surface:

$$\Delta V_{\text{subs}} = -\int_0^{\infty} 2\pi\rho\, u_z(\rho)\mathrm{d}\rho = 2c_{\text{m}}\,(1-\nu)\, V\alpha\Delta p_{\text{f}} \tag{12.41}$$

where ρ is the horizontal distance, that is, $R_1^2 = \rho^2 + D^2$, and we have used the integral

$$\int_0^{\infty} \frac{x}{\left(x^2 + a^2\right)^n}\mathrm{d}x = \frac{a^{2(1-n)}}{2n-2} \tag{12.42}$$

Eq. (12.41) shows that the volume of the subsidence bowl is related to the volume reduction of the compacting reservoir (ΔV_{comp}) by

$$\Delta V_{\text{subs}} = 2\,(1-\nu)\,\Delta V_{\text{comp}} \tag{12.43}$$

As an example, for $\nu = 0.25$, the volume of the subsidence bowl is 50% larger than the volumetric reservoir compaction.

Since the subsidence for an arbitrary shaped reservoir is found by integrating the result from many nuclei, it is clear that this ratio is generally valid (within the assumptions of the model i.e. elastic deformation and no contrast between the reservoir and the surroundings).

12.3.1.3 Subsidence above a disk-shaped reservoir

Given Eq. (12.40), it is in principle straightforward to compute the subsidence for an arbitrarily shaped reservoir by numerical methods. This is however of limited interest, since the method is based on constant elastic properties. When turning to numerical modelling, we should also include a more realistic model of the surrounding formations including the layers below the reservoir.

There exist, however, analytical solutions for a disk-shaped reservoir giving a general insight into the subsidence problem, which is useful even though the method is not suited for precise calculations.

We will first consider the vertical displacement along the centre line through a disk-shaped reservoir. From Eq. (12.39) we extract the vertical component of the displacement and integrate over the reservoir. In our calculations, we use a set of strain nuclei situated at $z = D$ and multiply by the reservoir height h to get the final result. The integrals involved are similar to Eq. (12.42), and the final result is

$$\begin{aligned} u_z = &\frac{1}{2} c_{\rm m} h \alpha \Delta p_{\rm f} \left(3 - 4\nu + \frac{D - z}{|D - z|} - \frac{D - z}{\sqrt{R^2 + (D - z)^2}} \right. \\ &\left. - \frac{(D + z)(3 - 4\nu)}{\sqrt{R^2 + (D + z)^2}} + \frac{2R^2 z}{\left(R^2 + (D + z)^2\right)^{3/2}} \right) \end{aligned} \tag{12.44}$$

The surface subsidence at the centre is found by letting $z = 0$:

$$u_z = w = 2 c_{\rm m} h \alpha \Delta p_{\rm f} (1 - \nu) \left(1 - \frac{D}{\sqrt{D^2 + R^2}} \right) \tag{12.45}$$

An important question to resolve is how much error we introduce by assuming the compaction to be localised at $z = D$. We recall from the discussion following Eq. (12.38) that for a sphere, we may assume all compaction to take place at the centre and still get the correct answer outside the reservoir. Thus, if we modelled a reservoir consisting of a distribution of spheres, then we would get the correct answer. Since a disk cannot be

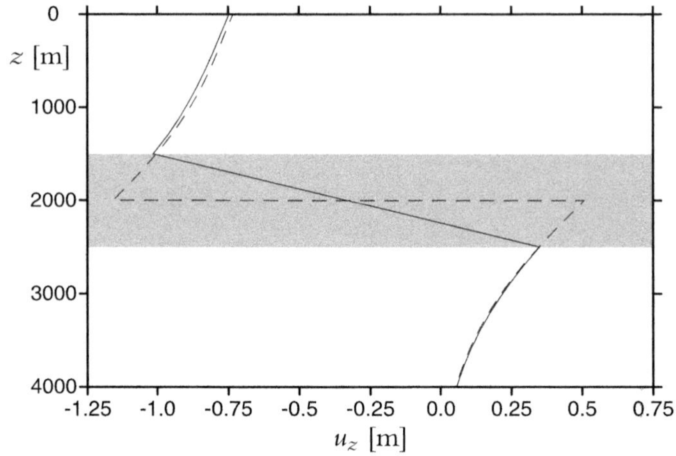

Figure 12.7 Comparison of the vertical displacement at the centre of a disk-shaped reservoir. The approximate solution (12.44) is shown by the dashed line, and the exact solution by the solid line.

fully approximated by spheres, we expect some deviation, but it might seem reasonable to expect that it will be small.

To address this question, we may divide Eq. (12.45) by h and integrate from $D - h/2$ to $D + h/2$, which gives the exact subsidence for a reservoir of thickness h:

$$u_z = w = 2c_m \alpha \Delta p_f (1 - \nu) \left(h - \sqrt{\left(D + \frac{h}{2}\right)^2 + R^2} + \sqrt{\left(D - \frac{h}{2}\right)^2 + R^2} \right) \tag{12.46}$$

By inserting numbers we find that the difference between the two calculations is only a couple of percent even when $h/D \sim 1/2$.

Eq. (12.44) may be integrated to give the exact result for a reservoir of a finite thickness. The result is shown in Fig. 12.7, where the exact result is compared with the approximate result, for reservoir properties given in Table 12.1 (p. 624) except for a large reservoir thickness of h =1000 m. Again, we see that outside the reservoir, the Geertsma solution is very close to the exact result.

Geertsma (1973a) also gave solutions for off-centre positions. He showed that the displacements and stresses can be expressed in terms of integrals of Bessel functions, which in turn can be expressed via elliptic integrals.

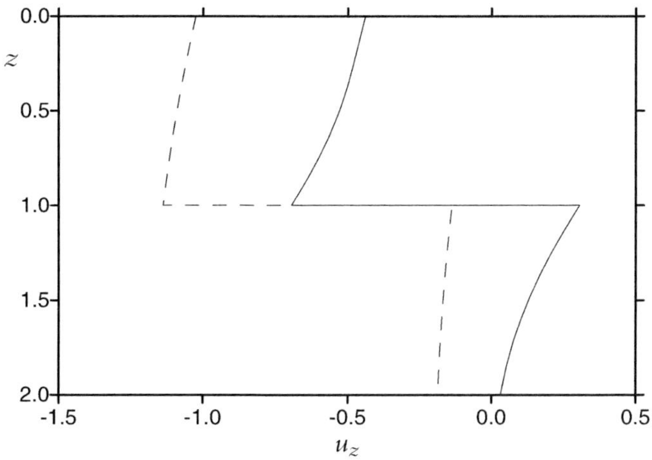

Figure 12.8 Vertical displacement as a function of depth along a vertical line through the centre of a disk-shaped reservoir with radius equal to the reservoir depth (full line) and radius equal to three times the reservoir depth (dashed line). The compaction has been normalised to 1.

We have included an overview of Geertsma's main formulas in Appendix D.5.

12.3.1.4 Some example results

The Geertsma solutions are useful for getting an overview of what happens during compaction and subsidence. From Eq. (12.44) we can compute the vertical displacement at the centre of a disk-shaped reservoir. Fig. 12.8 shows two examples for two different reservoir size-to-depth ratios R/D. The full line applies to a reservoir with radius equal to the depth. We observe that the reservoir top subsides while the reservoir bottom rises. This results in a subsidence that is smaller than the reservoir compaction.

On the other hand, for the case shown by the dashed line, which applies to a wide reservoir with radius three times the depth, both the reservoir top and bottom subside, and the Geertsma model predicts a subsidence that is larger than the reservoir compaction.

Fig. 12.9 shows the ratio of subsidence to reservoir compaction as a function of reservoir radius scaled by the reservoir depth. For very wide and/or shallow reservoirs, the ratio in this example eventually approaches 1.5.

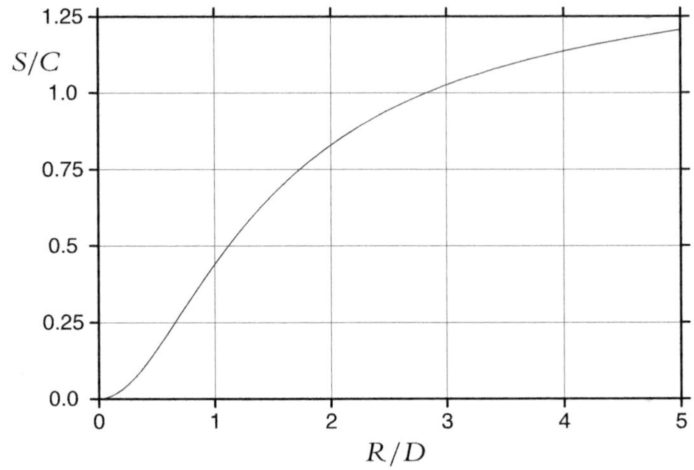

Figure 12.9 Ratio of subsidence S to compaction C as a function of R/D according to the Geertsma model for $\nu = 0.25$. See Eqs. (12.45) and (12.9).

Figs. 12.10 and 12.11 show the displacement field around depleting reservoirs with radius equal to the depth and three times the depth, respectively.

We observe that at the central parts of the reservoir, the displacement is essentially vertical, in accordance with the assumption of uniaxial compaction. However, at the outer radius the dominant displacement is horizontal. It is clear that the assumption of uniaxial vertical compaction is a simplification relative to the predictions of the purely elastic Geertsma model, which is of course again a simplification of the true behaviour.

Fig. 12.12 shows contour plots of the vertical and radial displacements for the parameters given in Table 12.1, which corresponds to a reservoir with radius equal to the depth.

Fig. 12.13 shows contour plots of the *change* in vertical and radial stresses for a depleting disk-shaped reservoir. The calculations have been made for the parameters given in Table 12.1.

Fig. 12.14 shows a plot of q/p' normalised to the value before depletion for the case in Figs. 12.12 and 12.13. The model does not give valid predictions inside the reservoir. We see that outside the perimeter of the reservoir, we have a region where there is a significant increase in q/p'.

The assumptions of Geertsma's model are rarely met in practice. The subsurface is not homogeneous, and in most cases, it is not linearly elastic and isotropic. Reservoirs are not disk-shaped nor ellipsoidal, and they

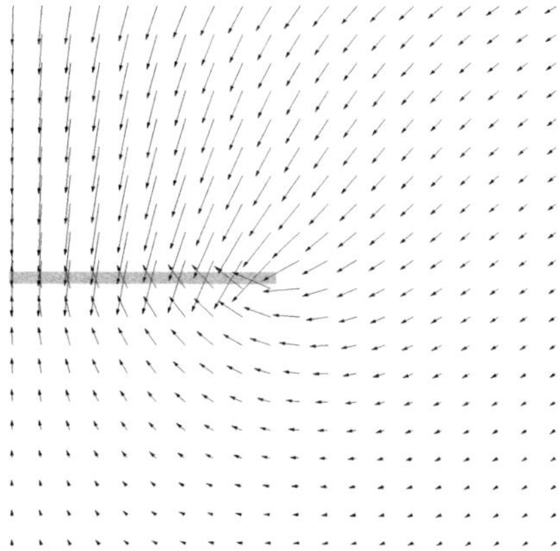

Figure 12.10 Displacement field around a circular reservoir with radius equal to the reservoir depth. The grey rectangle indicates the position of the reservoir. The centre line of the reservoir is to the left. (We here show the physical displacements, which point opposite of the formal definition in Section 1.2.)

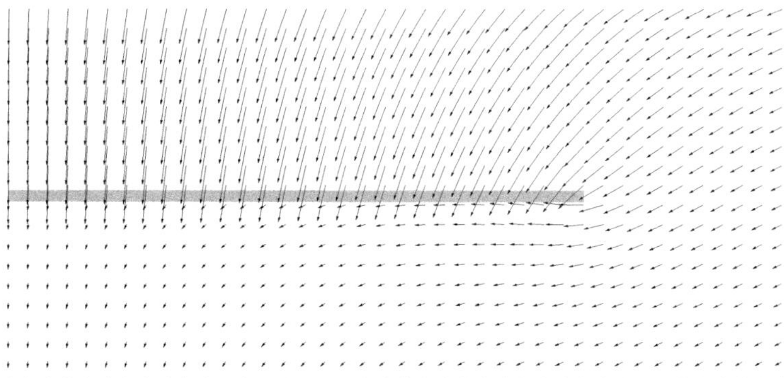

Figure 12.11 Displacement field around a reservoir with radius equal to three times the reservoir depth. The grey rectangle indicates the position of the reservoir. The centre line of the reservoir is to the left. (We here show the physical displacements, which point opposite of the formal definition in Section 1.2.)

are not necessarily embedded in the horizontal plane. Most of these issues can be treated numerically, but some analytical refinements have also been made.

In the early days of subsidence research, Gambolati (1972) generalised Geertsma's model to account for elastic contrast between depleting nuclei and surrounding rock. Van Opstal (1974) considered this indirectly by in-

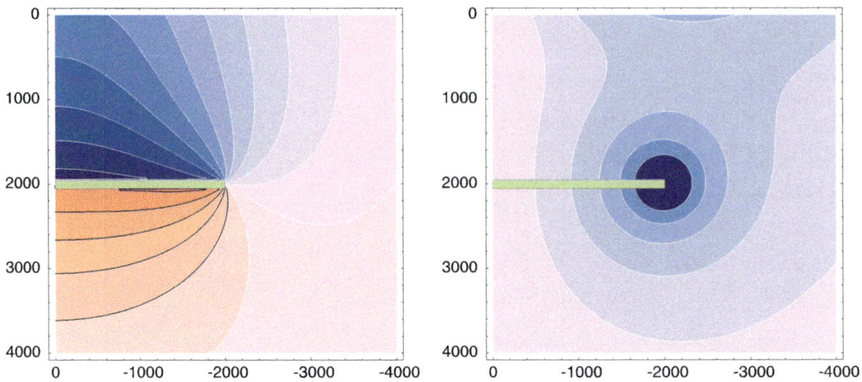

Figure 12.12 Contour plots of the vertical (left) and radial (right) displacements around a depleting disk-shaped reservoir with parameters given in Table 12.1. The contours for u_z run from -11 cm to 5 cm, in steps of 1 cm, whereas those for u_r run from -5 cm to -1 cm. Negative contours including 0 are in white, and positive contours in black.

Table 12.1 Reservoir properties used in the examples shown in Figs. 12.12–12.14.

Reservoir depth	D	2000 m
Reservoir radius	R	2000 m
Shear modulus	G	2 GPa
Poisson's ratio	ν	0.25
Reservoir height	h	100 m
Depletion	$\|\Delta p_f\|$	10 MPa
Density	ρ	2200 kg/m^3
Estimated compaction, assuming uniaxial deformation	$c_m h \|\Delta p_f\|$	0.17 m
Reservoir strain	$-\Delta h/h$	0.17 %
Subsidence at $r = 0$	From Eq. (12.45)	0.07 m

troducing a rigid bottom as a zero-strain boundary placed at some depth below the reservoir. He showed that this led to balance between compacted reservoir volume and volume of the subsidence bowl, independent of the position of the rigid bottom. This differs from the Geertsma solution, where the subsidence volume is $2(1 - \nu)$ times the compacted volume (Eq. (12.43)). The difference is manifested by small uplifts at the surface outside the radius of the depleting zone. The slopes of the subsidence bowl are steeper, and the vertical subsidence increases compared to the Geertsma case.

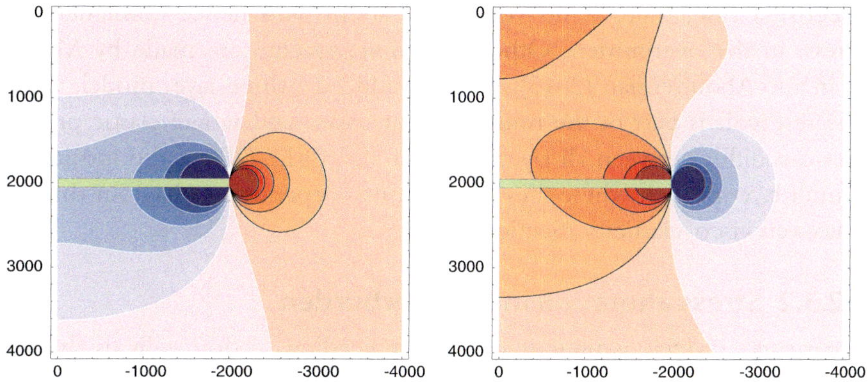

Figure 12.13 Contour plots of the *change* in the vertical (left) and radial (right) stresses around a depleting disk-shaped reservoir with parameters given in Table 12.1. The contours run from −0.25 MPa to 0.25 MPa in steps of 0.05 MPa. Negative contours including 0 are in white, and positive contours in black. The plot is not valid inside the reservoir and is expected to be inaccurate close to the reservoir edges.

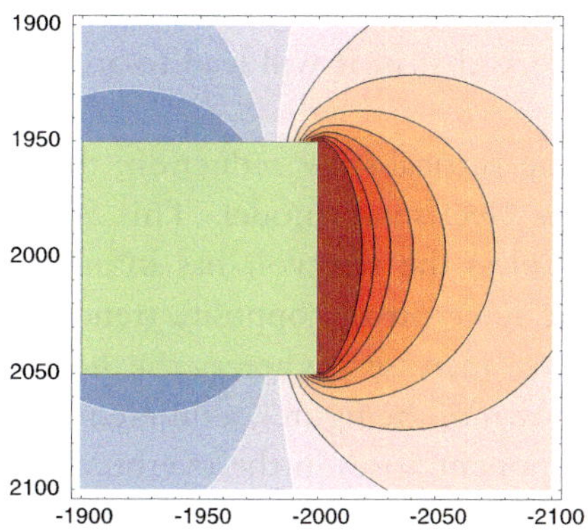

Figure 12.14 Contour plot of shear stress divided by mean effective stress q/p' normalised to the situation before depletion. The contours run from 0.8 to 1.8 in steps of 0.1. Contours from 0.8 to 1.0 are in white, and contours from 1.1 are in black. The model does not give valid predictions inside the reservoir, and it is expected to be inaccurate close to the reservoir edges.

Tempone et al. (2010) pointed out that the rigid bottom approach may represent a situation where there is crystalline bedrock or harder sedimentary rocks underneath the reservoir and demonstrated larger vertical stretching and larger horizontal compaction of the overburden compared to Geertsma's solution, resulting in a net volumetric compaction. In the

Geertsma case, there is no volumetric strain, and hence constant mean stress, in the overburden. Other related approaches are made by Mehrabian and Abousleiman (2015), who introduced a three-layer model, where the reservoir is part of the middle porous layer, and where elastic properties can differ between all layers. Fokker and Orlic (2006) also introduced a multilayer model, where they applied time-dependent behaviour through Maxwell visco-elasticity (Section 1.9.4).

12.3.2 Stress alteration in the overburden

Geertsma's model, as presented in the preceding section, tells us that the stress state will change within the rock masses surrounding a depleting reservoir. Since reservoir compaction leads to surface subsidence, and since the vertical displacement is usually smaller at the surface than at the top of the reservoir (see Figs. 12.7 and 12.8), the overburden above the central part of the reservoir will deform with an extensional vertical strain. Thus the vertical stress is reduced in this region, as illustrated in Fig. 12.13. Obviously, pore pressure increase due to injection will have an opposite (but not necessarily equal) effect, that is, it will lead to an increase in vertical stress in the overburden.

Accompanying the vertical stress reduction, the horizontal stress is increased according to Geertsma's model. This is an expression of stress arching, occurring unless the reservoir has infinite lateral extent ($e = 0$). Near the edge of the reservoir the opposite trend is evident from the figure: The vertical stress increases, whereas the horizontal stress decreases. Furthermore (not shown in the figure), a difference between the radial and tangential stress component arises in the overburden outside the depleting region.

We may extend the definition of the stress path coefficients introduced in Section 12.2.3 beyond the reservoir by considering the stress changes outside the reservoir while maintaining that the pore pressure change Δp_f in Eqs. (12.14)–(12.16) will be interpreted as the pore pressure change in the depleting reservoir zone. This is convenient, since we do not know the possible pore pressure changes in the overburden, and because it is the pore pressure in the reservoir that drives the process. We therefore define the ratio between the changes in horizontal and vertical stresses in the surroundings of the reservoir as

$$\kappa = \frac{\Delta\sigma_h}{\Delta\sigma_v} = \frac{\gamma_h}{\gamma_v} \tag{12.47}$$

The vertical stress is continuous across the boundary between top or bottom of the reservoir and the surroundings. Thus the vertical stress path coefficient is also continuous across these boundaries. The horizontal stresses may on the other hand be discontinuous at the same boundaries.

Segall and Fitzgerald (1998) found that the horizontal stress will increase above and below a depleting ellipsoidal inclusion with low aspect ratio e (having elastic properties identical to those of the surroundings, like in Geertsma's model) according to the following equation (valid for small e):

$$\Delta\sigma_{\mathrm{h}}^{\mathrm{exterior}} = -\alpha\,\Delta p_{\mathrm{f}}\frac{1-2\nu_{\mathrm{fr}}}{1-\nu_{\mathrm{fr}}}\frac{\pi}{4}e \tag{12.48}$$

At the edge of the depleting zone the horizontal stress in the radial direction is the same outside and inside the reservoir, whereas the vertical stress outside the reservoir increases by an amount of the order $\alpha(-\Delta p_{\mathrm{f}})$. This increase can be large, also for the effective vertical stress, since the change in pore pressure will most likely be small. This is a possible source of faulting or fault reactivation. (Recall that there is also a tangential horizontal stress component, which is not continuous at the edge of the reservoir.)

The stress evolution around a depleting zone can be calculated from the analytical model of Geertsma (1973a), as outlined in the preceding section. Figs. 12.15a and b show the resulting stress path coefficients along a vertical line through the centre of the depleting zone for depleting reservoir radius equal to 1000 m (a) and 500 m (b). The reservoir is disk-shaped and centred at 3000 m depth. The two plots illustrate that the arching effect decreases as the extent of the depleting zone increases. The region influenced by the depleting zone extends more than 1000 m above and below the depleting reservoir. The vertical extent increases with increasing radius of the depleting zone.

Furthermore, we see that $\gamma_{\mathrm{v}} + 2\gamma_{\mathrm{h}} \approx 0$ in Figs. 12.15a and b, which means that the mean stress remains constant, and hence there is no volumetric strain, only pure shear loading in the surrounding rock. That implies that there is no local pore pressure change, which means that the parameter κ (Eq. (12.47)) becomes equal to the ratio between γ_{h} and γ_{v}. Hence $\kappa \approx -1/2$ for the cases in Figs. 12.15a and b. This ratio is valid almost everywhere outside the depleting reservoir in these cases.

Fig. 12.15c shows a similar plot of the stress path coefficients for the case defined by the parameters in Table 12.1. In this case the reservoir is more shallow (centred at 2000 m), and the aspect ratio is relatively low (0.05). This leads to low values of the arching coefficients, but the vertical extent

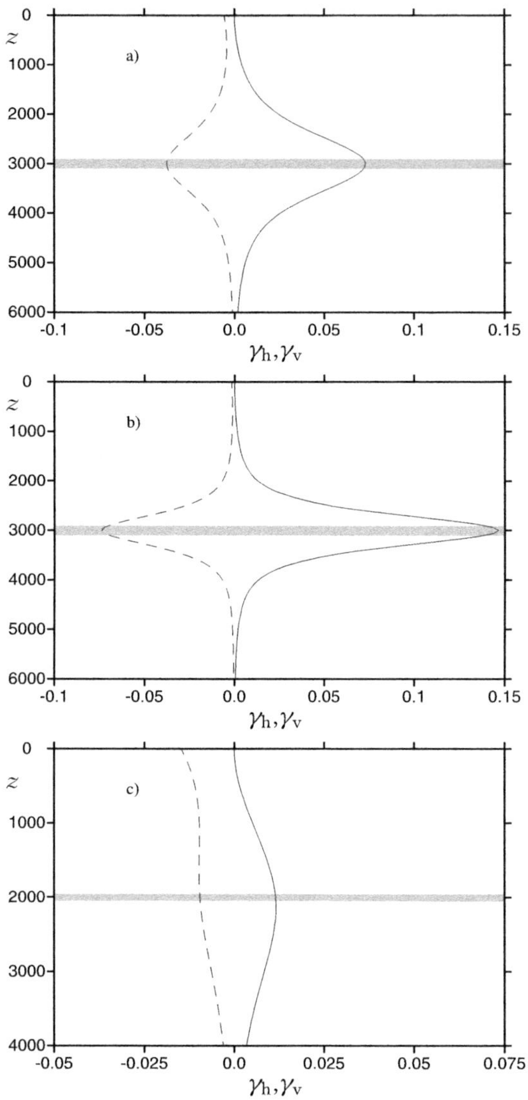

Figure 12.15 a) Vertical and horizontal stress path coefficients γ_v (full lines) and γ_h (dashed lines) along a vertical line through the centre of a depleting reservoir. The reservoir is 200 m thick and centred at 3000 m depth, with Young's modulus 12 GPa and Poisson's ratio 0.20. The calculations are performed with Geertsma's model, which implies that the drained reservoir properties are the same as the undrained surrounding rock properties. The radius of the depleting zone is 1000 m. b) As for (a), except the depleting zone radius is 500 m. c) Vertical and horizontal stress path coefficient along a vertical line through the centre of a depleting reservoir described by the parameters in Table 12.1. Note that the curves for γ_h are not valid within the reservoir.

of arching is large enough to cause influence by the free surface: We see that there is horizontal compression at the surface and that the mean stress is not constant in this case.

Fig. 12.16 shows the vertical and horizontal stress path coefficients as functions of the depleting zone radius for various depths above a reservoir at 2000 m depth, and with properties given by Table 12.1 (except the radius, which is varied in the plot).

Note that in the rock surrounding the reservoir (usually shale), the permeability is very low. There is hence limited possibility for fluid flow, so that the pore pressure response is that of an undrained case. The pore pressure change depends on the stress path and the constitutive behaviour of the shale, and is controlled by the Skempton coefficients (see Section 1.6.7). For an isotropic and linearly elastic medium, constant mean stress means constant volume; hence we would normally (as in Figs. 12.15a and b) not expect any pore pressure change in the overburden or anywhere else in the surroundings of the reservoir. This follows from the value of Skempton's A, which equals 1/3 in linear isotropic poroelasticity (Eqs. (1.215) and (1.216)). However, as shown by Skempton (1954), plasticity may change A drastically. Anisotropy, which is pronounced in most shales, also alters A (Cheng, 1997) and makes it depend on the orientation of the bedding plane with respect to the local stress field. Further, both plasticity and anisotropy may change the stress path. We may therefore expect pore pressure changes in the overburden as a result of depletion or inflation, and depending on the value of A, the pore pressure change may in both cases be either an increase or a decrease (Holt et al., 2018a).

Geertsma's analytical model, which we have used to exemplify the concepts of reservoir geomechanics, is based on an assumption of homogeneous elastic properties, that is, no elastic contrast between drained reservoir and undrained surrounding rocks. Furthermore, the reservoir is assumed to be disk-shaped, and there is no tilt. In a real case, we should apply numerical techniques, such as the finite element method (FEM) to obtain realistic answers. Adding elastic contrast by making the surroundings much stiffer than the reservoir rock promotes further arching. This leaves the rock above and below the reservoir in a state of shear loading and imposes spatially varying volume changes. If the reservoir also is tilted, then the arching effect can be very large.

In this section, we have addressed stress paths associated with depleting reservoirs. The analysis is general and thus relevant also for other cases of fluid withdrawal from underground strata. As long as the rocks behave

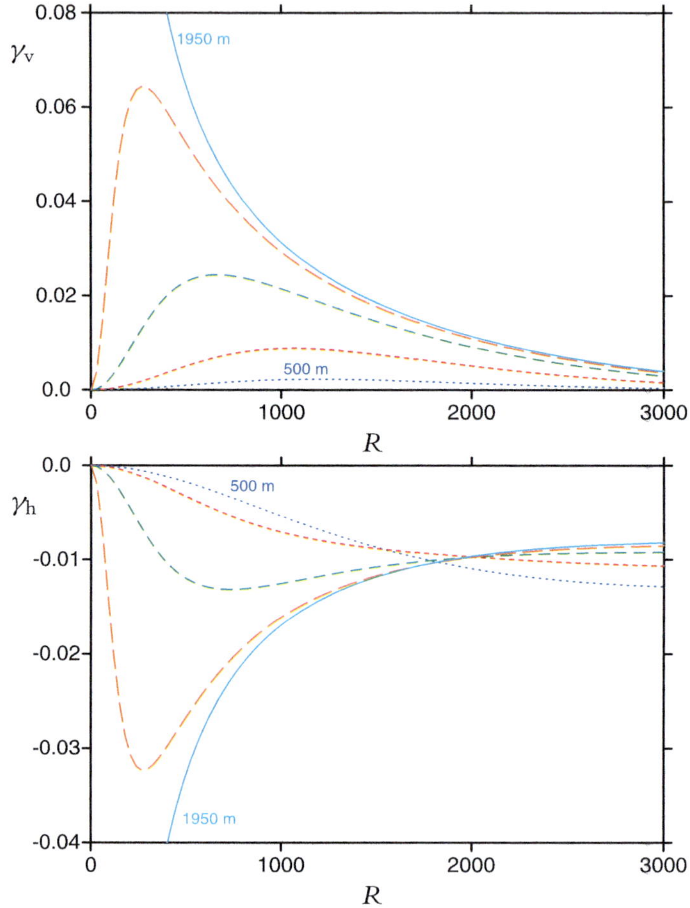

Figure 12.16 Vertical (upper) and horizontal (lower) stress path coefficients vs. depleting zone radius above a depleting reservoir described by the parameters in Table 12.1. The curves correspond to depths 500, 1000, 1500, 1800 and 1950 m. (The length of the dashes increases with depth.)

elastically, we may use the same formalism when an isolated layer within the Earth is pressurised by fluid (e.g. CO_2) injection, except that then the stress path will have opposite sign because Δp_f is positive. However, as commented in Section 12.2.4, any plasticity will make the response to inflation different from that of depletion.

Non-elastic effects may also be of significance in the rocks surrounding the reservoir. The strong stress concentration near the edges of a depleting reservoir may trigger pre-existing faults and be a source of seismicity. It may

Figure 12.17 Observed shear banding in a trap door experiment by Papamichos et al. (2001). The trap door at the bottom of the model is moved downwards, and the two figures represent snapshots of the overburden sand deformation at different times during the experiment. With permission from Elsevier.

also lead to faulting of intact rock itself. This mechanism is illustrated by physical modelling through the so-called trap-door experiment (Papamichos et al., 2001; based on an original experiment by Terzaghi, 1936). A container with a transparent front plate is filled with sand. At the bottom of the container, there is a trap door, which is lowered to mimic the compaction of a reservoir. Initially, selected flat layers of black painted sand were inserted for each 10 mm to visualise the displacement pattern during lowering of the trap door. Examples of observed patterns are shown in Fig. 12.17. Shear bands are seen to originate from the flanks of the "reservoir". When the reservoir is shallow, that is, when the trap door depth-to-width ratio is smaller than one, the shear bands immediately reach the surface, and compaction is directly transferred to subsidence. For deep reservoirs, the shear bands meet each other before they reach the surface. Additional shear band formation eventually leads to surface subsidence also in this case, with a time delay with respect to reservoir compaction. The subsidence bowl observed in these experiments is concentrated directly above the reservoir.

12.4 Geomechanical effects on reservoir performance

So far, we have focused on the purely mechanical aspects of reservoir geomechanics. Having the problems associated with subsidence and compaction in mind, it appears as if the geomechanical influence on the economy of the petroleum industry is largely negative. However, compaction may also act as a drive mechanism for petroleum production, just like water is expelled by squeezing a sponge. In addition, the stress changes associated with pore pressure depletion will also affect porosity, permeability and permeability anisotropy, which will have further effects on recovery.

12.4.1 Compaction drive

The basic definition of pore volume compressibility was discussed in Section 1.6.6. Eq. (1.206) gives the pore volume compressibility with respect to pore pressure, provided that the stresses are constant. However, it was seen in Section 12.2.3 that the stresses change during reservoir depletion. Hence the expression in Eq. (1.206) does not apply directly to a depleting reservoir.

To find the appropriate pore volume compressibility, we must take into account how $\overline{\sigma}$ in Eq. (1.204) depends on the pore pressure p_{f}.

From the basic definitions of the stress path parameters (12.14)–(12.16) we have (assuming that $\gamma_{\mathrm{h}} = \gamma_{\mathrm{H}}$)

$$\Delta\overline{\sigma} = \frac{1}{3}\left(\gamma_{\mathrm{v}} + 2\gamma_{\mathrm{h}}\right)\Delta p_{\mathrm{f}} = \overline{\gamma}\,\Delta p_{\mathrm{f}} \tag{12.49}$$

where $\overline{\gamma} = (\gamma_{\mathrm{v}} + 2\gamma_{\mathrm{h}})/3$.

Entering this expression into Eq. (1.204), we find the pore volume compressibility c_{pp}^{γ} for a given stress path as

$$c_{\mathrm{pp}}^{\gamma} = \frac{1}{V_{\mathrm{p}}}\frac{\Delta V_{\mathrm{p}}}{\Delta p_{\mathrm{f}}} = \frac{1-\overline{\gamma}}{\phi}\left(\frac{1}{K_{\mathrm{fr}}} - \frac{1}{K_{\mathrm{s}}}\right) - \frac{1}{K_{\mathrm{s}}} \tag{12.50}$$

From the discussion leading to Eq. (12.29) it is clear that for an ellipsoidal reservoir, we have

$$\overline{\gamma} = \frac{2}{3}\frac{1-2\nu_{\mathrm{fr}}}{1-\nu_{\mathrm{fr}}}\alpha \tag{12.51}$$

independent of the ellipticity as long as there is no elastic contrast between the reservoir and its surroundings. Thus, in this case, which includes the uniaxial compaction case, we have (Zimmerman, 2000)

$$c_{pp}^{\gamma} = \frac{1+\nu_{fr}}{3(1-\nu_{fr})}\frac{\alpha}{\phi}\frac{1}{K_{fr}} + \left[\frac{2(1-2\nu_{fr})\alpha}{3(1-\nu_{fr})\phi} - 1\right]\frac{1}{K_s} \tag{12.52}$$

If we may assume that $K_{fr} \ll K_s$, then the equation simplifies to

$$c_{pp}^{\gamma} = \frac{1+\nu_{fr}}{3(1-\nu_{fr})\phi K_{fr}} = \frac{c_m}{\phi} \tag{12.53}$$

which is a frequently used approximation.

As a result of a pore pressure decrease, the produced fluid volume (at reservoir conditions) is (Eq. (1.141))

$$\Delta V_{prod} = -V_p\left(c_f + c_{pp}^{\gamma}\right)\Delta p_f \tag{12.54}$$

The compressibility of the pore fluid is denoted by c_f and is the inverse of the bulk modulus ($c_f = 1/K_f$).

This illustrates that in particular for a soft rock, which has low stiffness and therefore high pore compressibility, production will be enhanced by compaction of the pores. This is what we call compaction drive. In case of high stiffness (low compressibility) the production drive will be dominated by the pore fluid compressibility.

As an example, if the pore fluid is oil with bulk modulus of 1 GPa, and the rock is soft with bulk modulus ~1 GPa, porosity ~25 % and Poisson's ratio ~0.30, then the compaction term in Eq. (12.54) contributes about 2.5 times as much as the fluid compressibility to the production. On the other hand, if the rock is stiff with bulk modulus ~10 GPa, porosity ~10 % and Poisson's ratio ~0.20, then the contribution from compaction is only about half of that of the fluid.

Gas has negligible compressibility above the bubble point when it is dissolved in the pore liquid. Below the bubble point, $c_{gas} \approx 1/p_f$, and in that case gas drive will completely dominate.

12.4.2 Stress effects on porosity

In this section, we describe how porosity, based on linear poroelastic theory, changes with stress and pore pressure. One objective is to evaluate porosity changes during depletion of a reservoir. Another objective is to be able to correct core-based porosity data to in situ conditions.

12.4.2.1 Porosity change during depletion

Porosity change associated with a stress change can be expressed in terms of pore and bulk strains as shown in Eq. (1.208):

$$\frac{\Delta\phi}{\phi} = \frac{\Delta V_{\mathrm{p}}}{V_{\mathrm{p}}} - \frac{\Delta V_{\mathrm{tot}}}{V_{\mathrm{tot}}} = c_{\mathrm{pp}}^{\gamma}\Delta p_{\mathrm{f}} + \varepsilon_{\mathrm{vol}} \tag{12.55}$$

Then inserting $\varepsilon_{\mathrm{vol}}$ from Eq. (12.30) and c_{pp}^{γ} from Eq. (12.50), we find

$$\frac{\Delta\phi}{\phi} = \frac{1}{K_{\mathrm{fr}}}\left(\frac{\alpha}{\phi} - 1\right)(1 - \overline{\gamma})\,\Delta p_{\mathrm{f}} \tag{12.56}$$

The latter part of this equation simply states the effective stress principle for porosity (cf. Eq. (1.209)).

For an ellipsoidal reservoir (including the uniaxial compaction case), we may replace $\overline{\gamma}$ from Eq. (12.51):

$$\frac{\Delta\phi}{\phi} = \frac{1+\nu_{\mathrm{fr}}}{3(1-\nu_{\mathrm{fr}})}\left(\frac{1}{\phi} - 1\right)\frac{\alpha\Delta p_{\mathrm{f}}}{K_{\mathrm{fr}}} + \left[\frac{2(1-2\nu_{\mathrm{fr}})\alpha}{3(1-\nu_{\mathrm{fr}})\phi} - 1\right]\frac{\Delta p_{\mathrm{f}}}{K_{\mathrm{s}}} \tag{12.57}$$

In the limit $K_{\mathrm{fr}} \ll K_{\mathrm{s}}$, the expression simplifies to

$$\Delta\phi = (1-\phi)\frac{1+\nu_{\mathrm{fr}}}{3(1-\nu_{\mathrm{fr}})K_{\mathrm{fr}}}\Delta p_{\mathrm{f}} \tag{12.58}$$

The maximum Δp_{f} is typically a few tens of MPa, whereas K_{fr} is in the GPa range. Hence, for most practical examples, porosity reduction as a result of depletion is small, typically less than one percent unit. In reality, porosity change would also be heterogeneous, partly because of rock heterogeneity and partly because of localised deformation.

12.4.2.2 Overburden correction of laboratory measured porosity

To estimate hydrocarbon reserves as accurately as possible, a reliable estimate of the in situ porosity is required. Cores represent a source for direct measurement of porosity. In principle, the core should be reloaded to the in situ stress state in a true triaxial setup, with pore pressure (and temperature) control. The test should be performed on a core properly oriented with respect to the Earth stresses. If both pore volume change (from the volume of expelled fluid from the core) and bulk volumetric strain are measured, then Eq. (12.55) directly gives the porosity change as the core is reloaded. Such tests are too complicated to be performed on a routine basis at an acceptable price, and thus simplified procedures are required (Teeuw, 1971; Nieto et al., 1994; Holt et al., 2003).

Core measurements of porosity are usually done in drained loading experiments, where the external stress is increased hydrostatically, and the pore pressure is kept at atmospheric level. Since porosity change is associated with change in pore and bulk volumes, it is, as long as the core behaves linearly elastic, responding to the mean stress: The confining pressure applied to the core should therefore be raised to the mean effective in situ stress. In standard Gassmann-based poroelasticity (Section 1.6.6), porosity is a function of an effective stress that does not depend on the Biot coefficient:

$$\phi = \phi(\overline{\sigma} - p_{\mathrm{f}}) \tag{12.59}$$

The core should thus be loaded to a confining pressure equal to

$$\overline{\sigma}' = \frac{\sigma_{\mathrm{h}} + \sigma_{\mathrm{H}} + \sigma_{\mathrm{v}}}{3} - p_{\mathrm{f}} \tag{12.60}$$

The porosity change as shown by Eq. (12.55) depends on the change in the bulk volume and the pore volume. Usually, however, we do not measure both pore and bulk volume changes.

Since the pore and bulk volume changes in a drained experiment are related through the Biot coefficient (see Eq. (1.194)) as

$$\frac{\Delta V_{\mathrm{p}}}{V_{\mathrm{p}}} = \frac{\alpha}{\phi} \frac{\Delta V_{\mathrm{tot}}}{V_{\mathrm{tot}}} \tag{12.61}$$

we can estimate porosity changes from pore or bulk strain measurements alone by making an appropriate guess for the Biot coefficient α:

$$\frac{\Delta \phi}{\phi} = \left(1 - \frac{\phi}{\alpha}\right) \frac{\Delta V_{\mathrm{p}}}{V_{\mathrm{p}}} \tag{12.62}$$

or

$$\frac{\Delta \phi}{\phi} = \left(\frac{\alpha}{\phi} - 1\right) \frac{\Delta V_{\mathrm{tot}}}{V_{\mathrm{tot}}} \tag{12.63}$$

Often, Biot's α is set to 1 in these equations. The error introduced by doing so is larger in the case where only bulk strain is measured than where pore strain is recorded. We may improve the reliability of the estimated porosity change by inserting a realistic value for α, preferably from a controlled core test (see e.g. Section 7.3.6) or by calculating the bulk frame

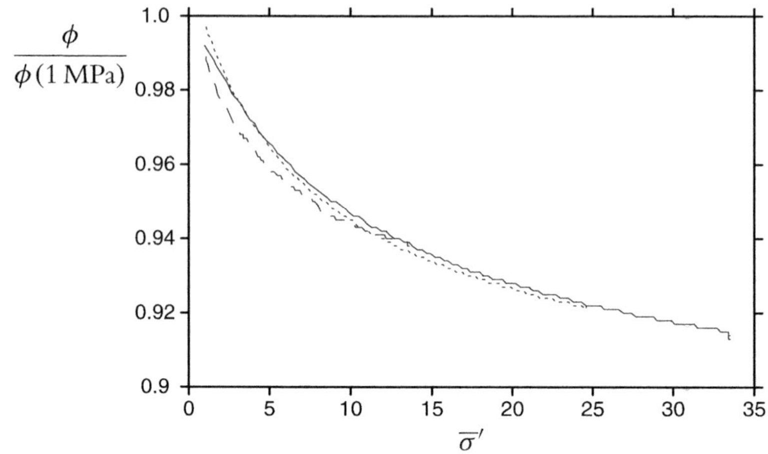

Figure 12.18 Porosity reduction for a field sandstone core as a result of mean effective stress increase, using three different stress paths, isotropic (full line), uniaxial compaction (dashed line) and proportional loading (dotted line). The resulting porosity is practically stress path independent (after Holt et al., 2003). The porosity reduction at in situ conditions is read at the mean in situ effective stress (which was close to 25 MPa).

modulus from the existing data and assuming a solid grain bulk modulus based on knowledge of the mineralogy.

The basis for the analysis above was linear poroelasticity. When a core is retrieved from the Earth, microcracks have developed as a result of stress release, and the subsequent core reloading is highly nonlinear, often with large strains (see Chapter 7). The underlying assumptions can therefore be discussed. Using physical model simulations (cementing synthetic rocks under stress) and discrete particle simulations plus experiments with natural field sandstone cores, Holt et al. (2003) demonstrated that the porosity reduction during reloading was largely stress path insensitive, that is, the use of an isotropic stress may still be valid in overburden correction. This is shown in Fig. 12.18. Cases where this is not likely to be valid are, for instance, in high-porosity rocks, where core damage may cause a permanent porosity reduction as a result of coring and subsequent reloading in the laboratory (Holt et al., 2000).

12.4.3 Stress effects on permeability

The rate of production from a reservoir as a result of a certain amount of depletion is largely controlled by reservoir rock permeability. Likewise, the ability of an injected fluid (water, gas) to displace hydrocarbons is also linked to permeability. Because of the heterogeneous distribution of rocks within a reservoir and because of fractures and faults, permeability may vary considerably over the reservoir volume. This variability, resulting in barriers, flow channels and compartments of rock bodies within the reservoir, controls how fluid is drained from the reservoir during depletion and during enhanced recovery operations. Furthermore, permeability anisotropy implies that there will be preferred flow directions in situ, even within homogeneous layers.

We know from the preceding sections of this chapter that stresses are changing during production. The question we address here is how Earth stress changes may affect permeability and permeability anisotropy and thereby reservoir drainage.

12.4.3.1 Flood directionality in the field

An extensive study of well test data was performed by Heffer and Dowokpor (1990) and Heffer et al. (1994). They showed (as can be seen in Fig. 12.19) that the preferred horizontal flow direction, seen from well tests in a majority of fractured and non-fractured reservoirs, is parallel to the maximum in situ horizontal stress. This implies that stress anisotropy leads to permeability anisotropy. The relevance of this to reservoir engineering can be illustrated by considering water injection from a vertical well: If the producing well is placed in the direction of maximum permeability, then the result would be early water-breakthrough and low recovery. The injector must hence be placed with respect to the producer by careful consideration of the in situ stress field. Thus identification of principal in situ stress directions may be of importance for the reservoir drainage strategy.

12.4.3.2 Permeability changes under isotropic stress conditions

Experimentally, we find that permeability is reduced with increasing isotropic stress. The relative reduction of permeability is generally larger in low-permeability and low-porosity rocks, where permeability is mainly due to cracks or narrow pores, which are easily closed by an applied stress, than in high-permeability and high-porosity rocks. As an example, data on sandstones (Yale, 1984; Kilmer et al., 1987; Holt, 1990; Schutjens et al., 2004) show that for milliDarcy or lower permeabilities, a confining pressure increase of 30 MPa may cause a permeability decrease of 50 % or more. For porosities above 15–20 % and permeabilities above 1 milliDarcy,

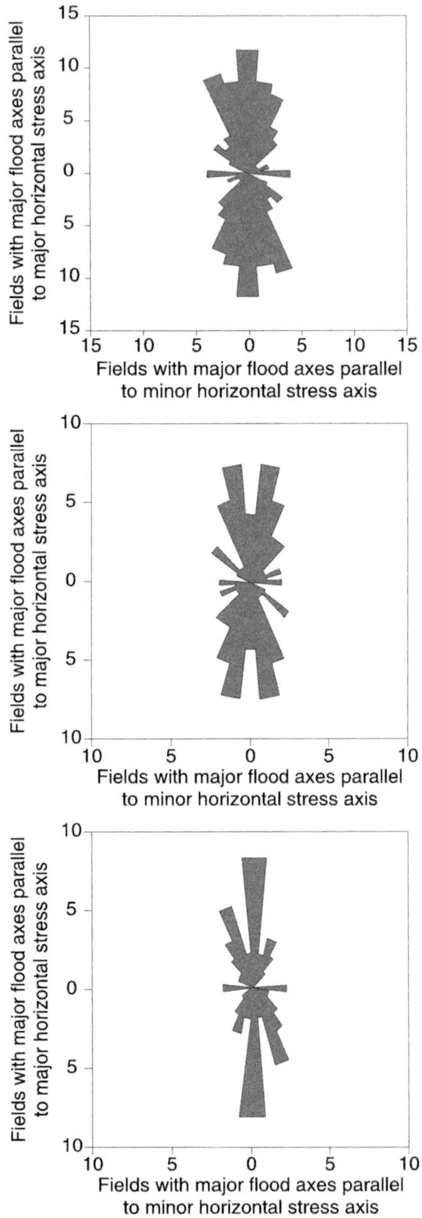

Figure 12.19 Preferred flow directions with respect to in situ stress directions in 80 petroleum reservoirs world wide (top), in the subset of non-fractured reservoirs (middle) and in naturally fractured reservoirs (bottom). After Heffer et al. (1994).

the permeability reduction caused by a similar stress increase is usually of the order 20 % or less.

From a microscopic viewpoint, permeability is controlled by the following parameters:

- porosity ϕ
- pore size/grain size (and size distribution)
- pore shape/grain shape
- tortuosity T

The Kozeny–Carman equation (see e.g. Dullien, 1992) is based on a simple network model, where the pores are considered as pipes:

$$k = \frac{d_g^2}{\kappa_0 T^2} \frac{\phi^3}{(1-\phi)^2} \tag{12.64}$$

where d_g is the grain diameter (assuming spherical grains), and κ_0 is a modifying factor introduced to account for realistic pore shapes. When stress is increased, as long as the rock remains elastic, the main factor changing in the equation above is the porosity. This is illustrated in Fig. 12.20, where measured permeability change during hydrostatic loading of a sandstone core (24 % porosity) from 4 to 35 MPa is compared to permeability change predicted from the Eq. (12.64) using strain recordings during the test to calculate porosity changes according to the theory presented in the preceding section. In this case a convincing agreement was obtained. Flornes (2005) found from hydrostatic laboratory experiments on several high-porosity sandstones (in overall agreement with a permeability–porosity relationship such as the Kozeny–Carman equation) a permeability change proportional to porosity change:

$$\frac{\Delta k}{k} = \chi \frac{\Delta \phi}{\phi} \tag{12.65}$$

The value of the proportionality factor χ was found to be between 4 and 10 for different rocks studied with an average near 7.

In low-permeability rocks the flow is more controlled by thin aspect ratio pores and cracks. In this case the stress sensitivity is much larger than expected from the porosity-based Kozeny–Carman model. This is primarily due to crack closure. Gangi (1978) derived a simple model for stress-dependent fracture permeability:

$$k = k_0 \left[1 - \left(\frac{\sigma}{\sigma_0} \right)^m \right]^3 \tag{12.66}$$

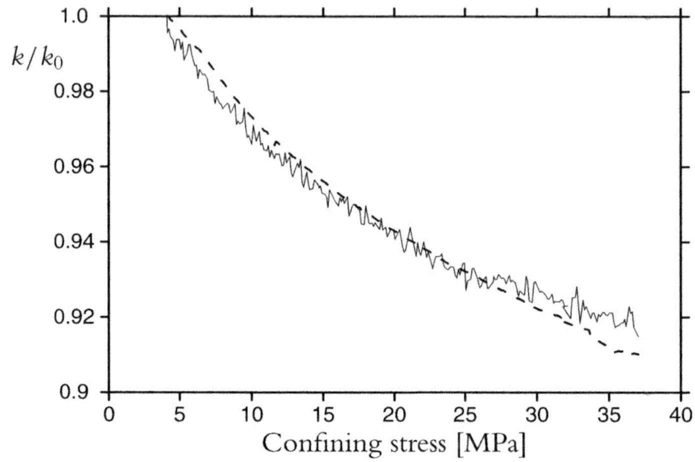

Figure 12.20 Permeability variation with confining pressure during drained hydrostatic loading of a 24% porosity outcrop sandstone. The variation of the factor $\phi^3/(1-\phi)^2$ is inserted into the Kozeny–Carman equation (12.64) derived from measurements of volumetric strain during the test (dashed line).

where σ_0 is a constant relating to the stiffness of the fracture, m is an exponent ($0 < m < 1$) that describes the asperity distribution within the fracture, and k_0 is the permeability at zero stress. Note that σ is the stress acting normal to the fracture plane. Fig. 12.21 shows the rapid decline of permeability according to this model. In case of shear stress applied to the fracture the permeability may increase slightly during initial shearing if the fracture dilates. If asperities are torn off and gouge produced within the fracture, then significant permeability reductions (2 or more orders of magnitude) have been observed (Makurat and Gutierrez, 1996).

The stress effect on permeability is a combined effect of increasing external stresses and decreasing pore pressure due to depletion. We may anticipate that permeability is functionally dependent on an effective stress

$$\sigma' = \sigma - \alpha_k p_f \tag{12.67}$$

This effective stress is not the same as the effective stress governing the elastic behaviour. The coefficient α_k is a poroelastic coefficient for permeability. Experiments with Berea sandstone (Zoback and Byerlee, 1975) gave α_k-values between 2 and 4, which of course is very different from the conventional assumption ($\alpha_k = 1$). Flornes (2005) found values close to 1 for high-porosity sandstones. This is in agreement with the observation

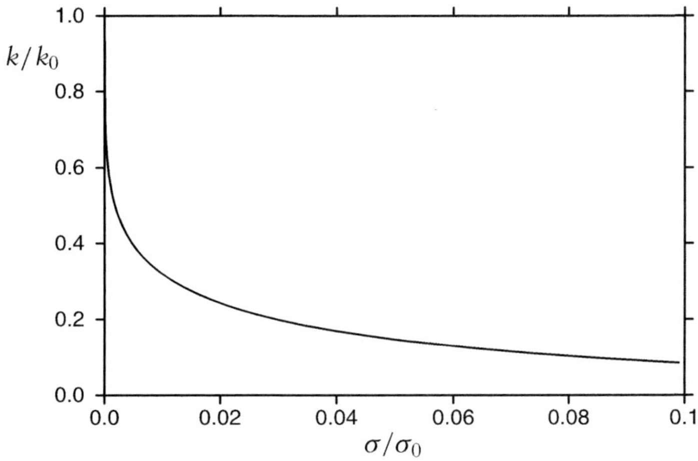

Figure 12.21 Predicted permeability drop of a fracture with increasing normal stress on the fracture, as derived from Gangi's (1978) model (12.66) with $m = 0.25$.

that permeability change under hydrostatic loading is primarily controlled by porosity change: The effective stress for porosity is equal to the net stress (confining minus pore pressure). However, if the solid phase is heterogeneous (as pointed out by Berryman, 1992), then the effective stress coefficient for permeability may deviate significantly from $\alpha_k = 1$. Zoback and Byerlee (1975) argued that if soft clay is coating the pore walls, then the pore pressure will have larger influence on the permeability than the external stress. Further work is required to reach a better understanding of the effective stress law for permeability.

12.4.3.3 Permeability changes under anisotropic stress conditions

Reservoir stresses and stress changes during production are anisotropic. We therefore also expect permeability and permeability changes to be anisotropic (cf. the field observation by Heffer et al. referred previously). There are few tests published in the literature where permeability anisotropy has been directly measured by monitoring both principal components of the permeability tensor in the same test. Morita et al. (1992) showed, using Berea sandstone, that permeability perpendicular to the direction of maximum stress would decrease, whereas permeability along this direction would remain constant during the main part of a triaxial test. Close to failure, the rock became dilatant, and both permeabilities showed slight increases.

Khan and Teufel (2000) performed experiments where permeability along the core axis was measured in comparative triaxial compression and extension tests performed on a low (12 %) porosity and low (0.1 milliDarcy) permeability sandstone, applying different stress paths. They concluded that for decreasing stress path coefficient κ' (increasing γ_h), the permeability anisotropy is increasing. The permeability was always higher in the direction along the maximum principal stress. Whereas in hydrostatic tests, the permeability evolution was essentially isotropic, the ratio between the maximum and minimum permeability was close to 3 when $\kappa' = 0$ (uniaxial load increase). The permeability along the maximum principal stress was seen (like in the tests of Morita et al.) to increase towards the end of the applied loading as a result of dilatancy.

There is evidence from borehole measurements in highly fractured and faulted crystalline rocks that the permeability of critically stressed faults can be much higher than that of faults being far from failure (Barton et al., 1995).

On the other hand, there are several observations in the literature that permeability drops during rock yielding and failure (Holt, 1990; Bruno, 1994; Sarda et al., 1998). This is most often seen in high-porosity and high-permeability materials, and at relatively high confining pressures. Fig. 12.22 shows an example where a triaxial test was performed with the same (24 % porosity) sandstone used for isotropic loading as shown in Fig. 12.20 at a confining pressure of 40 MPa. Again, the permeability evolution correlates extremely well with the porosity change as determined from the volumetric strain, but only until strain localisation takes place near rock failure. Then the measured permeability starts to drop at a much higher rate than before, whereas the core is actually dilating. This is probably explained by the formation of a shear band and associated crushing of grains, creating a barrier to fluid flow. A most dramatic example is the observation by Holcomb and Olsson (2003), who detected a permeability drop of 3–4 orders of magnitude when a Castlegate sandstone core at very high confinement was strained axially by 10 % in a triaxial test. They interpreted the permeability drop by the formation and propagation of a lateral compaction band containing crushed grains.

Boutéca et al. (2000) and Schutjens et al. (2004) both summarise the behaviour of permeability in stress space by distinguishing between an elastic or near-elastic regime, where permeability changes are small and probably related mainly to porosity change. Then as shear stress is increased, inelastic deformation takes over gradually, and the rock eventually fails. The per-

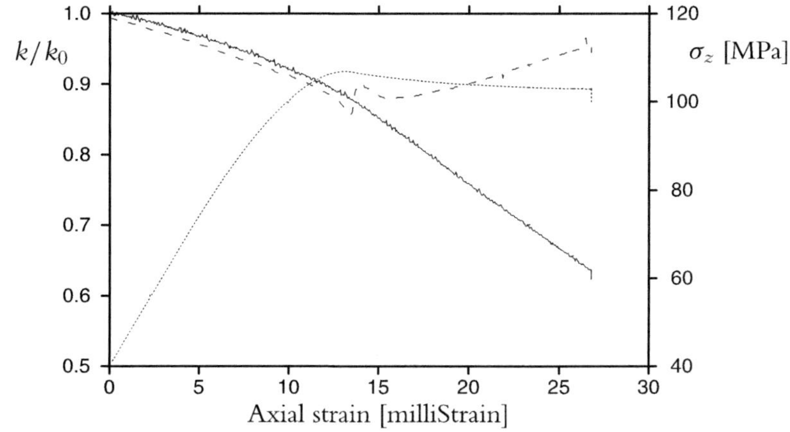

Figure 12.22 Axial permeability (full line) and axial stress (dotted line) vs. axial strain during a triaxial test with a 24% porosity sandstone at 40 MPa confining pressure. The variation of the factor $\phi^3/(1-\phi)^2$ is inserted into the Kozeny–Carman equation (Eq. (12.64)), derived from measurements of volumetric strain during the test (dashed line).

meability may then be enhanced or reduced, depending on the stress path, on the porosity and initial permeability of the rock and on its constitutive behaviour.

12.4.4 Geomechanics in reservoir simulation

The evolution of reservoir geomechanics calls for an implementation of geomechanical concepts into reservoir simulation. There are two fundamentally different ways of performing this:

12.4.4.1 Weakly coupled technique

A conventional reservoir simulator is used to calculate the pore pressure distribution within the reservoir. Such a model needs as input within each cell (10^5–10^6 cells in a total simulation): rock properties (porosity, the three components k_x, k_y and k_z of the permeability tensor plus pore compressibility), fluid properties (saturation of oil, gas and water, viscosities and densities for the three fluid components) and combined rock–fluid parameters (relative permeabilities, capillary pressure curve). In addition, well information must be included. Usually, the permeability and mechanical properties are thought to be constant over the lifetime of the reservoir.

A large-scale geomechanical (usually finite element) model is then used to calculate the stress changes and mechanical compaction and subsidence as a result of the pore pressure reductions found by the reservoir simulation. Permeability and pore compressibility changes can be calculated if their stress sensitivities are known, and these data can be fed back to the reservoir simulator, which is then updated, and so on. A challenge is that a geomechanical model needs to account for the overburden, plus the rock to the sides and underneath the reservoir. This in practice usually limits the reservoir to be represented by 10^4–10^5 elements, that is, on a much coarser grid than in common reservoir simulation.

Examples of this approach can be found, for example, in Bostrøm (2009), Renli et al. (2019), and Nasrollahzadeh et al. (2021).

12.4.4.2 Fully coupled simulation

The coupling between stresses and fluid flow goes back to Biot's poroelastic theory in Section 1.6. The idea is to use a single set of coupled governing equations that contain both mechanical and fluid flow aspects. The benefit is that a fully coupled approach is likely to be more correct. The problem is that such a simulator is computationally more complicated and hence more time-consuming. We should also keep in mind that the existing Biot theory might need modification to account correctly for multi-phase fluid saturation and flow.

More detailed descriptions of such models and their development can be found, for example, in Lewis and Schrefler (1998); Osorio et al. (1998); Kotsabeloulis et al. (1994); Gutierrez and Lewis (1998); Settari and Walters (2001); Stone et al. (2001).

12.4.5 Seismic reservoir monitoring

Seismic reservoir monitoring (time-lapse or "4D"-seismics) has become a powerful tool to monitor reservoir performance. The results of repeated seismic surveys are compared, and differences over time indicate the presence of dynamic processes in the subsurface. Such alterations may be observed as changes in two-way travel time or as a change in a reflectivity of a certain reflector (e.g. at bottom and/or top of the reservoir). Travel time is associated with seismic wave velocities (v_p, v_s), whereas seismic reflectivity (see Section 5.7) is related to contrast in acoustic impedance (ρv_p, ρv_s).

The most important sources of observable time lapse effects associated with producing reservoirs are:

- Fluid substitution, that is, hydrocarbons have been replaced by water due to drainage or replaced by an injected fluid.
- Changes in temperature as a result of a cold (water) or hot (steam) injection front.
- Changes in pore pressure and reservoir stresses.

12.4.5.1 Fluid substitution

The P-wave velocity of a fluid-saturated isotropic rock is given by the Biot theory. The result in the low-frequency limit, Eq. (5.39), can be written as

$$v_{\mathrm{p}} = \sqrt{\frac{H_{\mathrm{fr}} + \frac{K_{\mathrm{f}}}{\phi} \frac{\alpha^2}{1+\frac{K_{\mathrm{f}}}{\phi K_{\mathrm{s}}}(\alpha-\phi)}}{\rho_{\mathrm{dry}} + \rho_{\mathrm{f}}\phi}} \tag{12.68}$$

Pore fluid characteristics affect velocity and hence travel time through the fluid bulk modulus K_{f} and the fluid density ρ_{f}. If oil is replaced by water, then both the plane wave modulus in the numerator and the density in the denominator increase. The net effect is not obvious and depends on the other rock parameters involved, such as the plane wave modulus H_{fr} of the dry frame and the porosity. Typically, brine has a bulk modulus (K_{f}) of 2.5 GPa, whereas K_{f} for live oil is near 0.5 GPa. The densities are approximately 1.0 and 0.6 g/cm^3, respectively. In general, the fluid substitution effect will be more prominent in high- than in low-porosity rock, because a high-porosity rock has a low H_{fr}, making the fluid contribution to the plane wave modulus more significant. In such a case the P-wave velocity of the reservoir layer will increase if oil is replaced by water (see Fig. 5.2).

The effect of fluid substitution on the reflection coefficient (Eq. (5.84)) is larger and more consistent, because the acoustic impedance depends on the product of the modulus and the density, which both are increased as a result of water replacing oil:

$$Z_{\mathrm{p}} = \rho v_{\mathrm{p}} = \sqrt{\left(H_{\mathrm{fr}} + \frac{K_{\mathrm{f}}}{\phi} \frac{\alpha^2}{1+\frac{K_{\mathrm{f}}}{\phi K_{\mathrm{s}}}(\alpha-\phi)}\right)(\rho_{\mathrm{dry}} + \rho_{\mathrm{f}}\phi)} \tag{12.69}$$

Again, the effect of fluid substitution is larger in high-porosity than in low-porosity reservoir rocks.

12.4.5.2 Changes in temperature

The first published example of successful seismic reservoir monitoring was done in connection with a steam injection operation (Nur et al., 1984).

In this case the high temperature of the injection fluid compared to the reservoir rock strongly promoted the observed contrast. Also, when cold water is injected into a reservoir, thermal effects will influence the observed response.

The effects of a temperature change are two-fold: It affects rock and fluid properties, and it leads to thermal stresses and pore pressures, which alter the state of stress underground and thereby the velocities. These effects may be large: Water-injection often results in thermally induced fracturing, which is beneficial for the EOR operation. Clearly, such changes will lead to prominent changes in seismic behaviour. In most rocks, velocities are found in laboratory investigations to decrease with increasing temperature, typically, 5–10 % when the rock is heated from room temperature to around 100 °C. Because oil is more temperature sensitive than brine, the effect is expected to be larger in oil- than water-saturated regions.

12.4.5.3 Changes in pore pressure and reservoir stresses

When a reservoir is depleted, the effective stresses are increased, as outlined in Section 12.2.3. We should expect this to cause an increase in wave velocities in the reservoir zone, since velocities in most rocks are found from laboratory experiments to increase with increasing effective stress.

There is however a number of indications from 4D field data (see Rocks under stress: Special section in the December issue of The Leading Edge, 2005) that this effect is much smaller than anticipated. One probable reason for this is that reservoir rock in situ is less stress sensitive than it appears from core measurements, because cores have been altered as a result of the coring process (Section 7.1.2). Stress release during coring leads to development of microcracks that are closing during reloading. However, this process is not fully reversible, so that the rate of crack closure when the core reaches in situ conditions can be significantly different from that of the reservoir rock in situ. Hence perfect re-establishment of in situ behaviour cannot be expected.

Intuitively, we would anticipate that a well-cemented rock in the Earth is relatively stress insensitive. In situ stress dependence is then mainly associated with pre-existing or evolving cracks or fractures within the reservoir. If the rock is poorly cemented so that a number of grain contacts are not welded, then stress-dependence occurs as a result of the nonlinear (Hertzian) behaviour of these grain contacts. Also, a soft reservoir rock undergoing large depletion may compact with large porosity reduction, causing wave velocities to increase.

We should also remember that pore pressure itself affects the properties of the saturating fluid independent of stress changes. The bulk modulus of oil is reduced typically with 0.01 GPa per MPa pore pressure reduction, whereas the bulk modulus for brine is reduced slightly less, depending on temperature. This effect counteracts the increase in P-wave velocity caused by effective stress increase. For a stress-insensitive frame, the velocity reduction with reducing pore pressure is in the range 0.1–0.3% per MPa, being more significant for soft rock than for hard rock, and more significant for high than for low gas to oil ratios (GOR).

As was also seen in Section 12.2.3, the stress evolution as a result of pore pressure change is anisotropic and may cause non-elastic behaviour of the reservoir rock and even fault activation/initiation. This would clearly affect wave velocities, leading to stress-induced anisotropy. If the pore pressure is increased (e.g. as a result of water-injection), then the effective stress decreases anisotropically, and non-elastic behaviour may occur for relatively modest pore pressure changes: Initiation of microcracking by tensile bond breakages may happen and lead to velocity reduction, in accordance with field observations (Landrø et al., 2005).

12.4.5.4 Overburden 4D seismics

We saw in Section 12.3 that stress changes and strains associated with depletion also occur in the rock surrounding the reservoir. For the overburden above the centre of the reservoir, the vertical stress is reduced as a result of reservoir depletion. This may lead to a reduction of wave velocity for a vertically propagating seismic wave, increasing its two-way travel time and altering the reflectivity associated with top of the reservoir.

The influence zone of stress alteration is thick—in Fig. 12.15 close to 1000 m above the reservoir—which magnifies the effect on two-way travel time compared to that caused by a similar velocity change in the reservoir.

There is significant field evidence that depletion-induced stress changes in the overburden cause observable 4D effects (Hatchell and Bourne, 2005; Røste et al., 2006; MacBeth et al., 2018). Field data from Valhall (Barkved and Kristiansen, 2005) demonstrate increased S-wave splitting associated with the stress concentration above the edges of the reservoir, in agreement with predictions of the Geertsma model combined with a stress sensitive crack model (Fjær, 2006). Thus seismic observations of stress-arching can be used to identify depleting reservoir zones and to distinguish between sealing and non-sealing faults.

According to the Geertsma model, the decrease in vertical stress above the central part of the reservoir during depletion is accompanied by an increase in the horizontal stress, leaving the stress path κ close to the constant mean stress value of $-1/2$. In practice, elastic contrast between reservoir and overburden, presence of faults and fractures, tilted reservoirs, or heterogeneous lithology in the overburden may cause κ to be very different from the constant mean stress value. For instance, a stiff (undrained) overburden compared to a soft (drained) reservoir rock may cause both vertical and horizontal stresses to decrease above the reservoir centre, that is, a positive value of κ. Holt et al. (2018b) demonstrated through laboratory measurements that there may be significant stress path-dependent sensitivity of wave velocities in overburden shales. This implies that geomechanical modelling needs to accompany 4D seismic interpretation, but also that there is potential to explore the overburden stress path from 4D data. Quantitative prediction of stress and pore pressure evolution in the overburden is important for evaluation of cap rock integrity and stability during infill drilling.

12.4.5.5 Other monitoring methods

Reservoir monitoring can also be done in a passive mode by detecting seismicity originating from within the reservoir or in the cap rock immediately above or near the reservoir edge (Maxwell et al., 1998; Kristiansen et al., 2000). This type of monitoring is aimed at interpreting the compaction and subsidence mechanisms and does not directly give information about fluid movements. Another technique is to monitor surface deformation with a very high resolution using precision laser levelling. The result is inverted mathematically (e.g. using the Geertsma model described above) to find reservoir deformation and relate it to mechanisms responsible for the deformations seen, for example, thermal recovery, waste or massive fluid injection (Dusseault and Rothenburg, 2002).

12.5 Well problems and reservoir geomechanics

There is evidence from field operations worldwide that the stability of boreholes drilled or wells being produced are affected by stress changes described in the previous paragraphs. For example, as was pointed out in Chapter 9, stable drilling of infill wells may become difficult as the producing reservoir zones are depleted. This is related to the strong decrease of horizontal stress (and hence fracture pressure) in permeable layers at the same time as the pore pressure (and hence the collapse pressure) remains unchanged in ultralow-permeability shale layers within or just above the top of the reservoir. The mud weight window eventually closes at a critical amount of pore pressure reduction.

Sand failure is also affected by the changing stress field during depletion, as was discussed in Chapter 10. Onset of shear failure in the perforation cavities or in an open producing hole is controlled by the rock strength and by the far-field effective stress (Eq. (10.8)). The increase in far-field effective stress caused by pore pressure reduction enhances the risk of sand production. Pressure maintenance by water injection may result in reduction of rock strength and therefore also lead to increased risk of particle production as the reservoir is being produced.

12.5.1 Casing damage

Casing damage is one of the most significant well problems associated with compacting reservoirs. An extensive summary of several field cases worldwide is provided by Casero and Rylance (2020). Other examples documented in the literature are from the Ekofisk Field and the Valhall Field, North Sea, and from the Belridge Field and the Wilmington Field, California (Schwall et al., 1996; Kristiansen et al., 2000; Fredrich et al., 1998; Dusseault et al., 2001). Casing damage becomes an operational problem when the casing is deformed or bent (by creation of dog-legs) so much that it prevents workovers and recompletions (Cernocky and Scholibo, 1995). This necessitates costly drilling of a side-track well. Further possible consequences are loss of production and reduced ultimate recovery.

The main mechanisms responsible for casing damage are compressional, shear, tensile and bending deformations (Veeken et al., 1994; Cernocky and Scholibo, 1995; Fredrich et al., 1998; Dusseault et al., 2001). These mechanisms are illustrated in Fig. 12.23. Casing failure is influenced by several factors, as will be seen in the following. These include stress regime and rock properties, cement, tubulars and operational features (Casero and Rylance, 2020).

In a depleting and compacting zone the casing strain in a vertical well is assumed to be equal to the formation strain. This is based on the assumption of no axial slip and uniform reservoir compaction. If the well is deviated, then the casing strain equals the component of formation strain along the well direction. Thus a deviated well is preferable to a vertical well to reduce compressive casing damage. The risk of casing buckling is highest in unsupported intervals, for instance, near sand producing perforations where the pressure drawdown is also most severe, wherever the cement gives poor

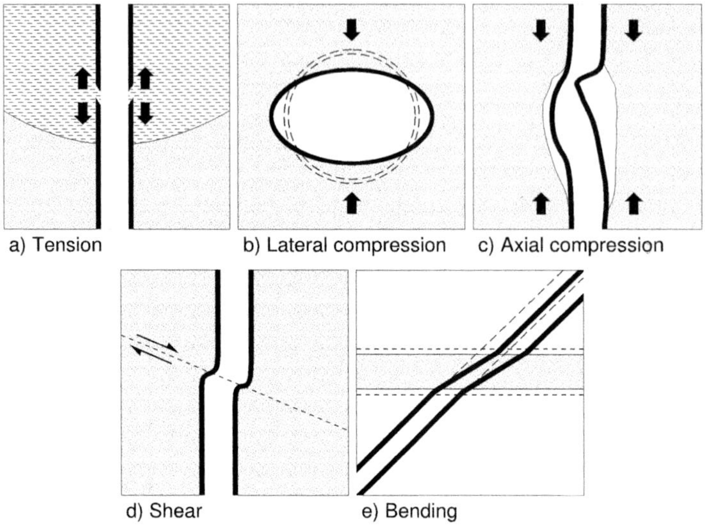

Figure 12.23 Casing damage scenarios (after Veeken et al., 1994).

support or at connections. Based on analytical and numerical modelling, Cernocky and Scholibo (1995) found that 0.5–0.75 % casing strain may be thought of as an upper limit for an unsupported 10-ft section. Bruno (2002) argued that casing compaction of 1 % or less is unlikely to impair casing integrity. Note that, as pointed out by Ewy (2021), compaction strains and hence the likelihood of casing damage will differ between reservoir zones of different stiffness. Obviously, strain localisation enhances the risk of casing collapse.

Deformation of the casing cross section due to non-uniform lateral loading is referred to as "crushing" (Cernocky and Scholibo, 1995). This is mainly a problem in reservoir zones with large depletion and where the formation is very soft particularly in horizontal wells. Formation (e.g. salt) creep against casing and cement may also in extreme conditions lead to point loading causing casing deformation and collapse (Casero and Rylance, 2020).

In their finite element simulations, Cernocky and Scholibo chose a "crushing limit" (i.e. the maximum tolerable decrease of the inner diameter) of 0.3″ (7.5 mm) based on field experience. They found that a low diameter-to-thickness ratio of the casing (say 10 or thereabout) is a key to reduce crushing risk in highly compactive sands. If a large diameter-to-thickness ratio is used (say 20 or so), then the crushing risk is much higher for a horizontal or deviated well than for a vertical one. For a low diameter-to-thickness ratio, the well angle plays a minor role.

Casing damage by shear is caused by high shear stresses within the rock (Dusseault et al., 2001). Casing shear occurs near lithology interfaces (shale-sand) along faults, fractures, or bedding planes. It is most common in the overburden, where shear damage to the casing most likely takes place near the flanks of the reservoir. Here the vertical stress is increased as a result of depletion, leading to possible fault reactivation. An example was given in Section 9.7.1 (Maury and Zurdo, 1996). Casing shear may also occur in the crestal region above the reservoir, where the vertical stress is reduced, but this reduction decreases towards the flanks. This lateral stress redistribution, which is linked to the variation in total compaction strain of the reservoir, promotes bending and flexing of the overburden (Ewy, 2021), leading to casing failure at weak points.

Casing shear may also occur at the top of production and injection intervals inside the reservoir. Note that stress changes during injection are of opposite signs but not necessarily the same as during depletion, so the two scenarios have to be treated differently. Within unconventional reservoirs, casing shear may happen during completion operations (Ewy, 2021), such as hydraulic fracturing.

The decrease in vertical stress above a depleting reservoir also may lead to tensional casing damage. As for compaction strain, tensile strain is assumed to be transferred directly to the casing. Several cases of tensile casing damages were observed in the Belridge field (Fredrich et al., 1998). Furui et al. (2012) describe casing and screen failures associated with compacting sandstone reservoirs. They point in particular to tensile failure caused by reducing vertical stress in a high-pressure gas sand layer as much as 300 m above the reservoir. Thermal expansion due to steam injection may also be a cause of tensile casing failures.

Casing bending may occur at locations of non-uniform compaction, such as in a deviated well near the top or bottom of a depleting reservoir zone.

Monitoring of casing deformation is regularly done to see if a workover tool can get through. There are several options, such as multifinger calipers (Ewy, 2021), which may provide 3D pictures that also demonstrate mechanisms involved.

Analysis and mitigation of casing failure first of all requires a good picture of the stress and strain evolution within and outside the reservoir, and it is particularly important to locate possibly reactivated faults and sites of massive particle production. Casero and Rylance (2020) point to the risk of using local experience uncritically in new areas, where stress regimes may differ. For instance, in strike-slip stress regimes where hydraulic fractures become horizontal, the risk of tensional casing failures increases. All in all, geomechanical simulations are necessary, and they need to be based on a solid foundation of input parameters. Simulations may be used to select well positions and well angles that reduce the risk of casing damage as much as possible. Dusseault et al. (2001) recommends in particular to increase the compliance of the wellbore–casing system (increased casing diameter, use of highly ductile cement) as a remedy against casing shear damage, whereas casing strength appears to be of minor importance. Kristiansen et al. (2000) found that events located from microseismic monitoring in the Valhall Field, North Sea, were spatially correlated with locations of casing collapse. They used this information to achieve optimum casing design in the highest risk areas.

12.5.2 Reservoir geomechanics as a tool to optimise drilling and production strategies

This section has demonstrated that well problems associated with drilling and production are strongly linked to the geomechanical response of the reservoir. Drilling instabilities, sand production and casing damage are triggered by depletion. This calls for a geomechanical analysis predicting stress changes and strain evolution within and outside the reservoir, based on different production scenarios. Such analyses will provide information like at what time and where infill wells have to be drilled to minimise stability problems and where casing damage mitigation is required.

Stress arching in general reduces reservoir compaction and associated well problems within the reservoir. On the other hand, the stress changes within and around the reservoir become more pronounced, which leads to more well failures in the overburden. Stress arching may however to some extent be controlled by the order in which reservoir compartments are produced, by pressure maintenance or by thermal stresses generated as a result of fluid injection. This opens up possibilities to use geomechanics actively and—through 4D and microseismic monitoring—interactively, as a part of reservoir management.

Figure 12.24 Three pictures of the Ekofisk tank. The one to the left was taken in the early phase of production (mid-1970s), the middle one at the time when subsidence was discovered (1984) and the one to the right in the late 1980s. Notice the number of slots visible above sea level, which indicates the amount of sea-floor subsidence (the distance between slots is about 1 m). Courtesy of Norwegian Petroleum Museum.

12.6 Some field cases: subsidence and induced seismicity

As mentioned in Section 12.1, subsidence as such represents a main operational challenge only in a few cases. These examples are to some extent based on an overview by Doornhof et al. (2006).

The Ekofisk field was discovered in 1969 as the first giant oil field in the North Sea. Subsidence was realised in 1984 (at that time it was around 3 m) by comparing the visible air gap between the platform and the sea with old photos from early field life (Nagel, 2001). Some pictures illustrating the ongoing subsidence in this period are shown in Fig. 12.24.

The subsidence rate was at this point about 30 cm/yr, which in 1987 led the operators to jack up the steel platforms by 6 m to keep the air gap larger than the estimated maximum height of ocean waves to maintain platform safety.

The reservoir is situated with its crest at 2900 m below the sea level (water depth is 78 m). It has an elliptical dome shape and lies above a salt diapir. The initial pore pressure was 49.6 MPa. The reservoir consists of two zones with average thickness of about 2–300 m, and the reservoir rock is chalk with porosities from 25 % to above 40 %. The reservoir is naturally fractured with a permeability of up to 200 milliDarcy, which is 10–100 times that of the intact chalk. The depletion driven compaction was boosted by inelastic deformation due to pore collapse. Water injection started in the late 1980s to enhance oil production. Until 1998 the pore pressure during water-flooding was more or less constant, but the subsidence rate had increased to around 40 cm/yr, indicating a water-weakening effect on the chalk. Only when new platforms were installed, and the pore pressure was increased back to 80–90 % of its initial value, the subsidence rate dropped significantly. In the later years, it has stayed at about 10 cm/yr, indicating a total subsidence (until 2020) of about 10 m. The costs involved in jack-up and replacement of platforms has by far been compensated by increased production due to compaction drive, extending field life by 50 years or more compared to early planning.

The neighbouring field Valhall, which has been produced since 1982, is also a chalk reservoir with even higher porosity (up to 50% or more). The reservoir zone is thinner than at Ekofisk, about 60 m on average, but highly variable, leading to variations in compaction and a complex overburden response. Although amounting to several metres, subsidence has been less problematic than at Ekofisk. Well problems (see Section 12.5) have however been frequent. Both Ekofisk and Valhall have installed life of field 4D seismic (LOFS) systems to monitor the reservoir performance (see Section 12.4.5).

The Groningen gas field onshore in The Netherlands was discovered in 1959 and produced from 1964. Groningen was indeed a precursor to the upcoming discoveries of offshore fields in the North Sea. It has been a major supplier of gas to Dutch and neighbouring households, largely replacing the need for coal as a source of energy in this region.

The reservoir is at about 3000 m depth with a diameter of about 30 km. It consists of relatively competent sandstone with porosities in the range 10–25% and a thickness of 100–200 m. The initial pore pressure was 32 MPa and has dropped to below 10 MPa. Due to the large extent of the reservoir, homogeneous depletion is expected to cause surface subsidence similar to the compaction. The fact that a significant part of the land is below sea level makes water management challenging, which means that very little subsidence can be tolerated. So, even if the measured subsidence is limited to about 30 cm (in 2020), a similar amount added over time would require strengthening of the dikes serving as protection against the sea. This calls for very detailed prediction, accounting for all mechanisms that could possibly contribute.

Observations show that subsidence has accelerated with time and with the level of depletion, implying effects of plasticity and possibly also creep. Time dependence is important in this case, since it may lead to continued subsidence after depletion has been terminated. This has in particular been addressed in the Wadden Sea area on the Dutch coast (outside the Groningen field), which is on the UNESCO World Heritage list as a protected area (de Waal et al., 2012; Fokker et al., 2018). Here subsidence due to gas production needs to be controlled not to exceed the acceptable limit set by authorities, which also includes tidal raise in water level and natural sand transport, deposition and erosion: This calls for prediction and monitoring within an accuracy of a few centimetres of subsidence over periods of 10–20 years.

Most subsidence modelling for field operation has been based on the theory of Geertsma and van Opstal (1973) (see Section 12.3), incorporating mimicked nonlinearity and creep (rate type compaction) of the reservoir sandstone, in addition to addressing possible creep in the salt overburden, and time delay due to slowly depleting aquifers beneath the reservoir (see further discussion by van Thienen-Visser and Fokker, 2017). Surface monitoring has with time become based on GPS data. In situ compaction monitoring has been performed by locating radioactive bullets placed in wells at an early stage of depletion. For the present and future, use of optical fibres significantly improves the accuracy of such measurements.

Besides subsidence, induced seismicity has been observed in the Groningen area since 1991. The number of observed earthquakes has increased with time, reaching an average of more than 100 per year in the period 2013–2017, and their magnitudes have also increased, with peaks between 3.5 and 4 on the Richter scale. Immediate consequences were damage to buildings in the area and public concern that seismicity could lead to fatalities. The operator developed a seismological model based on the apparent correlation between compaction and seismicity (Bourne et al., 2014), stating that the total seismic moment would be proportional to reservoir bulk volume change. With subsidence and pore pressure changes providing direct input, the model enables prediction with combined geomechanical modelling and reservoir simulation. Faults were not explicitly included, but their effect assumed to be accounted for through the empirical input. Implementing a probability distribution of seismic moment, numbers, locations and magnitudes of future earthquakes could be simulated. Each simulation contributed to an earthquake catalogue to be part of seismic risk and hazard assessment (van Elk et al., 2017) by using the Monte Carlo simulation method. For this, maps of expected peak ground motion could be developed based on the path for the seismic wave from the epicentre to the surface and the impact of the soil near the surface. Further, the response of buildings was included based on laboratory tests (including full-scale houses) and numerical simulations, together with an assessed risk of injury associated with the mode of building collapse and type of building. The production in Groningen was reduced to control the seismicity, and it has been decided to close the field from 2022.

Induced seismicity associated with petroleum production is not uncommon. Grasso (1992) described possible source mechanisms and summarised a number of field observations, covering depletion and injection scenarios. During depletion, stress alterations around, above and beneath the reservoir (as described in the previous parts of this chapter) may cause microseismicity or fault slip in these areas. Further, fault zones triggered by load removal due to massive oil extraction. Injection is thought to be responsible for magnitude 5–7 earthquakes amongst other in California and Uzbekistan.

Fluid injection may trigger fault activation because the effective stress is reduced by pore pressure increase. When injecting fluid into a depleted zone, we may expect seismicity after the pore pressure gets above its previous maximum value. It might however occur earlier or later, depending on the stress path during injection. It appears that all mechanisms acting during depletion and injection require relatively small stress changes (<1 MPa) to trigger seismic events.

It is however important to underline that earthquakes usually require sustained injection and are not occurring during, for example, hydraulic fracturing operations (Chapter 11). Exceptions may occur if one is injecting into a fault or if there is a natural pathway for the fluid to link up with a reactivating fault. It seems to be agreed that the large increase in seismic activity across the US interior during the last decades originates from disposal of wastewater in relation to natural gas production (Zoback, 2012; Ellsworth, 2013). Some of these earthquakes have exceeded magnitude 5. An example from the North Sea is the 2001 event (magnitude 4.1–4.4), which was strongly felt but caused no structural damage at the platforms in the Ekofisk area (Ottemöller et al., 2005). The source was located, and it was concluded that it resulted from unintentional water injection into the overburden.

Injection plays a central role in CO_2 storage and in geothermal energy recovery (Vilarrasa et al., 2019). In both cases, induced seismicity has been detected. To avoid leakage from storage reservoirs and damage to wells and other infrastructure, these events need to be understood to minimise their occurrence. Thermal effects appear to be particularly important to trigger seismicity for stimulation of geothermal energy production but may also be a concern in carbon storage. CO_2 injection however leads to a peak in pore pressure relatively soon after injection of gas has started, after which it reaches a plateau and a corresponding drop in microseismicity.

References

Addis, M.A., 1997. The stress-depletion response of reservoirs. SPE38720. In: SPE Annual Technical Conference and Exhibition. San Antonio, TX, USA, pp. 55–65.

Barkved, O., Kristiansen, T.G., 2005. Seismic time-lapse effects and stress changes: examples from a compacting reservoir. Lead. Edge 24 (12), 1244–1248. https://doi.org/10.1190/1.2149636.

Barton, C.A., Zoback, M.D., Moos, D., 1995. Fluid flow along potentially active faults in crystalline rock. Geology 23 (8), 683–686. https://doi.org/10.1130/0091-7613(1995)023<0683:FFAPAF>2.3.CO;2.

Berryman, J.G., 1992. Effective stress for transport properties of inhomogeneous porous rock. J. Geophys. Res. 97 (B12), 17409–17424. https://doi.org/10.1029/92JB01593.

Bostrøm, B., 2009. Development of a geomechanical reservoir modelling workflow and simulations. SPE124307. In: SPE Annual Technical Conference and Exhibition. New Orleans, LA, October 4–7, pp. 1–19.

Bourne, S.J., Oates, S.J., van Elk, J., Doornhof, D., 2014. A seismological model for earthquakes induced by fluid extraction from a subsurface reservoir. J. Geophys. Res. 119, 8991–9015. https://doi.org/10.1002/2014JB011663.

Boutéca, M.J., Sarda, J.-P., Vincké, O., 2000. Constitutive law for permeability evolution of sandstones during depletion. SPE58717. In: SPE International Symposium on Formation Damage. Lafayette, LA, USA.

Bruno, M.S., 1994. Micromechanics of stress-induced permeability anisotropy and damage in sedimentary rock. Mech. Mater. 18 (1), 31–48. https://doi.org/10.1016/0167-6636(94)90004-3.

Bruno, M.S., 2002. Geomechanical and decision analyses for mitigating compaction-related casing damage. SPE Drill. Completion 17 (3), 179–188. https://doi.org/10.2118/79519-PA.

Casero, A., Rylance, M., 2020. The unconventional unconventionals: tectonically influenced regions, stress states and casing failures. SPE199710. In: SPE Hydraulic Fracturing Technology Conference and Exhibition. The Woodlands, TX, USA.

Cernocky, E.P., Scholibo, F.C., 1995. Approach to casing design for service in compacting reservoirs. SPE30522. In: Proc. SPE Annual Technical Conference and Exhibition. Dallas, TX, USA.

Cheng, A.H.-D., 1997. Material coefficients of anisotropic poroelasticity. Int. J. Rock Mech. Min. Sci. 34 (2), 199–205. https://doi.org/10.1016/S0148-9062(96)00055-1.

Christensen, S.O., Janbu, N., Jones, M.E., 1988. Subsidence due to oil-gas production. In: Behavior of Offshore Struct. (BOSS '88), vol. V1. Trondheim, Norway, pp. 143–157.

de Waal, J.A., Smits, R.M.M., 1988. Prediction of reservoir compaction and surface subsidence: field application of a new model. SPE Form. Eval. 3 (2), 347–356. https://doi.org/10.2118/14214-PA.

de Waal, J.A., Roest, J.P.A., Fokker, P.A., Kroon, I.C., Breunese, J.N., Muntendam-Bos, A.G., Oost, A.P., van Wirdum, G., 2012. The effective subsidence capacity concept: how to assure that subsidence in the Wadden Sea remains within defined limits? Neth. J. Geosci. 91, 385–399. https://doi.org/10.1017/S0016774600000512.

Doornhof, D., Kristiansen, T.G., Nagel, N.B., Patillo, P.D., Sayers, C., 2006. Compaction and subsidence. Oilfield Rev. 18 (3), 50–68.

Dullien, F.A.L., 1992. Porous Media. Fluid Transport and Pore Structure, second edition. Academic Press, New York.

Dusseault, M.B., Rothenburg, L., 2002. Deformation analysis for reservoir management. Oil Gas Sci. Technol. – Rev. IFP 57 (5), 539–554. https://doi.org/10.2516/ogst:2002036.

Dusseault, M.B., Bruno, M.S., Barrera, J., 2001. Casing shear: causes, cases, cures. SPE Drill. Completion 16 (2), 98–107. https://doi.org/10.2118/72060-PA.

Ellsworth, W.L., 2013. Injection-induced earthquakes. Science 341, 1225942. https://doi.org/10.1126/science.1225942.

Eshelby, J.D., 1957. The determination of the elastic field of an ellipsoidal inclusion, and related problems. Proc. R. Soc. Lond. Ser. A 241, 376–396. https://doi.org/10.1098/rspa.1957.0133.

Ewy, R.T., 2021. Well shear associated with conventional and unconventional operations: diagnosis and mechanisms. SPE Drill. Completion 36 (2), 459–472. https://doi.org/10.2118/205007-PA.

Fjær, E., 2006. Geomechanics and geophysics for reservoir management. Rev. Eur. Génie Civ. 10, 703–730. https://doi.org/10.1080/17747120.2006.9692853.

Flornes, O., 2005. Stress Dependent Fluid Flow in Porous Rock: Experiments and Network Modelling. PhD thesis. NTNU.

Fokker, P.A., Orlic, B., 2006. Semi-analytical modelling of subsidence. Math. Geol. 38, 117–126. https://doi.org/10.1007/s11004-006-9034-z.

Fokker, P.A., van Leijeb, F.J., Orlic, B., van der Marel, H., Hanssen, R.F., 2018. Subsidence in the Dutch Wadden sea. Neth. J. Geosci. 97, 127–181. https://doi.org/10.1017/njg.2018.9.

Fredrich, J.T., Deitrick, G.L., Arguello, J.G., DeRouffignac, E.P., 1998. Reservoir compaction, surface subsidence, and casing damage: a geomechanics approach to mitigation and reservoir management. SPE/ISRM47284. In: Proc. Eurock '98. Trondheim, Norway.

Furui, K., Fuh, G.-F., Morita, N., 2012. Casing- and screen-failure analysis in highly compacting sandstone fields. SPE Drill. Completion 27 (2), 241–252. https://doi.org/10.2118/146231-PA.

Gambolati, G., 1972. A three-dimensional model to compute land subsidence. Bull. Int. Assoc. Sci. Hydrol. 17, 219–226. https://doi.org/10.1080/02626667209493823.

Gambolati, G., Teatini, P., 2015. Geomechanics of subsurface water withdrawal and injection. Water Resour. Res. 51, 3922–3955. https://doi.org/10.1002/2014WR016841.

Gangi, A.F., 1978. Variation of whole and fractured porous rock permeability with confining pressure. Int. J. Rock Mech. Min. Sci. Geomech. Abstr. 15 (5), 249–257. https://doi.org/10.1016/0148-9062(78)90957-9.

Geertsma, J., 1973a. A basic theory of subsidence due to reservoir compaction: the homogeneous case. Verh. K. Ned. Geol. Mijnbouwkd. Genoot. 28, 43–62.

Geertsma, J., 1973b. Land subsidence above compacting oil and gas reservoirs. J. Pet. Technol. 25 (6), 734–744. https://doi.org/10.2118/3730-PA.

Geertsma, J., van Opstal, G., 1973. A numerical technique for predicting subsidence above compacting reservoirs, based on the nucleus of strain concept. Verh. K. Ned. Geol. Mijnbouwkd. Genoot. 28, 63–78.

Grasso, J.R., 1992. Mechanics of seismic instabilities induced by the recovery of hydrocarbons. Pure Appl. Geophys. 139, 507–534.

Gutierrez, M., Lewis, R.W., 1998. The role of geomechanics in reservoir simulation. SPE/ISRM47392. In: Proc. Eurock '98, vol. 2. Trondheim, Norway, pp. 439–448.

Hatchell, P., Bourne, S., 2005. Rocks under strain: strain-induced time-lapse time shifts are observed for depleting reservoirs. Lead. Edge 24 (12), 1222–1225. https://doi.org/10.1190/1.2149624.

Heffer, K.J., Dowokpor, A.B., 1990. Relationship between azimuths of flood anisotropy and local earth stresses in oil reservoirs. In: North Sea Oil and Gas Reservoirs II.

Heffer, K.J., Last, N.C., Koutsabeloulis, N.C., Chan, H.C.M., Gutierrez, M., Makurat, A., 1994. The influence of natural fractures, faults and earth stresses on reservoir performance – geomechanical analysis by numerical modelling. In: Aasen, J.O., Berg, E., Buller, A.T.,

Hjelmeland, O., Holt, R.M., Kleppe, J., Torsæter, O. (Eds.), North Sea Oil and Gas Reservoirs — III. Kluwer Academic Publishers.

Hettema, M., Papamichos, E., Schutjens, P., 2002. Subsidence delay: field observations and analysis. Oil Gas Sci. Technol. – Rev. IFP 57 (5), 443–458. https://doi.org/10.2516/ogst:2002029.

Hettema, M.H.H., Schutjens, P.M.T.M., Verboom, B.J.M., Gussinklo, H.J., 2000. Production-induced compaction of a sandstone reservoir: the strong influence of stress path. SPE Reserv. Eval. Eng. 3 (4), 342–347. https://doi.org/10.2118/65410-PA.

Holcomb, D.J., Olsson, W.A., 2003. Compaction localization and fluid flow. J. Geophys. Res. 108 (B6), 2290. https://doi.org/10.1029/2001JB000813.

Holt, R.M., 1990. Permeability reduction induced by a nonhydrostatic stress field. SPE Form. Eval. 5 (4), 444–448. https://doi.org/10.2118/19595-PA.

Holt, R.M., Brignoli, M., Kenter, C.J., 2000. Core quality: quantification of coring-induced rock alteration. Int. J. Rock Mech. Min. Sci. 37 (6), 889–907. https://doi.org/10.1016/S1365-1609(00)00009-5.

Holt, R.M., Lehr, B.C., Kenter, C.J., Spiets, P., 2003. In-situ porosity from cores: poroelastic correction for stress relief during coring. Petrophysics 44 (4), 253–261.

Holt, R.M., Flornes, O., Li, L., Fjær, E., 2004. Consequences of depletion-induced stress changes on reservoir compaction and recovery. ARMA/NARMS04-589. In: Gulf Rocks. Houston, TX, USA.

Holt, R.M., Gheibi, S., Lavrov, A., 2016. Where does the stress path lead? Irreversibility and hysteresis in reservoir geomechanics. ARMA 16-496. In: 50th US Rock Mechanics/Geomechanics Symposium. Houston, TX, USA.

Holt, R.M., Bakk, A., Stenebråten, J.F., Bauer, A., Fjær, E., 2018a. Skempton's A – a key to man-induced subsurface pore pressure changes. ARMA18-0949. In: 52nd US Rock Mechanics/Geomechanics Symposium. Seattle, WA, USA.

Holt, R.M., Bauer, A., Bakk, A., 2018b. Stress-path dependent velocities in shales: impact on 4D seismic interpretation. Geophysics, MR353–MR367. https://doi.org/10.1190/GEO2017-0652.1.

Khan, M., Teufel, L.W., 2000. The effect of geological and geomechanical parameters on reservoir stress path and its importance in studying permeability anisotropy. SPE Reserv. Eval. Eng. 3 (5), 394–400. https://doi.org/10.2118/66184-PA.

Kilmer, N.H., Morrow, N.R., Pitman, J.K., 1987. Pressure sensitivity of low permeability sandstones. J. Pet. Sci. Eng. 1 (1), 65–81. https://doi.org/10.1016/0920-4105(87)90015-5.

Kotsabeloulis, N.C., Heffer, K.J., Wong, S., 1994. Numerical geomechanics in reservoir engineering. In: Siriwardane, H.J., Zaman, M.M. (Eds.), Proc. Computer Methods and Advances in Geomechanics, vol. 3. Balkema.

Kristiansen, T.G., Barkved, O., Pattillo, P.D., 2000. Use of passive seismic monitoring in well and casing design in the compacting and subsiding Valhall field, North Sea. SPE65134. In: Proc. SPE European Petroleum Conference. Paris, France.

Landrø, M., Digranes, P., Strønen, K., 2005. Pressure depletion measured by time-lapse VSP. Lead. Edge 24 (12), 1226–1232. https://doi.org/10.1190/1.2149628.

Lewis, R.W., Schrefler, B.A., 1998. The Finite Element Method in the Static and Dynamic Deformation and Consolidation of Porous Media, second edition. John Wiley & Sons, Chichester, England.

MacBeth, C., Kudarova, A., Hatchell, P., 2018. A semi-empirical model of strain sensitivity for 4D seismic interpretation. Geophys. Prospect. 66, 1327–1348. https://doi.org/10.1111/1365-2478.12648.

Makurat, A., Gutierrez, M., 1996. Fracture flow and fracture cross flow experiments. SPE36732. In: SPE Annual Technical Conference and Exhibition. Denver, CO, USA.

Maury, V., Zurdo, C., 1996. Drilling-induced lateral shifts along pre-existing fractures: a common cause of drilling problems. SPE Drill. Completion 11, 17–23. https://doi.org/10.2118/27492-PA.

Maxwell, S.C., Young, R.P., Bossu, R., Jupe, A., Dangerfeld, J., 1998. Microseismic logging of the Ekofisk reservoir. SPE/ISRM47276. In: Proc. Eurock '98. Trondheim, Norway.

Mehrabian, A., Abousleiman, Y.N., 2015. Geertsma's subsidence solution extended to layered stratigraphy. J. Pet. Sci. Eng. 130, 68–76. https://doi.org/10.1016/j.petrol.2015.03.007.

Mindlin, R.D., Cheng, D.H., 1950a. Nuclei of strain in the semi-infinite solid. J. Appl. Phys. 21, 926–930. https://doi.org/10.1063/1.1699785.

Mindlin, R.D., Cheng, D.H., 1950b. Thermoelastic stress in the semi-infinite solid. J. Appl. Phys. 21, 931–933. https://doi.org/10.1063/1.1699786.

Morita, N., Whitfill, D.L., Nygaard, O., Bale, A., 1989. A quick method to determine subsidence, reservoir compaction and in-situ stress induced by reservoir depletion. J. Pet. Technol. 41, 71–79. https://doi.org/10.2118/17150-PA.

Morita, N., Gray, K.E., Srouji, F.A.A., Jogi, P.N., 1992. Rock property change during reservoir compaction. SPE Form. Eval. 7 (3), 197–205. https://doi.org/10.2118/13099-PA.

Mulders, F.M.M., 2003. Modelling of stress development and fault slip in and around a producing gas reservoir. PhD thesis. TU Delft.

Nagel, N., 2001. Compaction and subsidence issues within the petroleum industry: from Wilmington to Ekofisk and beyond. Phys. Chem. Earth 26A, 3–14.

Nasrollahzadeh, B., Amiri, H.A.A., Ghabezloo, S., 2021. Field-scale fully coupled simulation of fluid flow and geomechanics: gas storage/recovery process in a depleted sandstone reservoir. J. Pet. Sci. Eng. 200. https://doi.org/10.1016/j.petrol.2021.108423.

Nieto, J.A., Yale, D.P., Evans, R.J., 1994. Improved methods for correcting core porosity to reservoir conditions. Log Analyst 35 (3), 21–30.

Nur, A., Tosaya, C., Vo-Thanh, D., 1984. Seismic monitoring of thermal enhanced oil recovery oil recovery processes. In: Proc. 54th Ann. Internat. Mtg. Soc. Explor. Geophys., pp. 337–340.

Osorio, J.G., Chen, H-Y., Teufel, L.W., Schaffer, S., 1998. A two-domain, 3D, fully coupled fluid-flow/geomechanical simulation model for reservoirs with stress-sensitive mechanical and fluid-flow properties. SPE/ISRM47397. In: Proc. Eurock '98, vol. 2. Trondheim, Norway, pp. 455–464.

Ottemöller, L., Nielsen, H.H., Atakan, K., Braunmiller, J., Havskov, J., 2005. The 7 May 2001 induced seismic event in the Ekofisk oil field, North Sea. J. Geophys. Res. 110, B10301. https://doi.org/10.1029/2004JB003374, pp. 1–15.

Papamichos, E., Heill, L.K., Vardoulakis, I., 2001. Overburden modeling above a compacting reservoir using a trap door apparatus. Phys. Chem. Earth 26A (1–2), 69–74. https://doi.org/10.1016/S1464-1895(01)00025-4.

Renli, L., Helstrup, O.A., Ke, G., Thompson, N., 2019. Fast 4D geomechanical modelling with multiple applications. SPE195605. In: SPE Norway One Day Seminar. Bergen, May 14, pp. 1–14.

Rudnicki, J.W., 1999. Alteration of regional stress by reservoirs and other inhomogeneities: stabilizing or destabilizing? In: Proceedings of the 9th International Congress on Rock Mechanics, pp. 1629–1637.

Røste, T., Stovas, A., Landrø, M., 2006. Estimation of layer thickness and velocity changes using 4D prestack seismic data. Geophysics 71, S219–S234. https://doi.org/10.1190/1.2335657.

Santarelli, F.J., Marsala, A.F., Brignoli, M., Rossi, E., Bona, N., 1998. Formation evaluation from logging on cuttings. SPE Reserv. Eval. Eng. 1 (3), 238–244. https://doi.org/10.2118/36851-PA.

Sarda, J.-P., Ferfera, F.M.R., Vincké, O., Boutéca, M., Longuemare, P., 1998. Experimental study of the stress path influence on monophasic permeability evolution. In: SCA Conferences Paper SCA 9827.
Schutjens, P.M.T.M., Hanssen, T.H., Hettema, M.H.H., Merour, J., de Bree, P., Coremans, J.W.A., Helliesen, G., 2004. Compaction-induced porosity/permeability reduction in sandstone reservoirs: data and model for elasticity-dominated deformation. SPE Reserv. Eval. Eng. 7, 202–216. https://doi.org/10.2118/88441-PA.
Schwall, G.H., Slack, M.V., Kaiser, T.M.V., 1996. Reservoir compaction well design for the Ekofisk field. SPE36621. In: SPE Annual Technical Conference and Exhibition. Denver, CO, USA.
Segall, P., Fitzgerald, S.D., 1998. A note on induced stress changes in hydrocarbon and geothermal reservoirs. Tectonophysics 289, 117–128. https://doi.org/10.1016/S0040-1951(97)00311-9.
Sen, B., 1951. Note on the stresses produced by nuclei of thermo-elastic strain in a semi-infinite elastic solid. Q. Appl. Math. 8, 365–369. https://doi.org/10.1090/qam/37717.
Settari, A., Walters, D.A., 2001. Advances in coupled geomechanical and reservoir modeling with applications to reservoir compaction. SPE J. 6 (3), 334–342. https://doi.org/10.2118/74142-PA.
Skempton, A.W., 1954. The pore pressure coefficients *A* and *B*. Géotechnique 4, 143–147. https://doi.org/10.1680/geot.1954.4.4.143.
Stone, T., Xian, C., Wan, J., 2001. Implementation of coupled geomechanics in the Eclipse reservoir simulator. In: Pres. at IFP Research Workshop on Geomechanics in Reservoir Simulation.
Søreide, O.K., Hansen, S., Stenebråten, J.F., 2014. Estimation of reservoir stress effects due to injection of cold fluids: an example from NCS. ARMA14-7394. In: 48th US Rock Mechanics/Geomechanics Symposium. Minneapolis, MN, USA.
Teeuw, D., 1971. Prediction of formation compaction from laboratory and compressibility data. Soc. Pet. Eng. J. 11, 263–271. https://doi.org/10.2118/2973-PA.
Tempone, P., Fjær, E., Landrø, M., 2010. Improved solution of displacements due to a compacting reservoir over a rigid basement. Appl. Math. Model. 34, 3352–3362. https://doi.org/10.1016/j.apm.2010.02.025.
Terzaghi, K., 1936. Stress distribution in dry and saturated sand above a yielding trap door. In: Proc. Int. Conference on Soil Mechanics and Foundation Engineering, vol. 1. Cambridge, Mass., pp. 307–311.
Teufel, L.W., Rhett, D.W., Farrell, H.E., 1991. Effect of reservoir depletion and pore pressure drawdown on in situ stress and deformation in the Ekofisk field, North Sea. In: Proc. 32nd U.S. Symp. on Rock Mechanics. Norman, OK, USA, pp. 62–73.
van Elk, J., Doornhof, D., Bommer, J.J., Bourne, S.J., Oates, S.J., Pinho, R., Crowley, H., 2017. Hazard and risk assessments for induced seismicity in Groningen. Neth. J. Geosci. 96, 259–269. https://doi.org/10.1017/njg.2017.37.
van Opstal, G.H.C., 1974. The effect of a base-rock rigidity on subsidence due to reservoir compaction. In: Proc. Third Congress ISRM, vol. 2, pp. 1102–1111.
van Thienen-Visser, K., Fokker, P.A., 2017. The future of subsidence modelling: compaction and subsidence due to depletion of the Groningen gas field in the Netherlands. Neth. J. Geosci. 96, 105–116. https://doi.org/10.1017/njg.2017.10.
Veeken, C.A.M., Wahleitner, J.P., Keedy, C.R., 1994. Experimental modelling of casing deformation in a compacting reservoir. SPE/ISRM28090. In: Proc. Eurock '94. Delft, Netherlands, pp. 497–506.
Vilarrasa, V., Carrera, J., Olivella, S., Rutqvist, J., Laloui, L., 2019. Induced seismicity in geologic carbon storage. Solid Earth 10, 871–892. https://doi.org/10.5194/se-10-871-2019.
Yale, D.P., 1984. Network Modelling of Flow, Storage and Deformation in Porous Rocks. PhD thesis. Stanford University.

Zimmerman, R.W., 2000. Implications of static poroelasticity for reservoir compaction. In: Girard, J., Liebman, M., Breeds, C., Doe, T. (Eds.), Proc. Pacific Rocks 2000. Rotterdam. Balkema, pp. 169–172. ISBN 90 5809 155 4.

Zoback, M.D., 2012. Managing the seismic risk posed by wastewater disposal. Earth Mag. 57, 38–43.

Zoback, M.D., Byerlee, J.D., 1975. Permeability and effective stress. AAPG Bull. 59, 154–158. https://doi.org/10.1306/83D91C40-16C7-11D7-8645000102C1865D.

APPENDIX A

Rock properties

In this Appendix, we include the following tables:

Table A.1 Static mechanical properties for some common rock types and materials.

Table A.2 Static mechanical properties for some specific rocks.

Table A.3 Density, elastic moduli and sound velocities for some common materials.

Table A.4 Sound velocities for some common rock types.

Table A.5 Thermal properties for some rocks and materials.

The data are compiled from the sources given in the list below. Elastic moduli are from tests under unconfined or near unconfined (low confining pressure) conditions.

Note: For most rocks, material properties such as elastic moduli, strength parameters and sound velocities vary from rock sample to rock sample, in addition to dependence on measurement conditions (stress, temperature, saturation). Also, material properties are often anisotropic. We have therefore chosen to indicate expected intervals (Tables A.1 and A.4) for the properties. In Tables A.2, A.3 and A.5, when some specific materials are referred to, the values are taken from a specific experiment and need not be representative for other experiments with the same type of rock.

There may also be slight variations in the numbers given by different sources.

Developments in Petroleum Science, Volume 72
ISSN 0376-7361, https://doi.org/10.1016/B978-0-12-822195-2.00022-X

Source	Used in Table
Bourbié, T., Coussy, O. and Zinsner, B., 1987: Acoustics of porous media. IFP publications, Editions Technip, Paris, France.	A.3, A.4
Carmichael, R.S., 1982: *Handbook of physical properties of rocks* (Vols. II & III), CRC Press, Boca Raton, Florida.	A.1, A.2, A.3, A.4
Deere, D.U. and Miller, R.P., 1966: Engineering classification and index properties for intact rock, Report no. AFWL-TR-65-116, Air Force Weapons Laboratory, Kirtland Air Force Base, New Mexico.	A.1, A.2
Gercek, H., 2007: Poisson's ratio values for rocks. *Int. J. Rock Mech. Min. Sci.*, **44**, 1–13.	A.1
Green, S.J. Griffin, R.M. and Pratt, H.R., 1973: Stress-strain and failure properties of a porous shale, SPE 4242, 6th Conf. Drilling and Rock Mechanics.	A.2
Head, K.H., 1984: *Soil laboratory testing*. Vols. I–III, ELE Intl., London.	A.1
SINTEF: Various inhouse tests, Trondheim.	A.1, A.2, A.4
Lide, D.R. (ed.), 1997: *Handbook of Chemistry and Physics*, 78th Ed., CRC Press, Boca Raton, Florida.	A.3, A.5
MacGillivray, D.A. and Dusseault, M.B., 1998: Thermal conduction and expansion of saturated quartz-illitic and smectitic shales as a function of stress, temperature, and bedding, anisotropy, *Int. J. Rock Mech. & Min. Sci.* **35: 4–5**, Paper No. 35.	A.5
Mavko, G., Mukerji, T. and Dvorkin, J., 1998: *The rock physics handbook*, Cambridge University Press, Cambridge.	A.2, A.3
Morita, N., Whitfill, D.L., Massie, I. and Knudsen, T.W., 1987: Realistic sand production prediction: Numerical approach, *SPE Production Eng.* **4, No. 1**, pp. 15–24.	A.2
Nur, A.M. and Wang, Z., 1989 (and references therein): Seismic and acoustic velocities in reservoir rocks, Geophysics Reprint Series, Soc. Expl. Geophysicists, Tulsa, Oklahoma.	A.3, A.4
Somerton, W.H., 1992: *Thermal properties and temperature-related behavior of rock/fluid systems*, Elsevier, Amsterdam.	A.5
TerraTek, 2001: Properties of Pierre shale, Personal communication.	A.2

Table A.1 Static mechanical properties for some common rock types and materials.

Material	**Density** ρ [10^3 **kg/m**^{3}]	**Young's modulus** E **[GPa]**	**Poisson's ratio** ν	**Unconf. compr. strength** C_0 **[MPa]**	**Tensile strength** T_0 **[MPa]**
Unconsolidated sands	1.5–1.7	0.01–0.1	~0.45		
Sandstone	2.0–2.65	0.1–30	0–0.45	1–250	
Clay	1.9–2.1	0.06–0.15	~0.40	0.2–0.5	
Shale	2.3–2.8	0.4–70	0–0.30	2–250	
High-porosity chalk	1.4–1.7	0.5–5	0.05–0.35	4–15	
Low-porosity chalk	1.7–2.0	5–30	0.05–0.30	10–40	
Basalt	2.7–2.9	50–100	0.2–0.3	200–350	10–15
Granite	2.6–2.8	5–85	0.1–0.34	50–350	5–15
Dolomite	2.4–3.2	10–100	0–0.5	40–350	
Limestone	1.4–2.9	2–100	0–0.3	5–250	
Gneiss	2.7–3.1	40–100	0.1–0.3	50–250	
Marble	2.7–3.2	5–90	0–0.3	50–250	
Ice	0.9	8	0.35		
Steel	7.9	200	0.28	220–900[a]	350–1200

[a] Yield strength.

Table A.2 Static mechanical properties for some specific rocks. E-modulus and Poisson's ratio are measured at 50 % of peak stress. Data are from unconfined tests, except for Castlegate sandstone and for Weak reservoir sandstone, where moduli and unconfined strength are based on triaxial tests at confining pressures up to 5 MPa. For Weak North Sea shale, Young's modulus and Poisson's ratio were measured at undrained conditions at an effective initial confining pressure of 5 MPa. Unconfined compressive strength is extrapolated from undrained triaxial tests at different confining pressures.

Material[a]		**Density** ρ [10^3 kg/m^3]	**Young's modulus** E [GPa]	**Poisson's ratio** ν	**Unconf. compr. strength** C_0 [MPa]	**Tensile strength** T_0 [MPa]
Red Wildmoor	⊥	1.9–2.0	1.8		14	0.4–0.7
sandstone	‖	1.9–2.0	1.3		7	0.4–0.7
Castlegate sandstone (dry)		1.93	4–7	0.2–0.3	15	0.5
Weak reservoir sandstone (North Sea, oil sat.)		2.03	1–3	0.1–0.2	5–7	
Berea sandstone		2.18	20	0.38	74	
St. Peter sandstone		2.34	4–10	0.05–0.10	37	
Weak shale	⊥	2.17	1.2	0.43	6	
(North Sea)	‖	2.17	2.1	0.49	8	
Pierre shale I		2.38	0.8–1.1	0.36	8–14	0.6–0.8
Shale (El Paso)		2.47	26	0.10	115	
Bedford limestone		2.21	29	0.29	51	1.6
Solenhofen limestone		2.62	64	0.29	245	4.0
Hackensack siltstone		2.59	26	0.22	123	3.0
Nevada Tuff		1.61	5.0	0.21	24	1.4
Schist	⊥	2.81	21	0.31	55	0.6
(Luther Falls)	‖	2.82	58	0.18	83	5.2

[a] ⊥ loaded perpendicular to bedding, ‖ loaded parallel to bedding.

Table A.3 Density, elastic moduli and sound velocities for some common materials.

Material	Density ρ [10^3 kg/m^3]	Bulk modulus K [GPa]	Shear modulus G [GPa]	P-wave velocity v_p [km/s]	S-wave velocity v_s [km/s]
Calcite	2.71	74.0	27.5	6.39	3.18
Quartz (α)[a]	2.65	37.7	46.9	5.90	3.94
Muscovite (mica)	2.79	52–62	31–41	5.78–6.46	3.33–3.84
Biotite	3.05–3.12	41–60	12–42	4.35–6.8	2.00–3.70
Feldspar (average)	2.63	76	26	4.68	2.39
Pyrite	4.93	147.5	132.5	8.10	5.18
Dolomite	2.87	76–95	45–52	6.93–7.34	3.96–4.23
Barite	4.50	53	22	4.29	2.22
Olivine	3.32	130	80	8.54	5.04
Steel	7.9	198–205	57–80	5.9–6.3	2.7–3.2
Aluminium	2.7	77–98	23–26	6.3–7.0	2.9–3.1
Water (fresh)	1.00	2.25		1.50	
Ice	0.9			1.3–1.7	
Crude oil (room temp.)	0.7–1.1	1.2–2.8		1.3–1.7	
Air (atmospheric press.)	$0.14 \cdot 10^{-3}$	$0.15 \cdot 10^{-3}$		0.33	
Plexiglass	1.2			2.55	1.28
Ethanol	0.79	0.89		1.06	

[a] The α-variant of quartz is stable up to 573 °C. Single crystals are anisotropic; see Section 6.5. The values for K and G were computed by Heyliger et al. (2003) assuming a randomly oriented aggregate of well bonded microcrystals, using formulas similar to those at the end of Appendix D.1.6.

Table A.4 Sound velocities for some common rock types.

Material	Density ρ [10^3 kg/m^3]	P-wave velocity v_P [km/s]	S-wave velocity v_S [km/s]	Conditions
Sand, dry, loose	1.5–1.7	0.3–1.0	0.05–0.4	dry, from surface to ≈ 50 m depth
Sand, dry	1.6–1.7	1.0–1.7	0.4–0.9	dry, loaded from ≈ 1 to ≈ 50 MPa
Sand, wet, loose	1.8–2.2	1.5–2.0	0.05–0.4	saturated, from surface to ≈ 50 m depth
Sandstone, competent	2.0–2.65	1.8–4.5	1.0–3.0	dry, various porosities
Berea sandstone	2.2	3.8–4.0	2.3–2.4	brine saturated, confined
Sandstone, weak	1.7–2.0	1.0–2.0	0.6–1.2	dry, various porosities
Red Wildmoor sst	2.0	1.7–2.0	1.1–1.3	dry, confined
Clay	1.9–2.1	1.5–1.6	0.1–0.3	saturated, from surface to ≈ 50 m depth
London Clay, deep	2.0	1.7–1.8	0.8–1.1	saturated
Shale	2.3–2.8	1.6–4.5	0.7–3.0	saturated, various porosities
Weak shale, North Sea	1.8–2.3	2.4–2.6	1.2–1.3	saturated, unconfined
Chalk, high porosity	1.4–1.7	1.8–2.6	1.0–1.5	saturated, field and lab. data
Chalk, low porosity	1.7–2.4	2.6–5.0	1.5–3.5	saturated, field and lab. data
Limestone	2.4–2.7	3.5–6.0	2.0–3.5	various
Basalt	2.5–2.9	3.5–5.5	1.7–3.4	dry & saturated; stress 0–100 MPa
Granite	2.6–2.7	5.5–6.5	3.0–3.5	dry; stress 0–100 MPa

Table A.5 Thermal properties for some rocks and materials.

Material[a]	Linear thermal exp. coeff. $[10^{-6}\,K^{-1}]$	Conditions	Thermal conductivity $[W\,m^{-1}\,K^{-1}]$	Conditions	Heat capacity $[J\,kg^{-1}\,K^{-1}]$	Conditions
Berea sst.	13	100–200 °C	2.34	20 °C, dry		
Bandera sst.	20	100–200 °C	1.70	20 °C, dry		
Boise sst.	17	100–200 °C	1.47	20 °C, dry		
Sandstone					824–1000	25 °C, dry
Pierre shale			1.30–1.70	35–75 °C, 1–24 MPa saturated		
Mancos shale	13–20	20–75 °C 0–7 MPa	1.50–2.25	35–80 °C, 1–24 MPa saturated		
Queenston shale	11–13	20–75 °C 0–3.5 MPa	1.74–1.95	35–90 °C, 1–24 MPa saturated		
Quartz ⊥c	18	20–100 °C				
‖c	10	20–100 °C	13	0 °C	735	0 °C
Calcite ⊥c	24	20–100 °C				
‖c	-5	20–100 °C				
Aluminium	23.1	25 °C	200	27 °C	900	25 °C
Water	70[b]	20 °C	0.6	20 °C	4182	20 °C

[a] ⊥c loaded perpendicular to crystallographic axis, ‖c loaded parallel to crystallographic axis.
[b] Computed from volumetric expansion.

Table A.6 Some examples of pore fluid properties (based on Batzle and Wang, 1992). The brine properties represent a salinity typical for sea water (3.5 % NaCl).

Fluid	Conditions	Density [g/cm^3]	Bulk modulus [GPa]
Air	25 °C, 0.1 MPa	0.00129	0.00014
Methane	25 °C, 0.1 MPa	0.00064	0.00013
Methane	50 °C, 15 MPa	0.10	0.024
Methane	100 °C, 30 MPa	0.16	0.061
Brine	25 °C, 0.1 MPa	1.05	2.40
Brine	50 °C, 15 MPa	1.02	2.60
Brine	100 °C, 30 MPa	1.00	2.65
Dead oil	25 °C, 0.1 MPa	0.74–0.82	1.11–1.47
Dead oil	50 °C, 15 MPa	0.70–0.80	0.90–1.30
Dead oil	100 °C, 30 MPa	0.65–0.75	0.75–1.10
Live oil	50 °C, 15 MPa	0.60–0.70	0.55–0.85
Live oil	100 °C, 30 MPa	0.60–0.70	0.45–0.70

Table A.7 Some empirical correlations for unconfined strength of shales.

Relation	Derived for	Reference
$C_0 = 10\left(\frac{304.8}{\Delta t_c} - 1\right)$	High porosity, tertiary shale	Lal (1999)
$C_0 = 0.77\left(\frac{304.8}{\Delta t_c}\right)^{2.93}$	High porosity, North Sea shales (normal to bedding)	Horsrud (2001)
$C_0 = 1.35\left(\frac{304.8}{\Delta t_c}\right)^{2.6}$	Global application	Chang et al. (2006)
$C_0 = 0.43\left(\frac{304.8}{\Delta t_c}\right)^{3.2}$	Gulf of Mexico	Chang et al. (2006)
$C_0 = 0.142\left(\frac{304.8}{\Delta t_c}\right)^{3.93} + 3.53$	Global application	Dewhurst et al. (2015)
$C_0 = 1.00\,\phi^{-1.14}$	$\phi < 0.1$	Lashkaripour and Dusseault (1993)
$C_0 = 2.93\,\phi^{-0.96}$	High porosity, North Sea shales	Horsrud (2001)
$C_0 = 0.28\,\phi^{-1.762}$	$\phi > 0.27$	Chang et al. (2006)
$C_0 = 1.98\,\phi^{-1.03}$	Global application	Dewhurst et al. (2015)

Units: C_0: [MPa], ϕ: [frac], Δt_c: [μs/ft].
Note: $304.8/\Delta t_c$ [μs/ft] $= v_p$ [km/s].

Table A.8 Some empirical correlations for unconfined strength of carbonates.

Relation	Derived for	Reference
$C_0 = 2.45\left(\frac{304.8}{\Delta t_c}\right)^{1.82}$	Limestone	Militzer and Stoll (1973)[a]
$C_0 = 1.9 \cdot 10^{0.358(304.8/\Delta t_c)}$	Limestone	Golubev and Rabinovich (1976)[a]
$C_0 = 31.5\left(\frac{304.8}{\Delta t_c}\right) - 63.7$	Limestone, dolomite and marble	Yasar and Erdogan (2004)
$C_0 = 276\,(1 - 3\phi)^2$	$\phi < 0.3$	Rzhevsky and Novick (1971)[a]
$C_0 = 174\,\mathrm{e}^{-7.57\phi}$	North Sea chalk $0.25 < \phi < 0.47$	Havmøller and Foged (1996)
$C_0 = 136\,\mathrm{e}^{-4.8\phi}$	Limestone & dolomite $\phi < 0.2$	Chang et al. (2006)

Units: C_0: [MPa], ϕ: [frac], Δt_c: [μs/ft].
Note: $304.8/\Delta t_c$ [μs/ft] $= v_p$ [km/s].
[a] Cited by Chang et al. (2006).

Table A.9 Some empirical correlations for unconfined strength of sandstones.

Relation	Derived for	Reference
$C_0 = 35\left(\frac{304.8}{\Delta t_c}\right) - 31.5$	Consolidated sandstone	Freyburg (1972)[a]
$C_0 = 1200\,e^{-11(\Delta t_c/304.8)}$	Fine/medium grained $60 < \Delta t_c < 140$	McNally (1987)
$C_0 = 0.50\left(\frac{304.8}{\Delta t_c}\right)^3$	Weak and unconsolidated sandstone	Chang et al. (2006)
$C_0 = 64.2\left(\frac{304.8}{\Delta t_c}\right) - 118$		Sharma and Singh (2008)
$C_0 = 0.40\left(\frac{304.8}{\Delta t_s}\right)^5$	Collection of rock types. Δt_s estimated from Δt_c	From Mason (1987)[b]
$C_0 = 258\,e^{-9\phi}$	$\phi < 0.30$	Sarda et al. (1993)
$C_0 = 254\,(1 - 2.7\phi)^2$	Clean sandstones $\phi < 0.30$	Vernik et al. (1993)
$C_0 = 357\,(1 - 2.8\phi)^2$	Clean sandstones $\phi < 0.357$	Plumb (1994)
$C_0 = 43 - 140\phi + 63\phi^2$	$0.2 < \phi < 0.35$	Raaen et al. (1996)

Units: C_0: [MPa], ϕ: [frac], Δt_c: [μs/ft], Δt_s: [μs/ft].
Note: $304.8/\Delta t_c$ [μs/ft] $= v_p$ [km/s], $304.8/\Delta t_s$ [μs/ft] $= v_s$ [km/s].
[a] Cited by Chang et al. (2006).
[b] The formula is not given in Mason's paper, but derived from data presented in the paper.

References

Batzle, M., Wang, Z., 1992. Seismic properties of pore fluids. Geophysics 57, 1396–1408. https://doi.org/10.1190/1.1443207.

Chang, C., Zoback, M.D., Khaksar, A., 2006. Empirical relations between rock strength and physical properties in sedimentary rocks. J. Pet. Sci. Eng. 51, 223–237. https://doi.org/10.1016/j.petrol.2006.01.003.

Dewhurst, D.N., Sarout, J., Delle Piane, C., Siggins, A.F., Raven, M.D., 2015. Empirical strength prediction for preserved shales. Mar. Pet. Geol. 67, 512–525. https://doi.org/10.1016/j.marpetgeo.2015.06.004.

Freyburg, E., 1972. Der Untere und Mittlere Buntsandstein SW-Thüringen in seinen gesteinstechnischen Eigenshaften. Ber. Dtsch. Ges. Geol. Wiss. A 17, 911–919.

Golubev, A.A., Rabinovich, G.Y., 1976. Resultaty primeneniya apparatury akusticheskogo karotasa dlya opredeleniya prochnostnykh svoistv gornykh porod na mestorozhdeniyakh tverdykh iskopaemykh. Prikl. Geofiz. Moskva 73, 109–116.

Havmøller, O., Foged, N., 1996. Review of rock mechanics data for chalk. In: Proc. 5th North Sea Chalk Symposium. Reims, France.

Heyliger, P., Ledbetter, H., Kim, S., 2003. Elastic constants of natural quartz. J. Acoust. Soc. Am. 114 (2), 644–650. https://doi.org/10.1121/1.1593063.

Horsrud, P., 2001. Estimating mechanical properties of shale from empirical correlations. SPE Drill. Completion 16 (2), 68–73. https://doi.org/10.2118/56017-PA.

Lal, M., 1999. Shale stability: drilling fluid interaction and shale strength. SPE54356. In: SPE Asia Pacific Oil and Gas Conference and Exhibition. Jakarta, Indonesia.

Lashkaripour, G.R., Dusseault, M.D., 1993. A statistical study of shale properties: relationships among principal shale properties. In: Li, K.S., Lo, S-C.R. (Eds.), Probabilistic Methods in Geotechnical Engineering. Balkema, pp. 195–200.

Mason, K.L., 1987. Three-cone bit selection with sonic logs. SPE Drill. Eng. 2, 135–142. https://doi.org/10.2118/13256-PA.

McNally, G.H., 1987. Estimation of coal measures rock strength using sonic and neutron logs. Geoexploration 24, 381–395. https://doi.org/10.1016/0016-7142(87)90008-1.

Militzer, H., Stoll, R., 1973. Einige Beiträge der Geophysik zur Primärdatenerfassung im Bergbau. Neue Bergbautechnik 3, 21–25.

Plumb, R.A., 1994. Influence of composition and texture on the failure properties of clastic rocks. SPE/ISRM28022. In: Proc. Eurock '94. Delft, Netherlands. Balkema, pp. 13–20.

Raaen, A.M., Hovem, K.A., Jøranson, H., Fjær, E., 1996. FORMEL: a step forward in strength logging. SPE36533. In: Proc. 1996 SPE Ann. Techn. Conf. & Exh. Denver, CO, USA, pp. 439–445.

Rzhevsky, V., Novick, G., 1971. The Physics of Rocks. MIR Publishers, Moscow.

Sarda, J.-P., Kessler, N., Wicquart, E., Hannaford, K., Deflandre, J.-P., 1993. Use of porosity as a strength indicator for sand production evaluation. SPE26454. In: SPE 68th Ann. Tech. Conf. & Exh. Houston, TX, USA.

Sharma, P.K., Singh, T.N., 2008. A correlation between P-wave velocity, impact strength index, slake durability index and uniaxial compressive strength. Bull. Eng. Geol. Environ. 67, 17–22. https://doi.org/10.1007/s10064-007-0109-y.

Vernik, L., Bruno, M., Bovberg, C., 1993. Empirical relations between compressive strength and porosity of siliciclastic rocks. Int. J. Rock Mech. Min. Sci. Geomech. Abstr. 30 (7), 677–680. https://doi.org/10.1016/0148-9062(93)90004-W.

Yasar, E., Erdogan, Y., 2004. Correlating sound velocity with the density, compressive strength and Young's modulus of carbonate rocks. Int. J. Rock Mech. Min. Sci. 41 (5), 871–875. https://doi.org/10.1016/j.ijrmms.2004.01.012.

APPENDIX B

SI metric conversion factors

To convert from	To	Multiply by
atm	MPa	0.101325*
bar	MPa	0.1*
bbl	m^3	0.1589873
cp	Pa·s	$1.0 \cdot 10^{-3}$*
Darcy	μm^2	0.9869233
dyne/cm^2	Pa	0.1*
ft	m	0.3048*
in	m	$2.54 \cdot 10^{-2}$*
lbf	N	4.44822
lbm	kg	0.4535924
lbm/USgal	g/cm^3	0.1198264
psi	kPa	6.894757
psi/ft	kPa/m	22.62059
psi$\sqrt{in}$	MPa$\sqrt{m}$	$1.099 \cdot 10^{-3}$

* Conversion factor is exact.

To convert from degrees Fahrenheit to degrees Celsius:

$$T_c = \frac{T_f - 32}{1.8} \tag{B.1}$$

where T_f is given in °F, and T_c is given in °C.

Developments in Petroleum Science, Volume 72
ISSN 0376-7361, https://doi.org/10.1016/B978-0-12-822195-2.00023-1

APPENDIX C

Mathematical background

C.1 Introduction

When using the first edition of this book as a basis for a course in Petroleum Related Rock Mechanics, we have repeatedly had the request for a refreshment of some mathematical background material, in particular, linear algebra. This appendix is an attempt to respond to these requests.

This appendix is not a full tutorial, and we do not attempt to provide full mathematical rigour in the presentation. Still, we hope that this appendix may be useful for readers that need a repetition of some central mathematical methods. We also hope that this appendix may make it easier to read some of the literature in the field.

C.2 Matrices

A *matrix* is a rectangular array of numbers or other mathematical objects. The elements of a matrix are normally referred to by two indices. The first index refers to the *rows*, and the second to the *columns*.

Thus an example of a 3 × 3 matrix is

$$\mathbf{a} = \begin{pmatrix} a_{11} & a_{12} & a_{13} \\ a_{21} & a_{22} & a_{23} \\ a_{31} & a_{32} & a_{33} \end{pmatrix} \tag{C.1}$$

C.2.1 The transpose of a matrix

The *transpose* of a matrix is the matrix found by interchanging the rows and columns of a matrix. The transpose of $\mathbf{a}$ is often denoted by $\mathbf{a}^{\mathsf{T}}$. The transpose of matrix (C.1) is

$$\mathbf{a}^{\mathsf{T}} = \begin{pmatrix} a_{11} & a_{21} & a_{31} \\ a_{12} & a_{22} & a_{32} \\ a_{13} & a_{23} & a_{33} \end{pmatrix} \tag{C.2}$$

Developments in Petroleum Science, Volume 72
ISSN 0376-7361, https://doi.org/10.1016/B978-0-12-822195-2.00024-3

C.2.2 Symmetric matrix

A *symmetric* matrix is equal to its transpose, that is, it is left unchanged when reflected about the diagonal:

$$\mathbf{a} = \mathbf{a}^{\mathsf{T}} \tag{C.3}$$

In component notation, this may be written as

$$a_{ij} = a_{ji} \tag{C.4}$$

C.2.3 Diagonal matrix

A *diagonal* matrix is a square matrix in which only the diagonal elements are non-zero. Sometimes the following notation is used:

$$\operatorname{diag}(a, b, c) = \begin{pmatrix} a & 0 & 0 \\ 0 & b & 0 \\ 0 & 0 & c \end{pmatrix} \tag{C.5}$$

C.2.4 Matrix addition

Two matrices of the same size are added by adding the corresponding elements. As an example, for two 2×2 matrices, we have

$$\mathbf{a} + \mathbf{b} = \begin{pmatrix} a_{11} + b_{11} & a_{12} + b_{12} \\ a_{21} + b_{21} & a_{22} + b_{22} \end{pmatrix} \tag{C.6}$$

C.2.5 Multiplication by a scalar

A matrix is multiplied by a scalar by multiplying all elements with the scalar. For example,

$$k\mathbf{a} = k \begin{pmatrix} a_{11} & a_{12} \\ a_{21} & a_{22} \end{pmatrix} = \begin{pmatrix} ka_{11} & ka_{12} \\ ka_{21} & ka_{22} \end{pmatrix} \tag{C.7}$$

C.2.6 Matrix multiplication

The product of an $n \times l$ matrix $\mathbf{a}$ and an $l \times m$ matrix $\mathbf{b}$ is the matrix in which the ij element is given by

$$\sum_k a_{ik} b_{kj} \tag{C.8}$$

Thus, for two 2 × 2 matrices, we get

$$\mathbf{ab} = \begin{pmatrix} a_{11}b_{11} + a_{12}b_{21} & a_{11}b_{12} + a_{12}b_{22} \\ a_{21}b_{11} + a_{22}b_{21} & a_{21}b_{12} + a_{22}b_{22} \end{pmatrix} \tag{C.9}$$

Matrix multiplication is non-commutative, that is, in general,

$$\mathbf{ab} \neq \mathbf{ba} \tag{C.10}$$

Matrix multiplication is associative, that is,

$$\mathbf{a}(\mathbf{bc}) = (\mathbf{ab})\mathbf{c} = \mathbf{abc} \tag{C.11}$$

Matrix multiplication is distributive, that is,

$$\mathbf{a}(\mathbf{b} + \mathbf{c}) = \mathbf{ab} + \mathbf{ac} \tag{C.12}$$

C.2.7 The identity matrix

A square matrix with all diagonal elements equal to 1 and all other elements equal to 0 is called the *identity* matrix and is often denoted by **I**.

We see that

$$\mathbf{aI} = \mathbf{Ia} = \mathbf{a} \tag{C.13}$$

Note that in component form the identity matrix can be written using the Kronecker δ (see Section C.12.2):

$$I_{ij} = \delta_{ij} \tag{C.14}$$

C.2.8 The inverse matrix

A square ($n \times n$) matrix **a** is said to be *non-singular* if an inverse matrix, denoted by $\mathbf{a}^{-1}$, exists. The inverse matrix is defined by the relations

$$\mathbf{aa}^{-1} = \mathbf{a}^{-1}\mathbf{a} = \mathbf{I} \tag{C.15}$$

The inverse of a 2 × 2 matrix is

$$\mathbf{a}^{-1} = \frac{1}{a_{11}a_{22} - a_{12}a_{21}} \begin{pmatrix} a_{22} & -a_{12} \\ -a_{21} & a_{11} \end{pmatrix} \tag{C.16}$$

and that of a 3 × 3 matrix is

$$\mathbf{a}^{-1} = \frac{1}{|\mathbf{a}|}\begin{pmatrix} a_{22}a_{33} - a_{23}a_{32} & a_{13}a_{32} - a_{12}a_{33} & a_{12}a_{23} - a_{13}a_{22} \\ a_{23}a_{31} - a_{21}a_{33} & a_{11}a_{33} - a_{13}a_{31} & a_{13}a_{21} - a_{11}a_{23} \\ a_{21}a_{32} - a_{22}a_{31} & a_{12}a_{31} - a_{11}a_{32} & a_{11}a_{22} - a_{12}a_{21} \end{pmatrix} \tag{C.17}$$

where the *determinant* $|\mathbf{a}|$ of the matrix is defined by Eq. (C.21) below.

C.2.9 The trace of a matrix

The *trace* of a square matrix is defined as the sum of its diagonal elements. In component form, we may write

$$\mathrm{Tr}(\mathbf{a}) = \sum_i a_{ii} \tag{C.18}$$

The trace of a matrix product is independent of the order of the factors:

$$\mathrm{Tr}(\mathbf{ab}) = \mathrm{Tr}(\mathbf{ba}) \tag{C.19}$$

Sometimes, the trace of a matrix is denoted by $\mathrm{Sp}(\mathbf{a})$, from the German word Spur.

C.2.10 Determinants

The *determinant* is another characteristic number associated with a square matrix $\mathbf{a}$ and is denoted by $|\mathbf{a}|$ or $\det(\mathbf{a})$. Here we will not discuss determinants in general, but note that the determinant of a 2 × 2 matrix is given by

$$\begin{vmatrix} a_{11} & a_{12} \\ a_{21} & a_{22} \end{vmatrix} = a_{11}a_{22} - a_{21}a_{12} \tag{C.20}$$

and of a 3 × 3 matrix by

$$\begin{vmatrix} a_{11} & a_{12} & a_{13} \\ a_{21} & a_{22} & a_{23} \\ a_{31} & a_{32} & a_{33} \end{vmatrix} = \begin{matrix} a_{11}a_{22}a_{33} + a_{21}a_{32}a_{13} + a_{31}a_{12}a_{23} \\ -a_{11}a_{32}a_{23} - a_{21}a_{12}a_{33} - a_{31}a_{22}a_{13} \end{matrix} \tag{C.21}$$

The determinant of the product of two matrices is equal to the product of the determinants, that is,

$$|\mathbf{ab}| = |\mathbf{a}|\,|\mathbf{b}| \tag{C.22}$$

From this equation and Eq. (C.15) it follows that

$$\left|\mathbf{a}^{-1}\right| = \frac{1}{|\mathbf{a}|} \tag{C.23}$$

since $|\mathbf{I}| = 1$.

It is clear from Eqs. (C.20) and (C.21) that

$$\left|\mathbf{a}^{\mathrm{T}}\right| = |\mathbf{a}| \tag{C.24}$$

for 2×2 and 3×3 matrices. This relation is valid in general.

From Eq. (C.17) we see that a 3×3 matrix is singular (the inverse does not exist) if the determinant is zero. This is also a general property.

C.2.11 Systems of linear equations

Matrices form an efficient way of writing a system of linear equations. For instance, we may write the system

$$a_{11}x_1 + a_{12}x_2 = b_1 \tag{C.25}$$

$$a_{21}x_1 + a_{22}x_2 = b_2 \tag{C.26}$$

as

$$\mathbf{ax} = \mathbf{b} \tag{C.27}$$

where $\mathbf{a}$ is the 2×2 matrix

$$\mathbf{a} = \begin{pmatrix} a_{11} & a_{12} \\ a_{21} & a_{22} \end{pmatrix} \tag{C.28}$$

$\mathbf{x}$ and $\mathbf{b}$ are the column vectors (or 2×1 matrices)

$$\mathbf{x} = \begin{pmatrix} x_1 \\ x_2 \end{pmatrix}, \quad \mathbf{b} = \begin{pmatrix} b_1 \\ b_2 \end{pmatrix} \tag{C.29}$$

Assuming $\mathbf{a}$ to have an inverse, the solution of Eq. (C.27) may be written as

$$\mathbf{x} = \mathbf{a}^{-1}\mathbf{b} \tag{C.30}$$

C.2.11.1 Homogeneous systems

If all b_i are zero, we have a *homogeneous* system

$$\mathbf{ax} = \mathbf{0} \tag{C.31}$$

Then the formal solution (C.30) assuming $\mathbf{a}$ to be invertible only allows the trivial solution $\mathbf{x} = \mathbf{0}$, which is normally not useful.

To get non-trivial solutions for a homogeneous system, the matrix $\mathbf{a}$ must thus be singular, that is its determinant must be zero:

$$|\mathbf{a}| = 0 \tag{C.32}$$

C.2.12 Eigenvalues and eigenvectors

The equation

$$\mathbf{ax} = \lambda\mathbf{x} \tag{C.33}$$

where $\mathbf{a}$ is an $n \times n$ matrix, $\mathbf{x}$ is a column vector (or $n \times 1$ matrix) and λ is a scalar (i.e. a number, in general, complex), is called the *eigenvalue equation* for the matrix $\mathbf{a}$. The values of λ for which Eq. (C.33) has non-trivial solutions for $\mathbf{x}$ are called the *eigenvalues* of the matrix.

We see from Eq. (C.32) that we may find the eigenvalues by solving the determinant equation

$$|\mathbf{a} - \lambda\mathbf{I}| = 0 \tag{C.34}$$

Eq. (C.34) is used to determine the principal stresses and strains in Eqs. (1.31) and (1.84).

An $n \times n$ matrix has n eigenvalues λ_i, some or all of which may coincide. If we insert, for example, an eigenvalue λ_1 into (C.33), we get the equation

$$\mathbf{ax} = \lambda_1\mathbf{x} \tag{C.35}$$

which has a non-trivial solution for $\mathbf{x}$. The solution is called the *eigenvector* corresponding to λ_1. There are n eigenvectors for an $n \times n$ matrix. The eigenvectors are (or can be chosen to be) linearly independent (orthogonal). Eqs. (1.32)–(1.34) are an example of the use of Eq. (C.35).

It can be shown that the eigenvalues of a *symmetric* matrix with real elements are real.

C.2.13 Similarity transforms and orthogonal transforms

If we have a non-singular matrix **S**, then the transform

$$\mathbf{SaS}^{-1} \tag{C.36}$$

is called a *similarity* transform of **a**.

We see that since

$$\left|\mathbf{SaS}^{-1} - \lambda\mathbf{I}\right| = \left|\mathbf{SaS}^{-1} - \mathbf{S}\lambda\mathbf{IS}^{-1}\right| = |\mathbf{S}|\,|\mathbf{a} - \lambda\mathbf{I}|\,\left|\mathbf{S}^{-1}\right| = |\mathbf{a} - \lambda\mathbf{I}| \tag{C.37}$$

the eigenvalues are unchanged by a similarity transform. It can also be shown that the trace and determinant of the matrix are left unchanged by the similarity transform. (We leave this to the reader as a simple exercise. Use Eqs. (C.19) and (C.22).)

A matrix **O** that fulfils

$$\mathbf{O}^{-1} = \mathbf{O}^{\mathsf{T}} \tag{C.38}$$

is called an *orthogonal* matrix.

It follows from $\mathbf{I} = \mathbf{OO}^{\mathsf{T}}$ that

$$\delta_{ik} = \sum_j O_{ij}(\mathbf{O}^{\mathsf{T}})_{jk} = \sum_j O_{ij}O_{kj} \tag{C.39}$$

which shows that the row vectors of **O** (i.e. the vectors formed by the rows of **O**) are orthogonal and that their length is 1. From $\mathbf{I} = \mathbf{O}^{\mathsf{T}}\mathbf{O}$ it follows that the same applies to the column vectors (the vectors formed by the columns of **O**).

It is clear that *orthogonal transforms* defined by

$$\mathbf{OaO}^{\mathsf{T}} \tag{C.40}$$

are similarity transforms and thus preserve the trace, determinant and eigenvalues.

Any *symmetric* matrix with real elements can be *diagonalised*, that is, transformed to a diagonal matrix by an orthogonal transformation.

From $\mathbf{I} = \mathbf{OO}^{\mathsf{T}}$ and Eq. (C.24) it follows that $|\mathbf{O}| = \pm 1$. We will see below that a rotation of the coordinate system corresponds to an orthogonal transform. A matrix with $|\mathbf{O}| = 1$ corresponds to a *proper* rotation, which preserves the "handedness" of the coordinate system. If $|\mathbf{O}| = -1$, then a reflection is involved, indicating that the transform is taking a right-handed coordinate system into a left-handed one, or vice versa.

C.3 Vectors and coordinate transforms

A vector is a geometric object independent of the coordinate system we use. We can visualise a vector as an arrow with given direction and length, which makes the previous statement obvious.

Let us assume that we have a vector in some coordinate system xyz and want to express it in some other coordinate system $x'y'z'$ rotated relative to the first one. This may be written as a matrix equation

$$\begin{pmatrix} v'_x \\ v'_y \\ v'_z \end{pmatrix} = \begin{pmatrix} a_{11} & a_{12} & a_{13} \\ a_{21} & a_{22} & a_{23} \\ a_{31} & a_{32} & a_{33} \end{pmatrix} \begin{pmatrix} v_x \\ v_y \\ v_z \end{pmatrix} \tag{C.41}$$

Since both the starting coordinate system and the final coordinate system are orthogonal, and the length of a vector is the same in both systems, we may infer some conditions on the transformation matrix **a**. We see that the column vectors must all have length unity and that they are orthogonal. The same applies to the row vectors.

Thus the matrix **a** is an *orthogonal* matrix. The conditions imposed reduce the number of independent elements in **a** to 3, which means that it takes three parameters to specify an arbitrary rotation in space.

To transform from the new coordinate system back to the old, we must use the inverse matrix. Since the matrix is orthogonal, we have

$$\begin{pmatrix} v_x \\ v_y \\ v_z \end{pmatrix} = \begin{pmatrix} a_{11} & a_{21} & a_{31} \\ a_{12} & a_{22} & a_{32} \\ a_{13} & a_{23} & a_{33} \end{pmatrix} \begin{pmatrix} v'_x \\ v'_y \\ v'_z \end{pmatrix} \tag{C.42}$$

It is now easy to find the coordinate axes of the new system in terms of the old one. We see that for the x-direction,

$$\begin{pmatrix} a_{11} \\ a_{12} \\ a_{13} \end{pmatrix} = \begin{pmatrix} a_{11} & a_{21} & a_{31} \\ a_{12} & a_{22} & a_{32} \\ a_{13} & a_{23} & a_{33} \end{pmatrix} \begin{pmatrix} 1 \\ 0 \\ 0 \end{pmatrix} \tag{C.43}$$

Thus we see that we can identify the elements of the orthogonal transformation matrix with the direction cosines used e.g. in Sections 1.1.5 and 4.3.1:

$$l_{x'x} = a_{11} \tag{C.44}$$

$$l_{x'y} = a_{12} \tag{C.45}$$

$$l_{x'z} = a_{13} \tag{C.46}$$

In general, we get

$$\begin{pmatrix} a_{11} & a_{12} & a_{13} \\ a_{21} & a_{22} & a_{23} \\ a_{31} & a_{32} & a_{33} \end{pmatrix} = \begin{pmatrix} l_{x'x} & l_{x'y} & l_{x'z} \\ l_{y'x} & l_{y'y} & l_{y'z} \\ l_{z'x} & l_{z'y} & l_{z'z} \end{pmatrix} \tag{C.47}$$

The scalar (or inner) product of two unit vectors is equal to the cosine of the angle between them. This means that we may write

$$a_{ij} = \hat{\mathbf{x}}'_i \cdot \hat{\mathbf{x}}_j \tag{C.48}$$

where the unit vector $\hat{\mathbf{x}}'_i$ represents the direction of the ith axis of the new coordinate system, and the unit vector $\hat{\mathbf{x}}_j$ represents the direction of the jth axis of the current coordinate system. Clearly, all six unit vectors must be referred to the same coordinate system. (This may be the unprimed system, the primed system or any other system.)

C.4 Tensors and coordinate transforms

Like vectors, tensors are geometric objects that have a meaning independent of coordinate systems. Unlike vectors, we have no intuitive picture of what the tensor object is.

One possible picture is to view a second-order tensor as a machine that accepts one vector as input and returns another. For example, the stress tensor may be seen as an object that given a unit vector in space, computes the force per unit area in that direction.

Let us denote a second-order tensor by $\mathbf{T}$ and the vectors it relates by $\mathbf{f}$ and $\mathbf{r}$. We thus have

$$\mathbf{f} = \mathbf{T}\,\mathbf{r} \tag{C.49}$$

If we now transform the coordinate system according to an orthogonal matrix $\mathbf{a}$, then we have

$$\mathbf{f}' = \mathbf{a}\mathbf{f} = \mathbf{a}\,(\mathbf{T}\,\mathbf{r}) = \mathbf{a}\mathbf{T}\left(\mathbf{a}^{\mathsf{T}}\mathbf{a}\right)\mathbf{r} = \left(\mathbf{a}\mathbf{T}\mathbf{a}^{\mathsf{T}}\right)\mathbf{r}' \tag{C.50}$$

which shows that the second-order tensor transforms according to

$$\mathbf{T}' = \mathbf{a}\,\mathbf{T}\,\mathbf{a}^{\mathsf{T}} \tag{C.51}$$

which is an orthogonal transform.

Writing this equation in component form, we find

$$T'_{ij} = \sum_l \sum_k a_{il} T_{lk} \left(a^{\mathrm{T}}\right)_{kj} = \sum_l \sum_k a_{il} a_{jk} T_{lk} \tag{C.52}$$

This equation allows us to give a possible definition of a tensor:

A tensor is an object that transforms as a vector in each of its indices.

Without further discussion, we state that higher-order tensors transform according to the same rule. For example, the fourth-order stiffness tensor transforms according to

$$C'_{ijkl} = \sum_r \sum_s \sum_t \sum_u a_{ir} a_{js} a_{kt} a_{lu} C_{rstu} \tag{C.53}$$

C.5 Eigenvalues, eigenvectors and diagonalisation

We have seen that a real symmetric matrix has real eigenvalues and that any real symmetric matrix can be diagonalised by an orthogonal transformation. We have further seen that an orthogonal transformation of a second-order tensor may be seen as a rotation of the coordinate system.

We have also observed that an orthogonal transformation does not change the eigenvalues of a matrix. Since it is easy to see that the eigenvalues of a diagonal matrix are the diagonal elements themselves, it is clear that there is a connection between the eigenvalues and the diagonalisation process: By diagonalising a matrix we get the eigenvalues as the diagonal elements.

To make the picture complete, we now need to clarify the connection between the eigenvectors and the orthogonal transformation matrix.

By diagonalising a second-order tensor we change to a new coordinate system in which it is diagonal. The eigenvectors give the directions of the new coordinate axes referred to the initial coordinate system. The length of the eigenvectors is not specified by the eigenvalue equation (multiplying $\mathbf{x}$ by a number obviously does not invalidate Eq. (C.33)). However, by scaling the eigenvector to unit length we see that its components can be identified with the direction cosines. Writing

$$\mathbf{v}_1 = \begin{pmatrix} v_{1x} \\ v_{1y} \\ v_{1z} \end{pmatrix} \tag{C.54}$$

for the normalised eigenvector associated with the first eigenvalue, and similarly for the other eigenvectors, we may use Eq. (C.47) to write down the orthogonal transformation matrix as

$$\mathbf{a} = \begin{pmatrix} v_{1x} & v_{1y} & v_{1z} \\ v_{2x} & v_{2y} & v_{2z} \\ v_{3x} & v_{3y} & v_{3z} \end{pmatrix} \tag{C.55}$$

C.6 Rotation of the coordinate system: the Euler angles

We discussed above that to specify an arbitrary rotation of the coordinate system in space, three parameters are required.

There are many ways to specify a rotation of the coordinate system. One option is the direction cosines used, for instance, in Sections 1.1.5 and 4.3.1. We saw from Eq. (C.47) how the direction cosines, specifying the orientation of the new coordinate axes, constitute the elements of the orthogonal transformation matrix. Since there are nine direction cosines, they are obviously not independent.

A convenient choice of three independent parameters is the Euler angles ϕ, θ and β.

There are several conventions for the definition of the Euler angles. Here we will use the one adopted by Rose (1957), which defines the angles in the following way:

- a rotation ϕ about the original z-axis,
- a rotation θ about the *new* y-axis,
- a rotation β about the *new* z-axis.

The angles ϕ and θ specify the direction of the new z-axis in conventional spherical coordinates, whereas β specifies the rotation of the (x, y) axes relative to the starting position resulting from the rotations given by ϕ and θ. The convention used in Section 4.3.1 (see Fig. 4.6) is thus consistent with the Rose (1957) convention.

The same final rotation may be obtained by the following rotations about the initial axes:

- a rotation β about the original z-axis,
- a rotation θ about the original y-axis,
- a rotation ϕ about the original z-axis.

To build the rotation matrix for a full Euler rotation, we thus need the matrices for rotations about the y- and z-axes. These are

$$\mathbf{R}_y(\theta) = \begin{pmatrix} \cos\theta & 0 & -\sin\theta \\ 0 & 1 & 0 \\ \sin\theta & 0 & \cos\theta \end{pmatrix} \tag{C.56}$$

$$\mathbf{R}_z(\phi) = \begin{pmatrix} \cos\phi & \sin\phi & 0 \\ -\sin\phi & \cos\phi & 0 \\ 0 & 0 & 1 \end{pmatrix} \tag{C.57}$$

The matrix $\mathbf{R}_z(\beta)$ is found by exchanging ϕ with β in $\mathbf{R}_z(\phi)$.

The signs have been chosen such that multiplying these matrices with a vector in the old system, we find the coordinates of that vector in the rotated system. To find the direction of the axes of the new system expressed in the old system, we must use the inverse matrices, which are the same as the transposed matrices since the matrices are orthogonal.

By using the matrices in sequence, it is now simple to find the rotation matrix corresponding to the rotation specified by the three Euler angles. The result is

$$\mathbf{R}(\phi,\theta,\beta) = \mathbf{R}_z(\beta)\mathbf{R}_y(\theta)\mathbf{R}_z(\phi) = \begin{pmatrix} \cos\theta\cos\phi\cos\beta - \sin\phi\sin\beta & \cos\theta\sin\phi\cos\beta + \cos\phi\sin\beta & -\cos\beta\sin\theta \\ -\cos\theta\cos\phi\sin\beta - \sin\phi\cos\beta & -\cos\theta\sin\phi\sin\beta + \cos\phi\cos\beta & \sin\beta\sin\theta \\ \sin\theta\cos\phi & \sin\theta\sin\phi & \cos\theta \end{pmatrix} \tag{C.58}$$

For $\beta = 0$, this equation reduces to

$$\mathbf{R}(\phi,\theta,0) = \begin{pmatrix} \cos\theta\cos\phi & \cos\theta\sin\phi & -\sin\theta \\ -\sin\phi & \cos\phi & 0 \\ \sin\theta\cos\phi & \sin\theta\sin\phi & \cos\theta \end{pmatrix} \tag{C.59}$$

which should be compared to Eq. (4.78).

C.7 Examples

In this section, we give a few examples of the use of the formalism discussed in this appendix.

C.7.1 Rotation of the stress tensor

Let us now use Eq. (C.51) and the Euler matrices to derive an expression for the stress tensor in a coordinate system that is rotated by an angle θ around the z-axis:

$$\mathbf{R}_z(\theta)\begin{pmatrix}\sigma_x & \tau_{xy} & \tau_{xz}\\ \tau_{xy} & \sigma_y & \tau_{yz}\\ \tau_{xz} & \tau_{yz} & \sigma_z\end{pmatrix}(\mathbf{R}_z(\theta))^{\mathsf{T}} \tag{C.60}$$

which written in full is

$$\begin{pmatrix}\cos\theta & \sin\theta & 0\\ -\sin\theta & \cos\theta & 0\\ 0 & 0 & 1\end{pmatrix}\begin{pmatrix}\sigma_x & \tau_{xy} & \tau_{xz}\\ \tau_{xy} & \sigma_y & \tau_{yz}\\ \tau_{xz} & \tau_{yz} & \sigma_z\end{pmatrix}\begin{pmatrix}\cos\theta & -\sin\theta & 0\\ \sin\theta & \cos\theta & 0\\ 0 & 0 & 1\end{pmatrix} \tag{C.61}$$

By performing the matrix multiplications we end up with

$$\begin{pmatrix}\begin{array}{c}\sigma_x\cos^2\theta+\sigma_y\sin^2\theta\\ +2\tau_{xy}\sin\theta\cos\theta\end{array} & \begin{array}{c}(\sigma_y-\sigma_x)\sin\theta\cos\theta\\ +\tau_{xy}(\cos^2\theta-\sin^2\theta)\end{array} & \tau_{xz}\cos\theta+\tau_{yz}\sin\theta\\ \begin{array}{c}(\sigma_y-\sigma_x)\sin\theta\cos\theta\\ +\tau_{xy}(\cos^2\theta-\sin^2\theta)\end{array} & \begin{array}{c}\sigma_x\sin^2\theta+\sigma_y\cos^2\theta\\ -2\tau_{xy}\sin\theta\cos\theta\end{array} & \tau_{yz}\cos\theta-\tau_{xz}\sin\theta\\ \tau_{xz}\cos\theta+\tau_{yz}\sin\theta & \tau_{yz}\cos\theta-\tau_{xz}\sin\theta & \sigma_z\end{pmatrix} \tag{C.62}$$

This is exactly the operation needed to switch from Cartesian coordinates to cylindrical coordinates, and thus Eq. (C.62) is equivalent to Eqs. (4.1)–(4.6). Note that the sum of the diagonal elements is unchanged as $\sigma_x+\sigma_y+\sigma_z$ (the trace of the matrix) is invariant under an orthogonal transform.

C.7.2 Inversion of an axis

Let us see how the strain tensor changes as we invert the z-axis as in Section 1.7.1. We must calculate

$$\begin{pmatrix}1 & 0 & 0\\ 0 & 1 & 0\\ 0 & 0 & -1\end{pmatrix}\begin{pmatrix}\varepsilon_{11} & \varepsilon_{12} & \varepsilon_{13}\\ \varepsilon_{12} & \varepsilon_{22} & \varepsilon_{23}\\ \varepsilon_{13} & \varepsilon_{23} & \varepsilon_{33}\end{pmatrix}\begin{pmatrix}1 & 0 & 0\\ 0 & 1 & 0\\ 0 & 0 & -1\end{pmatrix} \tag{C.63}$$

which gives the following result:

$$\begin{pmatrix} \varepsilon_{11} & \varepsilon_{12} & -\varepsilon_{13} \\ \varepsilon_{12} & \varepsilon_{22} & -\varepsilon_{23} \\ -\varepsilon_{13} & -\varepsilon_{23} & \varepsilon_{33} \end{pmatrix} \tag{C.64}$$

C.7.3 Rotation and strain

It is obvious that a rigid rotation of a body gives no deformation and hence no strain. However, if we use Eq. (C.57) to transform the coordinates, then compute the displacements according to Eqs. (1.57)–(1.58) and the strain from Eq. (1.77), we end up with the strain tensor

$$\begin{pmatrix} 1-\cos\phi & 0 & 0 \\ 0 & 1-\cos\phi & 0 \\ 0 & 0 & 0 \end{pmatrix} \tag{C.65}$$

This is zero only to first order in ϕ. When ϕ increases and second-order terms become important, the result is clearly unphysical. This is related to the assumptions made when deriving Eq. (1.77).

It is thus clear that a condition for the validity of Eq. (1.77) is not only small deformations, but also small rotations. In rock mechanics, large rotations are rarely an issue, but in other fields, this must be taken into account, and a more complicated definition of strain is required.

We leave it as an exercise to show that for the Green–Lagrange strain (see any advanced textbook on elasticity, for example, Rice, 1999) defined by

$$\varepsilon_{ij} = \frac{1}{2}\left(\frac{\partial u_i}{\partial x_j} + \frac{\partial u_j}{\partial x_i} + \sum_{l=1}^{3} \frac{\partial u_l}{\partial x_i}\frac{\partial u_l}{\partial x_j}\right) \tag{C.66}$$

an arbitrary rotation gives zero strain.

C.7.4 Lode coordinates

Lode coordinates, introduced in Section 1.1.8, are cylindrical coordinates in which the z-axis points along the hydrostatic axis in principal stress space.

The transformation matrix to go from the basic principal stress coordinate system to the Cartesian Lode system is

$$\mathbf{a} = \begin{pmatrix} \sqrt{\frac{2}{3}} & -\sqrt{\frac{1}{6}} & -\sqrt{\frac{1}{6}} \\ 0 & \sqrt{\frac{1}{2}} & -\sqrt{\frac{1}{2}} \\ \sqrt{\frac{1}{3}} & \sqrt{\frac{1}{3}} & \sqrt{\frac{1}{3}} \end{pmatrix} \tag{C.67}$$

This matrix corresponds to the Euler angles $\phi = 45°$, $\theta = \arccos(1/\sqrt{3}) \approx 54.7°$ and $\beta = -60°$. A moment's reflection will show that this must be correct: The first two rotations leave the z-axis in the right direction, but the x-axis then points exactly between the original x- and y-axes, and hence we must rotate backwards 60° about the new z-axis to align it correctly.

From Fig. 1.10 and the discussion in Section 1.1.8 it is clear that the point $(\sigma_1, \sigma_2, \sigma_3)$ may be written in Cartesian Lode coordinates as

$$\mathbf{v} = \begin{pmatrix} \sqrt{\frac{2}{3}}\, q \cos \vartheta_{\mathrm{L}} \\ \sqrt{\frac{2}{3}}\, q \sin \vartheta_{\mathrm{L}} \\ \sqrt{3}\,\overline{\sigma} \end{pmatrix} \tag{C.68}$$

We find $(\sigma_1, \sigma_2, \sigma_3)$ by transforming back according to $\mathbf{a}^{\mathrm{T}} \cdot \mathbf{v}$, and the formulas given in Eqs. (D.1)–(D.3) result.

Note that in this case, we transform in the abstract principal stress space, not in 3D coordinate space, but the formalism is the same.

C.8 Matrix invariants

Let us investigate how the expression $\sum_i \sum_j B_{ij} B_{ji}$ transforms under an orthogonal transformation. We find

$$\sum_i \sum_j B'_{ij} B'_{ji} = \sum_i \sum_j \left(\sum_l \sum_m a_{il} a_{jm} B_{lm} \right) \left(\sum_r \sum_s a_{jr} a_{is} B_{rs} \right) \tag{C.69}$$

Now, from the relation (see Eq. (C.39))

$$\sum_i a_{il} a_{is} = \delta_{ls} \tag{C.70}$$

we derive

$$\sum_i \sum_j B'_{ij} B'_{ji} = \sum_l \sum_s \sum_r \sum_m \delta_{ls} \delta_{rm} B_{lm} B_{rs} = \sum_s \sum_r B_{sr} B_{rs} \tag{C.71}$$

This tells us that $\sum_i \sum_j B_{ij} B_{ji}$ is invariant under an orthogonal transformation.

By similar calculations it is easy to show that

$$\sum_k B_{kk} \tag{C.72}$$

$$\sum_i \sum_j B_{ij} B_{ji} \tag{C.73}$$

$$\sum_i \sum_j \sum_k B_{ij} B_{jk} B_{ki} \tag{C.74}$$

$$\sum_i \sum_j \sum_k \sum_l B_{ij} B_{jk} B_{kl} B_{li} \tag{C.75}$$

are invariants of linear, quadratic, cubic and quartic degree. It is obvious how fifth- and higher-order invariants are constructed in the same way. It should however be noted that only the first three invariants are truly independent; the higher-order invariants may be expressed as sums of combinations of the first three.

In rock mechanics, these invariants are not used directly, except for the first. The standard rock mechanical invariants are defined by writing the third-degree equation resulting from Eq. (1.31) as

$$\sigma^3 - I_1 \sigma^2 - I_2 \sigma - I_3 = 0 \tag{C.76}$$

Since the solutions of the equation are the principal stresses, it is obvious that the coefficients are invariants. (Beware that sometimes Eq. (C.76) is written with a +-sign in front of the I_2-term, leading to the opposite sign for I_2.)

By expanding Eq. (1.31) and comparing with Eq. (C.76) expressions (1.41)–(1.43) follow. The relations between the rock mechanical invariants and those derived in this section are

$$\sum_k \sigma_{kk} = I_1 \tag{C.77}$$

$$\sum_i \sum_j \sigma_{ij} \sigma_{ji} = I_1^2 + 2I_2 \tag{C.78}$$

$$\sum_i \sum_j \sum_k \sigma_{ij}\sigma_{jk}\sigma_{ki} = I_1^3 + 3I_2I_1 + 3I_3 \tag{C.79}$$

Observe that for a symmetric matrix,

$$\sum_i \sum_j B_{ij}B_{ji} = \sum_i \sum_j (B_{ij})^2 \tag{C.80}$$

that is, it is equal to the sum of the squares of the elements.

This means that the Frobenius matrix *norm* defined by

$$\mathcal{N}_\mathrm{F} = \sqrt{\sum_i \sum_j (B_{ij})^2} \tag{C.81}$$

is also an invariant for symmetric entities like stress and strain.

C.9 Some trigonometric formulas

In computations with stresses and strains, trigonometric formulas are used extensively. Here we repeat some of the basic formulas:

$$\cos^2\theta + \sin^2\theta = 1 \tag{C.82}$$

$$\cos(\alpha \pm \beta) = \cos\alpha\cos\beta \mp \sin\alpha\sin\beta \tag{C.83}$$

$$\sin(\alpha \pm \beta) = \sin\alpha\cos\beta \pm \sin\beta\cos\alpha \tag{C.84}$$

$$\cos 2\theta = \cos^2\theta - \sin^2\theta = 2\cos^2\theta - 1 = 1 - 2\sin^2\theta \tag{C.85}$$

$$\sin 2\theta = 2\sin\theta\cos\theta \tag{C.86}$$

from which we derive

$$\cos^2\theta = \frac{1}{2} + \frac{1}{2}\cos 2\theta \tag{C.87}$$

$$\sin^2\theta = \frac{1}{2} - \frac{1}{2}\cos 2\theta \tag{C.88}$$

which are useful, for example, when deriving Eq. (1.19) from Eq. (1.18).

From $\tan\theta = \sin\theta/\cos\theta$ and the above we find

$$\tan 2\theta = \frac{2\tan\theta}{1 - \tan^2\theta} \tag{C.89}$$

$$\tan^2\theta = \frac{1-\cos^2\theta}{\cos^2\theta} = \frac{\sin^2\theta}{1-\sin^2\theta} \tag{C.90}$$

$$\cos^2\theta = \frac{1}{1+\tan^2\theta} \tag{C.91}$$

$$\sin^2\theta = \frac{\tan^2\theta}{1+\tan^2\theta} \tag{C.92}$$

The last two equations are useful, for example, when deriving Eqs. (1.23) and (1.24) from Eqs. (1.19) and (1.22).

C.10 The Voigt notation spelled out

Since the stress and strain tensors are symmetrical, they may be mapped into six-component vectors. Similarly, because of the symmetries of the stiffness and compliance tensors, they may be mapped into symmetric 6×6 matrices.

One way to do this is to use the Voigt notation (Voigt, 1910). This method is commonly adopted, in spite of the fact that it has some theoretical shortcomings (stiffness and compliance do not transform as a tensor; see e.g. Mehrabadi and Cowin, 1990; Helbig, 1994, 1996, 2013).

The Voigt mapping for the stiffness tensor was introduced in Section 1.7.1. In this section, we treat it in more detail.

The Voigt notation is the mapping in which pairs of indices are mapped into a single index according to

$$11 \to 1 \tag{C.93}$$

$$22 \to 2 \tag{C.94}$$

$$33 \to 3 \tag{C.95}$$

$$23 \to 4 \tag{C.96}$$

$$31 \to 5 \tag{C.97}$$

$$12 \to 6 \tag{C.98}$$

These relations may be summarised by

$$I = i\delta_{ij} + (9-i-j)(1-\delta_{ij}) \tag{C.99}$$

where I is the Voigt index, i and j are the tensor indices, and δ_{ij} is the Kronecker symbol. Beware that sometimes a different order of the shear components is used. An example is Cheng (1997).

For the stress and stiffness tensors, the mapping is direct, whereas for the strain and compliance tensors, some numerical constants are necessary, as shown below.

C.10.1 The Voigt mapping for the stress tensor

$$\begin{pmatrix} \sigma_1 \\ \sigma_2 \\ \sigma_3 \\ \sigma_4 \\ \sigma_5 \\ \sigma_6 \end{pmatrix} = \begin{pmatrix} \sigma_{11} \\ \sigma_{22} \\ \sigma_{33} \\ \sigma_{23} \\ \sigma_{13} \\ \sigma_{12} \end{pmatrix} \tag{C.100}$$

C.10.2 The Voigt mapping for the strain tensor

$$\begin{pmatrix} \varepsilon_1 \\ \varepsilon_2 \\ \varepsilon_3 \\ \varepsilon_4 \\ \varepsilon_5 \\ \varepsilon_6 \end{pmatrix} = \begin{pmatrix} \varepsilon_{11} \\ \varepsilon_{22} \\ \varepsilon_{33} \\ 2\,\varepsilon_{23} \\ 2\,\varepsilon_{13} \\ 2\,\varepsilon_{12} \end{pmatrix} \tag{C.101}$$

Note the factor 2 for the shear strains.

C.10.3 The Voigt mapping for the stiffness tensor

$$\mathbf{C} = \begin{pmatrix} C_{1111} & C_{1122} & C_{1133} & C_{1123} & C_{1113} & C_{1112} \\ C_{1122} & C_{2222} & C_{2233} & C_{2223} & C_{1322} & C_{1222} \\ C_{1133} & C_{2233} & C_{3333} & C_{2333} & C_{1333} & C_{1233} \\ C_{1123} & C_{2223} & C_{2333} & C_{2323} & C_{1323} & C_{1223} \\ C_{1113} & C_{1322} & C_{1333} & C_{1323} & C_{1313} & C_{1213} \\ C_{1112} & C_{1222} & C_{1233} & C_{1223} & C_{1213} & C_{1212} \end{pmatrix} \tag{C.102}$$

C.10.4 The Voigt mapping for the compliance tensor

$$\mathbf{S} = \begin{pmatrix} S_{1111} & S_{1122} & S_{1133} & 2\,S_{1123} & 2\,S_{1113} & 2\,S_{1112} \\ S_{1122} & S_{2222} & S_{2233} & 2\,S_{2223} & 2\,S_{1322} & 2\,S_{1222} \\ S_{1133} & S_{2233} & S_{3333} & 2\,S_{2333} & 2\,S_{1333} & 2\,S_{1233} \\ 2\,S_{1123} & 2\,S_{2223} & 2\,S_{2333} & 4\,S_{2323} & 4\,S_{1323} & 4\,S_{1223} \\ 2\,S_{1113} & 2\,S_{1322} & 2\,S_{1333} & 4\,S_{1323} & 4\,S_{1313} & 4\,S_{1213} \\ 2\,S_{1112} & 2\,S_{1222} & 2\,S_{1233} & 4\,S_{1223} & 4\,S_{1213} & 4\,S_{1212} \end{pmatrix} \tag{C.103}$$

Note the numerical factors 2 and 4.

C.10.5 The Voigt mapping for the Biot tensor

$$\begin{pmatrix} \alpha_1 \\ \alpha_2 \\ \alpha_3 \\ \alpha_4 \\ \alpha_5 \\ \alpha_6 \end{pmatrix} = \begin{pmatrix} \alpha_{11} \\ \alpha_{22} \\ \alpha_{33} \\ \alpha_{23} \\ \alpha_{13} \\ \alpha_{12} \end{pmatrix} \tag{C.104}$$

C.10.6 The Voigt mapping for the Skempton tensor

$$\begin{pmatrix} B_1 \\ B_2 \\ B_3 \\ B_4 \\ B_5 \\ B_6 \end{pmatrix} = \begin{pmatrix} B_{11} \\ B_{22} \\ B_{33} \\ 2\,B_{23} \\ 2\,B_{13} \\ 2\,B_{12} \end{pmatrix} \tag{C.105}$$

Note the factor 2.

C.10.7 Direct rotation of the Voigt matrices

To rotate the Voigt vectors and matrices, we must convert to tensor form, rotate and transform back. It is easy to do this symbolically for stress and strain, and come up with formulas that work directly on the Voigt vectors:

$$\boldsymbol{\sigma}' = \mathbf{M}\boldsymbol{\sigma} \tag{C.106}$$

$$\boldsymbol{\varepsilon}' = \mathbf{N}\boldsymbol{\varepsilon} \tag{C.107}$$

The matrices **M** and **N** are often referred to as the *Bond matrices*, after Bond (1943).

In terms of the components of a generic orthogonal matrix a_{ij} (Section C.3), the matrices are given by

$$\mathbf{M} = \begin{pmatrix} a_{11}^2 & a_{12}^2 & a_{13}^2 & 2a_{12}a_{13} & 2a_{11}a_{13} & 2a_{11}a_{12} \\ a_{21}^2 & a_{22}^2 & a_{23}^2 & 2a_{22}a_{23} & 2a_{21}a_{23} & 2a_{21}a_{22} \\ a_{31}^2 & a_{32}^2 & a_{33}^2 & 2a_{32}a_{33} & 2a_{31}a_{33} & 2a_{31}a_{32} \\ a_{21}a_{31} & a_{22}a_{32} & a_{23}a_{33} & a_{23}a_{32} + a_{22}a_{33} & a_{23}a_{31} + a_{21}a_{33} & a_{22}a_{31} + a_{21}a_{32} \\ a_{11}a_{31} & a_{12}a_{32} & a_{13}a_{33} & a_{13}a_{32} + a_{12}a_{33} & a_{13}a_{31} + a_{11}a_{33} & a_{12}a_{31} + a_{11}a_{32} \\ a_{11}a_{21} & a_{12}a_{22} & a_{13}a_{23} & a_{13}a_{22} + a_{12}a_{23} & a_{13}a_{21} + a_{11}a_{23} & a_{12}a_{21} + a_{11}a_{22} \end{pmatrix} \tag{C.108}$$

$$\mathbf{N}=\begin{pmatrix} a_{11}^2 & a_{12}^2 & a_{13}^2 & a_{12}a_{13} & a_{11}a_{13} & a_{11}a_{12} \\ a_{21}^2 & a_{22}^2 & a_{23}^2 & a_{22}a_{23} & a_{21}a_{23} & a_{21}a_{22} \\ a_{31}^2 & a_{32}^2 & a_{33}^2 & a_{32}a_{33} & a_{31}a_{33} & a_{31}a_{32} \\ 2a_{21}a_{31} & 2a_{22}a_{32} & 2a_{23}a_{33} & a_{23}a_{32}+a_{22}a_{33} & a_{23}a_{31}+a_{21}a_{33} & a_{22}a_{31}+a_{21}a_{32} \\ 2a_{11}a_{31} & 2a_{12}a_{32} & 2a_{13}a_{33} & a_{13}a_{32}+a_{12}a_{33} & a_{13}a_{31}+a_{11}a_{33} & a_{12}a_{31}+a_{11}a_{32} \\ 2a_{11}a_{21} & 2a_{12}a_{22} & 2a_{13}a_{23} & a_{13}a_{22}+a_{12}a_{23} & a_{13}a_{21}+a_{11}a_{23} & a_{12}a_{21}+a_{11}a_{22} \end{pmatrix} \quad \text{(C.109)}$$

It can be shown that $\mathbf{M}^{-1}=\mathbf{N}^{\mathrm{T}}$ and $\mathbf{N}^{-1}=\mathbf{M}^{\mathrm{T}}$. This allows us to write

$$\boldsymbol{\sigma}'=\mathbf{MC}\boldsymbol{\varepsilon}=\mathbf{MCN}^{-1}\mathbf{N}\boldsymbol{\varepsilon}=\mathbf{MCM}^{\mathrm{T}}\boldsymbol{\varepsilon}' \quad \text{(C.110)}$$

which shows that

$$\mathbf{C}'=\mathbf{MCM}^{\mathrm{T}} \quad \text{(C.111)}$$

and similarly

$$\mathbf{S}'=\mathbf{NSN}^{\mathrm{T}} \quad \text{(C.112)}$$

Note that the Bond transformations do not map a 6×6 unit matrix onto itself, since this is not a proper tensor rotation.

If we shared the factor 2 equally between the shear stresses and shear strains (a factor $\sqrt{2}$ on each), then we would get only one Bond matrix, a 6×6 orthogonal matrix. See Helbig (1994, 1996, 2013).

C.11 Elastic stability

In Section 1.4, we discussed that K and G must be non-negative for isotropic media. The general criterion is that any deformation of an elastic material should lead to a positive strain energy.

This immediately tells us that the diagonal elements of the stiffness matrix must be positive, which follows by letting one of the elements of the strain vector be non-zero at the time.

This is however not sufficient, as other requirements may follow from various combinations of the strains. It is however not necessary to check all possible strain combinations due to mathematical theorems established in the 19th century (see e.g. Born and Huang, 1954; Backus, 1962; Mouhat and Coudert, 2014),

A matrix with the desired properties is referred to as a *positive definite* matrix. It has been proven that the positive definiteness of a matrix is equivalent to other sets of criteria, such as

- All the eigenvalues of the matrix are positive;

- All the principal minors of the matrix are positive (Sylvester's criterion). The principal minors are the determinants of all sub-matrices that include the upper left corner, that is, matrices of size 1×1 up to 6×6.

Although equivalent, the methods may give the results in different forms, which are not necessarily straightforward to deduce from each other.

As an example, let us consider transverse isotropy. The stiffness matrix may be written (see Eq. (1.264))

$$\mathbf{C} = \begin{pmatrix} C_{11} & C_{12} & C_{13} & 0 & 0 & 0 \\ C_{12} & C_{11} & C_{13} & 0 & 0 & 0 \\ C_{13} & C_{13} & C_{33} & 0 & 0 & 0 \\ 0 & 0 & 0 & C_{44} & 0 & 0 \\ 0 & 0 & 0 & 0 & C_{44} & 0 \\ 0 & 0 & 0 & 0 & 0 & \frac{C_{11}-C_{12}}{2} \end{pmatrix} \tag{C.113}$$

The first principal minor gives the trivial result $C_{11} > 0$. The second is $C_{11}^2 - C_{12}^2$, which means $C_{11} > |C_{12}|$. The third is $(C_{11} - C_{12})((C_{11} + C_{12})C_{33} - 2C_{13}^2)$, which, combined with the previous, gives $(C_{11} + C_{12})C_{33} > 2C_{13}^2$. Together with $C_{44} > 0$, these are the necessary and sufficient stability conditions for TI (see e.g. Mouhat and Coudert, 2014).

The eigenvalues of the stiffness matrix are

$$C_{11} - C_{12} \tag{C.114}$$

$$\frac{1}{2}\left(C_{11} + C_{12} + C_{33} - \sqrt{8C_{13}^2 + (C_{11} + C_{12} - C_{33})^2}\right) \tag{C.115}$$

$$\frac{1}{2}\left(C_{11} + C_{12} + C_{33} + \sqrt{8C_{13}^2 + (C_{11} + C_{12} - C_{33})^2}\right) \tag{C.116}$$

$$C_{44} \tag{C.117}$$

$$C_{44} \tag{C.118}$$

$$(C_{11} - C_{12})/2 \tag{C.119}$$

from which the above criteria may also be derived.

By investigating the eigenvectors we find that four of the eigenvalues are associated with pure shear, whereas the other two have volumetric and shear components. (This is in contrast to isotropy, where there are five pure shear eigenvalues and one pure volumetric.) See Helbig (1996) for an in-depth discussion.

The volumetric eigenvalues represent the maximum and minimum generalised bulk moduli

$$K^{\min} = \frac{1}{6}\left(C_{11} + C_{12} + C_{33} - \sqrt{8C_{13}^2 + (C_{11} + C_{12} - C_{33})^2}\right) \quad \text{(C.120)}$$

$$K^{\max} = \frac{1}{6}\left(C_{11} + C_{12} + C_{33} + \sqrt{8C_{13}^2 + (C_{11} + C_{12} - C_{33})^2}\right) \quad \text{(C.121)}$$

Each of these results from loading according to the specific eigenvector, whereas all values in between may be reached by loading according to linear combinations of the eigenvectors. The isostrain and isostress bulk moduli discussed in Section 1.7.3 are particular cases, which are clearly within these bounds.

See Holt (2016) for an extended discussion for transverse isotropy and references to further relevant literature.

Note that eigenvalues (C.114) and (C.119) differ by the factor 2. This is because the Voigt matrix is not a tensor; in a theoretically stringent formulation, they would be equal.

C.12 The Einstein summing convention and other notation conventions

In this book, we have attempted to use as explicit notation as possible, to minimise the risk for confusion.

There exist, however, a number of notation conventions that are so widely used that they are not always defined in books and papers that use them. Such conventions make equations more compact (and in fact easier to read when one is accustomed to them) and include the Einstein summing convention and operator and comma notation for partial derivatives. Here we include a short overview of these notation conventions.

C.12.1 The Einstein summing convention

The Einstein summing convention states that when an index is *repeated* in certain expressions, a summation is implied. For example, if we assume that the index runs from 1 to 3, then we have

$$a_{ii} \overset{\text{def}}{=} \sum_{i=1}^{3} a_{ii} = a_{11} + a_{22} + a_{33} \quad \text{(C.122)}$$

and

$$a_{ij}b_{jk} \stackrel{\text{def}}{=} \sum_{j=1}^{3} a_{ij}b_{jk} = a_{i1}b_{1k} + a_{i2}b_{2k} + a_{i3}b_{3k} \tag{C.123}$$

The expression may be a single quantity with multiple indices or a product. A summation over i is thus *not* implied in sums such as

$$a_i = b_i + c_i \tag{C.124}$$

C.12.2 Kronecker's delta

A symbol that is often encountered, both in conjunction with the summing convention and otherwise, is Kronecker's δ defined by

$$\delta_{ij} \stackrel{\text{def}}{=} \begin{cases} 1 \text{ if } i = j \\ 0 \text{ if } i \neq j \end{cases} \tag{C.125}$$

It is easy to see that

$$\sum_i \delta_{ii} = \delta_{11} + \delta_{22} + \delta_{33} = 3 \tag{C.126}$$

C.12.3 Comma notation for partial derivatives

A partial derivative is sometimes denoted by a subscript preceded by a comma, that is (f is a scalar),

$$f_{,j} \stackrel{\text{def}}{=} \frac{\partial f}{\partial x_j} \tag{C.127}$$

A second-order derivative is indicated in a similar way (g_i is here component i of a vector):

$$g_{i,jk} \stackrel{\text{def}}{=} \frac{\partial^2 g_i}{\partial x_j \partial x_k} \tag{C.128}$$

In this notation the volumetric strain (see Eq. (1.79)) is written as (here we also use the summing convention)

$$\varepsilon_{\text{vol}} = u_{i,i} \tag{C.129}$$

and the equation of equilibrium (Eq. (1.17)) as

$$\sigma_{ji,j} + \rho X_i = 0 \tag{C.130}$$

Sometimes, differentiation with respect to time is shown in a similar way, for example,

$$u_{i,tt} \stackrel{\text{def}}{=} \frac{\partial^2 u_i}{\partial t^2} \tag{C.131}$$

An example of a central reference using this notation is Detournay and Cheng (1993).

C.12.4 Operator notation for partial derivatives

Sometimes, an operator notation is used for partial derivatives:

$$\partial_i f \stackrel{\text{def}}{=} \partial_{x_i} f \stackrel{\text{def}}{=} \frac{\partial f}{\partial x_i} \tag{C.132}$$

The volumetric strain becomes

$$\varepsilon_{\text{vol}} = \partial_i u_i \tag{C.133}$$

and the equation of equilibrium is

$$\partial_j \sigma_{ji} + \rho X_i = 0 \tag{C.134}$$

References

Backus, G.E., 1962. Long-wave elastic anisotropy produced by horizontal layering. J. Geophys. Res. 67 (11), 4427–4440. https://doi.org/10.1029/JZ067i011p04427.

Bond, W.L., 1943. The mathematics of the physical properties of crystals. Bell Syst. Tech. J. 22 (1), 1–72.

Born, M., Huang, K., 1954. Dynamical Theory of Crystal Lattices. Clarendon Press, Oxford.

Cheng, A.H.-D., 1997. Material coefficients of anisotropic poroelasticity. Int. J. Rock Mech. Min. Sci. 34 (2), 199–205. https://doi.org/10.1016/S0148-9062(96)00055-1.

Detournay, E., Cheng, A.H.-D., 1993. Fundamentals of poroelasticity. In: Hudson, J.A. (Ed.), Comprehensive Rock Engineering, vol. 2. Pergamon Press, Oxford, pp. 113–171.

Helbig, K., 1994. Foundation of Anisotropy for Exploration Geophysicists. Pergamon, Oxford.

Helbig, K., 1996. Kelvin and the early history of seismic anisotropy. In: Fjær, E., Holt, R.M., Rathore, J.S. (Eds.), Seismic Anisotropy. Society of Exploration Geophysicists, pp. 15–36.

Helbig, K., 2013. Review paper: what Kelvin might have written about elasticity. Geophys. Prospect. 61, 1–20. https://doi.org/10.1111/j.1365-2478.2011.01049.x.

Holt, R.M., 2016. Bounds of elastic parameters characterizing transversely isotropic media: application to shales. Geophysics 81 (5), C243–C252. https://doi.org/10.1190/GEO2015-0485.1.

Mehrabadi, M.M., Cowin, S.C., 1990. Eigentensors of linear anisotropic elastic materials. Q. J. Mech. Appl. Math. 43 (1), 15–41. https://doi.org/10.1093/qjmam/43.1.15.

Mouhat, F., Coudert, F.-C., 2014. Necessary and sufficient elastic stability conditions in various crystal systems. Phys. Rev. B 90, 224104. https://doi.org/10.1103/PhysRevB.90.224104.

Rice, J.R., 1999. Foundations of solid mechanics. In: Meyers, M.A., Armstrong, R.W., Kirchner, H. (Eds.), Mechanics and Materials: Fundamentals and Linkages. Wiley, New York.

Rose, M.E., 1957. Elementary Theory of Angular Momentum. John Wiley & Sons, New York.

Voigt, W., 1910. Lehrbuch der Kristallphysik. Teubner, Leipzig.

APPENDIX D

Some relevant formulas

In this appendix, we give a few formulas that may be useful but are not essential for the main text.

D.1 Elasticity

D.1.1 Stress invariants

Principal stresses ($\overline{\sigma}$ = mean stress; q = von Mises stress; ϑ_{L} = Lode angle):

$$\sigma_A = \overline{\sigma} + \frac{2}{3} q \cos \vartheta_{\mathrm{L}} \tag{D.1}$$

$$\sigma_B = \overline{\sigma} - \frac{1}{3} q (\cos \vartheta_{\mathrm{L}} - \sqrt{3} \sin \vartheta_{\mathrm{L}}) \tag{D.2}$$

$$\sigma_C = \overline{\sigma} - \frac{1}{3} q (\cos \vartheta_{\mathrm{L}} + \sqrt{3} \sin \vartheta_{\mathrm{L}}) \tag{D.3}$$

The numerical order depends on ϑ_{L}. For $0 \leq \vartheta_{\mathrm{L}} \leq \pi/3$, we have $\sigma_A \geq \sigma_B \geq \sigma_C$, and so on.

Invariants of stress deviation (recall that $J_1 = s_1 + s_2 + s_3 = 0$):

$$J_2 = \frac{1}{2} \sum_{i,j} s_{ij} s_{ji} = \frac{1}{2} \sum_{i,j} \sigma_{ij} \sigma_{ji} - \frac{1}{6} \sum_{k} \sigma_{kk}^2 \tag{D.4}$$

$$= \frac{1}{3} I_1^2 + I_2 = \frac{q^2}{3} \tag{D.5}$$

$$= -\left(s_x s_y + s_y s_z + s_z s_x\right) + s_{xy}^2 + s_{yz}^2 + s_{xz}^2 \tag{D.6}$$

$$= \frac{1}{2} \left(s_x^2 + s_y^2 + s_z^2 + 2 s_{yz}^2 + 2 s_{xz}^2 + 2 s_{xy}^2\right) \tag{D.7}$$

$$= \frac{1}{2} \left(s_1^2 + s_2^2 + s_3^2\right) \tag{D.8}$$

$$= \frac{1}{2} \left[(\sigma_1 - \overline{\sigma})^2 + (\sigma_2 - \overline{\sigma})^2 + (\sigma_3 - \overline{\sigma})^2\right] \tag{D.9}$$

$$= \frac{1}{6} \left[\left(\sigma_y - \sigma_z\right)^2 + (\sigma_z - \sigma_x)^2 + \left(\sigma_x - \sigma_y\right)^2\right] + \tau_{yz}^2 + \tau_{xz}^2 + \tau_{xy}^2 \tag{D.10}$$

$$= \frac{1}{6} \left[(\sigma_2 - \sigma_3)^2 + (\sigma_3 - \sigma_1)^2 + (\sigma_1 - \sigma_2)^2\right] \tag{D.11}$$

Developments in Petroleum Science, Volume 72
ISSN 0376-7361, https://doi.org/10.1016/B978-0-12-822195-2.00025-5

$$= \frac{1}{3}\left(\sigma_1^2 + \sigma_2^2 + \sigma_3^2 - \sigma_2\sigma_3 - \sigma_3\sigma_1 - \sigma_1\sigma_2\right) \tag{D.12}$$

Eq. (D.7) shows that $2J_2$ is equal to the sum of the squares of the elements of the deviatoric stress tensor.

$$J_3 = \frac{1}{3}\sum_{i,j,k} s_{ij}s_{jk}s_{ki} \tag{D.13}$$

$$= \frac{1}{3}\left(\sum_{i,j,k}\sigma_{ij}\sigma_{jk}\sigma_{ki} - \sum_{i,j,k}\sigma_{ij}\sigma_{ji}\sigma_{kk} + \frac{2}{9}\sum_{k}\sigma_{kk}^3\right) \tag{D.14}$$

$$= I_3 + \frac{1}{3}I_2 I_1 + \frac{2}{27}I_1^3 \tag{D.15}$$

$$= I_3 + \frac{1}{3}J_2 I_1 - \frac{1}{27}I_1^3 \tag{D.16}$$

$$= s_x s_y s_z + 2 s_{xy} s_{yz} s_{xz} - s_x s_{yz}^2 - s_y s_{xz}^2 - s_z s_{xy}^2 \tag{D.17}$$

$$= \frac{1}{27}[54\tau_{xy}\tau_{yz}\tau_{xz} \tag{D.18}$$

$$+ 9\tau_{xy}^2(\sigma_x + \sigma_y - 2\sigma_z) - 9\tau_{yz}^2(2\sigma_x - \sigma_y - \sigma_z) \tag{D.19}$$

$$+ 9\tau_{xz}^2(\sigma_x - 2\sigma_y + \sigma_z) \tag{D.20}$$

$$+ (\sigma_x + \sigma_y - 2\sigma_z)(2\sigma_x - \sigma_y - \sigma_z)(\sigma_x - 2\sigma_y + \sigma_z)] \tag{D.21}$$

$$= \frac{1}{3}\left(s_1^3 + s_2^3 + s_3^3\right) \tag{D.22}$$

$$= s_1 s_2 s_3 \tag{D.23}$$

$$= (\sigma_1 - \overline{\sigma})(\sigma_2 - \overline{\sigma})(\sigma_3 - \overline{\sigma}) \tag{D.24}$$

$$= \frac{1}{27}(\sigma_1 + \sigma_2 - 2\sigma_3)(2\sigma_1 - \sigma_2 - \sigma_3)(\sigma_1 - 2\sigma_2 + \sigma_3) \tag{D.25}$$

Also note that J_3 is equal to the determinant of the deviatoric stress tensor (compare Eqs. (C.21) and (D.17)).

D.1.2 Strain in spherical coordinates

$$\varepsilon_r = \frac{\partial u_r}{\partial r} \tag{D.26}$$

$$\varepsilon_\theta = \frac{u_r}{r} + \frac{1}{r}\frac{\partial u_\theta}{\partial \theta} \tag{D.27}$$

$$\varepsilon_\phi = \frac{1}{r\sin\theta}\frac{\partial u_\phi}{\partial \phi} + \frac{u_\theta}{r}\cot\theta + \frac{u_r}{r} \tag{D.28}$$

$$\Gamma_{r\theta} = \frac{1}{2r}\frac{\partial u_r}{\partial \theta} + \frac{1}{2}\frac{\partial u_\theta}{\partial r} - \frac{u_\theta}{2r} \tag{D.29}$$

$$\Gamma_{\theta\phi} = \frac{1}{2r}\left(\frac{\partial u_\phi}{\partial \theta} - u_\phi \cot\theta\right) + \frac{1}{2r\sin\theta}\frac{\partial u_\theta}{\partial \phi} \tag{D.30}$$

$$\Gamma_{\phi r} = \frac{1}{2}\left(\frac{1}{r\sin\theta}\frac{\partial u_r}{\partial \phi} + \frac{\partial u_\phi}{\partial r} - \frac{u_\phi}{r}\right) \tag{D.31}$$

D.1.3 Isotropic linear elastic stiffness tensor

$$C_{ijkl} = \lambda\delta_{ij}\delta_{kl} + G\left(\delta_{ik}\delta_{jl} + \delta_{il}\delta_{jk}\right) \tag{D.32}$$

D.1.4 Isotropic linear thermoporoelastic stress–strain law

Solved for stresses:

$$\sigma_x = (\lambda_{\text{fr}} + 2G_{\text{fr}})\varepsilon_x + \lambda_{\text{fr}}\varepsilon_y + \lambda_{\text{fr}}\varepsilon_z + \alpha_{\text{T}}(3\lambda_{\text{fr}} + 2G_{\text{fr}})(T - T_0) + \alpha p_{\text{f}} \tag{D.33}$$

$$= \lambda_{\text{fr}}\varepsilon_{\text{vol}} + 2G_{\text{fr}}\varepsilon_x + \alpha_{\text{T}}(3\lambda_{\text{fr}} + 2G_{\text{fr}})(T - T_0) + \alpha p_{\text{f}} \tag{D.34}$$

$$\sigma_y = \lambda_{\text{fr}}\varepsilon_x + (\lambda + 2G_{\text{fr}})\varepsilon_y + \lambda_{\text{fr}}\varepsilon_z + \alpha_{\text{T}}(3\lambda_{\text{fr}} + 2G_{\text{fr}})(T - T_0) + \alpha p_{\text{f}} \tag{D.35}$$

$$= \lambda_{\text{fr}}\varepsilon_{\text{vol}} + 2G_{\text{fr}}\varepsilon_y + \alpha_{\text{T}}(3\lambda_{\text{fr}} + 2G_{\text{fr}})(T - T_0) + \alpha p_{\text{f}} \tag{D.36}$$

$$\sigma_z = \lambda_{\text{fr}}\varepsilon_x + \lambda_{\text{fr}}\varepsilon_y + (\lambda_{\text{fr}} + 2G_{\text{fr}})\varepsilon_z + \alpha_{\text{T}}(3\lambda_{\text{fr}} + 2G_{\text{fr}})(T - T_0) + \alpha p_{\text{f}} \tag{D.37}$$

$$= \lambda_{\text{fr}}\varepsilon_{\text{vol}} + 2G_{\text{fr}}\varepsilon_z + \alpha_{\text{T}}(3\lambda_{\text{fr}} + 2G_{\text{fr}})(T - T_0) + \alpha p_{\text{f}} \tag{D.38}$$

$$\tau_{yz} = 2G_{\text{fr}}\Gamma_{yz} \tag{D.39}$$

$$\tau_{xz} = 2G_{\text{fr}}\Gamma_{xz} \tag{D.40}$$

$$\tau_{xy} = 2G_{\text{fr}}\Gamma_{xy} \tag{D.41}$$

Solved for strains:

$$E_{\text{fr}}\varepsilon_x = \sigma_x - \nu_{\text{fr}}(\sigma_y + \sigma_z) - E_{\text{fr}}\alpha_{\text{T}}(T - T_0) - (1 - 2\nu_{\text{fr}})\alpha p_{\text{f}} \tag{D.42}$$

$$E_{\text{fr}}\varepsilon_y = \sigma_y - \nu_{\text{fr}}(\sigma_x + \sigma_z) - E_{\text{fr}}\alpha_{\text{T}}(T - T_0) - (1 - 2\nu_{\text{fr}})\alpha p_{\text{f}} \tag{D.43}$$

$$E_{\text{fr}}\varepsilon_z = \sigma_z - \nu_{\text{fr}}(\sigma_x + \sigma_y) - E_{\text{fr}}\alpha_{\text{T}}(T - T_0) - (1 - 2\nu_{\text{fr}})\alpha p_{\text{f}} \tag{D.44}$$

Various forms on compact notation:

$$\sigma_{ij} = \lambda_{\text{fr}}\varepsilon_{\text{vol}}\delta_{ij} + 2G_{\text{fr}}\varepsilon_{ij} + 3\alpha_{\text{T}}K_{\text{fr}}(T - T_0)\delta_{ij} + \alpha p_{\text{f}}\delta_{ij} \tag{D.45}$$

$$= \frac{3K_{\text{fr}}\nu_{\text{fr}}}{1 + \nu_{\text{fr}}}\varepsilon_{\text{vol}}\delta_{ij} + 3K_{\text{fr}}\frac{1 - 2\nu_{\text{fr}}}{1 + \nu_{\text{fr}}}\varepsilon_{ij} + 3\alpha_{\text{T}}K_{\text{fr}}(T - T_0)\delta_{ij} + \alpha p_{\text{f}}\delta_{ij} \tag{D.46}$$

$$= \frac{E_{\text{fr}}\nu_{\text{fr}}}{(1 + \nu_{\text{fr}})(1 - 2\nu_{\text{fr}})}\varepsilon_{\text{vol}}\delta_{ij} + \frac{E_{\text{fr}}}{1 + \nu_{\text{fr}}}\varepsilon_{ij} + \frac{E_{\text{fr}}}{1 - 2\nu_{\text{fr}}}\alpha_{\text{T}}(T - T_0)\delta_{ij} + \alpha p_{\text{f}}\delta_{ij} \tag{D.47}$$

$$= 2\frac{G_{\text{fr}}\nu_{\text{fr}}}{1-2\nu_{\text{fr}}}\varepsilon_{\text{vol}}\delta_{ij} + 2G_{\text{fr}}\varepsilon_{ij} + 2G_{\text{fr}}\frac{1+\nu_{\text{fr}}}{1-2\nu_{\text{fr}}}\alpha_{\text{T}}\left(T-T_0\right)\delta_{ij} + \alpha p_{\text{f}}\delta_{ij} \quad \text{(D.48)}$$

$$2G_{\text{fr}}\varepsilon_{ij} = \sigma_{ij} - 3\frac{\nu_{\text{fr}}}{1+\nu_{\text{fr}}}\overline{\sigma}\delta_{ij} - 2G_{\text{fr}}\alpha_{\text{T}}\left(T-T_0\right)\delta_{ij} - \frac{1-2\nu_{\text{fr}}}{1+\nu_{\text{fr}}}\alpha p_{\text{f}}\delta_{ij} \quad \text{(D.49)}$$

D.1.5 The force balance equation

The basic force balance equation is

$$\sum_j \frac{\partial \sigma_{ji}}{\partial x_j} + \rho f_i = 0 \quad \text{(D.50)}$$

Assuming homogeneous isotropic thermoporoelasticity and neglecting body forces, the equation may be written in terms of the displacements as

$$\frac{1}{1-2\nu_{\text{fr}}}\sum_k \frac{\partial^2 u_k}{\partial x_j \partial x_k} + \sum_i \frac{\partial^2 u_j}{\partial x_i^2} + \frac{\alpha}{G_{\text{fr}}}\frac{\partial p_{\text{f}}}{\partial x_j} + 2\frac{1+\nu_{\text{fr}}}{1-2\nu_{\text{fr}}}\alpha_{\text{T}}\frac{\partial T}{\partial x_j} = 0 \quad \text{(D.51)}$$

In coordinate-independent form, this is

$$\frac{1}{1-2\nu_{\text{fr}}}\nabla\left(\nabla\cdot\vec{u}\right) + \nabla^2\vec{u} + \frac{\alpha}{G_{\text{fr}}}\nabla p_{\text{f}} + 2\frac{1+\nu_{\text{fr}}}{1-2\nu_{\text{fr}}}\alpha_{\text{T}}\nabla T = 0 \quad \text{(D.52)}$$

or, alternatively,

$$\nabla\left(\nabla\cdot\vec{u}\right) - \frac{1-2\nu_{\text{fr}}}{2\left(1-\nu_{\text{fr}}\right)}\nabla\times\nabla\times\vec{u} + \frac{1-2\nu_{\text{fr}}}{2\left(1-\nu_{\text{fr}}\right)}\frac{\alpha}{G_{\text{fr}}}\nabla p_{\text{f}} + \frac{1+\nu_{\text{fr}}}{1-\nu_{\text{fr}}}\alpha_{\text{T}}\nabla T = 0 \quad \text{(D.53)}$$

D.1.6 Transverse isotropy

See Section 1.7.2 for an explanation of the notation.

$$E_{\text{v}} = C_{33} - \frac{C_{13}^2}{C_{11}-C_{66}} \quad \text{(D.54)}$$

$$E_{\text{h}} = 4C_{66}\frac{(C_{11}-C_{66})C_{33}-C_{13}^2}{C_{11}C_{33}-C_{13}^2} \quad \text{(D.55)}$$

$$\nu_{\text{vh}} = \frac{C_{13}}{2(C_{11}-C_{66})} = \frac{C_{13}}{C_{11}+C_{12}} \quad \text{(D.56)}$$

$$\nu_{\text{hh}} = \frac{(C_{11}-2C_{66})C_{33}-C_{13}^2}{C_{11}C_{33}-C_{13}^2} = \frac{C_{12}C_{33}-C_{13}^2}{C_{11}C_{33}-C_{13}^2} \quad \text{(D.57)}$$

$$\nu_{hv} = \frac{2C_{66}C_{13}}{C_{11}C_{33} - C_{13}^2} \tag{D.58}$$

$$\frac{E_h}{E_v} = \frac{\nu_{hv}}{\nu_{vh}} = 4C_{66}\frac{C_{11} - C_{66}}{C_{11}C_{33} - C_{13}} \tag{D.59}$$

$$\frac{\nu_{hv}}{E_h} = \frac{\nu_{vh}}{E_v} = \frac{C_{13}}{2\left[(C_{11} - C_{66})C_{33} - C_{13}^2\right]} \tag{D.60}$$

$$\frac{1}{E(\theta)} = \frac{\cos^4\theta}{E_v} + \frac{\sin^4\theta}{E_h} + \left(\frac{1}{G_v} - 2\frac{\nu_{vh}}{E_v}\right)\cos^2\theta\sin^2\theta \tag{D.61}$$

$$C_{11} = \frac{E_h}{1+\nu_{hh}}\frac{E_v - \nu_{vh}^2 E_h}{(1-\nu_{hh})E_v - 2\nu_{vh}^2 E_h} \tag{D.62}$$

$$C_{33} = \frac{E_v}{1 - 2\frac{\nu_{vh}^2}{1-\nu_{hh}}\frac{E_h}{E_v}} \tag{D.63}$$

$$C_{12} = C_{11} - 2C_{66} = \frac{E_h}{1+\nu_{hh}}\frac{\nu_{hh}E_v + \nu_{vh}^2 E_h}{(1-\nu_{hh})E_v - 2\nu_{vh}^2 E_h} \tag{D.64}$$

$$C_{13} = \frac{\nu_{vh}E_v E_h}{(1-\nu_{hh})E_v - 2\nu_{vh}^2 E_h} \tag{D.65}$$

$$C_{44} = G_v \tag{D.66}$$

$$C_{66} = \frac{1}{2}(C_{11} - C_{12}) = \frac{E_h}{2(1+\nu_{hh})} \tag{D.67}$$

$$C_{11} + C_{12} = 2(C_{11} - C_{66}) \tag{D.68}$$

The isostrain and isostress bulk moduli defined in Section 1.7.3 are given by

$$\begin{aligned} K^{\varepsilon} &= \frac{1}{9}\left[C_{33} + 4(C_{11} - C_{66} + C_{13})\right] = \frac{1}{9}\left[2C_{11} + C_{33} + 2C_{12} + 4C_{13}\right] \\ &= \frac{1}{9}\frac{(1-\nu_{hh})E_v + 2E_h(1+2\nu_{vh})}{1 - \nu_{hh} - 2\nu_{vh}^2\frac{E_h}{E_v}} \end{aligned} \tag{D.69}$$

$$\begin{aligned} K^{\sigma} &= \frac{(C_{11} - C_{66})C_{33} - C_{13}^2}{C_{11} + C_{33} - C_{66} - 2C_{13}} = \frac{C_{33}(C_{11} + C_{12}) - 2C_{13}^2}{C_{11} + C_{12} + 2C_{33} - 4C_{13}} \\ &= \frac{E_v}{1 + 2(1-\nu_{hh})\frac{E_v}{E_h} - 4\nu_{vh}} \end{aligned} \tag{D.70}$$

A polycrystalline aggregate of randomly oriented TI elements appears as isotropic on length scales large relative to the individual elements. The Voigt and Reuss bounds (see Section 6.2) give upper and lower bounds for the isotropic properties.

For the bulk modulus, the Voigt and Reuss bounds are equal to the isostrain and isostress bulk moduli above, respectively. For the shear modulus, the bounds are (e.g. Antonangeli et al., 2005)

$$G_V = \frac{1}{30}(7C_{11} + 2C_{33} - 5C_{12} - 4C_{13} + 12C_{44}) \tag{D.71}$$

$$G_R = \frac{15A}{4(C_{11} + C_{12}) + 8C_{13} + \frac{6A}{C_{44}} + \frac{12A}{C_{11}-C_{12}} + 2C_{33}} \tag{D.72}$$

where

$$A = C_{33}(C_{11} + C_{12}) - 2C_{13}^2 \tag{D.73}$$

These formulas must clearly be used with caution. They assume completely random orientation of the elements and also perfect bonding between them: no weaknesses at the element boundaries.

Note that $A > 0$ according to the stability criteria (Section C.11).

D.2 Elastic wave propagation in rocks

D.2.1 Correction for non-laminar flow

The function $F(\kappa)$ introduced in Section 5.3.1 is defined as

$$F(\kappa) = \frac{1}{4}\frac{\kappa\, T(\kappa)}{1 - \frac{2}{\mathrm{j}\kappa}T(\kappa)} \tag{D.74}$$

with

$$T(\kappa) = \frac{\mathrm{ber}'(\kappa) + \mathrm{j}\,\mathrm{bei}'(\kappa)}{\mathrm{ber}(\kappa) + \mathrm{j}\,\mathrm{bei}(\kappa)} \tag{D.75}$$

where ber and bei are the so-called Kelvin functions defined by

$$\mathrm{ber}(\kappa) + \mathrm{j}\,\mathrm{bei}(\kappa) = J_0(\mathrm{j}\sqrt{\mathrm{j}}\,\kappa) \tag{D.76}$$

J_0 is the zeroth-order Bessel function of the first kind.

D.2.2 Reflection, transmission and conversion coefficients at non-normal incidence

The coefficients for reflection, transmission and conversion for an incoming P-wave are given as solutions of the equation (Aki and Richards, 1980; Berkhout, 1987)

$$\bar{M}\vec{X} = \vec{B} \tag{D.77}$$

where

$$\vec{X} = \begin{pmatrix} r_{\mathrm{pp}} \\ t_{\mathrm{pp}} \\ r_{\mathrm{sp}} \\ t_{\mathrm{sp}} \end{pmatrix} \tag{D.78}$$

$$\vec{B} = \begin{pmatrix} -(1-2\sin^2\theta_{\mathrm{rs}}) \\ 2\left(\frac{v_{\mathrm{s1}}}{v_{\mathrm{p1}}}\right)^2 \sin\theta_{\mathrm{i}}\cos\theta_{\mathrm{i}} \\ -\sin\theta_{\mathrm{rs}} \\ \rho_2 v_{\mathrm{p2}}\cos\theta_{\mathrm{i}} \end{pmatrix} \tag{D.79}$$

$$\bar{M} = \begin{pmatrix} 1-2\sin^2\theta_{\mathrm{rs}} & -(1-2\sin^2\theta_{\mathrm{ts}}) & -2\sin\theta_{\mathrm{rs}}\cos\theta_{\mathrm{rs}} & -2\sin\theta_{\mathrm{ts}}\cos\theta_{\mathrm{ts}} \\ 2\left(\frac{v_{\mathrm{s1}}}{v_{\mathrm{p1}}}\right)^2 \sin\theta_{\mathrm{i}}\cos\theta_{\mathrm{i}} & 2\left(\frac{v_{\mathrm{s2}}}{v_{\mathrm{p2}}}\right)^2 \sin\theta_{\mathrm{t}}\cos\theta_{\mathrm{t}} & 1-2\sin^2\theta_{\mathrm{rs}} & -(1-2\sin^2\theta_{\mathrm{ts}}) \\ \sin\theta_{\mathrm{rs}} & -\frac{\rho_1 v_{\mathrm{s1}}}{\rho_2 v_{\mathrm{s2}}}\sin\theta_{\mathrm{ts}} & \cos\theta_{\mathrm{rs}} & \frac{\rho_1 v_{\mathrm{s1}}}{\rho_2 v_{\mathrm{s2}}}\cos\theta_{\mathrm{ts}} \\ \rho_2 v_{\mathrm{p2}}\cos\theta_{\mathrm{i}} & \rho_1 v_{\mathrm{p1}}\cos\theta_{\mathrm{t}} & -\rho_2 v_{\mathrm{p2}}\sin\theta_{\mathrm{i}} & \rho_1 v_{\mathrm{p1}}\sin\theta_{\mathrm{t}} \end{pmatrix} \tag{D.80}$$

The incoming P-wave is entering through medium 1. The amplitude of the reflected P-wave is r_{pp}, the amplitude of the transmitted P-wave is t_{pp}, the amplitude of the reflected and converted wave is r_{sp}, and the amplitude of the transmitted and converted wave is t_{sp}. The other parameters are defined in Section 5.7.

D.3 Rock models

D.3.1 Stresses at a crack tip

A crack tip may experience three types of loading, commonly denoted Mode I (opening), Mode II (in-plane shear) and Mode III (out-of-plane shear), respectively. The three loading modes are illustrated in Fig. D.1.

For an elliptical crack, the stresses at the crack tip can be expressed as

$$\sigma_{ij} = \frac{K_{\mathrm{M}}}{\sqrt{2\pi r}} f_{ij}^{(\mathrm{M})}(\theta) \tag{D.81}$$

where M = I, II or III. K_{M} represents the stress intensity factor of mode M and depends on the far-field stress and the crack length (as shown in Eq. (6.71) for K_{I}).

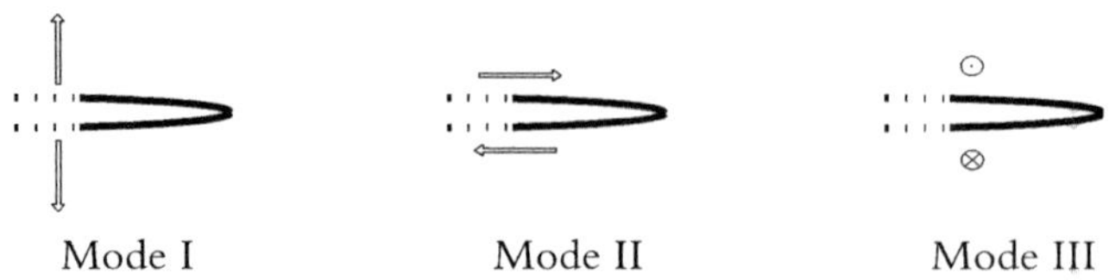

Figure D.1 The three modes of loading for a crack tip.

The functions $f_{ij}^{(M)}(\theta)$ define the angular dependence of the stress at the crack tip for Mode M. The angle θ, the distance r and the coordinate axes are defined in Fig. 6.10. Axis 3 is the out-of-plane axis. The non-zero components of $f_{ij}^{(M)}(\theta)$ are as follows.

Mode I:

$$f_{11}^{(I)}(\theta) = \cos\frac{\theta}{2}\left[1 - \sin\frac{\theta}{2}\sin\frac{3\theta}{2}\right] = \frac{3}{4}\cos\frac{\theta}{2} + \frac{1}{4}\cos\frac{5\theta}{2} \tag{D.82}$$

$$f_{22}^{(I)}(\theta) = \cos\frac{\theta}{2}\left[1 + \sin\frac{\theta}{2}\sin\frac{3\theta}{2}\right] = \frac{5}{4}\cos\frac{\theta}{2} - \frac{1}{4}\cos\frac{5\theta}{2} \tag{D.83}$$

$$f_{12}^{(I)}(\theta) = \sin\frac{\theta}{2}\cos\frac{\theta}{2}\cos\frac{3\theta}{2} = \frac{1}{4}\sin\frac{5\theta}{2} - \frac{1}{4}\sin\frac{\theta}{2} \tag{D.84}$$

For plane strain conditions, $f_{33}^{(I)}(\theta) = \nu(f_{11}^{(I)}(\theta) + f_{22}^{(I)}(\theta))$, whereas for plane stress conditions, $f_{33}^{(I)}(\theta) = 0$ (ν is the Poisson ratio).

Mode II:

$$f_{11}^{(II)}(\theta) = -\sin\frac{\theta}{2}\left[2 + \cos\frac{\theta}{2}\cos\frac{3\theta}{2}\right] = -\frac{7}{4}\sin\frac{\theta}{2} - \frac{1}{4}\sin\frac{5\theta}{2} \tag{D.85}$$

$$f_{22}^{(II)}(\theta) = \sin\frac{\theta}{2}\cos\frac{\theta}{2}\cos\frac{3\theta}{2} \tag{D.86}$$

$$f_{12}^{(II)}(\theta) = \cos\frac{\theta}{2}\left[1 - \sin\frac{\theta}{2}\sin\frac{3\theta}{2}\right] \tag{D.87}$$

For plane strain conditions, $f_{33}^{(II)}(\theta) = \nu(f_{11}^{(II)}(\theta) + f_{22}^{(II)}(\theta))$, whereas for plane stress conditions, $f_{33}^{(II)}(\theta) = 0$.

Mode III:

$$f_{13}^{(III)}(\theta) = -\sin\frac{\theta}{2} \tag{D.88}$$

$$f_{23}^{(III)}(\theta) = \cos\frac{\theta}{2} \tag{D.89}$$

For a more comprehensive description, see, for instance, Anderson (2005).

D.3.2 Self-consistent model for composite media

Berryman (1980) introduced self-consistency into the dynamic theory of Kuster and Toksöz (1974) and derived the following set of implicit equations for the effective moduli:

$$\sum_{i=1}^{N} c_i(K_i - K^*)P^{*i} = 0 \tag{D.90}$$

$$\sum_{i=1}^{N} c_i(G_i - G^*)Q^{*i} = 0 \tag{D.91}$$

$$\sum_{i=1}^{N} c_i(\rho_i - \rho^*) = 0 \tag{D.92}$$

where c_i is the volumetric concentration of component i in the material, ρ^* is the total density, and the coefficients P^{*i} and Q^{*i} are given as follows:

For spherical inclusions,

$$P^{*i} = \frac{3K^* + 4G^*}{3K_i + 4G^*} \tag{D.93}$$

$$Q^{*i} = \frac{G^* + F^*}{G_i + F^*} \tag{D.94}$$

where

$$F^* = \frac{G^*(9K^* + 8G^*)}{6(K^* + 2G^*)} \tag{D.95}$$

For penny-shaped inclusions (resembling e.g. flat cracks),

$$P^{*i} = \frac{3K^* + 4G_i}{3K_i + 4G_i + 3\pi\gamma\beta^*} \tag{D.96}$$

$$Q^{*i} = \frac{1}{5}\left[1 + \frac{8G^*}{4G_i + \pi\gamma(G^* + 2\beta^*)} + 2\frac{3K_i + 2G_i + 2G^*}{3K_i + 4G_i + 3\pi\gamma\beta^*}\right] \tag{D.97}$$

where

$$\beta^* = \frac{G^*(3K^* + G^*)}{3K^* + 4G^*} \tag{D.98}$$

Although the solution for ρ^* is trivial, explicit solutions for K^* and G^* are not known; however, solutions can be found by numerical iteration.

D.4 Solids production

D.4.1 Critical drawdown for turbulent flow

The critical drawdown for a semispherical sand arch in a non-Darcy flow regime is given in implicit form by (Ong et al., 2000)

$$\frac{S_1 + 3S_2}{m+1}\left(q_a\right)^{-\frac{m}{m+1}} = \frac{4\sin\varphi}{1-\sin\varphi} \tag{D.99}$$

where

$$S_1 = \frac{2(\sqrt{1+h_s(q_b - q_a)} - 1)}{h_s\left(1 - \frac{R_c}{R_e}\right)} \tag{D.100}$$

$$S_2 = \frac{(\sqrt{1+h_s(q_b - q_a)} - 1)^2}{h_s\left(1 - \left(\frac{R_c}{R_e}\right)^3\right)} \tag{D.101}$$

$$h_s = \frac{4k^2\beta\gamma\left(1 - \left(\frac{R_c}{R_e}\right)^3\right)}{3R_c(m+1)\left(\eta_f\left(1 - \frac{R_c}{R_e}\right)\right)^2}\left(\frac{S_0}{\tan\varphi}\right)^{m+1} \tag{D.102}$$

$$q_a = \left(\frac{(p_{fo} - \mathcal{D}_c)\tan\varphi}{S_0}\right)^{m+1} \tag{D.103}$$

$$q_b = \left(\frac{p_{fo}\tan\varphi}{S_0}\right)^{m+1} \tag{D.104}$$

and R_e is the radius of the drainage area. Eq. (D.99) is derived from the assumption that the fluid flow follows the Forchheimer equation (Forchheimer, 1901)

$$-\nabla p_f = \frac{\eta_f Q}{kA} + \beta\rho_f\left(\frac{Q}{A}\right)^2 \tag{D.105}$$

where the non-Darcy flow coefficient β is a constant of dimension m^{-1}. For $\beta = 0$, Eq. (D.105) is reduced to the Darcy equation (Eq. (1.308)), and Eq. (D.99) is reduced to Eq. (10.37).

D.5 Subsidence

Geertsma's formulas for the deformation and stress fields *outside* a disk-shaped reservoir:

$$u_r = \frac{c_m R h \Delta p_f}{2}\left[\mathrm{I}_1(|z-D|) + (3-4\nu)\mathrm{I}_1(z+D) - 2z\mathrm{I}_2(z+D)\right] \tag{D.106}$$

$$u_z = \frac{c_m R h \Delta p_f}{2}\left[\frac{z-D}{|z-D|}\mathrm{I}_3(|z-D|) - (3-4\nu)\mathrm{I}_3(z+D) - 2z\mathrm{I}_4(z+D)\right] \tag{D.107}$$

$$\sigma_r = G c_m R h \Delta p_f\left\{\left[\mathrm{I}_4(|z-D|) + 3\mathrm{I}_4(z+D) - 2z\mathrm{I}_6(z+D)\right] - \frac{1}{r}\left[\mathrm{I}_1(|z-D|) + (3-4\nu)\mathrm{I}_1(z+D) - 2z\mathrm{I}_2(z+D)\right]\right\} \tag{D.108}$$

$$\sigma_\theta = G c_m R h \Delta p_f\left\{4\nu\mathrm{I}_4(z+D) + \frac{1}{r}\left[\mathrm{I}_1(|z-D|) + (3-4\nu)\mathrm{I}_1(z+D) - 2z\mathrm{I}_2(z+D)\right]\right\} \tag{D.109}$$

$$\sigma_z = G c_m R h \Delta p_f\left[-\mathrm{I}_4(|z-D|) + \mathrm{I}_4(z+D) + 2z\mathrm{I}_6(z+D)\right] \tag{D.110}$$

$$\tau_{rz} = G c_m R h \Delta p_f\left[-\frac{z-D}{|z-D|}\mathrm{I}_2(|z-D|) - \mathrm{I}_2(z+D) + z\mathrm{I}_7(z+D)\right] \tag{D.111}$$

where

$$\mathrm{I}_1(q) = \frac{2}{\pi\sqrt{mrR}}\left[\left(1-\frac{m}{2}\right)K(m) - E(m)\right] \tag{D.112}$$

$$\mathrm{I}_2(q) = \frac{q\sqrt{m}}{2\pi(rR)^{3/2}}\left[\frac{1-\frac{m}{2}}{1-m}E(m) - K(m)\right] \tag{D.113}$$

$$\mathrm{I}_3(q) = -\frac{q\sqrt{m}K(m)}{2\pi\sqrt{rR}R} + (\mathrm{U}(r-R)-\mathrm{U}(R-r))\frac{\Lambda_0(\beta|m)}{2R} + \frac{1}{R}\mathrm{U}(R-r) \tag{D.114}$$

$$\mathrm{I}_4(q) = \frac{m^{3/2}\left(R^2-r^2-q^2\right)E(m)}{8\pi(rR)^{3/2}R(1-m)} + \frac{\sqrt{m}K(m)}{2\pi\sqrt{rR}R} \tag{D.115}$$

$$\mathrm{I}_6(q) = \frac{q m^{3/2}}{8\pi R(1-m)(rR)^{3/2}} \times \left\{ 3E(m) + m\frac{R^2 - r^2 - q^2}{rR} \left[\frac{1-\frac{m}{2}}{1-m} E(m) - \frac{1}{4} K(m) \right] \right\} \quad \text{(D.116)}$$

$$\mathrm{I}_7(q) = -\frac{1}{q}\mathrm{I}_2(q) + \frac{q^2 m^{3/2}}{8\pi(1-m)(rR)^{5/2}} \times \left[\frac{1-m+m^2}{1-m} E(m) - \left(1 - \frac{1}{2}m\right) K(m) \right] \quad \text{(D.117)}$$

$$\Lambda_0(\beta|m) = \frac{2}{\pi}\left[E(m)F(\beta|1-m) + K(m)E(\beta|1-m) - K(m)F(\beta|1-m)\right]$$

$$m = \frac{4Rr}{q^2 + (r+R)^2}$$

$$\sin\beta = \frac{q}{\sqrt{q^2 + (R-r)^2}}$$

where $F(\alpha|m)$ and $E(\alpha|m)$ are elliptic integrals of the first and second kind of *parameter* m, respectively, $K(m) = F(\frac{\pi}{2}|m)$ and $E(m) = E(\frac{\pi}{2}|m)$ are complete elliptic integrals of the first and second kind, and U is the unit step function ($\mathrm{U}(x \geq 0) = 1$, $\mathrm{U}(x < 0) = 0$). The function $\Lambda_0(\beta|m)$ is called the Heuman lambda function; in particular, $\Lambda_0(\frac{\pi}{2}|m) = 1$.

When working with elliptic integrals, it is important to be aware that there exist many different conventions for the definitions of the arguments, even for the order of the arguments. In the convention used here the integrals are given by

$$F(\alpha|m) = \int_0^{\alpha} \frac{1}{\sqrt{1 + m\sin^2\theta}} \mathrm{d}\theta \quad \text{(D.118)}$$

and

$$E(\alpha|m) = \int_0^{\alpha} \sqrt{1 + m\sin^2\theta}\,\mathrm{d}\theta \quad \text{(D.119)}$$

The elliptic integrals are available in modern integrated mathematics computer programs, like Maple, Mathematica and others. Given the availability of such a program, it is an easy and quick task to code and study the consequences of Geertsma's equations.

In Table D.1, we give some numerical values for the integrals to make it easy to check an implementation. Full tables of the integrals are found in Geertsma (1973).

Table D.1 Some values for the Geertsma integrals $I_i(q)$.

r, q, R	I_1	I_2	I_3	I_4	I_6	I_7
0.2, 0.4, 1.2	0.0595689	0.0461035	0.565282	0.598431	0.477736	-0.0544387
0.4, 0.4, 1	0.162188	0.197103	0.592418	0.818683	1.15435	-0.0445251

D.6 Permeability of tubes

Laminar flow through tubes and slits may be described by an equivalent permeability. We give some basic formulas and refer to fluid mechanics textbooks such as Landau and Lifshitz (1987) for context and derivations.

Permeability of an elliptical tube with half-axes a and b:

$$k = \frac{a^2 b^2}{4\left(a^2 + b^2\right)} = \frac{b^2}{4} \frac{1}{1 + \frac{b^2}{a^2}} \tag{D.120}$$

Circular tube ($a = b$):

$$k = \frac{b^2}{8} \tag{D.121}$$

Thin elliptical fracture ($a \gg b$):

$$k = \frac{b^2}{4} \tag{D.122}$$

Thin rectangular slit of width $w = 2b$:

$$k = \frac{w^2}{12} = \frac{b^2}{3} \tag{D.123}$$

Circular annulus with outer radius r_o and inner radius r_i:

$$k = \frac{1}{8}\left(r_o{}^2 + r_i{}^2 - \frac{r_o{}^2 - r_i{}^2}{\ln \frac{r_o}{r_i}}\right) \tag{D.124}$$

By series expansion we find that for a thin annulus,

$$k = \frac{(r_o - r_i)^2}{12} \tag{D.125}$$

in accordance with Eq. (D.123).

As a numerical example, consider a circular tube with diameter 1 μm. Then $k = 31$ milliDarcy. A tube with diameter equal to typical human hair thickness of 75 μm gives $k = 180$ Darcy.

D.7 Vector operators in cylindrical coordinates

The gradient of a scalar function:

$$\nabla f = \frac{\partial f}{\partial r}\vec{e}_r + \frac{1}{r}\frac{\partial f}{\partial \theta}\vec{e}_\theta + \frac{\partial f}{\partial z}\vec{e}_z \tag{D.126}$$

The Laplacian of a scalar function:

$$\nabla^2 f = \frac{1}{r}\frac{\partial f}{\partial r} + \frac{\partial^2 f}{\partial r^2} + \frac{1}{r^2}\frac{\partial^2 f}{\partial \theta^2} + \frac{\partial^2 f}{\partial z^2} \tag{D.127}$$

$$= \frac{1}{r}\frac{\partial}{\partial r}r\frac{\partial f}{\partial r} + \frac{1}{r^2}\frac{\partial^2 f}{\partial \theta^2} + \frac{\partial^2 f}{\partial z^2} \tag{D.128}$$

The divergence of a vectorial function:

$$\nabla \cdot \vec{u} = \frac{u_r}{r} + \frac{\partial u_r}{\partial r} + \frac{1}{r}\frac{\partial u_\theta}{\partial \theta} + \frac{\partial u_z}{\partial z} \tag{D.129}$$

$$= \frac{1}{r}\frac{\partial}{\partial r}(r u_r) + \frac{1}{r}\frac{\partial u_\theta}{\partial \theta} + \frac{\partial u_z}{\partial z} \tag{D.130}$$

The Laplacian of a vectorial function:

$$(\nabla^2 \vec{u})_r = \frac{\partial^2 u_r}{\partial r^2} + \frac{1}{r}\frac{\partial u_r}{\partial r} - \frac{u_r}{r^2} + \frac{1}{r^2}\frac{\partial^2 u_r}{\partial \theta^2} - \frac{2}{r^2}\frac{\partial u_\theta}{\partial \theta} + \frac{\partial^2 u_r}{\partial z^2} \tag{D.131}$$

$$(\nabla^2 \vec{u})_\theta = \frac{\partial^2 u_\theta}{\partial r^2} + \frac{1}{r}\frac{\partial u_\theta}{\partial r} - \frac{u_\theta}{r^2} + \frac{1}{r^2}\frac{\partial^2 u_\theta}{\partial \theta^2} + \frac{2}{r^2}\frac{\partial u_r}{\partial \theta} + \frac{\partial^2 u_\theta}{\partial z^2} \tag{D.132}$$

$$(\nabla^2 \vec{u})_z = \frac{\partial^2 u_z}{\partial r^2} + \frac{1}{r}\frac{\partial u_z}{\partial r} + \frac{1}{r^2}\frac{\partial^2 u_z}{\partial \theta^2} + \frac{\partial^2 u_z}{\partial z^2} \tag{D.133}$$

The curl of a vectorial function:

$$(\nabla \times \vec{u})_r = \frac{1}{r}\frac{\partial u_z}{\partial \theta} - \frac{\partial u_\theta}{\partial z} \tag{D.134}$$

$$(\nabla \times \vec{u})_\theta = \frac{\partial u_r}{\partial z} - \frac{\partial u_z}{\partial r} \tag{D.135}$$

$$(\nabla \times \vec{u})_z = \frac{u_\theta}{r} + \frac{\partial u_\theta}{\partial r} - \frac{1}{r}\frac{\partial u_r}{\partial \theta} \tag{D.136}$$

$$= \frac{1}{r}\frac{\partial (r u_\theta)}{\partial r} - \frac{1}{r}\frac{\partial u_r}{\partial \theta} \tag{D.137}$$

The gradient of the divergence of a vectorial function:

$$(\nabla(\nabla\cdot\vec{u}))_r = -\frac{u_r}{r^2} + \frac{1}{r}\frac{\partial u_r}{\partial r} + \frac{\partial^2 u_r}{\partial r^2} - \frac{1}{r^2}\frac{\partial u_\theta}{\partial \theta} + \frac{1}{r}\frac{\partial^2 u_\theta}{\partial r\partial\theta} + \frac{\partial^2 u_z}{\partial r\partial z} \tag{D.138}$$

$$= \frac{\partial}{\partial r}\frac{1}{r}\frac{\partial}{\partial r}(r u_r) - \frac{1}{r^2}\frac{\partial u_\theta}{\partial\theta} + \frac{1}{r}\frac{\partial^2 u_\theta}{\partial r\partial\theta} + \frac{\partial^2 u_z}{\partial r\partial z} \tag{D.139}$$

$$(\nabla(\nabla\cdot\vec{u}))_\theta = \frac{1}{r^2}\frac{\partial u_r}{\partial\theta} + \frac{1}{r}\frac{\partial^2 u_r}{\partial r\partial\theta} + \frac{1}{r^2}\frac{\partial^2 u_\theta}{\partial\theta^2} + \frac{1}{r}\frac{\partial^2 u_z}{\partial\theta\partial z} \tag{D.140}$$

$$(\nabla(\nabla\cdot\vec{u}))_z = \frac{1}{r}\frac{\partial u_r}{\partial z} + \frac{\partial^2 u_r}{\partial r\partial z} + \frac{1}{r}\frac{\partial^2 u_\theta}{\partial\theta\partial z} + \frac{\partial^2 u_z}{\partial z^2} \tag{D.141}$$

The curl of the curl of a vectorial function:

$$(\nabla\times\nabla\times\vec{u})_r = \frac{1}{r^2}\frac{\partial u_\theta}{\partial\theta} + \frac{1}{r}\frac{\partial^2 u_\theta}{\partial r\partial\theta} - \frac{1}{r^2}\frac{\partial^2 u_r}{\partial\theta^2} - \frac{\partial^2 u_r}{\partial z^2} + \frac{\partial^2 u_z}{\partial r\partial z} \tag{D.142}$$

$$(\nabla\times\nabla\times\vec{u})_\theta = \frac{u_\theta}{r^2} - \frac{1}{r}\frac{\partial u_\theta}{\partial r} - \frac{\partial^2 u_\theta}{\partial r^2} - \frac{1}{r^2}\frac{\partial u_r}{\partial\theta} + \frac{1}{r}\frac{\partial^2 u_r}{\partial r\partial\theta} \tag{D.143}$$

$$+ \frac{1}{r}\frac{\partial^2 u_z}{\partial\theta\partial z} - \frac{\partial^2 u_\theta}{\partial z^2} \tag{D.144}$$

$$(\nabla\times\nabla\times\vec{u})_z = \frac{1}{r}\frac{\partial u_r}{\partial z} + \frac{\partial^2 u_r}{\partial r\partial z} - \frac{1}{r}\frac{\partial u_z}{\partial r} - \frac{\partial^2 u_z}{\partial r^2} \tag{D.145}$$

$$- \frac{1}{r^2}\frac{\partial^2 u_z}{\partial\theta^2} + \frac{1}{r}\frac{\partial^2 u_\theta}{\partial\theta\partial z} \tag{D.146}$$

References

Aki, K., Richards, P.G., 1980. Quantitative Seismology. W.H. Freeman and Co., New York.

Anderson, T.L., 2005. Fracture Mechanics. Fundamentals and Applications, third edition. Taylor & Francis. (Note: Fourth edition 2017).

Antonangeli, D., Krisch, M., Fiquet, G., Badro, J., Farber, D.L., Bossak, A., Merkel, S., 2005. Aggregate and single-crystalline elasticity of hcp cobalt at high pressure. Phys. Rev. B 72, 134303. https://doi.org/10.1103/PhysRevB.72.134303.

Berkhout, A.J., 1987. Applied Seismic Wave Theory. Elsevier, Amsterdam.

Berryman, J.G., 1980. Long-wavelength propagation in composite elastic media. II. Ellipsoidal inclusions. J. Acoust. Soc. Am. 68, 1820–1831. https://doi.org/10.1121/1.385172.

Forchheimer, P., 1901. Wasserbewegung durch Boden. Z. Ver. Dtsch. Ing. 45, 1782–1788.

Geertsma, J., 1973. A basic theory of subsidence due to reservoir compaction: the homogeneous case. Verh. K. Ned. Geol. Mijnbouwkd. Genoot. 28, 43–62.

Kuster, G.T., Toksöz, M.N., 1974. Velocity and attenuation of seismic waves in two-phase media: Part 1. Theoretical formulations. Geophysics 39 (5), 587–606. https://doi.org/10.1190/1.1440450.

Landau, L.D., Lifshitz, E.M., 1987. Fluid Mechanics, second edition. Pergamon Press.

Ong, S., Ramos, G.G., Zheng, Z., 2000. Sand production prediction in high rate, perforated and open-hole gas wells. SPE58721. In: SPE International Symposium on Formation Damage Control. Lafayette, LA, USA.

APPENDIX E

Abbreviations

AE	Acoustic Emission
API	American Petroleum Institute
ASR	Anelastic Strain Recovery
AV	Annular Velocity
AVO	Amplitude Versus Offset
BHA	Bottom Hole Assembly
BHTV	Borehole TeleViewer
CEC	Cation Exchange Capacity
CHOPS	Cold Heavy Oil Production with Sand
CMP	Common Midpoint Gather
CMS	Constant Mean Stress test
CSL	Critical State Line
CT	Computer Tomography
CWT	Continuous Wave Technique
CU	Consolidated Undrained
DEM	Discrete Element Method
DIF	Drilling Induced Fracture
DWVA	Differential Wave Velocity Analysis
ECD	Equivalent Circulating Density
EGS	Enhanced Geothermal Systems
EOR	Enhanced Oil Recovery
ESD	Equivalent Static Density
FBP	Fracture Breakdown Pressure
FCP	Fracture Closure Pressure
FEM	Finite Element Method
FIP	Fracture Initiation Pressure
FIT	Formation Integrity Test
FPP	Fracture Propagation Pressure
GOR	Gas Oil Ratio
GPS	Global Positioning System
HPHT	High Pressure High Temperature
ISIP	Instantaneous Shut-In Pressure
ISRM	International Society for Rock Mechanics
KGD	Khristianovitch–Geertsma–de Klerk

continued on next page

Developments in Petroleum Science, Volume 72
ISSN 0376-7361, https://doi.org/10.1016/B978-0-12-822195-2.00026-7

LCM	Lost Circulation Material
LEDO	Low Energy Drilling Operations
LEFM	Linear Elastic Fracture Mechanics
LOP	Leak Off Pressure
LOT	Leak Off Test
LVDT	Linear Variable Differential Transformer
MHF	Massive Hydraulic Fracturing
MPD	Managed Pressure Drilling
MSL	Mean Sea Level
MWD	Measurement While Drilling
NMO	Normal MoveOut
PFC	Particle Flow Code
PKN	Perkins–Kern–Nordgren
PWD	Pressure While Drilling
REV	Representative Elementary Volume
ROP	Rate Of Penetration
RMS	Root Mean Square
RT	Rotary Table (drillfloor level)
SEM	Scanning Electron Microscope
SG	Specific Gravity; density relative to water
SRV	Stimulated Reservoir Volume
sst	Sandstone
TI	Transverse Isotropy
TIF	Thermally Induced Fracturing
TWC	Thick-Walled Cylinder
TWT	Two-Way Traveltime
UCS	Uniaxial (Unconfined) Compressive Strength
UST	Uniaxial Strain Test
VSP	Vertical Seismic Profiling
WIFF	Wave-Induced Fluid Flow
XLOT	Extended Leak-Off Test
XRD	X-Ray Diffraction
YP	Yield Point (mud property)

APPENDIX F

List of symbols

This is a list of most of the symbols used in this book. General mathematical symbols, some basic symbols such as x, r and so on, and a few symbols used only where they are defined are not included.

The order is alphabetical, in the order of symbols, script letters, Greek letters and Latin letters. Capital letters precede lowercase letters. Where applicable, the list shows the number of the first equation where a symbol is used.

$\stackrel{\text{def}}{=}$	(1.159)	Define something
$\sim$	(3.15)	Roughly equal to, coarser than $\approx$
$\mathcal{A}$	(2.9)	Attraction
$\mathcal{D}$	(10.4)	Drawdown pressure
$\mathcal{D}_c$	(10.7)	Critical drawdown pressure
$\mathcal{I}_2$	(1.75)	Second invariant of strain
$\mathcal{I}_3$	(1.76)	Third invariant of strain
$\mathcal{J}_2$	(1.118)	Second deviatoric invariant of strain
$\Delta\Pi$	(9.14)	Osmotic potential
Δt_c	(5.26)	Compressional transit time
Δt_{cr}	(10.47)	Sand production cycle time
Δt_s		Shear transit time
Ψ	(2.76)	Angle of dilatancy
α	(5.43)	Acoustic attenuation
α	(1.168)	Biot's poroelastic parameter
$\overline{\alpha}$	(1.279)	Biot's poroelastic parameter, mean
α_{SF}	(5.48)	Acoustic attenuation, solid friction
α_T	(1.121)	Coefficient of linear thermal expansion
α_{TV}		Coefficient of volumetric thermal expansion
α_{Tf}	(1.231)	Coefficient of linear thermal expansion, fluid
α_{Th}	(5.67)	Thomsen's alpha
α_{Tu}	(1.234)	Undrained coefficient of linear thermal expansion
$\alpha_{scatt.}$	(5.49)	Acoustic attenuation, scattering
β	(6.74)	Brown Korringa factor
β	(2.10)	Failure angle
β	(D.102)	Non-Darcy flow coefficient

continued on next page

Developments in Petroleum Science, Volume 72
ISSN 0376-7361, https://doi.org/10.1016/B978-0-12-822195-2.00027-9

β_{Th}	(5.68)	Thomsen's beta
β_{w}	(2.84)	Failure angle, weak plane
χ	(6.79)	Effective stress coefficient for porosity
χ	(1.339)	Visco-elastic coefficient
δ	(5.46)	Acoustic loss tangent
δ_{Th}	(5.66)	Thomsen's delta
$\boldsymbol{\varepsilon}$	(1.244)	Strain matrix
$\varepsilon_1^{\mathrm{p}}$	(2.69)	First principal plastic strain
$\varepsilon_2^{\mathrm{p}}$		Second principal plastic strain
$\varepsilon_3^{\mathrm{p}}$	(2.70)	Third principal plastic strain
ε^{F}	(6.84)	Fracture strain
ε_{H}	(12.3)	Horizontal strain
$\varepsilon_{\mathrm{Th}}$	(5.64)	Thomsen's epsilon
$\varepsilon_{\mathrm{cr}}$	(1.331)	Creep strain
$\dot{\varepsilon}_{\mathrm{cr}}$	(1.335)	Creep strain rate
ε_{h}	(12.3)	Horizontal strain
ε_{ij}	(1.77)	Strain, general form
$\varepsilon_{ij}^{\mathrm{e}}$	(2.62)	Elastic strain, general form
$\varepsilon_{ij}^{\mathrm{p}}$	(2.62)	Plastic strain, general form
ε_{v}	(12.3)	Vertical strain
$\varepsilon_{\mathrm{vol}}^{\mathrm{p}}$	(2.72)	Volumetric plastic strain
$\varepsilon_{\mathrm{vol}}$	(1.74)	Volumetric strain
$\varepsilon_{\mathrm{vol,f}}$	(1.133)	Volumetric strain of fluid
$\varepsilon_{\mathrm{vol,p}}$	(1.142)	Volumetric strain of pore
$\varepsilon_{\mathrm{vol,s}}$	(1.133)	Volumetric strain of solid
$\varepsilon_x^{\mathrm{o}}$	(5.33)	Strain amplitude
ϕ	(1.134)	Porosity
φ	(2.8)	Angle of internal friction
φ_{w}	(2.84)	Angle of internal friction, weak plane
γ	(6.41)	Crack aspect ratio
γ	(2.19)	Angle describing the slope of the failure line in the principal stress plot
γ	(10.38)	Gas density coefficient
γ	(2.52)	Surface tension
γ_{Th}	(5.65)	Thomsen's gamma
γ^{p}	(2.77)	Generalised plastic shear strain
γ	(12.17)	Stress path parameter
γ'	(12.17)	Effective stress path parameter
$\overline{\gamma}$	(12.49)	Mean stress path parameter
$\overline{\gamma'}$	(12.30)	Mean effective stress path parameter
γ_{H}	(12.16)	Stress path parameter
γ_{h}	(12.16)	Stress path parameter
γ_{h}'	(12.19)	Effective stress path parameter, horizontal
γ_{v}	(12.16)	Stress path parameter

continued on next page

γ'_v	(12.19)	Effective stress path parameter, vertical
η_p	(2.87)	Patchiness
η	(4.53)	The poroelastic stress coefficient
η_L	(2.37)	Lade criterion parameter
η_f	(1.308)	Dynamic viscosity
η_w		Dynamic viscosity of water
κ	(2.78)	Hardening parameter
κ	(1.338)	Creep parameter
κ_T	(1.330)	Thermal conductivity
κ	(12.47)	Overburden stress path parameter
κ'	(7.14)	Effective stress path parameter
λ	(2.64)	Plastic strain parameter
λ	(1.101)	Lamé's modulus
λ_{fr}	(1.184)	Lamé modulus, frame (drained)
λ_p	(4.167)	Plasticity lambda parameter
λ_{sand}	(10.41)	Sand production coefficient
λ_u	(1.156)	Lamé modulus, undrained
λ_w	(5.7)	Wavelength
μ	(1.305)	Coefficient of internal friction
μ		Creep parameter
$\nu_{h,fr}$	(1.293)	TI: in plane Poisson's ratio, frame (drained)
$\nu_{v,fr}$	(1.293)	TI: out of plane Poisson's ratio, frame (drained)
ν	(1.95)	Poisson's ratio
ν_{dyn}	(8.2)	Dynamic Poisson's ratio
ν_{fr}	(1.221)	Drained Poisson's ratio
ν_{hh}	(1.265)	TI: in plane Poisson's ratio
ν_{hv}	(1.266)	TI: Poisson's ratio for in-plane loading out-of-plane deformation
ν_s	(6.22)	Poisson's ratio, solid material
ν_u	(1.221)	Undrained Poisson's ratio
ν_{vh}	(1.265)	TI: out of plane Poisson's ratio
θ_i	(5.86)	Angle of incidence
θ_r		Angle of reflection
θ_{rs}	(5.88)	Angle of "reflected" S-wave
θ_t	(5.86)	Angle of refracted P-wave
θ_{ts}	(5.89)	Angle of "refracted" S-wave
ϑ_L	(1.54)	The Lode angle
ρ	(1.13)	Density
ρ_{dry}	(12.68)	Density of dry rock
ρ_f	(5.27)	Density, fluid
ρ_g	(5.74)	Density of gas
ρ_s	(5.27)	Density, solid material
ρ_w	(5.94)	Density of borehole fluid
ρ_w	(5.74)	Density of water

continued on next page

$\overline{\sigma}$	(1.40)	Mean stress
$\boldsymbol{\sigma}$	(1.244)	Stress matrix
σ	(1.3)	Normal stress
σ^*	(1.341)	Creep parameter
σ	(9.18)	Membrane efficiency
σ'_{ij}	(1.191)	Effective stress in ij-direction
σ'	(2.2)	Effective stress
$\overline{\sigma}'$	(1.187)	Mean effective stress
σ'_1	(2.1)	Major effective principal stress
σ'_2	(2.1)	Intermediate effective principal stress
σ'_3	(2.1)	Minor effective principal stress
σ'_{v}	(3.3)	Effective vertical stress
$\sigma_{\theta \mathrm{i}}$	(7.16)	Hollow cylinder inner tang. stress
$\sigma_{\theta,\max}$	(4.107)	Tangential stress, maximum
$\sigma_{\theta,\min}$	(4.108)	Tangential stress, minimum
σ_1	(1.24)	Major principal stress
σ_2	(1.24)	Intermediate principal stress
σ_3	(1.37)	Minor principal stress
σ_{H}	(3.12)	Major horizontal principal stress
σ'_{H}	(9.4)	Effective major horizontal principal stress
σ_{c}	(4.178)	Stress on casing
σ_{d}	(1.207)	Stress difference $\sigma_p - p_{\mathrm{f}}$
σ_{h}	(4.45)	Minor horizontal principal stress
σ'_{h}	(3.3)	Effective minor horizontal principal stress
σ_{ij}	(1.123)	Stress in ij-direction
$\sigma_{\max\perp}$		Maximum stress normal to borehole
$\sigma'_{\max\perp}$	(9.7)	Effective maximum stress normal to borehole
$\sigma_{\min\perp}$		Minimum stress normal to borehole
$\sigma'_{\min\perp}$	(9.7)	Effective minimum stress normal to borehole
σ_{o}	(4.197)	Hydrostatic far-field stress
σ_{oct}	(1.55)	Octahedral normal stress
σ_p	(1.102)	External hydrostatic stress
σ'_p		Effective external hydrostatic stress
$\sigma'_{\parallel}$	(9.7)	Effective stress parallel to borehole axis
$\sigma_{r\mathrm{i}}$	(7.17)	Hollow cylinder inner radial stress
$\sigma_{r\mathrm{o}}$	(4.38)	Radial stress, outer
$\overline{\sigma}_{\mathrm{s}}$	(1.212)	Mean grain stress
σ_{v}	(3.1)	Vertical stress
σ_{y}	(1.340)	Yield stress
$\sigma_{z\mathrm{i}}$	(7.15)	Hollow cylinder inner axial stress
τ	(1.4)	Shear stress
τ_{D}	(1.326)	Diffusion time
τ_{c}	(1.305)	Critical shear stress
$\tau_{\max}$		Maximum shear stress

continued on next page

τ_{oct}	(1.56)	Octahedral shear stress
τ_{s}	(10.47)	Characteristic sand erosion time
υ	(2.81)	Specific volume
ω	(5.5)	Angular frequency
ξ	(1.304)	Crack density
ξ_{incl}	(6.33)	Density of inclusions
ζ	(1.139)	Biot's strain parameter
ζ^{o}	(5.34)	Biot's strain parameter, amplitude
A	(1.1)	Area
A	(1.215)	Skempton's A parameter
B	(1.214)	Skempton's B parameter
$\overline{B}$	(1.283)	Skempton's B parameter, mean
CEC	(3.26)	Cation exchange capacity
$\mathbf{C}_{\text{g}}$	(6.98)	Global stiffness matrix
$\mathbf{C}$	(1.245)	Stiffness matrix
C	(1.156)	Biot's C-parameter
C_0	(2.21)	Uniaxial compressive strength
C_{TWC}	(10.18)	Thick-walled cylinder strength
C_{dr}	(10.61)	Dimensionless drag coefficient
C_{ij}	(6.42)	Elastic coefficient, Voigt notation, $i, j = 1, \ldots, 6$
C^*_{ij}	(6.42)	Effective elastic coefficient, Voigt notation, $i, j = 1, \ldots, 6$
C^{fr}_{ij}	(1.292)	Elastic coefficient, drained, Voigt notation, $i, j = 1, \ldots, 6$
C^{u}_{ij}	(1.292)	Elastic coefficient, undrained, Voigt notation, $i, j = 1, \ldots, 6$
C_{ijkl}	(1.236)	Elastic coefficient, tensor notation, $i, j, k, l = 1, 2, 3$
C^{fr}_{ijkl}	(1.276)	Elastic coefficient, drained, tensor notation, $i, j, k, l = 1, 2, 3$
C^{u}_{ijkl}	(1.286)	Elastic coefficient, undrained, tensor notation, $i, j, k, l = 1, 2, 3$
C_{lo}	(11.24)	Fracture leak-off coefficient
C_{r}	(1.330)	Specific heat capacity, (saturated) rock
C_{s}	(11.46)	Specific heat capacity, rock matrix
C_{w}	(11.46)	Specific heat capacity, water
D	(3.1)	Depth
D_{T}	(1.329)	Thermal diffusivity
D_{c}	(6.39)	Drainage parameter for cracks
D_{cp}	(6.45)	Drainage parameter for cracks and pores (low freq)
D_{h}	(5.75)	Hydraulic diffusivity
D_{n}	(6.25)	Normal force constant (grain pack models)
D_{p}	(6.48)	Drainage parameter for pores
D_{t}	(6.26)	Tangential force constant (grain pack models)
E	(D.117)	Incomplete elliptic integral of the first kind
E	(1.94)	Young's modulus
E_{h}	(1.265)	TI: in plane Young's modulus
E_{v}	(1.265)	TI: out of plane Young's modulus
$E_{\text{h,fr}}$	(1.293)	TI: in plane Young's modulus, frame (drained)

continued on next page

$E_{v,fr}$	(1.293)	TI: out of plane Young's modulus, frame (drained)
E'	(1.112)	Plane strain modulus
E_{dyn}	(5.71)	Dynamic Young's modulus
E_{eff}	(1.304)	Effective Young's modulus
E_{fr}	(4.73)	Drained Young's modulus
E_s	(6.22)	Young's modulus, solid material
E_{sec}	(1.301)	Secant Young's modulus
E_{stat}	(5.71)	Static Young's modulus
E_{tan}	(1.302)	Tangential Young's modulus
F	(D.118)	Elliptic integral of the second kind
F	(1.1)	Force
F_d	(10.61)	Drag force
F_g	(10.60)	Net gravity force
F_h	(10.3)	Hydrodynamic force on a grain
F_n	(1.3)	Normal force
F_p	(1.4)	Parallel force
F_r	(10.1)	Force needed to remove a grain from the rock
G	(1.101)	Shear modulus
G_{fr}	(1.160)	Shear modulus of the frame (drained)
G_u		Shear modulus, undrained
G_s	(6.17)	Shear modulus, solid material
G_v	(1.265)	TI: out of plane shear modulus
H	(11.20)	Fracture height
H	(3.17)	Uniaxial compaction modulus
H_{fr}	(1.202)	Uniaxial compaction modulus
H_u	(1.316)	Undrained plane wave modulus
I_1	(1.43)	1st invariant of stress
I_1'	(2.37)	1st invariant of effective stress
I_2	(1.43)	2nd invariant of stress
I_3	(1.43)	3rd invariant of stress
I_3'	(2.37)	3rd invariant of effective stress
J_1	(1.47)	1st invariant of deviatoric stress
J_2	(1.47)	2nd invariant of deviatoric stress
J_3	(1.47)	3rd invariant of deviatoric stress
JCS	(6.90)	Wall strength
JRC	(6.90)	Joint roughness coefficient
K	(D.117)	Elliptic integral of the first kind
K'	(3.3)	Principal stress ratio
K	(1.102)	Bulk modulus
K^{ε}	(1.269)	Isostrain bulk modulus
K^{ε}_{fr}	(1.279)	Isostrain bulk modulus, drained (fr)
K^{ε}_{u}	(1.287)	Isostrain bulk modulus, undrained
K_{ϕ}	(1.207)	Inverse pore compressibility w.r.t. pore pressure at constant stress difference

continued on next page

K^{σ}	(1.272)	Isostress bulk modulus
K^{σ}_{fr}	(1.282)	Isostress bulk modulus, drained (fr)
K^{σ}_{u}	(1.290)	Isostress bulk modulus, undrained
K_{I}	(6.69)	Stress intensity factor, mode I
K_{IC}	(7.35)	Fracture toughness
K_{M}	(6.73)	Inverse total compressibility w.r.t. pore pressure at constant stress difference
$K_{\mathrm{b}p}$	(1.195)	Bulk modulus with respect to pore pressure
K_{dyn}	(5.70)	Dynamic bulk modulus
K_{eff}	(1.132)	Effective bulk modulus
K_{f}	(1.136)	Bulk modulus, fluid
$K_{\mathrm{f}i}$	(5.76)	Bulk modulus, fluid component i
$K_{\mathrm{f,eff}}$	(5.77)	Effective bulk modulus, fluid
K_{fr}	(1.159)	Drained bulk modulus
K_{g}	(5.73)	Bulk modulus of gas
K_{o}	(3.19)	Compaction modulus
K_{p}	(1.205)	Inverse pore compressibility w.r.t. confining pressure
$K_{\mathrm{p}p}$	(1.206)	Inverse pore compressibility w.r.t. pore pressure
K_{s}	(1.136)	Bulk modulus, solid material
K_{stat}	(5.70)	Static bulk modulus
K_{u}	(1.137)	Undrained bulk modulus
K_{w}	(5.73)	Bulk modulus of water
L	(11.14)	Fracture half length
$\mathbf{M}$	(C.106)	Bond M matrix
M	(1.156)	Biot's M-parameter (storage modulus)
$M^{\mathrm{T}}_{\mathrm{sand}}$	(10.59)	Total mass of produced sand in one event
$M^{\mathrm{o}}_{\mathrm{sand}}$	(10.57)	Initial mass of removed sand
M_{sand}	(10.51)	Total mass of produced sand
$\dot{M}_{\mathrm{sand}}$	(10.51)	Sand production rate
$\mathbf{N}$	(C.107)	Bond N matrix
N_{c}	(6.27)	Coordination number
Q	(1.335)	Activation energy
Q_{ij}	(6.42)	Inclusion impact parameter
Q	(6.43)	Inclusion impact matrix
Q_K	(6.33)	Inclusion impact parameter, bulk modulus
Q_G	(6.34)	Inclusion impact parameter, shear modulus
$\vec{Q}$	(1.308)	Flow rate
Q	(4.200)	Flow rate (scalar)
Q	(5.45)	Acoustic Q-factor
Q_{o}	(4.63)	Oil flow rate, scalar
R	(12.44)	Reservoir radius
R	(1.335)	Molar gas constant
R_{c}	(4.178)	Casing radius
R_{c}	(4.197)	Cavity radius

continued on next page

R_{co}	(10.53)	Initial cavity radius
R_{cool}	(11.46)	Radius of cooled region
R_{cr}	(10.45)	Critical radius for sand erosion
R_e	(4.70)	Outer radius for flow
R_{flood}	(11.46)	Radius of flooded region
R_i	(7.15)	Inner radius
R_o	(4.38)	Outer radius
R_p	(4.147)	Radius of the plastified zone
R_w	(4.45)	Well radius
Re	(10.62)	Reynolds number
$\mathbf{S}$	(1.252)	Compliance matrix
S	(1.199)	Specific storage
S_ε	(1.200)	Specific storage, constant strain
S_σ	(1.201)	Specific storage, constant stress
S_0	(1.202)	Specific storage, uniaxial strain
S_0	(1.305)	Cohesion
S_{0w}	(2.86)	Cohesion, weak plane
S_{ijkl}	(1.239)	Elastic compliance, tensor notation, $i, j, k, l = 1, 2, 3$
S^{fr}_{ijkl}	(1.289)	Elastic compliance, drained, tensor notation, $i, j, k, l = 1, 2, 3$
S^{u}_{ijkl}	(1.282)	Elastic compliance, undrained, tensor notation, $i, j, k, l = 1, 2, 3$
S_L	(2.38)	Lade criterion parameter
S_p	(11.26)	Spurt loss
S_w	(5.73)	Water saturation
S_{we}	(2.53)	Saturation of wetting fluid
S_i	(5.76)	Saturation of fluid component i
T	(1.121)	Temperature
T	(5.37)	Tortuosity (Biot theory)
T_i	(1.121)	Initial temperature
T_0	(2.2)	Tensile strength
T_a	(1.335)	Activation temperature
T_{cool}	(11.42)	Temperature of a cooled region
T_o	(4.73)	Far-field temperature
T_w	(4.73)	Well temperature
Tr	(C.18)	Trace of a matrix
U	(D.117)	Unit step function
U	(10.61)	Fluid velocity
U_t	(10.64)	Terminal settling velocity
V	(1.79)	Volume
V_{cl}	(3.26)	Volume fraction of clay
V_f	(1.133)	Fluid volume
V_{frac}	(11.21)	Fracture volume
V_{inj}	(11.21)	Fracture injected volume

continued on next page

V_{lo}	(11.21)	Fracture leak-off volume
V_{p}	(1.139)	Pore volume
V_{prod}	(12.54)	Produced volume
V_{s}	(1.133)	Solid volume
V_{sp}	(10.46)	Volume of sand producing zone
V_{tot}	(1.133)	Total volume
V_{w}	(9.14)	Molar volume of water
W	(1.114)	Strain energy
W_{c}	(6.65)	Strain energy due to a crack
W_{s}	(6.66)	Surface energy of a crack
a	(10.49)	Cavity growth parameter
a	(4.129)	Ellipse major semiaxis
a	(4.78)	Azimuth angle
a_{w}	(2.54)	Water activity
$a_{\mathrm{w,df}}$	(9.14)	Water activity in drilling fluid
$a_{\mathrm{w,sh}}$	(9.14)	Water activity in shale
b	(5.101)	Gutenberg–Richter *b*-value
b	(10.53)	Erosion-inducing fluid flux parameter
b	(4.129)	Ellipse minor semiaxis
c_{b}		Bulk compressibility
c_{ϕ}	(1.207)	Pore compressibility w.r.t. pore pressure at constant stress difference
c_{M}	(6.78)	Total compressibility w.r.t. pore pressure at constant stress difference
$c_{\mathrm{b}p}$	(1.196)	Bulk compressibility w.r.t. pore pressure
c_{f}	(1.138)	Fluid compressibility
c_{fr}	(1.174)	Drained compressibility, K_{fr}^{-1}
c_{pp}^{γ}	(12.50)	Pore compressibility w.r.t. pore pressure for a stress path
c_{gas}		Gas compressibility
c_{m}	(12.9)	Coefficient of uniaxial compaction
c_{p}	(1.205)	Pore compressibility w.r.t. confining pressure
$c_{\mathrm{p}p}$	(1.206)	Pore compressibility w.r.t. pore pressure
c_{s}	(1.138)	Compressibility, solid material, K_{s}^{-1}
c_{u}	(1.138)	Undrained compressibility, K_{u}^{-1}
d_{g}	(10.1)	Grain diameter
e	(1.335)	Euler's number e=2.71828...
e	(12.26)	Aspect ratio
e	(2.80)	Void ratio
f	(4.129)	Ellipse flattening
f	(5.6)	Frequency
f_i	(1.17)	Body force in the *i*-direction
g	(3.1)	Acceleration of gravity
g	(2.66)	Plastic potential
g_{pn}^{c}	(10.23)	Critical drawdown pressure gradient

continued on next page

g_{pn}	(10.22)	Normalised drawdown pressure gradient
h	(12.4)	Reservoir thickness
i	(4.78)	Inclination angle
j	(5.5)	Imaginary unit, $\sqrt{-1}$
k	(1.308)	Permeability
k	(4.149)	Triaxial strength factor
$\hat{l}$	(5.13)	Direction cosine vector
l_D	(1.326)	Diffusion length
l_i		Direction cosine i-axis
m	(1.336)	Creep exponent
m_b	(2.89)	Hoek–Brown constant
m_{sand}	(10.40)	Mass of produced sand from a volume element
p'	(2.56)	Mean effective stress
p^*	(2.28)	Hydrostatic compressive strength
p_{cp}	(2.51)	Capillary suction
p_{frac}	(11.13)	Fracturing pressure
p_f	(1.140)	Fluid pressure
p_{fn}	(3.2)	Normal fluid pressure i.e. corresponding to a water column
p_{fo}	(4.50)	Far-field fluid pressure
$p_{w,max}^{frac}$	(4.118)	Maximum well pressure, fracture initiation
$p_{w,max}^{fracprop}$	(9.9)	Maximum well pressure, fracture propagation
p_{net}	(11.13)	Net fracturing pressure
p_{nw}	(2.51)	Fluid pressure, non-wetting fluid
p_{tip}	(11.40)	Fracture tip pressure drop
p_{tip}^{cr}	(11.41)	Critical fracture tip pressure drop
p_w	(4.38)	Well pressure
p_{we}	(2.51)	Fluid pressure, wetting fluid
$p_{w,min}$	(4.116)	Minimum well pressure
q	(1.48)	Generalised shear stress
q	(5.5)	Wavenumber
q_{fl}^{cr}	(10.41)	Critical fluid flux
$\vec{q}_{fl}$	(1.148)	Fluid flux vector
q_{fl}	(10.41)	Fluid flux
r	(1.49)	Stress invariant
r_L		The Lode r-coordinate
r_{pp}	(5.82)	pp reflection coefficient
r_{sp}	(D.78)	ps "reflection" coefficient
t'	(4.109)	Dimensionless time
t_T	(5.95)	Two-way travel time
t_a	(10.55)	Accumulated cavity growth parameter
t^*	(6.52)	Relaxation time for flow between cracks and pores
t_{inj}	(11.21)	Fracture pumping time
Q_{inj}	(11.21)	Fracture pumping rate

continued on next page

t_{pp}	(5.83)	pp transmission coefficient
t_{sp}	(D.78)	pp "refraction" coefficient
u	(1.59)	Displacement in the negative x (or r) direction
u^F	(6.81)	Relative shear displacement of fracture surfaces
$\vec{u}_f^{fr}$	(1.147)	Fluid "frame" displacement vector
u_f^{fr}	(5.27)	Fluid "frame" displacement
$\vec{u}_s^{fr}$	(1.146)	Solid frame displacement vector
u_s^{fr}	(5.27)	Solid frame displacement
$\vec{d}_g$	(6.98)	Global displacement vector
$\vec{F}_g$	(6.98)	Global force vector
v	(1.59)	Displacement in the negative y (or θ) direction
v	(5.8)	Sound velocity
v_{group}	(5.10)	Group velocity
v_p	(3.21)	P-wave velocity
v_{pa}		P-wave velocity of altered medium
v_s	(5.19)	S-wave velocity
v_w	(5.93)	Velocity of borehole fluid
v_{sa}		S-wave velocity of altered medium
w	(1.59)	Displacement in the negative z direction
$\overline{w}$	(11.27)	Mean fracture width
w	(11.28)	Fracture width
w^F	(6.81)	Relative normal displacement of fracture surfaces
w_o^F		Initial fracture opening
w_{max}^F	(6.82)	Fracture opening at maximum closure
w_{max}	(11.29)	Maximum fracture width
w_s	(6.66)	Surface energy per unit area
w_{sw}		Surface energy per unit area, solid water interface
z_L		The Lode z-coordinate

Index

Symbols

π-plane, 16, 94, 111

A

Abnormal pressures, 172, 501
Abnormally pressured zone, 171
Absorption, 272
Acoustic
  emission, 314, 413
  impedance, 295
  logging, 429
    while drilling, 433
  measurements, 386
    drill cuttings, 433
    small samples, 408
  transducer, 387
  wave, 264
Activation energy, 82
Active solids control, 520
Activity, *see* Chemical activity
AE (Acoustic Emission), 314
Amontons' law, 97
Anelastic strain recovery, 164, 412
Angle
  contact, 378
  critical, 296
  dilatancy, 132
  Euler, 216
  failure, 99
  friction, 98, 175, 179, 182, 186, 188
  Lode, 15, 16, 111
  of incidence, 296
  of internal friction, 98
Angular frequency (def.), 261
Anisotropic
  horizontal stresses, 251, 481, 563
  material, 54, 56, 61, 143
  models, 339
  poroelasticity, 62
  rock, 59
  shale, 489
  Skempton, 237
  stress, 61, 237, 641
  field, 210, 391
Anisotropy, 30, 54, 55, 62, 182, 183, 190, 194, 240, 277, 378, 489
  elliptical, 281
  intrinsic, 55, 143
  lithological, 55
  rock, 487, 489
  shale, 195
  stress induced, 55, 70
  structural, 143
  weak, 280
Antiplane strain, 23
Arching, 604, 607, 610
  coefficient, 606, 610
Argillaceous mineral, *see* Clay minerals
Aspect ratio, 338, 609
Associated flow, 129
Attenuation, 271
Attraction, 98, 246
Average modulus
  definition of, 393
Average velocity, 313
AVO (Amplitude Versus Offset), 313
Axial stress, 391

B

Backus average, 321
Barefoot completion, 520
Baushinger effect, 134
Bed of nails, 332, 359, 360
Biaxial testing, 391
Bifurcation, 142, 394
  point, 142
Biot coefficient
  anisotropic, 63
  definition of, 44
  determination, 397
Biot flow, 270, 274
Biot tensor, 63, 696
Biot theory, 35, 37, 319
  stiff frame limit (def.), 42
  wave equation, *see* Wave equation Biot theory

weak frame limit (def.), 43
Biot–Gassmann equation, 42, 326, 605
Blowout, 555
BO (breakouts), 463
Body force, 6
Bolivar, 602
Bond matrices, 696
Borehole
acoustics, 299
alteration, 308
collapse, 407, 461, 464
elliptical, 232
elongation, 233, 238, 440
failure, 224–226, 238, 240, 471
criteria, 224
time-delayed, 223, 478
instability, 461, 463, 474
logs, 439
modes, 302
permeable, 560
stability, 217, 391, 461, 462, 469
analysis, 487, 489
problem, 461, 495
televiewer (BHTV), 441
vertical, 217, 224, 230
Bound water, 122, 177, 190, 356, 377
Bounds, 324
Hashin–Shtrikman, 324
Reuss, 35, 324, 353
Voigt, 324, 353
Brazilian test, 415
Breakout, 238, 240–243, 439, 463
Breckels–van Eekelen correlation, 453
Brie model, 290
Brinell hardness, 416
Brittle deformation, 90
Brittleness, 591
Brown–Korringa model, 33, 40, 48, 355
Buckling, 142
Bulk modulus, 392, 395
definition of, 26
generalised, 61, 699
isostrain, 61
isostress, 62
undrained, 54
Burgers substance, 83

C

Caliper log, 440
Cantilever, 386
Capillary
effect, 118
forces, 289
pressure, 119, 378
suction, 119, 380
Carbonate, 175, 182, 186, 503
Carter leak-off, 576
Cased borehole, 250
Casing, 250, 491, 509
collapse, 495
damage, 649, 651
failure, 649, 651
shear, 651
Catastrophic sand production, 522, 546
Cation exchange capacity, 420, 421
Causality, 276
Cementation, 179–181
processes, 181, 327
Chalk, 178, 186, 188, 551, 552
fluid effects, 380
North Sea, 188
porosity, 187
preparation, 379
production, 519, 551
Chemical
activity, 121, 421, 485
consolidation, 521
effects, 293
CHOPS (Cold Heavy Oil Production with Sand), 520
Christoffel
equation, 277
matrix, 278
Clastic sediment, 178
Clay, 175
minerals, 176, 177, 181, 190, 191, 377, 462, 484, 489
normally consolidated, 137
strongly overconsolidated, 137
Closure pressure, 582
CMP gather, 312
CMP stack, 312
CO_2, 288, 289, 291, 293, 601, 630, 656
Coal, 506

Coefficient
- arching, 606, 610
- Biot, 44
- compaction, 604
- consolidation, 76
- drag, 548
- fluid loss, 576
- gas density, 539
- joint roughness, 361
- leak-off, 576
- of consolidation, 402
- of friction, 70
- of internal friction, 97
- poroelastic stress, 208, 228, 606
- reflection, 295
- sand production, 541
- Skempton, 50, 140, 172, 396
- stress path, 606
- thermal expansion, 31, 589
- transmission, 295

Cohesion, 97

Collapse, 461
- pressure, 531, 551

Common midpoint, 312

Compaction, 103, 179, 601
- band, 142, 186, 536
- coefficient, 604
- criterion, 104
- delayed, 615
- drive, 632, 633
- ellipsoidal reservoir, 608
- failure, 103
- reservoir, 601, 603
- uniaxial, 603, 604, 608
- volumetric, 610
- well problems, 648

Compatibility conditions, 22

Complex moduli, 276

Compliance
- matrix, 58
- tensor, 56

Compressibility
- definition, 26
- pore volume, 48
- uniaxial, 604

Compressional stress, 416

Compressional wave, 263, 431

Compton scattering, 438

Confining fluid, 384

Confining pressure, 391

Consolidated-undrained test, 399

Consolidation, 72, 73, 180, 383, 399, 479, 482, 614
- coefficient of, 76

Constant mean stress, 404
- test, 403

Constant net stress test, 397

Constant stress ratio tests, 403

Contact angle, 378

Continuous failure state test, 404

Continuous wave technique, 408, 433

Converted waves, 297

Cooled vs. flooded zone, 589

Coordination number, 329

Core
- alteration, 375
- depth, 381
- discing, 376
- freezing, 377
- handling, 377
- orientation, 381
- preparation, 378
- representativeness, 374
- sandstone, 377
- shale, 377
- size, 374
- wrapping, 377

Crack density, 69, 338

Crack growth, 348–350
- subcritical, 80

Cracks, 69, 94, 101, 115, 271, 275, 285, 335, 375, 413

Creep, 72, 80, 150, 159, 276, 383, 483
- accelerating, 80
- primary, 80
- process, 504, 511
- rate, 82, 504
- rheological models, 83
- secondary, 80
- steady state, 80
- tertiary, 80
- transient, 80

Criterion
  borehole failure, 224
  compaction, 104
  Drucker–Prager, 109, 124
  Griffith, 101
    extended, 110, 124
    modified, 102
  Hoek–Brown, 148
  modified Lade, 110, 124
  Mogi–Coulomb, 110, 489
  Mohr–Coulomb, 97, 225, 244, 404
  tensile failure, 95, 230, 334, 472, 524
  Tresca, 97, 98, 108, 131, 243, 530
  von Mises, 109, 124
  yield, 127
Critical angle, 296
Critical drawdown, 526, 527
  pressure gradient, 533
Critical effective pressure, 104
Critical porosity, 325
Critical state line, 139
Critical state soil mechanics, 135
Cross-dipole, 432
Crushing, 650
  limit, 650
  pressure, 104
Cusiana, 495
Cuttings, 501

D

Damage surface, 414
Darcy, definition of unit, 73
Darcy's law, 73, 210
Deep water drilling, 462, 497
Deformation
  ductile, 90
  plastic, 126, 128, 512, 530, 551, 611
Dehydration, 377
Delayed compaction, 615
Delayed subsidence, 615
Density log, 438
Depleting sphere, 605
Depletion, 527
Depth correction, 493
Depth reference, 493
Determinant, 680
Deviatoric stress, 13–15
Diagenesis, 180
DIF (drilling induced fractures), 439, 463
Differential effective medium theory, 346
Differential sticking, 464
Differential strain curve analysis, 410
Differential stress, 46
Differential wave velocity analysis, 413
Diffusion
  constant, 76
  equation, 76
  length, 77
Dilatancy, 70, 130
Dilatancy angle, 132
Dip-angle, 164
Dipole log, 307
Dipole transducer, 432
Direction cosines, 11, 214
Discrete particle modelling, 333
Dispersion, 262
  acoustic, 389
  relation, 268
Displacement, 16
  control, 383
  discontinuity, 343
  radial, 605
  rate, 399
Dix inversion, 312
Dog leg, 466
Drag coefficient, 548
Drained stiffness, 75
Drained test, 39, 135, 138, 395
  triaxial, 392
Drawdown, 526
  critical, 526, 527
Drill cuttings, 408, 417, 433, 435, 501
Drilling, 461
  data, 436, 455
  fluid additives, 484
  induced fractures, 441, 463
  infill, 506, 602, 648
  stable, 462, 466
  underbalanced, 472, 480
Drucker–Prager criterion, 109
  in the (p', q) plane, 124
Dry rock, 380
Ductile, 68, 126
  deformation, 90
Dynamic moduli (def.), 281

Dynamic vs. static moduli, *see* Static vs. dynamic moduli
Dynamic well pressure, 470

E

Earthquakes, 314
ECD (equivalent circulating density), 463, 468, 470, 480, 498, 500
Effective medium, 319
  self-consistent, 343
Effective stress, 14, 92, 117, 170
  definition, 44
  generalised, 379
  history, 86
  principle
    acoustic velocities, 287
  Terzaghi, 46
Eigenvalue, 682
Einstein summing convention, 699
Ekofisk, 163, 164, 188, 196, 602, 607, 649, 653, 656
Elastic moduli, 429
  anisotropy, 56
  definition, 25
  frame modulus, 39
  relations between, 27
  secant modulus, 68
  second-order, 29
  suspension, 34
  tangent modulus, 68
  third-order, 29, 67, 444
  unloading, 68
Elastic stability, 30, 56, 697
Elastic tensor, *see* Stiffness tensor
Elastic wave, 259, 264
Elasticity, 1
  linear, 1, 24
  non-linear, 66
  perfect, 67
Elements, 365
Ellipse flattening, 232
Ellipsoidal reservoir, 608
Elliptical
  anisotropy, 281
  borehole, 232
  microcracks, 101
ELOT, *see* Extended leak-off test
End cap, 104
  in the (p', q) plane, 125
Engineering shear strain, 20
Equation
  Biot–Gassmann, 42, 326, 605
  Christoffel, 277
  diffusion, 76
  Gassmann, 42
  Kirsch, 214
  Kozeny–Carman, 524, 639
  wave, 260
  Wood's, 35
  Wyllie, 266
  Young–Laplace, 378
Equation of equilibrium, 7
  Cartesian coordinates, 8
  cylindrical coordinates, 203
Equivalent circulating density (ECD), 468, 470
Equivalent core diameter, 415
ESD (equivalent static density), 463, 468, 480
Euler angles, 216, 687
  conventions, 687
Extended Griffith criterion, 110
  in the (p', q) plane, 124
Extended leak-off test, 445, 476, 557
Extension test, 404

F

Failure, 89
  angle, 99
  envelope, 96, 391, 404
  line, 96
  rock, 89, 92
  shear, 92, 95, 236, 471, 526
  surface, 93
Failure criteria
  compaction, *see* Compaction criterion
  Drucker–Prager, *see* Drucker–Prager criterion
  extended Griffith, *see* Extended Griffith criterion
  Griffith, *see* Griffith criterion
  Hoek–Brown, *see* Hoek–Brown criterion
  modified Griffith, *see* Modified Griffith criterion

modified Lade, *see* Modified Lade criterion
Mogi–Coulomb, *see* Mogi–Coulomb criterion
Mohr–Coulomb, *see* Mohr–Coulomb criterion
Tresca, *see* Tresca criterion
von Mises, *see* von Mises criterion
Fatigue failure, 151
Fault, 164
Faulting, 612
FBP (fracture breakdown pressure), 478, 507
FCP (fracture closure pressure), 447
FEM (Finite Element Method), 365
Filter cake, 563
Finite element method, 365, 629
FIP (fracture initiation pressure), 447, 478
FIT, *see* Formation integrity test
Flexural wave, 307, 432, 443
Flood directionality, 637
Flow rule, 127
Fluid framework, 37
Fluid loss coefficient, 576
Fluid properties, 670
Forchheimer equation, 712
Formation breakdown, 558
Formation integrity test, 451
Formation strength, 373
FPP (fracture propagation pressure), 447
Frac pack, 522, 556
Fracture, 357, 374
aperture, 568
closure, 582
pressure, 447, 455, 477
confinement, 565, 567
elliptical hole, 235
extension pressure, 568
geometry, 574
gradient, 462, 476, 557
growth, 565
initiation, 235, 558
pressure, 444, 447, 478
orientation, 565
penny-shaped, 567
pressure, 569
propagation pressure, 447
radial, 580
test, 444
wireline tools, 452
tip, 568
toughness, 350, 419, 420, 564, 565, 572, 580
width, 577, 578
Fracture test, 444
wireline tools, 452
Fractured
reservoir, 361, 364, 569, 613
reservoirs, 569
rocks, 148, 357
Fracturing, 557, 567, 574
hydraulic, 226, 555, 556
shale, 563, 585, 590
wedge effect, 572
Frame, 35
Frame modulus, 319
definition of, 39
Framework, 35, 37
Free water, 377
Friction
angle, 175, 179, 182, 186, 188
definition of, 98
coefficient of, 70

G

Gamma ray, 381, 438
Gas density coefficient, 539
Gassmann equation, *see* Biot–Gassmann equation
Gassmann–Hill model, 291
Gassmann–Wood model, 289
Geertsma nucleus of strain model, 616
Generalised bulk modulus, 61
Generalised plane strain, 23
Generalised shear stress, 14, 124
Global stiffness matrix, 365
Goose Creek, 602
Grain
compressibility, 41, 398
material, 35
mean stress, 49
salt, 196
shape, 175
size, 175

distribution, 381
sorting, 175
Gravel packing, 520
Green–Lagrange strain, 690
Griffith crack, 349
Griffith criterion, 101
Groningen, 602, 654, 655
Group velocity, 262
Gullfaks, 173
Gumbo shale, 465
Gutenberg–Richter law, 314

H

Haigh–Westergaard coordinates, *see* Lode coordinates
Haigh–Westergaard stress space, *see* Principal stress space
Hard rock, 42
Hardening rule, 127
Hardness tests, 416
Hashin–Shtrikman bounds, 324
Heidrun, 497
Heim's rule, 159
Hertz–Mindlin theory, 329
Hertzian contact, 328
Hod, 173
Hoek cell, 384
Hoek–Brown criterion, 148
Hole cleaning, 464
Hollow cylinder, 203
test, 405
Hooke's law, 25, 603
cylindrical coordinates, 202
Hoop stress, 206
Horizontal stress, 158, 410
direction, 439
empirical relations, 453
magnitude, 442, 444
HPHT (high pressure, high temperature), 539
Hvorslev surface, 140
Hydraulic diffusivity, 76
Hydraulic fracturing, 226, 555, 556
massive, 555
Hydrodynamic forces, 524
Hydrogeology, 47
Hydrostat, 94
Hydrostatic pressure, 35
Hydrostatic stress state, 26
Hydrostatic test, 395
Hysteresis, 68

I

Ideally plastic material, 127
Identity matrix, 679
Image log
acoustical, 441
electrical, 441
Impermeable borehole wall, 471, 558
Increment of fluid content, 35
Indentation tests, 417
Index tests, 414
Infill drilling, 506, 602, 648
Infinitesimal strain, 18
Inherent shear strength, 97
Initial modulus
definition of, 393
Injection, 611
In situ stress, *see* Stress in situ
Interface wave, 298
Interfacial tension, 378
Intermediate principal stress, 96, 106, 108, 405
Interval transit time, 266
Invariant
strain, 21
stress, 13
stress deviation, 14
ISIP (Instantaneous Shut-In Pressure), 583
Isostrain, 324
Isostrain bulk modulus, 61
Isostress, 35, 292, 324
Isostress bulk modulus, 62
Isotropic
hardening, 134
materials
definition of, 25
rock, 59
ISRM
standards, 380

J

Jacketed test, 39
Joint, 164, 357
Joint roughness coefficient, 361

K

Kaiser effect, 315
Kelvin substance, 84
Keyseat, 440, 466
KGD (Khristianovic–Geertsma–de Klerk) model, 574, 586
Kick, 467
Kinematic hardening, 134
Kirsch equations, 214
Kozeny–Carman equation, 524, 639
Kronecker's delta, 700

L

Laboratory testing, 373
Lamé modulus, 26, 39
Largest horizontal stress, 438
LCM (Lost Circulation Material), 503
Leak-off
  coefficient, 576
  point, 445
  pressure, 445, 476
  test, 445, 557
Limestone, 178
Linear elastic fracture mechanics (LEFM), 350, 420
Linear isotropic models, 336
Liquefaction, 143, 552
Liquefied rock, 143
Lithification, 180
Lithostatic stress, 159, 567
Load cell, 385
Load frame, 382
Loading pistons, 384
Loading rate, 151
Lobo Carneiro test, 415
Local flow, 271, 274
Localisation, 141
Lode angle, 15, 16, 111, 703
Lode coordinates, 15, 690
Longitudinal wave, 263
Lost circulation, 462, 464, 467, 555
LOT, *see* Leak-off test
LVDT, 386

M

Macroscopic stress, 327
Macroscopic theory, 37, 319
Managed pressure drilling (MPD), 498, 503, 509
Mandel–Cryer effect, 78
Mandel's problem, *see* Mandel–Cryer effect
Massive hydraulic fracturing (MHF), 555
Matrix
  addition, 678
  Christoffel, 278
  compliance, 58
  compressibility, 398
  definition, 677
  diagonal, 678
  diagonalisation, 683
  eigenvalue, 682
  global stiffness, 365
  identity, 679
  invariants
    general, 691
  inverse, 679
  multiplication, 678
  norm, 693
  orthogonal, 683
  similarity transform, 683
  singular, 679
  spur, 680
  stiffness, 58, 60, 61, 65, 278, 285
  symmetric, 678
  trace, 680
  transpose, 677
Maxwell substance, 84, 626
Mechanical properties
  drilling data, 436
  empirical correlations, 435
Mesh, 365
Methylene blue test, 421
Micro-frac test, 452
Microcracks, 55, 101, 115, 120, 184, 285, 388, 409
Microseismic monitoring, 314, 591, 593, 652, 656
Mineralogy, 381, 421
Mini-frac test, 451, 556
Mode I loading, 349, 709
Mode II loading, 709
Mode III loading, 709
Model
  Brie, 290

Brown–Korringa, 33, 40, 48, 355
Gassmann–Hill, 291
Gassmann–Wood, 289
Geertsma nucleus of strain, 616
Khristianovic–Geertsma–de Klerk, 574, 586
linear isotropic, 336
Perkins–Kern–Nordgren, 574, 578, 586
Modified Griffith criterion, 102
Modified Lade criterion, 110
in the (p', q) plane, 124
Modulus
average, 393
bulk, 26
generalised, 61
isostrain, 61
isostress, 62
undrained, 54
complex, 276
frame, 39, 319
initial, 393
Lamé, 26, 39
P-wave, 26, 265
plane strain, 28, 327, 577
plane wave, 26, 265
secant, 68, 393
shear, 26, 393
static, 282, 283, 285, 398
elastic, 394
storage, 47, 398
tangent, 393
uniaxial compaction, 26, 265, 402, 605
unloading, 68
Young's, 25
Mogi–Coulomb criterion, 110, 489
Mohr–Coulomb criterion, 97, 225, 244, 404
Mohr's circle, 10
3D, 12
Mohr's hypothesis, 96
Moment tensor inversion, 315
Monopole transducer, 432
MPD (Managed Pressure Drilling), 498, 503
Mud
filtrate, 375
loss, 464
oil-based, 450, 484
water-based, 378, 450, 483
weight, 462
Mud weight calibration, 502
Mud weight window, 462
Multiple failure state test, 404

N

Natural completion, 520
Net fracturing pressure, 564, 568
Net stress, 46
Node, 365
Non-associated plastic flow, 129, 132
Non-Darcy flow coefficient, 712
Non-laminar flow, 268, 708
Noordbergum effect, 78
Normal fault, 166, 438
Normal moveout, 312
velocity, 312
Normal strain, 17
Normalised drawdown pressure gradient, 533
Normally consolidated, 136, 400
North Sea chalk, 188
Nucleus of strain model, 616

O

Octahedral normal stress, 16
Octahedral plane, 16
Octahedral shear stress, 16
Octahedral stresses, 16
Oedometer
modulus, 26, 402
test, 400
Offset, 311
Oil-based mud, 450, 484
Oilfield, 2
Open hole completion, 520
Oriented core, 381, 409
Orthogonal matrix, 683
Orthogonal transform, 683
Osmosis, 173, 484
Overbalance, 472, 503
Overburden stress, 437, 438
Overconsolidated, 136, 139, 163, 400
Overconsolidation ratio, 136, 592
Overpressured zone, 171
Overthrust, 166

P

P-wave, 264
  modulus, 26, 265
  velocity, 265, 387
P&A (Plug and Abandonment), 451
Paleosols, 506
Paleostress, 163
Partial saturation, 118, 288, 537
Patchiness, 148
  parameter, 148
Patchy saturation, 290
Patchy weakness, 147
Penny-shaped fracture, 567
Perforation, 521
  oriented, 438, 529
Permanent deformation, 127
Permanent overpressure, 512
Permeability, 73
  effective stress for, 640
  shale, 171, 172, 399
  stress effects on, 636
Permeable borehole wall, 220, 228, 479
Phase
  of elastic wave, 261
  velocity, 261
PKN (Perkins–Kern–Nordgren) model, 574, 578, 586
Plane of weakness model, 144
Plane strain, 23, 204
  modulus, 28, 327, 577
Plane stress, 23, 24, 211
Plane wave modulus, 26, 265
Plastic
  deformation, 126, 128, 512, 530, 551, 611
  potential, 128
  strain, 126, 248, 530
  zone, 243
Plasticity, 126
  theory, 126
Point load strength, 415
Poisson's ratio, 30, 392
  definition of, 25
  dynamic, 265, 430
  undrained, 51
Polarisation, 264
Polyaxial, 391
Poorly consolidated, 180
Pore collapse, 92, 103, 188, 395
Pore pressure, 35, 158, 170, 601
  drillout-induced, 218
Pore volume compressibility, 48
  stress path, 632
  uniaxial, 633
Poroelastic stress coefficient, 208, 228, 606
Poroelasticity, 33
  analogy to thermoelasticity, 52
Porosity, 34, 35
  chalk, 187
  change during depletion, 634
  critical, 325
  effective stress for, 48, 635
  overburden correction, 634
  stress dependent, 633
Post failure, 383, 474, 523
Preconsolidation stress, 136, 163, 402
Pressure
  abnormal, 172, 501
  critical effective, 104
  fracture, 569
  gradient, 492
  hydrostatic, 35
  net fracturing, 564, 568
  pore, 35, 158, 170, 601
  rebound, 587
  solution, 181
  static, 470
Pressure transmission test, 422
Pressurisation rate, 560
Primary wave, 263
Principal axes
  of strain, 22
  of stress, 9, 10, 12
    orthogonal, 10
Principal strain, 22
Principal stress
  intermediate, 96, 106, 108, 405
  space, 93
Proper rotation, 683
Proppants, 522
Pseudo-Rayleigh wave, 303
PTT (Pressure Transmission Test), 422
Pump rate, 561

Pump-in/flowback test, 447, 585
Punch test, 418

Q

Quality factor (Q-factor), 272
Quartz, 62, 183, 354
Quasi-waves, 278

R

Radial
  displacement, 605
  fracture, 580
Radial flow
  stationary, 210, 249
Radius of curvature, 272
Rayleigh scattering, 275
Rayleigh wave, 298
Reflection coefficient, 295
Refraction, 296
Representative elementary volume (REV), 353
Reservoir
  compaction, 601, 603
  ellipsoidal, 608
  fractured, 361, 364, 569, 613
  geomechanics, 601
  monitoring, 648
    4D seismic, 644
  simulation, 643
    geomechanics, 643
  simulator, 643
  stress path, *see* Stress path
Reuss bound, 35, 324, 353
Reynolds number, 548
Rigid body
  rotation, 17
  translation, 17
Rock
  anisotropic, 59
  anisotropy, 487, 489
  creeps, 150
  dry, 380
  failure, 89, 92
  isotropic, 59
  liquefied, 143
  mass, 361
  salt, 196, 197, 504
  sedimentary, 55, 94, 134, 158, 175, 277, 319, 327, 353
  specimen, 375
Rock mechanical tests, 389
Root-mean-square velocity, 312
ROP (rate of penetration), 436
Roscoe surface, 139
Rotation matrix
  Euler angles, 688
Rotation of a rigid body, 17

S

S-wave, 264
  velocity, 387
Salt, 196, 504
  grains, 196
  rock, 196, 197, 504
Sample
  diameter, 380
  size, 380
  surface, 380
Sand, 175
  prediction, 391, 550
  production, 522, 549
    catastrophic, 522
    coefficient, 541
    continuous, 522
    rate, 541, 544, 545
    transient, 522, 531
  transport, 547
  unconsolidated, 117, 167
Sandstone, 178, 179, 182, 183, 186, 502
  fluid effects, 380
  fracturing, 563
  preparation, 379
  synthetic, 612
Scattering
  Compton, 438
  loss, 275
  Rayleigh, 275
Scholte wave, 299
Scratch test, 417
Screens, 521
Screw mode, 307
Secant modulus, 68, 393
Secondary wave, 264
Sedimentary rocks, 55, 94, 134, 158, 175, 277, 319, 327, 353

Seismic reservoir monitoring, 644, 645
Seismic resolution, 313
Seismicity, induced, 601, 603, 630, 648, 655, 656
Self-consistent, 343, 711
Shale, 178, 179, 182, 189, 435, 461, 483
  anisotropic, 489
  anisotropy, 195
  barrier, 509–511
  fracturing, 563, 585, 590
  gas, 556, 590, 593
  oil, 556, 590, 593
  permeability, 171, 172, 399
  preparation, 379
  saturation, 379
  sloughing, 464
  testing of, 391, 399
Shallow water flow, 498
Shear band, 141, 631
Shear eigenmodes wave, 303
Shear failure, 92, 95, 236, 471, 526
Shear modulus, 393
  definition of, 26
Shear strain, 18
Shear wave, 264, 431
Shear-enhanced compaction, 104
Shut-in/decline test, 446, 583
Sign convention
  strain, 18
  stress, 2
Skempton coefficients, 140, 172, 396
  anisotropic, 63, 237
  definition of, 50
Skempton tensor, 63, 696
Sleeve, 384
  fracturing, 452
Sloughing shale, 464
Slow formations, 307, 431
Slow wave, 271
Smallest horizontal stress, 437
  estimation of, 445, 453, 582
Soil mechanics, 135
Solid friction loss, 273
Solids production, 519, 551
Sonic pulse technique, 408, 433
Sonic tool
  cross-dipole, 432
  full waveform, 431
  long-spaced, 432
  multipole, 432
Sound velocity, 262, 387
  effect of temperature, 288
Spallings, 464
Specific storage, 47, 77
Specific surface area, 177, 421
Specific volume, 135
Spreading ridge, 161
Spurt loss, 576
Squirt flow, 274
Stable drilling, 462, 466
Stacking, 312
Static
  moduli, 282, 283, 285, 398
  pressure, 470
Static elastic moduli, 394
Static moduli (def.), 281
Static vs. dynamic moduli, 281, 282, 387, 430, 605
Static well pressure, 470
Stiff frame, 43
  limit, 77, 576
Stiff frame limit (def.), 42
Stiffness
  drained, 75
  system, 448, 586
  undrained, 75
Stiffness matrix, 58, 60, 61, 65, 278, 285, 697
Stiffness tensor, 56, 395, 686, 695
  inverse relationship, 56
Stimulated reservoir volume (SRV), 591
Stimulation, 555
Stokes' law, 548
Stoneley wave, 299, 302, 431
Storage modulus, 47, 398
Strain, 16
  amplitude, 430
  antiplane, 23
  cylindrical coordinates, 201
  definition of, 17
  energy, 28, 349
  finite, 71, 250
  from displacements
    cylindrical coordinates, 202

gauge, 386
generalised plane (def.), 23
hardening, 133
infinitesimal, 18
invariant, 21
natural, 72
normal
definition of, 17
plane, 23, 204
plastic, 126, 248, 530
principal, 22
rate, 383
shear, 18
tangential, 205
tensor, 20, 410
thermal, 31
true, 72
volumetric, 21
Strength, 89
anisotropy, 143
inherent shear, 70
parameters, 429, 434
Stress
anisotropic, 61, 237, 641
around borehole, 206, 469
general orientation, 216
in principal stress direction, 217
non-linear, 212
plastic, 243
time effects, 219
varying pore pressure, 208
at borehole wall
temperature effects, 212
varying pore pressure, 209
compressional, 416
concentration, 207, 222, 226, 284, 469, 523, 559, 606, 612, 630, 647
corrosion, 80
index, 151
cylindrical coordinates, 201, 689
definition of, 2
deviation, 14
deviator, 14
deviatoric, 13–15
deviatoric (def.), 13
differential, 46
effective, *see* Effective stress
evolution, 627, 647
generalised plane (def.), 24
history, 84, 94, 149, 160, 163, 286, 348, 358, 413
hoop, 206
horizontal, 158, 410
hydrostatic, 35
definition of, 26
in situ, 157, 409, 429, 437, 582
determination, field tests, 437
from laboratory tests, 409
intensity factor, 349, 350, 709
invariant, 13, 14, 703
level, 89
lithostatic, 159, 567
macroscopic, 327
memory, 315
net, 46
normal
definition of, 4
octahedral, 16
normal, 16
shear, 16
overburden, 437, 438
path, 93, 123, 137, 402
coefficient, 606
overburden, 626, 627
reservoir, 606
plane, 23, 24, 211
plane (def.), 24
polygon, 168
principal
from invariants, 703
principal (def.), 9
relaxation, 80
residual, 162
shadow, 570
shadowing, 570
shear (def.), 4
space, 93
structural, 162
tectonic, 161, 495
tensor, 5, 13, 689
thermal, *see* Thermoelastic stress, 31
thermoelastic, 31, 212, 482, 589
transformation formulas, 214, 689
vertical, 157, 437, 438

Stress–strain relation
anisotropic, 56
Biot theory, 38, 43
isotropic, 27, 28
isotropic linear elasticity, 25
linear thermoelasticity, 31
Strike, 164
Strike-slip
fault, 166
stress regime, 444
Stuck pipe, 461, 464, 467, 478, 495
Subcritical crack growth, 80, 81, 151
Subduction zone, 161
Subsidence, 601
bowl, 618
delayed, 615
disk shaped reservoir, 619
nucleus of strain model, 616
some examples, 621
Suction, 119
Summing convention, 699
Superposition principle, 209
Surface energy, 349, 377
Surge and swab, 499
Suspension, 34, 271, 288, 324
Swelling, 176, 422, 462, 483
clay, 177, 465
minerals, 462, 483
System stiffness, 448, 586

T

Tangent modulus
definition of, 393
Tangential strains, 205
Tectonic forces, 157
Tectonic plates, 161
Tectonic stress, *see* Stress tectonic
Tensile failure, 92, 94, 555, 557
criterion, 95, 230, 334, 472, 524
Tensile strength, 94, 185, 192, 197, 226, 415
chalk, 188
Tensor, 685
Terminal settling velocity, 548
Terzaghi effective stress, *see* Effective stress
Test
Brazilian, 415
consolidated-undrained, 399
constant mean stress, 403
constant net stress, 397
constant stress ratio, 403
continuous failure state, 404
drained, 39, 135, 138, 395
triaxial, 392
extended leak-off, 445, 476, 557
extension, 404
formation integrity, 451
fracture, 444
hardness, 416
hollow cylinder, 405
hydrostatic, 395
indentation, 417
jacketed, 39
leak-off, 445, 557
Lobo Carneiro, 415
methylene blue, 421
micro-frac, 452
mini-frac, 451, 556
multiple failure state, 404
pressure transmission, 422
pump-in/flowback, 447, 585
punch, 418
rock mechanical, 389
scratch, 417
shut-in/decline, 446, 583
triaxial, 89, 90, 391
true triaxial, 391, 405
undrained, 135, 137, 139, 140, 395
uniaxial, 89, 90, 394
unjacketed, 40, 41, 354, 397
Theory
Biot, 35, 37, 319
Hertz–Mindlin, 329
macroscopic, 37, 319
plasticity, 126
Thermal diffusivity, 78
Thermal expansion
coefficient, 31, 589
undrained, 54
volumetric, 31
Thermal strain, 31
Thermal stress, 31
Thermally induced fracturing (TIF), 587
Thermoelastic stress, 31, 212, 482, 589

Thermoelasticity, 31
  analogy to poroelasticity, 52
Thermoporoelasticity, 53, 705
Thick-walled cylinder
  strength, 531
  test, 530
Thickness-to-diameter ratio, 609
Thomsen parameters, 280
Thrust fault, 166
Tight hole, 461, 464, 470, 478, 495
Time average equation, 266
Time-delayed borehole failure, 223, 478
Tortuosity, 268, 270, 639
Total reflection, 296
Translation of a rigid body, 17
Transmission coefficient, 295
Transversal wave, 264
Transverse isotropy, 60, 62, 182, 279, 321, 698, 706
Transversely isotropic, 60, 65, 195, 321, 443
Tresca criterion, 97, 98, 108, 131, 243, 530
Triaxial
  cell, 384
  stress states, 64, 414
  test, 89, 90, 391
Trigonometric identities, 693
True triaxial, 391
  test, 391, 405
Tube wave, 302
Turbidity current, 178
TWT (Two-Way Traveltime), 311

U

Unconfined
  compression test, *see* Uniaxial test
  strength, *see* Uniaxial compr. strength
Unconfined strength, 197
Unconsolidated, 180
  sand, 117, 167
Unconventionals, 556, 590
Underbalance, 521
Underbalanced drilling, 472, 480
Undrained
  bulk modulus, 54
  conditions, 49, 53, 121, 395
  stiffness, 75
    matrix, 65
  test, 135, 137, 139, 140, 395
Undrained (def.), 39
Uniaxial
  compaction, 603, 604, 608
  compressibility, 604
  compression, 400
  reservoir compaction, 603, 606
  tests, 89
Uniaxial compaction modulus, 26, 265, 402, 605
  chalk, 188
Uniaxial compressive strength, 90, 100, 403
  chalk, 188
  sandstone, 184
  shale, 192
Uniaxial test, 90, 394
Unjacketed loading, 37
Unjacketed test, 40, 41, 354, 397
Unloading moduli, 68
UST (Uniaxial Strain Test), 401

V

Valhall, 173, 188, 602, 647, 649, 652, 654
Velocity
  average, 313
  group, 262
  phase, 261
  root-mean-square, 312
  sound, 262, 387
Vertical borehole, 217, 224, 230
Vertical stress, 157, 437, 438
Void ratio, 135
Voigt bound, 324, 353
Voigt notation, 57, 278, 694
Volumetric
  compaction, 610
  deformations, 610
  strain, 21
von Mises criterion, 109
  in the (p', q) plane, 124
von Mises stress, 14, 703
  effective, 14
  equivalent, 14
VSP (Vertical Seismic Profiling), 195

W

Washout, 465, 500

Waste disposal, 557
Water
  bound, *see* Bound water
  content, 420
  free, 377
Water injection, 556, 587, 637, 646, 649, 653
Water weakening, 121
Water-based mud, 378, 450, 483
Wave
  acoustic, 264
  compressional, 263, 431
  converted, 297
  elastic, 259, 264
  flexural, 307, 432, 443
  interface, 298
  longitudinal, 263
  primary, 263
  pseudo-Rayleigh, 303
  Rayleigh, 298
  Scholte, 299
  secondary, 264
  shear, 264, 431
    eigenmodes, 303
  slow, 271
  Stoneley, 299, 302, 431
  transversal, 264
  tube, 302
Wave equation, 260
  Biot theory, 267
Wave-induced fluid flow, 276, 291
Wavelength (def.), 261
Wavenumber (def.), 261
Weak anisotropy, 280
Weak frame, 43, 399
Weak frame limit (def.), 43
Wedge effect, 572
Well completion, 520
Wellbore, *see* Borehole
Wilmington, 602, 649
Wing cracks, 285, 351
Wood's equation, 35
Work hardening, 134
Wormhole, 536
Wyllie equation, 266

X

X-ray diffraction, 421
XLOT, *see* Extended leak-off test

Y

Yield criterion (def), 127
Yield point, 68, 90
Young–Laplace equation, 378
Young's modulus
  chalk, 188
  correlation with strength, 185
  definition of, 25
  dynamic, 430
    and Brinell hardness, 417
  lab. test, 392
  salt, 196
  stress dependent, 212
  transverse isotropy, 60

Z

Zone
  abnormally pressured, 171
  cooled vs. flooded, 589
  overpressured, 171
  plastic, 243
  subduction, 161